Introduction to Fluid Mechanics and Fluid Machines

Third Edition

INTRODUCTION TO FLUID MECHANICS AND FLUID MACHINES

THIRD EDITION

S K Som
Professor
Department of Mechanical Engineering
Indian Institute of Technology Kharagpur, India

Gautam Biswas
Director and JC Bose National Fellow
Indian Institute of Technology Guwahati
and
Professor
Department of Mechanical Engineering
Indian Institute of technology Kanpur

Suman Chakraborty
Professor
Department of Mechanical Engineering
Indian Institute of Technology Kharagpur, India

McGraw Hill Education (India) Private Limited

Published by McGraw Hill Education (India) Private Limited
Registered Office: Anjana Complex No: 5/90 A, Butt Road, St. Thomas Mount, Chennai 600016, Tamil Nadu

Introduction to Fluid Mechanics and Fluid Machines, 3e

22nd reprint 2024

ISBN-13: 978-0-07-132919-4
ISBN-10: 0-07-132919-6

23 24 25 26 27 ADA 28 27 26 25 24

Typeset at Text-O-Graphics, B-1/56, Aravali Apartments, Sector 34, Noida 201301, and Printed and bound in India at Adarsh Printers, Jhilmil Industrial Area, Delhi 110095.

Visit us at: www.mheducation.co.in
Toll free in India: 1800 103 5875
Write to us at: support.india@mheducation.com
CIN: U80302TN2010PTC111532

Contents

Preface

The continued use of successive editions of this book by a large number of readers is a testament to the book having core strength in explaining the principles of fluid mechanics. At the suggestions of the publisher, Tata McGraw-Hill, and peer groups, the authors agreed to pen a third edition. The authors of the first two editions, S K Som and G Biswas, are glad to be joined by Prof. S Chakraborty in the authorship of this edition.

This edition of the book is rearranged and rejuvenated in such a way that each chapter introduces the topic and then familiarises the readers with all the associated principles and applications in a systematic manner. This edition also provides the reader with a good foundation to understand fluid mechanics and apply that knowledge in the proliferating world of engineering science ranging from micro-nano-world to the regime of spacecrafts. The book provides overall exposure to the subject, highlighting its importance in creative learning of science and engineering. The book appropriately covers all prerequisites of higher level courses in Aerospace Engineering, Civil Engineering, Chemical Engineering, Mechanical Engineering, Metallurgy and Materials Engineering, Biological Engineering and Applied Mathematics, and it will be especially useful for students of these subjects.

This book has the following salient features:

- Introduction to fundamental concepts through elementary principles of continuum mechanics
- A lucid treatment of fundamental interfacial phenomena in fluid mechanics
- Thoroghly revised treatment of fundamental equations of conservation which includes both differential and integral approach based on control mass system and control volume formulation
- Expanded coverage of Fluid Kinematics with detailed description (both analytical and physical) of irrotational flows, circulation, velocity potential and vortex flows
- Momentum conservation theorem in a non-inertial frame of reference, encompassing both rectilinear acceleration and rotational motion, has been derived and discussed in detail using suitable illustrative examples
- Complete derivation of the Navier Stokes equation in a concise and lucid manner
- Enhanced coverage of Boundary Layer, Turbulent Flows, Unsteady and Compressible Flows
- Inclusion of chapter-end summary
- Comprehensive selection of solved examples and exercise problems, especially chosen to explain the nuances of basic principles

We would like to thank numerous individuals from various colleges, universities and institutes who provided inputs for improvements. We are grateful to Prof. Shaligram Tiwari, Prof. Sarit K Das and Prof. K Arul Prakash of IIT Madras, Prof. Nishith Verma and Prof. P K Panigarhi of IIT Kanpur for their valuable suggestions. Mr Ajoy Kuchlyan of CSIR-CMERI Durgapur has also helped us in miscellaneous ways. We would also like to express our special appreciation for Mr Sukumar Pati, a senior research fellow of IIT Kharagpur, for his indispensable help in checking the technical contents and providing valuable suggestions.

The publishers would like to thank with appreciation the efforts of our reviewers.

Debasish Roy
Jadavpur University, Kolkata

Basudeb Bhattacharya
Bengal Engineering College and Science University, Shibpur, West Bengal

A S Dhoble
Visvesvaraya National Institute of Technology, Nagpur, Maharashtra

D C Mahale
B S Deore College of Engineering, Dhule, Maharashtra

S Jayaraj
National Institue of Technology Calicut, Kerala

K R Balasubramanian
National Institute of Technology Tiruchirappalli, Tamil Nadu

S Suresh
Sona College of Technology, Salem, Tamil Nadu

Shaligram Tiwari
Indian Institute of Technology Madras, Chennai

Umamahesh Nanduri
National Institute of Technology Warangal, Andhra Pradesh

T P Ashok Babu
National Institute of Technology Karanataka, Surathkal

Finally we are hopeful that this book will help the students to analyse and design various flow systems. We also hope that the book will be a source to learning through reasoning and inference. We cordially welcome suggestions and comments from our readers for further improvements. These can be sent to us at sksom@mech.iitkgp.ernet.in, gtm@iitk.ac.in, suman@mech.iitkgp.ernet.in

S K Som
G Biswas
S Chakraborty

Publisher's Note

Remember to write to us. We look forward to receiving your feedback, comments and ideas to enhance the quality of this book. You can reach us at *tmh.elefeedback@gmail.com*. Please mention the title and authors' name as the subject.

In case you spot piracy of this book, please do let us know.

1

INTRODUCTION AND FUNDAMENTAL CONCEPTS

1.1 A BROAD PERSPECTIVE OF FLUID MECHANICS

Fluid mechanics deals with the behaviour of liquids and gases in rest or in motion. Numerous intriguing questions can be answered using fundamental concepts of fluid mechanics. Some such questions are the following:

- How does a rocket go up?
- How should one design the shape of a car to minimise the wind resistances against its motion?
- How does the roughness of a solid body influence its resistance against relative motion with a fluid?
- How do insects fly?
- How does blood flow through arteries and veins?
- How are nutrients supplied from the ground up into the tall branches of a tree, against gravity?
- How does the human heart act like a pump?
- What are the factors that influence sports ball (such as golf ball or cricket ball) dynamics?
- How should one design model airplanes so that results on their testing may be utilised to predict the performance of real prototype aircrafts?
- How are ocean currents formed?

Such a list of questions goes on and on, but one aspect looks obvious–the mechanics of fluids is very interesting, interdisciplinary in appeal, and practical. Scientists and engineers often need to develop a fundamental understanding of this subject, in order to address and tackle challenging problems in real-life design and analysis. The aim of this introductory text is, accordingly, to provide a foundation of the basic principles of fluid mechanics and its applications.

1.2 DEFINITION OF A FLUID

A fluid is a substance that deforms continuously when subjected to a tangential or shear stress, however small the shear stress may be. As such, this continuous deformation under the application of shear stress constitutes a flow. For example (Fig.

1.1), if a shear stress τ is applied at any location in a fluid, the element $011'$ which is initially at rest, will move to $022'$, then to $033'$ and to $044'$ and so on. In other words, the tangential stress in a fluid body depends on the velocity of deformation, and vanishes as this velocity approaches zero.

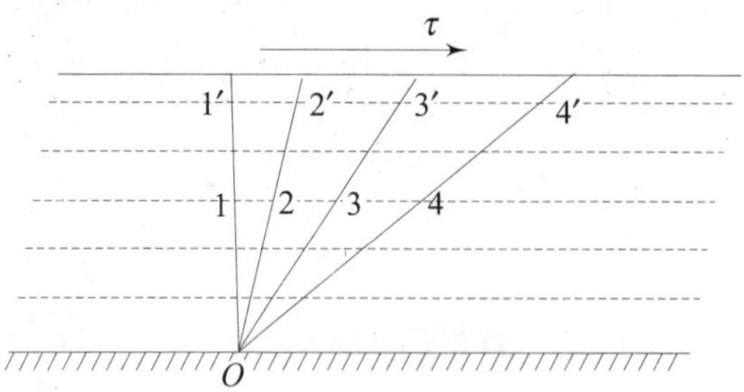

Fig. 1.1 Shear stress on a fluid body

1.3 DISTINCTION BETWEEN A SOLID AND A FLUID

The molecules of a solid are more closely packed as compared to that of a fluid. This may be attributed to the fact that attractive forces between the molecules of a solid are much larger than those of a fluid.

A deformable solid body undergoes either a definite (say, α) angular deformation (Fig. 1.2) or breaks completely when shear stress is applied on it. The amount of deformation (denoted by the angle α) is proportional to the magnitude of applied stress up to some limiting condition.

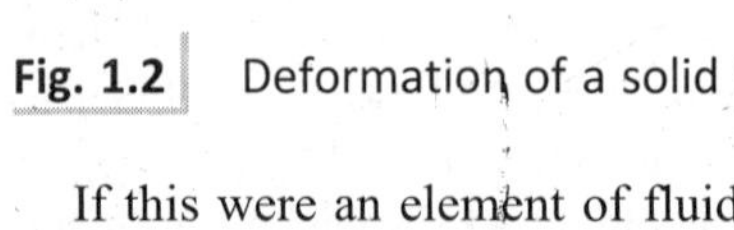

Fig. 1.2 Deformation of a solid body

If this were an element of fluid, there would have been no fixed α even for an infinitesimally small shear stress. Instead, a continuous deformation would have persisted as long as the shear stress was applied. It can be simply said, in other words, that while solids can resist tangential stress under static conditions, fluids can do it only under dynamic situation. For fluids, therefore, the rate of deformation rather than the absolute deformation is a more interesting parameter by itself, since a fluid will any way go on deforming more and more as time progresses, if subjected to a shear stress.

In the subsequent section, we will first consider different approaches for analysing the mechanics of fluids, keeping in view the characteristic physical scales that the analyser is interested to address.

1.4 MACROSCOPIC AND MICROSCOPIC POINTS OF VIEW: THE CONCEPT OF A CONTINUUM

The most fundamental approach for analysing the mechanical behaviour of a fluidic system may be a 'deterministic' molecular approach in which the dynamics of individual molecules is investigated by writing their respective equations of motion. While this is fundamentally appealing and may be suitable for certain cases, several practical constraints also seem to be inevitable. In order to appreciate the underlying consequences, consider the molar volume of a gas at normal temperature and pressure, which, by Avogadro's hypothesis, contains 6.023×10^{23} number of molecules. To describe motion of each of these molecules, three translational velocity components (along three mutually perpendicular coordinate directions) and three rotational components (along the same coordinate directions as above) need to be specified. Therefore, one has to deal with $6 \times 6.023 \times 10^{23}$ number of equations of motion, even for an elementary molar volume, which is an extremely demanding computational task even today, in spite of the advent of high-speed supercomputers. Hence, from a practical point of view, there must be certain approaches that can reduce the number of variables to a figure that can be handled conveniently for practical computations.

In particular, there are two specific approaches that can be introduced in this context. In one approach, we deal with 'statistically averaged' behaviour of many molecules constituting the matter under investigation. This is exactly the approach followed in kinetic theories of matter and statistical mechanics, which in general is termed as the '**microscopic**' point of view, since the primary focus of attention is on the averaged behaviour of individual microscopic constituents of matter.

The second approach reduces the number of variables even further, by considering the gross effect of many molecules that can be captured by direct measuring instruments and can be perceived by our senses. Such an approach is the so-called '**macroscopic**' approach. For a clearer distinction between macroscopic and microscopic approaches, we can refer to a very simple example as follows: When we conceive the term 'pressure' of a gas (we will deal with this term more formally later) in the microscopic point of view, it originates out of the rate of change of momentum of molecules as a consequence of a collision. On the other hand, from a macroscopic point of view, we describe the same quantity in terms of time-averaged force over a given area, which can be measured by a pressure gauge.

Though the macroscopic approach provides us with a more practical method of analysis, one should not presume that it can give the correct solution in all situations. In order to assess the underlying implications, let us consider certain aggregations of molecules, such as a set of widely spaced molecules for a gas and a set of closely spaced molecules for a liquid. If we consider a very small elementary volume within the medium, then numerous molecules may come into it or move out of it during some specified interval of time. For someone interested in calculating density of the medium, it may apparently be sufficient to calculate the number of molecules (n) in that elementary volume, and compute a 'locally averaged' density based on that information. However, the number of molecules residing over that elemental volume at any instant is rather uncertain. Now, if that elemental volume is too small, then the number of molecules residing in that volume may be very few, and hence, even an uncertainty in a few number of molecules residing in that volume may result in serious uncertainties of the computed density based on that information. On the other hand, if that elemental volume is sufficiently large (say, the characteristic linear dimension of the volume is significantly large in comparison to the mean free path of the molecules, the later being nothing but the distance traversed by a molecule between two successive collisions), then the number of molecules contained in that volume may be large enough, so that it hardly matters whether there is an uncertainty over a few number of molecules contained in that volume. Therefore, for each situation, there is a limiting volume, above which we can treat the substance as being a continuous medium without any significant uncertainty in prediction of the averaged behaviour of the molecules constituting the medium (However, the chosen elemental volume should not be too large either, in which case we shall not be able to capture the local variation of properties). Such a medium is known as a **continuum** (continuous medium), which basically implies that variation of properties within the medium is smooth enough (i.e., randomness due to uncertainties in molecular behaviour is virtually ruled out) so that differential calculus can be used to analyse the averaged physical behaviour of the medium mathematically, without showing any additional concern on the behaviour of individual molecules. Therefore, validity of continuum assumption is a basic prerequisite to the validity of the macroscopic point of view. In essence, the macroscopic point of view gives a direct approach of analysis if *(i) there is a large number of molecules and (ii) the dimensions under consideration are extremely large as compared to the dimensions in atomic/ molecular scale (typically, the 'mean free path')*. This approach, however, fails, when the mean free path approaches the characteristic linear dimension (the so-called 'length scale') of the system, such as for the flow of rarefied gases at low pressure (i.e., high vacuum). The microscopic approach turns out to be essential for addressing such specific situations. It can be noted here that in the present text, we will present the subject matter primarily from a macroscopic point of view.

In the context of the use of the microscopic description of matter, it may be instructive to look into some specific contexts in which employment of such approaches may become inevitable. To give the issue a technological perspective, one may refer to micro electromechanical systems (or MEMS)-based appliances, which find their applications in a wide variety of emerging technologies, ranging from microactuators, microsensors, microreactors to thermo-mechanical data storage systems. In such devices the flow physics may be faceted by the fact that the classical continuum hypothesis may cease to work as the distance traversed by molecules between successive collisions (i.e., the mean free path, λ) becomes comparable with the characteristic length scale of the system (L) over which characteristic changes in the flow are expected to occur. The ratio of these two quantities, known as the

Knudsen number ($Kn = \lambda/L$), is an indicator of the degree of rarefaction of the system, which determines the extent of deviation from a possible continuum behaviour. For $0 < Kn < 0.01$, the flow domain can be treated as a continuum (see Fig. 1.3), in which the continuum equations with the classical boundary conditions of no relative slip between the fluid and the solid boundary remains applicable (This is known as no-slip boundary condition, which we will discuss in detail in Subsection 1.6.4.1). On the other extreme, when $Kn > 10$, the flow becomes free molecular in nature, because of negligible molecular collisions. The range of Kn spanning from 0.01 to 0.1 is known as the slip flow regime, over which the no-slip boundary condition becomes invalid, although continuum conservation equations can still be used for characterising the bulk flow. However, over the Kn range of 0.1 to 10 (the so-called transitional regime), the continuum hypothesis progressively ceases to work altogether, thereby necessitating a shift of paradigm from the continuum-based analysis to particle-based analysis (following microscopic point of view) for this range of Kn and beyond.

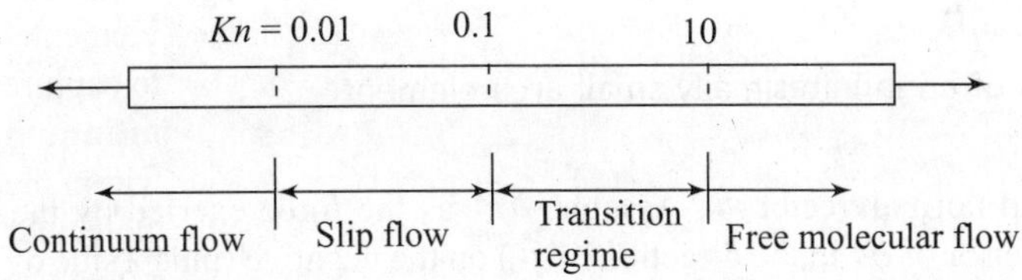

Fig. 1.3 Flow regimes for gases

In the subsequent discussions, we will consider that the range of Kn is such that the continuum hypothesis is valid. Accordingly, we will first develop a formalism to describe the state of stress in a fluid medium considering the macroscopic viewpoint.

1.5 CONCEPTS OF PRESSURE AND STRESS IN A FLUID

As discussed earlier, a fluid is basically a substance that undergoes continuous deformation when subjected to a shear (tangential), no matter how small the shear stress is. The fluid moves and deforms continuously as long as the shear stress is applied. Therefore, fluid at rest must be in a state of zero shear stress, a state often called 'hydrostatic state of stress'. It has to be noted here that the hydrostatic state of stress basically originates out of the physical quantity known as pressure (p), which acts equally from all directions, and is inherently compressive in sense.

At this juncture, it is important to distinguish between two frequently confused terms, namely, pressure and stress. Towards that, we first develop the basic concept of stress at a point in a fluid. Consider an arbitrary volume of a fluid, which can be subjected to basically two types of forces, namely, (i) **surface forces** (i.e., forces that act at surfaces and are proportional to the surface area) and (ii) **body forces** (i.e., forces acting throughout the volume of the fluid). Typical examples of surface forces and body forces in a fluid are forces due to pressure and gravity, respectively. Further, assume that the fluid volume element occupies a surface area S. Forces acting on surface of this element may be due to interactions from surrounding fluid elements or a solid boundary in contact. Let 'A' be any closed surface within the volume, as depicted in Fig. 1.4. Also, let δA be an infinitesimal area segment of the surface,

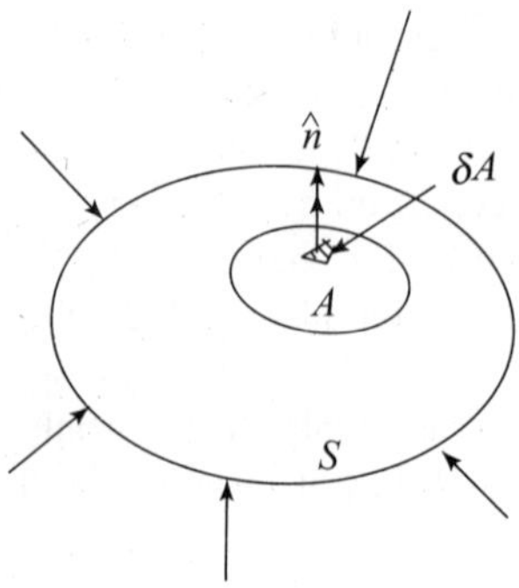

Fig. 1.4 Definition of an infinitesimally small area element

having a unit outward normal vector $\hat{n}$. Assume $\delta\bar{F}$ as the force exerted by the positive side (i.e., exterior of δA in the direction of $\hat{n}$) on the negative (inner) side of the infinitesimal area δA. Note that $\delta\bar{F}$ depends on the location of the area δA, magnitude of the area δA and orientation of the area δA (specified by the vector $\hat{n}$). Based on this information, we can define a quantity called **traction vector** as $\bar{T}^n = \underset{\delta A \to 0}{Lt} \dfrac{\delta \bar{F}}{\delta A} = \dfrac{d\bar{F}}{dA}$. Physically, $\bar{T}^n$ is the force per unit area outward at a point on an elemental surface with outward direction normal $\hat{n}$; the force being exerted by the material located on the positive side of the area by that located on the negative side of the area.

Let T_j^n be the jth component of the vector $\bar{T}^n$ (note that $j = 1$ means component along x, $j = 2$ means component along y, and $j = 3$ means component along z, in a Cartesian indexing system). Further, we define a notation $\tau_{ij} \equiv T_j^i$, only for special cases in which the directions i and j correspond to directions conforming the Cartesian coordinate axes. This is known as the Cartesian index notation and the components τ_{ij} are known as the Cartesian stress tensor components. Here τ_{ij} represents the jth component of the traction vector on a plane whose outward normal is parallel to the ith direction. Note that each of the indices i and j may vary between 1 to 3. Some alternative notations are also followed in many texts, such as $\sigma_x = \tau_{11}$, $\sigma_y = \tau_{22}$, $\sigma_z = \tau_{33}$, $\tau_{xy} = \tau_{12}$, $\tau_{yx} = \tau_{21}$, $\tau_{xz} = \tau_{13}$, $\tau_{zx} = \tau_{31}$, $\tau_{yz} = \tau_{23}$, $\tau_{zy} = \tau_{32}$. For an illustration of the notation presented above, one may refer to Fig. 1.5, in which the τ_{ij} components are shown for the top, bottom and the two side faces for a rectangular-parallelepiped shaped element. Referring to Fig. 1.6, for the right-side face, outward normal vector is directed along x, and hence the index $i = 1$ for all τ_{ij}'s acting on that face. The other index j signifies the direction of action of the stress component. For instance, $j = 1$ means that it acts along the x direction, $j = 2$ means it acts along the y direction, and $j = 3$ means it acts along the z direction. One can easily observe that τ_{11} acts normal to the surface under consideration (right face), and is therefore termed as a 'normal component of stress', whereas τ_{12} and τ_{13}, for example, act tangential to the surface under consideration, and are therefore termed as 'shear components of stress'. On the left-side face, the positive senses of action of these components are assumed to be opposite to that for the right face, since outward normal vector for the left face is directed along the negative x direction. More precisely, we develop our sign convention such that we take positive sense of τ_{ij} in the positive j direction, in case the outward normal vector for the surface is oriented along the positive i direction. Accordingly, we take positive sense of τ_{ij} in the negative j direction, in case the outward normal vector for the surface is oriented along the negative i direction. For example, one can see that a positive sense of τ_{11} is directed along the negative x_1, direction on the left face of the parallelepiped depicted in Fig. 1.5, since that face itself has an outward normal directed along the negative x_1 direction. For the same reason, the positive sense of τ_{12} is directed along the negative x_2 direction of the same face and similarly for τ_{13}.

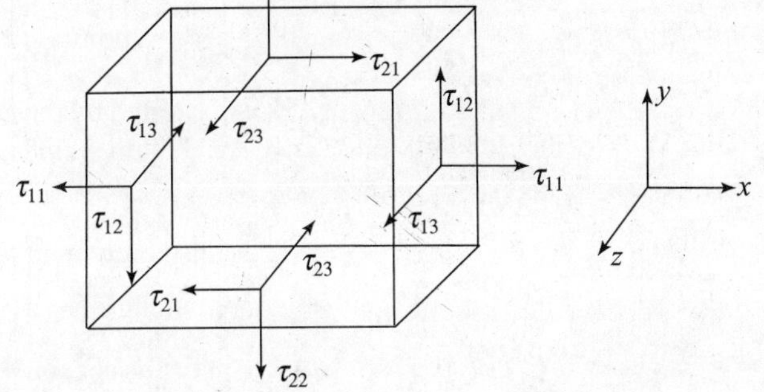

Fig. 1.5 Components of the stress tensor demonstrated in a cuboid

*Next, we will try to utilise the concept developed so far to express traction vector on an arbitrarily oriented surface in terms of the stress components. For that purpose, consider a infinitesimal tetrahedral fluid element formed by three surfaces dS_1, dS_2, and dS_3, such that dS_1 is parallel to the yz plane, dS_2 is parallel to the xz plane and dS_3

*This portion may be omitted without loss of continuity

is parallel to the *x-y* plane, and a surface dS (surface *ABC*) whose outward unit normal vector is $\hat{n}$ (refer to Fig. 1.6). Let $\overline{T}^n$ be the traction vector on the surface dS. Also, let h be the perpendicular distance from the vertex O to the surface dS. It is apparent that dS_1 is nothing but the projection of dS on the *y-z* plane. Now, the angle between dS_1 and dS (say, θ) is basically the angle between the directions of their respective normals, i.e., angle between unit vectors $\hat{i}$ and $\hat{n}$ (where $\hat{n} = n_1\hat{i} + n_2\hat{j} + n_3\hat{k}$, in terms of its components, i.e., n_i represents the component of vector $\hat{n}$ along the ith direction). Therefore, from the definition of dot product of two vectors we have, $\hat{i}.\hat{n} = |\hat{i}||\hat{n}|\cos\theta$. Now, since $|\hat{i}| = |\hat{n}| = 1$, we have $\cos\theta = \hat{i}.\hat{n} = n_1$ (using the components of the vector $\hat{n}$). Thus, $dS_1 = dS\cos\theta = dSn_1$. Similarly, $dS_2 = dSn_2$, and $dS_3 = dSn_3$. Also, volume of the element is given by $dV = \frac{1}{3} \times h \times dS$ (recall formula for volume of a tetrahedron). With this information, let us now try and apply Newton's 2nd law of motion on the fluid element for motion along the x direction. For that purpose, we first identify forces on each of the surfaces along the x direction. For the surface dS_1, the outward normal vector is along the negative x direction, and therefore the corresponding force is $-\tau_{11}dS_1$ (we are neglecting a small variation in τ_{11} between the point O and any other point on the surface dS, which is perfectly legitimate as $h \to 0$). Similarly, the force on dS_2 along the x direction is $-\tau_{21}dS_2$ (the minus sign originates out of the fact that outward normal to dS_2 is directed along the negative

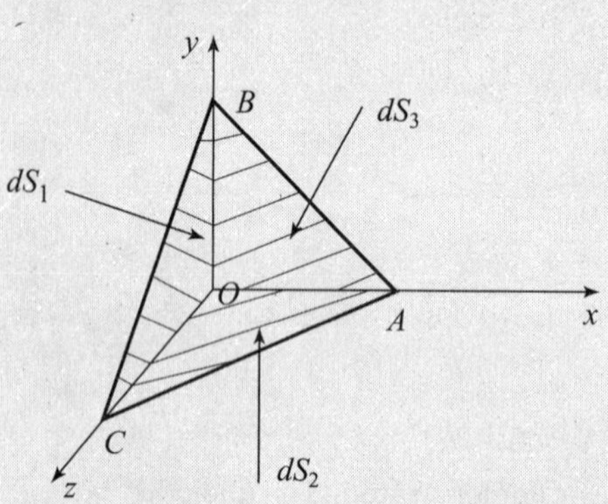

Fig. 1.6 A tetrahedral element of fluid

y direction) and the force acting on dS_3 along the x direction is $-\tau_{31}dS_3$ (the minus sign originates out of the fact that outward normal to dS_3 is directed along the negative z direction). Also, the force acting on dS along the x direction is $T_1^n dS$. Therefore, the net force acting on the element along x direction is given by

$$\sum F_x = \text{Net surface force along } x + \text{Net body force along } x$$

$$= \left(-\tau_{11}dS_1 - \tau_{21}dS_2 - \tau_{31}dS_3 + T_1^n dS\right) + \left(b_1 \frac{1}{3} \times h \times dS\right) \quad (1.1)$$

where b_1 is the body force per unit volume. Now, if ρ be the density of the element of mass dm, then $dm = \rho \times \frac{1}{3} \times h \times dS$. Also, as per Newton's 2nd law of motion, $\sum F_x = dm(a_1)$, where a_1 is acceleration of the element along x. Substituting $dS_1 = dSn_1$, $dS_2 = dSn_2$, $dS_3 = dSn_3$, we have

$$\left(-\tau_{11}n_1 - \tau_{21}n_2 - \tau_{31}n_3 + T_1^n\right) + b_1 \frac{h}{3} = \rho \frac{h}{3} a_1 \quad (1.2)$$

Finally, since $h \to 0$ (we need to evaluate traction vector at a point O, and therefore we need to shrink the tetrahedron to virtually a single point in the limiting sense), we have

$$T_1^n = \tau_{11}n_1 + \tau_{21}n_2 + \tau_{31}n_3 = \sum_{j=1}^{3} \tau_{j1}n_j \quad (1.3)$$

Identically, for any general coordinate direction i (i = 1, 2 or 3)

$$T_i^n = \sum_{j=1}^{3} \tau_{ji}n_j \quad (1.4)$$

The above result is an outcome of the so-called **Cauchy's theorem** relating traction vector components acting on a plane having unit normal vector $\hat{n}$ to Cartesian stress components τ_{ij}. The above expression may also be conveniently written in an equivalent matrix form as

$$= \underbrace{\begin{bmatrix} T_1^n \\ T_2^n \\ T_3^n \end{bmatrix}}_{\substack{\text{traction} \\ \text{vector} \\ (\bar{T}^n)}} = \underbrace{\begin{bmatrix} \tau_{11} & \tau_{21} & \tau_{31} \\ \tau_{12} & \tau_{22} & \tau_{32} \\ \tau_{13} & \tau_{23} & \tau_{33} \end{bmatrix}}_{\substack{\text{stress} \\ \text{components}}} \underbrace{\begin{bmatrix} n_1 \\ n_2 \\ n_3 \end{bmatrix}}_{\substack{\text{unit} \\ \text{normal} \\ \text{vector}(\hat{n})}} \quad (1.5)$$

We can, therefore, see that the Cartesian components of stress tensor, altogether, map a vector $\hat{n}$ into another vector $\bar{T}^n$. Therefore, from a fundamental mathematical definition which states that *the linear transformation that maps a vector into a vector is nothing but a second order tensor*, we can conclude that **stress is a**

second order tensor quantity. At this stage, it may not be too comfortable to proceed deeper into the mathematics of tensor, and we would neither make any effort of that kind here. However, it is very easy to appreciate from our previous discussions that tensor is a quantity much more general than a vector.

For the stress tensor, there are actually nine components (since $i\ \varepsilon[1,3]$ and $j\ \varepsilon[1,3]$, out of which six are independent. This may be attributed to the fact that from angular momentum conservation, $\tau_{ji} = \tau_{ij}$. This may be illustrated as follows:

Let us consider a two-dimensional fluid element, as shown in Fig 1.7

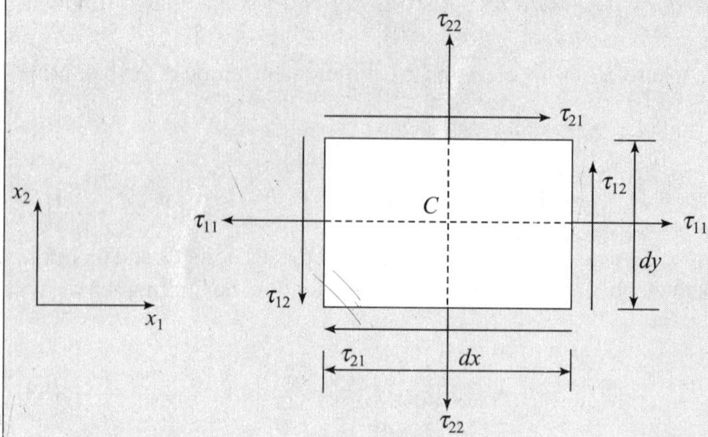

Fig. 1.7

For rotation of the fluid element with respect to its centroid, net moment of all forces about C may be expressed as $\sum M_c = I_c\ddot{\theta}$, where I_c is the moment of inertia of the element with respect to its centroidal axis, and $\ddot{\theta}$ is its angular acceleration.

Thus $(\tau_{12}dy)dx - (\tau_{21}dx)dy = (k\rho dxdy)(dx^2 + dy^2)\ddot{\theta}$

where k is a constant, ρ is the fluid density. In the limit as $dx, dy \to 0$, the above yields

$$\tau_{12} = \tau_{21}$$

This, interestingly, is true even if the fluid element has angular acceleration. To generalize,

$$\tau_{ji} = \tau_{ij}$$

The only restriction to the applicability of this equation is that there is no body couple acting on the fluid element.

At this stage, it may be imperative to demarcate a second order tensor such as stress from a vector such as force. Importantly, a second order tensor component τ_{ij}, for its specification, requires two indices i and j, the first index i specifying the direction normal of the face on which it is acting, and the second index j specifying the direction of action of the stress component itself. Therefore, unlike force, stress at a point is not unique, and is strongly dependent on orientation of the chosen area based on which it is calculated. A vector (such as force, for example) is the special case of a tensor; in fact it is a tensor of order one, since it requires only one index (say, i) to specify its ith component. Scalar, in that sense, is a tensor of order zero.

Next, let us formally investigate the nature of the quantity pressure. For that purpose, let us refer to Fig. 1.8, in which a differentially small fluid element in the form of a triangular prism is supposed to be in static condition. Since it is not in motion, it is not under the action of any tangential (shear) force exerted by the surrounding fluid elements (we have already learnt that fluid cannot resist even an infinitesimally small shear without being deformed, and therefore a static fluid element must be subjected to a zero-shear condition), and forces exerted by the surrounding fluid elements (surface forces) are only normal to the respective surfaces under consideration. Such forces are always compressive in nature (usually expressed in terms of force per unit area). For the situation shown in Fig. 1.8, let us assume that p_1, p_2, and p_3 are the pressures acting on the three surfaces. Let b_i be the component of body force (per

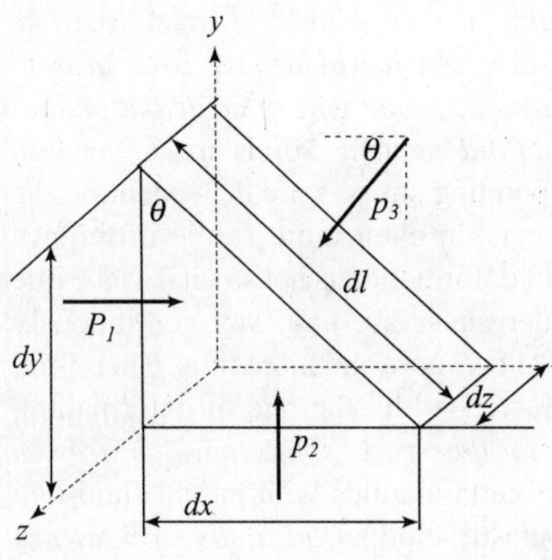

Fig. 1.8 A triangular wedge-shaped fluid element in equilibrium

unit volume) along the ith direction. Now, for equilibrium of the element,

$$\sum F_x = 0$$

$$p_1 dydz - p_3 dldz \cos\theta + (1/2)dl(dy \sin\theta)dzb_x = 0 \qquad (1.6)$$

Now, since $dy = dl\cos\theta$, we have

$$p_1 - p_3 + (1/2)(dl\sin\theta)b_x = 0 \tag{1.7}$$

Finally, since $dl \to 0$, we have

$$p_1 = p_3 \tag{1.8}$$

Also, for equilibrium.

$$\sum F_y = 0$$

$$p_2 dxdz - p_3 dldz\sin\theta + (1/2)dl(dy\sin\theta)dzb_y = 0 \tag{1.9}$$

Since $dx = dl\sin\theta$, we have

$$p_2 - p_3 + (1/2)(dl\cos\theta)b_y = 0 \tag{1.10}$$

As $dl \to 0$, we have

$$p_2 = p_3 \tag{1.11}$$

Therefore, $p_1 = p_2 = p_3$, i.e., pressure acts equally from all possible directions, no matter whether any body force is acting or not.

We may extend our previous discussions to the case of a general fluid element in motion, for which the *traction vector on any surface should clearly have two separate contributions; the first one is pressure, which would have been present even if the fluid element was non-deforming, and a second type of contribution that is related to the deformation behaviour of the fluid element.* Solids are, in general, not pressure-sensitive as such, and the corresponding stresses are dependent on the nature of deformation alone. However, there is an even more subtle difference between these two. For fluids, the net amount of deformation is not so vital as the rate of deformation. This is because, fluids under shear are any way continuously deforming, and the deformation is more as we allow more and more time. Therefore, resistance towards deformation is better characterised by the rate of deformation, rather than the deformation itself. Thus, *unlike for the case of solids in which stresses are mostly related to strain alone* (except for certain solids with partial fluid-like characteristics, namely, the so-called 'visco-elastic' solids), *in fluids stresses are predominantly related to the rate of strain* (or, rate of deformation). In fact, in fluids, the force necessary to produce deformation approaches zero as the rate of deformation is reduced.

Such clear demarcations between fluids (liquids and gases, in totality) and solids, however may encounter serious concerns in some borderline cases. For example, some apparently solid substance such as asphalt or lead resist shear stress for short periods but actually deform slowly and exhibit definite fluid behaviour for long periods. Other substances such as colloid and slurry mixtures or even metals at semi-solid (mushy) state can resist small shear stresses but 'yield' at larger shear stresses and begin to flow as fluids do. Such cases demand a more general description of material deformation, in a specialised field of science known as rheology of substances. For most engineering problems, however, a more or less clear decision about fluid or solid states can be made in terms of constitutive behaviour of the substance (i.e., stress response of the material as a function of internal deformation and/or rate of internal deformation). For example, within proportional limit, elastic solids have stress linearly proportional to strain and analogously, Newtonian fluids (such as water; we will define Newtonian fluids more formally later) have shear stress linearly proportional to the rate of shear strain.

1.6 FLUID PROPERTIES

Certain characteristics of a continuous fluid are independent of the motion of the fluid. These characteristics are called basic properties of the fluid. We shall discuss a few such basic properties here.

1.6.1 Specific Volume (v) and Density (ρ)

Specific volume of a substance is formally defined as

$$v = \lim_{\delta \mathcal{V} \to \delta \mathcal{V}'} \frac{\delta \mathcal{V}}{\delta m} \tag{1.12}$$

where $\delta \mathcal{V}$ is the volume of a small elemental mass δm, and $\delta \mathcal{V}'$ is the smallest volume for which the system can be considered as a continuum. If $\delta \mathcal{V}$ is substantially less than $\delta V'$, it falls in the domain of 'microscopic uncertainty' (as discussed earlier), in which the whole continuum assumption fails altogether. Therefore, $\delta \mathcal{V} \to 0$ (in place of $\delta \mathcal{V} \to \delta \mathcal{V}'$) would be a wrong limiting assumption, and should never be used casually in such formal definitions. As easily understandable, specific volume is the inverse of a term more familiar to you, namely the density ($v = 1/\rho$). For two-phase mixtures of simple compressible substances, specific volume of a mixture can simply be obtained as a weighted average of specific volume of individual constituent phases, and therefore, in thermodynamics, specific volume is a more popular terminology used in preference over density. In mechanics of fluids, on the other hand, the concept of density is more commonly used, which can formally be defined as

$$\rho = \lim_{\delta \mathcal{V} \to \delta \mathcal{V}'} \frac{\delta m}{\delta \mathcal{V}} \tag{1.13}$$

with symbols having usual meaning in which they were used to define specific volume.

Regarding the choice of the elemental volume ($\delta \mathcal{V}$) for calculating the 'local' properties, certain important points need to be carefully noted. Such infinitesimal elements must be chosen small enough to be considered uniform (i.e., any spatial variations in properties inside the volume element itself are negligible); and at the same time are large enough in size to contain a statistically large number of molecules. Materials obeying this 'hypothesis' are said to behave as a continuum, as mentioned earlier. Thus, the continuum hypothesis for estimation of local variations in properties works well provided all the dimensions of the system are large compared to the molecular size. In Fig. 1.9, is shown an optimal range of elementary volume that ensures capturing the local density variations and at the same time is not too small to depict fluctuations originating out of molecular scale uncertainties.

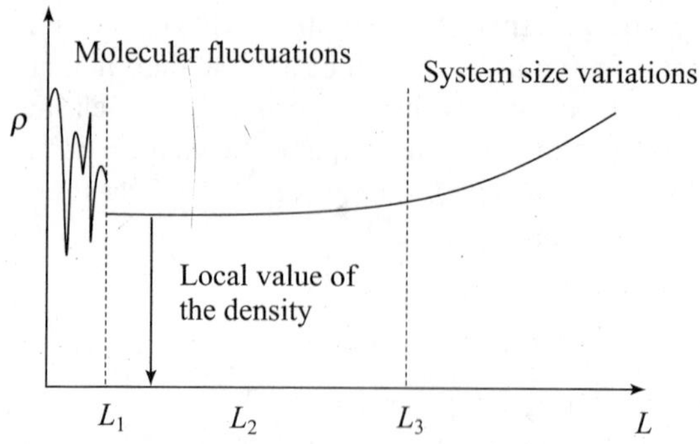

Fig. 1.9 Density variations as function of the length scale of the chosen elemental volume

1.6.2 Specific Weight (γ)

Specific weight is the weight of fluid per unit volume. Specific weight is given by

$$\gamma = \rho g$$

where g is the gravitational acceleration. Just as weight must be clearly distinguished from mass, so must the specific weight be distinguished from density. In SI units, γ will be expressed in N/m^3.

1.6.3 Specific Gravity (s)

For liquids, it is the ratio of density of a liquid at actual conditions to the density of pure water at 101 kN/m^2 and at 4°C. The specific gravity of a gas is the ratio of its density to that of either hydrogen or air at some specified temperature or pressure. However, there is no general standard, so the conditions must be stated while referring to the specific gravity of a gas.

1.6.4 Viscosity

1.6.4.1 *The No-slip Boundary Condition*

Consider a flat plate of large width being kept in a 'free stream' flow, so that the fluid just before encountering with the plate is having a uniform velocity of U (see Fig. 1.10). However, as the first set of fluid molecules comes in contact with the plate,

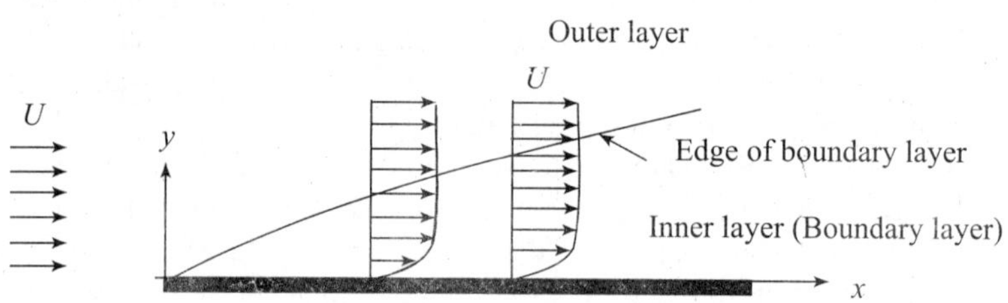

Fig. 1.10 Free stream flow of a viscous fluid over a flat plate

these molecules tend to stick to the solid. In other words, *it is commonly expected that there should be no relative tangential velocity component between the fluid and the solid boundary at their contact points. This conceptual paradigm is known as 'no-slip' boundary condition,* which has been successfully employed in explaining the experimentally observed fluid dynamics over most length scales addressing common engineering problems. However, the physical origin of the no-slip boundary condition over a solid boundary has not been established with certainty, and has been a matter of strong debate in the research community for long. One theory argues that the molecules of a fluid next to a solid surface are adsorbed onto the surface for a short period of time, and are then desorbed and ejected into the fluid. This process slows down the fluid and renders the tangential component of the fluid velocity equal to the corresponding component of the boundary velocity.

*The above consideration remains valid only if the fluid adjacent to the solid wall is in thermodynamic equilibrium. The achievement of thermodynamic equilibrium, in turn, requires an infinitely large number of collisions between the fluid molecules and the solid surface, which may not be possible for a 'rarified' or a less dense medium. This may result in a 'slip' between the fluid and the solid boundary in small channels where the mean free path may be of comparable order as that of the channel dimension. This phenomenon may be more aggravated by the presence of strong local gradients of temperature and/or density, because of which the molecules tending to 'slip' on the walls experience a net driving force. Such phenomena are usually termed as 'thermophoresis' and 'diffusophoresis', respectively.

Slip in liquids is something much less intuitive. Due to sufficient intermolecular forces of attraction between the molecules of the solid surface and a dense medium such as the liquid, it is expected that the liquid molecules would remain stationary relative to the solid boundary at their points of contact. Only at very high shear rates (typically realisable only in extremely narrow confinements of size, roughly a few molecular diameters), the straining may be sufficient enough in moving the fluid molecules adhering to the solid by overcoming the van der Waals forces of attraction. Another theory argues that the no-slip boundary condition arises due to microscopic boundary roughness, since fluid elements may get locally trapped within the surface asperities. If the fluid is a liquid then it may not be possible for the molecules to escape from that trapping, because of an otherwise compact molecular packing. Following this argument, it may be conjectured that a molecularly smooth boundary would allow the liquid to slip, because of the non-existence of the surface asperity barriers.

The phenomenon of 'true' slip, mentioned above, is not very common in most practical liquid flows. Rather, an 'apparent' slip phenomenon may be a more probable feature, especially if the fluid is flowing through a very narrow confinement. Recent studies have demonstrated that the intuitive assumption of 'no slip at the boundary' can fail greatly not only when the solid surfaces are sufficiently smooth,

---

*This portion may be omitted without loss of continuity

but also when they are sufficiently rough. Towards this, research investigations have attempted to resolve the apparent anomaly of 'reduced' fluid friction in the presence of 'rough' surface elements under specific conditions. These studies have conjectured that while the microscopic roughness of the solid surface impedes the motion of the adhering fluid a thin layer of vapour formed on the surface roughness elements tends to augment the level of slippage. This thin layer of vapour (typically, of nanometer length scale) may be spontaneously formed on hydrophobic surfaces (i.e., surfaces which have a phobia of water and therefore do not easily get wet), because of complicated thermodynamic interactions that are mostly triggered on account of the narrowness of the confinement (such interactions, therefore, are existent only in case of very narrow channels but not in large pipes or other conduits). This vapour layer, in effect, acts like a shield, preventing the liquid from being directly exposed to the irregularities of the channel surface. In such cases, the liquid is not likely to feel the presence of the wall directly and may smoothly sail over the intervening vapour layers, instead of being in direct contact with the wall roughness elements. Utilising this concept, researchers may potentially develop engineered rough surfaces with triggered hydrophobic interactions and consequent inception of friction-reducing vapour phases. Narrow confinements capable of mimicking the selective and rapid fluidic transport attainable in biological cellular channels but designed on the basis of such newly discovered surface roughness-hydrophobicity coupling would open up a wide range of new applications, such as transdermal rapid drug delivery systems, selective chemical sensing and mixing in nano-scales and several other new biological applications.

It is now clear from the above physical description that the existence of a thin vapour layer (of a few nm thickness) adhering the solid substrate leads to an occurrence of the so-called 'apparent slip' behaviour. This is illustrated in Fig. 1.11. Usually, the vapour layer is too thin to be resolved in experimental details, and therefore the velocity profile (i.e., the locus of the tip of the velocity vectors at any section) in the liquid layer (shown in the magnified near-wall image) provides the only experimentally observable continuum picture. That profile, when extrapolated (in practice, tangent to the profile at the liquid interface is only extrapolated) to the wall, signifies a slip velocity, u_{slip}, which is different from zero. However, when this profile continues to be extended by a length of l_s further away from the wall in its normal direction, it will show a zero velocity. The length l_s is typically known as the slip length. From simple trigonometry, it follows that

$$u_{slip} = l_s \left.\frac{\partial u}{\partial y}\right|_{\text{interface}} \tag{1.14}$$

where $\left.\frac{\partial u}{\partial y}\right|_{\text{interface}}$ represents the interfacial slope of the liquid velocity profile. Thus, the no-slip boundary condition is a special case of this conjecture, with $l_s = 0$

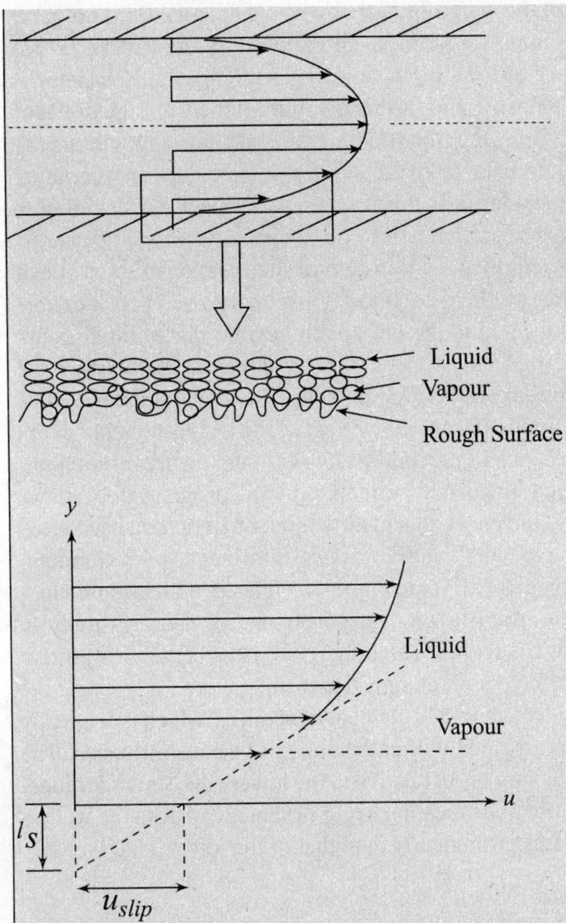

Fig. 1.11 A mechanism of apparent slip

Despite interesting exceptions, the no-slip boundary condition has been confirmed in overwhelming majority of applications, and is the standard choice in mainstream fluid dynamics. Accordingly, throughout this book we shall follow the no-slip boundary condition for our analysis. However, we shall keep in mind that it is a mere paradigm rather than a ritual, and may be easily violated in certain applications that are not so familiar with the classical engineering community but are emerging in the present day technological scenario.

1.6.4.2 *Viscous Effects and Newtonian Fluids*

Following the no-slip boundary condition, let us revisit the description of flow over a flat plate as an example, as depicted in Fig 1.10. The first layer of fluid molecules adhering to the plate feels the presence of the wall directly, and tends to stick to the same with zero relative velocity. However, the layer of fluid molecules located immediately above the wall adhering layer does not directly feel the presence of the wall.

Nevertheless, it feels the effect of the wall implicitly by virtue of some fluid property that enables the effect of the momentum disturbance imposed by the wall to propagate from the wall adjacent layer to the further outer layers. This fluid property is known as **viscosity**. As we move further and further away from the wall, the effect of the wall is less prominently felt. Thus, the free stream effect becomes more and more successful in forcing an outer fluid layer to move at a faster pace than another layer that is located somewhat closer to the wall. In this way, one may reach a distance from the solid boundary at which the free stream condition is virtually reached (this distance is quite short for reasonably high-speed flow, as we shall demonstrate in one of our later chapters), and no further gradient of velocity occurs beyond that location. This implicitly means that fluid located in the far-stream beyond this point does not feel the presence of the wall at all! As one progress further along the axis of the plate and considers another section, similar effects are felt. Only, the effect of the wall now penetrates deeper into the bulk, simply because the effect of the wall is now being felt more strongly because of the presence of preceding slow-moving upstream sections. In this way, the zone of influence of viscous effects 'grows' progressively as we march along the axis of the plate. One may conceptually think of an imaginary demarcating boundary between the 'active' and 'inactive' zones of influence of the momentum disturbance effect induced by the walls (see Fig. 1.10). This demarcating boundary is known as the 'edge' of a **boundary layer**. Thus, the boundary layer is simply an imaginary layer adjacent to the walls within which the viscous effects are important and outside which the viscous effects are not significant.

To proceed further, let us consider a thin fluid element of width Δy which is originally rectangular, as shown in Fig. 1.12, located within the zone of viscous influence. The upper edge of the element has a velocity of Δu relative to the lower edge (this additional velocity of the upper edge is because of its location being further away from the wall so that it feels the effect of the wall less prominently than that of the lower edge).

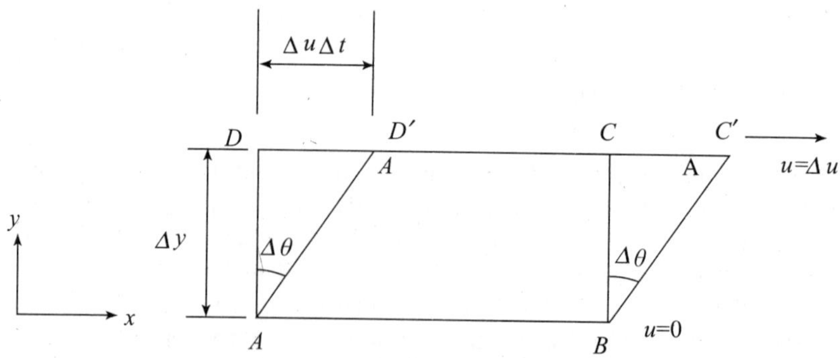

Fig. 1.12 Angular deformation of a rectangular fluid element under unidirectional flow

Therefore, in a small time interval of Δt, the upper layer displaces by an incremental amount of $\Delta u \Delta t$, as relative to the lower one. This distorts the original rectangular element to a parallelogram shape ($ABC'D'$), with an angular strain. The extent of this angular strain is measured by the shear angle $\Delta\theta$, which from trigonometry can be expressed as $\tan \Delta\theta = \frac{\Delta u \Delta t}{\Delta y}$. Since Δt is chosen as small, $\Delta\theta$ is also small, which implies that $\tan \Delta\theta \approx \Delta\theta$. The rate of angular deformation (or, equivalently the rate of shear strain), $\dot{\gamma}$, may be then estimated as $\dot{\gamma} = \lim_{\Delta t \to 0} \frac{\Delta\theta}{\Delta t} = \lim_{\Delta y \to 0} \frac{\Delta u}{\Delta y} = \frac{du}{dy}$. This estimation is strictly valid only when the fluid has a unidirectional motion (i.e., the fluid has only one velocity component that is directed along the x direction). Thus, for fluid flows with more than one velocity component, the expression of $\dot{\gamma}$ will be more involved (which will be derived in a subsequent chapter).

Notwithstanding the underlying kinematic details, we may say that the shear stress (τ_{xy}, or simply in a generic notation τ) in the fluid will be some function of the rate of shear deformation (we already mentioned that one important characteristic of a fluid is that its shear stress is related to the rate of shear deformation), $\dot{\gamma}$. In general, this relationship (known as constitutive behaviour, since it originates from the constitution or internal molecular structure of the concerned fluid) may be somewhat complicated and involved. However, fortunately, for many common engineering fluids (including air and water), this relationship is linear, i.e.,

$$\tau \, \alpha \, \dot{\gamma}$$

or

$$\tau = \mu\dot{\gamma} \tag{1.15}$$

The proportionality factor μ in the above equation is known as the coefficient of viscosity or simply viscosity, and the above constitutive relationship is known as Newton's law of viscosity. Fluids obeying this law are known as Newtonian fluids. *Physically, the quantity of viscosity is an indicator of the resistance in the fluid against the relative motion between adjacent layers.* Viscosity has a dimension of $[M][L^{-1}][T^{-1}]$ (which is evident by dividing the dimension of shear stress with that of the shear rate or velocity gradient). Its SI unit is kg-m^{-1}s^{-1} or equivalently, Pa-s. In CGS units, it is expressed in the unit of Poise (1 Pa-s = 10 Poise), in the honour of the legendary scientist Poiseuille who had pioneering contributions towards the fundamental understanding of viscous flows.

It is interesting in many cases to compare the viscous forces with other forces acting in the system, by expressing their relative strengths in terms of pertinent dimensionless numbers. One such important and widely used dimensionless number, for example, is the **Reynolds number** (Re), which may be an indicator of the ratio of inertia force to viscous force. For a fluid element with a length scale of l and velocity scale u, the inertia force (mass × acceleration) may be expressed as $\rho l^3 u^2/l$. Viscous force on the same element, estimated through Newton's law of viscosity, may be expressed as (shear stress × area), or in terms of the pertinent scales: $\mu(u/l)l^2$. The ratio of inertia to viscous forces thus scales as $\rho ul/\mu$. Here l is a characteristic length (or length scale) corresponding to the chosen physical problem (it is simply a representative linear dimension of the system over which characteristic changes in properties or velocity occur). As an example, for flow over a flat plate the appropriate length scale may be the axial length of the plate, whereas for flow through a pipe the appropriate length scale may be its diameter. It is also important to note here that the Reynolds number is merely a collection of physical parameters and remains defined as non-zero even when the inertia force acting on the fluid is zero (i.e., the fluid elements are non-accelerating). In that case, the inertia force in the Reynolds number interpretation may be interpreted as an equivalent hypothetical inertia force acting on the fluid element to accelerate the same from zero velocity to the velocity under concern (u).

Closely related to the term viscosity is another property known as **kinematic viscosity** (ν), which is the ratio of viscosity to the density of a fluid ($\nu = \mu/\rho$). Kinematic viscosity has an SI unit of m^2-s^{-1}, and a CGS unit of cm^2-s^{-1}. The CGS unit is also known as Stokes, in honour of the famous mathematician Stokes, who contributed immensely in the early development of viscous flows. The kinematic viscosity is considered as a kinematic quantity, since its unit does not contain any unit of mass. Physically, the kinematic viscosity represents the relative ability of a fluid to diffuse a disturbance in momentum as compared to its ability of sustaining the original momentum. This interpretation stems from the fact that mass is a measure of inertia, which in turn is proportional to the density of the fluid (that appears in the denominator of the expression of kinematic viscosity). In other words, kinematic viscosity may also be qualified as momentum diffusivity. In that sense, the shear stress (which is shear force per unit area) is also considered as momentum flux.

Although the detailed molecular picture of viscosity is somewhat involved, an elementary consideration of the same may be presented in the purview of this fundamental text. Interestingly, the scenario is grossly different for liquids and gases. In liquids, viscosity primarily originates because of intermolecular forces of attraction. Accordingly, if the temperature of a liquid is increased then its viscosity usually decreases. This may be attributed to the fact that a higher temperature implies a more vigorous random motion with respect to their mean position, thereby weakening the effective intermolecular attraction. For gases, the intermolecular force of attraction may not be dominating in most cases (primarily because of a lesser molecular density), and the viscosity originates mainly because of the transfer and exchange of molecular momentum. In order to appreciate the underlying implications from both qualitative and quantitative aspects, let us consider two adjacent layers of ideal gas molecules that are separated by one mean free path (λ) distance. The lower layer moves to the right with an average translational speed of u, whereas the upper layer moves in the same direction with an average translational speed of $u + \Delta u$. When a molecule from the upper layer joins the lower layer by virtue of its thermal energy, it shares some of its excess energy with the upper layer by virtue of a collision with one of the molecules belonging to the upper layer. In other words, the slower moving layer tends to move faster. On the other hand, a molecule joining the upper layer from the lower one tends to slow the upper layer down. This is known as exchange or transfer of molecular momentum, which tends to resist a relative motion between adjacent fluid layers and therefore is physically responsible behind the viscous behaviour of gases. Since the exchange of molecular momentum becomes more vigorous at elevated temperatures (because of higher levels of thermal agitation of the individual molecules), the viscosity of gases commonly increases with increase in temperature. Fig. 1.13 shows the typical variation of viscosity with temperature for some commonly used liquids and gases.

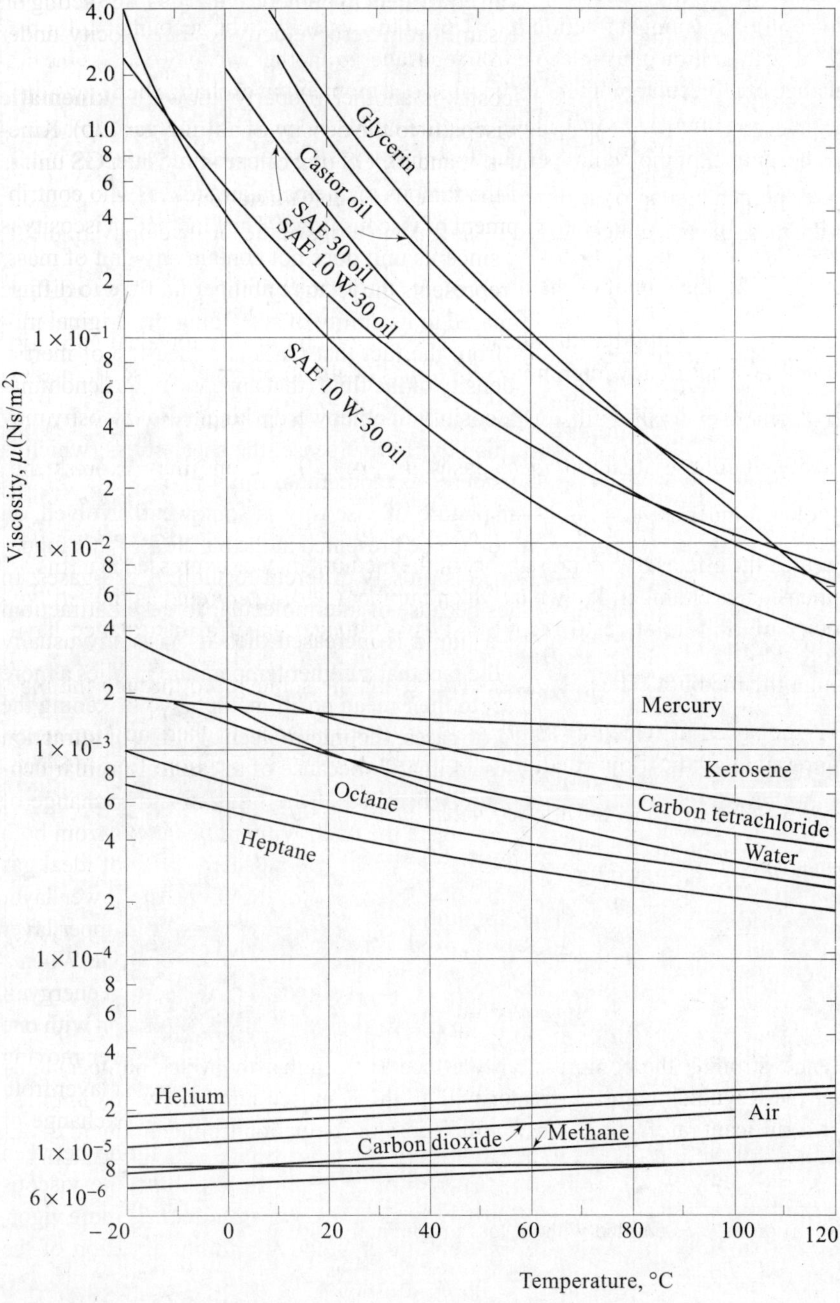

Fig. 1.13 Viscosity of common fluids as a function of temperature

*Consider the above example of layered motion of molecules of an ideal gas. Because of a thermal agitation, molecules in each layer posses a transverse mean velocity as given by $v_m = \sqrt{\frac{8RT}{\pi}}$ in addition to the axial velocity (u), where R is the gas constant (refer to any text book detailing the kinetic theory of gases for its derivation). Assuming a fraction α ($\approx 1/6$) of this velocity to be accountable for the transverse migration of a molecule from one layer to another, we may express the net exchange of molecular momentum because of a momentum exchange between adjacent layers as $nv\Delta t\Delta Am\Delta u$, where n is the number density of the gas molecules (i.e., the number of molecules per unit volume), Δt is the elapsed time interval, ΔA is the area perpendicular to the direction of the transverse migration (y), $v = \alpha v_m$, and m is the mass of each molecule. Expressing Δu in terms of the local velocity gradient as $\Delta u = \frac{\partial u}{\partial y}\lambda$, the rate of exchange of molecular momentum per unit area per unit time (which is nothing but the shear stress or momentum flux for ideal gases, in complete absence of any inter molecular forces of attraction) may be ascertained as $\rho v\lambda \frac{\partial u}{\partial y}$, where $\rho = nm$. Utilising the analogy of this form with Newton's law of viscosity, it follows that for ideal gases $\mu = \rho v\lambda$. Thus, one may express the Reynolds number as, $\mathrm{Re} = \frac{\rho ul}{\mu} = \frac{\rho ul}{\alpha\rho\lambda\sqrt{\gamma RT/\pi}}$.

Further, the effects of compressibility of the medium may be expressed in terms of a dimensionless number, known as Mach number (Ma), interpreted as the ratio of velocity of flow relative to the velocity of propagation of a disturbance have through the medium. Thus, $Ma = \frac{u}{a}$, where a is the sonic speed through the medium; the speed at which a disturbance wave propagates. Higher the Ma, more compressible is the flow medium (for more details, see Section 1.6.5). Since $a = \sqrt{\gamma RT}$ for an ideal gas, it follows that $Ma = \frac{u}{a} = \frac{u}{\sqrt{\gamma RT}}$.

Thus we get,

$$\frac{Ma}{\mathrm{Re}} = \alpha\sqrt{\frac{8}{\pi\gamma}}Kn \tag{1.16}$$

where $$Kn = \frac{\lambda}{l}.$$

Hence, stronger the compressibility effects (i.e., higher the Mach number) and lower the Reynolds number, greater will be the Knudsen number (which implies more significant rarefaction effects with a possible deviation from continuum).

*This portion may be omitted without loss of continuity

The viscosity variation of real gases with temperature may be well described by the Sutherland formula as

$$\mu = \mu_0 \frac{T_0 + C}{T + C}\left(\frac{T}{T_0}\right)^{3/2} \tag{1.17}$$

where μ = viscosity in (Pa-s) at input temperature T, μ_0 = reference viscosity in (Pa-s) at reference temperature T_0, T = input temperature in Kelvin, T_0 = reference temperature in Kelvin, and C = Sutherland's constant for the gaseous material in question. The above formula is valid for temperatures between $0 < T < 555$ K, with an error due to pressure being less than 10% below 3.45 MPa. Sutherland's constant and the reference temperature for some gases are mentioned in Table 1.1.

Table 1.1 Viscosity variation data with temperature for comman gases (see www.complore.com/viscosity-gases)

Gas	*C [K]*	T_0 *[K]*	μ_0 *[10^{-6} Pa s]*
Air	120	291.15	18.27
Nitrogen	111	300.55	17.81
Oxygen	127	292.25	20.18
Carbon dioxide	240	293.15	14.8
Carbon monoxide	118	288.15	17.2
Hydrogen	72	293.85	8.76
Ammonia	370	293.15	9.82
Sulphur dioxide	416	293.65	12.54

Importantly, the law of viscosity variation cannot be used for liquids. For many liquids, the temperature dependence of viscosity can be represented reasonably well by the Arrhenius equation:

$$\mu = A\exp\left(B/T\right) \tag{1.18}$$

where T is the absolute temperature. If the viscosity of a liquid is known at two different temperatures, this information can be used to evaluate the parameters A and B, which then permits the calculation of the viscosity at any other temperature.

1.6.4.3 *Non-Newtonian Behaviour*

An important class of fluids exists that differs from the Newtonian fluids in a sense that the constitutive relationship between the shear stress and the shear deformation rate is more complicated than a simple linear one through the origin, i.e., shear stress is zero when shear rate is zero. Such fluids are called **non-Newtonian** or rheological fluids (rheology is a branch of science that deals with the constitutive behaviour of fluids). It needs to be mentioned here that the knowledge of non-Newtonian fluid mechanics is still in an early stage and many aspects of the same are yet to be fundamentally resolved. Although non-Newtonian fluids do not have the property of viscosity (since viscosity is defined through the Newton's law of viscosity only), their characteristics may be cast in a Newtonian form by introducing an apparent viscosity, which is the ratio of the local shear stress to the shear rate at that point. The apparent viscosity is not a true property for non-Newtonian fluids, since its value depends upon the flow field, or shear rate.

*Purely viscous non-Newtonian fluids are those in which the shear stress is a function only of the shear rate but in a more complicated manner than that described by the Newton's law. Figure 1.14 illustrates the characteristics of purely viscous time-independent fluids. In Fig. 1.14, (a) and (b) are fluids where the shear stress depends only on the shear rate but in a non-linear way. Fluid (a) is called pseudoplastic (or shear thinning); and fluid (b) is called dilatant (or shear thickening). Curve (c) is one which has an initial yield stress after which it acts as a Newtonian fluid, called Bingham plastic; and curve (d) is called Hershel-Buckley, also has a yield stress after which it becomes pseudoplastic. Curve (e) depicts a Newtonian fluid. The reason for which the apparent viscosity of a pseudoplastic fluid decreases with increase in the shear rate may be qualitatively attributed to a breakdown of loosely bonded aggregates by the shearing effect of flow. Examples of such fluid include aqueous or non-aqueous suspensions of polymers, etc. On the other hand, the increment of apparent viscosity with increases in the shear rate for dilatant fluids may be due to the shift of a closely packed particulate system to a more open arrangement under shear, which may entrap some of the liquid. Examples include aqueous suspensions of magnetite, galena, and ferro-silicons. The limiting case of a 'plastic' fluid is one that requires a finite yield stress before beginning to flow. Such fluids are known as Bingham plastic fluids such as sewage sludge, mud, clay, etc. A typical example is toothpaste, which does not flow out of a tube until a finite shear stress is applied by squeezing.

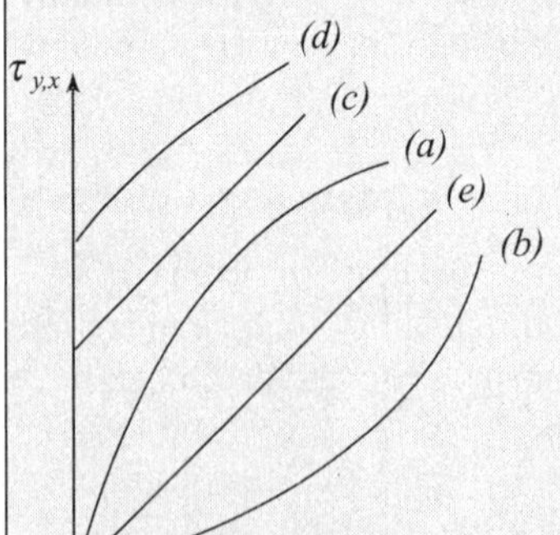

Fig. 1.14 Flow curves for viscous, time-dependent fluids: (a) Pseudoplastic, (b) Dilatant, (c) Bingham plastic, (d) Hershel-Buckley, and (e) Newtonian.

*This portion may be omitted without loss of continuity

Figure 1.15 shows flow curves for two common classes of purely viscous time dependent non-Newtonian fluids. It is seen that such fluids have a hysteresis loop or memory whose shape depends upon the time-dependent rate at which the shear stress is applied. Curve (a) illustrates a pseudoplastic time-dependent fluid and curve (b) a dilatant time-dependent fluid. Examples shown in Fig. 1.15 (a) and 1.15 (b) are called thixotropic and rheopectic fluids, respectively. Apparent viscosity of a thixotropic fluid decreases with time under constant shear. Classical example is water suspension in bentonitic clay, which is a typical drilling fluid used in the petroleum industry. Fluids for which the apparent viscosity increases with increase in time for a given shear rate are called rheopectic fluids. Examples of rheopectic fluids include gypsum pastes and printers inks.

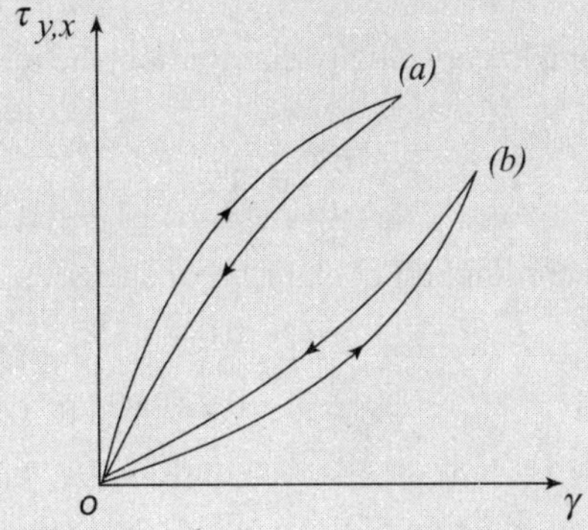

Fig. 1.15 Flow curves for viscous, time-dependent fluids: (a) Thixotropic and (b) Rheopectic

Example 1.1

A fluid has a solute viscosity of 0.048 Pas and a specific gravity of 0.913. For the flow of such a fluid over a flat solid surface, the velocity at a point 75 mm away from the surface is 1.125 m/s. Calculate the shear stresses at the solid boundary, at points 25 mm, 50 mm, and 75 mm away from the boundary surface. Assume (i) a linear velocity distribution and (ii) a parabolic velocity distribution with the vertex at the point 75 mm away from the surface where the velocity is 1.125 m/s.

Solution

Consider a two-dimensional Cartesian coordinate system with the velocity of fluid *V* as abscisa and the normal distance *Y* from the surface as the ordinate with the origin *O* at the solid surface (Fig. 1.16)

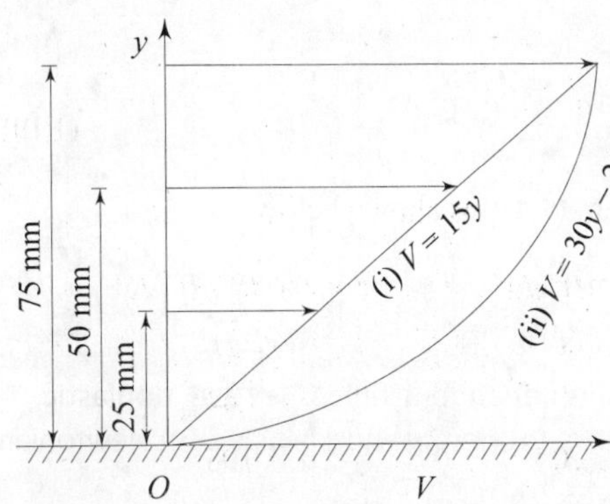

Fig. 1.16 Velocity distribution in the flow of lucid as described in Example 1.1.

According to the no-slip condition at the solid surface.

$$V = 0 \text{ at } y = 0$$

$$\text{Again, } V = 1.125 \text{ m/s at } y = 0.075 \text{ m (given in the problem)}$$

(i) For a linear velocity distribution, the relation between V and y is $V = \frac{1.125}{0.075} y = 15y$

Hence $$\frac{dV}{dy} = 15 \text{ s}^{-1}$$

According to Eq. (1.1), shear stress $\tau = \mu \, dV/dy$

$$= 0.048 \times 15$$

$$= 0.72 \text{ Pa (N/m}^2\text{)}$$

In this case, shear stress is uniform throughout.

(ii) The equation of the parabolic velocity distribution is considered to be given by

$$V = A + By + Cy^2$$

where the constants A, B and C are to be determined from the boundary conditions given in the problem as

$$V = 0 \text{ at } y = 0 \text{ (No-slip at the plate surface)}$$

$$V = 1.125 \text{ at } y = 0.075$$

$$\frac{dV}{dy} = 0 \text{ at } y = 0.075 \text{ (the condition for the vertex of the parabola)}$$

Substitution of the boundary conditions in the expression of velocity profile we get

$$A = 0$$

$$1.125 = 0.075\,B + (0.075)^2\,C$$

$$O = B + 0.15\,C$$

which give $$B = 30, C = -200$$

Therefore the expression of velocity profile becomes

$$V = 30y - 200y^2$$

Hence, $$\frac{dV}{dy} = 30 - 400y \quad (1.19)$$

Tabulation of results with the help of Eq. (1.19) is shown below:

y (m)	V (m/s)	dV/dy (s^{-1})	$\tau = 0.048\ (dV/dy)$ (Pa)
0	0	30	1.44
0.025	0.625	20	0.96
0.050	0.880	10	0.48
0.075	1.125	0	0

It is observed that the shear stress decreases as the velocity gradient decreases with the distance y from the plate and becomes zero where the velocity gradient is zero.

Example 1.2

A cylinder of 0.12 m radius rotates concentrically inside a fixed hollow cylinder of 0.13 m radius. Both the cylinders are 0.3 m long. Determine the viscosity of the liquid which fills the space between the cylinders if a torque of 0.88 Nm is required to maintain an angular velocity of 2π rad/s.

Solution

The torque applied = The resisting torque by the fluid

= (Shear stress) × (Surface area) × (Torque arm)

Hence, at any radial location r from the axis of rotation.

$$0.88 = \tau(2\pi r \times 0.3)r$$

$$\tau = \frac{0.467}{r^2}$$

Now, according to Eq. (1.1),

$$\frac{dV}{dy} = \frac{\tau}{\mu} = \frac{0.467}{\mu r^2}$$

Rearranging the above expression and substituting – dr for dy (the minus sign indicates that r, the radial distance, decreases as V increases), we obtain

$$\int_{V_{outer}}^{V_{inner}} dV = \frac{0.467}{\mu}\int_{0.13}^{0.12} -\frac{dr}{r^2}$$

Hence $$V_{inner} - V_{outer} = \frac{0.467}{\mu}\left[\frac{1}{r}\right]_{0.13}^{0.12}$$

The velocity of the inner cylinder,

$$V_{inner} = 2\pi \times 0.12 = 0.754 \text{ m/s}$$

Hence, $$(0.754 - 0) = \frac{0.467}{\mu}\left(\frac{1}{0.12} - \frac{1}{0.13}\right)$$

From which $$\mu = 0.397 \text{ Pa s}$$

Example 1.3

The velocity profile in laminar flow through a round pipe is expressed as

$$u = 2U(1 - r^2/r_0^2)$$

where U is the average velocity, r is the radial distance from the centre line of the pipe, and r_0 is the pipe radius. Draw the dimensionless shear stress profile τ/τ_0 against

r/r_0, where τ_0 is the wall shear stress. Find the value of τ_0, when fuel oil having an absolute viscosity $\mu = 4 \times 10^{-2}$ N^{-s}/m^2 flows with an average velocity of 4 m/s in a pipe of diameter 150 mm.

Solution

The given velocity profile is

$$u = 2U(1 - r^2/r^2{}_0)$$

$$\frac{du}{dr} = -\frac{4Ur}{r_0^2}$$

Shear stress at any radial location r can be written as

$$\tau = -\mu \frac{du}{dr}$$

Hence,
$$\frac{\tau}{\tau_0} = \frac{du/dr}{(du/dr)_{r=r_0}} = \frac{r}{r_0}$$

Figure 1.17 shows the shear stress distribution

Wall shear stress $\tau_0 = -\mu \left(\frac{du}{dr}\right)_{r=r_0}$

$$= \mu \left(\frac{4U}{r_0}\right)$$

$$= (4 \times 10^{-2}) \left(\frac{4 \times 4}{0.075}\right)$$

$$= 8.533 \text{ N/m}^2$$

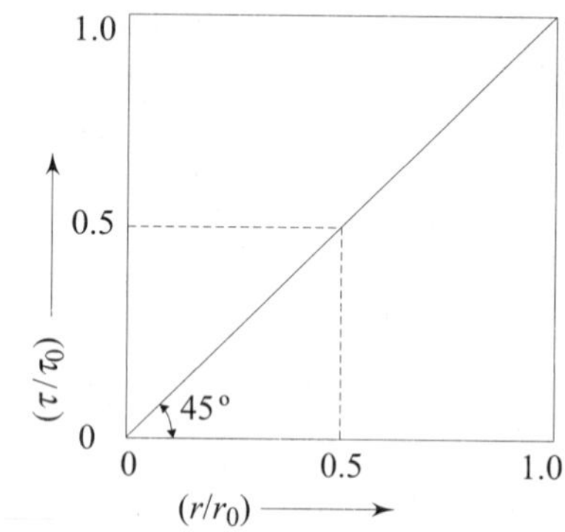

Fig. 1.17 Shear stress distribution in the pipe flow problem of Example 1.3

The shear stress distibution in the pipe flow problem is depicted in Fig. 1.17.

1.6.5 Compressibility

Compressibility of any substance is the measure of its change in volume under the action of external forces, namely, the normal compressive forces. The normal compressive stress of any fluid element at rest is known as hydrostatic pressure p and arises as a result of innumerable molecular collisions in the entire fluid. The degree of compressibility of a substance is characterised by the bulk modulus of elasticity E defined as

$$E = \lim_{\Delta V \to \Delta V'} \frac{-\Delta p}{\Delta V / V} \tag{1.20}$$

where ΔV and Δp are the changes in the volume and pressure respectively, V is the initial volume, and $\Delta V'$ is the smallest possible volume over which continuum hypothesis remains valid. The negative sign in Eq. (1.20) indicates that an increase in pressure is associated with a decrease in volume. For a given mass of a substance, the change in its volume and density satisfies the relation

$$\frac{\Delta \mathcal{V}}{\mathcal{V}} = -\frac{\Delta \rho}{\rho} \tag{1.21}$$

With the help of Eq. (1.20), E can be expressed in differential form as

$$E = \rho \frac{dp}{d\rho} \tag{1.22}$$

Values of E for liquids are very high as compared with those of gases (except at very high pressures). Therefore, liquids are usually termed as incompressible fluids though, in fact, no substance is theoretically incompressible with a value of E as ∞. For example, the bulk modulus of elasticity for water and air at atmospheric pressure are approximately 2×10^6 kN/m^2 and 101 kN/m^2 respectively. It indicates that air is about 20,000 times more compressible than liquid water. Hence liquid water can be treated as incompressible. Another characteristic parameter, known as compressibility K, is usually defined for gases. It is the reciprocal of E as

$$K = \frac{1}{\rho}\frac{d\rho}{dp} = -\frac{1}{\mathcal{V}}\left(\frac{d\mathcal{V}}{dp}\right) \tag{1.23}$$

K is often expressed in terms of specific volume $\mathcal{V}$. For any gaseous substance, a change in pressure is generally associated with a change in volume and a change in temperature simultaneously. A functional relationship between the pressure, volume and temperature at any equilibrium state is known as thermodynamic equation of state for the gas. For an ideal gas, the thermodynamic equation of state is given by

$$p = \rho R T \tag{1.24}$$

where T is the temperature in absolute thermodynamic or gas temperature scale (which are, in fact, identical), and R is known as the characteristic gas constant, the value of which depends upon a particular gas. However, this equation is also valid for real gases which are thermodynamically far from their liquid phase. For air, the value of R is 287 J/kg K. The relationship between the pressure p and the volume $\mathcal{V}$ for any process undergone by a gas depends upon the nature of the process. A general relationship is usually expressed in the form of

$$p\mathcal{V}^x = \text{constant} \tag{1.25}$$

For a constant temperature (isothermal) process of an ideal gas, $x = 1$. If there is no heat transfer to or from the gas, the process is known as *adiabatic*. A frictionless adiabatic process is called an *isentropic* process and x equals to the ratio of specific heat at constant pressure to that at constant volume. The Eq. (1.25) can be written in a differential form as

$$\frac{d\mathcal{V}}{dp} = -\frac{\mathcal{V}}{xp} \tag{1.26}$$

Using the relation (1.26), Eqs (1.22) and (1.23) yield

$$E = xp \tag{1.27a}$$

or $$K = \frac{1}{xp} \tag{1.27b}$$

Therefore, the compressibility K, or bulk modulus of elasticity E for gases depends on the nature of the process through which the pressure and volume change. For an isothermal process of an ideal gas ($x = 1$), $E = p$ or $K = 1/p$. The value of E for air quoted earlier is the isothermal bulk modulus of elasticity at normal atmospheric pressure and hence the value equals to the normal atmospheric pressure.

1.6.6 Distinction between an Incompressible and a Compressible Fluid

In order to know whether it is necessary to take into account the compressibility of gases in fluid flow problems, we have to consider whether the change in pressure brought about by the fluid motion causes large change in volume or density.

From Bernoulli's equation (to be discussed in a subsequent chapter), $p + \frac{1}{2}\rho V^2 =$ constant (V being the velocity of flow), and therefore the change in pressure, Δp, in a flow field, is of the order of $\frac{1}{2}\rho V^2$ (dynamic head). Invoking this relationship into Eq. (1.22) we get,

$$\frac{\Delta\rho}{\rho} \approx \frac{1}{2}\frac{\rho V^2}{E} \tag{1.28}$$

Now, we can say that if $(\Delta\rho/\rho)$ is very small, the flow of gases can be treated as incompressible with a good degree of approximation. According to Laplace's equation, the velocity of sound is given by $a = \sqrt{E/\rho}$. Hence,

$$\frac{\Delta\rho}{\rho} \approx \frac{1}{2}\frac{V^2}{a^2} \approx \frac{1}{2} Ma^2 \tag{1.29}$$

where *Ma* is the ratio of the velocity of flow to the acoustic velocity in the flowing medium at the condition and is known as *Mach number*.

From the aforesaid argument, it is concluded that the compressibility of gas in a flow can be neglected if $\Delta\rho/\rho$ is considerably less than unity, i.e., $\frac{1}{2} Ma^2 = 1$. In other words, if the flow velocity is small as compared to the local acoustic velocity, compressibility of gases can be neglected. Considering a maximum relative change in density of five per cent as the criterion of an incompressible flow, the upper limit of Mach number becomes approximately 0.33. In case of air at standard pressure and temperature, the acoustic velocity is about 335.28 m/s. Hence a Mach number of 0.33 corresponds to a velocity of about 110 m/s. Therefore flow of air up to a velocity of 110 m/s under standard condition can be considered as incompressible flow.

Example 1.4

If we neglect the temperature effect, an empirical pressure-density relation for water is $p/p_a = 3001 \times (\rho/\rho_a)^7 - 3000$, where subscript a refers to atmospheric conditions. Determine the isothermal bulk modulus of elasticity and compressibility of water at 1, 10 and 100 atmospheric pressure.

Solution

Pressure-density relationship is given as

$$\frac{p}{p_a} = 3001 \times \left(\frac{\rho}{\rho_a}\right)^7 - 3000 \tag{1.30}$$

$$\frac{1}{p_a}\frac{dp}{d\rho} = 7 \times 3001 \frac{\rho^6}{\rho_a^7}$$

Hence

$$\frac{dp}{d\rho} = 7 \times 3001\, p_a \frac{\rho^6}{\rho_a^7}$$

$$\rho\frac{dp}{d\rho} = 7 \times 3001\, p_a \left(\frac{\rho}{\rho_a}\right)^7$$

According to Eq. (1.22),

$$E = \rho\frac{dp}{d\rho} = 7 \times 3001\, p_a \left(\frac{\rho}{\rho_a}\right)^7 \tag{1.31}$$

Substituting the value of $\left(\frac{\rho}{\rho_a}\right)^7$ from Eq. (1.30) to Eq. (1.31), we get

$$E = \frac{7 \times 3001}{3001} \cdot p_a \left[\frac{p}{p_a} + 3000\right]$$

$$= 7p_a \left[\frac{p}{p_a} + 3000\right]$$

Therefore,

$$(E)_{1\text{ atm pressure}} = 7 \times 3001\, p_a$$
$$= 2.128 \times 10^6\ \text{kN/m}^2$$

(The atmospheric pressure p_a is taken as that at the sea level and equals to 1.0132×10^5 N/m^2)

$$(E)_{10\text{ atm pressure}} = 7 \times 3010\, p_a$$
$$= 2.135 \times 10^6\ \text{kN/m}^2$$

$$(E)_{100\text{ atm pressure}} = 7 \times 3100\, p_a$$

$$= 2.198 \times 10^6 \text{ kN/m}^2$$

Respective compressibilities are

$$(K)_{1 \text{ atm pressure}} = \frac{1}{(E)_{1 \text{ atm pressure}}} = 0.47 \times 10^{-6} \text{ m}^3\text{/kN}$$

$$(K)_{10 \text{ atm pressure}} = \frac{1}{(E)_{10 \text{ atm pressure}}} = 0.468 \times 10^{-6} \text{ m}^2\text{/kN}$$

$$(K)_{100 \text{ atm pressure}} = \frac{1}{(E)_{100 \text{ atm pressure}}} = 0.455 \times 10^{-6} \text{ m}^3\text{/kN}$$

It is found from the above example that the bulk modulus of elasticity or compressibility of water is almost independent of pressure.

Example 1.5

A cylinder contains 0.35 m^3 of air at 50 °C and 276 kN/m^2 absolute. The air is compressed to 0.071 m^3. (a) Assuming isothermal conditions, what is the pressure at the new volume and what is the isothermal bulk modulus of elasticity at the new state. (b) Assuming isentropic conditions, what is the pressure and what is the isentropic bulk modulus of elasticity? (Take the ratio of specific heats of air $\gamma = 1.4$)

Solution

(a) For isothermal conditions,

$$p_1 V_1 = p_2 V_2$$

Then, $(2.76 \times 10^5)0.35 = (p_2).071$

which gives,

$$p_2 = 13.6 \times 10^5 \text{ N/m}^2$$
$$= 1.36 \text{ MN/m}^2$$

The isothermal bulk modulus of elasticity at any state of an ideal gas equals to its pressure at that state. Hence $E = p_2 = 1.36$ MN/m^2.

(b) For isentropic conditions

$$P_1 V_1^{1.4} = P_2 V_2^{1.4}$$

Then, $(2.76 \times 10^5)(0.35)^{1.4} = (p_2)(.071)^{1.4}$

From which,

$$p_2 = 25.8 \times 10^5 \text{ N/m}^2$$
$$= 2.58 \text{ MN/m}^2$$

The isentropic bulk modulus of elasticity

$$E = \gamma p = 1.40 \times 25.8 \times 10^5 \text{ N/m}^2$$
$$= 3.61 \text{ MN/m}^2$$

1.6.7 Interfacial Phenomenon and Surface Tension

1.6.7.1 *The Molecular Origin*

Consider an interface between two fluid phases (say, phase 1 is liquid and phase 2 is gas, for example). Molecules located in the bulk of the liquid have identical interactions with all the neighbouring molecules (mostly through van der Waals attractive interactions for organic liquids and through hydrogen bonds for polar liquids like water). However, molecules on the interface have liquid molecules on one side and gas molecules on the other. Because of the distinctive compactness of molecules in these two phases, the interfacial molecules may have a net attractive 'pull' towards the liquid. If the interfacial molecules do not possess sufficient energy to overcome this net attraction and remain at the interface, they would ultimately get dissolved in the liquid and get lost! However, when an interface is formed, this does not occur. This implies that the interface has sufficient energy to overcome any net driving force on its molecules and allow them to be located on the interface. This energy is qualitatively known as interfacial energy or surface energy.

The interfacial phenomenon elucidated above is quite generic in nature, and takes place for an interface formed by any two distinctive phases (such as between a solid and a liquid phase, or even between two liquid phases such as oil and water). For example, let us consider a solid-liquid interface. Molecules in the liquid (say, water) are attracted towards the interface by van der Waals' forces. However, these molecules usually do not stick to the wall because of Brownian motion. At the same time, impurities contained in the liquid, such as dust particles or biological polymers (such as proteins) may adhere permanently to the solid surface since they may experience stronger attractions to the solid. This can be attributed to the fact that the size of polymer molecules is typically much larger than that of the water molecules, giving rise to more number of contacts for the polymer with the solid and hence a strengthened van der Waals interaction.

The description of the physics in the smaller scale may get too involved, and one may be interested to represent a gross manifestation of the underlying interactions rather than presenting a detailed molecular picture. This gross representation is achieved by introducing a macroscopic physical property known as surface tension (σ), which may also be interpreted as the interfacial energy per unit area (which is equivalently force per unit length). Thus, the SI unit of surface tension is J-m^{-2} or N-m^{-1}. Typical surface tension values of some common fluids at standard temperature are given in Table 1.2. The force-interpretation of surface tension may be visualised by a very simple experiment. Take a solid frame and a solid tube that can roll on this frame. If we form a liquid film of soap between the frame and the tube by plunging one side of the structure in a soap-water solution, we see that the tube immediately starts to move towards the region where there is the liquid film. This is because of the fact that the surface tension of the liquid film exerts a net force on its free boundary. Thus, additional work needs to be done to overcome this force and stretch the film to a desired extent.

Table 1.2. Surface tension data for some representative fluids (see http://web.mit.edu/hml/ncfmf/04STFM.pdf)

Fluid combination	*Surface tension (in mN-m^{-1})*
Water/air	73
Salt water/oil	75
Ether/air	17
Alcohol/air	23
Carbon tetrachloride/air	27
Mercury/air	480

It is due to surface tension that a curved liquid interface in equilibrium results in a greater pressure at the concave side of the surface than that at its convex side.

Consider an elemental curved liquid surface (Fig. 1.18) separating the bulk of liquid in its concave side and a gaseous substance or another immiscible liquid on the convex side. The surface is assumed to be curved on both the sides with radii of curvature as r_1 and r_2 and with the length of the surfaces subtending angles of $d\theta_1$ and $d\theta_2$, respectively, at the centre of curvature as shown in Fig. 1.18. Let the surface be subjected to the uniform pressure p_i and p_o at its concave and convex sides respectively acting perpendicular to the elemental surface. The surface tension forces across the boundary lines of the surface appear to be the external forces acting on the surface. Considering the equilibrium of this small elemental surface, a force balance in the direction perpendicular to the surface results.

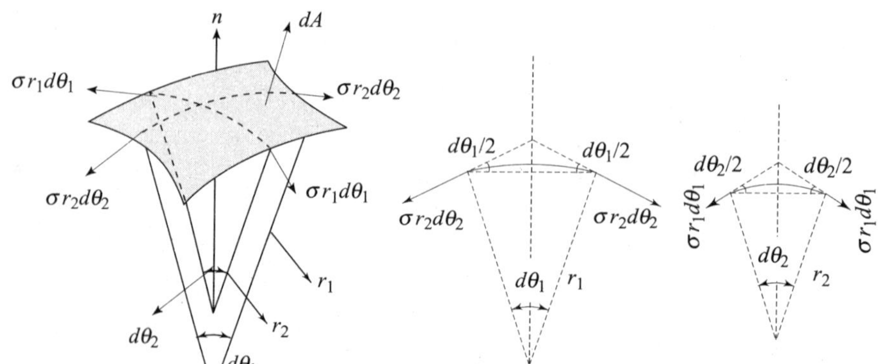

Fig. 1.18 State of stress and force balance on a curved liquid interface in equilibrium with surrounding due to surface tension

$$2\,\sigma r_2\, d\theta_2 \sin\left(\frac{d\theta_1}{2}\right) + 2\,\sigma r_1\, d\theta_1 \sin\left(\frac{d\theta_2}{2}\right) = (p_i - p_o)\, r_1 r_2\, d\theta_1\, d\theta_2$$

For small angles

$$\sin\left(\frac{d\theta_1}{2}\right) \approx \frac{d\theta_1}{2}, \; \sin\left(\frac{d\theta_2}{2}\right) \approx \frac{d\theta_2}{2}$$

Hence, from the above equation of force balance we can write

$$\frac{\sigma}{r_1} + \frac{\sigma}{r_2} = (p_i - p_o)$$

or
$$\Delta p = \frac{\sigma}{r_1} + \frac{\sigma}{r_2} \tag{1.32}$$

where
$$\Delta p = p_i - p_o$$

and σ is the surface tension of the liquid in contact with the specified fluid at its convex side. Equation (1.32) is also known as Young–Laplace equation. If the liquid surface coexists with another immiscible fluid, usually gas, on both the sides, the surface tension force appears on both the concave and convex interfaces and the net surface tension force on the surface will be twice as that described by Eq. (1.32). Hence the equation for pressure difference in this case becomes

$$\Delta p = 2\left(\frac{\sigma}{r_1} + \frac{\sigma}{r_2}\right) \tag{1.33}$$

Special Cases For a spherical liquid drop, Eq. (1.32) is applicable with $r_1 = r_2 = r$ (the radius of the drop) to determine the difference between the pressure inside and outside the drop as

$$\Delta p = 2\sigma/r \tag{1.34}$$

The excess pressure in a cylindrical liquid jet over the pressure of the surrounding atmosphere can be found from Eq. (1.32) with $r_1 \Rightarrow \alpha$ and $r_2 = r$ (the radius of the jet) as

$$\Delta p = \sigma/r \tag{1.35}$$

In case of the spherical bubble, the Eq. (1.33) is applicable with $r_1 = r_2 = r$ (radius of the bubble), which gives

$$\Delta p = 4\sigma/r \tag{1.36}$$

Alternatively, Eq. (1.32) may also be derived from work energy principle, as follows.

Consider a small section of an arbitrarily curved surface, as shown in Fig. 1.19.

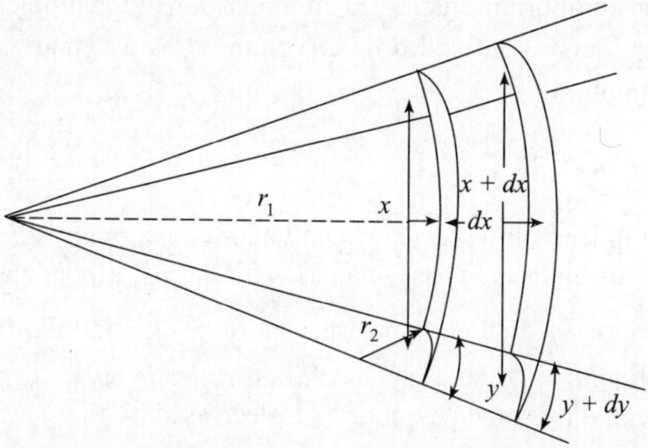

Fig.1.19 Geometry of a stretching surface element

The section taken is small enough so that R_1 and R_2 are approximately constants within the chosen section. If the elemental surface is displaced by a small amount outwards, the net change in its area will be $\Delta A = (x+dx)(y+dy) - xy = xdy + ydx$. The work done in forming this additional surface is $\Delta W = \sigma_{lv}(xdy + ydx)$, where σ_{lv} represents the liquid vapour surface tension coefficient. If the interfacial pressure difference is Δp, it acts on an area *dxdy* and is associated with a displacement of *dz*. The corresponding work is equivalent to $\Delta W = \Delta p dxdydz$. Equating the two expressions for work, it follows that

$$\Delta p dxdydz = \sigma_{lv}(xdy + ydx) \tag{1.37}$$

Further, from the geometry of the figure (similar triangles), it follows that $\frac{x+dx}{r_1+dz} = \frac{x}{r_1}$, or equivalently $dx = \frac{xdz}{r_1}$. Similarly, $dy = \frac{ydz}{r_2}$. Utilising these in Eq. (1.37), we get an expression identical to Eq. (1.32).

Interfacial equilibrium accounted by surface tension may also be analysed from energy minimisation principle as applied to an interface of any arbitrary shape. For simplicity in illustration, we consider a droplet of the shape of a part of sphere, in equilibrium, as shown in Fig. 1.20. The droplet tends to minimise its net surface

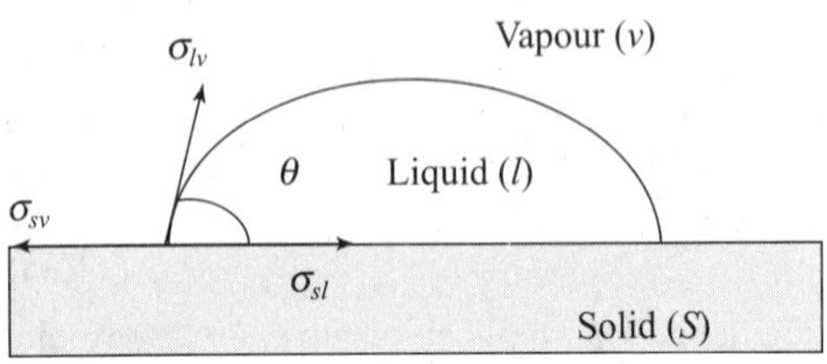

Fig.1.20 Equilibrium of a three-phase contact line

energy in an effort to come to an equilibrium shape. The minimisation of the droplet surface energy under the constraints of a fixed droplet volume $\forall$ is as good as minimising the following function:

$$E = \sum_{i \neq j} A_{ij}\sigma_{ij} - \lambda\forall \tag{1.38}$$

where and λ is a Lagrange multiplier to enforce a constant volume constraint. Here A_{ij} is the interfacial area that demarcates the phases *i* and *j*, with the corresponding surface energy being designated as σ_{ij}. In this chapter, the subscripts *l*, *s* and *v* will be employed to represent the liquid, solid and vapour phases, respectively. It can also be noted that if A_{ls} is increased by some amount, A_{sv} is decreased by the same amount. From the geometry of the above figure, one may write: $A_{sl} = \pi R^2 \sin^2\theta$ $A_{lv} = 2\pi R^2(1-\cos\theta)$, $\forall = \pi R^3\left(\frac{2}{3}-\frac{3}{4}\cos\theta+\frac{\cos 3\theta}{12}\right)$.

Thus, for a spherical droplet

$$E = \underbrace{\pi R^2 \sin^2\theta(\sigma_{ls}-\sigma_{sv})+\sigma_{lv}2\pi R^2(1-\cos\theta)}_{f} - \lambda\left\{\underbrace{\pi R^3\left(\frac{2}{3}-\frac{3}{4}\cos\theta+\frac{\cos 3\theta}{12}\right)}_{g}\right\} \tag{1.39}$$

where k is the volume of the droplet. For minimisation of E, one must have $\frac{\partial E}{\partial \theta}=0$ and $\frac{\partial E}{\partial R}=0$, which implies

$$\frac{\partial f}{\partial \theta}\frac{\partial g}{\partial R} = \frac{\partial f}{\partial R}\frac{\partial g}{\partial \theta} \tag{1.40}$$

Performing the necessary algebra and simplifying, it follows from Eqs. (1.39) and (1.40) that

$$\cos\theta = \frac{\sigma_{sv}-\sigma_{sl}}{\sigma_{lv}} \tag{1.41}$$

When $0 \le \theta \le 90°$, the liquid is termed as partially wetting, or equivalently, the solid substrate is characterised as hydrophilic (the word 'hydro' is somewhat specific to water, whereas the terminology 'wetting' is more applicable for any general liquid). When $\theta > 90°$, the liquid is non-wetting and the substrate is termed as hydrophobic. The case $\theta = 0$, a very special one, represents a theoretically complete wetting. Eq. (1.41), also known as 'Young's Law', and can be interpreted as a balance between the horizontal components of all the forces that act on the three-phase contact line (refer to Fig. 1.20). The vertical component of this resultant force, on the other hand, is balanced by the normal stress in the rigid solid substrate. Further, substituting the value of $\cos\theta$ from Eq. (1.41), one may calculate a value of the parameter λ, as

$$\lambda = \frac{\partial f / \partial \theta}{\partial g / \partial \theta} = \frac{2\sigma_{lv}}{R} \tag{1.42}$$

Physically, the Lagrange multiplier (λ) may be interpreted as the differential pressure, Δp, across the two sides (liquid and vapour sides in this specific example) of the droplet.

It is important to note here that Equations (1.32) and (1.41) are the two necessary conditions for equilibrium but not sufficient, since in addition, the second variation of E must also be positive for a minimisation of the same. In fact, in the presence of complex surfaces, certain morphologies may, indeed, result in unstable droplets, even though the necessary conditions of equilibrium are satisfied.

*The celebrated Young's equation, as described by Eq. (1.32), is somewhat restricted in nature in a strict sense, since it neglects the bulk internal forces within the droplet that might generate due to gravity, electric field, etc. In an effort to generalize the underlying mathematical description, one may note that, for equilibrium of a spherical droplet

$$dE(R,\theta) = 0 \tag{1.43}$$

which implies

$$\frac{\partial E}{\partial R} dR + \frac{\partial E}{\partial \theta} d\theta = 0 \tag{1.44}$$

Further, since $\forall = \pi R^3 \left(\frac{2}{3} - \frac{3}{4}\cos\theta + \frac{\cos 3\theta}{12} \right)$, the condition of $d\forall = 0$ implies that

$$dR - Rf_1(\theta)d\theta = 0 \tag{1.45}$$

where

$$f_1(\theta) = \frac{-2\cos^2\dfrac{\theta}{2}\cot\dfrac{\theta}{2}}{2 + \cot\theta} \tag{1.46}$$

Substituting Eq. (1.46) in Eq. (1.44), one finally gets

$$\frac{\partial E}{\partial R} Rf_1(\theta) + \frac{\partial E}{\partial \theta} = 0 \tag{1.47}$$

Incorporating the effects of gravitational potential energy in the expression E, for illustration, one can write

$$E = \underbrace{R^2\left[2\pi\sigma_{lv}(1-\cos\theta) + \pi\sin^2\theta(\sigma_{ls} - \sigma_{sv})\right]}_{\text{contribution from interfacial terms}} + \underbrace{R^4 \frac{2\pi}{3}\rho g(3+\cos\theta)\sin^6\left(\frac{\theta}{2}\right)}_{\text{contribution from gravity}} \tag{1.48}$$

Utilising the above expression for E in Eq. (1.47), one can arrive at a more generalized form of Young's equation incorporating gravity effects as

$$\cos\theta - \frac{\sigma_{sv} - \sigma_{sl}}{\sigma_{lv}} - \frac{\rho g R^2}{\sigma_{lv}} \left[\frac{\cos\dfrac{\theta}{3} - \cos 2\theta}{12} - \frac{1}{4} \right] = 0 \tag{1.49}$$

From Eq. (1.49), following two important observations may be carefully noted: (i) With gravity effects, θ depends on R, unlike the case without gravity.

*This portion may be omitted without loss of continuity

(ii) If $\frac{\rho g R^2}{\sigma_{lv}}$ is small, gravity effects can safely be neglected. This ratio is known as the Bond number (Bo), which essentially compares the square of the system length scale (R) with the square of a characteristic length scale, l_s, that depicts the relative contributions of surface tension and gravity influences, as $l_s = \frac{\sigma_{lv}}{\rho g}$. For $Bo \leq 10^{-3}$, gravity effects can safely be neglected, as appropriate to many of problems involving tiny droplets.

1.6.7.2 *Capillary Rise*

As a simple demonstration of Young's equation, we may take up the example of **capillary rise** or **capillary depression** phenomenon. Consider a narrow glass tube dipped in water. It may be observed that water spontaneously rises to a new level (see Fig. 1.21) in the capillary tube. This phenomenon is called as capillary rise. Depending on the fluid-solid combination, it could also be capillary depression (instead of rise). as guided by the contact angle.

The capillary rise phenomenon may be mathematically analysed following a simplistic approach, by assuming the meniscus as hemispherical (which is approximately the case if the capillary tube is very narrow). The pressure differential across the meniscus (see Fig. 1.21) can then be ascertained by the Young–Laplace equation with $R_1 = R_2 = R$, as $\Delta p = \frac{2\sigma}{R}$ (where $\sigma = \sigma_{lv}$). Also, if h is the height of the capillary rise, then the pressure difference between the inner and outer (atmospheric) side of the meniscus is given by $\Delta p = h\rho g$. Equating these two and noting that $R = r/\cos\theta$ (see Fig. 1.21), where r is the radius of the capillary, it follows that

$$h = \frac{2\sigma\cos\theta}{\rho g r}. \tag{1.50}$$

Since $h \,\alpha\, 1/r$, an ultra-narrow radius may ensure a large capillary rise against the gravity, despite the fact that no external pumping effort is deployed. This remarkable phenomenon is one of the key affects observed in fluid mechanics of small-scale systems, allowing a very efficient capillary transport with the aid of surface tension forces alone. This kind of physical principle also plays a key role in transmitting water from the ground level to the upper extremities of trees through microcapillaries, a phenomenon known as ascent of sap.

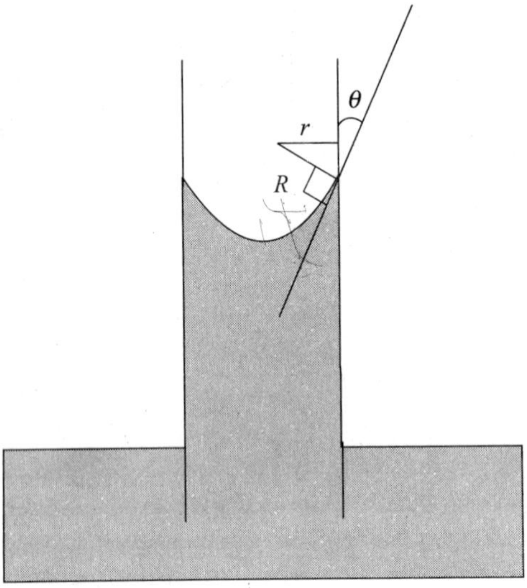

Fig.1.21 Demonstration of the capillary rise phenomenon

Marangoni Effect Surface tension is not a constant in general, as it depends on the temperature and/or concentration of chemical species at the surface. Temperature dependence of surface tension influences many practical processes, including the fusion welding of materials. Typically, many metals have an increasing surface tension with a decreasing temperature. Thus, if a material is bombarded with a high energy beam, the molten region that is created surrounding the central point of heating is subjected to distinctive surface tension forces at different locations. To understand the underlying physical consequences, we consider two different fluid elements in the top layer of the molten pool. The fluid element with a higher temperature (close to the heat source) has less surface tension than that with a lower temperature (away from the heat source). In an effort to minimise the surface energy, the fluid element with lower surface tension stretches itself towards the fluid element with a higher surface tension, thereby inducing a radially outward fluid flow (see the schematic depicted in Fig. 1.22). This is an example of temperature-gradient induced surface tension driven flow, or Marangoni flow.

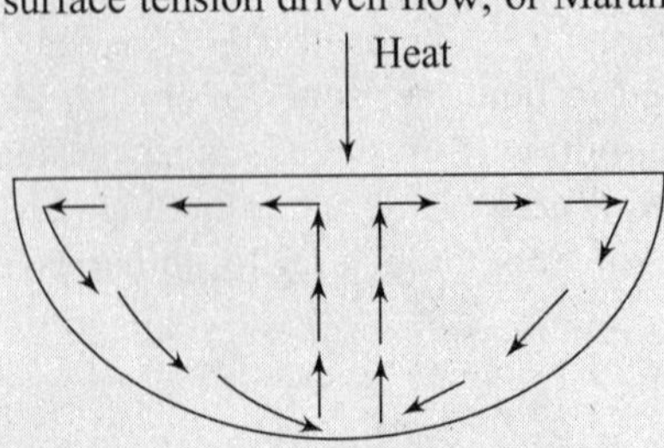

Fig. 1.22 Schematic of Marangoni flow in a weldpool

*This portion may be omitted without loss of continuity

Surface tension may also be strongly influenced by the solutal concentration at an interface. Surfactants, also known as tensides, are wetting agents that lower the surface tension of a liquid, allowing easier spreading, and lower the interfacial tension between two liquids. The term surfactant is a blend of 'surface acting agent'. Surfactants are usually organic compounds that are amphipathic, meaning they contain both hydrophobic groups (their 'tails') and hydrophilic groups (their 'heads'). Therefore, they are soluble in both organic solvents and water. The most common biological example of surfactant is that coating the surfaces of the alveoli, the small air sacs of the lungs that serve as the site of gas exchange. Surfactants reduce the surface tension of water by absorbing at the liquid-gas interface. They also reduce the interfacial tension between oil and water by absorbing at the liquid-liquid interface. Many surfactants can also assemble in the bulk solution into aggregates. Some of these aggregates are known as micelles. Thermodynamics of the surfactant systems are of great importance, theoretically and practically. This is because surfactant systems represent systems between ordered and disordered states of matter. Surfactant solutions may contain an ordered phase (micelles) and a disordered phase (free surfactant molecules and/or ions in the solution).

Example 1.6

Analyse the shape of the water-air interface near a plane wall, as shown in Fig. 1.23, assuming that the slope is small, $1/R \approx d^2\eta/\mathrm{d}x^2$ (where R is the radius of curvature of the interface) and the pressure difference across the interface is balanced by the product of specific weight and interface height as $\Delta p = \rho g h$. Boundary conditions: area wetting contact angle $\theta = \theta_0$ at $x = 0$, and $\theta = 90°$ as $x \to \infty$. What is the height h at the wall?

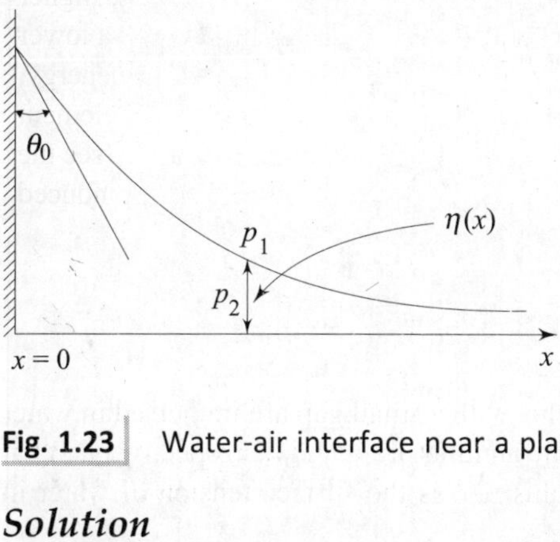

Fig. 1.23 Water-air interface near a plane wall

Solution

The curved interface is plane in the other direction. Hence the pressure difference across the interface can be written according to Eq. (1.32) as

$$\Delta p = p_1 - p_2 = \sigma\left(\frac{1}{R}\right) \tag{1.51}$$

From the given data

$$\frac{1}{R} = \frac{d^2\eta}{dx^2}$$

and $$\Delta p = \rho g \eta$$

Substituting the values of $1/R$ and Δp in Eq. (1.51), we get

$$\frac{d^2\eta}{dx^2} - \frac{\rho g}{\sigma}\eta = 0 \tag{1.52}$$

The solution of η from the above Eq. (1.52) is,

$$\eta = \text{A}e^{-\sqrt{\frac{\rho g}{\sigma}}x} + \text{B}e^{\sqrt{\frac{\rho g}{\sigma}}x} \tag{1.53}$$

where A and B are parametric constants. The value of A and B are found out using he boundary conditions as follows:

$$\text{at } x = 0, \frac{d\eta}{dx} = -\cot\theta_0$$

and $$\text{at } x \to \infty \quad \frac{d\eta}{dx} = 0$$

which give,

$$\text{A} = \sqrt{\frac{\sigma}{\rho g}}\cot\theta_0$$

$$\text{B} = 0$$

Hence Eq. (1.53) becomes

$$\eta = \sqrt{\frac{\sigma}{\rho g}}\cot\theta_0\, e^{-\sqrt{\frac{\sigma}{\rho g}}x}$$

which defines the shape of the interface

$$(\eta)_{x=0} = \sqrt{\frac{\sigma}{\rho g}}\cot\theta_0$$

Example 1.7

Two coaxial glass tubes forming an annulus with a small gap are immersed in water in a trough. The inner and outer radii of the annulus are r_i and r_0, respectively. What is the capillary rise of water in the annulus if σ is the surface tension of water in contact with air?

Solution

The area wetting contact angle for air-water interface in a glass tube is 0° (Fig. 1.24). Therefore, equating the weight of water column in the annulus with the total surface tension force, we get,

$$W = T_i + T_o \quad (1.54)$$

Again, $T_i = \sigma(2\pi r_i)$

$T_o = \sigma(2\pi r_o)$

and $W = \pi(r_o^2 - r_i^2)h\rho g$

Substitution of these values of T_i, T_o and W in Eq. (1.54) gives,

$$\pi(r_o^2 - r_i^2)h\rho g = 2\pi\sigma(r_o + r_i)$$

from which

$$h = \frac{2\sigma}{\rho g(r_o - r_i)}$$

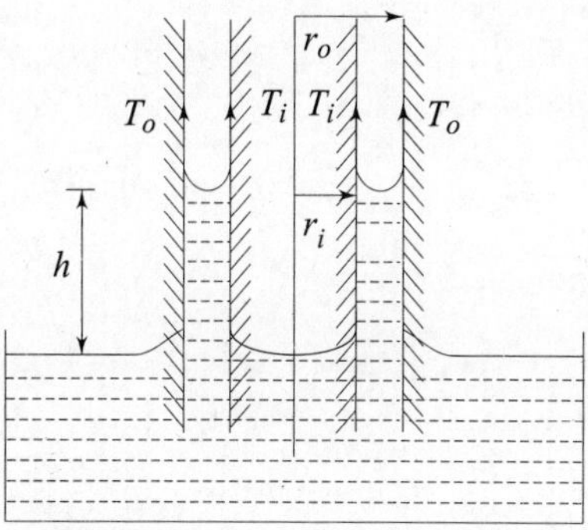

Fig. 1.24 Capillary rise of water in the annulus of two coaxial glass tubes

Example 1.8

What is the pressure within a 1 mm diameter spherical droplet of water relative to the atmospheric pressure outside? Assume σ for pure water to be 0.073 N/m.

Solution Equation (1.34) is used to determine the pressure difference Δp (= $p_2 - p_1$; refer to Fig. 1.25) as

$$\Delta p = 2\sigma/R$$

or $\Delta p = 2 \times 7.3 \times 10^{-2}/(0.5 \times 10^{-3}) = 292 \text{ N/m}^2$

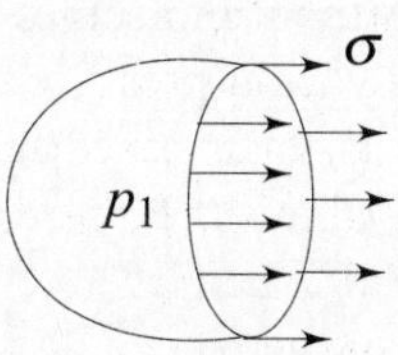

Fig. 1.25 Surface tension force on a spherical water droplet

Example 1.9

A spherical water drop of 1 mm in diameter splits up in air into 64 smaller drops of equal size. Find the work required in splitting up the drop. The surface tension coefficient of water in air = 0.073 N/m.

Solution

An increase in the surface area out of a given mass takes place when a bigger drop splits up into a number of smaller drops, and the work required is given by the product of surface tension coefficient and the increase in surface area.

Let d be the diameter of the smaller drops.

From conservation of mass

$$64 \times \pi \times \frac{d^3}{6} = \pi \times \frac{(0.001)^3}{6}$$

which gives $$d = \frac{0.001}{4} = 0.25 \times 10^{-3} \text{ m}$$

Initial surface area (due to the single drop)

$$= \pi \times (0.001)^2$$
$$= \pi \times 10^{-6} \text{ m}^2$$

Final surface area (due to 64 smaller drops)

$$= 64 \times \pi (0.25 \times 10^{-3})^2$$
$$= 4\pi \times 10^{-6} \text{ m}^2$$

Hence, the increase in surface area

$$= (4 - 1)\pi \times 10^{-6}$$
$$= 3\,\pi \times 10^{-6} \text{ m}^2$$

Therefore, the required work

$$= 0.073 \times 3\pi \times 10^{-6} \text{ J}$$
$$= 0.69 \times 10^{-6} \text{ J}$$

1.6.8 Vapour Pressure

All liquids have a tendency to evaporate when exposed to a gaseous atmosphere. The rate of evaporation depends upon the molecular energy of the liquid, which in turn depends upon the type of liquid and its temperature. The vapour molecules exert a partial pressure in the space above the liquid, known as vapour pressure. If the space above the liquid is confined (Fig. 1.26) and the liquid is maintained at constant temperature, after sufficient time, the confined space above the liquid will contain vapour molecules to the extent that some of them will be forced to enter the liquid. Eventually an equilibrium condition will evolve when the rate at which the number of vapour molecules striking back the liquid surface and condensing is just equal to the rate at which they leave from the surface. The space above the liquid then becomes saturated with vapour. The vapour pressure of a given liquid is a function of temperature only and is equal to the saturation pressure for boiling corresponding to that temperature. Hence, the vapour pressure increases with the increase in temperature. Therefore, the phenomenon of boiling of a liquid is closely related to the vapour pressure. In fact, when the vapour pressure of a liquid becomes equal to the total pressure impressed on its surface, the liquid starts boiling. This concludes that boiling can be achieved either by raising the temperature of the liquid, so that its vapour pressure is elevated to the ambient pressure, or by lowering the pressure of the ambience (surrounding gas) to the liquid's vapour pressure at the existing temperature.

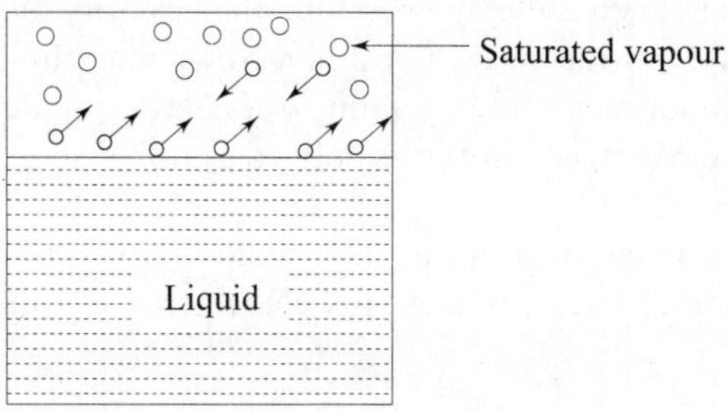

Fig. 1.26 To and fro movement of liquid molecules from an interface in a confined space as a closed surrounding

SUMMARY

- A fluid is a substance that deforms continuously when subjected to even an infinitesimal shear stress. Solids can resist tangential stress at static conditions undergoing a definite deformation while a fluid can do it only at dynamic conditions undergoing a continuous deformation as long as the shear stress is applied.
- The concept of a continuum assumes a continuous distribution of mass within the matter or system with no empty space. In the continuum approach, properties of a system can be expressed as continuous functions of space and time. A dimensionless parameter known as the Knudsen number, $Kn = \lambda/L$, where λ is the mean free path and L is the characteristic length, aptly describes the degree of departure from a continuum. The concept of a continuum usually holds good when $Kn < 0.01$.
- Stress at a point is essentially a surface traction force per unit area that sensitively depends on the orientation of the area chosen to calculate the stress. Unlike force, it requires two indices for its specification, one for the direction normal of the chosen area and the other for the direction of action of the force itself. The state of stress for non-deforming fluids (fluids at rest) may be solely represented by a normal inward force per unit area, which acts equally from all directions. This quantity is called pressure.

- Viscosity is a property of a fluid by virtue of which it offers resistance to flow. The shear stress at a point in a moving fluid is directly proportional to the rate of shear strain. For one-dimensional flow, $\tau = \mu(du/dy)$. The constant of proportionality μ is known as the coefficient of viscosity of simple the viscosity. The relationship is known as the Newton's law of viscosity and the fluids which obey this law are known as *Newtonian fluids.*
- The relationship between the shear stress and the rate of shear strain is known as the constitutive equation. The fluids whose constitutive equations are not linear through origin (i.e., do not obey Newton's law of viscosity) are known as *non-Newtonian fluids*. For a Newtonian fluid, viscosity is a function of temperature only. With an increase in temperature, the viscosity of a liquid decreases, while that of a gas increases. For a non-Newtonian fluid, the apparent viscosity depends not only on temperature but also on the deformation rate of the fluid. Kinematic viscosity, ν, is defined as μ/ρ, where ρ is the density.
- Compressibility of a substance is the measure of its change in volume or density under the action of external forces. It is usually characterised by the *bulk modulus of elasticity*

  $$E = \lim_{\Delta \forall \to \Delta \forall'} \frac{-\Delta p}{\Delta \forall / \forall}$$

  where $\Delta \forall'$ is the smallest possible volume over which the continuum hypothesis remains valid.
- A fluid is said to be incompressible when the change in its density due to the change in pressure brought about by the fluid motion is negligibly small. When the flow velocity is equal to or less then 0.33 times of the local acoustic speed, the relative change in density of the fluid, due to flow, becomes equal to or less than 5 per cent respectively, and hence the flow is considered to be incompressible.
- The force of attraction between the molecules of a fluid is known as cohesion, while that between the molecules of a fluid and of a solid is known as cohesion. The interplay of these two intermolecular forces explains the phenomena of surface tension and capillary rise or depression. A free surface of the liquid is always under stretched condition implying the existence of a tensile force on the surface. The magnitude of this force per unit length of an imaginary line drawn along the liquid surface is known as the surface tension coefficient of simply, the surface tension.
- It is due to surface tension that a curved liquid interface, in equilibrium, results in a greater pressure at the concave side than that at its convex side.

  The pressure difference Δp is given by $\Delta p = \sigma\left(\frac{1}{r_1} + \frac{1}{r_2}\right)$.
- A liquid wets a solid surface and results in a capillary rise when the forces of cohesion between the liquid molecules are lower than the forces of adhesion between the molecules of the liquid and the solid in contact. Non-wettability

of solid surfaces and capillary depression are exhibited by the liquids for which forces of cohesion are more than the forces of adhesion. The capillary rise (or depression) in a tube of radius r may be estimated by the expression,

$h = \dfrac{2\sigma \cos\theta}{\rho g r}$, where θ is the area wetted contact angle.

EXERCISES

1.1 Choose the correct answer:

(i) A fluid is a substance that
- (a) always expands until it fills any container
- (b) is practically incompressible
- (c) cannot withstand any shear force
- (d) cannot remain at rest under the action of any shear force
- (e) obeys Newton's law of viscosity
- (f) None of the above

(ii) Newton's law of viscosity relates to
- (a) pressure, velocity and viscosity
- (b) shear stress and rate of angular deformation in a fluid
- (c) shear stress, temperature, viscosity and velocity
- (d) pressure, viscosity and rate of angular deformation
- (e) None of the above

(iii) The bulk modulus of elasticity
- (a) is independent of temperature
- (b) increases with the pressure
- (c) has the dimensions of 1/P
- (d) is larger when the fluid is more compressible
- (e) is independent of pressure and viscosity

(iv) The phenomenon of capillary rise or depression
- (a) is observed only in vertical tubes
- (b) depends solely upon the surface tension of the liquid
- (c) depends upon the surface tension of the liquid, material of the tube and the surrounding gas in contact of the liquid
- (d) depends upon the pressure difference between the liquid and the environment
- (e) is influenced by the viscosity of the liquid

1.2 One measure as to a gas is in continuum, is the size of its mean free path. According to the kinetic theory of gas, the mean free path is given by

$$\lambda = 1.26\, \mu/\rho\, (RT)^{1/2}$$

What will be the density of air when its mean free path is 10 mm. The temperature is 20 °C, $\mu = 1.8 \times 10^{-5}$ kg/ms, $R = 287$ J/kg K.

Ans. (0.782×10^{-5} kg/m^3)

1.3 A thin film of liquid flows down an inclined channel. The velocity distribution in the flow is given by

$$u = \frac{1}{2\mu}(h^2 - y^2)\,\rho g \sin \alpha$$

where h is the depth of flow, α is the angle of inclination of the channel to the horizontal, ρ is the density of liquid, μ is the dynamic viscosity of the liquid and the coordinate y is measured from the free surface along its normal. Calculate the shear stress: (a) at the bottom of the channel (b) at mid-depth and (c) at the free surface.

Ans. ((a) $\rho gh \sin \alpha$ (b) $\frac{\rho gh}{2} \sin \alpha$ (c) 0)

1.4 A shaft 80 mm in diameter is being pushed through a bearing sleeve 80.2 mm in diameter and 0.3 m long. The clearance, assumed uniform, is flooded with lubricating oil of viscosity 0.1 kg/ms and specific gravity 0.9. (a) If the shaft moves axially at 0.8 m/s, estimate the resistance force exerted by the oil on the shaft, and (b) If the shaft is axially fixed and rotated at 1800 rpm, estimate the resisting torque exerted by the oil and the power required to rotate the shaft.

Ans. (60.32N, 22.74 Nm, 4.29 kW)

1.5 A body weighing 1000 N slides down at a uniform speed of 1 m/s along a lubricated inclined plane making a 30° angle with the horizontal. The viscosity of lubricant is 0.1 kg/ms and contact area of the body is 0.25 m^2. Determine the lubricant thickness assuming linear velocity distribution.

Ans. (0.05 mm)

1.6 The space between two large flat and parallel walls 20 mm apart is filled with a liquid of absolute viscosity 0.8 Pas. Within this space, a thin flat plate 200 mm × 200 mm is towed at a velocity of 200 mm/s at a distance of 5 mm from one wall. The plate and its movement are parallel to the walls. Assuming a linear velocity distribution between the plate and the walls, determine the force exerted by the liquid on the plate.

Ans. (1.71 N)

1.7 A uniform film of oil 0.13 mm thick separates two discs, each of 200 mm diameter, mounted coaxially. Ignoring the edge effects, calculate the torque necessary to rotate one disc relative to other at a speed of 7 rev/s, if the oil has a viscosity of 0.14 Pas.

Ans. (7.43 Nm)

1.8 Two discs of 250 mm diameter are placed 1.5 mm apart and the gap is filled with oil. A power of 500 W is required to rotate the upper disc at 500 rpm while keeping the lower one stationary. Determine the viscosity of the oil. Ignore the edge effects.

Ans. (0.71 kg/ms)

1.9 A piston 79.6 mm diameter and 210 mm long works in a cylinder 80 mm diameter. If the annular space is filled with a lubricating oil having a viscosity of 0.065 kg/ms, calculate the speed with which the piston will move through the cylinder when an axial load of 85.6 N is applied. Neglect the inertia of the piston.

Ans. (5.01 m/s)

1.10 (a) Find the change in volume of 1.00 m^3 of water at 26.7 °C when subjected to a pressure increase of 2 MN/m^2 (The bulk modulus of elasticity of water at 26.7 °C is 2.24×10^9 N/m^2). *Ans.* (0.89×10^{-3} m^3)

(b) From the following test data, determine the bulk modulus of elasticity of water: at 3.5 MN/m^2, the volume was 1.000 m^3 and at 24 MN/m^2, the volume was 0.990 m^3.

Ans. (2.05×10^9 N/m^2)

1.11 A pressure vessel has an internal volume of 0.5 m^3 at atmospheric pressure. It is desired to test the vessel at 300 bar by pumping water into it. The estimated variation in the change of the empty volume of the container due to pressurisation to 300 bar is 6 per cent. Calculate the mass of water to be pumped into the vessel to attain the desired pressure level. Given the bulk modulus of elasticity of water as 2×10^9 N/m^2 and at atmospheric pressure, P = 1000 kg/m^3.

Ans. (538 kg)

1.12 Find an expression for the isothermal bulk modulus of elasticity for a gas which obeys van der Waals law of state according to the equation

$$P = \rho RT\left(\frac{1}{1-b\rho} - \frac{a\rho}{RT}\right),$$

where a and b are constants.

1.13 What is the approximate capillary rise of water in contact with air (surface tension 0.073 N/m) in a clean glass tube of 5 mm in diameter?

Ans. (5.95 mm)

1.14 An atomiser forms water droplets with a diameter of 5×10^{-5} m. What is the pressure within the droplets at 20 °C, if the pressure outside the droplets is 101 kN/m^2? Assume the surface tension of water at 20 °C is 0.0718 N/m.

Ans. (106.74 kN/m^2)

1.15 A spherical soap bubble of diameter d_1 coalesces with another bubble of diameter d_2 to form a single bubble of diameter d_3 containing the same amount of air. Assuming an isothermal process, derive an analytical expression for d_3 as a function of d_1, d_2, the ambient pressure p_0 and the surface tension of soap solution σ. If d_1 = 20 mm, d_2 = 40 mm p_0 = 101 kN/m^2 and σ = 0.09 N/m, determine d_3.

Ans. $(P_0 + 8\sigma/d_3)d_3^3 = (P_0 + 8\sigma/d_1)d_1^3 + (P_0 + 8\sigma/d_2)d_2^3$; $d_3 = 41.60$ mm)

1.16 By how much does the pressure in a cylindrical jet of water 4 mm in diameter exceed the pressure of the surrounding atmosphere if the surface tension of water is 0.0718 N/m?

Ans. (35.9 N/m^2)

1.17 Calculate the capillary depression of mercury at 20 °C (contact angle θ = 140°) to be expected in a 2.5 mm diameter tube. The surface tension of mercury at 20 °C is 0.4541 N/m.

Ans. (4.2 mm)

2

FLUIDS UNDER REST/RIGID BODY MOTION

2.1 INTRODUCTION

In this chapter, we shall be describing situations involving fluids under rest/rigid body motion. These two apparently unconnected situations fall in the same paradigm of mathematical analysis, since in both cases, there is no relative deformation between the adjacent fluid layers. Thus, in this chapter, we consider that fluid is under zero shear and only normal component of stress, manifested in terms of negative of the thermodynamic pressure, acts on the fluid.

2.2 FUNDAMENTAL EQUATION OF FLUIDS AT REST

Referring to Fig. 2.1, let T^n be the traction vector acting on a plane having the unit normal vector $\hat{n}$. Let T_i^n be the ith component of the traction vector T^n (note that i = 1 means component along x, i = 2 means component along y, and i = 3 means component along z in a Cartesian indexing system). In special cases when the direction $\hat{n}$ coincides with any of the coordinate directions (j), the notations T_i^j and τ_{ji} may be used equivalently. The later ones are also known to constitute the stress tensor components (see Chapter 1). Note that each of the indices i and j may vary between 1 and 3.

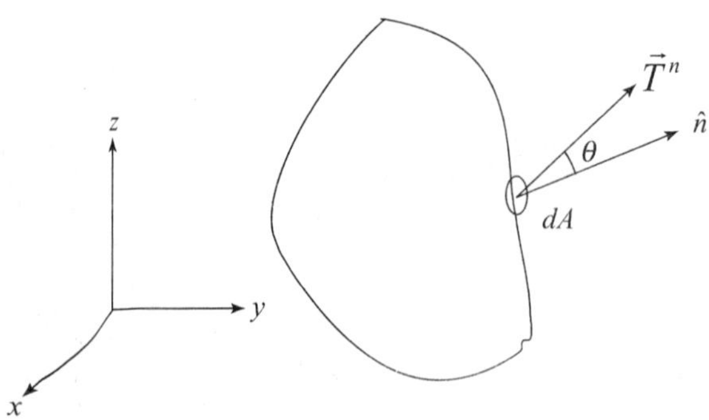

Fig. 2.1

For any arbitrary orientation $\hat{n}$, the traction vector at a point may be expressed in terms of the stress tensor components, by utilising Cauchy's theorem, as follows (see Chapter 1 for detailed derivation):

$$T_i^n = \sum_{j=1}^{3} \tau_{ji} n_j = \sum_{j=1}^{3} \tau_{ij} n_j \quad \text{(since, from angular momentum conservation, } \tau_{ij} = \tau_{ij}\text{)}$$

where, $\hat{n} = n_1\hat{i} + n_2\hat{j} + n_3\hat{k}$ and $\tau_{ij} n_j = \tau_{i1} n_1 + \tau_{i2} n_2 + \tau_{i3} n_3 = \vec{\tau}_i.\hat{n}$ (defining $\vec{\tau}_i$ as $\vec{\tau}_i = \tau_{i1}\hat{i} + \tau_{i2}\hat{j} + \tau_{i3}\hat{k}$)

The net surface force acting on the fluid element along the ith direction is given by

$$F_{\text{surface,i}} = \int_{CS} T_i^n \, dA = \int (\vec{\tau}_i.\hat{n}) dA \tag{2.1}$$

Now, using the divergence theorem ($\int_{CS} \vec{F}.\hat{n} dA = \int_{CV} \nabla.\vec{F} d\forall$), the area integrals appearing in Eq. (2.1) may be converted into volume integrals, to yield

$$F_{\text{surface,i}} = \int_{CV} \nabla.\vec{\tau}_i d\forall$$

$$= \int_{CV} \frac{\partial \tau_{ij}}{\partial x_j} d\forall \tag{2.2}$$

It is important to note here that in a fluid at rest, there are only normal components of stress on a surface that are independent of the orientation of the surface. In other words, the stress tensor is isotropic or spherically symmetrical. Because the stress in a static fluid is isotropic, it must be of the form

$$\tau_{ij} = -p\delta_{ij} \tag{2.3}$$

where p is the thermodynamic pressure (which acts equally from all directions; see Chapter 1), which may be related to density and temperature by an equation of state and δ_{ij} is the Kronecker delta, and is given by

$$\delta_{ij} = 1 \text{ if } i = j$$

$$= 0 \text{ if } i \neq j$$

A negative sign is introduced in Eq. (2.3) because of the fact that the normal components of stress are regarded as positive if they indicate tension, whereas pressure by definition is compressive in nature.

Thus, Eq. (2.2) can be written as

$$F_{\text{surface,i}} = \int_{CV} -\frac{\partial p}{\partial x_i} d\forall \tag{2.4}$$

The net surface force acting on the fluid element is

$$F_{\text{surface}} = \int_{CV} -\nabla p d\forall \tag{2.5}$$

Let F_{body} be the resultant force on the fluid and $\vec{b}$ be the body force per unit mass. Thus, one may write,

$$F_{\text{body}} = \int_{CV} \vec{b}\rho d\forall \qquad (2.6)$$

where $d\forall$ is an element of volume whose mass is $\rho d\forall$.

For equilibrium of the fluid element, we have

$$F_{\text{body}} + F_{\text{surface}} = \int_{CV} \left(\vec{b}\rho - \nabla p\right) d\forall = 0 \qquad (2.7)$$

Equation (2.7) is valid for any arbitrary choice of $\forall$, and hence

$$\vec{b}\rho - \nabla p = 0$$

or

$$\nabla p = \vec{b}\rho \qquad (2.8)$$

Equation (2.8) is the fundamental equation of a fluid at rest. If gravity is considered to be the only external body force acting on the fluid, the vector form of Eq. (2.8) can be expressed in its scalar components with respect to a Cartesian coordinate system (Fig. 2.1) as

$$\frac{\partial p}{\partial x} = 0 \qquad (2.8a)$$

$$\frac{\partial p}{\partial y} = 0 \qquad (2.8b)$$

$$\frac{\partial p}{\partial z} = b_z \rho = -g\rho \qquad (2.8c)$$

where b_z, the external body force per unit mass in the positive direction of z (vertically upward), equals the negative value of g, the acceleration due to gravity. From Eqs. (2.8a) to (2.8c), it can be concluded that the pressure p is a function of z only. Therefore, Eq. (2.8c) can be written as

$$\frac{dp}{dz} = -\rho g \qquad (2.9)$$

The explicit functional relationship of hydrostatic pressure p with z can be obtained by integrating the Eq. (2.9). However, this integration is not possible unless the variation of ρ with p and z is known.

Constant Density Solution

For a constant density (ρ) fluid, Eq. (2.9) can be integrated as

$$p = -\rho g z + C \qquad (2.10)$$

where C is the integration constant.

If we consider an expanse of fluid with a free surface, where the pressure is defined as $p = p_0$ (Fig. 2.2), Eq. (2.10) can be written as

$$p - p_0 = \rho g \left(z_0 - z_1\right) = \rho g h \qquad (2.11)$$

Therefore, Eq. (2.11) gives the expression of hydrostatic pressure p at a point whose vertical depression from the free surface is h. Thus, the difference in pressure between two points in a constant density fluid at rest can be expressed in terms of the vertical distance between the points. This result is known as *Toricelli's principle* which is the basis for differential pressure measuring devices. The pressure $p = p_0$ at free surface is the local atmospheric pressure. Therefore, it can be stated from Eq. (2.11), that the pressure at any point in an expanse of a stagnant fluid with a free surface exceeds that of the local atmospheric by an amount ρgh, where h is the vertical depth of the point from the free surface.

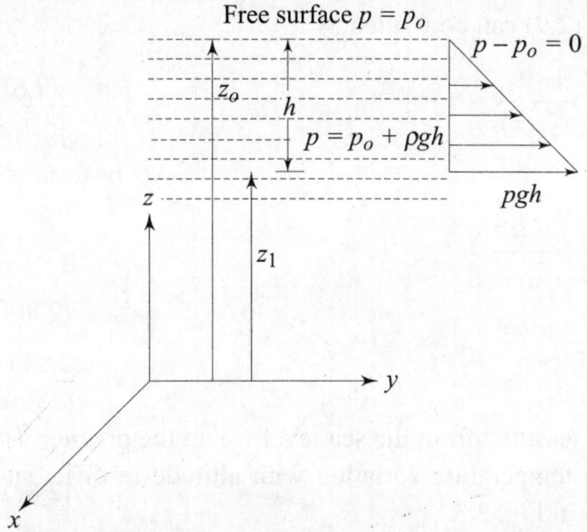

Fig. 2.2 Pressure variation in an incompressible fluid at rest with a free surface

Variable Density Solution (Pressure Variation in a Compressible Fluid)

The pressure variation in a compressible fluid at rest depends on how the fluid density changes with height z and pressure p.

Constant Temperature Solution (Isothermal Fluid)

The equation of state for a compressible system generally relates its density to its pressure and temperature. If the fluid is a perfect gas at rest at constant temperature, it can be written from Eq. (1.24) as

$$\frac{p}{\rho} = \frac{p_0}{\rho_0} \tag{2.12}$$

where p_0 and ρ_0 are the pressure and density at some reference horizontal plane. With the help of Eq. (2.12), Eq. (2.9) becomes

$$\frac{dp}{p} = -\frac{\rho_0}{p_0} g \, dz \tag{2.13}$$

$$p = p_0 \exp\left[-\frac{\rho_0 g}{p_0}(z - z_0)\right] \tag{2.14}$$

where z and z_0 are the vertical coordinates of the plane concerned for pressure p and the reference plane respectively from any fixed datum.

Non-isothermal Fluid

The temperature of the atmosphere up to a certain altitude is frequently assumed to decrease linearly with the altitude z as given by

$$T = T_0 - \alpha z \tag{2.15}$$

where T_0 is the absolute temperature at sea level and the constant α is known as *lapse rate*. For the standard atmosphere, $\alpha = 6.5$ K/km and $T_0 = 288$ K. With the help of Eq. (1.24) and (2.15), Eq. (2.9) can be written as

$$\frac{dp}{p} = \frac{-g}{R} \frac{dz}{(T_0 - \alpha z)} \tag{2.16}$$

Integration of Eq. (2.16) yields

$$\ln \frac{p}{p_0} = \frac{g}{R\alpha} \ln \frac{T_0 - \alpha z}{T_0}$$

Hence,

$$\frac{p}{p_0} = \left(1 - \frac{\alpha z}{T_0}\right)^{g/R\alpha} \tag{2.17}$$

The altitude z in Eq. (2.17) is measured from the sea level where the pressure is p_0. Experimental evidence of the temperature variation with altitude in different layers of the atmosphere is shown in Fig. 2.3.

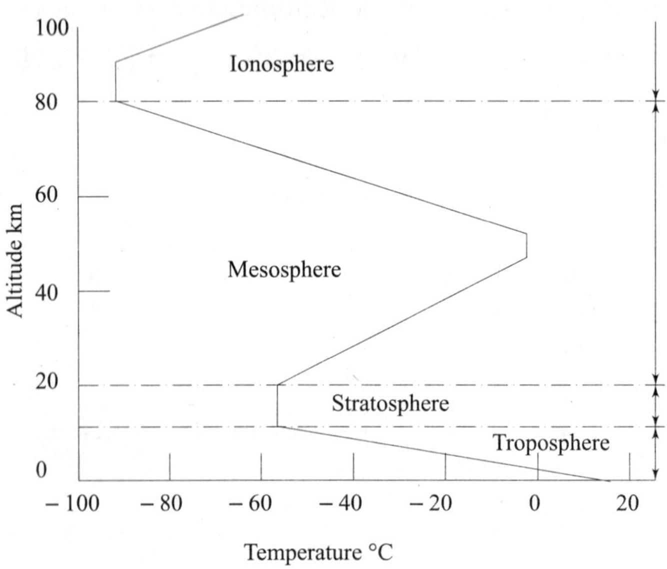

Fig. 2.3 Temperature variation in atmosphere

Example 2.1

What is the intensity of pressure in the ocean at a depth of 1500 m, assuming (a) salt water is incompressible with a specific weight of 10050 N/m^3 and (b) salt water is compressible and weighs 10050 N/m^3 at the free surface? E (bulk modulus of elasticity of salt water) = 2070 MN/m^2 (constant).

Solution

(a) For an incompressible fluid, the intensity of pressure at a depth, according to Eq. (2.11), is

p (pressure in gauge) = $\rho g h$ = 10050 (1500) N/m^2 = 15.08 MN/m^2 gauge

(b) The change in pressure with the depth of liquid h from free surface can be written according to Eq. (2.19) as

$$\frac{\mathrm{d}p}{\mathrm{d}h} = \rho g \tag{2.18}$$

Again from the definition of bulk modulus of elasticity E (Eq. (1.20)),

$$\mathrm{d}p = E\,\frac{\mathrm{d}\rho}{\rho} \tag{2.19}$$

Integrating equation (2.19), for a constant value of E, we get

$$p = E \ln \rho + C \tag{2.20}$$

The integration constant C can be found out by considering $p = p_0$ and $\rho = \rho_0$ at the free surface.

Therefore, Eq. (2.20) becomes

$$p - p_0 = E \ln\left(\frac{\rho}{\rho_0}\right) \tag{2.21}$$

Substitution of dp from Eq. (2.18) into Eq. (2.19) yields

$$\mathrm{d}h = \frac{E\,d\rho}{g\rho^2}$$

After integration

$$h = -\frac{E}{g\rho} + C_1$$

The constant C_1 is found out from the condition that, $\rho = \rho_0$ at $h = 0$ (free surface)

Hence, $$h = \frac{E}{g}\left(\frac{1}{\rho_0} - \frac{1}{\rho}\right)$$

from which $$\frac{\rho}{\rho_0} = \frac{E}{E - h\rho_0 g}$$

Substituting this value of ρ/ρ_0 in Eq. (2.21), we have

$$p - p_0 = E \ln\left(\frac{E}{E - h\rho_0 g}\right)$$

Therefore,

$$p \text{ (in gauge)} = 2.07 \times 10^9 \ln\left[\frac{2.07\times10^9}{2.07\times10^9 - (10050)(1500)}\right] \text{N/m}^2 \text{ gauge}$$

$$= 15.13 \text{ MN/m}^2 \text{ gauge}$$

2.3 UNITS AND SCALES OF PRESSURE MEASUREMENT

The unit of pressure is N/m^2 and is known as *Pascal*. Pressure is usually expressed with reference to either absolute zero pressure (a complete vacuum) or local atmospheric pressure. *Absolute pressure* is the pressure expressed as a difference between its value and the absolute zero pressure. When a pressure is expressed as a diference between its value and the local atmospheric pressure, it is known as *gauge pressure* (Fig. 2.4).

Therefore,

$$p_{abs} = p - 0 = p \tag{2.22a}$$

$$p_{gauge} = p - p_{atm} \tag{2.22b}$$

If the pressure p is less than the local atmospheric pressure, the gauge pressure p_{gauge}, defined by the Eq. (2.22b), becomes negative and is called *vacuum pressure*.

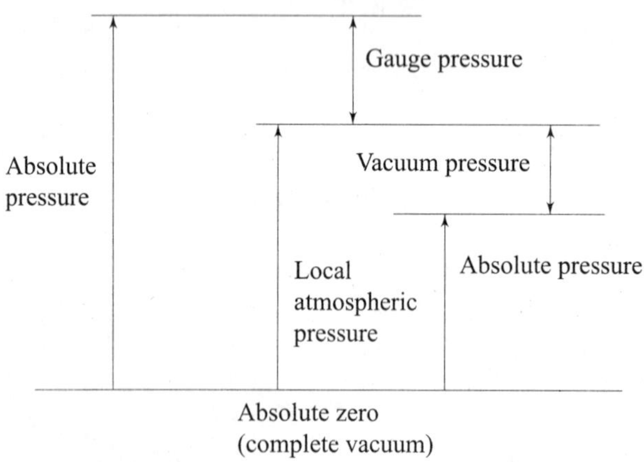

Fig. 2.4 The scale of pressure

At sea level, the international standard atmosphere has been chosen as

$$p_{atm} = 101.32 \text{ kN/m}^2$$

2.4 BAROMETER

It is already established that there is a simple relation (Eq. 2.16) between the height of a column of liquid and the pressure at its base. The direct proportionality between gauge pressure and the height h for a fluid of constant density enables the pressure to be simply visualised in terms of the vertical height, $h = p/\rho g$. The height h is termed as *pressure head* corresponding to pressure p. For a liquid without a free surface in a closed pipe, the pressure head $p/\rho g$ at a point corresponds to the vertical height above the point to which a free surface would rise, if a small tube of sufficient length and open to atmosphere is connected to the pipe (Fig. 2.5).

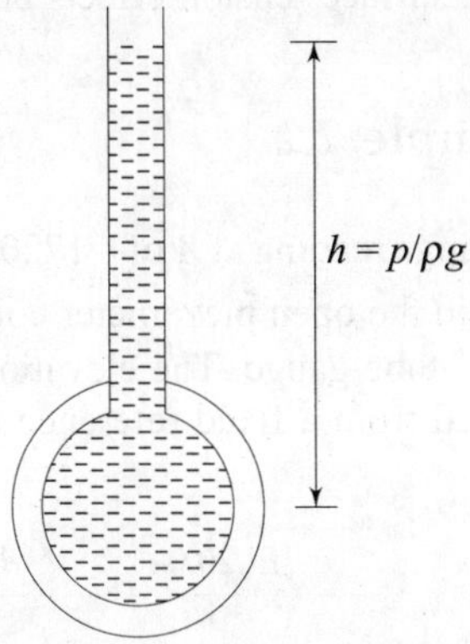

Fig. 2.5 A piezometer tube

Such a tube is called a *piezometer tube*, and the height h is the measure of the gauge pressure of the fluid in the pipe. If such a piezometer tube of sufficient length were closed at the top and the space above the liquid surface were a perfect vacuum, the height of the column would then correspond to the absolute pressure of the liquid at the base. This principle is used in the well-known mercury barometer to determine the local atmospheric pressure. Mercury is employed because its density is sufficiently high for a relative short column to be obtained, and also because it has very small vapour pressure at normal temperature. A perfect vacuum at the top of the tube (Fig. 2.6) is never possible; even if no air is present, the space would be occupied by the mercury vapour and the pressure would equal to the vapour pressure of mercury at its existing temperature. This almost vacuum condition above the mercury in the barometer is known as *Torricellian vacuum*. The pressure at A is equal to that at B (Fig. 2.6) which is the atmospheric pressure p_{atm} since A and B lie on the same horizontal plane. Therefore, we can write

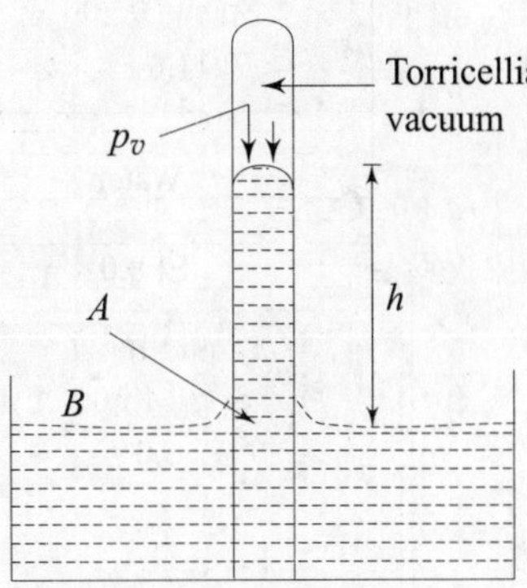

Fig. 2.6 A barometer

$$p_B = p_{atm} = p_v + \rho g h \tag{2.23}$$

The vapour pressure of mercury p_v, can normally be neglected in comparison to p_{atm}. At 20 °C, p_v is only 0.16 p_{atm}, where $p_{atm} = 1.0132 \times 10^5$ Pa at sea level. Then we get from Eq. (2.23)

$$h = p_{atm}/\rho g = \frac{1.0132 \times 10^5\ \text{N/m}^2}{(13560\ \text{kg/m}^3)\,(9.81\ \text{N/kg})} = 0.762\ \text{m of Hg}$$

For accurate work, small corrections are necessary to allow for the variation of ρ with temperature, the thermal expansion of the scale (usually made of brass), and surface tension effects. If water was used instead of mercury, the corresponding height of the column would be about 10.4 m provided that a perfect vacuum could be achieved above the water. However, the vapour pressure of water at ordinary temperature is appreciable and so the actual height at, say, 15 °C would be about 180 mm less than this value. Moreover, with a tube smaller in diameter than about 15 mm, surface tension effects become significant.

Example 2.2

For a gauge reading at *A* of – 17200 Pa (Fig. 2.7), determine: (a) The elevation of the liquids in the open piezometer columns *E*, *F*, *G*, and (b) The deflection of mercury in the U-tube gauge. The elevations EL of the interfaces, as shown in Fig. 2.7, are measured from a fixed reference datum.

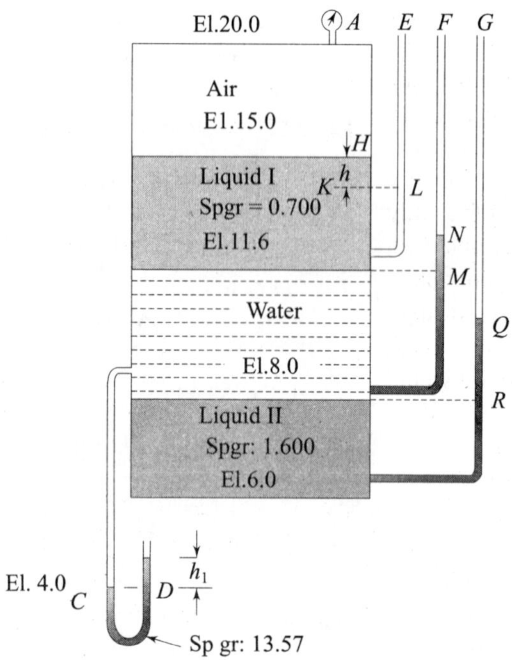

Fig. 2.7 Piezometer tubes connected to a tank containing different liquids

Solution

(a) Since the specific weight of air (= 12 N/m^3) is very small compared to that of the liquids, the pressure at elevation 15.0 may be considered to be –17200 Pa gauge by neglecting the weight of air above it without introducing any significant error in the calculations.

For column E: Since the pressure at H is below the atmospheric pressure, the elevation of liquid in the piezometer E will be below H, and assume this elevation is L as shown in Fig. 2.7.

From the principle of hydrostatics, $p_K = p_L$

Then $p_{atm} - 17200 + (0.700 \times 9.81 \times 10^3)h = p_{atm}$

(where p_{atm} is the atmospheric pressure)

or $$h = 2.5 \text{ m}$$

Hence the elevation at L is $15 - 2.5 = 12.5$ m

For column F: Pressure at EL11.6 = Pressure at EL15.0 + Pressure of the liquid I

$$= -17200 + (0.7 \times 9.81 \times 10^3)(15 - 11.6)$$
$$= 6148 \text{ Pa gauge}$$

which must equal the pressure at M.

The height of the water column corresponding to this pressure is $\dfrac{6148}{9810} = 0.63$ m, and therefore the water column in the piezometer F will rise 0.63 m above M.

Hence the elevation at N is $(11.6 + 0.63) = 12.23$ m

For column G : Pressure at

$$\text{EL8.0} = \text{Pressure at EL11.6} + \text{Pressure of 3.6 m of water}$$
$$= 6148 + 9.81 \times 3.6 \times 10^3 = 41464 \text{ Pa}$$

which must be the pressure at R and equals to a column of

$$\frac{41464}{1.6 \times 9810} = 2.64 \text{ m of liquid II}$$

Therefore, the liquid column in piezometer G will rise 2.64 m above R and elevation at Q is $(8.0 + 2.64) = 10.64$ m.

(b) For the U-tube gauge,

$$\text{Pressure at } D = \text{Pressure at } C$$
$$9810 \times 13.57\, h_1 = \text{Pressure at EL11.6} + \text{Pressure of 7.6 m of water}$$

or $$13.57\, h_1 = 0.63 + 7.6$$

from which $$h_1 = 0.61 \text{ m}$$

2.5 MANOMETERS

Manometers are devices in which columns of a suitable liquid are used to measure the difference in pressure between two points or between a certain point and the atmosphere. For measuring very small gauge pressures of liquids, simple piezometer tube (Fig. 2.5) may be adequate, but for larger gauge pressures, some modifications of the tube are necessary and this modified tube is known as a *manometer*. A common type manometer is like a transparent 'U-tube' as shown in Fig. 2.8(a). One of its ends is connected to a pipe or a container having a fluid (A) whose pressure is to be measured while the other end is open to the atmosphere. The lower part of the U-tube contains a liquid immiscible with the fluid A and is of greater density than that of A. This fluid is called *manometric fluid*. The pressures at the two points P and Q (Fig. 2.8(a)) in a horizontal plane within the continuous expanse of the same fluid (liquid B in this case) must be equal. Then equating the pressures at P and Q in terms of the heights of the fluids above those points, with the aid of the fundamental equation of hydrostatics (Eq. 2.11), we have

$$p_1 + \rho_A g (y + x) = p_{\text{atm}} + \rho_B g x$$

Hence, $$p_1 - p_{\text{atm}} = (\rho_B - \rho_A) g x - \rho_A g y \tag{2.24}$$

where p_1 is the absolute pressure of the fluid A in the pipe or container at its centre line, and p_{atm} is the local atmospheric pressure. When the pressure of the fluid in the container is lower than the atmospheric pressure, the liquid levels in the manometer would be adjusted as shown in Fig. 2.8(b). Hence it becomes,

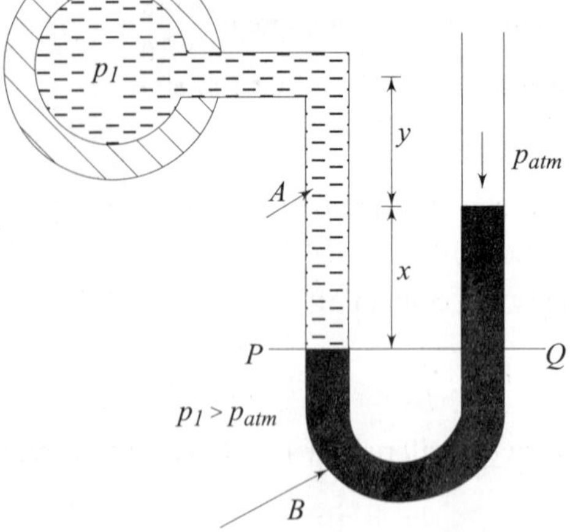

Fig. 2.8(a) A simple manometer to measure gauge pressure

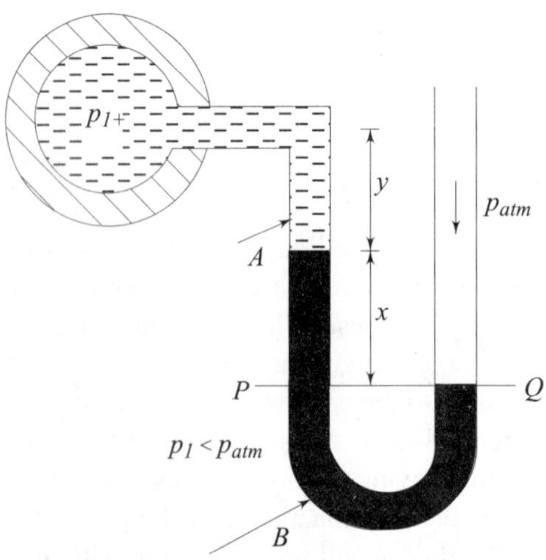

Fig. 2.8(b) A simple manometer measuring vacuum pressure

$$p_1 + \rho_A g y + \rho_B g x = p_{atm}$$

$$p_{atm} - p_1 = (\rho_A y + \rho_B x)g \tag{2.25}$$

In the similar fashion, a manometer is frequently used to measure the pressure difference, in course of flow, across a restriction in a horizontal pipe (Fig. 2.9).

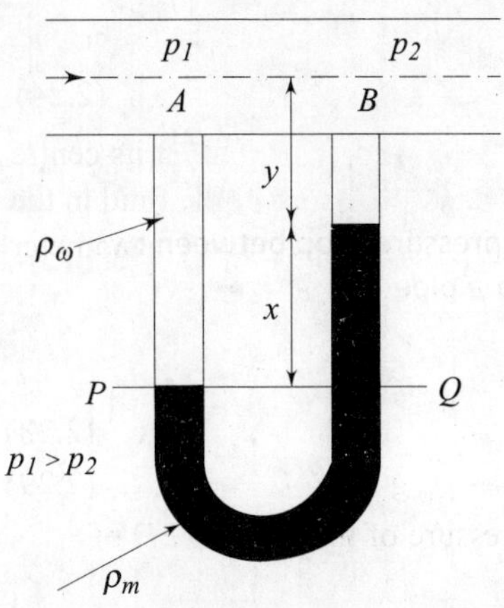

Fig. 2.9 A manometer measuring pressure differential

It is very important that the axis of each connecting tube at *A* and *B* to be perpendicular to the direction of flow and also for the edges of the connections to be smooth. Applying the principle of hydrostatics at *P* and *Q* we have,

$$p_1 + (y + x)\rho_w g = p_2 + y\,\rho_w g + x\,\rho_m g$$

$$p_1 - p_2 = (\rho_m - \rho_w)gx \tag{2.26}$$

where, ρ_m is the density of manometric fluid and ρ_w is the density of the working fluid flowing through the pipe. Sometimes it is desired to express this difference of pressure in terms of the difference of heads (height of the working fluid at equilibrium).

Thus, $$h_1 - h_2 = \frac{p_1 - p_2}{\rho_w g} = \left(\frac{\rho_m}{\rho_w} - 1\right)x \tag{2.27}$$

Example 2.3

A typical differential manometer is attached to two sections *A* and *B* in a horizontal pipe through which water is flowing at a steady rate (Fig. 2.10). The deflection of mercury in the manometer is 0.6 m with the level nearer *A* being the lower one as shown in the figure. Calculate the difference in pressure between Sections *A* and *B*. Take the densities of water and mercury as 1000 kg/m^3 and 13570 kg/m^3, respectively.

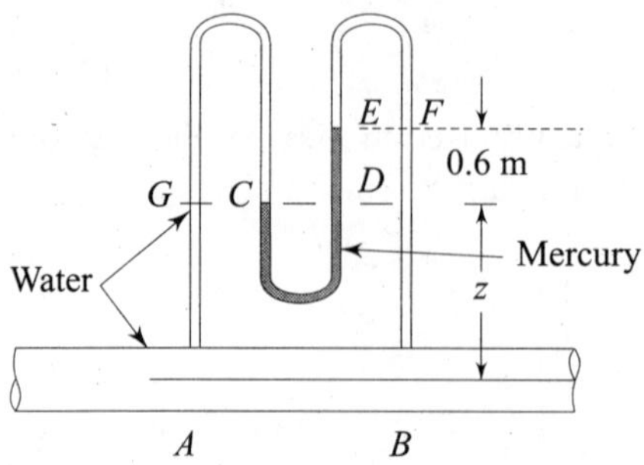

Fig. 2.10 A differential manometer measuring pressure drop between two sections in the flow of water through a pipe

Solution

$$p_C \text{ (Pressure at } C) = p_D \text{ (Pressure at } D) \tag{2.28}$$

Again $\quad p_C = p_G \text{ (Pressure at } G) = p_A - \rho_w g z \quad (2.29)$

and $\quad p_D = p_E$ (Pressure at E) + Pressure of the column ED of mercury

$$= p_F \text{ (Pressure at } F) + \rho_m g\,(0.6)$$

$$= p_B - (z + 0.6)\,\rho_w\, g + 0.6\,\rho_m\, g \tag{2.30}$$

With the help of Eq. (2.29) and (2.30), the Eq. (2.28) can be written as

$$p_A - \rho_w g z = p_B - (z + 0.6)\,\rho_w g + 0.6\,\rho_m g$$

or

$$p_A - p_B = 0.6g\,(\rho_m - \rho_w) = 0.6 \times 9.81\,(13.57 - 1) \times 10^3 \text{ Pa}$$

$$= 74 \text{ kPa}$$

2.5.1 Inclined Tube Manometer

To obtain a reasonable value of x [Eq. (2.27)] for accurate measurement of small pressure differences by a ordinary U-tube manometer, it is essential that the ratio ρ_m/ρ_w should be close to unity. If the working fluid is a gas, this is not possible. Moreover, it may not be always possible to have a manometric liquid of density very close to that of the working liquid and giving at the same time a well defined meniscus at the interface. For this purpose, an inclined tube manometer is used. For example, if the transparent tube of a manometer instead of being vertical is set at an angle θ to the horizontal (Fig. 2.11), then a pressure difference corresponding to a vertical difference of levels x gives a movement of the meniscus $s = x/\sin\theta$ along the slope (Fig. 2.11).

If θ is small, a considerable magnification of the movement of the meniscus may be achieved. Angles less than 5°, however, are not usually satisfactory, because it becomes difficult to determine the exact position of the meniscus. One limb is usually made very much greater in cross section than the other. When a pressure difference is applied across the manometer, the movement of the liquid surface in the wider limb is practically negligible compared to that occurring in the narrower limb. If the level of the surface in the wider limb is assumed constant, the displacement of the meniscus in the narrower limb needs only to be measured, and therefore only this limb is required to be transparent.

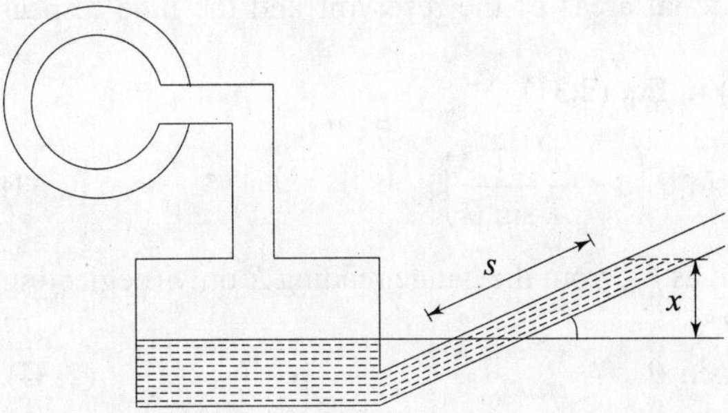

Fig. 2.11 An inclined tube manometer

Example 2.4

An inclined tube manometer measures the gauge pressure p_s of a system S (Fig. 2.12). The reservoir and tube diameters of the manometer are 50 mm and 5 mm respectively. The inclination angle of the tube is 30°. What will be the percentage error in measuring p_S if the reservoir deflection is neglected.

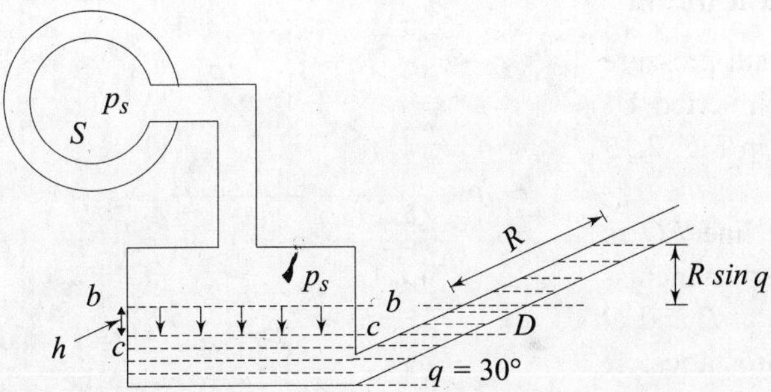

Fig. 2.12 An inclined tube manometer measuring gauge pressure of a system

Solution

Let, with the application of pressure p_S, the level of gauge fluid in the reservoir lowers down from *bb* to *cc*

Now, pressure at c = Pressure at D

or $$p_S = \rho_g \cdot g\,(R \sin\theta + h) \tag{2.31}$$

where ρ_g is the density of the gauge fluid. From continuity of the fluid in both the limbs,

$$A \cdot h = a \cdot R$$

or $$h = \frac{aR}{A} \tag{2.32}$$

where A and a are the cross-sectional areas of the reservoir and the tube respectively.
Substituting for h from Eq. (2.32) in Eq. (2.31)

$$p_S = \rho_g g R \sin\theta \left(1 + \frac{a}{A}\frac{1}{\sin\theta}\right) \tag{2.33}$$

Let the pressure p_S be measured as p_S' from the gauge reading R only (neglecting the reservoir deflection h).

Then
$$p_S' = \rho_g g R \sin\theta \tag{2.34}$$

The percentage error in measuring p_S as p_S' can now be calculated with the help of Eqs. (2.33) and (2.34) as

$$e = \frac{(p_S - p_S') \times 100}{p_S} = \frac{1}{\left(1 + \frac{A}{a}\sin\theta\right)} \times 100$$

$$= \frac{1}{\left[1 + \left(\frac{50}{5}\right)^2 \frac{1}{2}\right]} \times 100 = 1.96\%$$

2.5.2 Inverted Tube Manometer

For the measurement of small pressure differences in liquids, an inverted U-tube manometer as shown in Fig. 2.13 is often used.

Here $\rho_m < \rho_w$, and the line PQ is taken at the level of the higher meniscus to equate the pressures at P and Q from the principle of hydrostatics. If may be written that

$$p_1^* - p_2^* = (\rho_w - \rho_m) g x \tag{2.35}$$

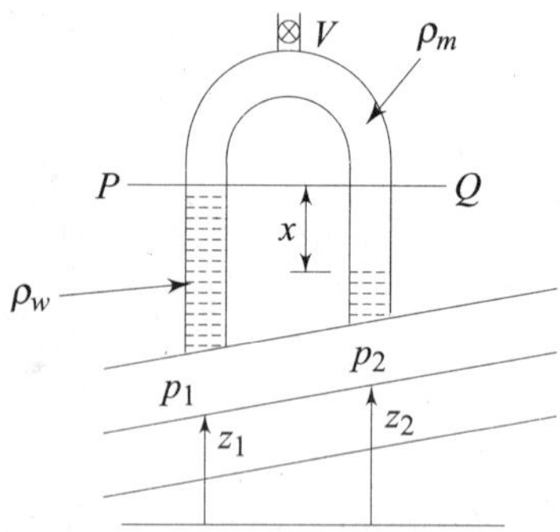

Fig. 2.13 An inverted tube manometer

where p^* represents the *piezometric pressure* $p + \rho g z$ (z being the vertical height of the point concerned from any reference datum). In case of a horizontal pipe ($z_1 = z_2$), the difference in piezometric pressure becomes equal to the difference in the static pressure. If $(\rho_w - \rho_m)$ is sufficiently small, a large value of x may be obtained for a small value of $p_1^* - p_2^*$. Air is used as the manometric fluid. Therefore, ρ_m is negligible compared with ρ_w and hence,

$$p_1^* - p_2^* \approx \rho_w g x \tag{2.36}$$

Air may be pumped through a value V at the top of the manometer until the liquid menisci are at a suitable level.

2.5.3 Micromanometer

When an additional gauge liquid is used in a U-tube manometer, a large difference in meniscus levels may be obtained for a very small pressure difference. The typical arrangement is shown in Fig. 2.14.

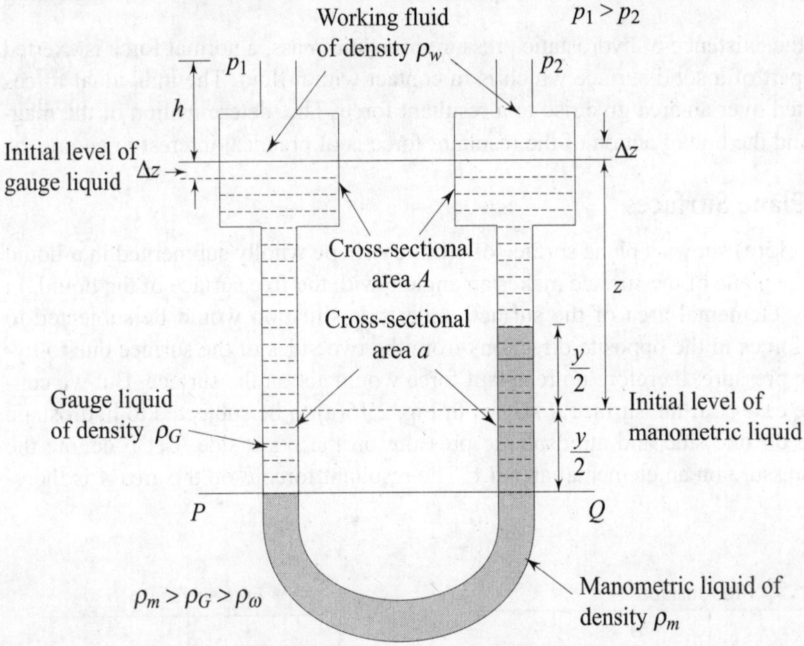

Fig. 2.14 A micromanometer

The equation of hydrostatic equilibrium at PQ can be written as

$$p_1 + \rho_w\, g(h + \Delta z) + \rho_G\, g\left(z - \Delta z + \frac{y}{2}\right) = p_2 + \rho_w\, g(h - \Delta z)$$

$$+ \rho_G\, g\left(z + \Delta z - \frac{y}{2}\right) + \rho_m\, g\, y \tag{2.37}$$

where ρ_w, ρ_G and ρ_m are the densities of working fluid, gauge liquid and manometric liquid respectively.

From continuity of gauge liquid,

$$A\,\Delta z = a\,\frac{y}{2} \tag{2.38}$$

Substituting for Δz from Eq. (2.38) in Eq. (2.37), we have

$$p_1 - p_2 = gy\left\{\rho_m - \rho_G\left(1 - \frac{a}{A}\right) - \rho_w\,\frac{a}{A}\right\} \tag{2.39}$$

If a is very small compared to A,

$$p_1 - p_2 \approx (\rho_m - \rho_G)\, gy \tag{2.40}$$

With a suitable choice for the manometric and gauge liquids so that their densities are close ($\rho_m \approx \rho_G$), a reasonable value of y may be achieved for a small pressure difference.

2.6 HYDROSTATIC THRUSTS ON SUBMERGED SURFACES

Due to the existence of hydrostatic pressure in a fluid mass, a normal force is exerted on any part of a solid surface which is in contact with a fluid. The individual forces distributed over an area give rise to a resultant force. The determination of the magnitude and the line of action of the resultant force is of practical interest to engineers.

2.6.1 Plane Surfaces

Figure 2.15(a) shows a plane surface of arbitrary shape wholly submerged in a liquid so that the plane of the surface makes an angle θ with the free surface of the liquid. In fact, any elemental area of the surface under this situation would be subjected to normal forces in the opposite directions from the two sides of the surface due to hydrostatic pressure; therefore no resultant force would act on the surface. But we consider the case as if the surface A shown in Fig. 2.15(a) to be subjected to hydrostatic pressure on one side and atmospheric pressure on the other side. Let p denote the gauge pressure on an elemental area dA. The resultant force F on the area A is therefore,

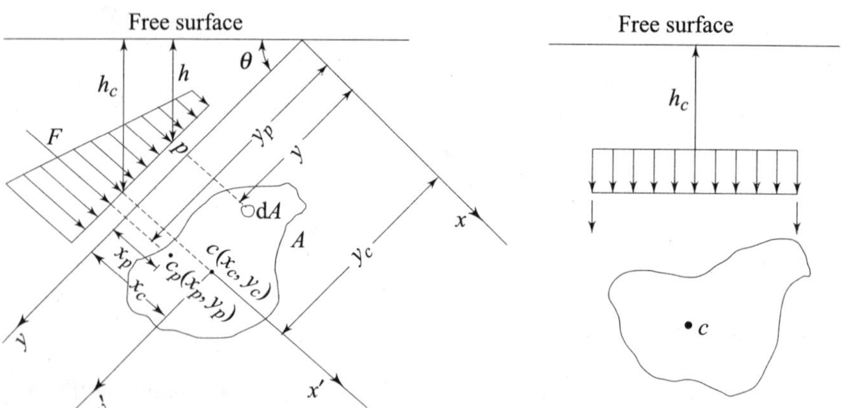

(a) Inclined surface (b) Horizontal surface

Fig. 2.15 Hydrostatic thrust on submerged plane surface

$$F = \iint_A p\, \mathrm{d}A \tag{2.41}$$

According to Eq. (2.11), Eq. (2.41) reduces to

$$F = \rho g \iint h \, \mathrm{d}A = \rho g \sin\theta \iint y \, \mathrm{d}A \tag{2.42}$$

where h is the vertical depth of the elemental area dA from the free surface and the distance y is measured from the x axis, the line of intersection between the extension of the inclined plane and the free surface (Fig. 2.15(a)). The ordinate of the centre of area of the plane surface A is defined as

$$y_c = \frac{1}{A} \iint y \, \mathrm{d}A \tag{2.43}$$

Hence from Eqs (2.42) and (2.43), we get

$$F = \rho g y_c \sin\theta A = \rho g h_c A \tag{2.44}$$

where h_c (= y_c sin θ) is the vertical depth (from free surface) of centre of area c.

Equation (2.44) implies that the hydrostatic thrust on an inclined plane is equal to the pressure at its centroid times the total area of the surface, i.e., the force that would have been experienced by the surface if placed horizontally at a depth h_c from the free surface (Fig. 2.15(b)).

The point of action of the resultant force on the plane surface is called the centre of pressure c_p. Let x_p and y_p be the distances of the centre of pressure from the y and x axes respectively. Equating the moment of the resultant force about the x axis to the summation of the moments of the component forces, we have

$$y_p F = \int y \, \mathrm{d}F = \rho g \sin\theta \iint y^2 \, \mathrm{d}A \tag{2.45}$$

Solving for y_p from Eq. (2.45) and replacing F from Eq. (2.42), we can write

$$y_p = \frac{\iint_A y^2 \, \mathrm{d}A}{\iint_A y \, \mathrm{d}A} \tag{2.46}$$

In the same manner, the x coordinate of the centre of pressure can be obtained by taking moment about the y axis. Therefore,

$$x_p F = \int x \, \mathrm{d}F = \rho g \sin\theta \iint xy \, \mathrm{d}A$$

from which,

$$x_p = \frac{\iint_A xy \, \mathrm{d}A}{\iint_A y \, \mathrm{d}A} \tag{2.47}$$

The two double integrals in the numerators of Eqs. (2.46) and (2.47) are the moment of inertia about the x axis I_{xx} and the product of inertia I_{xy} about the x and y axes of the plane area, respectively. By applying the theorem of parallel axis,

$$I_{xx} = \iint y^2 \, \mathrm{d}A = I_{x'x'} + A y_c^2 \tag{2.48}$$

$$I_{xy} = \iint xy \, \mathrm{d}A = I_{x'y'} + A x_c y_c \tag{2.49}$$

where, $I_{x'x'}$ and $I_{x'y'}$ are the moment of inertia and the product of inertia of the surface about the centroidal axes $(x' - y')$, x_c and y_c are the coordinates of the centre of area c with respect to the x and y axes.

With the help of Eqs. (2.48), (2.49) and (2.43), Eqs. (2.46) and (2.47) can be written as

$$y_p = \frac{I_{x'x'}}{A\,y_c} + y_c \tag{2.50a}$$

$$x_p = \frac{I_{x'y'}}{A\,y_c} + x_c \tag{2.50b}$$

The first term on the right-hand side of Eq. (2.50a) is always positive. Hence, the centre of pressure is always at a higher depth from the free surface than that at which the centre of area lies. This is obvious because of the typical variation of hydrostatic pressure with the depth from the free surface. When the plane area is symmetrical about the y' axis, $I_{x'y'} = 0$, and $x_p = x_c$.

Example 2.5

Oil of specific gravity 0.800 acts on a vertical triangular area whose apex is in the oil surface. The triangle is isosceles of 3 m high and 4 m wide. A vertical rectangular area of 2 m high is attached to the 4 m base of the triangle and is acted upon by water. Find the magnitude and point of action of the resultant hydrostatic force on the entire area.

Solution

The submerged area under oil and water is shown in Fig. 2.16.

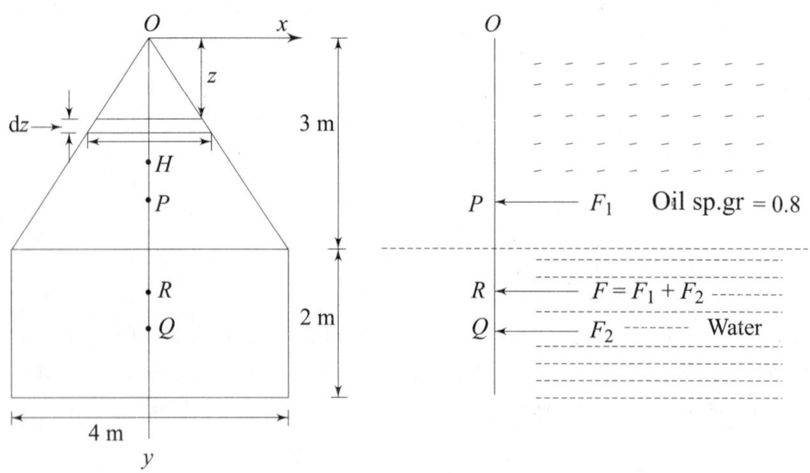

Fig. 2.16 The submerged surface under oil and water as described in Example 2.5

The hydrostatic force F_1 on the triangular area

$$= 9.81\times 0.8\times 10^3\times\left(\frac{2}{3}\times 3\right)\times\left(\frac{1}{2}\times 3\times 4\right)N = 94.18\text{ kN}$$

The hydrostatic force F_2 on the rectangular area

$$= 9.81\times 10^3\,(3\times 0.8+1)\times(2\times 4)\,N = 266.83\text{ kN}$$

Therefore the resultant force on the entire area

$$F = F_1 + F_2 = 94.18 + 266.83 = 361\text{ kN}$$

Since the vertical line through the apex 0 is the axis of symmetry of the entire area, the hydrostatic forces will always act through this line. To find the points of action of the forces F_1 and F_2 on this line, the axes *Ox* and *Oy* are taken as shown in Fig. 2.16.

For the triangular area, moments of forces on the elemental strips of thickness *dz* about *Ox* give

$$F_1 \cdot OP = \int_0^3 9.81\times 10^3\,(0.8z)\,(Hdz.z)$$

Again from geometry, $H = \frac{4}{3}z$

Hence,
$$OP = \frac{\int_0^3 9.81\times 10^3\times 0.8\left(\frac{4}{3}\right)z^3\,dz}{94.18\times 10^3} = 2.25\text{ m}$$

In a similar way, the point of action of the force F_2 on the rectangular area is found out as

$$OQ = \frac{\int_3^5 9.81\times 10^3\,\{(3\times 0.8)+(z-3)\}\,(4\,dz)\,z}{266.83\times 10^3} = 4.1\text{ m}$$

Finally the point of action *R* (Fig. 2.16) of the resultant force *F* is found out by taking moments of the forces F_1 and F_2 about *O* as

$$OR = \frac{94.18\times 2.25 + 266.83\times 4.1}{361} = 3.62\text{ m}$$

Example 2.6

Figure 2.17 shows a flash board. Find the depth of water *h* at the instant when the water is just ready to tip the flash board.

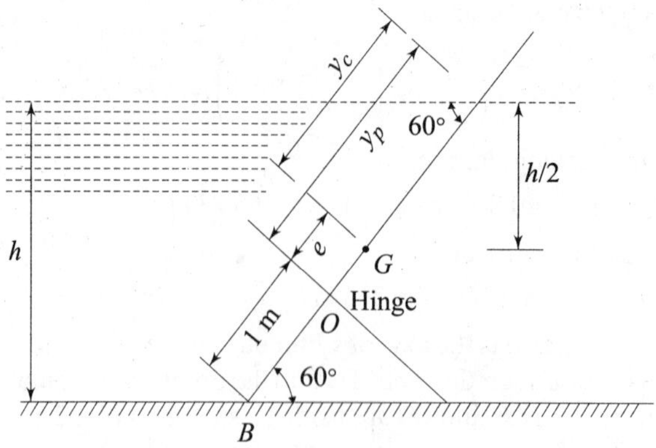

Fig. 2.17 A flash board in water

Solution

The flash board will tip if the hydrostatic force on the board acts at a point away from the hinge towards the free surface. Therefore, the depth of water h for which the hydrostatic force F_p passes through the hinge point O is the required depth when water is just ready to tip the board. Let G be the centre of gravity of the submerged part of the board (Fig. 2.17).

Then $$BG = \frac{h/2}{\sin 60°} = \frac{h}{\sqrt{3}}$$

If y_p and y_c are the distances of the pressure centre (point of application of the hydrostatic force F_p) and the centre of gravity respectively from the free surface along the board, then from Eq. (2.50a)

$$e = y_p - y_c = \frac{(2h/\sqrt{3})^3}{12\left(\frac{2h}{\sqrt{3}}\right)\frac{h}{\sqrt{3}}} = h/\left(3\sqrt{3}\right) \tag{2.51}$$

(considering unit length of the board)

Again from the geometry,

$$e = BG - BO = \left(h/\sqrt{3}\right) - 1 \tag{2.52}$$

Equating the two expressions of e from Eqs. (2.51) and (2.52), we have

$$h/\left(3\sqrt{3}\right) = h/\sqrt{3} - 1$$

from which $$h = \frac{3\sqrt{3}}{2} = 2.6 \text{ m}$$

Example 2.7

The plane gate (Fig. 2.18) weighs 2000 N/m length normal to the plane of the figure, with its centre of gravity 2 m from the hinge O. Find h as a function of θ for equilibrium of the gate.

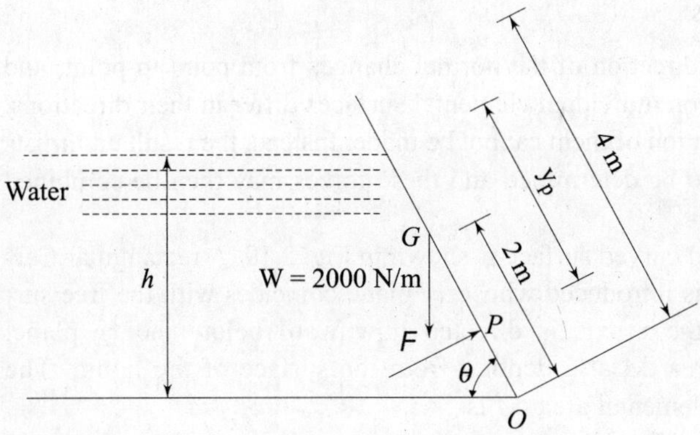

Fig. 2.18 A plain gate in equilibrium under hydrostatic force due to water

Solution

Let F be the hydrostatic force acting on the gate at point P

Then $\quad F =$ Pressure at the centroid of the submerged portion of gate × submerged area of the gate

$$= \frac{h}{2} \times 9.81 \times 10^3 \times \frac{h}{\sin\theta} \times 1 = \frac{4905\, h^2}{\sin\theta} \tag{2.53}$$

The distance of the pressure centre P from the free surface along the gate is found out, according to Eq. (2.50), as

$$y_p = \frac{h}{2\sin\theta} + \frac{1 \times (h/\sin\theta)^3}{12 \times 1 \times \dfrac{h}{\sin\theta}\left(\dfrac{h}{2\sin\theta}\right)} = \frac{h}{\sin\theta}\left(\frac{1}{2} + \frac{1}{6}\right) = \frac{2}{3}\frac{h}{\sin\theta}$$

Now, $\quad OP = \dfrac{h}{\sin\theta} - \dfrac{2}{3}\dfrac{h}{\sin\theta} = \dfrac{1}{3}\dfrac{h}{\sin\theta}$

For equilibrium of the gate, moment of all the forces about the hinge O will be zero.

Hence, $\quad F\left(\dfrac{1}{3}\dfrac{h}{\sin\theta}\right) - 2000\,(2\cos\theta) = 0$

Substituting F from Eq. (2.53),

$$\frac{4905\, h^2}{\sin\theta}\left(\frac{1}{3}\frac{h}{\sin\theta}\right) - 4000\cos\theta = 0$$

from which $\quad h = 1.347\,(\sin^2\theta\cos\theta)^{1/3}$

2.6.2 Curved Surfaces

On a curved surface, the direction of the normal changes from point to point, and hence the pressure forces on individual elemental surfaces differ in their directions. Therefore, a scalar summation of them cannot be made. Instead, the resultant thrusts in certain directions are to be determined and these forces may then be combined vectorially.

An arbitrary submerged curved surface is shown in Fig. 2.19. A rectangular Cartesian coordinate system is introduced whose xy plane coincides with the free surface of the liquid and the z axis is directed downward below the xy plane. Consider an elemental area dA at a depth z from the surface of the liquid. The hydrostatic force on the elemental area dA is

$$\mathrm{d}F = \rho g z \,\mathrm{d}A \tag{2.54}$$

and the force acts in a direction normal to the area dA. The components of the force dF in x, y and z directions are

$$\mathrm{d}F_x = l\,\mathrm{d}F = l\,\rho gz\,\mathrm{d}A \tag{2.55a}$$
$$\mathrm{d}F_y = m\,\mathrm{d}F = m\,\rho gz\,\mathrm{d}A \tag{2.55b}$$
$$\mathrm{d}F_z = n\,\mathrm{d}F = n\,\rho gz\,\mathrm{d}A \tag{2.55c}$$

where l, m and n are the direction cosines of the normal to d A.

The components of the surface element d A projected on the yz, xz and xy planes are, respectively

$$\mathrm{d}A_x = l\,\mathrm{d}A \tag{2.56a}$$
$$\mathrm{d}A_y = m\,\mathrm{d}A \tag{2.56b}$$
$$\mathrm{d}A_z = n\,\mathrm{d}A \tag{2.56c}$$

Substituting Eqs (2.56a–2.56c) into (2.55), we can write

$$\mathrm{d}F_x = \rho g\, z\,\mathrm{d}A_x \tag{2.57a}$$
$$\mathrm{d}F_y = \rho g\, z\,\mathrm{d}A_y \tag{2.57b}$$
$$\mathrm{d}F_z = \rho g\, z\,\mathrm{d}A_z \tag{2.57c}$$

Therefore, the components of the total hydrostatic force along the coordinate axes are

$$F_x = \iint_A \rho g\, z\,\mathrm{d}A_x = \rho g\, z_c A_x \tag{2.58a}$$

$$F_y = \iint_A \rho g\, z\,\mathrm{d}A_y = \rho g\, z_c A_y \tag{2.58b}$$

$$F_z = \iint_A \rho g\, z\,\mathrm{d}A_z \tag{2.58c}$$

where z_c is the z coordinate of the centroid of area A_x and A_y (the projected areas of curved surface on yz and xz plane, respectively). If z_p and y_p are taken to be the coordinates of the point of action of F_x on the projected area A_x on the yz plane, following the method discussed in 2.6.1, we can write

$$z_p = \frac{1}{A_x z_c}\iint z^2\,\mathrm{d}A_x = \frac{I_{yy}}{A_x z_c} \tag{2.59a}$$

$$y_p = \frac{1}{A_x z_c} \iint yz \, dA_x = \frac{I_{yz}}{A_x z_c} \tag{2.59b}$$

where I_{yy} is the moment of inertia of area A_x about the y axis and I_{yz} is the product of inertia of A_x with respect to the axes y and z. In the similar fashion, z'_p and x'_p, the coordinates of the point of action of the force F_y on area A_y, can be written as

$$z'_p = \frac{1}{A_y z_c} \iint z^2 \, dA_y = \frac{I_{xx}}{A_y z_c} \tag{2.60a}$$

$$x'_p = \frac{1}{A_y z_c} \iint xz \, dA_y = \frac{I_{xz}}{A_y z_c} \tag{2.60b}$$

where I_{xx} is the moment of inertia of area A_y about the x axis and I_{xz} is the product of inertia of A_y about the axes x and z.

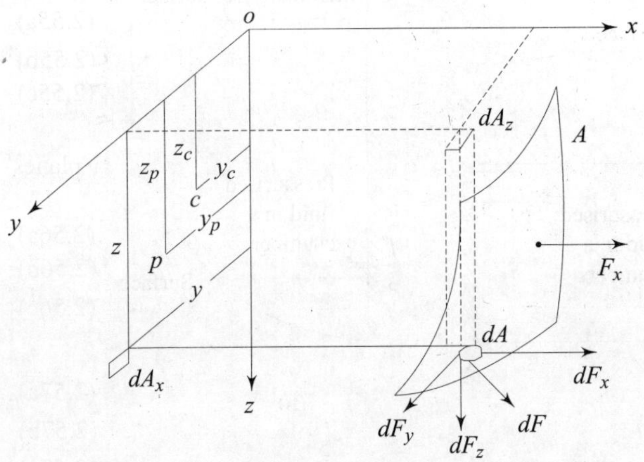

Fig. 2.19 Hydrostatic thrust on a submerged curved surface

We can conclude from Eqs (2.58), (2.59) and (2.60) that for a curved surface, the component of hydrostatic force in a horizontal direction is equal to the hydrostatic force on the projected plane surface perpendicular to that direction and acts through the centre of pressure of the projected area. From Eq. (2.58c), the vertical component of the hyrostatic force on the curved surface can be written as

$$F_z = \rho g \iint z \, dA_z = \rho g \forall \tag{2.61}$$

where $\forall$ is the volume of the body of liquid within the region extending vertically above the submerged surface to the free surface of the liquid. Therefore, the vertical component of hydrostatic force on a submerged curved surface is equal to the weight of the liquid volume vertically above the solid surface to the free surface of the liquid and acts through the centre of gravity of the liquid in that volume.

In some instances (Fig. 2.20), it is only the underside of a curved surface which is subjected to hydrostatic pressure. The vertical component of the hydrostatic thrust on the surface in this case acts upward and is equal, in magnitude, to the weight of an imaginary volume of liquid extending from the surface up to the level of the free surface. If a free surface does not exist in practice, an imaginary free surface may be considered (Fig. 2.21(a), 2.21(b)) at a height $p/\rho g$ above any point where the pressure p is known. The hydrostatic forces on the surface can then be calculated by considering the surface as a submerged one in the same fluid with an imaginary free surface as shown.

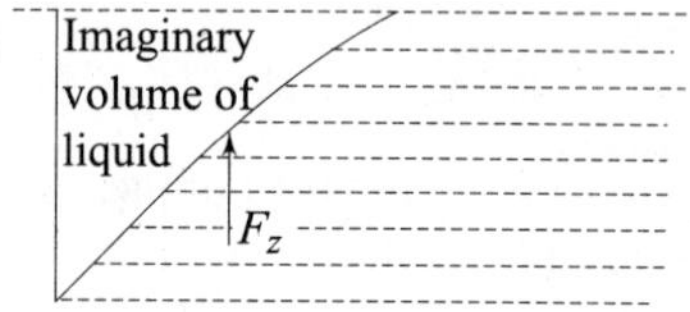

Fig. 2.20 Hydrostatic thrust on the underside of a curved surface

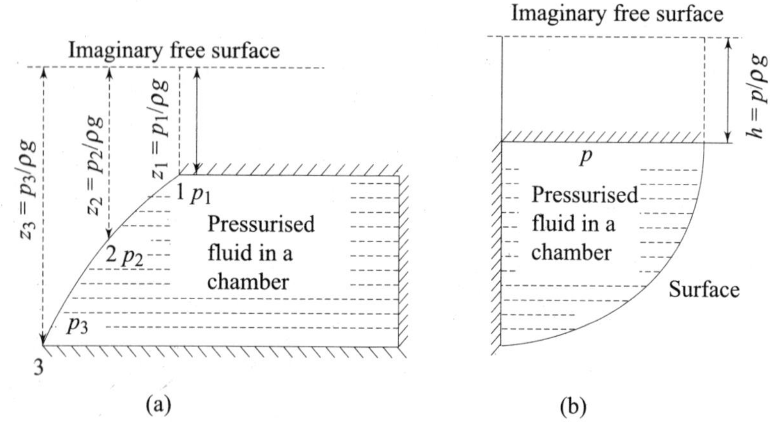

Fig. 2.21 Hydrostatic force exerted on a curved surface by a fluid without a free surface

Example 2.8

A circular cylinder of 1.8 m diameter and 2.0 m long is acted upon by water in a tank as shown in Fig. 2.22(a). Determine the horizontal and vertical components of hydrostatic force on the cylinder.

Solution

Let us consider, at a depth z from the free surface, an elemental surface on the cylinder that subtends an angle $d\theta$ at the centre. The horizontal and vertical components of hydrostatic force on the elemental area can be written as

$$dF_H = 9.81 \times 10^3 \{0.9 (1 + \cos\theta)\} (0.9\, d\theta \times 2) \sin\theta$$

and

$$dF_V = 9.81 \times 10^3 \{0.9 (1 + \cos\theta)\} (0.9\, d\theta \times 2) \cos\theta$$

Therefore, the horizontal and vertical components of the net force on the entire cylindrical surface in contact with water are given by

$$F_H = \int_0^{\pi} 9.81 \times 10^3 \{0.9(1 + \cos\theta)\}\, 1.8 \sin\theta \, d\theta \text{ N} = 31.78 \text{ kN}$$

$$F_V = \int_0^{\pi} 9.81 \times 10^3 \{0.9(1 + \cos\theta)\}\, 1.8 \cos\theta \, d\theta \text{ N}$$

$$= 9.81 \times 10^3 \times 0.9 \times 1.8 \left[\int_0^{\pi} \cos\theta \, d\theta + \int_0^{\pi} \cos^2\theta \, d\theta\right]$$

$$= 9.81 \times 10^3 \times 0.9 \times 1.8 \left[0 + \frac{\pi}{2}\right] \text{N}$$

$$= 24.96 \text{ kN}$$

Alternative method:
The horizontal component of the hydrostatic force on surface *ACB* (Fig. 2.22 (b)) is equal to the hydrostatic force on a projected plane area of 1.8 m high and 2 m long.

Therefore, $F_H = 9.81 \times 10^3 \times 0.9 \times (1.8 \times 2) \text{ N} = 31.78 \text{ kN}$

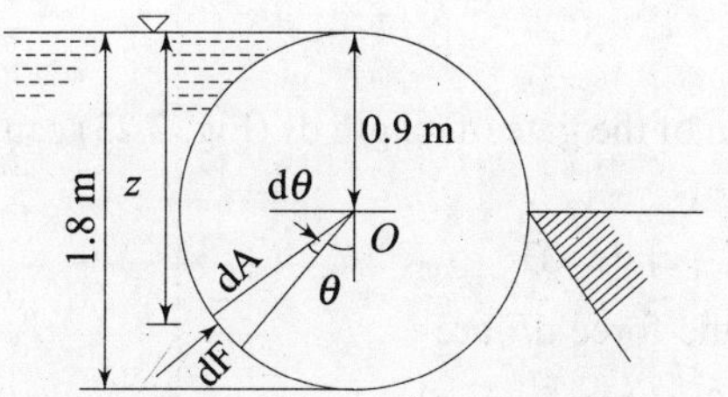

Fig. 2.22(a) A circular cylinder in a tank of water

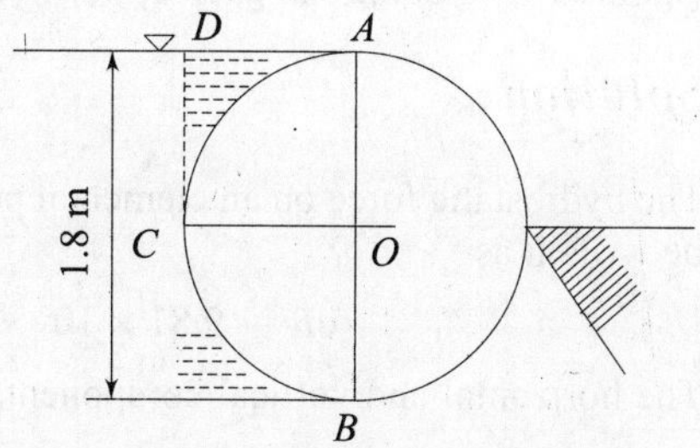

Fig. 2.22(b) A circular cylinder in a tank of water

The downward vertical force acting on surface *AC* is equal to the weight of water contained in the volume *CDAC*. The upward vertical force acting on surface *CB* is equal to the weight of water corresponding to a volume *BCDAB*.

Therefore the net upward vertical force on surface *ACB*

= Weight of water corresponding to volume of *BCDAB* – Weight of water in volume *CDAC*

= Weight of water corresponding to a volume of *BCAB* (half of the cylinder volume)

Hence, $F_V = 9.81 \times 10^3 \times \frac{1}{2} \{3.14 \times (0.9)^2 \times 2\} \text{N} = 24.96 \text{ kN}$

Example 2.9

A parabolic gate AB is hinged at A and latched at B as shown in Fig. 2.23. The gate is 3 m wide. Determine the components of net hydrostatic force on the gate exerted by water.

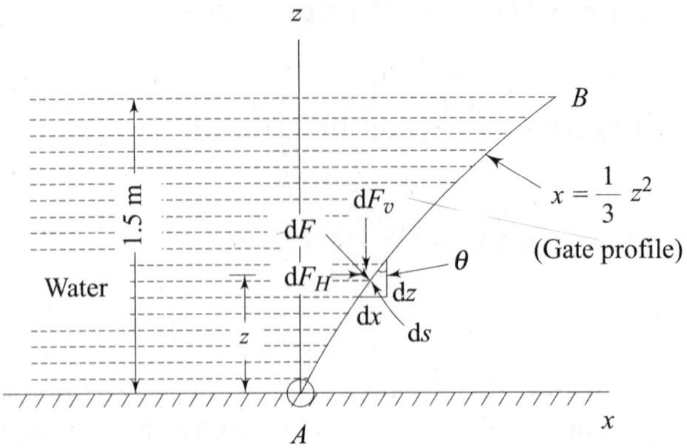

Fig. 2.23 A parabolic gate under hydrostatic pressure

Solution

The hydrostatic force on an elemental portion of the gate of length ds (Fig. 2.23) can be written as

$$dF = 9.81 \times 10^3 \times (1.5 - z)\, ds \times 3$$

The horizontal and vertical components of the force dF are

$$dF_H = 9.81 \times 10^3 \times 3(1.5 - z) \times ds \cos\theta$$
$$= 9.81 \times 3 \times (1.5 - z) \times 10^3\, dz$$

and

$$dF_V = 9.81 \times 10^3 \times 3(1.5 - z) \times ds \sin\theta$$
$$= 9.81 \times 3 \times (1.5 - z) \times 10^3\, dx$$

Therefore, the horizontal component of hydrostatic force on the entire gate

$$F_H = \int_0^{1.5} 9.81 \times 3 \times (1.5 - z) \times 10^3\, dz$$

$$= 9.81 \times 10^3 \times \frac{1.5 \times 1.5}{2} \times 3 \text{ N} = 33.11 \text{ kN}$$

The vertical component of force on the entire gate

$$F_V = \int_0^{1.5} 9.81 \times 3 \times (1.5 - z) \times 10^3 \left(\frac{2}{3}z\right) dz$$

$$\left(\text{Since } x = \frac{1}{3}z^2 \text{ for the gate profile, } dx = \frac{2}{3}z\ dz\right)$$

$$= \frac{2}{3} \times 9.81 \times 10^3 \times \frac{(1.5)^3}{6} \times 3 \text{ N} = 11.04 \text{ kN}$$

Example 2.10

A sector gate, of radius 4 m and length 5 m, controls the flow of water in a horizontal channel. For the equilibrium condition shown in Fig. 2.24, determine the total thrust on the gate.

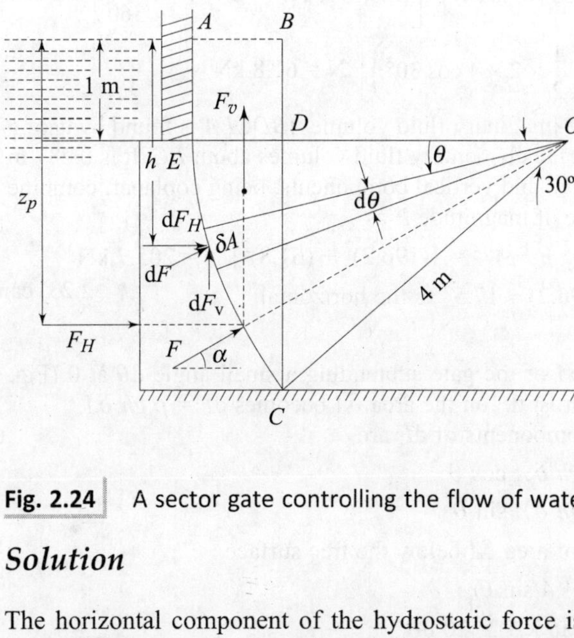

Fig. 2.24 A sector gate controlling the flow of water in a channel

Solution

The horizontal component of the hydrostatic force is the thrust which would be exerted by the water on a projected plane surface in a vertical plane. The height of this projected surface is 4 sin 30° m = 2 m and its centroid is (1 + 2/2) m = 2 m below the free surface.

Therefore, the horizontal component of hydrostatic thrust

$$F_H = \rho g \bar{h} A = 1000 \times 9.81 \times 2 \times (5 \times 2) \text{ N} = 196.2 \text{ kN}$$

The line of action of F_H passes through the centre of pressure which is at a distance z_p below the free surface, given by (see Eq. 2.50a)

$$z_P = 2 + \frac{5(2)^3}{12 \times (5 \times 2) \times 2} = 2.167 \text{ m}$$

The vertical component of the hydrostatic thrust F_V

= Weight of imaginary water contained in the volume *ABDCEA*

Now, Volume $ABDCEA =$ Volume $ABDEA$ + Volume $OECO$ – Volume $ODCO$

Volume $ABDEA = 5 \times AB \times BD = 5 \times (4 - 4\cos 30°) \times 1$

$= 5 \times 0.536$

Volume $OECO = 5 \times \pi \times (OC)^2 \times 30/360$

$= 5 \times \pi \times (4)^2 \times (30/360)$

Volume $ODCO = 5 \times \frac{1}{2} \times 4 \sin 30° \times 4\cos 30°$

$= 5 \times \frac{1}{2} \times 2 \times 4 \cos 30°$

Therefore,

$$F_V = 1000 \times 9.81 \times 5 \left[(0.536 \times 1) + \left(\pi \times 4^2 \times \frac{30}{360}\right) - \left(\frac{1}{2} \times 2 \times 4\cos 30°\right)\right] \text{N} = 61.8 \text{ kN}$$

The centre of gravity of the imaginary fluid volume *ABDCEA* is found by taking moments of the weights of all the elementary fluid volumes about *BC*. It is 0.237 m to the left of *BC*. The horizontal and vertical components, being coplanar, combine to give a single resultant force of magnitude *F* as

$$F = (F_H^2 + F_V^2)^{1/2} = \{(196.2)^2 + (61.8)^2\}^{1/2} = 205.7 \text{ kN}$$

at an angle $\alpha = \tan^{-1}(61.8/196.2) \approx 17.5°$ to the horizontal.

Alternative method:

Consider an elemental area δA of the gate subtending a small angle $d\theta$ at 0 (Fig. 2.24). Then the hydrostatic thrust dF on the area δA becomes $dF = \rho g h\, \delta A$. The horizontal and vertical components of dF are

$$dF_H = \rho g h\, \delta A \cos\theta$$

$$dF_y = \rho g h\, \delta A \sin\theta$$

where *h* is the vertical depth of area δA below the free surface.

Now, $h = (1 + 4\sin\theta)$

and $\delta A = (4\, d\theta \times 5) = 20\, d\theta$

Therefore, the total horizontal and vertical components are

$$F_H = \int dF_H = 1000 \times 9.81 \times 20 \int_0^{\pi/6} (1 + 4\sin\theta)\cos\theta\, d\theta \text{N} = 196.2 \text{ kN}$$

$$F_V = 1000 \times 9.81 \times 20 \int_0^{\pi/6} (1 + 4\sin\theta)\sin\theta\, d\theta \text{N} = 61.8 \text{ kN}$$

Since all the elemental thrusts are perpendicular to the surface, their lines of action pass through *O* and that of the resultant force therefore also passes through *O*.

2.7 BUOYANCY

When a body is either wholly or partially immersed in a fluid, the hydrostatic lift due to the net vertical component of hydrostatic pressure forces experienced by the body is called the *buoyant force* and the phenomenon is called *buoyancy*. Consider a solid body of arbitrary shape completely submerged in a homogeneous liquid as shown in Fig. 2.25. Hydrostatic pressure forces act on the entire surface of the body.

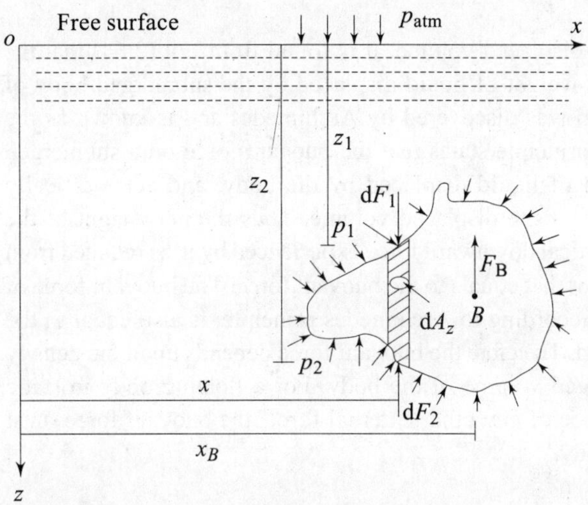

Fig. 2.25 Buoyant force on a submerged body

It is evident according to the earlier discussion in Section 2.6, that the resultant horizontal force in any direction for such a closed surface is always zero. To calculate the vertical component of the resultant hydrostatic force, the body is considered to be divided into a number of elementary vertical prisms. The vertical forces acting on the two ends of such a prism of cross section dA_z (Fig. 2.25) are respectively

$$dF_1 = (p_{atm} + p_1)\, dA_z = (p_{atm} + \rho g z_1) dA_z \tag{2.62a}$$

$$dF_2 = (p_{atm} + p_2)\, dA_z = (p_{atm} + \rho g z_2)\, dA_z \tag{2.62b}$$

Therefore, the buoyant force (the net vertically upward force) acting on the elemental prism is

$$dF_B = dF_2 - dF_1 = \rho g\, (z_2 - z_1)\, dA_z = \rho g d\forall \tag{2.63}$$

where $d\forall$ is the volume of the prism.

Hence, the buoyant force F_B on the entire submerged body is obtained as

$$F_B = \iiint_{\forall} \rho g\, d\forall = \rho g \forall \tag{2.64}$$

where $\forall$ is the total volume of the submerged body. The line of action of the force F_B can be found by taking moment of the force with respect to the z axis. Thus

$$x_B F_B = \int x \, dF_B \tag{2.65}$$

Substituting for dF_B and F_B from Eqs. (2.63) and (2.64) respectively into Eq. (2.65), the x coordinate of the centre of buoyancy is obtained as

$$x_B = \frac{1}{\forall} \iiint_{\forall} x \, d\forall \tag{2.66}$$

which is the centroid of the displaced volume. It is found from Eq. (2.64) that the buoyant force F_B equals the weight of liquid displaced by the submerged body of volume $\forall$. This phenomenon was discovered by Archimedes and is known as the *Archimedes principle.* This principle states that the buoyant force on a submerged body is equal to the weight of liquid displaced by the body, and acts vertically upward through the centroid of the displaced volume. Thus the net weight of the submerged body, (the net vertical downward force experienced by it) is reduced from its actual weight by an amount that equals to the buoyant force. The buoyant force of a partially immersed body, according to Archimedes principle, is also equal to the weight of the displaced liquid. Therefore the buoyant force depends upon the density of the fluid and the submerged volume of the body. For a floating body in static equilibrium and in the absence of any other external force, the buoyant force must balance the weight of the body.

Example 2.11

A block of steel (sp. gr. 7.85) floats at a mercury water interface as in Fig. 2.26. What is the ratio of a and b for this condition? (sp. gr. of mercury is 13.57).

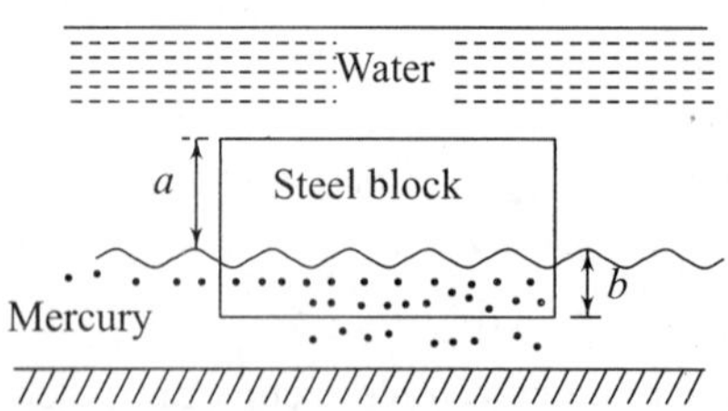

Fig. 2.26 A steel block floating at mercury water interface

Solution

Let the block have a uniform cross-sectional area A.

Under the condition of floating equilibrium as shown in Fig. 2.26,

Weight of the body = Total buoyancy force acting on it

$$A \times (a + b)(7850) \times g = (b \times 13.57 + a) \times A \times g \times 10^3$$

Hence $\quad 7.85\,(a+b) = 13.57b + a$

or $\quad \dfrac{a}{b} = \dfrac{5.72}{6.85} = 0.835$

Example 2.12

An aluminium cube 150 mm on a side is suspended by a string in oil and water as shown in Fig. 2.27. The cube is submerged with half of it being in oil and the other half in water. Find the tension in the string if the specific gravity of oil is 0.8 and the specific weight of aluminium is 25.93 kN/m^3.

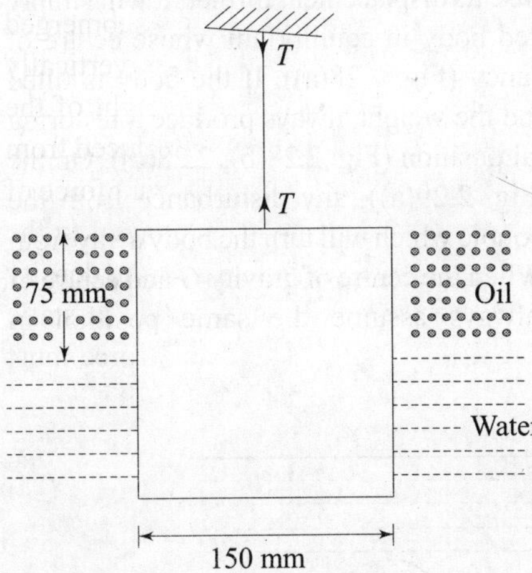

Fig. 2.27 An aluminium cube suspended in an oil and water system

Solution

Tension T in the string can be written in consideration of the equilibrium of the cube as

$$\begin{aligned} T &= W - F_B \\ &= 25.93 \times 10^3 \times (.15)^3 - 9.81 \times 10^3\,[(.15^3 \times 0.5 \times 0.8 \\ &\quad + (.15)^3 \times 0.5 \times 1]\ \text{N} \\ &= 57.71\ \text{N} \end{aligned}$$

(W = Weight of the cube and F_B = Total buoyancy force on the cube)

2.8 STABILITY OF UNCONSTRAINED BODIES IN FLUIDS

2.8.1 Submerged Bodies

For a body not otherwise restrained, it is important to know whether it will rise or fall in a fluid, and also whether an orginally vertical axis in the body will remain vertical.

When a body is submerged in a liquid, the equilibrium requires that the weight of the body acting through its cetre of gravity should be colinear with an equal hydrostatic lift acting through the centre of buoyancy. In general, if the body is not homogeneous in its distribution of mass over the entire volume, the location of centre of gravity G does not coincide with the centre of volume, i.e., the centre of buoyancy B. Depending upon the relative locations of G and B, a floating or submerged body attains different states of equilibrium, namely, (i) *stable equilibrium* (ii) *unstable equilibrium* and (iii) *neutral equilibrium.*

A body is said to be in stable equilibrium, if it, being given a small angular displacement and hence released, returns to its original position by retaining the originally vertical axis as vertical. If, on the other hand, the body does not return to its original position but moves further from it, the equilibrium is unstable. In neutral equilibrium, the body having been given a small displacement and then released will neither return to its original position nor increase its displacement further, it will simply adopt its new position. Consider a submerged body in equilibrium whose centre of gravity is located below the centre of buoyancy (Fig. 2.28(a)). If the body is tilted slightly in any direction, the buoyant force and the weight always produce a restoring couple trying to return the body to its original position (Fig. 2.28(b), 2.28(c)). On the other hand, if point G is above point B (Fig. 2.29(a)), any disturbance from the equilibrium position will create a destroying couple which will turn the body away from its original position (Figs. 2.29(b), 2.29(c)). When the centre of gravity G and centre of buoyancy B coincides, the body will always assume the same position in

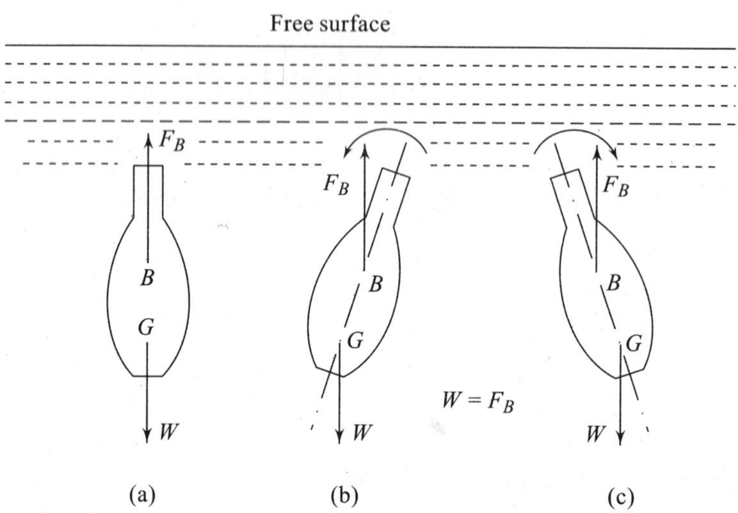

Fig. 2.28 A submerged body in stable equilibrium

which it is placed (Fig. 2.30) and hence it is in neutral equilibrium. Therefore, it can be concluded from the above discussion that a submerged body will be in stable, unstable or neutral equilibrium if its centre of gravity is below, above or coincident with the centre of buoyancy respectively (Fig. 2.31).

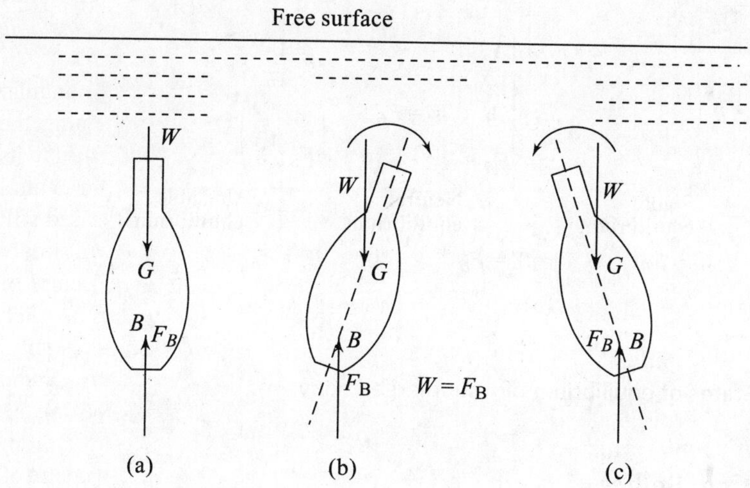

Fig. 2.29 A submerged body in unstable equilibrium

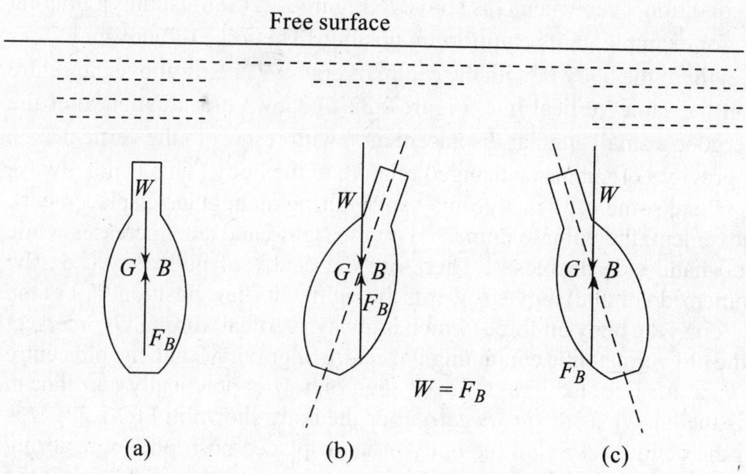

Fig. 2.30 A submerged body in neutral equilibrium

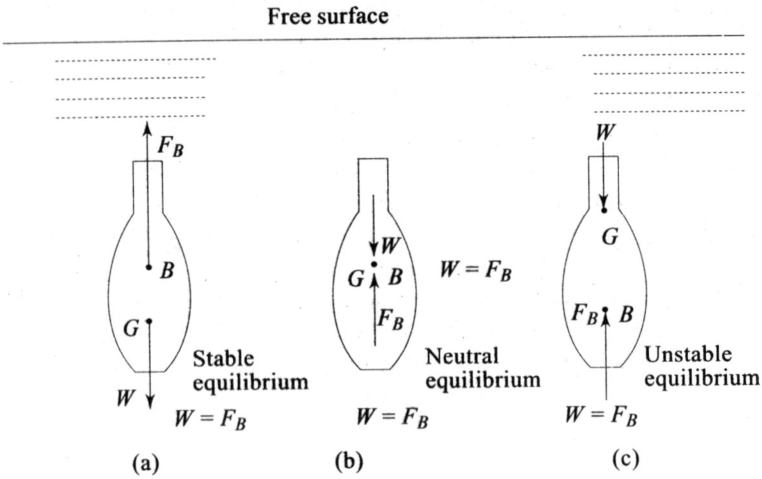

Fig. 2.31 States of equilibrium of a submerged body

2.8.2 Floating Bodies

The condition for angular stability of a floating body is a little more complicated. This is because, when the body undergoes an angular displacement about a horizontal axis, the shape of the immersed volume changes and so the centre of buoyancy moves relative to the body. As a result, stable equlibrium can be achieved, under certain condition, even when G is above B. Figure 2.32(a) illustrates a floating body—a boat, for example, in its equilibrium position. The force of buoyancy F_B is equal to the weight of the body W with the centre of gravity G being above the centre of buoyancy in the same vertical line. Figure 2.32(b) shows the situation after the body has undergone a small angular displacement θ with respect to the vertical axis. The centre of gravity G remains unchanged relative to the body (This is not always true for ships where some of the cargo may shift during an angular displacement). During the movement, the volume immersed on the right-hand side increases while that on the left-hand side decreases. Therefore the centre of buoyancy (i.e., the centroid of immersed volume) moves towards the right to its new position B'. Let the new line of action of the buoyant force (which is always vertical) through B' intersect the axis BG (the old vertical line containing the centre of gravity G and the old centre of buoyancy B) at M. For small values of θ, the point M is practically constant in position and is the known as the *metacentre*. For the body shown in Fig. 2.32, M is above G, and the couple acting on the body in its displaced position is a restoring couple which tends to turn the body to its original position. If M were below G, the couple would be an overturning couple and the original equilibrium would have been unstable. When M coincides with G, the body will assume its new position without any further movement and thus will be in neutral equilibrium. Therefore, for a floating body, the stability is determined not simply by the relative position of B and G, but rather by the relative position of M and G. The distance of the metacentre above G along the line BG is known as metacentric height GM which can be written as

$$GM = BM - BG$$

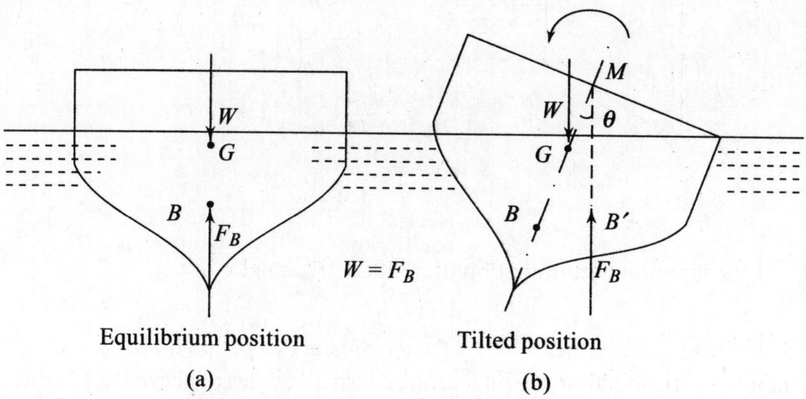

Fig. 2.32 A floating body in stable equilibrium

Hence the condition of stable equilibrium for a floating body can be expressed in terms of metacentric height as follows:

$GM > 0$	(M is above G)	Stable equilibrium
$GM = 0$	(M coinciding with G)	Neutral equilibrium
$GM < 0$	(M is below G)	Unstable equilibrium

The angular displacement of a boat or ship about its longitudinal axis is known as 'rolling' while that about its transverse axis is known as 'pitching'.

2.8.3 Experimental Determination of Metacentric Height

A simple experiment is usually conducted to determine the metacentric height. Suppose that for the boat, shown in Fig. 2.33, the metacentric height corresponding to 'roll' about the longitudinal axis (the axis perpendicular to the plane of the figure) is required. Let a weight P be moved transversely across the deck (which was initially horizontal) so that the boat heels through a small angle θ and comes to rest at this new position of equilibrium. The new centres of gravity and buoyancy are therefore again vertically in line. The movement of the weight P through a distance x in fact causes a parallel shift of the centre of gravity (centre of gravity of the boat including P) from G to G'.

Hence, $$P \cdot x = W GG'$$

Again, $$GG' = GM \tan\theta$$

Therefore, $$GM = \frac{P \cdot x}{W} \cot\theta \tag{2.67}$$

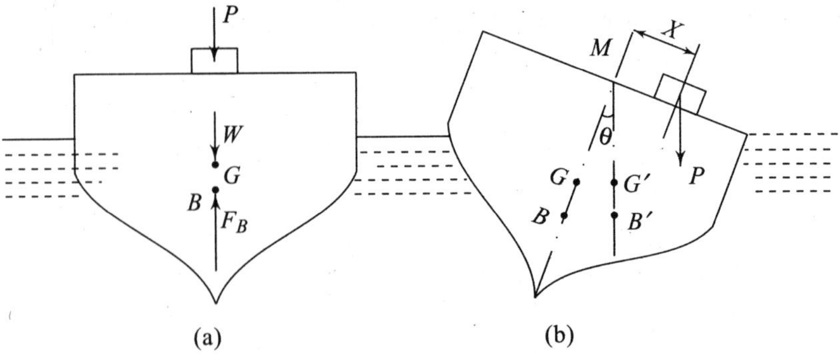

Fig. 2.33 Experimental determination of metacentric height

The angle of heel θ can be measured by the movement of a plumb line over a scale. Since the point M corresponds to the metacentre for small angles of heel only, the true metacentric height is the limiting value of GM as $\theta \to 0$. This may be determined from a graph of nominal values of GM calculated from Eq. (2.67) for various values of θ (positive and negative).

It is well understood that the metacentric height serves as the criterion of stability for a floating body. Therefore it is desirable to establish a relation between the metacentric height and the geometrical shape and dimensions of a body so that one can determine the position of the metacentre beforehand and then construct the boat or the ship accordingly. This may be done simply by considering the shape of the hull. Figure 2.34(a) shows the cross section, perpendicular to the axis of rotation, in which the centre of buoyancy B lies at the initial equilibrium position. The position of the body after a small angular displacement is shown in Fig. 2.34(b). The section on the left, indicated by cross-hatching, has emerged from the liquid, whereas the cross-hatched section on the right has moved down into the liquid. It is assumed that there is no overall vertical movement; thus the vertical equilibrium is undisturbed. As the total weight of the body remains unaltered so does the volume immersed, and therefore the volumes corresponding to the cross-hatched sections are equal. This is so if the planes of flotation for the equilibrium and displaced positions intersect along the centroidal axes of the planes. The coordinate axes are chosen through O as origin. OY is perpendicular to the plane of Fig 2.34(a) and 2.34(b), OY lies in the original plane of flotation (Fig. 2.34(c)) and OZ is vertically downwards in the original equilibrium position. The total immersed volume is considered to be made up of elements each underneath an area dA in the plane of flotation as shown in Figs 2.34(a) and 2.34(c). The by definition is the centroid of the immersed volume (the liquid being assumed homogeneous). The x coordinate x_B of the centre of buoyancy may therefore be determined by taking moments of elemental volumes about the yz plane as

$$\forall x_B = \int (z\, \mathrm{d}A)\, x \tag{2.68}$$

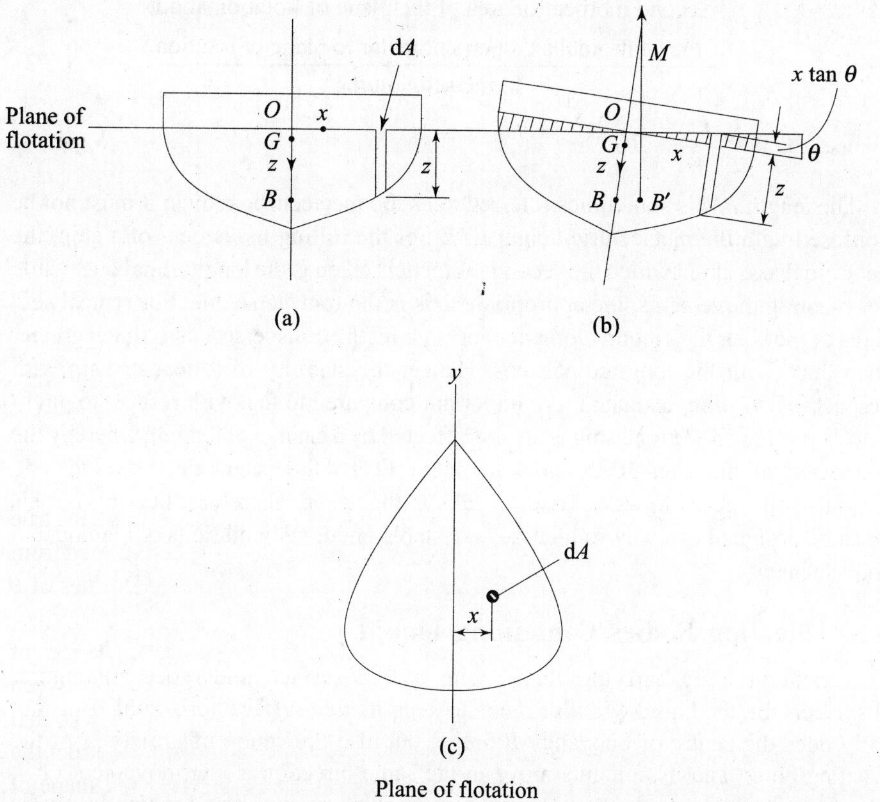

Fig. 2.34 Analysis of metacentric height

After displacement, the depth of each elemental volume immersed is $z + x \tan \theta$ and hence the new centre of buoyancy x'_B can be written as

$$\not{V} x'_B = \int (z + x \tan \theta)\, \mathrm{d}A\, x \tag{2.69}$$

Subtracting Eq. (2.68) from Eq. (2.69), we get

$$\not{V}(x'_B - x_B) = \int x^2 \tan \theta\, \mathrm{d}A = \tan \theta \int x^2\, \mathrm{d}A \tag{2.70}$$

The second moment of area of the plane of flotation about the axis Oy is defined as

$$I_{yy} = \int x^2\, \mathrm{d}A \tag{2.71}$$

Again, for small angular displacements,

$$x'_B - x_B = BM \tan \theta \tag{2.72}$$

With the help of Eqs (2.71) and (2.72), Eq. (2.70) can be written as

$$BM = \frac{I_{yy}}{\not{V}}$$

$$= \frac{\text{Second moment of area of the plane of flotation about the centroidal axis perpendicular to plane of rotation}}{\text{Immersed volume}} \quad (2.73)$$

Hence, $$GM = \frac{I_{yy}}{\forall} - BG \quad (2.74)$$

The length *BM* is sometimes referred to as the metacentric radius; it must not be confused with the metacentric height *GM*. For the rolling movement of a ship, the centroidal axis about which the second moment is taken is the longitudinal one, while for pitching movements, the appropriate axis is the transverse one. For typical sections of the boat, the second moment of area about the transverse axis is much greater than that about the longitudinal axis. Hence, the stability of a boat or ship with respect to its rolling is much more important compared to that with respect to pitching. The value of *BM* for a ship is always affected by a change of loading whereby the immersed volume alters. If the sides are not vertical at the water-line, the value of I_{yy} may also change as the vessel rises or falls in the water. Therefore, floating vessels must be designed in a way so that they are stable under all conditions of loading and movement.

2.8.4 Floating Bodies Containing Liquid

If a floating body carrying liquid with a free surface undergoes an angular displacement, the liquid will also move to keep its free surface horizontal. Thus not only does the centre of buoyancy *B* move, but also the centre of gravity *G* of the floating body and its contents move in the same direction as the movement of *B*. Hence the stability of the body is reduced. For this reason, liquid which has to be carried in a ship is put into a number of separate compartments so as to minimise its movement within the ship.

2.8.5 Period of Oscillation

It is observed from the foregoing discussion that the restoring couple caused by the buoyant force and gravity force acting on a floating body displaced from its equilibrium position is $W \cdot GM \sin\theta$ (Fig. 2.32). Since the torque equals the mass moment of inertia (i.e., second moment of mass) multiplied by angular acceleration, it can be written as

$$W(GM)\sin\theta = -I_M\,(\mathrm{d}^2\theta/\mathrm{d}t^2) \quad (2.75)$$

where I_M represents the mass moment of inertia of the body about its axis of rotation. The minus sign in the RHS of Eq. (2.75) arises since the torque is a retarding one and decreases the angular acceleration. If θ is small, $\sin\theta \approx \theta$ and hence Eq. (2.75) can be written as

$$\frac{\mathrm{d}^2\theta}{\mathrm{d}t^2} + \frac{W \cdot GM}{I_M}\theta = 0 \quad (2.76)$$

Equation (2.76) represents a simple harmonic motion. The time period (i.e., the time of a complete oscillation from one side to the other and back again) equals to $2\pi(I_M/W \cdot GM)^{1/2}$. The oscillation of the body results in a flow of the liquid around it and this flow has been disregarded here. In practice, of course, viscosity in the liquid introduces a damping action which quickly suppresses the oscillation unless further disturbances such as waves cause new angular displacements.

The metacentric height of ocean-going vessel is usually of the order of 0.3 m to 1.2 m. An increase in the metacentric height results in a better stability but reduces the period of roll, and so the vessel is less comfortable for passengers. In cargo vessels the metacentric height and the period of roll are adjusted by changing the position of the cargo. If the cargo is placed further from the centre-line, the moment of inertia of the vessel and consequently the period may be increased with little sacrifice of stability. On the other hand, in warships and racing yachts, stability is more important than comfort, and such vessels have larger metacentric heights.

Example 2.13

A cube of side a floats with one of its axes vertical in a liquid of specific gravity S_L. If the specific gravity of the cube material is S_c, find the values of S_L/S_c for the metacentric height to be zero.

Solution

Let the cube float with h as the submerged depth, as shown in Fig. 2.35.

For equilibrium of the cube,

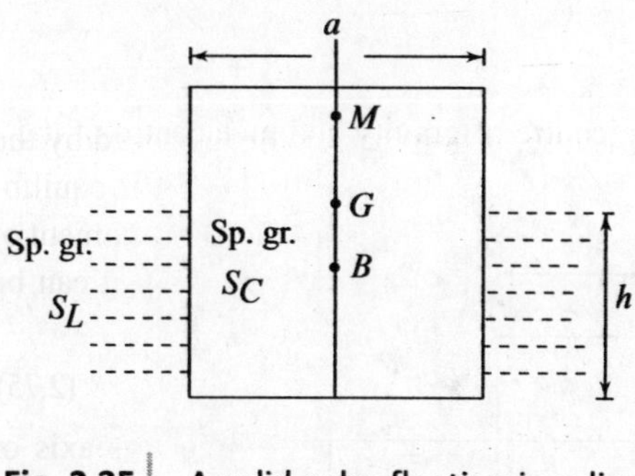

Fig. 2.35 A solid cube floating in a liquid

$$\text{Weight} = \text{Buoyant force}$$

$$a^3 S_c \times 10^3 \times 9.81 = h a^2 \times S_L \times 10^3 \times 9.81$$

or,
$$h = a\,(S_c/S_L) = a/x$$

where
$$S_L/S_c = x$$

The distance between the centre of buoyancy B and centre of gravity G becomes

$$BG = \frac{a}{2} - \frac{h}{2} = \frac{a}{2}\left(1 - \frac{1}{x}\right)$$

Let M be the metacentre, then

$$BM = \frac{I}{V\!\!\!-} = \frac{a\left(\frac{a^3}{12}\right)}{a^2 h} = \frac{a^4}{12\, a^2\left(\frac{a}{x}\right)} = \frac{ax}{12}$$

The metacentric height $MG = BM - BG = \frac{ax}{12} - \frac{a}{2}\left(1 - \frac{1}{x}\right)$

According to the given condition,

$$MG = \frac{ax}{12} - \frac{a}{2}\left(1 - \frac{1}{x}\right) = 0$$

or $$x^2 - 6x + 6 = 0$$

which gives $$x = \frac{6 \pm \sqrt{12}}{2} = 4.732, 1.268$$

Hence, $$S_L/S_c = 4.732 \quad \text{or} \quad 1.268$$

Example 2.14

A rectangular barge of width b and a submerged depth of H has its centre of gravity at the waterline. Find the metacentric height in terms of b/H, and hence show that for stable equilibrium of the burge $b/H \geq \sqrt{6}$.

Solution

Let B, G and M be the centre of buoyancy, centre of gravity and metacentre of the burge (Fig. 2.36), respectively.

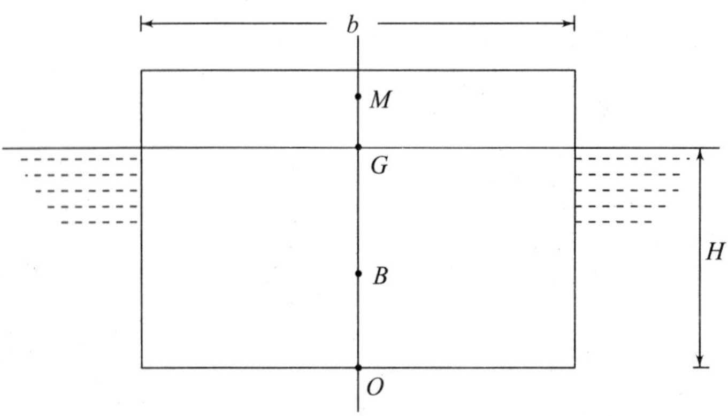

Fig. 2.36 A rectangular barge in water

Now, $OB = H/2$

and $OG = H$ (as given in the problem)

Hence, $$BG = OG - OB = H - \frac{H}{2} = \frac{H}{2}$$

Again, $$BM = \frac{I}{V} = \frac{L\,b^3}{12 \times L \times b \times H} = \frac{b^2}{12H}$$

where, L is the length of the barge in a direction perpendicular to the plane of the Fig. 2.36.

Therefore, $$MG = BM - BG = \frac{b^2}{12H} - \frac{H}{2} = \frac{H}{2}\left\{\frac{1}{6}\left(\frac{b}{H}\right)^2 - 1\right\}$$

For stable equilibrium of the barge, $MG \geq 0$

Hence, $$\frac{H}{2}\left\{\frac{1}{6}\left(\frac{b}{H}\right)^2 - 1\right\} \geq 0$$

which gives $$b/H \geq \sqrt{6}$$

Example 2.15

A solid hemisphere of density ρ and radius r floats with its plane base immersed in a liquid of density ρ_l ($\rho_l > \rho$). Show that the equilibrium is stable and the metacentric height is

$$\frac{3}{8} r \left(\frac{\rho_l}{\rho} - 1\right)$$

Solution

The hemisphere in its floating condition is shown in Fig. 2.37. Let $\forall$ be the submerged volume. Then from equilibrium under floating condition,

$$\frac{2}{3}\pi r^3 \times \rho = \forall \times \rho_l$$

or $$\forall = \frac{2}{3}\pi r^3 \times \frac{\rho}{\rho_l}$$

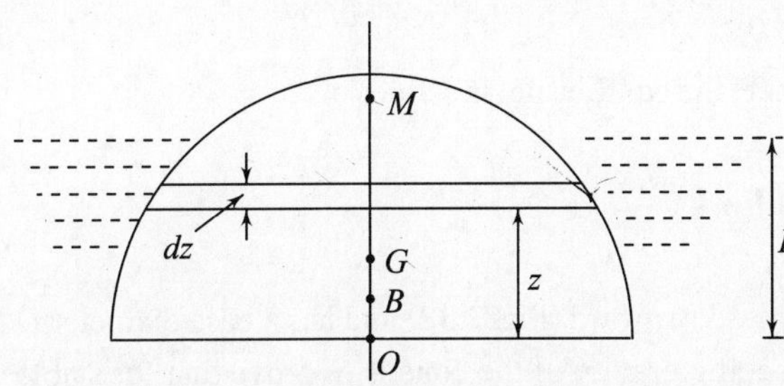

Fig. 2.37 A solid hemisphere floating in a liquid

The centre of gravity G will lie on the axis of symmetry of the hemisphere. The distance of G along this line from the base of the hemisphere can be found by taking moments of elemental circular strips (Fig. 2.37) about the base as

$$OG = \frac{\int_0^r \pi (r^2 - z^2)\, z\, dz}{\frac{2}{3}\pi r^3} = \frac{3}{8} r$$

In a similar way, the location of centre of buoyancy which is the centre of immersed volume $\forall$ is found as

$$OB = \frac{\int_0^H \pi (r^2 - z^2)\, z\, dz}{\frac{2}{3}\pi r^3 \frac{\rho}{\rho_l}} = \frac{3}{8}\frac{\rho_l}{\rho} r \frac{H^2}{r^2}\left(2 - \frac{H^2}{r^2}\right) \quad (2.77)$$

where H is the depth of immersed volume as shown in Fig. 2.37.

If r_h is the radius of cross section of the hemisphere at water line, then we can write

$$H^2 = r^2 - r_h^2$$

Substituting the value of H in Eq. (2.77), we have

$$OB = \frac{3}{8}\frac{\rho_l}{\rho} r \left(1 - \frac{r_h^4}{r^4}\right)$$

The height of the metacentre M above the centre of buoyancy B is given by

$$BM = \frac{I}{V} = \frac{\pi r_h^4}{4\left\{\left(\frac{2}{3}\pi r^3\right)\frac{\rho}{\rho_l}\right\}} = \frac{3}{8}\frac{\rho_l}{\rho} r \frac{r_h^4}{r^4}$$

Therefore, the metacentric height MG becomes

$$MG = MB - BG = MB - (OG - OB)$$

$$= \frac{3}{8}\frac{\rho_l}{\rho} r \frac{r_h^4}{r^4} - \frac{3}{8} r + \left[\frac{3}{8}\frac{\rho_l}{\rho} r \left(1 - \frac{r_h^4}{r^4}\right)\right]$$

$$= \frac{3}{8} r \left(\frac{\rho_l}{\rho} - 1\right)$$

Since $\rho_l > \rho$, $MG > 0$, and hence, the equilibrium is stable.

Example 2.16

A cone floats in water with its apex downward (Fig. 2.38) and has a base diameter D and a vertical height H. If the specific gravity of the cone is S, prove that for stable equilibrium,

$$H^2 < \frac{1}{4}\left(\frac{D^2 S^{1/3}}{1 - S^{1/3}}\right)$$

Solution

Let the submerged height of the cone under floating condition be h, and the diameter of the cross section at the plane of flotation be d (Fig. 2.38).

For the equilibrium,

Weight of the cone = Total buoyancy force

$$\frac{1}{3}\pi\left(\frac{D^2}{4}\right)H \cdot S = \frac{1}{3}\pi\left(\frac{d^2}{4}\right)\cdot h \tag{2.78}$$

Again from geometry,

$$d = D\frac{h}{H} \tag{2.79}$$

Using the value of d from Eq. (2.79) in Eq. (2.78), we get

$$h = H S^{1/3} \tag{2.80}$$

The centre of gravity G of the cone is found out by considering the mass of cylindrical element of height dz and diameter Dz/H, and its moment about the apex 0 in the following way:

$$OG = \frac{\int_0^H \pi \frac{D^2 z^2}{4H^2} z\,dz}{\frac{1}{3}\pi\frac{D^2}{4}H} = \frac{3}{4}H$$

The centre of buoyancy B is the centre of volume of the submerged conical part and hence $OB = \frac{3}{4}h$.

Therefore, $BG = OG - OB = \frac{3}{4}(H - h)$

Substituting h from Eq. (2.80) we can write

$$BG = \frac{3}{4}H(1 - S^{1/3}) \tag{2.81}$$

If M is the metacentre, the metacentric radius BM can be written according to Eq. (2.73) as

$$BM = \frac{1}{\forall} = \frac{\pi d^4}{64 \times \frac{1}{3}\pi(d^2/4)h} = \frac{3}{16}\frac{d^2}{h}$$

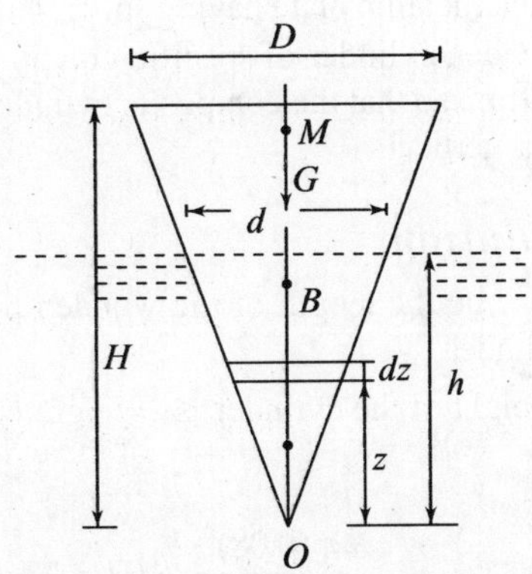

Fig. 2.38 A solid cone floating in water

Substituting d from Eq. (2.79) and h from Eq. (2.80), we can write

$$BM = \frac{3}{16}\frac{D^2}{H} S^{1/3}$$

The metacentric height

$$MG = BM - BG$$

$$= \frac{3}{16}\frac{D^2}{H} S^{1/3} - \frac{3}{4} H(1 - S^{1/3})$$

For stable equilibrium, $MG > 0$

Hence,
$$\frac{3}{16}\frac{D^2}{H} S^{1/3} - \frac{3}{4} H(1 - S^{1/3}) > 0$$

or
$$\frac{D^2}{H^2} S^{1/3} - 4(1 - S^{1/3}) > 0$$

or
$$\frac{D^2}{H^2} S^{1/3} > 4(1 - S^{1/3})$$

or
$$\frac{H^2}{D^2 S^{1/3}} < \frac{1}{4(1 - S^{1/3})}$$

Hence,
$$H^2 < \frac{D^2 S^{1/3}}{4(1 - S^{1/3})}$$

Example 2.17

An 80 mm diameter composite solid cylinder consists of an 80 mm diameter 20 mm thick metallic plate having sp. gr. 4.0 attached at the lower end of an 80 mm diameter wooden cylinder of specific gravity 0.8. Find the limits of the length of the wooden portion so that the composite cylinder can float in stable equilibrium in water with its axis vertical.

Solution

Let l be the length of the wooden piece. For floating equilibrium of the composite cylinder,

Weight of the cylinder ≤ Weight of the liquid of the same volume as that of the cylinder

Hence,
$$\frac{\pi (0.08)^2}{4}\{0.02 \times 4 + 0.8l\} \le \frac{\pi (0.08)^2}{4}\{0.02 + l\}$$

From which
$$l \ge 0.3\,\text{m}$$

Hence, the minimum length of the wooden portion $l_{minimum} = 0.3 \text{ m} = 300 \text{ mm}$.

The minimum length corresponds to the situation when the cylinder will just float with its top edge at the free surface (Fig. 2.39a). For any length l greater than 300 mm, the cylinder will always float in equilibrium with a part of its length submerged as shown in (Fig. 2.39b). The upper limit of l would be decided from the consideration of stable equilibrium (angular stability) of the cylinder.

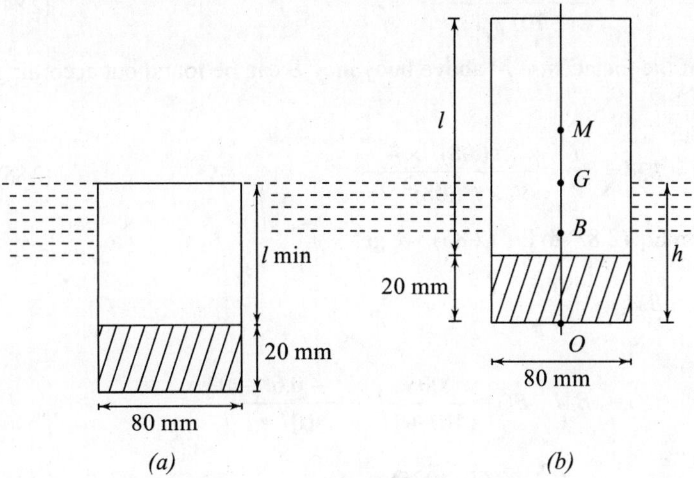

Fig. 2.39 A composite cylinder floating in water

For stable equilibrium,

Metacentric height > 0 (2.82)

The location of centre of gravity G of the composite cylinder can be found as

$$OG = \frac{\pi \dfrac{(.08)^2}{4}[.02 \times 4 \times .01 + l \times .8(0.5l + 0.02)]}{\pi \dfrac{(.08)^2}{4}(.08 + .8l)}$$

$$= \frac{5l^2 + 0.2l + 0.01}{10l + 1}$$

The submerged length h of the wooden cylinder is found from the consideration of floating equilibrium as

Weight of the cylinder = Buoyancy force

$$\frac{\pi(.08)^2}{4}(.02 \times 4 + .8l\} = \frac{\pi(.08)^2}{4} \times h$$

or

$$h = 0.08\,(10l + 1) \tag{2.83}$$

The location of the centre of buoyancy B can therefore be expressed as $OB = h/2 = 0.04\,(10\,l + 1)$

Now
$$BG = OG - OB = \frac{5l^2 + 0.2l + 0.01}{10l + 1} - 0.04(10l + 1)$$

$$= \frac{l^2 - 0.6l - .03}{10l + 1} \qquad (2.84)$$

The location of the metacentre M above buoyancy B can be found out according to Eq. (2.73) as

$$BM = \frac{I}{V} = \frac{\pi(.08)^4 \times 4}{64 \times \pi(.08)^2 \times h} \qquad (2.85)$$

Substituting h from Eq. (2.83) to Eq. (2.85), we get

$$BM = \frac{.005}{10l + 1}$$

Therefore,
$$MG = BM - BG = \frac{.005}{10l + 1} - \frac{l^2 - 0.6l - .03}{10l + 1}$$

$$= \frac{-(l^2 - 0.6l - .035)}{10l + 1}$$

Using the criterion for stable equilibrium as $MG > 0$ we have,

$$\frac{-(l^2 - 0.6l - .035)}{10l + 1} > 0$$

or
$$l^2 - 0.6\,l - .035 < 0$$

or
$$(l - 0.653)\,(l + 0.053) < 0$$

The length l can never be negative. Hence, the physically possible condition is

$$l - 0.653 < 0$$

or
$$l < 0.653$$

2.9 FLUIDS UNDER RIGID BODY MOTION (RELATIVE EQUILIBRIUM)

In certain instances of fluid flow, the behaviour of fluids in motion can be found from the principles of hydrostatics. Fluids in such motions are said to be in relative equilibrium or in relative rest. These situations arise when a fluid flows with uniform velocity without any acceleration or with uniform acceleration.

2.9.1 Flow with Constant Acceleration

When a fluid moves uniformly in a straight line without any acceleration, there is neither shear force nor inertia force acting on the fluid particle which maintains its motion simply due to inertia. The weight of a fluid particle is balanced by the pressure-force as it happens in case of a fluid mass at absolute rest, and therefore the hydrostatic equations can be applied without change. If all the fluid concerned now undergo a uniform acceleration in a straight line without any layer moving relative to another, there are still no shear forces, but an additional force acts to cause the acceleration. Nevertheless, provided that due allowance is made for the additional force, the system may be studied by the methods of hydrostatics.

Let us consider a rectangular fluid element in a three-dimensional rectangular Cartesian coordinate system, as shown in Fig. 2.40. The pressure in the centre of the element is p. The fluid element is moving with a constant acceleration whose components along the coordinate axes x, y, and z are a_x, a_y and a_z, respectively. The force acting on the fluid element in the x direction is

$$\left[\left(p - \frac{\partial p}{\partial x}\frac{1}{2}\,\mathrm{d}x\right) - \left(p + \frac{\partial p}{\partial x}\frac{1}{2}\,\mathrm{d}x\right)\right] \mathrm{d}y\,\mathrm{d}z$$

$$= \frac{\partial p}{\partial x}\, dx\, dy\, \mathrm{d}z$$

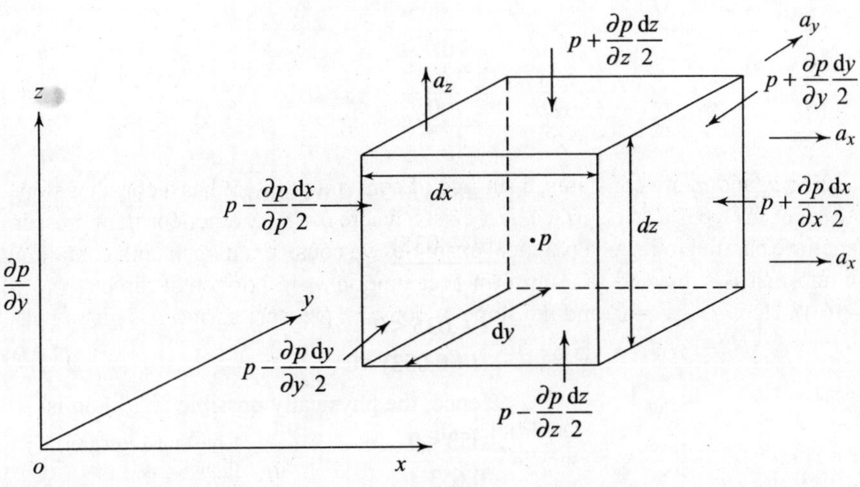

Fig. 2.40 Equilibrium of fluid element moving with constant acceleration

Therefore the equation of motion in the x direction can be written as

$$\rho\,\mathrm{d}x\,\mathrm{d}y\,\mathrm{d}z\, a_x = -\frac{\partial p}{\partial x}\,\mathrm{d}x\,\mathrm{d}y\,\mathrm{d}z$$

or
$$\frac{\partial p}{\partial x} = -\rho\, a_x \tag{2.86a}$$

where ρ is the density of the fluid. In a similar fashion, the equation of motion in the y direction can be written as

$$\frac{\partial p}{\partial y} = -\rho a_y \tag{2.86b}$$

The net force on the fluid element in z direction is the difference of pressure force and the weight. Therefore the equation of motion in the z direction is written as

$$\left(-\frac{\partial p}{\partial z} - \rho g\right) \mathrm{d}x\,\mathrm{d}y\,\mathrm{d}z = \rho a_z \,\mathrm{d}x\,\mathrm{d}y\,\mathrm{d}z$$

or

$$\frac{\partial p}{\partial z} = -\rho (g + a_z) \tag{2.86c}$$

It is observed that the governing equations of pressure distribution [Eqs (2.86a), (2.86b) and (2.86c)] are similar to the pressure distribution equation of hydrostatics.

If we consider, for simplicity, a two-dimensional case where the y component of the acceleration a_y is zero, then a surface of constant pressure in the fluid will be one along which

$$\mathrm{d}p = \frac{\partial p}{\partial x}\mathrm{d}x + \frac{\partial p}{\partial z}\mathrm{d}z = 0$$

or

$$\frac{\mathrm{d}z}{\mathrm{d}x} = -\frac{\frac{\partial p}{\partial x}}{\frac{\partial p}{\partial z}} = -\frac{a_x}{g + a_z} \tag{2.87}$$

Since a_x and a_z are constants, a surface of constant pressure has a constant slope. One such surface is a free surface, if it exists, where $p = p_{atm}$; other constant pressure planes are parallel to it. As a practical example, we consider an open tank containing a liquid that is subjected to a uniform acceleration a_x in horizontal direction (Fig. 2.41(a)). Here $a_y = a_z = 0$, and the slope of constant pressure surfaces is given by

$$\tan\theta = \mathrm{d}z/\mathrm{d}x = -a_x/g \tag{2.87a}$$

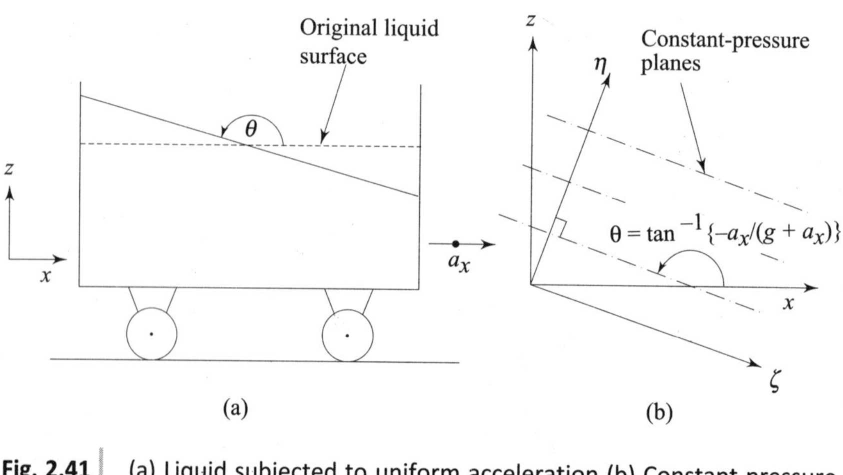

Fig. 2.41 (a) Liquid subjected to uniform acceleration (b) Constant pressure planes

If the tank is uniformly accelerated only in the vertical direction, then from the Eq. (2.87), $dz/dx = 0$ and planes of constant pressure are horizontal. Therefore, when a container with a liquid in it is allowed to fall freely under gravity, then the free surface remains horizontal. Moreover, from the Eq. (2.86(c)), $dp/dz = -\rho(g-g) = 0$. This implies that a point in the liquid under this situation experiences no hydrostatic pressure due to the column of liquid above it. Therefore pressure is throughout atmospheric, provided a free surface exists, for example, if the container is open. From the above discussion, an interesting fact can be concluded, that if there is a hole on the base of a container with an open top, liquid will not leak through it during the free fall of the container.

For a two-dimensional system in a vertical plane, pressure at a point in the fluid may be determined from Eq. (2.86a) and (2.86c) as

$$p = \int dp = \int \frac{\partial p}{\partial x}\, dx + \int \frac{\partial p}{\partial x}\, dz$$

$$= -\rho\, a_x x - \rho\,(g + a_z)\, z + \text{constant} \tag{2.88}$$

The flow is considered to be of constant density and the integration constant is determined by any given condition of the problem. An alternative expression for pressure distribution can be obtained with respect to a frame of coordinates with ζ and η axes (Fig. 2.41b), parallel and perpendicular to the constant-pressure planes respectively. Then $dp/d\zeta = 0$ and

$$\frac{\partial p}{\partial \eta} = \frac{\partial p}{\partial x} \bigg/ \frac{\partial \eta}{\partial x} = \frac{-\rho a_x}{\sin\theta} \tag{2.89}$$

Again from Eq. (2.87),

$$\frac{dz}{dx} = \tan\theta = \frac{-a_x}{g + a_z}$$

which gives,

$$\sin\theta = \frac{a_x}{(a_x^2 + (g + a_z)^2)^{1/2}} \tag{2.90}$$

Since p is a function of η only, $\partial p/\partial \eta$ can be written as $dp/d\eta$. Hence the Eq. (2.89) can be written with the help of the Eq. (2.90) as

$$\frac{dp}{d\eta} = -\rho\left(a_x^2 + (g + a_z)^2\right)^{1/2} \tag{2.91}$$

A comparison of the Eq. (2.91) with the pressure distribution equation in hydrostatics [Eq. (2.9)] shows that pressure in case of fluid motions with uniform acceleration may be calculated by the hydrostatic principle provided that $\left(a_x^2 + (g + a_z)^2\right)^{1/2}$ takes the place of g, and η the place of vertical coordinate.

Discussion on fluids under rigid body rotation (forced vortex flow) has been made in Chapter 4.

Example 2.18

Determine the equation of free surface of water in a tank 4 m long, moving with a constant acceleration of 0.5 g along the *x* axis as shown in Fig. 2.42.

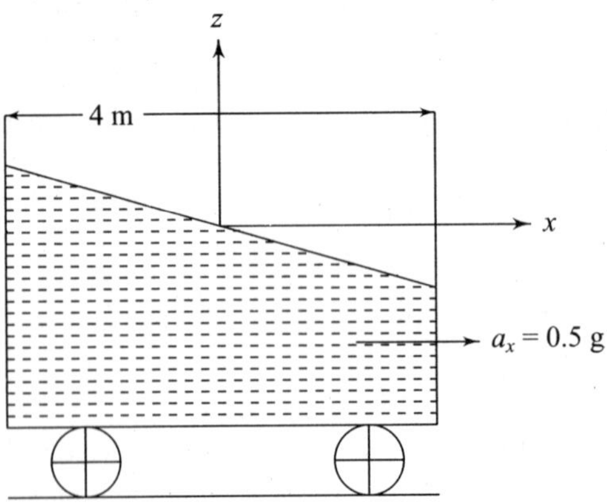

Fig. 2.42 Liquid in a tank under uniform acceleration

Solution

Let us consider the pressure p at a point to be a function of x and z.

Hence, $$\mathrm{d}p = \frac{\partial p}{\partial x}\,\mathrm{d}x + \frac{\partial p}{\partial z}\,\mathrm{d}z \tag{2.92}$$

From Eqs. (2.86a) and (2.86c),

$$\partial p/\partial x = -\rho\, a_x$$

$$\partial p/\partial z = -\rho\,(g + a_z)$$

where a_x and a_z are the accelerations in x and z directions, respectively.

Here, $a_x = 0.5$ g

and $a_z = 0$

Therefore, Eq. (2.92) becomes,

$$\mathrm{d}p = -\rho\,(0.5\mathrm{g}\,\mathrm{d}x + g\,\mathrm{d}z)$$

Integrating the equation, we obtain

$$p = -\rho g\,(0.5x + z) + c \tag{2.93}$$

where c is a constant. Considering the origin of the coordinate axes at free surface, we have

$p = p_{\mathrm{atm}}$ (atmospheric pressure), at $x = 0$ and $z = 0$

Therefore, Eq. (2.93) becomes

$$-\rho g\,(0.5x + z) = p - p_{\mathrm{atm}} \tag{2.94}$$

The equation of free surface can be obtained by putting $p = p_{atm}$ in Eq. (2.94) as

$$-\rho g(0.5x + z) = 0$$

or

$$z + 0.5x = 0$$

Example 2.19

A rectangular tank of length $L = 10$ m, height $h = 2$ m and width (perpendicular to the plane of Fig. 2.43) 1 m is initially half-filled with water. The tank suddenly accelerates along the horizontal direction with an acceleration = g (acceleration due to gravity). Will any water spill out of the tank?

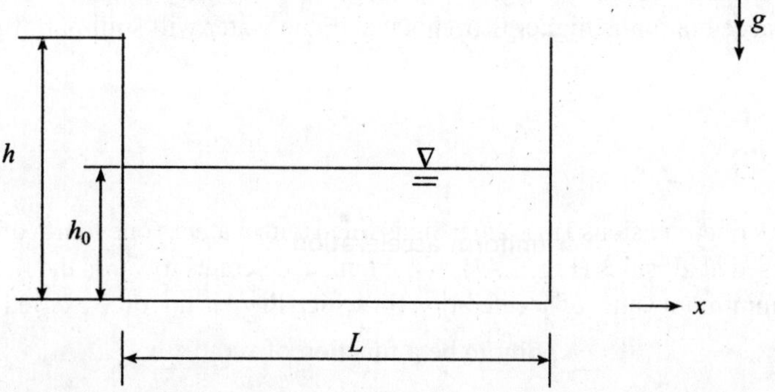

Fig. 2.43

Solution

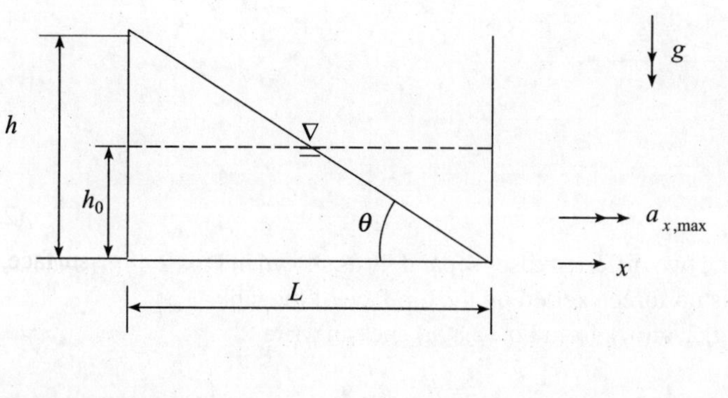

Fig. 2.43(a)

Let us consider the case that the tank accelerates with a maximum acceleration a_x along the horizontal direction without spilling the water. Since the volume of the water in the tank remains unchanged, the free surface takes the shape as shown in Fig. 2.43(a).

From Eq. (2.87a), we have

$$\tan\theta = \frac{h}{L} = \frac{a_{x,\max}}{g}$$

or

$$\frac{a_{x,\max}}{g} = \frac{2}{10} = 0.2$$

or

$$a_{x,\max} = 0.2g$$

This is the maximum acceleration that can be given without spilling the water. Since the given acceleration is higher than this value, the water will spill out of the tank.

Example 2.20

A rectangular tank of dimensions 1m × 2m × 3m is filled with water to one fourth of its height and is closed at all sides (Fig. 2.44). If the tank accelerates towards the right, then what is the minimum value of acceleration for which there is no force exerted on the top face?

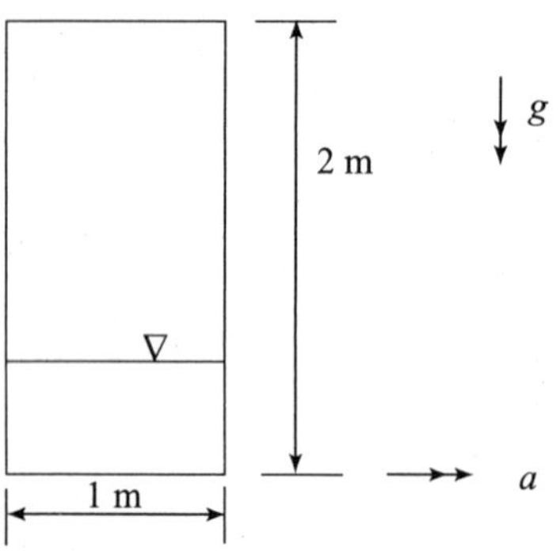

Fig. 2.44

Solution

Let us consider the two different cases separately as shown in Figs 2.44(a) and 2.44(b) for which there is no force exerted on the top face of the tank.

For case 1 (Fig. 2.44(a)), from Eq. (2.87a) one can write

$$\tan\theta_1 = \frac{2-h}{1} = \frac{a_x}{g} \tag{2.95}$$

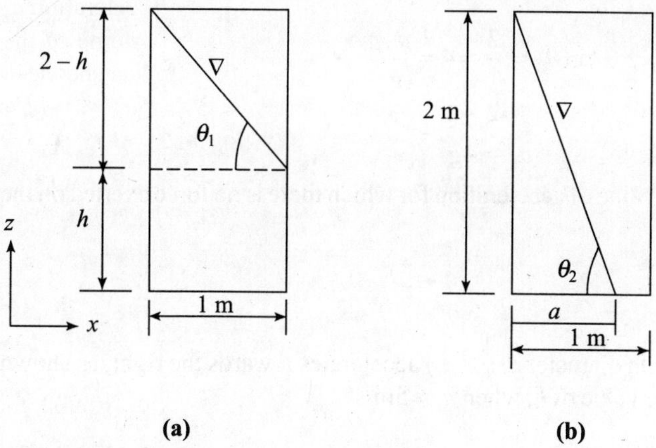

Fig. 2.44

Since the volume of the water in the tank remains unchanged, we have

$$\forall_1 = \forall_0$$

or
$$\frac{1}{2}\times 1\times \tan\theta_1 + h\times 1 = \frac{1}{4}\times 1\times 2 \quad (2.96)$$

From Eqs (2.95) and (2.96), we get

$$\frac{2-h}{2} + h = \frac{1}{2}$$

or
$$h = -1$$

Therefore, this case is not possible.

For case 2 (Fig. 2.44b), from Eq. (2.87a) one can write

$$\tan\theta_2 = \frac{2}{a} = \frac{a_x}{g} \quad (2.97)$$

Since the volume of the water in the tank remains unchanged, we have

$$\forall_2 = \forall_0$$

or
$$\frac{1}{2}\times a\times a\tan\theta_2 = \frac{1}{4}\times 1\times 2 \quad (2.98)$$

From Eqs (2.97) and (2.98), we get

$$\frac{a^2}{2}\times\frac{2}{a} = \frac{1}{2}$$

or
$$a = \frac{1}{2}$$

Now, $$\tan\theta_2 = \frac{2}{\frac{1}{2}} = 4 = \frac{a_x}{g}$$

or $$a_x = 4g$$

This is the minimum value of acceleration for which there is no force exerted on the top face of the tank.

Example 2.21

A U-tube manometer with diameter D ($<< L$) accelerates towards the right, as shown in Fig. 2.45. Find out the value of h, when $a_x = 5\text{m/s}^2$.

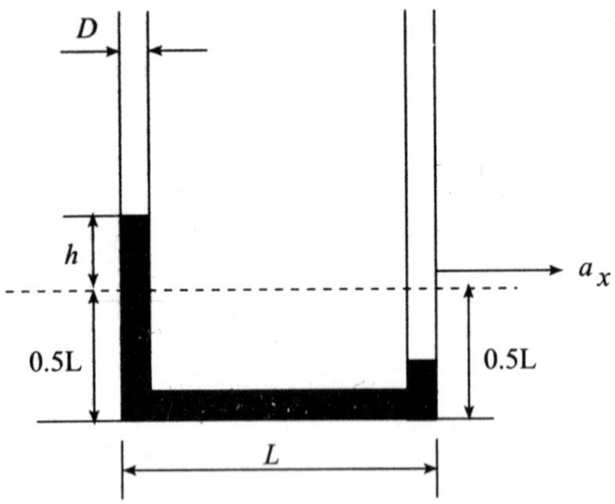

Fig. 2.45

Solution

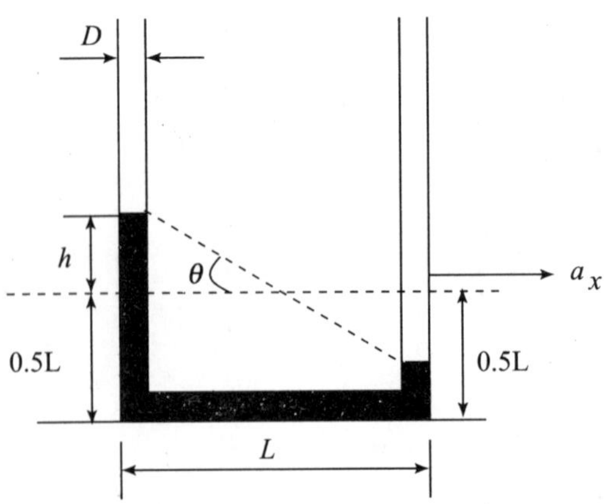

Fig. 2.45(a)

The free surface (see Fig. 2.45(a)) can be considered as a point because the diameter of the manometer is so small. With this consideration, from Eq. (2.87a), one can write

$$\tan\theta = \frac{a_x}{g} = \frac{5}{9.81} = 0.51$$

or
$$\theta = 27.02°$$

Then, h can be found from the geometry of the Fig. (2.45(a)) as

$$h = 0.5L\tan\theta = 0.5L\times 0.51 = 0.255L$$

Example 2.22

Water containing in a cylindrical tank, which is initially rotated at a constant angular speed, ω, about its axis as shown in Fig. 2.46. After a short time duration, there is no relative motion between the water and the tank. Assuming the water to be undergoing a rigid body-like rotation, determine the equation of free surface of the water.

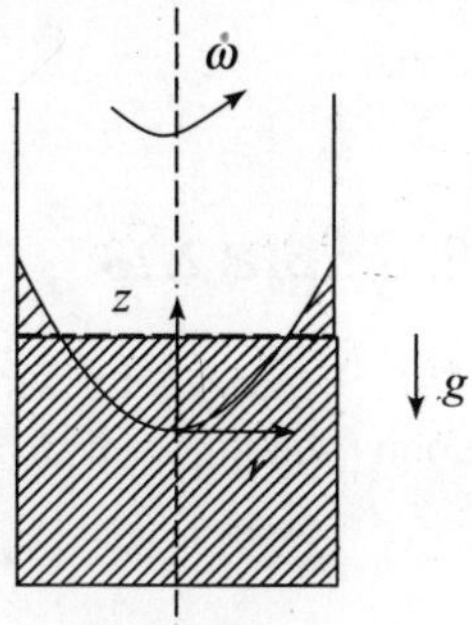

Fig. 2.46

Solution

Let us consider the pressure p at a point to be a function of r and z.

Hence,
$$dp = \frac{\partial p}{\partial r}dr + \frac{\partial p}{\partial z}dz \tag{2.99}$$

From Eqs (2.86a) and (2.86c),

$$\frac{\partial p}{\partial r} = -\rho a_r$$

$$\frac{\partial p}{\partial z} = -\rho(g + a_z)$$

Here,
$$a_r = -\omega^2 r \qquad \text{(centripetal acceleration)}$$

and
$$a_z = 0$$

Therefore, Eq. (2.99) becomes

$$dp = \rho\omega^2 r dr - \rho g dz \tag{2.99a}$$

Integrating the Eq. (2.99a), we obtain

$$p = \frac{\rho\omega^2 r^2}{2} - \rho g z + c \tag{2.100}$$

where c is an integration constant. Considering the origin of the coordinate axes at free surface, we have

$$p = p_{atm}, \text{ at } z = 0 \text{ and } r = 0$$

Therefore, Eq. (2.100) becomes

$$p - p_{atm} = \frac{\rho\omega^2 r^2}{2} - \rho g z \tag{2.101}$$

The equation of free surface can be obtained by putting $p = p_{atm}$ in Eq. (2.101) as

$$\rho g z = \frac{\rho\omega^2 r^2}{2}$$

or

$$z = \frac{\omega^2 r^2}{2g} \tag{2.102}$$

Example 2.23

A U-tube manometer with diameter D ($<<L$) is rotated about its right leg with angular velocity ω without translation, as shown in Fig. 2.47. Find out the value of h.

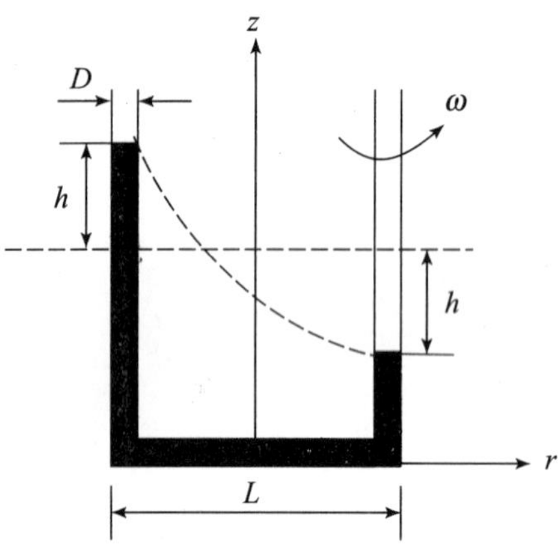

Fig. 2.47

Solution

The free surface of the fluid is shown in Fig. 2.47 for the case of rotation without translation.

Since $D << L$, from Eq. (2.102) the pressure difference between the two legs can be calculated as

$$2h = \frac{\omega^2 r^2}{2g} = \frac{\omega^2 L^2}{2g}$$

or

$$h = \frac{\omega^2 L^2}{4g}$$

SUMMARY

- Normal stresses at any point in a fluid at rest, being directed towards the point from all directions, are of equal magnitude. The scalar magnitude of the stress is known as hydrostatic or thermodynamic pressure.
- The fundamental equations of fluid statics are written as $\partial p/\partial x = 0$, $\partial p/\partial y = 0$ and $\partial p/\partial z = -\rho g$ with respect to a Cartesian frame of reference with $x-y$ plane as horizontal and axis z being directed vertically upwards. For an incompressible fluid, pressure p at a depth h below the free surface can be written as $p = p_0 + \rho g h$, where p_0 is the local atmospheric pressure.
- At sea level, the international standard atmospheric pressure has been chosen as $p_{atm} = 101.32$ kN/m^2. The pressure expressed as the difference between its value and the local atmospheric pressure is known as gauge pressure.
- Piezometer tube measures the gauge pressure of a flowing liquid in terms of the height of liquid column. Manometers are devices in which columns of a suitable liquid are used to measure the difference in pressure between two points or between a certain point and the atmosphere. A simple U-tube manometer can be modified as an inclined tube manometer, an inverted tube manometer and a micro manometer to measure a small difference in pressure through a relatively large deflection of liquid columns.
- The hydrostatic force on any one side of a submerged plane surface is equal to the product of the area and the pressure at the centre of area. The force acts in a direction perpendicular to the surface and its point of action, known as pressure centre, is always at a higher depth than that at which the centre of area lies. The distance of centre of pressure from the centre of area along the axis of symmetry is given by $y_p - y_C = I_{x'x'}/Ay_c$.
- For a curved surface, the component of hydrostatic force in any horizontal direction is equal to the hydrostatic force on the projected plane surface on a vertical plane perpendicular to that direction and acts through the centre of pressure for the projected plane area. The vertical component of hydrostatic force on a submerged curved surface is equal to the weight of the liquid volume vertically above the submerged surface to the level of the free surface of liquid and acts through the centre of gravity of the liquid in that volume.
- When a solid body is either wholly or partially immersed in a fluid, the hydrostatic lift due to the net vertical component of the hydrostatic pressure forces experienced by the body is called the *buoyant force*. The buoyant force on a submerged or floating body is equal to the weight of liquid displaced by the body and acts vertically upward through the centroid of displaced volume known as the *centre of buoyancy*.
- The equilibrium of floating or submerged bodies requires that the weight of the body acting through its centre of gravity has to be colinear with an equal buoyant force acting through the centre of buoyancy. A submerged body will be in stable, unstable or neutral equilibrium if its centre of gravity is below, above or coincident with the centre of buoyancy respectively. The metacentre of a floating body is defined as the point of intersection of the centre line of cross section containing the centre of gravity and centre of buoyancy with the vertical line through new centre of buoyancy due to any

small angular displacement of the body. For stable equilibrium of floating bodies, metacentre M has to be above the centre of gravity G. M coinciding with G or lying below G refers to the situation of neutral and unstable equilibrium respectively. The distance of metacentre from centre of gravity along the centre line of cross section is known as metacentric height and is given by $MG = (I_{yy}/\forall) - BG$.

- Fluids moving with a uniform velocity or uniform acceleration develop no shear stress in the flow field. The weight of the fluid particle is balanced by the pressure force and a constant inertia force (zero in the case of uniform velocity). The pressure distribution equations under the situations are similar to those in hydrostatics in a sense that the pressure gradients in space coordinates are constants. The fluids in such motions are said to be in relative equilibrium.

EXERCISES

2.1 Choose the correct answer:

(i) The normal stress is the same in all directions at a point in a fluid
   (a) only when the fluid is frictionless
   (b) only when the fluid is frictionless and incompressible
   (c) in a liquid at rest
   (d) when the fluid is at rest, regardless of its nature

(ii) The magnitude of hydrostatic force on one side of a circular surface of unit area, with the centroid 10 m below a free water (density ρ) surface is
   (a) less than 10 ρg
   (b) equals to 10 ρg
   (c) greater than 10 ρg
   (d) the product of ρg and the vertical distance from the free surface to pressure centre
   (e) None of the above

(iii) The line of action of the buoyancy force acts through the
   (a) centre of gravity of any submerged body
   (b) centroid of the volume of any floating body
   (c) centroid of the displaced volume of fluid
   (d) centroid of the volume of fluid vertically above the body
   (e) centroid of the horizontal projection of the body

(iv) For stable equilibrium of floating bodies, the centre of gravity has to:
   (a) be always below the centre of buoyancy
   (b) be always above the centre of buoyancy
   (c) be always above the metacentre
   (d) be always below the metacentre
   (e) coincide with metacentre

(v) A fish tank is being carried on a car moving with constant horizontal acceleration. The level of water will
   (a) remain unchanged
   (b) rise on the front side of the tank only

(c) rise on the front side of the tank and fall on the back side
(d) rise on the back side of the tank and fall on the front side
(e) None of the above

2.2 In construction, a barometer is a graduated inverted tube with its open end dipped in the measuring liquid contained in a trough opened to the atmosphere.

Estimate the height of the liquid column in a barometer where the atmospheric pressure is 100 kN/m^2 (a) When the liquid is mercury and (b) When the liquid is water. The measuring temperature is 50 °C, the vapour pressures of mercury and water at this temperature are respectively 0.015×10^4 N/m^2 and 1.23×10^4 N/m^2, and the densities are 13500 and 980 kg/m^3, respectively. What would be the percentage error if the effect of vapour pressure is neglected.

Ans. (0.754 m, 9.12 m, 0.14%, 14.05%)

2.3 The density of a fluid mixture ρ (in kg/m^3) in a chemical reactor varies with the vertical distance z (in metre) above the bottom of the reactor according to the relation

$$\rho = 10.1\left[1 - \frac{z}{500} + \left(\frac{z}{1000}\right)^2\right]$$

Assuming the mixture to be stationary, determine the pressure difference between the bottom and top of a 60 m tall reactor.

Ans. (5.59 kN/m^2)

2.4 The intensity of pressure in the ocean at a depth of eight kilometers is 100 MPa. Determine the specific weight of sea water at this depth if the specific weight at the surface is 10 kN/m^3 and the average bulk modulus of elasticity of sea water is 2.30 GPa. Neglect the variation of g.

Ans. (10.44 kN/m^3)

2.5 Find the atmospheric pressure just at the end of troposphere which extends up to a height of 11.02 km from sea level. Consider a temperature variation in the troposphere as $T = 288.16 - 6.49 \times 10^{-3} z$, where z is in metres and T in Kelvin. The atmospheric pressure at sea level is 101.32 kN/m^2.

Ans. (22.55 kN/m^2)

2.6 Find the pressure at an elevation of 3000 m above the sea level by assuming (a) an isothermal condition of air and (b) an isentropic condition of air. Pressure and temperature at sea level are 101.32 kN/m^2 and 293.15 K. Consider air to be an ideal gas with R (characteristic gas constant) = 287 J/kg K, and γ (ratio of specific heats) = 1.4.

Ans. (71.41 kN/m^2, 70.08 kN/m^2)

2.7 For the system shown in Fig 2.48, determine the air pressure p_A which will make the pressure at N one fourth of that at M.

Ans. (3.33 kPa)

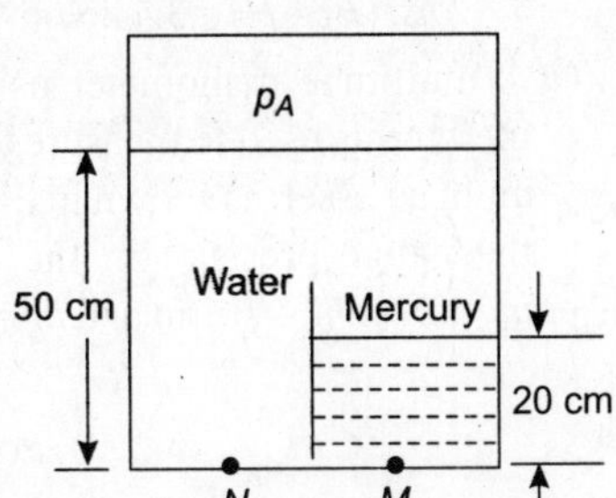

Fig. 2.48

2.8 Consider the pipe and manometer system as shown in Fig 2.49. The pipe contains water. Find the value of manometer reading h, and the difference in pressure between A and B if there is no flow. If there is a flow from A towards B and the manometer reading is $h = 60$ mm, then determine the static pressure difference $p_A - p_B$.

Ans. (0, 2.94 kPa; 3.53 kPa)

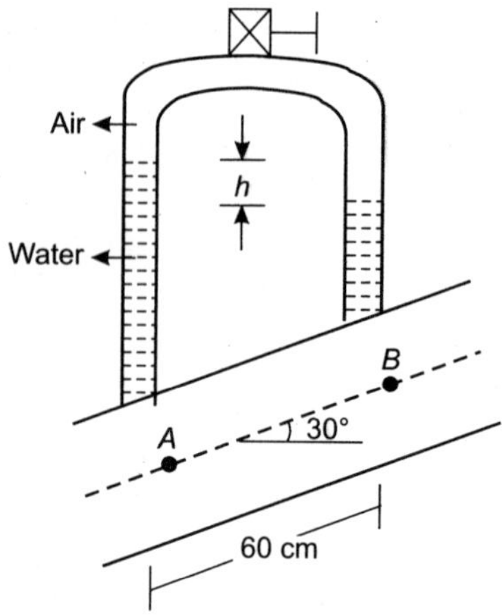

Fig. 2.49 An inclined pipe and inverted U-tube manometer

2.9 Two pipes A and B (Fig. 2.50) are in the same elevation. Water is contained in A and rises to a level of 1.8 m above it. Carbon tetrachloride (Sp. gr. = 1.59) is contained in B. The inverted U-tube is filled with compressed air at 300 kN/m^2 and 30 °C. The barometer reads 760 mm of mercury. Determine:

(a) The pressure difference in kN/m^2 between A and B if $z = 0.45$ m.

(b) The absolute pressure in B in mm of mercury.

Ans. ($P_B - P_A = 3.4$ kN/m^2, 2408.26 mm)

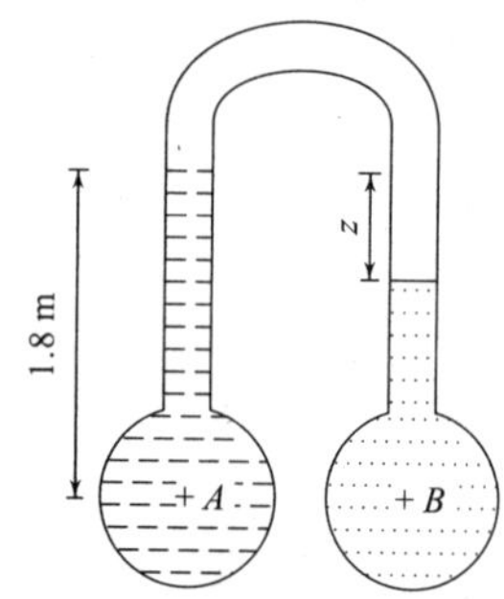

Fig. 2.50 Pipes with water and carbon tetrachloride

2.10 A multitube manometer using water and mercury is used to measure the pressure of air in a vessel, as shown in Fig. 2.51. For the given values of heights, calculate the gauge pressure in the vessel. h_1 = 0.4 m, h_2 = 0.5 m, h_3 = 0.3 m, $h_4 = 0.7$ mh_5 = 0.1 m and $h_6 = 0.5$ m.

Ans. (190.31 kN/m^2 gauge)

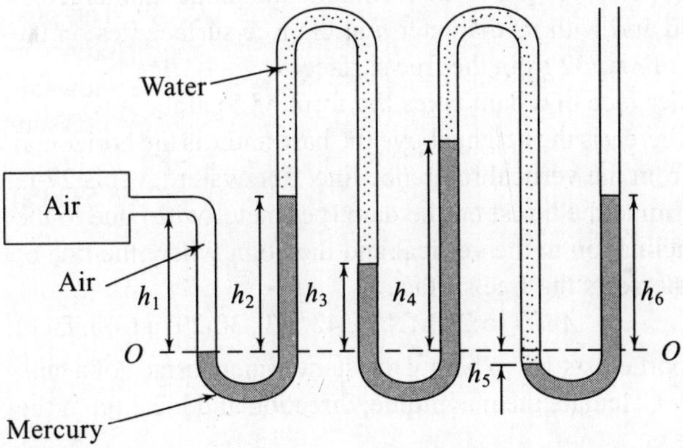

Fig. 2.51 A multitube manometer measuring air pressure in a vessel

2.11 Determine the air pressure above the water surface in the tank if a force of 8 kN is required to hold the hinge door in position as shown in Fig. 2.52.

Ans. (10.76 kPa)

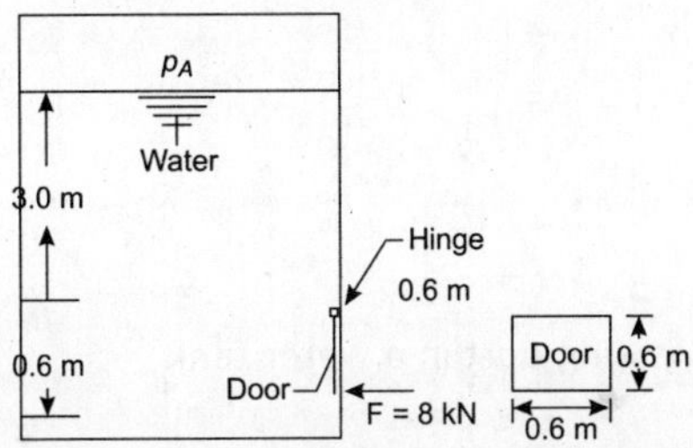

Fig. 2.52 A tank containing water and air with a door on one side

2.12 Gate *AB* in Fig. 2.53 is 1.2 m wide (in a direction perpendicular to the plane of the figure) and is hinged at *A*. Gauge *G* reads – 0.147 bar and oil in the right hand tank is having a relative density 0.75. What horizontal force must be applied at *B* for equilibrium of gate *AB*?

Ans. (3.66 kN)

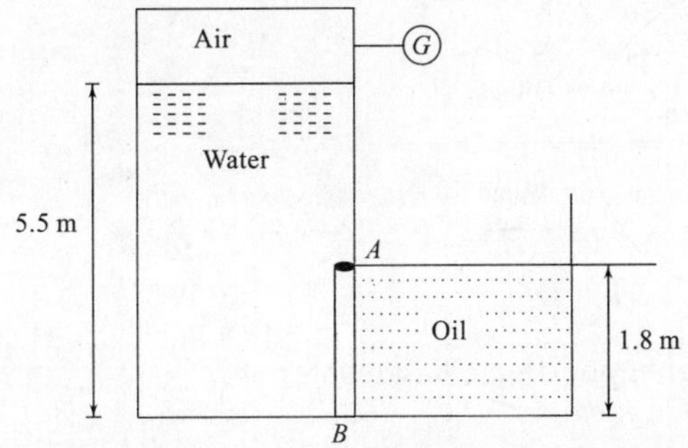

Fig. 2.53 A plane gate with water on one side and oil on the other side

2.13 Show that the centre of pressure for a vertical semicircular plane submerged in a homogeneous liquid and with its diameter d at the free surface lies on the centre line at a depth of $3\pi d/32$ from the free surface.

2.14 The profile of the inner face of a dam takes the form of a parabola with the equation $18y = x^2$, where y is the height above the base and x is the horizontal distance of the face from the vertical reference line. The water level is 27 m above the base. Determine the thrust on the dam (per meter with) due to the water pressure, its inclination to the vertical and the point where the line of action of this force intersects the free surface.

Ans. (15.28 MN/m, 42° 33', 30.29 m from face)

2.15 A spherical viewing port exists 1.5 m below the static water surface of a tank as shown in Fig. 2.54. Calculate the magnitude, direction and location of the thrust on the viewing port.

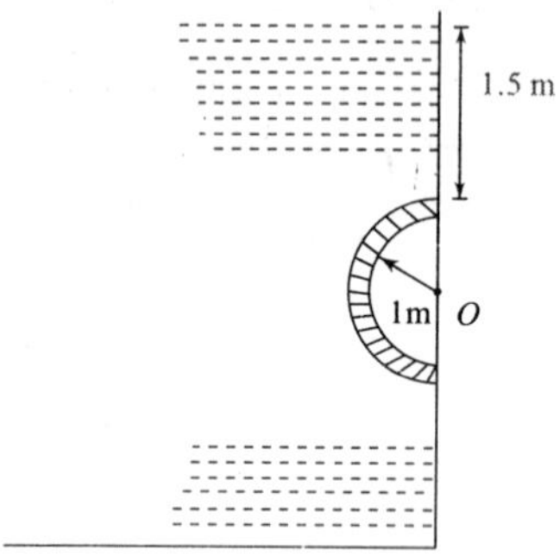

Fig. 2.54 A spherical viewing port in a water tank

Ans. (79.74 kN, 75° in a direction 75° clockwise from a vertically upward line and passes through the centre *O*)

2.16 Find the weight of the cylinder (dia = 2 m) per metre length if it supports water and oil (Sp. gr. 0.82) as shown in Fig. 2.55. Assume contact with wall as frictionless.

Ans. (14.02 kN)

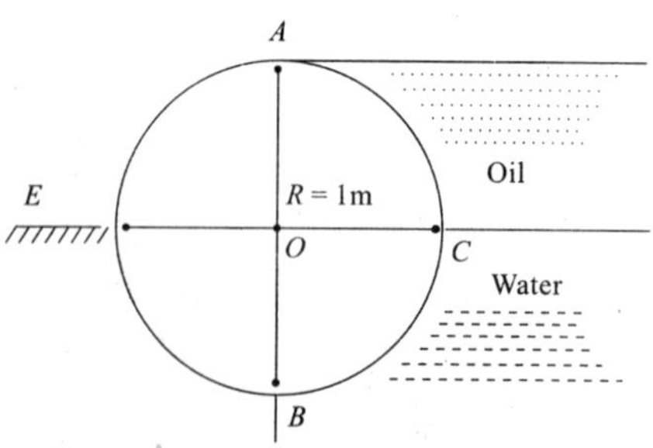

Fig. 2.55 A cylinder supporting oil and water

2.17 Calculate the force F required to hold the gate in a closed position (Fig. 2.56), if $R = 0.6$ m.

Ans. (46.02 kN)

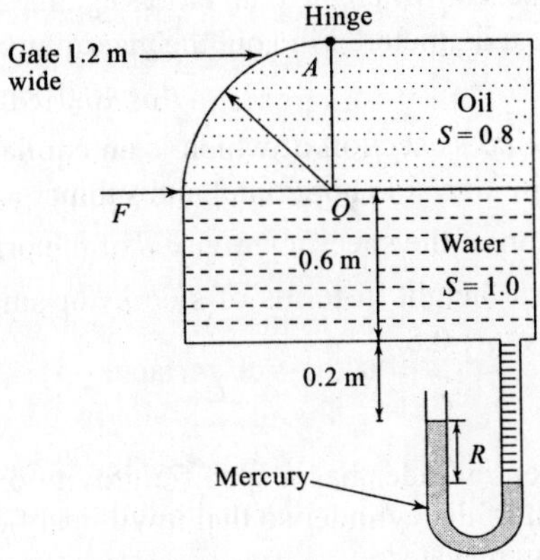

Fig. 2.56 A gate in closed position supporting oil and water in a tank

2.18 A cylindrical log of specific gravity 0.425 is 5 m long and 2 m in diameter. To what depth the log will sink in fresh water with its axis being horizontal?

Ans. (0.882 m)

2.19 A sphere of 1219 mm diameter floats half submerged in salt water (ρ = 1025 kg/m^3). What minimum mass of concrete (ρ = 2403 kg/m^3) has to be used as an anchor to submerge the sphere completely?

Ans. (848.47 kg)

2.20 A metal sphere of volume $\forall_m$ = 0.1 m^3, specific gravity S_m = 2 and fully immersed in water is attached by a flexible wire to a buoy of volume $\forall_B$ = 1 m^3 and specific gravity S_m = 0.1 (Fig. 2.57). Calculate the tension T in the wire and volume of the buoy that is submerged. Refer to Fig 5.13.

Ans. (981 N, 0.2 m^3)

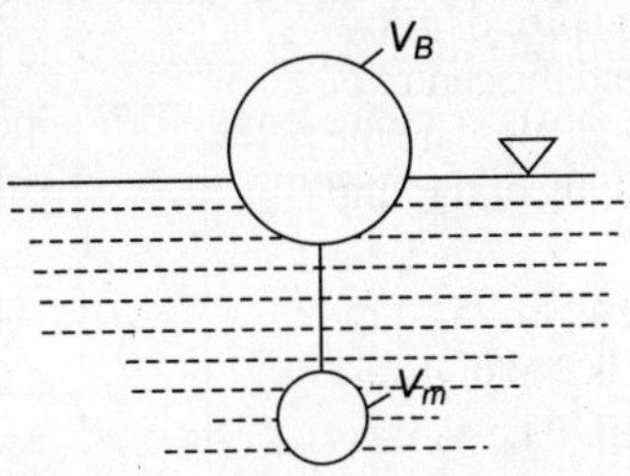

Fig. 2.57

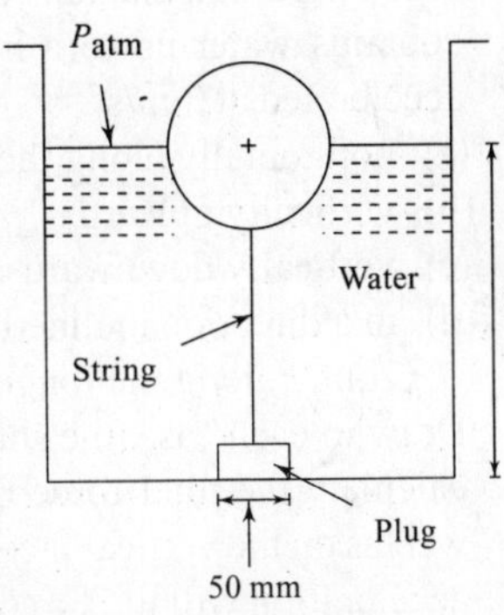

Fig. 2.58 A typical drain plug

2.21 The drain plug shown in Fig. 2.58 is closed initially. As the water fills up and the level reaches 2 m, the buoyancy force on the float opens the plug. Find the volume of the spherical weight if the total mass of the plug and the weight is 5 kg. As soon as the plug opens it is observed that the plug-float assembly jumps upward and attains a floating position. Explain why. Determine the level in the reservoir when the plug closes again. Can the plug diameter be larger than the float diameter? Find out the maximum possible plug diameter.

Ans. (0.018 m^3, 1.95 m, No, 87.5 mm)

2.22 A long prism, the cross section of which is an equilateral triangle of side a, floats in water with one side horizontal and submerged to a depth h. Find

(a) h/a as a function of the specific gravity S of the prism.

(b) The metacentric height in terms of side a for small angle of rotation if specific gravity, $S = 0.8$.

Ans. ($\sqrt{3s}/2, 0.11a$)

2.23 A uniform wooden cylinder has a specific gravity of 0.6. Find the ratio of diameter to length of the cylinder so that it will just float upright in a state of neutral equilibrium in water.

Ans. (1.386)

2.24 A solid uniform cylinder of length 150 mm and diameter 75 mm is to float upright in water. Determine the limits within which its mass should lie.

Ans. (0.641 kg and 0.663 kg)

2.25 Find the minimum apex angle of a solid cone of specific gravity 0.8 so that it can float in stable equilbrium in fresh water with its axis vertical and vertex downward.

Ans. (31.12°)

2.26 A ship weighing 25 MN floats in sea water with its axis vertical. A pendulum 2 m long is observed to have a horizontal displacement of 20 mm when a weight of 40 kN is moved 5 m across the deck. Find the metacentric height of the ship.

Ans. (0.8 m)

2.27 A ship of mass 2×10^6 kg has a cross section at the waterline as shown in Fig. 2.59. The centre of buoyancy is 1.5 m below the free surface, and the centre of gravity is 0.6 m above the free surface. Calculate the metacentric height for rolling and pitching of the ship with a small angle of tilt.

Ans. (0.42 m, 25.41 m)

2.28 An open rectangular tank of 5 m × 4 m is 3 m high and contains water up to a height of 2 m. The tank is accelerated at 3 m/s^2

(a) horizontally along the longer side

(b) vertically upwards

(c) vertically downwards and

(d) in a direction inclined at 30° upwards to the hori zontal along the longer side.

Draw in each case, the shape of the free surface and calculate the total force on the base of the tank as well as on the vertical faces of the container. At what acceleration will the force on each face be zero?

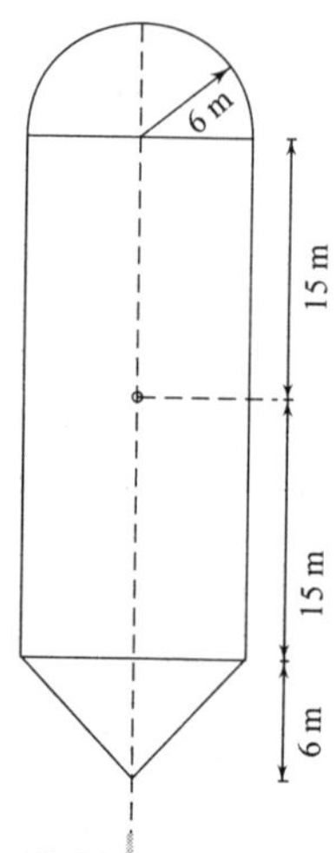

Fig. 2.59 Cross section of a ship at the waterline

Ans. [(a) base: 392.40 kN, leading face: 29.97 kN, trailing face: 149.89 kN, other two faces: 102.74 kN

(b) base: 512.40 kN, faces with longer side: 128.10 kN, other two faces: 102.48 kN

(c) base: 272.40 kN, faces with longer side: 68.10 kN, other two faces: 54.48 kN

(d) base: 452.40 kN, leading face: 45.93 kN, trailing face: 149.98 kN, other two faces: 116.22 kN; downward acceleration of 9.81 m/s^2].

2.29 An open-topped tank in the form of a cube of 900 mm side, has a mass of 340 kg. It contains 0.405 m^3 of oil of specific gravity 0.85 and is accelerated uniformly up along a slope at $\tan^{-1}$ (1/3) to the horizontal, the base of the tank remains parallel to the slope, and the side faces are parallel to the direction of motion. Neglecting the thickness of the walls of the tank, estimate the net force (parallel to the slope) accelerating the tank if the oil is just on the point of spilling.

Ans. (3538 N)

2.30 An open rectangular tank of 5 m × 4 m is 3 m high. It contains water up to a height of 2 m and is accelerated horizontally along the longer side. Determine the maximum acceleration that can be given without spilling the water and also calculate the percentage of water spilt over, if this acceleration is increased by 20%.

Ans. (3.92 m/s^2, 10%)

3

KINEMATICS OF FLUID FLOW

3.1 INTRODUCTION

Kinematics of fluid flow is that branch of fluid mechanics which describes the fluid motion and its consequences without consideration of the nature of forces causing the motion. The basic understanding of the fluid kinematics forms the groundwork for the studies on dynamical behaviour of fluids in consideration of the forces accompanying the motion. The subject has the following three main aspects:

(a) The development of methods and techniques for describing and specifying the motions of fluids.

(b) Characterisation of different types of motion and associated deformation rates of any fluid element.

(c) The determination of the conditions for the kinematic possibility of fluid motions, i.e., the exploration of the consequences of continuity in the motion.

3.2 SCALAR AND VECTOR FIELDS

Scalar A quantity which has only magnitude is defined to be a scalar. A scalar quantity can be completely specified by a single number representing its magnitude. Typical scalar quantities are mass, density and temperature. The magnitude of a scalar (a real number) will change when the units expressing the scalar are changed, but the physical entity remains the same.

Vector A quantity which is specified by both magnitude and direction is known to be a vector. Force, velocity and displacement are typical vector quantities. The magnitude of a vector is a scalar.

Scalar Field If at every point in a region, a scalar function has a defined value, the region is called a scalar field. The temperature distribution in a rod is an example of a scalar field.

Vector Field If at every point in a region, a vector function has a defined value, the region is called a vector field. Force and velocity fields are typical examples of vector fields.

3.3 FLOW FIELD AND DESCRIPTION OF FLUID MOTION

A flow field is a region in which the flow velocity is defined at each and every point in space, at any instant of time. Usually, the velocity describes the flow. In other words, a flow field is specified by the velocities at different points in the region at different times. A fluid mass can be conceived of consisting of a number of fluid particles. Hence the instantaneous velocity at any point in a fluid region is actually the velocity of a particle (of the same density as that of the fluid) that exists at that point at that instant. In order to obtain a complete picture of the flow, the fluid motion is described by two methods discussed as follows:

1. *Lagrangian Method* In this method, the fluid motion is described by tracing the kinematic behaviour of each and every individual particle constituting the flow. Identities of the particles are made by specifying their initial position (spatial location) at a given time. The position of a particle at any other instant of time then becomes a function of its identity and time. This statement can be analytically expressed as

$$\vec{S} = S(\vec{S}_0, t) \tag{3.1}$$

where $\vec{S}$ is the position vector of a particle (with respect to a fixed point of reference) at a time t. $\vec{S}_0$ is its initial position at a given time $t = t_0$, and thus specifies the identity of the particle. Equation (3.1) can be written into scalar components with respect to a rectangular Cartesian frame of coordinates as

$$x = x(x_0, y_0, z_0, t) \tag{3.1a}$$

$$y = y(x_0, y_0, z_0, t) \tag{3.1b}$$

$$z = z(x_0, y_0, z_0, t) \tag{3.1c}$$

where x_0, y_0, z_0 are the initial coordinates and x, y, z are the coordinates at a time t of the particle. Hence $\vec{S}$ in Eq. (3.1) can be expressed as

$$\vec{S} = \hat{i}x + \hat{j}y + \hat{k}z$$

where $\hat{i}$, $\hat{j}$ and $\hat{k}$ are the unit vectors along the x, y and z axes, respectively. The velocity $\vec{V}$ and acceleration $\vec{a}$ of the fluid particle can be obtained from the material derivatives of the position of the particle with respect to time. Therefore,

$$\vec{V} = \left[\frac{\mathrm{d}\vec{S}}{\mathrm{d}t}\right]_{S_0} \tag{3.2a}$$

or, in terms of scalar components,

$$u = \left[\frac{\mathrm{d}x}{\mathrm{d}t}\right]_{x_0,\, y_0\, z_0} \tag{3.2b}$$

$$v = \left[\frac{\mathrm{d}y}{\mathrm{d}t}\right]_{x_0,\, y_0\, z_0} \tag{3.2c}$$

$$w = \left[\frac{dz}{dt}\right]_{x_0,\, y_0,\, z_0} \tag{3.2d}$$

u, v, w are the components of velocity in x, y and z directions, respectively. For the acceleration,

$$\vec{a} = \left[\frac{d^2\vec{S}}{dt^2}\right]_{S_0} \tag{3.3a}$$

and hence,

$$a_x = \left[\frac{d^2x}{dt^2}\right]_{x_0,\, y_0,\, z_0} \tag{3.3b}$$

$$a_y = \left[\frac{d^2y}{dt^2}\right]_{x_0,\, y_0,\, z_0} \tag{3.3c}$$

$$a_z = \left[\frac{d^2z}{dt^2}\right]_{x_0,\, y_0,\, z_0} \tag{3.3d}$$

The subscripts in Eqs (3.2) and (3.3) represent the initial (at $t = t_0$) position of the particle and thus specify the particle identity. The favourable aspect of the method lies in the information about the motion and trajectory of each and every particle of the fluid so that at any time it is possible to trace the history of each fluid particle. In addition, by virtue of the fact that particles are initially identified and traced through their motion, conservation of mass is inherent. However, the serious drawback of this method is that the solution of the equations (Eqs (3.2) and (3.3)) presents appreciable mathematical difficulties except certain special cases and therefore, the method is rarely suitable for practical applications.

2. *Eulerian Method* The method given by Leonhard Euler is of greater advantage since it avoids the determination of the movement of each individual fluid particle in all details. Instead it seeks the velocity $\vec{V}$ and its variation with time t at each and every location ($\vec{S}$) in the flow field. In the Lagrangian view, all hydrodynamic parameters are tied to the particles or elements, whereas in the Eulerian view, they are functions of location and time. Mathematically, the flow field in the Eulerian method is described as

$$\vec{V} = \vec{V}(\vec{S}, t) \tag{3.4}$$

where,

$$\vec{V} = \hat{i}u + \hat{j}v + \hat{k}w$$

and

$$\vec{S} = \hat{i}x + \hat{j}y + \hat{k}z$$

Therefore,

$$u = u(x, y, z, t) \tag{3.4a}$$

$$v = v(x, y, z, t) \tag{3.4b}$$

$$w = w(x, y, z, t) \tag{3.4c}$$

The relationship between the Eulerian and the Lagrangian method can now be shown. Equation (3.4) of Eulerian description can be written as

$$\frac{d\vec{S}}{dt} = \vec{V}(\vec{S}, t) \tag{3.5}$$

or
$$\frac{dx}{dt} = u(x, y, z, t) \tag{3.5a}$$

$$\frac{dy}{dt} = v(x, y, z, t) \tag{3.5b}$$

$$\frac{dz}{dt} = w(x, y, z, t) \tag{3.5c}$$

The integration of Eq. (3.5) yields the constants of integration which are to be found from the initial coordinates of the fluid particles. Hence, the solution of Eq. (3.5) gives the equations of Lagrange as

$$\vec{S} = S(\vec{S}_0, t)$$

or
$$x = x(x_0, y_0, z_0, t)$$
$$y = y(x_0, y_0, z_0, t)$$
$$z = z(x_0, y_0, z_0, t)$$

Therefore, it is evident that, in principle, the Lagrangian method of description can always be derived from the Eulerian method. But the solution of the set of three simultaneous differential equations is generally very difficult.

Example 3.1

In a one-dimensional flow field, the velocity at a point may be given in the Eulerian system by $u = x + t$. Determine the displacement of a fluid particle whose initial position is x_0 at initial time t_0 in the Lagrangian system.

Solution

$$u = x + t$$

or
$$dx/dt = x + t$$

Using D as the operator d/dt, the above equation can be written as

$$(D - 1)x = t$$

The solution of the equation is

$$x = Ae^t - t - 1$$

The constant A is found from the initial condition as follows:

$$x_0 = Ae^{t0} - t_0 - 1$$

Hence,
$$A = \frac{x_0 + t_0 + 1}{e^{t_0}}$$

Substituting the value of A in the solution, we get

$$x = (x_0 + t_0 + 1)\, e^{(t - t0)} - t - 1$$

This equation is the required Lagrangian version of the fluid particle having the identity $x = x_0$ at $t = t_0$.

Example 3.2

A two-dimensional flow is described in the Lagrangian system as

$$x = x_0 e^{-kt} + y_0\,(1 - e^{-2kt})$$

and
$$y = y_0\, e^{kt}$$

Find (a) the equation of a fluid particle in the flow field and (b) the velocity components in the Eulerian system.

Solution

(a) Trajectory of fluid particle in the flow field is found by eliminating t from the equations describing its motion as follows:

$$e^{kt} = y/y_0$$

Hence,
$$x = x_0(y_0/y) + y_0(1 - y_0^2/y^2)$$

which finally gives after some arrangement

$$(x - y_0)y^2 - x_0 y_0 y + y_0^3 = 0$$

This is the required equation

(b) u (the x component of velocity)

$$= \frac{dx}{dt}$$

$$= \frac{d}{dt}\left[x_0 e^{-kt} + y_0(1 - e^{-2kt})\right]$$

$$= -kx_0\, e^{-kt} + 2\, ky_0\, e^{-2kt}$$

$$= -k\,[x - y_0(1 - e^{-2kt})] + 2\, ky_0\, e^{-2kt}$$

$$= -kx + ky_0\,(1 + e^{-2kt})$$

$$= -kx + ky\,(e^{-kt} + e^{-3kt})$$

v (the y component of velocity)

$$= \frac{dy}{dt} = \frac{d}{dt}\left(y_0\, e^{kt}\right)$$

$$= y_0\, ke^{kt} = ky$$

Example 3.3

The velocities at a point in a fluid in the Eulerian system are given by

$$u = x + y + z + t, \quad v = 2(x + y + z) + t,\ w = 3\,(x + y + z) + t$$

Show that the displacements of a fluid particle in the Lagrangian system are

$$x = \frac{5}{6}x_0 - \frac{1}{6}y_0 - \frac{1}{6}z_0 + \frac{1}{6}\left(x_0 + y_0 + z_0 + \frac{1}{12}\right)e^{6t} - \frac{1}{12}t + \frac{1}{4}t^2 - \frac{1}{72}$$

$$y = -\frac{1}{3}x_0 + \frac{2}{3}y_0 - \frac{1}{3}z_0 + \frac{1}{3}\left(x_0 + y_0 + z_0 + \frac{1}{12}\right)e^{6t} - \frac{1}{6}t - \frac{1}{36}$$

$$z = -\frac{1}{2}x_0 - \frac{1}{2}y_0 + \frac{1}{2}z_0 + \frac{1}{2}\left(x_0 + y_0 + z_0 + \frac{1}{12}\right)e^{6t} - \frac{1}{4}t + \frac{1}{4}t^2 - \frac{1}{24}$$

where x_0, y_0, and z_0 are the initial position in the Lagrangian coordinates when $t = t_0 = 0$.

Solution

$$u + v + w = 6(x + y + z) + 3t$$

or
$$\frac{dx}{dt} + \frac{dy}{dt} + \frac{dz}{dt} = 6(x + y + z) + 3t$$

or
$$\frac{d}{dt}(x + y + z) = 6(x + y + z) + 3t$$

$$x + y + z = c\,e^{6t} + \frac{3}{36}(-6t - 1) = ce^{6t} - \frac{1}{12}(6t + 1)$$

or
$$x + y + z = ce^{6t} - \frac{1}{12}(6t + 1)$$

At $t = t_0$, $x = x_0$, $y = y_0$, and $z = z_0$, which implies

$$x_0 + y_0 + z_0 = c - \frac{1}{12}$$

or
$$c = x_0 + y_0 + z_0 + \frac{1}{12}$$

Thus,
$$x + y + z = \left(x_0 + y_0 + z_0 + \frac{1}{12}\right)e^{6t} - \frac{1}{12}(6t + 1)$$

Now,

$$u = \frac{dx}{dt} = x + y + z + t$$

or
$$\frac{dx}{dt} = \left(x_0 + y_0 + z_0 + \frac{1}{12}\right)e^{6t} - \frac{1}{12}(6t + 1) + t$$

Integrating, we get

$$x = \left(x_0 + y_0 + z_0 + \frac{1}{12}\right)\frac{e^{6t}}{6} - \frac{1}{12}\left(6\frac{t^2}{2} + t\right) + \frac{t^2}{2} + c_1$$

The integration constant c_1 can be found from the initial conditions as $t = t_0 = 0, x = x_0$

$$x_0 = \left(x_0 + y_0 + z_0 + \frac{1}{12}\right)\frac{1}{6} + c_1$$

or
$$c_1 = \frac{5}{6}x_0 - \frac{1}{6}y_0 - \frac{1}{6}z_0 - \frac{1}{72}$$

Putting the value of c_1, we obtain

$$x = \frac{5}{6}x_0 - \frac{1}{6}y_0 - \frac{1}{6}z_0 + \frac{1}{6}\left(x_0 + y_0 + z_0 + \frac{1}{12}\right)e^{6t} - \frac{1}{12}t + \frac{1}{4}t^2 - \frac{1}{72}$$

Again
$$v = \frac{dy}{dt} = 2(x + y + z) + t$$

or
$$\frac{dy}{dt} = 2\left(x_0 + y_0 + z_0 + \frac{1}{12}\right)e^{6t} - \frac{1}{6}(6t + 1) + t$$

Integrating, we obtain

$$v = \left(x_0 + y_0 + z_0 + \frac{1}{12}\right)\frac{e^{6t}}{3} - \frac{1}{6}\left(6\frac{t^2}{2} + t\right) + \frac{t^2}{2} + c_2$$

The integration constant c_2 can be found from the initial conditions as $t = t_0 = 0, y = y_0$

$$y_0 = \left(x_0 + y_0 + z_0 + \frac{1}{12}\right)\frac{1}{3} + c_2$$

or
$$c_2 = -\frac{1}{3}x_0 + \frac{2}{3}y_0 - \frac{1}{3}z_0 - \frac{1}{36}$$

Putting the value of c_2, we obtain

$$y = -\frac{1}{3}x_0 + \frac{2}{3}y_0 - \frac{1}{3}z_0 + \frac{1}{3}\left(x_0 + y_0 + z_0 + \frac{1}{12}\right)e^{6t} - \frac{1}{6}t - \frac{1}{36}$$

Again
$$w = \frac{dz}{dt} = 3(x + y + z) + t$$

or
$$\frac{dz}{dt} = 3\left(x_0 + y_0 + z_0 + \frac{1}{12}\right)e^{6t} - \frac{1}{4}(6t + 1) + t$$

Integrating, we obtain

$$z = \left(x_0 + y_0 + z_0 + \frac{1}{12}\right)\frac{e^{6t}}{2} - \frac{1}{4}\left(6\frac{t^2}{2} + t\right) + \frac{t^2}{2} + c_3$$

The integration constant c_3 can be found from the initial conditions as $t = t_0, z = z_0$

$$z_0 = \left(x_0 + y_0 + z_0 + \frac{1}{12}\right)\frac{1}{2} + c_3$$

or
$$c_3 = -\frac{1}{2}x_0 - \frac{1}{2}y_0 + \frac{1}{2}z_0 - \frac{1}{24}$$

Putting the value of c_3, we obtain

$$z = -\frac{1}{2}x_0 - \frac{1}{2}y_0 + \frac{1}{2}z_0 + \frac{1}{2}\left(x_0 + y_0 + z_0 + \frac{1}{12}\right)e^{6t} - \frac{1}{4}t + \frac{1}{4}t^2 - \frac{1}{24}$$

3.3.1 Variation of Flow Parameters in Time and Space

In general, the flow velocity and other hydrodynamic parameters like pressure and density may vary from one point to another at any instant, and also from one instant to another at a fixed point. According to the type of variations, different categories of flow are described as follows:

3.3.1.1 *Steady and Unsteady Flows*

A steady flow is defined as a flow in which the flow velocities and fluid properties at any point do not change with time. The flow in which any of these parameters change with time is termed as unsteady flow. In the Eulerian approach, a steady flow is described as

$$\vec{V} = \vec{V}(\vec{S})$$

which means that the flow velocity is a function of space coordinates only. This implies that, in a steady flow, the flow velocity and fluid properties may vary with location, but the spatial distribution of any such parameter essentially remains invariant with time.

In the Lagrangian approach, time is inherent in describing the trajectory of any particle (Eq. (3.1)). But in steady flow, the velocities of all particles passing through any fixed point at different times will be same. In other words, the description of velocity as a function of time for a given particle will simply show the velocities at different points through which the particle has passed and thus furnish the information of velocity as a function of spatial location as described by the Eulerian method. Therefore, the Eulerian and the Lagrangian approaches of describing fluid motion become identical under this situation.

In practice, absolute steady flow is the exception rather than the rule, but many problems may be studied effectively by assuming that the flow is steady. Though minor fluctuations of velocity and other quantities with time occur in reality, the average value of any quantity over a reasonable interval of time remains unchanged. Moreover, a particular flow may appear steady to one observer but unsteady to another. This is because all movement is relative. The motion of a body or a fluid element is described with respect to a reference frame. Therefore, the flow may appear unsteady depending upon the choice of the reference frame. For example, the movement of water past the sides of a motorboat traveling at constant velocity would (apart from small fluctuations) appear steady to an observer in the boat. He/she would compare the water flow with an imaginary set of reference frame fixed to the boat. To an observer on a bridge, however, the same flow would appear to change with time as the boat passes underneath him/her. He/she would be comparing the flow with reference axes fixed to the bridge. Since the examination of steady flow is usually much simpler that that of unsteady flow, reference frames are chosen, where possible, so that flow with respect to the reference frame becomes steady .

3.3.1.2 *Uniform and Non-uniform Flow*

When velocity and fluid properties at any instant of time, do not change from point to point in flow field, the flow is said to be uniform. If, however, changes do occur from one point to another, the flow is non-uniform. Hence, for a uniform flow, the velocity is a function of time only, which can be expressed in the Eulerian description as

$$\vec{V} = \vec{V}(t)$$

This implies that for a uniform flow, there will be no spatial distribution of fluid velocities and flow properties. Any such parameter will have a unique value in the entire field, which of course, may change with time if the flow is unsteady.

For non-uniform flow, the changes with position may be found either in the direction of flow or in directions perpendicular to it. The latter kind of non-uniformity is always encountered near solid boundaries past which the fluids flow. This is because all fluids possess viscosity which reduces the relative velocity (velocity relative to solid boundary) to zero at a solid boundary (no-slip condition as described in Chapter 1).

For a steady and uniform flow, velocity is neither a function of time nor of space coordinates, and hence it assumes a constant value throughout the region of flow at all times. Steadiness of flow and uniformity of flow do not necessarily go together. Any of the four combinations as shown in Table 3.1 are possible.

Table 3.1

Type	*Example*
1. Steady uniform flow	Flow at constant rate through a duct of uniform cross section. (The region close to the walls of the duct is however disregarded.)
2. Steady non-uniform flow	Flow at constant rate through a duct of non-uniform cross section (tapering pipe.)
3. Unsteady uniform flow	Flow at varying rates through a long straight pipe of uniform cross section. (Again the region close to the walls is ignored.)
4. Unsteady non-uniform flow	Flow at varying rates through a duct flow of non-uniform cross section.

3.3.2 Material Derivative and Acceleration

Let the position of a particle at any instant t in a flow field be given by the space coordinates (x, y, z) with respect to a rectangular Cartesian frame of reference. The velocity components u, v, w of the particle along the x, y and z directions, respectively, can then be written in the Eulerian form as

$$\begin{aligned} u &= u(x, y, z, t) \\ v &= v(x, y, z, t) \\ w &= w(x, y, z, t) \end{aligned}$$

At an infinitesimal time interval Δt later, let the particle move to a new position given by the coordinates $(x + \Delta x, y + \Delta y, z + \Delta z)$, and its velocity components at this new position be $u + \Delta u$, $v + \Delta v$ and $w + \Delta w$. Therefore, we can write

$$u + \Delta u = u\,(x + \Delta x, y + \Delta y, z + \Delta z, t + \Delta t) \tag{3.6a}$$

$$v + \Delta v = v\,(x + \Delta x, y + \Delta y, z + \Delta z, t + \Delta t) \tag{3.6b}$$

$$w + \Delta w = w\,(x + \Delta x, y + \Delta y, z + \Delta z, t + \Delta t) \tag{3.6c}$$

The expansion of the right-hand side of the Eqs (3.6a) to (3.6c) in the form of Taylor's series gives

$$u + \Delta u = u(x, y, z, t) + \frac{\partial u}{\partial x}\Delta x + \frac{\partial u}{\partial y}\Delta y + \frac{\partial u}{\partial z}\Delta z + \frac{\partial u}{\partial t}\Delta t$$
$$+ \text{ higher order terms in } \Delta x, \Delta y, \Delta z \text{ and } \Delta t \tag{3.7a}$$

$$v + \Delta v = v(x, y, z, t) + \frac{\partial v}{\partial x}\Delta x + \frac{\partial v}{\partial y}\Delta y + \frac{\partial v}{\partial z}\Delta z + \frac{\partial v}{\partial t}\Delta t$$
$$+ \text{ higher order terms in } \Delta x, \Delta y, \Delta z \text{ and } \Delta t \tag{3.7b}$$

$$w + \Delta w = w(x, y, z, t) + \frac{\partial w}{\partial x}\Delta x + \frac{\partial w}{\partial y}\Delta y + \frac{\partial w}{\partial z}\Delta z + \frac{\partial w}{\partial t}\Delta t$$
$$+ \text{ higher order terms in } \Delta x, \Delta y, \Delta z \text{ and } \Delta t \tag{3.7c}$$

The increment in space coordinates can be written as

$$\Delta x = u\,\Delta t, \qquad \Delta y = v\,\Delta t, \qquad \Delta z = w\,\Delta t$$

Substituting the values of Δx, Δy and Δz in Eqs (3.7a) to (3.7c), we have

$$\frac{\Delta u}{\Delta t} = u\frac{\partial u}{\partial x} + v\frac{\partial u}{\partial y} + w\frac{\partial u}{\partial z} + \frac{\partial u}{\partial t}$$
$$+ \text{ terms containing } \Delta t \text{ and its higher orders}$$

$$\frac{\Delta v}{\Delta t} = u\frac{\partial v}{\partial x} + v\frac{\partial v}{\partial y} + w\frac{\partial v}{\partial z} + \frac{\partial v}{\partial t}$$
$$+ \text{ terms containing } \Delta t \text{ and its higher orders}$$

$$\frac{\Delta w}{\Delta t} = u\frac{\partial w}{\partial x} + v\frac{\partial w}{\partial y} + w\frac{\partial w}{\partial z} + \frac{\partial w}{\partial t}$$
$$+ \text{ terms containing } \Delta t \text{ and its higher orders}$$

The limiting forms of the equations as $\Delta t \to 0$ become

$$\frac{\mathrm{D}u}{\mathrm{D}t} = \frac{\partial u}{\partial t} + u\frac{\partial u}{\partial x} + v\frac{\partial u}{\partial y} + w\frac{\partial u}{\partial z} \tag{3.8a}$$

$$\frac{\mathrm{D}v}{\mathrm{D}t} = \frac{\partial v}{\partial t} + u\frac{\partial v}{\partial x} + v\frac{\partial v}{\partial y} + w\frac{\partial v}{\partial z} \tag{3.8b}$$

$$\frac{\mathrm{D}w}{\mathrm{D}t} = \frac{\partial w}{\partial t} + u\frac{\partial w}{\partial x} + v\frac{\partial w}{\partial y} + w\frac{\partial w}{\partial z} \tag{3.8c}$$

$$\left[\text{Since } \lim_{\Delta t \to 0}\frac{\Delta u}{\Delta t} = \frac{\mathrm{D}u}{\mathrm{D}t}, \quad \lim_{\Delta t \to 0}\frac{\Delta v}{\Delta t} = \frac{\mathrm{D}v}{\mathrm{D}t}, \quad \lim_{\Delta t \to 0}\frac{\Delta w}{\Delta t} = \frac{\mathrm{D}w}{\mathrm{D}t},\right.$$

$$\lim_{\Delta t \to 0} \text{(terms containing } \Delta t \text{ and its higher orders)} = 0 \Bigg]$$

It is important to note in this context that in the limit as $\Delta t \to 0$, the Eulerian and the Lagrangian description merge (i.e., the velocity of flow at a point at any instant is the velocity of a fluid particle that exists at that point at that instant), so that $\lim_{\Delta t \to 0} \frac{\Delta x}{\Delta t} = u$, which is the x component of velocity at the point of concern, and similarly for the components v and w.

It is evident from the above equations that the operator for total differential with respect to time, D/Dt in a convective field is related to the partial differential $\partial/\partial t$ as

$$\frac{\mathrm{D}}{\mathrm{D}t} = \frac{\partial}{\partial t} + u\frac{\partial}{\partial x} + v\frac{\partial}{\partial y} + w\frac{\partial}{\partial z} \tag{3.9}$$

The total differential D/Dt is known as the *material* or *substantial derivative* with respect to time. The first term $\partial/\partial t$ on the right-hand side of Eq. (3.9) is known as temporal or local derivative which expresses the rate of change with time, at a fixed position. The last three terms on the right-hand side of Eq. (3.9) are together known as *convective derivative* which represent the time rate of change due to change in position in the field. Therefore, the terms on the left-hand sides of Eqs (3.8a) to (3.8c) are defined the as the x, y and z components of substantial or material acceleration. The first terms on the right-hand sides of Eqs (3.8a) to (3.8c) represent the respective local or temporal accelerations, while the other terms are convective accelerations. Thus we can write,

$$a_x = \frac{\mathrm{D}u}{\mathrm{D}t} = \frac{\partial u}{\partial t} + u\frac{\partial u}{\partial x} + v\frac{\partial u}{\partial y} + w\frac{\partial u}{\partial z} \tag{3.9a}$$

$$a_y = \frac{\mathrm{D}v}{\mathrm{D}t} = \frac{\partial v}{\partial t} + u\frac{\partial v}{\partial x} + v\frac{\partial v}{\partial y} + w\frac{\partial v}{\partial z} \tag{3.9b}$$

$$a_z = \frac{\mathrm{D}w}{\mathrm{D}t} = \frac{\partial w}{\partial t} + u\frac{\partial w}{\partial x} + v\frac{\partial w}{\partial y} + w\frac{\partial w}{\partial z} \tag{3.9c}$$

(Material or substantial acceleration) = (Temporal or local acceleration) + (Convective acceleration)

In a steady flow, the temporal acceleration is zero, since the velocity at any point is invariant with time. In a uniform flow, on the other hand, the convective acceleration is zero, since the velocity components are not the functions of space coordinates. In a steady and uniform flow, both the temporal and convective acceleration vanish and hence there exists no material acceleration. Existence of the components of acceleration for different types of flow, as described in Table 3.1, is shown in Table 3.2.

Table 3.2

Type of flow	*Material acceleration*	
	Temporal	*Convective*
Steady and uniform	0	0
Steady and non-uniform	0	Exists
Unsteady and uniform	Exists	0
Unsteady and non-uniform	Exists	Exists

3.3.2.1 *Components of Acceleration in Other Coordinate Systems*

In a cylindrical polar coordinate system (Fig. 3.1 (a), the components of acceleration in r, θ and z directions can be written as

$$a_r = \frac{\mathrm{D}v_r}{\mathrm{D}t} - \frac{v_\theta^2}{r} = \frac{\partial v_r}{\partial t} + v_r \frac{\partial v_r}{\partial r} + \frac{v_\theta}{r}\frac{\partial v_r}{\partial \theta} + v_z \frac{\partial v_r}{\partial z} - \frac{v_\theta^2}{r} \tag{3.10a}$$

$$a_\theta = \frac{\mathrm{D}v_\theta}{\mathrm{D}t} + \frac{v_r v_\theta}{r} = \frac{\partial v_\theta}{\partial t} + v_r \frac{\partial v_\theta}{\partial r} + \frac{v_\theta}{r}\frac{\partial v_\theta}{\partial \theta} + v_z \frac{\partial v_\theta}{\partial z} + \frac{v_r v_\theta}{r} \tag{3.10b}$$

$$a_z = \frac{\mathrm{D}v_z}{\mathrm{D}t} = \frac{\partial v_z}{\partial t} + V_r \frac{\partial v_z}{\partial r} + \frac{v_\theta}{r}\frac{\partial v_z}{\partial \theta} + v_z \frac{\partial v_z}{\partial z} \tag{3.10c}$$

The term $-v_\theta^2/r$ in the Eq. (3.10a) appears due to an inward radial acceleration arising from a change in the direction of v_θ (velocity component in the azimuthal direction θ) with θ as shown in Fig. 3.1 (a). This is typically known as centripetal acceleration. In a similar fashion, the term $v_r\, v_\theta/r$ represents a component of acceleration in azimuthal direction caused by a change in the direction of v_r with θ (Fig. 3.1(a).

The acceleration components in a spherical polar coordinate system (Fig. 3.1(b) can be expressed as

$$a_R = \frac{\partial v_r}{\partial t} + v_r \frac{\partial v_r}{\partial R} + \frac{v_\phi}{R}\frac{\partial v_r}{\partial \phi} + \frac{v_\theta}{R\sin\phi}\frac{\partial v_r}{\partial \theta} - \frac{v_\phi^2 + v_\theta^2}{R} \tag{3.11a}$$

$$a_\phi = \frac{\partial v_\phi}{\partial t} + v_R \frac{\partial v_\phi}{\partial R} + \frac{v_\phi}{R}\frac{\partial v_\phi}{\partial_\phi} + \frac{v_\theta}{R\sin\phi}\frac{\partial v_\phi}{\partial \theta}$$

$$- \frac{v_R v_\phi}{R} - \frac{v_\theta^2 \cot\phi}{R} \tag{3.11b}$$

$$a_\theta = \frac{\partial v_\theta}{\partial t} + v_R \frac{\partial v_\theta}{\partial R} + \frac{v_\phi}{R}\frac{\partial v_\theta}{\partial \phi} + \frac{v_\theta}{R\sin\phi}\frac{\partial v_\theta}{\partial \theta}$$

$$+ \frac{v_\theta v_R}{R} + \frac{v_\theta\, v_\phi \cot\phi}{R} \tag{3.11c}$$

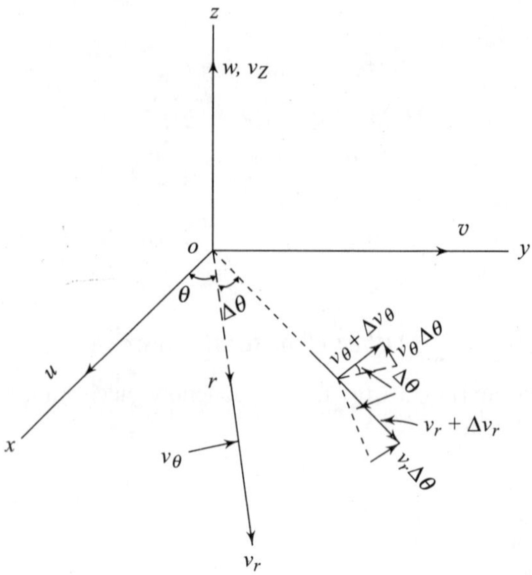

Fig. 3.1(a) Velocity components in a cylindrical polar coordinate system

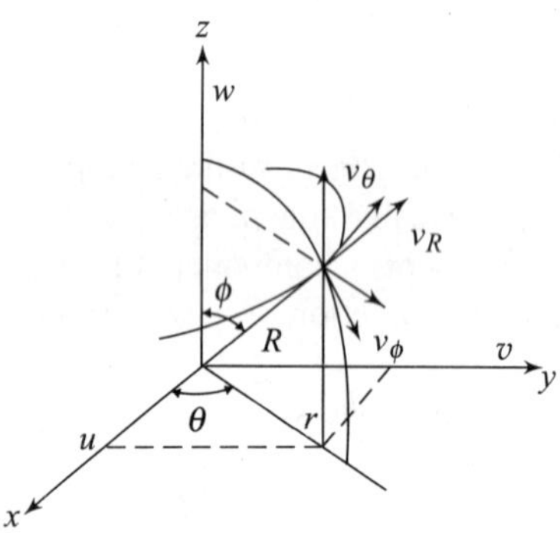

Fig. 3.1(b) Velocity components in a spherical polar coordinate system

Example 3.4

Given a velocity field $\vec{V} = (4 + xy + 2t)\hat{i} + 6x^3 \hat{j} + (2xt^2 + z)\hat{k}$. Find the acceleration of a fluid particle at (2, 4, – 4) and time $t = 3$.

Solution

$$\vec{a} = \frac{D\vec{V}}{Dt} = \frac{\partial \vec{V}}{\partial t} + (\vec{V} \cdot \nabla)\vec{V}$$

$$\frac{\partial \vec{V}}{\partial t} = 2\hat{i} + 6xt\hat{k}$$

$$(\vec{V}.\nabla)\vec{V} = \left(u\frac{\partial}{\partial x} + v\frac{\partial}{\partial y} + w\frac{\partial}{\partial z}\right)[(4 + xy + 2t)\hat{i} + 6x^3\hat{j} + (2xt^2 + z)\hat{k}]$$

$$= u(y\hat{i} + 18x^2\hat{j} + 3t^2\hat{k}) + v(x\hat{i}) + w(\hat{k})$$

$$= (4 + xy + 2t)\ (y\hat{i} + 18x^2\hat{j} + 3t^2\hat{k}) + 6x^3(x\hat{i}) + (2xt^2 + z)\hat{k}$$

$$= (4y + xy^2 + 2ty + 6x^4)\ \hat{i}\ + (72x^2 + 18x^3y + 36tx^2)\ \hat{j}$$

$$+ (12t^2 + 3xyt^2 + 6t^3 + 3xt^2 + z)\ \hat{k}$$

Finally, the acceleration field can be expressed as

$$\vec{a} = (2 + 4y + xy^2 + 2ty + 6x^4)\,\hat{i}\ + (72x^2 + 18x^3y + 36tx^2)\,\hat{j}$$

$$+ (6xt + 12t^2 + 3xyt^2 + 6t^3 + 3xt^2 + z)\,\hat{k}$$

The acceleration vector at the point (2, 4, –4) and at time $t = 3$ can be found out by substituting the values of x, y, z and t in the Eq. (3.36) as

$$\vec{a} = 170\,\hat{i}\ + 1296\,\hat{j}\ + 572\,\hat{k}$$

Hence, x component of acceleration a_x = 170 units

y component of acceleration a_y = 1296 units

z component of acceleration a_z = 572 units

Magnitude of resultant acceleration

$$|\vec{a}| = [(170)^2 + (1296)^2 + (572)^2]^{1/2}$$

$$= 1375.39 \text{ units}$$

Example 3.5

Find the acceleration components at a point (1, 1, 1) for the following flow field:

$$u = 2x^2 + 3y, \quad v = -2xy + 3y^2 + 3zy, \quad w = -\frac{3}{2}z^2 + 2xz - 9y^2z$$

Solution

x component of acceleration

$$a_x = \frac{\partial u}{\partial t} + u\frac{\partial u}{\partial x} + v\frac{\partial u}{\partial y} + w\frac{\partial u}{\partial z}$$

$$= 0 + (2x^2 + 3y)\,4x + (-2xy + 3y^2 + 3zy)3 + 0$$

Therefore, $(a_x)_{\text{at }(1,\,1,\,1)} = 0 + 5 \times 4 + 4 \times 3 + 0 = 32$ units

y component of acceleration

$$a_y = \frac{\partial v}{\partial t} + u\frac{\partial v}{\partial x} + v\frac{\partial v}{\partial y} + w\frac{\partial v}{\partial z}$$

$$= 0 + (2x^2 + 3y)\,(-2y) + (-2xy + 3y^2 + 3zy)\,(-2x + 6y + 3z)$$

$$+\left(-\frac{3}{2}z^2 + 2xz - 9y^2z\right)3y$$

Therefore, $(a_y)_{\text{at}\,(1,\,1,\,1)} = 5\times(-2)+4\times7+(-8.5)\times3 = -7.5$ units

z component of acceleration

$$a_z = \frac{\partial w}{\partial t} + u\frac{\partial w}{\partial x} + \text{v}\frac{\partial w}{\partial y} + w\frac{\partial w}{\partial z}$$

$$= 0+(2x^2+3y)\,2z+(-2xy+3y^2+3zy)\,(-18yz)$$

$$+\left(-\frac{3}{2}z^2 + 2xz - 9y^2z\right)(-3z+2x-9y^2)$$

Therefore $\quad (a_z)_{\text{at}\,(1,\,1,\,1)} = 0+(2+3)\times2-(-2+3+3)18$

$$+\left(-\frac{3}{2}+2-9\right)(-3+2-9)$$

$$= 23 \text{ units}$$

Example 3.6

The velocity and density fields in a diffuser are given by

$$u = u_0\,e^{-2x/L} \qquad \text{and} \qquad \rho = \rho_0\,e^{-x/L}$$

Find the rate of change of density at $x = L$.

Solution

The rate of change of density in this case can be written as

$$\frac{\mathrm{D}\rho}{\mathrm{D}t} = \frac{\partial \rho}{\partial t} + u\frac{\partial \rho}{\partial x}$$

$$= 0 + u_0\,e^{-2x/L}\,\frac{\partial}{\partial x}\left(\rho_0\,e^{-x/L}\right)$$

$$= u_0\,e^{-2x/L}\left(-\frac{\rho_0}{L}\right)e^{-x/L}$$

$$= -\frac{\rho_0\,u_0}{L}e^{-3x/L}$$

$$\text{at } x = L,\quad \mathrm{D}\rho/\mathrm{D}t = -\frac{\rho_0\,u_0}{L}e^{-3}$$

Example 3.7

Find the acceleration of a fluid particle at the point $r = 2a$, $\theta = \pi/2$ for a two-dimensional flow given by

$$v_r = -u\left(1-\frac{a^2}{r^2}\right)\cos\theta\,, \quad v_\theta = u\left(1+\frac{a^2}{r^2}\right)\sin\theta$$

Solution

$$\frac{\partial v_r}{\partial \theta} = u \sin\theta\left(1-\frac{a^2}{r^2}\right), \quad \frac{\partial v_\theta}{\partial \theta} = u\cos\theta\left(1+\frac{a^2}{r^2}\right)$$

$$\frac{\partial v_r}{\partial r} = \frac{-2ua^2}{r^3}\cos\theta, \quad \frac{\partial v_\theta}{\partial r} = \frac{-2ua^2}{r^3}\sin\theta,$$

Acceleration in the radial direction

$$a_r = v_r\frac{\partial v_r}{\partial r} + \frac{v_\theta}{r}\frac{\partial v_r}{\partial \theta} - \frac{v_\theta^2}{r}$$

$$= \frac{2u^2a^2}{r^3}\left(1-\frac{a^2}{r^2}\right)\cos^2\theta + \frac{u^2}{r}\left(1-\frac{a^4}{r^4}\right)\sin^2\theta - \frac{u^2}{r}\left(1+\frac{a^2}{r^2}\right)^2\sin^2\theta$$

Hence, (a_r) at $r = 2a$, $\theta = \pi/2$

$$= 0 + \frac{u^2}{2a}\left(1-\frac{1}{16}\right) - \frac{u^2}{2a}\left(1+\frac{1}{4}\right)^2$$

$$= -\frac{5u^2}{16a}$$

Acceleration in the azimuthal direction

$$a_\theta = v_r\frac{\partial v_\theta}{\partial r} + \frac{v_\theta}{r}\frac{\partial v_\theta}{\partial \theta} + \frac{v_r v_\theta}{r}$$

$$= \frac{2u^2a^2}{r^3}\left(1-\frac{a^2}{r^2}\right)\sin\theta\cos\theta + \frac{u^2}{r}\left(1+\frac{a^2}{r^2}\right)^2\sin\theta\cos\theta$$

$$-\frac{u^2}{r}\left(1-\frac{a^4}{r^4}\right)\sin\theta\cos\theta$$

Hence, $(a_\theta)_{(\text{at } r = 2a,\ \theta = \pi/2)} = 0 + 0 + 0 = 0$

Therefore, at $r = 2a$, $\theta = \pi/2$

$$a_r = -5u^2/16, \qquad a_\theta = 0$$

Example 3.8

A fluid is flowing at a constant volume flow rate of Q through a divergent pipe having inlet and outlet diameters of D_1 and D_2, respectively, and a length of L. Assuming the velocity to be axial and uniform at any section, show that the accelerations at the inlet and outlet of the pipe are given by $-\frac{32Q^2(D_2 - D_1)}{\pi^2 L D_1^5}$ and $-\frac{32Q^2(D_2 - D_1)}{\pi^2 L D_2^5}$, respectively.

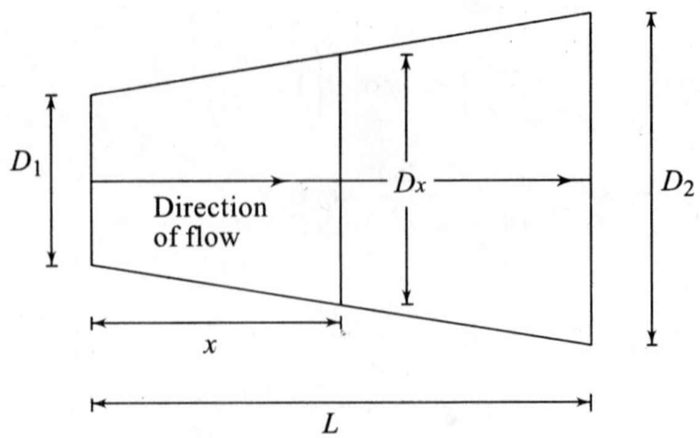

Fig. 3.2 Flow through a divergent duct

Solution

The diameter of the duct at an axial distance x from the inlet plane (Fig. 3.2) is given by

$$D_x = D_1 + \frac{x}{L}\cdot(D_2 - D_1)$$

Therefore, the velocity at this section can be written as

$$u = \frac{4Q}{\pi\left[D_1 + \frac{x}{L}(D_2 \bullet D_1)\right]^2}$$

Acceleration at this section can be written as

$$a = u\,\frac{\partial u}{\partial x}$$

$$= \frac{4Q}{\pi\left[D_1 + \frac{x}{L}(D_2 \bullet D_1)\right]^2} \times \frac{\bullet 8Q}{\pi\left[D_1 + \frac{x}{L}(D_2 \bullet D_1)\right]^3}\,\frac{(D_2 \bullet D_1)}{L}$$

$$= \frac{\bullet 32\,Q^2(D_2 \bullet D_1)}{\pi^2 L\left[D_1 + \frac{x}{L}(D_2 \bullet D_1)\right]^5}$$

This is the general expression of acceleration at any section at a distance x from the inlet of the pipe. Substituting the values of $x = 0$ (for inlet) and $x = L$ (for outlet) in the above equation, we have

$$\text{Acceleration at the inlet} = \frac{-32\,Q^2(D_2 - D_1)}{\pi^2 L D_1^5}$$

$$\text{and acceleration at the outlet} = \frac{-32\,Q^2(D_2 - D_1)}{\pi^2 L D_2^5}$$

3.3.3 Streamlines, Path Lines and Streak Lines

3.3.3.1 *Streamlines*

The analytical description of flow velocities by the Eulerian approach is geometrically depicted through the concept of streamlines. In the Eulerian method, the velocity vector is defined as a function of time and space coordinates. *If for a fixed instant of time, a space curve is drawn so that it is tangent everywhere to the velocity vector, then this curve is called a streamline. Therefore, the Eulerian method gives a series of instantaneous streamlines of the state of motion (Fig. 3.3(a)). In other words, a streamline at any instant can be defined as an imaginary curve or line in the flow field so that the tangent to the curve at any point represents the direction of the instantaneous velocity at that point.* In an unsteady flow where the velocity vector changes with time, the pattern of streamlines also changes from instant to instant. In a steady flow, the orientation or the pattern of streamlines will be fixed. From the above definition of streamline, it can be written

$$\vec{V} \times \mathrm{d}\vec{S} = 0 \tag{3.12}$$

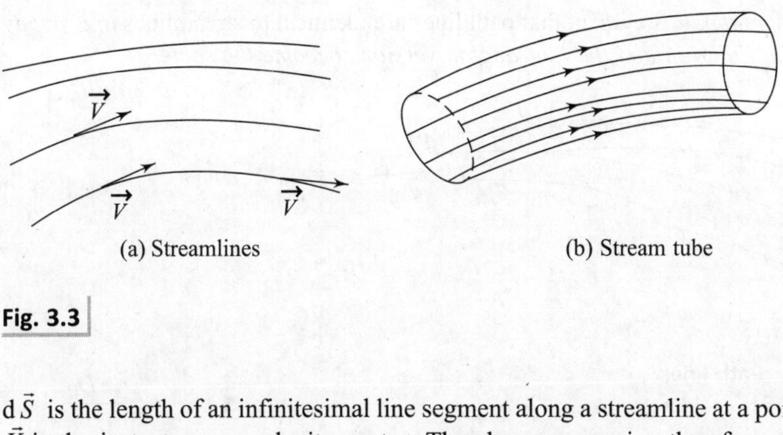

(a) Streamlines (b) Stream tube

Fig. 3.3

$\mathrm{d}\vec{S}$ is the length of an infinitesimal line segment along a streamline at a point where $\vec{V}$ is the instantaneous velocity vector. The above expression therefore represents the differential equation of a streamline. In a Cartesian coordinate system, the vectors $\vec{V}$ and $\mathrm{d}\vec{S}$ can be written in terms of their components along the coordinate axes as $\vec{V} = \hat{i}u + \hat{j}v + \hat{k}w$ and $\mathrm{d}\vec{S} = \hat{i}\,\mathrm{d}x + \hat{j}\,\mathrm{d}y + \hat{k}\,\mathrm{d}z$. Then Eq. (3.12) gives

$$\frac{\mathrm{d}x}{u} = \frac{\mathrm{d}y}{v} = \frac{\mathrm{d}z}{w} \tag{3.13}$$

and thus describes the differential equation of streamlines in a cartesian frame of reference.

A bundle of neighbouring streamlines may be imagined to form a passage through which the fluid flows (Fig. 3.3(b)). This passage (not necessarily circular in cross section) is known as a stream tube. A stream tube with a cross section small enough for the variation of velocity over it to be negligible is sometimes termed as a stream filament. Since the stream tube is bounded on all sides by streamlines, and by definition, velocity does not exist across a streamline, no fluid may enter or leave a stream tube except through its ends. The entire flow in a flow field may be imagined to be composed of flows through stream tubes arranged in some arbitrary positions.

3.3.3.2 *Path Lines*

Path lines are the outcome of the Lagrangian method in describing fluid flow and show the paths of different fluid particles as a function of time. In other words, *a path line is the trajectory of a fluid particle of fixed identity as defined by Eq. (3.1). Therefore a family of path lines represents the trajectories of different particles, say,* P_1*,* P_2*,* P_3*, etc. (Fig. 3.4)*. It can be mentioned in this context that while stream lines are referred to a particular instant of time, the description of path lines *inherently* involves *the variation of time*, since a fluid particle takes time to move from one point to another. Two path lines can intersect with one another or *a single path line itself can form a loop*. This is quite possible in a sense that, under certain conditions of flow, different particles *or even a same particle can arrive at same location at different instants of time*. The two stream lines, on the other hand, can never intersect each other since *the instantaneous velocity vector at a given location is always unique*. It is evident that path lines are identical to streamlines *in a steady flow as the Eulerian and the Lagrangian versions become the same*.

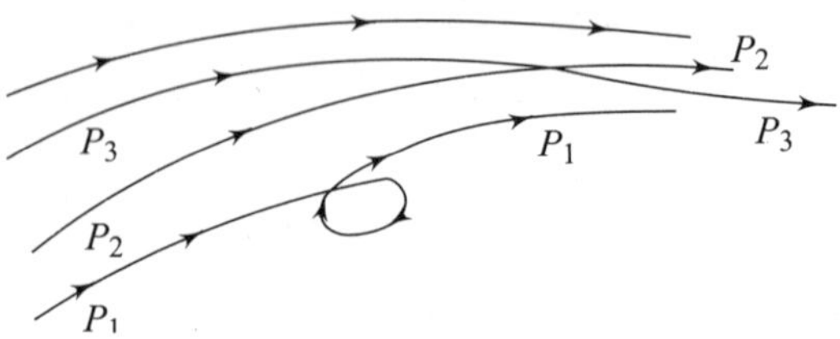

Fig. 3.4 Path lines

3.3.3.3 *Streak Lines*

A streak line at any instant of time is the locus of the temporary locations of all particles that have passed though a fixed point in the flow field. While a path line refers to the identity of a fluid particle, a streak line is specified by a fixed point in the flow field. This line is of particular interest in experimental flow visualisation. If dye is injected into a liquid at a fixed point in the flow field, then at a later time t, the dye will indicate the end points of the path lines of particles which have passed through the injection point. The equation of a streak line at time t can be derived by the Lagrangian method. If a fluid particle ($\vec{S}_0$) passes through a fixed point ($\vec{S}_1$) in a course of time t, then the Lagrangian method of description gives the equation

$$S(\vec{S}_0, t) = \vec{S}_1 \tag{3.14}$$

or solving for $\vec{S}_0$,

$$\vec{S}_0 = F(\vec{S}_1, t) \tag{3.15}$$

If the positions ($\vec{S}$) of the particles which have passed through the fixed point ($\vec{S}_1$) are determined, then a streak line can be drawn through these points. The equation of the streak line at a time t is given by

$$\vec{S} = f(\vec{S}_0, t) \tag{3.16}$$

Upon substitution of Eq. (3.15) into Eq. (3.16) we obtain,

$$\vec{S} = f[F(\vec{S}_1, t), t] \tag{3.17}$$

This is the final form of the equation of a streak line referred to a fixed point S_1. Figure 3.5 describe the difference between streak lines and path lines. Let P be a fixed point in space through which particles of different identities pass at different times. In an unsteady flow, the velocity vector at P will change with time and hence the particles arriving at P at different times will traverse different paths like PAQ, PBR, and PCS which represent the path lines of the particles. Let at any instant t_1, these particles arrive at points Q, R, and S. Thus, Q, R, and S represent the end points of the trajectories of these three particles at the instant t_1. Therefore, the curve joining the points S, R, Q, and the fixed point P will define the streak line at that instant t_1. The fixed point P will also lie on the line, since at any instant, there will be always a particle of some identity at that point. For a steady flow, the velocity vector at any point is invariant with time and hence the path lines of the particles with different identities passing through P at different times will not differ, rather would coincide with one another in a single curve which will indicate the streak line too. Therefore, in a steady flow, the path lines, streak lines and streamlines are identical.

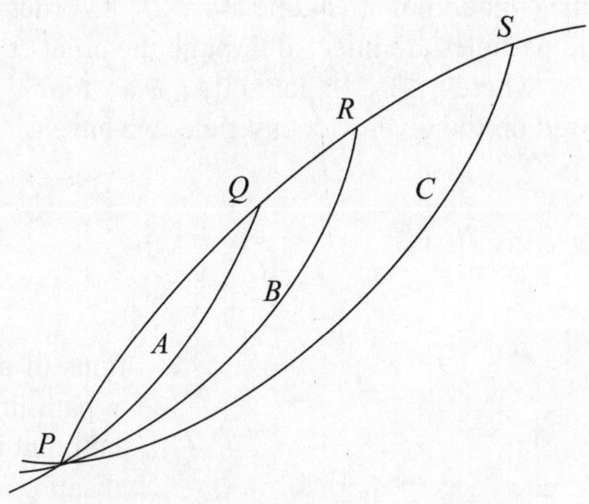

Fig. 3.5 Description of a streakline

Example 3.9

A two-dimensional flow field is defined as $\vec{V} = \hat{i}x - \hat{j}y$. Derive the equation of stream line passing through the point (1, 1).

Solution

The equation of stream line is

$$\vec{V} \times d\vec{s} = 0$$

or, $$udy - vdx = 0$$

Hence, $$dy/dx = v/u = -y/x$$

or, $$dy/y + dx/x = 0$$

Integration of the above equation gives $xy = C$, where C is constant

For the stream line passing through (1,1), the value of the constant C is 1. Hence, the required equation of stream line passing through (1,1) is $xy - 1 = 0$.

Example 3.10

The velocity components in a flow field are given as follows:

$u = \dfrac{x}{t + t_0}$, $v = v_0$, $w = 0$, where t_0 and v_0 are constants. A coloured dye is injected at the point A (x_0, y_0, z_0) in the flow field. Find (i) the equation of a coloured line visible in the flow field at $t = 2t_0$, as a consequence of the dye injection, given that the dye injection starts at $t = t_0$; and (ii) Find the locus of a fluid particle that passes through the point A at $t = t_0$.

Solution

(i) Here, essentially, we need to obtain the equation of streakline at $t = 2t_0$. In order to obtain the same, we consider that fluid particles are injected through the point at instants of time symbolically denoted by t_i, where $t_i \le 2t_0$. Importantly t_i is a variable and not a fixed instant of time. Now, based on the given velocity field, we have

$$u = \frac{dx}{dt} = \frac{x}{t + t_0}$$

$$\Rightarrow \quad \int_{x_0}^{x} \frac{dx}{x} = \int_{t_i}^{2t_0} \frac{dt}{t + t_0}$$

$$\Rightarrow \quad \ln \frac{x}{x_0} = \ln \frac{2t_0 + t_0}{t_i + t_0}$$

$$\Rightarrow \quad \frac{x}{x_0} = \frac{3t_0}{t_i + t_0} \tag{3.18}$$

Similarly, $$v = \frac{dy}{dt} = v_0$$

$$\Rightarrow \quad \int_{y_0}^{y} dy = \int_{t_i}^{2t_0} v_0 \, dt$$

$$\Rightarrow \qquad y - y_0 = -v_0\,(2t_0 - t_i)$$

$$\Rightarrow \qquad t_i = 2t_0 - \frac{y - y_0}{v_0} \tag{3.19}$$

Eliminating t_i from Eqs (3.18) and (3.19), one can write

$$\frac{x}{x_0} = \frac{3t_0}{2t_0 - \dfrac{y - y_0}{v_0} + t_0} = \frac{3t_0}{3t_0 - \dfrac{y - y_0}{v_0}} = \frac{1}{1 - \dfrac{y - y_0}{3t_0 v_0}}$$

$$\frac{x}{x_0}\left(1 - \frac{y - y_0}{3t_0 v_0}\right) = 1 \tag{3.20}$$

Equation. (3.20) is the equation of a coloured line visible in the flow field at $t = 2t_0$.

(ii) Here, essentially, we need to obtain the equation of the pathline, corresponding to a fluid particle that at $t = t_0$ passed through (x_0, y_0)

Now, $$u = \frac{dx}{dt} = \frac{x}{t + t_0}$$

$$\Rightarrow \qquad \int_{x_0}^{x} \frac{dx}{x} = \int_{t_0}^{t} \frac{dt}{t + t_0}$$

$$\Rightarrow \qquad \ln\frac{x}{x_0} = \ln\frac{t + t_0}{t_0 + t_0} = \ln\frac{t + t_0}{2t_0}$$

$$\Rightarrow \qquad \frac{x}{x_0} = \frac{t + t_0}{2t_0} \tag{3.21}$$

Similarly, $$v = \frac{dy}{dt} = v_0$$

$$\int_{y_0}^{y} dy = \int_{t_0}^{t} v_0\,dt$$

$$y - y_0 = v_0\,(t - t_0)$$

$$t - t_0 = +\frac{y - y_0}{v_0} \tag{3.22}$$

Eliminating t from Eqs (3.21) and (3.22), one can write

$$\frac{x}{x_0} = \frac{t_0 + \dfrac{y - y_0}{v_0} + t_0}{2t_0} = 1 + \frac{y - y_0}{2t_0 v_0}$$

$$\frac{x}{x_0} - \frac{y - y_0}{2t_0 v_0} = 1 \tag{3.23}$$

Equation. (3.23) is the equation of the locus of a fluid particle that passes through the point A at $t = t_0$.

3.3.4 One-, Two- and Three-Dimensional Flows

In general, fluid flow is three dimensional. This means that the flow parameters like velocity, pressure and so on vary in all the three coordinate directions. Sometimes simplification is made in the analysis of different fluid flow problems by selecting the coordinate directions so that appreciable variation of the hydrodynamic parameters take place in only two directions or even in only one.

So *one-dimensional flow* is that in which all the flow parameters may be expressed as functions of time and one space coordinate only. This single space coordinate is usually the distance measured along the centre line (not necessarily straight) of some conduit in which the fluid is flowing. For instance, the flow in a pipe is considered one-dimensional when variations of pressure and velocity occur along the length of the pipe, but any variation over the cross section is assumed negligible. In reality the flow is never one dimensional because viscosity causes the velocity to decrease to zero at the solid boundaries. If, however, the non-uniformity of the actual flow is not too great, valuable results may often be obtained from 'one-dimensional analysis'. Under this situation, the average values of the flow parameters at any given section (perpendicular to the flow) are assumed to be applied to the entire flow at that section. In two-dimensional flow, the flow parameters are functions of time and two space coordinates (say, x and y). There is no variation in the z direction, and therefore the same streamline pattern could, at any instant, be found in all planes perpendicular to the z direction. In three-dimensional flow, the hydrodynamic parameters are functions of three space coordinates and time.

3.3.5 Translation, Rate of Deformation and Rotation

The movement of a fluid element in space has three distinct features, namely, translation, rate of deformation, and rotation. A fluid motion, in general, consists of these three features simultaneously. Translation and rotation without deformation represent rigid-body displacements which do not induce any strain in the body. Figure. 3.6 shows the picture of a pure translation in absence of rotation and deformation of a fluid element in a three-dimensional flow described by a rectangular Cartesian coordinate system. In absence of deformation and rotation, there will be no change in the length of the sides or in the included angles made by the sides of the fluid element. The sides are displaced parallelly. This is possible when the flow velocities u (the x component velocity), v (the y component velocity), and w (the z component velocity) are neither a function of x, y, nor of z, in other words, the flow field is totally uniform.

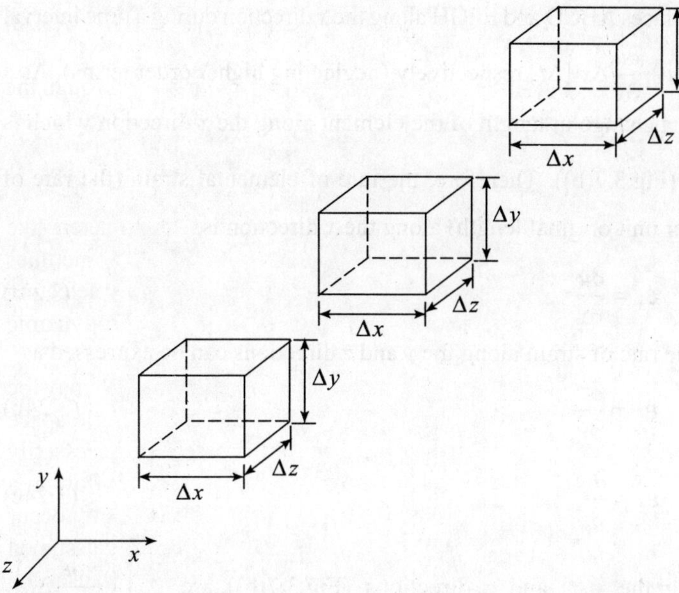

Fig. 3.6

3.3.5.1 *Rate of Linear Deformation*

Now consider a situation where a component of flow velocity becomes the function of only one space coordinate along which that velocity component is defined, for example, $u = u(x)$, $v = v(y)$ and $w = w(z)$. Let us consider a fluid element, as shown in Fig. 3.7 (a), which in course of its translation suffers a change in its linear dimensions without any change in the included angle by the sides.

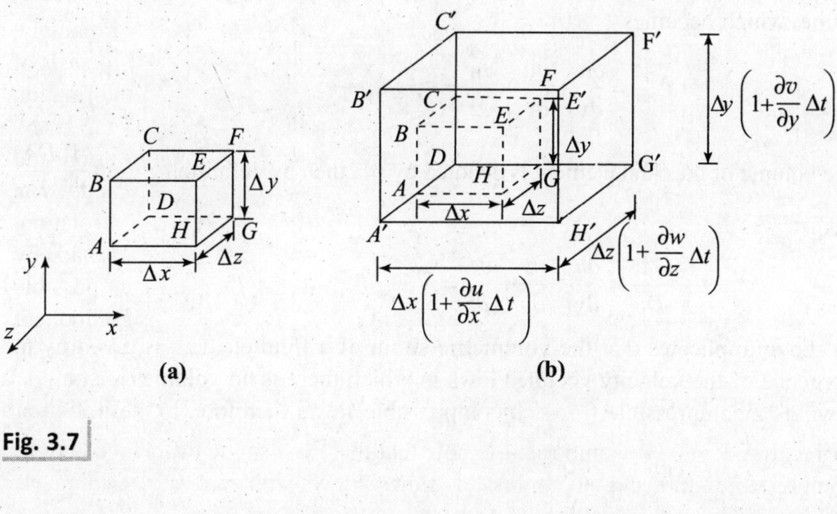

Fig. 3.7

Displacement of the faces ABCD and EFGH along the x direction during a time interval of Δt are $u\Delta t$ and $\left(u+\frac{\partial u}{\partial x}\Delta x\right)\Delta t$, respectively (neglecting higher order terms). As a result, there occurs a change in length of the element along the x direction which is equal to $\frac{\partial u}{\partial x}\Delta x\Delta t$ (Fig.3.7(b)). Therefore, the rate of elemental strain (the rate of change of length per unit original length) along the x direction is

$$\dot{\varepsilon}_x = \frac{\partial u}{\partial x} \tag{3.24a}$$

In a similar way, the rate of strain along the y and z directions can be expressed as

$$\dot{\varepsilon}_y = \frac{\partial v}{\partial y} \tag{3.24b}$$

$$\dot{\varepsilon}_z = \frac{\partial w}{\partial z} \tag{3.24c}$$

New lengths along the x, y and z directions (Fig.3.7(b)) are $\Delta x\left(1+\frac{\partial u}{\partial x}\Delta t\right)$, $\Delta y\left(1+\frac{\partial v}{\partial y}\Delta t\right)$ and $\Delta z\left(1+\frac{\partial w}{\partial z}\Delta t\right)$, respectively. Therefore, the change in volume of the fluid element (Fig. 3.7 (b)) is

$$\Delta x\Delta y\Delta z\left(1+\frac{\partial u}{\partial x}\Delta t\right)\left(1+\frac{\partial v}{\partial y}\Delta t\right)\left(1+\frac{\partial w}{\partial z}\Delta t\right) - \Delta x\Delta y\Delta z$$

$$= \Delta x\Delta y\Delta z\left[\left(\frac{\partial u}{\partial x}+\frac{\partial v}{\partial y}+\frac{\partial w}{\partial z}\right)\Delta t\right] \text{ (neglecting higher order terms)}$$

Rate of volumetric strain is given by the rate of change of volume per unit original volume, which becomes

$$\dot{\varepsilon}_{\text{vol}} = \frac{\partial u}{\partial x}+\frac{\partial v}{\partial y}+\frac{\partial w}{\partial z} = \dot{\varepsilon}_x + \dot{\varepsilon}_y + \dot{\varepsilon}_z$$

If the volume of the fluid element is denoted by $\forall$, then by definition $\dot{\varepsilon}_{\text{vol}} = \frac{1}{\forall}\frac{D\forall}{Dt}$.

Thus,

$$\frac{1}{\forall}\frac{D\forall}{Dt} = \frac{\partial u}{\partial x}+\frac{\partial v}{\partial y}+\frac{\partial w}{\partial z} = \nabla.\vec{V} \tag{3.24d}$$

The above implicates that the volumetric strain of a fluid element is given by the divergence of the velocity vector. Flows in which there is no volumetric strain are known as incompressible flows. Incompressible flows therefore, by definition are flows with $\nabla.\vec{V} = 0$. It is important to note that there is a subtle difference between incompressible fluid, and incompressible flows. Flows with zero rate of volumetric strain are termed as incompressible flows. On contrary, a fluid is called incompressible if its density does not change significantly with change in pressure.

It is important to mention here that the condition for incompressibility is intimately related to the consideration of conservation of mass for incompressible flows. In Section 3.4 we will revisit this issue with more elaborate details, starting from the general principles of conservation of mass applied to fluid flow.

3.3.5.2 *Rate of Angular Deformation and Rotation of Fluid Elements*

For simplicity, let us consider a fluid element ABCD (Fig. 3.8) in a two-dimensional velocity field, where both the velocities u and v become functions of x and y, i.e.,

$$u = u(x, y)$$
$$v = v(x, y)$$

In the expansion of u and v about x and y in a Taylor series function, the higher order terms are neglected for any derivation as follows. Displacement of A and B along the x direction during a time duration Δt are $u\Delta t$ and $\left(u+\frac{\partial u}{\partial x}\Delta x\right)\Delta t$, respectively. Similarly, the displacement of A and B along the y direction during a time duration Δt are $v\Delta t$ and $\left(v+\frac{\partial v}{\partial x}\Delta x\right)\Delta t$, respectively. Therefore, the point B has relative displacements in both x and y directions with respect to point A, whose magnitudes are $\frac{\partial u}{\partial x}\Delta x\Delta t$ and $\frac{\partial v}{\partial x}\Delta x\Delta t$, respectively. Similarly, point D has relative displacements in both x and y directionss with respect to point A, whose magnitudes are $\frac{\partial u}{\partial y}\Delta y\Delta t$ and $\frac{\partial v}{\partial y}\Delta y\Delta t$, respectively. Because of the relative displacement of B in the y direction with respect to A and the relative displacement of D in the x direction with respect to A, the included angle between AB and AD changes, and the fluid element suffers a continuous angular deformation along with the linear deformations in course of its motion. The rate of angular strain or angular deformation $\dot{\varepsilon}_{xy}$ is defined as the rate of change of the angle between two line elements in the fluids which were originally perpendicular to each other. Consider AB and AD as two such line elements, for example. On deformation, the angle between them is, $< \mathrm{B'A'D'} = \frac{\pi}{2} - (\Delta\alpha + \Delta\beta)$.

The change in the angle is $= \frac{\pi}{2} - \left[\frac{\pi}{2} - (\Delta\alpha + \Delta\beta)\right] = (\Delta\alpha + \Delta\beta)$

Rate of change of the angle, or rate of angular deformation is

$$\dot{\varepsilon}_{xy} = \lim_{\Delta t \to 0}\left(\frac{\Delta\alpha + \Delta\beta}{\Delta t}\right)$$

or

$$\dot{\varepsilon}_{xy} = \frac{d\alpha}{dt} + \frac{d\beta}{dt}$$

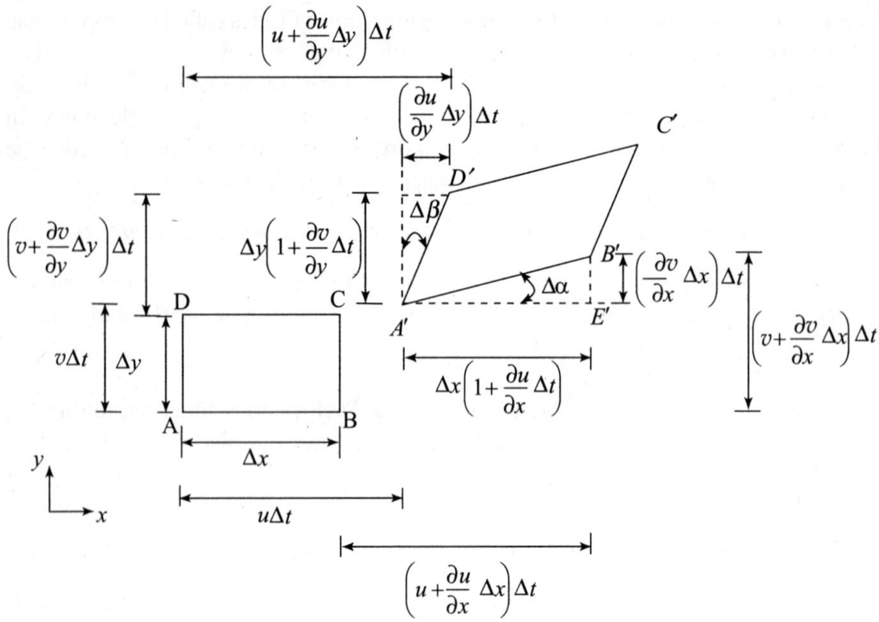

Fig. 3.8

From the geometry, we obtain

$$\tan \Delta\alpha = \frac{B'E'}{A'E'} = \frac{\frac{\partial v}{\partial x}\Delta x\Delta t}{\Delta x\left(1+\frac{\partial u}{\partial x}\Delta t\right)}$$

For small values of $\Delta\alpha$ (which occurs at small Δt), $\Delta\alpha \approx \Delta\alpha$

Hence,
$$\Delta\alpha = \frac{\frac{\partial v}{\partial x}\Delta x\Delta t}{\Delta x\left(1+\frac{\partial u}{\partial x}\Delta t\right)}$$

$$\frac{d\alpha}{dt} = \lim_{\Delta t\to 0}\frac{\Delta\alpha}{\Delta t} = \lim_{\Delta t\to 0}\frac{\frac{\partial v}{\partial x}\Delta x}{\Delta x\left(1+\frac{\partial u}{\partial x}\Delta t\right)}$$

$$= \frac{\partial v}{\partial x}$$

Similarly,
$$\frac{d\beta}{dt} = \lim_{\Delta t\to 0}\frac{\Delta\beta}{\Delta t} = \lim_{\Delta t\to 0}\frac{\frac{\partial u}{\partial y}\Delta y}{\Delta x\left(1+\frac{\partial v}{\partial y}\Delta t\right)}$$

$$= \frac{\partial u}{\partial y}$$

Thus, $$\dot{\varepsilon}_{xy} = \frac{d\alpha}{dt} + \frac{d\beta}{dt} = \frac{\partial v}{\partial x} + \frac{\partial u}{\partial y} \tag{3.25}$$

Similarly,

$$\dot{\varepsilon}_{yz} = \frac{\partial v}{\partial z} + \frac{\partial w}{\partial y} \tag{3.25a}$$

$$\dot{\varepsilon}_{xz} = \frac{\partial w}{\partial x} + \frac{\partial u}{\partial z} \tag{3.25b}$$

In index notation, the rate of angular deformation can be expressed as

$$\dot{\varepsilon}_{ij} = \frac{\partial u_j}{\partial x_i} + \frac{\partial u_i}{\partial x_j} \tag{3.26}$$

Specification of $\dot{\varepsilon}$, thus requires two indices; and like stress, it is also a second order tensor. It may be easily derived that the strain rate tensor is symmetric, i.e., $\dot{\varepsilon}_{ij} = \dot{\varepsilon}_{ji}$.

It is important to note in this context that because of relative motion between fluid layers, deformable fluids do not rotate like rigid bodies. However, an angular velocity for such fluid flow may be artificially defined in such a way that in the limiting case of non-deformable fluids, it conforms to the motion of rigid body rotation. Accordingly, the angular velocity at a point is defined as the arithmetic mean of angular velocities of two line elements at that point, which were originally perpendicular to each other. Referring to Fig. 3.8, the angular velocities of two originally perpendicular line elements AB and AD in the *xy* plane are $\frac{d\alpha}{dt}$ and $\frac{d\beta}{dt}$ respectively, but oriented in the opposite sense. Considering the anticlockwise direction as positive, the angular velocity of the fluid element at A thus can be written as,

$$\omega_{xy} = \frac{1}{2}\left(\dot{\alpha} - \dot{\beta}\right) = \frac{1}{2}\left(\frac{\partial v}{\partial x} - \frac{\partial u}{\partial y}\right) \tag{3.27}$$

Instead of ω_{xy}, one may also use an alternative notation of ω_z, considering that rotation in the *xy* plane and rotation with respect to the *z* axis is essentially the same thing.

In three-dimensional flow, the components of rotation are defined as

$$\omega_{xy} = \omega_z = \frac{1}{2}\left(\frac{\partial v}{\partial x} - \frac{\partial u}{\partial y}\right) \tag{3.27a}$$

$$\omega_{yz} = \omega_x = \frac{1}{2}\left(\frac{\partial w}{\partial y} - \frac{\partial v}{\partial z}\right) \tag{3.27b}$$

$$\omega_{zx} = \omega_y = \frac{1}{2}\left(\frac{\partial u}{\partial z} - \frac{\partial w}{\partial x}\right) \tag{3.27c}$$

In index notation, the angular velocity can be expressed as

$$\omega_{ij} = \frac{1}{2}\left(\frac{\partial u_j}{\partial x_i} - \frac{\partial u_i}{\partial x_j}\right) \tag{3.28}$$

Obviously the angular velocity tensor is skew-symmetric as $\omega_{ij} = -\omega_{ji}$. It is interesting to note in this context that the velocity gradient tensor $\frac{\partial u_i}{\partial x_j}$ can be decomposed into two components, viz., one symmetric and another skew-symmetric, as follows:

$$\frac{\partial u_i}{\partial x_j} = \underbrace{\frac{1}{2}\left(\frac{\partial u_i}{\partial x_j} + \frac{\partial u_j}{\partial x_i}\right)}_{\dot{\varepsilon}_{ij}} + \underbrace{\frac{1}{2}\left(\frac{\partial u_i}{\partial x_j} - \frac{\partial u_j}{\partial x_i}\right)}_{-\omega_{ij}} \tag{3.29}$$

Here the symmetric component represents the rate of shear strain tensor and the skew-symmetric part represents the angular velocity tensor.

The angular velocity of fluid flow has further mathematical implications, which may be inferred by referring to the fact that following Eqs (3.27a) to (3.27c), rotation in a flow field can be expressed in a vector form as

$$\vec{\omega} = \frac{1}{2}\left(\nabla \times \vec{V}\right) \tag{3.30}$$

The quantity $\vec{\Omega} = \nabla \times \vec{V}$ is defined as the **vorticity of flow,** which is a mathematical measure of rotationalities in the flow field. In Cartesian coordinate systems

$$\vec{\Omega} = \nabla \times \vec{V} = \begin{vmatrix} \hat{i} & \hat{j} & \hat{k} \\ \frac{\partial}{\partial x} & \frac{\partial}{\partial y} & \frac{\partial}{\partial z} \\ u & v & w \end{vmatrix}$$

$$= \left(\frac{\partial w}{\partial y} - \frac{\partial v}{\partial z}\right)\hat{i} + \left(\frac{\partial u}{\partial z} - \frac{\partial w}{\partial x}\right)\hat{j} + \left(\frac{\partial v}{\partial x} - \frac{\partial u}{\partial y}\right)\hat{k} = 2\vec{\omega} \tag{3.31}$$

If an imaginary line is drawn in the fluid so that the tangent to it at each point is in the direction of the vorticity vector $\vec{\Omega}$ at that point, the line is called a vortex line. Therefore, the general equation of the vortex line can be written as

$$\vec{\Omega} \times d\vec{s} = 0 \tag{3.32}$$

Let us now consider some special cases of angular deformation of fluid elements

$$\frac{\partial v}{\partial x} = -\frac{\partial u}{\partial y},$$

$$\dot{\varepsilon}_{xy} = 0 \text{ (from Eq. 3.25)} \tag{3.33a}$$

and
$$\omega_z = \frac{\partial v}{\partial x} = -\frac{\partial u}{\partial y} \quad \text{(from Eq. 3.27)} \tag{3.33b}$$

This implies that the linear segments *AB* and *AD* move with the same angular velocity (both in magnitude and direction) and hence the included angle between them remains the same and no angular deformation takes place. This situation is known as pure rotation (Fig. 3.9(a). In another special case,

when
$$\frac{\partial v}{\partial x} = \frac{\partial u}{\partial y}$$

$$\dot{\gamma}_{xy} = 2\frac{\partial v}{\partial x} = 2\frac{\partial u}{\partial y} \quad \text{(from Eq. 3.25)} \tag{3.34a}$$

and
$$\omega_z = 0 \text{ (from Eq. (3.27))} \tag{3.34b}$$

This implies that the fluid element has an angular deformation rate but no rotation about the *z* axis (Fig. 3.9(b)

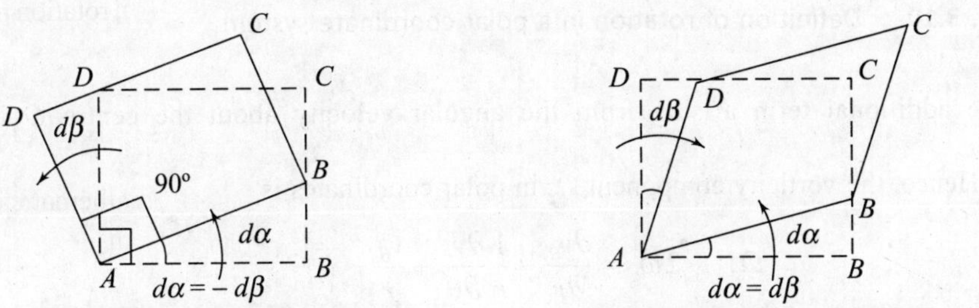

(a) Fluid element in pure rotation in absence of angular deformation

(b) Fluid element with angular deformation in absence of rotation

Fig. 3.9

When the components of rotation at all points in a flow field become zero, the flow is said to be irrotational. Therefore, the necessary and sufficient condition for a flow field to be irrotational is

$$\nabla \times \vec{V} = 0 \tag{3.35}$$

3.3.5.3 *Vorticity in Polar Coordinates*

In a two-dimensional polar coordinate system (Fig. 3.10), the angular velocity of segment Δr can be written as

$$\lim_{\Delta t \to 0}\left[\frac{(v_\theta + (\partial v_\theta/\partial r)\,\Delta r - v_\theta)\,\Delta t}{\Delta r\,\Delta t}\right] = \frac{\partial v_\theta}{\partial r}$$

Also, the angular velocity of segment $r\Delta\theta$ becomes

$$\lim_{\Delta t \to 0}\left[-\frac{(v_r + (\partial v_r/\partial \theta)\,\Delta \theta - v_r)\,\Delta t}{r\,\Delta\theta\,\Delta t}\right] = -\frac{1}{r}\frac{\partial v_r}{\partial \theta}$$

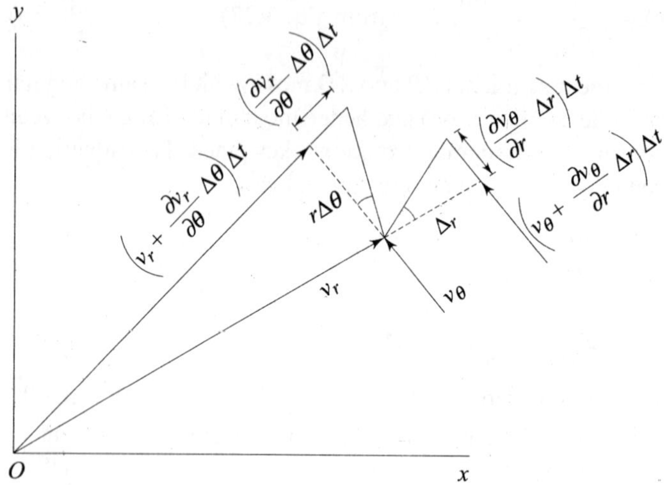

Fig. 3.10 Definition of rotation in a polar coordinate system

The additional term arising from the angular velocity about the centre O is V_θ/r.

Hence, the vorticity component Ω_z in polar coordinates is

$$\Omega_z = 2\omega_z = \frac{\partial v_\theta}{\partial r} - \frac{1}{r}\frac{\partial v_r}{\partial \theta} + \frac{v_\theta}{r}$$

Therefore, in a three-dimensional cylindrical polar coordinate system, the vorticity components can be written as

$$\Omega_z = \frac{\partial v_\theta}{\partial r} - \frac{1}{r}\frac{\partial v_r}{\partial \theta} + \frac{v_\theta}{r} \tag{3.36a}$$

$$\Omega_r = \frac{1}{r}\frac{\partial v_z}{\partial \theta} - \frac{\partial v_\theta}{\partial z} \tag{3.36b}$$

$$\Omega_\theta = \frac{\partial v_r}{\partial z} - \frac{\partial v_z}{\partial r} \tag{3.36c}$$

In a spherical polar coordinate system (Fig. 3.1(b)), the vorticity components are defined as

$$\Omega_R = \frac{1}{R}\frac{\partial v_\theta}{\partial \phi} - \frac{1}{R\sin\phi}\frac{\partial v_\phi}{\partial \theta} + \frac{v_\theta}{R}\cot\phi \qquad (3.37a)$$

$$\Omega_\phi = \frac{1}{R\sin\phi}\frac{\partial v_R}{\partial \theta} - \frac{\partial v_\theta}{\partial R} - \frac{v_\theta}{R} \qquad (3.37b)$$

$$\Omega_\theta = \frac{\partial v_\phi}{\partial R} + \frac{v_\phi}{R} - \frac{1}{R}\frac{\partial v_R}{\partial \phi} \qquad (3.37c)$$

3.3.5.4 *Circulation*

The circulation Γ around a closed contour is defined as the line integral of the tangential component of velocity and is given by (Fig. 3.11)

$$\Gamma = \oint \vec{V}.d\vec{s} \qquad (3.38)$$

where $d\vec{s}$ is a line element of the contour. Circulation is traditionally taken to be positive in the anticlockwise direction (not a mandatory but a general convention). The circulation can be expressed in terms of area integral using Stokes theorem as

$$\Gamma = \oint \vec{V}.d\vec{s} = \int_A \left(\nabla \times \vec{V}\right).\hat{n}dA \qquad (3.39)$$

where dA is the elemental area of the surface and A is the total surface area bounded by the close curve around which circulation is evaluated. In Eq. (3.39), $\hat{n}$ is a unit vector outward normal to dA.

Using the definition of vorticity, Eq. (3.39) becomes

$$\Gamma = \int_A \vec{\Omega}.\hat{n}dA \qquad (3.40)$$

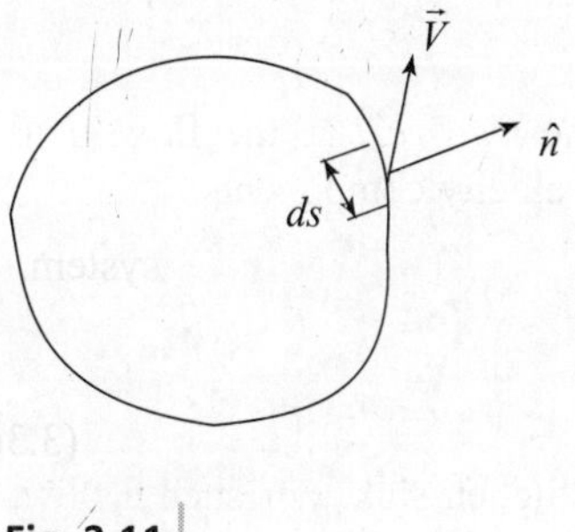

Fig. 3.11

It is interesting to illustrate the concept of circulation through some special types of flows known as vortex flows. Vortex flows are flows for which the radial component of velocity, v_r, is zero; but the cross-radial component of velocity, v_θ, is non-zero (in fact, a function of r).

The first example of vortex flow that we introduce here is called *forced vortex*, which is defined by the following velocity components:

$$v_\theta = \omega r$$

and $$v_r = 0$$

Consider a close contour *ABCD* in the above flow field shown in Fig. 3.12. Circulation along the close contour *ABCD* is given by

$$\Gamma_{ABCD} = \oint \vec{V}.d\vec{s} = -v_{\theta_{AB}} r d\theta + v_{r_{BC}} dr + v_{\theta_{CD}} (r+dr) d\theta - v_{r_{DA}} r d\theta$$

$$= -(\omega r) r d\theta + v_{r_{BC}}^{\;\;0} dr + \omega (r+dr)(r+dr)\, d\theta - v_{r_{DA}}^{\;\;0} dr$$

$$\Gamma_{ABCD} = 2\omega r dr d\theta \tag{3.41}$$

Circulation per unit area is given by

$$\frac{\Gamma_{ABCD}}{r dr d\theta} = 2\omega \tag{3.41a}$$

Interestingly, here 2ω is the vorticity of flow. Thus, circulation per unit area is the vorticity of the flow.

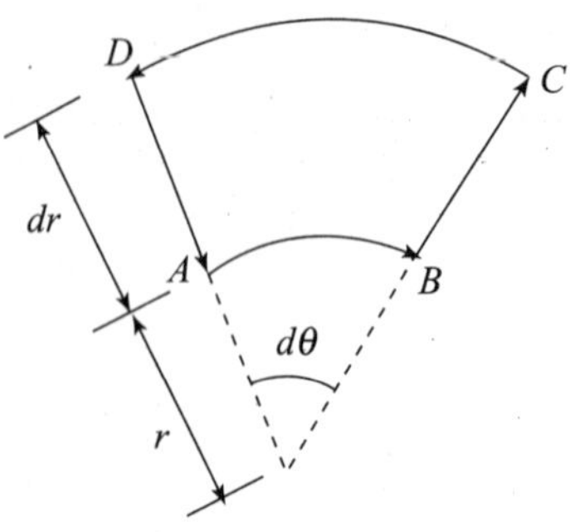

Fig. 3.12

As a next example of vortex flow, we consider the *free vortex,* the flow field corresponding to which is defined by the following velocity components:

$$v_\theta = \frac{C}{r}$$

and $$v_r = 0$$

Close to such a situation is the flow observed in a kitchen sink. For such a flow, circulation along the closed contour *ABCD* is given by

$$\Gamma_{ABCD} = \oint \vec{V}.d\vec{s} = -v_{\theta_{AB}} r d\theta - v_{r_{BC}} dr + v_{\theta_{CD}} (r+dr)\, d\theta + v_{r_{DA}} r d\theta$$

$$= -\frac{C}{r} r d\theta - v_{r_{BC}}^{\;\;0} dr + \frac{C}{r+dr}(r+dr) d\theta - v_{r_{DA}}^{\;\;0} r d\theta = 0$$

It is important to mention here that free vortex flow at the origin is impossible because of mathematical singularity (v_θ tends to infinity as r tends to zero). In this context, one may also note that although free vortex is an irrotational motion, the circulation around a path containing any singular point (such as the origin) is non-zero. However, that is not a technically correct way of evaluating the circulation, since the contour chosen must not include any point of singularity. Thus, if the circulation is calculated for a free vortex flow along any closed contour excluding the singular point (the origin), it should be zero, consistent with the conditions to be satisfied by an irrotational flow field.

A classical example of vortex flow in nature is a tornado. Up to the eye of a tornado, it is a forced vortex, having a strong element of rotationality. Beyond a critical radius it is like a free vortex. Thus, a tornado can be represented by a *Rankine vortex* which is a combination of forced and free vortex (Fig. 3.13). At the critical radius, v_θ evaluated from free vortex zone must be equal to that evaluated from the forced vortex zone.

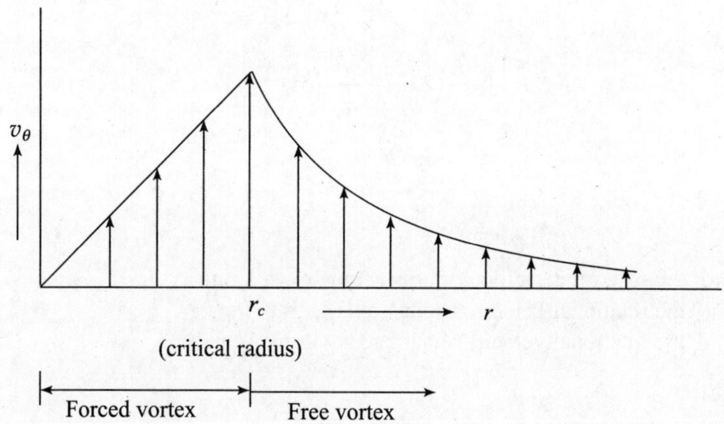

Fig. 3.13 Velocity distribution in a Rankine vortex

Example 3.11

The velocity field in a fluid medium is given by

$$\vec{V} = 3xy^2\,\hat{i} + 2xy\,\hat{j} + (2zy + 3t)\,\hat{k}$$

Find the magnitudes and directions of (i) translational velocity, (ii) rotational velocity, and (iii) the vorticity of a fluid element at (1, 2, 1) and at time $t = 3$.

Solution

(i) Translational velocity vector at (1, 2, 1) and at $t = 3$ can be written as

$$\vec{V} = 3(2)(4)\,\hat{i} + 2(1)(2)\,\hat{j} + [2(1)(2) + 3(3)]\,\hat{k} = 12\,\hat{i} + 4\,\hat{j} + 13\,\hat{k}$$

Hence, x component of translational velocity $u = 12$ units
y component of translational velocity $v = 4$ units
z component of translational velocity $w = 13$ units

(ii) Rotational velocity vector is found as

$$\vec{\omega} = \frac{1}{2}(\nabla \times \vec{V}) = \frac{1}{2}\begin{vmatrix} \hat{i} & \hat{j} & \hat{k} \\ \dfrac{\partial}{\partial x} & \dfrac{\partial}{\partial y} & \dfrac{\partial}{\partial z} \\ u & v & w \end{vmatrix}$$

$$= \frac{\hat{i}}{2}\left\{\frac{\partial w}{\partial y} - \frac{\partial v}{\partial z}\right\} + \frac{\hat{j}}{2}\left\{\frac{\partial u}{\partial z} - \frac{\partial w}{\partial x}\right\} + \frac{\hat{k}}{2}\left\{\frac{\partial v}{\partial x} - \frac{\partial u}{\partial y}\right\}$$

$$= \hat{i}\frac{1}{2}\left[\frac{\partial}{\partial y}(2zy + 3t) - \frac{\partial}{\partial z}(2xy)\right]$$

$$+ \hat{j}\frac{1}{2}\left[\frac{\partial}{\partial z}(3xy^2) - \frac{\partial}{\partial x}(2zy + 3t)\right]$$

$$+ \hat{k}\frac{1}{2}\left[\frac{\partial}{\partial x}(2xy) - \frac{\partial}{\partial y}\left(3xy^2\right)\right]$$

$$= z\hat{i} + (y - 3xy)\hat{k}$$

at (1, 2, 1) and $t = 3$,

$$\vec{\omega} = \hat{i} - 4\hat{k}$$

Therefore, the rotational velocity about x axis $\omega_x = 1$ unit
the rotational velocity about y axis $\omega_y = 0$ unit
the rotational velocity about z axis $\omega_z = -4$ units

(iii) The vorticity $\vec{\Omega} = 2\vec{\omega}$

Hence, $\vec{\Omega} = 2\hat{i} - 8\hat{k}$

Example 3.12

Find the vorticity components at a point (1,1,1) for the following flow field:

$$u = 2x^2 + 3y, \quad v = -2xy + 3y^2 + 3zy, \quad w = \frac{3}{2}z^2 + 2xz - 9y^2z$$

Solution

$$\Omega_x = \frac{\partial w}{\partial y} - \frac{\partial v}{\partial z} = 18yz - 3y = -(18yz + 3y)$$

$$\Omega_y = \frac{\partial u}{\partial z} - \frac{\partial w}{\partial x} = 0 - 2z = -2z$$

$$\Omega_z = \frac{\partial v}{\partial x} - \frac{\partial u}{\partial y} = -2y - 3 = -(2y + 3)$$

at the point (1, 1, 1)

$$\Omega_x = -(18 + 3) = -21 \text{ units}$$

$$\Omega_y = -2 \text{ units}$$

$$\Omega_z = -(2 + 3) = -5 \text{ units}$$

3.3.5.5 *Generalised Expression of the Movement of a Fluid Element*

An analytical expression to represent the most general form of the movement of a fluid element consisting of translation, rotation and deformation can be developed as follows.

Consider the movement of a fluid element in a fluid continuum as shown in Fig. 3.14.

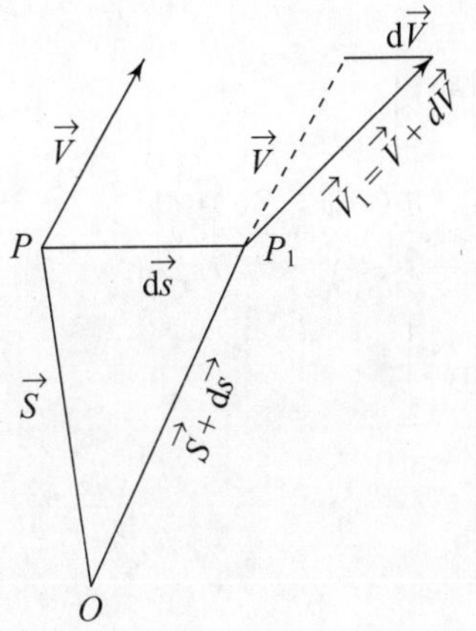

Fig. 3.14 General representation of fluid motion

The velocity at a point $P(x, y, z)$ is $\vec{V}$ and at point $P_1(x_1, y_1, z_1)$, a small distance $d\vec{S}$ from P, is $\vec{V}_1$.

The velocity vector $\vec{V}_1$ can be written as

$$\vec{V}_1 = \hat{i}u_1 + \hat{j}\mathrm{v}_1 + \hat{k}\mathrm{w}_1 = \vec{V} + d\vec{V}$$

$$= \hat{i}u + \hat{j}\mathrm{v} + \hat{k}w + \frac{\partial \vec{V}}{\partial x}\mathrm{dx} + \frac{\partial \vec{V}}{\partial y}\mathrm{dy} + \frac{\partial \vec{V}}{\partial z}\mathrm{dz}$$

$$= \hat{i}\left[u + \frac{\partial u}{\partial x}\mathrm{dx} + \frac{\partial u}{\partial y}\mathrm{dy} + \frac{\partial u}{\partial z}\mathrm{dz}\right]$$

$$+ \hat{j}\left[v + \frac{\partial v}{\partial x}\mathrm{d}x + \frac{\partial v}{\partial y}\mathrm{d}y + \frac{\partial v}{\partial z}\mathrm{d}z\right]$$

$$+ \vec{k}\left[w + \frac{\partial w}{\partial x}\mathrm{d}x + \frac{\partial w}{\partial y}\mathrm{d}y + \frac{\partial w}{\partial z}\mathrm{d}z\right] \quad (3.42)$$

where u_1, v_1, w_1 are the respective x, y and z components of $\vec{V}_1$ and u, v, w are those of $\vec{V}$.

Equation (3.42) can be rearranged as

$$\vec{V}_1 = \hat{i}\left[u + \left\{\frac{\partial u}{\partial x}\mathrm{d}x + \frac{1}{2}\left(\frac{\partial v}{\partial x} + \frac{\partial u}{\partial y}\right)\mathrm{d}y + \frac{1}{2}\left(\frac{\partial u}{\partial z} + \frac{\partial w}{\partial x}\right)\mathrm{d}z\right\}\right.$$

$$\left. + \frac{1}{2}\left\{\left(\frac{\partial u}{\partial z} - \frac{\partial w}{\partial x}\right)\mathrm{d}z - \left(\frac{\partial v}{\partial x} - \frac{\partial u}{\partial y}\right)\mathrm{d}y\right\}\right]$$

$$+ \hat{j}\left[v + \left\{\frac{\partial v}{\partial y}\mathrm{d}y + \frac{1}{2}\left(\frac{\partial v}{\partial x} + \frac{\partial u}{\partial y}\right)\mathrm{d}x + \frac{1}{2}\left(\frac{\partial w}{\partial y} + \frac{\partial v}{\partial z}\right)\mathrm{d}z\right\}\right.$$

$$\left. + \frac{1}{2}\left\{\left(\frac{\partial v}{\partial x} - \frac{\partial u}{\partial y}\right)\mathrm{d}x - \left(\frac{\partial w}{\partial y} - \frac{\partial v}{\partial z}\right)\mathrm{d}z\right\}\right]$$

$$+ \hat{k}\left[w + \left\{\frac{\partial w}{\partial z}\mathrm{d}z + \frac{1}{2}\left(\frac{\partial u}{\partial z} + \frac{\partial w}{\partial x}\right)\mathrm{d}x + \frac{1}{2}\left(\frac{\partial w}{\partial y} + \frac{\partial v}{\partial z}\right)\mathrm{d}y\right\}\right.$$

$$\left. + \frac{1}{2}\left\{\left(\frac{\partial w}{\partial y} - \frac{\partial v}{\partial z}\right)\mathrm{d}y - \left(\frac{\partial u}{\partial z} - \frac{\partial w}{\partial x}\right)\mathrm{d}x\right\}\right]$$

$$= \vec{V} + \frac{1}{2}\vec{\Omega} \times \mathrm{d}\vec{s} + \vec{D} \quad (3.43)$$

where, $\mathrm{d}\vec{S} = \hat{i}\,\mathrm{d}x + \hat{j}\,\mathrm{d}y + \hat{k}\,\mathrm{d}z$,

$\vec{\Omega}$ is the vorticity vector as defined by Eq. (3.31)

and

$$\vec{D} = \hat{i}\left[\frac{\partial u}{\partial x}\mathrm{d}x + \frac{1}{2}\left(\frac{\partial v}{\partial x} + \frac{\partial u}{\partial y}\right)\mathrm{d}y + \frac{1}{2}\left(\frac{\partial u}{\partial z} + \frac{\partial w}{\partial x}\right)\mathrm{d}z\right]$$

$$+ \hat{j}\left[\frac{1}{2}\left(\frac{\partial v}{\partial x} + \frac{\partial u}{\partial y}\right)\mathrm{d}x + \frac{\partial v}{\partial y}\mathrm{d}y + \frac{1}{2}\left(\frac{\partial w}{\partial y} + \frac{\partial v}{\partial z}\right)\mathrm{d}z\right]$$

$$+ \hat{k}\left[\frac{1}{2}\left(\frac{\partial u}{\partial z} + \frac{\partial w}{\partial x}\right)\mathrm{d}x + \frac{1}{2}\left(\frac{\partial w}{\partial y} + \frac{\partial v}{\partial z}\right)\mathrm{d}y + \frac{\partial w}{\partial z}\mathrm{d}z\right] \quad (3.44)$$

Equation (3.43) represents the most general form of the movement of a fluid element. The first term represents the translational velocity which indicates linear motion without any change of shape of the fluid body. The second term represents a rigid body rotation of the fluid element, while the third term $\vec{D}$ represents the rate of deformation.

3.4 CONSERVATION OF MASS FOR FLUID FLOW

A fluid being a material body, must obey the law of conservation of mass in the course of its flow. In other words, if a velocity field $\vec{V}$ has to exist in a fluid continuum, the velocity components must obey the mass conservation principle. Velocity components in accordance with the mass conservation principle are said to constitute a possible fluid flow, whereas these in violation of this principle, are said to constitute an impossible flow. Therefore, the existence of a physically possible flow field is verified from the principle of conservation of mass.

In Section (3.3.5), we have shown that condition for incompressibility of fluid flow leads to the kinematic constraint $\nabla.\vec{V} = 0$. Interestingly, this consideration is intimately related to the consideration of the conservation of mass for incompressible flows. In this section, we will emphasise on the mathematical description of conservation of mass for general fluid flows from different perspectives, and eventually arrive at certain specific conclusions related to the kinematic constraints of incompressible flows.

3.4.1 Conservation of Mass for a Fluid Element (Lagrangian Point of View: A Control Mass System Approach)

Let us consider a fluid element of mass m, density ρ, and volume $\mathcal{V}$. Mass of a fluid element can be expressed in terms of elemental density and elemental volume as

$$m = \rho \mathcal{V}$$

or

$$\text{In}\, m = \text{In}\, \rho + \text{In}\, \mathcal{V}$$

Differentiating (obtaining total derivative) with respect to time t, we obtain

$$\frac{1}{m}\frac{Dm}{Dt} = \frac{1}{\rho}\frac{D\rho}{Dt} + \frac{1}{\mathcal{V}}\frac{D\mathcal{V}}{Dt}$$

Since mass of the element is conserved, $\frac{Dm}{Dt} = 0$, which implies that we can write in a Cartesian system (noting that $\frac{1}{\mathcal{V}}\frac{D\mathcal{V}}{Dt}$ is the rate of volumetric strain; see Eq. 3.24d)

$$0 = \frac{1}{\rho}\frac{D\rho}{Dt} + \frac{\partial u}{\partial x} + \frac{\partial v}{\partial y} + \frac{\partial w}{\partial z}$$

$$\Rightarrow \quad 0 = \frac{1}{\rho}\left[\frac{\partial \rho}{\partial t} + u\frac{\partial \rho}{\partial x} + v\frac{\partial \rho}{\partial y} + w\frac{\partial \rho}{\partial z}\right] + \frac{\partial u}{\partial x} + \frac{\partial v}{\partial y} + \frac{\partial w}{\partial z}$$

$$\Rightarrow \quad \frac{\partial \rho}{\partial t} + u\frac{\partial \rho}{\partial x} + v\frac{\partial \rho}{\partial y} + w\frac{\partial \rho}{\partial z} + \rho\frac{\partial u}{\partial x} + \rho\frac{\partial v}{\partial y} + \rho\frac{\partial w}{\partial z} = 0$$

$$\frac{\partial \rho}{\partial t} + \frac{\partial}{\partial x}(\rho u) + \frac{\partial}{\partial y}(\rho v) + \frac{\partial}{\partial z}(\rho w) = 0 \qquad (3.45)$$

Equation. (3.45) is the well-known equation of continuity in a rectangular Cartesian coordinate system. The equation can be written in a vector form as

$$\frac{\partial \rho}{\partial t} + \nabla.\left(\rho \vec{V}\right) = 0 \tag{3.46}$$

where $\vec{V}$ represents the velocity vector. Equation. (3.46) is the celebrated continuity equation in fluid mechanics, which mathematically represents the conservating mass for fluid flow.

In case of steady flow,

$$\frac{\partial \rho}{\partial t} = 0$$

Hence, Eq. (3.46) becomes

$$\nabla.\left(\rho \vec{V}\right) = 0 \tag{3.47}$$

or, in a rectangular Cartesian coordinate system

$$\frac{\partial}{\partial x}(\rho u) + \frac{\partial}{\partial y}(\rho v) + \frac{\partial}{\partial z}(\rho w) = 0 \tag{3.48}$$

In case of incompressible flow, we have earlier seen by definition that

$$\nabla.\vec{V} = 0 \tag{3.49}$$

or, in a Cartesian system $\dfrac{\partial u}{\partial x} + \dfrac{\partial v}{\partial y} + \dfrac{\partial w}{\partial z} = 0$ (3.50)

Substituting Eq. (3.49) in Eq. (3.46), we thus get, for incompressible flow

$$\frac{\partial \rho}{\partial t} + \vec{V}.\nabla \rho = 0 \tag{3.51}$$

or

$$\frac{D\rho}{Dt} = 0 \tag{3.51a}$$

Equation. (3.51a) implies that for an incompressible flow, the total derivative of density is zero (in a Cartesian system: $\dfrac{D\rho}{Dt} = \dfrac{\partial \rho}{\partial t} + u\dfrac{\partial \rho}{\partial x} + v\dfrac{\partial \rho}{\partial y} + w\dfrac{\partial \rho}{\partial z} = 0$). *However, this does not necessarily imply that the density has to be a constant for incompressible flow.* In fact, the density can be functions of position and time in Eq. (3.50) in a manner such that all terms in the expression for $\dfrac{D\rho}{Dt}$ cancel each other, leading to $\dfrac{D\rho}{Dt} = 0$. Such flows are known as variable density, incompressible flows. Nevertheless, for the special case of a constant density flow, $\dfrac{D\rho}{Dt}$ is trivially zero. Thus, constant density flow is a special outclass of incompressible flow.

3.4.2 Alternate Derivation of Continuity Equation (Eulerian Point of View: A Control Volume Approach)

In order to derive the continuity equation from an Eulerian perspective, one may consider a differentially small rectangular control volume (i.e., an identified volume in space across which fluid flows) with coordinate axes as shown in Fig. 3.15. Importantly, fluid velocity components normal to the faces of the control volume solely contribute towards the volume flow rates across those respective faces. For illustration, we consider first the flow along the x direction, and subsequently extend similar inferences to y and z directions. Let the fluid enter across the face ABCD with a velocity u and density ρ. Therefore, the rate of mass entering the control volume (CV) along the x direction through face ABCD is given by

$$(\dot{m}_{in})_x = \rho u \Delta y \Delta z$$

The rate of mass leaving the control volume along the x direction through face EFGH is

$$(\dot{m}_{out})_{x+\Delta x} = (\dot{m}_{in})_x + \frac{\partial}{\partial x}(\dot{m}_{in})_x \Delta x + \text{higher order terms (h.o.t.)}$$

$$= \rho u \Delta y \Delta z + \frac{\partial}{\partial x}(\rho u) \Delta x \Delta y \Delta z + \text{h.o.t.}$$

Hence, the net rate of mass entering the control volume along the x direction is

$$(\dot{m}_{in})_x - (\dot{m}_{out})_{x+\Delta x} = -\frac{\partial}{\partial x}(\rho u) \Delta x \Delta y \Delta z + \text{h.o.t.}$$

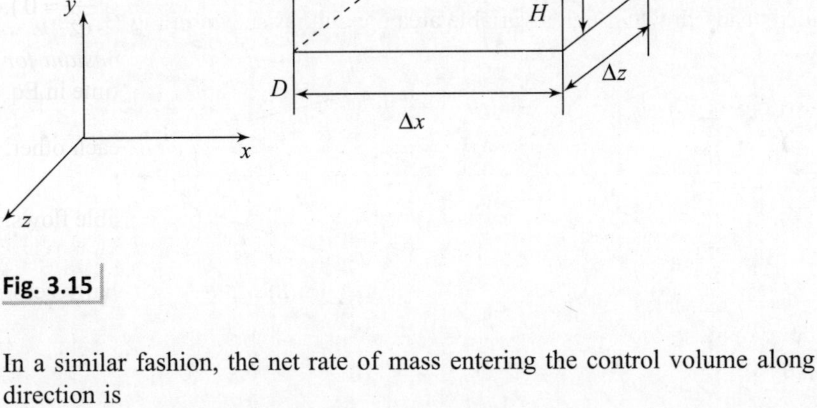

Fig. 3.15

In a similar fashion, the net rate of mass entering the control volume along the y direction is

$$(\dot{m}_{in})_y - (\dot{m}_{out})_{y+\Delta y} = -\frac{\partial}{\partial y}(\rho v)\Delta x \Delta y \Delta z + \text{h.o.t.}$$

and the net rate of mass entering the control volume along the z direction is

$$(\dot{m}_{in})_z - (\dot{m}_{out})_{z+\Delta z} = -\frac{\partial}{\partial z}(\rho w)\Delta x \Delta y \Delta z + \text{h.o.t.}$$

The conservation of mass for the control volume gives

$$\dot{m}_{in} - \dot{m}_{out} = \frac{\partial}{\partial t}(m_{CV}) = \frac{\partial}{\partial t}(\rho \Delta x \Delta y \Delta z) = \frac{\partial \rho}{\partial t}\Delta x \Delta y \Delta z \tag{3.52}$$

(Since $\Delta x \Delta y \Delta z$ is invariant with time).
Taking Δx, Δy, $\Delta z \to 0$ so that higher order term $\to 0$, and substituting expressions for $\dot{m}_{in} - \dot{m}_{out}$ different terms in Eq. (3.52), we obtain

$$-\frac{\partial}{\partial x}(\rho u)\Delta x \Delta y \Delta z - \frac{\partial}{\partial y}(\rho v)\Delta x \Delta y \Delta z - \frac{\partial}{\partial z}(\rho w)\Delta x \Delta y \Delta z = \frac{\partial \rho}{\partial t}\Delta x \Delta y \Delta z$$

$$\left[\frac{\partial \rho}{\partial t} + \frac{\partial}{\partial x}(\rho u) + \frac{\partial}{\partial y}(\rho v) + \frac{\partial}{\partial z}(\rho w)\right]\Delta x \Delta y \Delta z = 0$$

Since the equation is valid irrespective of the size $\Delta x\ \Delta y\ \Delta z$ of the control volume, we can write

$$\frac{\partial \rho}{\partial t} + \frac{\partial}{\partial x}(\rho u) + \frac{\partial}{\partial y}(\rho v) + \frac{\partial}{\partial z}(\rho w) = 0$$

or,
$$\frac{\partial \rho}{\partial t} + \nabla.(\rho \vec{V}) = 0$$

3.4.3 One-dimensional Cross-sectionally Averaged Form of the Continuity Equation

Consider steady flow through a variable area conduit, as shown in Fig. 3.16(a).

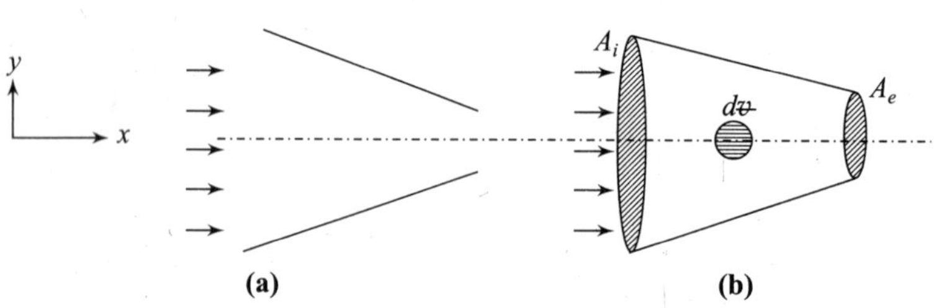

Fig. 3.16

The corresponding continuity equation reads

$$\nabla.\left(\rho\vec{V}\right)=0$$

Integrating the same over the domain (an elemental volume $d\forall$ is depicted in Fig. 3.16b), we get

$$\int_{\forall}\nabla.\left(\rho\vec{V}\right)d\forall=0$$

Using the divergence theorem, the above integral may be converted into an integral over an area that bounds the volume $\forall$, so that one may write

$$\int_{A}\left(\rho\vec{V}\right).\hat{n}dA=0$$

where $\hat{n}$ is a unit vector normal to dA. Since the velocity vector is zero except for the inflow (A_i) and outflow (A_e) boundaries, the above integral may be written as

$$\int_{A_i}\left(\rho\vec{V}\right).\hat{n}dA+\int_{A_e}\left(\rho\vec{V}\right).\hat{n}dA=0$$

Since $\hat{n}=-\hat{i}$ over A_i and $\hat{n}=\hat{i}$ over A_e, one may write the above as

$$-\int_{A_i}\rho u dA+\int_{A_e}\rho u dA=0 \tag{3.53}$$

where u is the x component of flow velocity.

For steady flow, thus the mass flow rate at section i is equal to the mass flow rate at section e. Many times it is convenient to express the above in terms of average velocity. For example, consider a situation in which density does not vary over a given section. Further, we note that the cross-sectionally average velocity is defined as

$$\bar{u}=\frac{\int_A u dA}{A} \tag{3.54}$$

Physically, the average velocity is an equivalent uniform velocity that could have given rise to the same volumetric flow rate as that induced by the variable velocity under consideration. Combining Eqs (3.53) and (3.54), one can write

$$\rho_i\bar{u}_iA_i=\rho_e\bar{u}_eA_e \tag{3.55}$$

where the subscript i and e indicate the inlet and exit conditions respectively. If density is not varying spatially, i.e., $\rho_i=\rho_e$, Eq. (3.55) becomes

$$\bar{u}_iA_i=\bar{u}_eA_e \tag{3.56}$$

3.4.4 Continuity Equation in a Cylindrical Polar Coordinate System

The continuity equation in any coordinate system can be derived in two ways, viz., (i) either by expanding the vectorial form of general Eq. (3.46) with respect to the particular coordinate system, or (ii) by considering an elemental control volume appropriate to the reference frame of coordinates and then by applying the fundamental principle of conservation of mass. The term $\nabla \cdot (\rho \vec{V})$ in a cylindrical polar coordinate system (Fig. 3.17) can be written as

$$\nabla \cdot (\rho \vec{V}) = \frac{\partial}{\partial r}(\rho v_r) + \frac{\rho v_r}{r} + \frac{1}{r}\frac{\partial (\rho v_\theta)}{\partial \theta} + \frac{\partial}{\partial z}(\rho v_z) \qquad (3.57)$$

Therefore, the equation of continuity in a cylindrical polar coordinate system can be written as

$$\frac{\partial \rho}{\partial t} + \frac{\partial}{\partial r}(\rho v_r) + \frac{\rho v_r}{r} + \frac{1}{r}\frac{\partial (\rho v_\theta)}{\partial \theta} + \frac{\partial}{\partial z}(\rho v_z) = 0 \qquad (3.58)$$

The above equation can also be derived by considering the mass fluxes in the control volume shown in Fig. 3.18.

Rate of mass entering the control volume through face *ABCD*

$$= \rho\, v_r\, r\, \mathrm{d}\theta\, \mathrm{d}z$$

Rate of mass leaving the control volume through the face *EFGH*

$$= \rho\, v_r\, r\, \mathrm{d}\theta\, \mathrm{d}z + \frac{\partial}{\partial r}(\rho v_r r\, \mathrm{d}\theta\, \mathrm{d}z)\mathrm{d}r$$

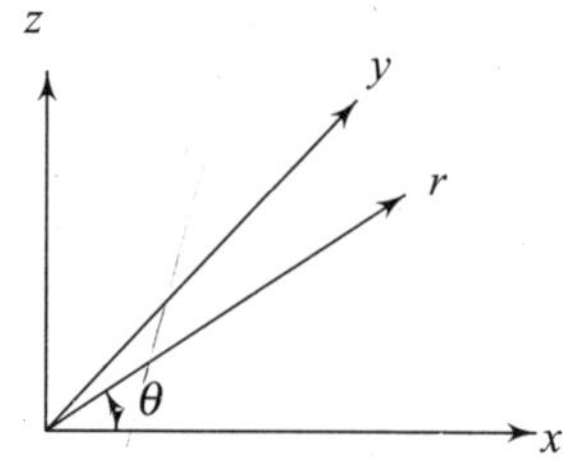

Fig. 3.17 A cylindeical polar coordinate system

Hence, the net rate of mass efflux in the r direction $= \frac{1}{r}\frac{\partial}{\partial r}(\rho v_r\, r)\,\mathrm{d}\forall$

where, $\mathrm{d}\forall = r\, \mathrm{d}r\, \mathrm{d}\theta\, \mathrm{d}z$ (the elemental volume)

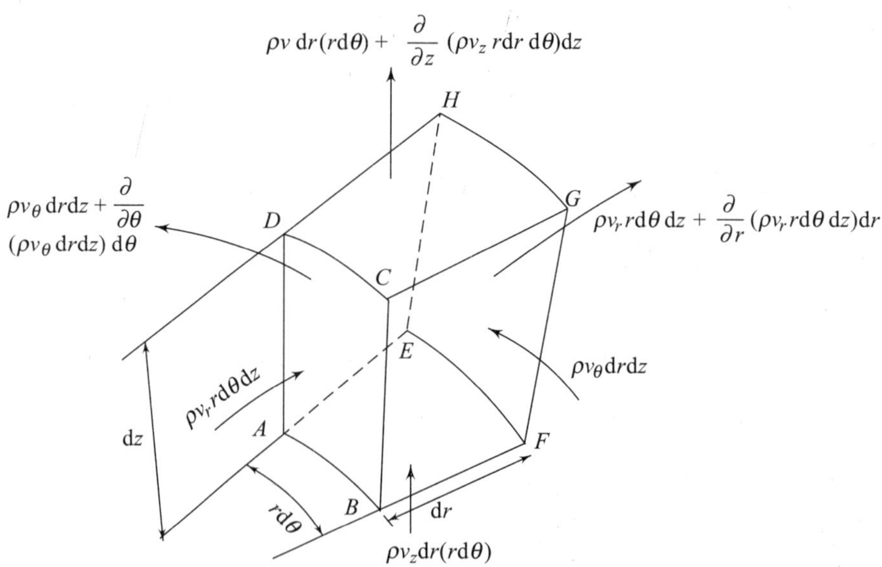

Fig. 3.18 A control volume appropriate to a cylindrical polar coordinate system

The net rate of mass efflux from the control volume, in θ direction, is the difference of mass leaving through face *ADHE* and the mass entering through face *BCGF* and can be written as

$$\frac{1}{r}\frac{\partial}{\partial \theta}(\rho v_\theta)\,\mathrm{d}\forall\,.$$

The net rate of mass efflux in the *z* direction can be written in a similar fashion as

$$\frac{\partial}{\partial z}(\rho v_z)\,\mathrm{d}\forall$$

The rate of increase of mass within the control volume becomes

$$\frac{\partial}{\partial t}(\rho\,\mathrm{d}\forall) = \frac{\partial \rho}{\partial t}(\mathrm{d}\forall)$$

Hence, following the fundamental principle of conservation of mass (rate of accumulation of mass in the control volume + net rate of mass efflux from the control volume = 0), the final form of continuity equation in a cylindrical polar coordinate system becomes

$$\frac{\partial \rho}{\partial t} + \frac{1}{r}\frac{\partial}{\partial r}(\rho v_r r) + \frac{1}{r}\frac{\partial}{\partial \theta}(\rho v_\theta) + \frac{\partial}{\partial z}(\rho v_z) = 0$$

or, $$\frac{\partial \rho}{\partial t} + \frac{\partial}{\partial r}(\rho v_r) + \frac{\rho v_r}{r} + \frac{1}{r}\frac{\partial}{\partial \theta}(\rho v_\theta) + \frac{\partial}{\partial z}(\rho v_z) = 0$$

In case of an incompressible flow,

$$\frac{\partial v_r}{\partial r} + \frac{v_r}{r} + \frac{1}{r}\frac{\partial v_\theta}{\partial \theta} + \frac{\partial v_z}{\partial z} = 0 \tag{3.59}$$

The equation of continuity in a spherical polar coordinate system (Fig. 3.1) can be written by expanding the term $\nabla\cdot(\rho\vec{V})$ of Eq. (3.46) as

$$\frac{\partial \rho}{\partial t} + \frac{1}{R^2}\frac{\partial}{\partial R}(R^2\rho v_R) + \frac{1}{R\sin\phi}\frac{\partial(\rho v_\theta)}{\partial \theta} + \frac{1}{R\sin\phi}\frac{\partial(\rho v_\phi \sin\phi)}{\partial \phi} = 0 \tag{3.60}$$

For an incompressible flow, Eq. (3.60) reduces to

$$\frac{1}{R}\frac{\partial}{\partial R}(R^2 v_R) + \frac{1}{\sin\phi}\frac{\partial v_\theta}{\partial \theta} + \frac{1}{\sin\phi}\frac{\partial(v\phi\sin\phi)}{\partial \phi} = 0 \tag{3.61}$$

The derivation of Eq. (3.60) by considering an elemental control volume appropriate to a spherical polar coordinate system is left as an exercise for the readers.

Example 3.13

Does a velocity field given by

$$\vec{V} = 5x^3\,\hat{i} - 15x^2\,y\,\hat{j} + t\,\hat{k}$$

represent a possible incompressible flow of fluid?

Solution

In order to check for a physically possible incompressible fluid flow, one has to look for its compliance with the equation of continuity.

The continuity equation (in differential form) for a three-dimensional incompressible flow can be written as

$$\nabla \cdot \vec{V} = \frac{\partial u}{\partial x} + \frac{\partial v}{\partial y} + \frac{\partial w}{\partial z} = 0$$

Here, $u = 5x^3, v = -15x^2y$ and $w = t,$

Hence, $\frac{\partial u}{\partial x} = 15x^2, \frac{\partial v}{\partial y} = -15x^2$ and $\frac{\partial w}{\partial z} = 0$

which, on substitution in the continuity equation satisfies it for all x, y, z and t values. This shows that the above velocity field represents a physically possible incompressible flow.

Example 3.14

Which of the following sets of equations represent possible two-dimensional incompressible flows?

(i) $u = x + y;\quad v = x - y$
(ii) $u = x + 2y;\quad v = x^2 - y^2$
(iii) $u = 4x + y;\quad v = x - y^2$
(iv) $u = xt + 2y;\quad v = x^2 - yt^2$
(v) $u = xt^2;\quad v = xyt + y^2$

Solution

The continuity equation (in differential form) for a two-dimensional incompressible flow is $\frac{\partial u}{\partial x} + \frac{\partial v}{\partial y} = 0$

(i) $u = x + y \quad v = x - y; \quad \frac{\partial u}{\partial x} + \frac{\partial v}{\partial y} = 1 + (-1) = 0$

$\therefore$ two-dimensional incompressible flow is possible.

(ii) $u = x + 2y, \quad v = x^2 - y^2; \quad \frac{\partial u}{\partial x} + \frac{\partial v}{\partial y} = 1 - 2y$

$\therefore$ two-dimensional incompressible flow is not possible.

(iii) $u = 4x + y, \quad v = x - y^2; \quad \frac{\partial u}{\partial x} + \frac{\partial v}{\partial y} = 4 - 2y$

$\therefore$ two-dimensional incompressible flow is not possible.

(iv) $u = xt + 2y, \quad v = x^2 - yt^2; \quad \dfrac{\partial u}{\partial x} + \dfrac{\partial v}{\partial y} = t - t^2$

$\therefore$ two-dimensional incompressible flow is not possible.

(v) $u = xt^2, v = xyt + y^2; \quad \dfrac{\partial u}{\partial x} + \dfrac{\partial v}{\partial y} = t^2 + xt + 2y$

two-dimensional incompressible flow is not possible.

Example 3.15

For a flow in the *xy* plane, the *y* component of velocity is given by

$$v = y^2 - 2x + 2y$$

Determine a possible *x* component for a steady incompressible flow. How many possible *x* components are there?

Solution

The flow field is steady and incompressible. Therefore, from continuity

$$\frac{\partial u}{\partial x} + \frac{\partial v}{\partial y} = 0$$

or

$$\frac{\partial u}{\partial x} = -\frac{\partial v}{\partial y}$$

Now,

$$-\frac{\partial v}{\partial y} = -\frac{\partial}{\partial y}(y^2 - 2x + 2y) = -(2y + 2) = -2y - 2$$

Hence,

$$u = \int \frac{\partial u}{\partial x}\,dx = \int -\frac{\partial v}{\partial y}\,dx = \int -(2y + 2)\,dx$$

$$= -2yx - 2x + f(y)$$

There are infinite number of possible *x* components, since $f(y)$ is arbitrary. The simplest one would be found by setting $f(y) = 0$.

Example 3.16

A heated rectangular electronic chip floats on the top of a thin layer of air, above a bottom plate. Air is blown at a uniform velocity v_0 through holes in the bottom plate. For steady, inviscid, and constant density flow, find out the components of velocity and acceleration. Width of the chip perpendicular to the plane of the figure is *b*.

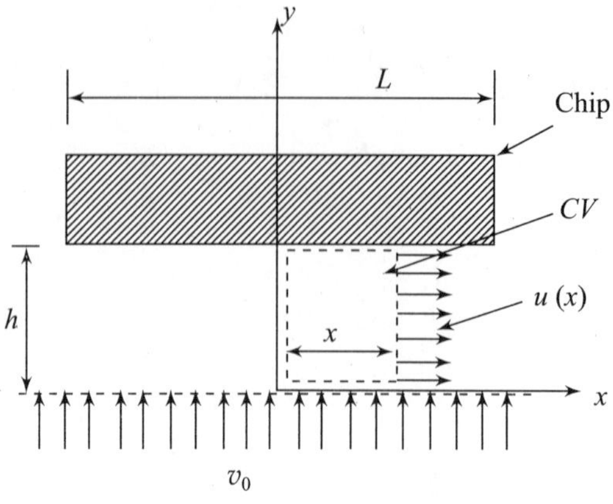

Fig. 3.19

Solution

Choose a fixed control volume as shown by the dashed line in Fig. 3.19.

The rate at which air enters this control volume through the bottom is the same rate at which it leaves the control volume across the gap, to maintain continuity. Thus, one may write

$$v_0\, xb = u(x)\, hb$$

or
$$u(x) = \frac{v_0 x}{h}$$

Continuity equation becomes

$$\frac{\partial u}{\partial x} + \frac{\partial v}{\partial y} = 0$$

or
$$\frac{v_0}{h} + \frac{\partial v}{\partial y} = 0$$

or
$$\partial v = -\frac{v_0}{h}\partial y$$

$$\therefore v = -\frac{v_0}{h}y + f(x) \tag{3.62}$$

The integration constant $f(x)$ can be found by applying an appropriate boundary condition as follows:

$$\text{At } y = h, \quad v = 0$$

$$\therefore f(x) = \frac{v_0}{h}h = v_0$$

Thus,
$$v = v_0\left(1 - \frac{y}{h}\right) \tag{3.62a}$$

$$a_x = \cancelto{0}{\frac{\partial u}{\partial t}} + u\frac{\partial u}{\partial x} + v\frac{\partial u}{\partial y}$$

$$= \frac{v_0 x}{h}\frac{v_0}{h} = \frac{{v_0}^2 x}{h^2}$$

$$a_y = \frac{\partial v}{\partial t}^{0} + u\frac{\partial v}{\partial x} + v\frac{\partial v}{\partial y}$$

$$= v_0\left(1-\frac{y}{h}\right)\left(-\frac{v_0}{h}\right) = -\frac{{v_0}^2}{h}\left(1-\frac{y}{h}\right)$$

The acceleration field can be expressed as

$$\vec{a} = \frac{{v_0}^2 x}{h^2}\hat{i} - \frac{{v_0}^2}{h}\left(1-\frac{y}{h}\right)\hat{j}$$

This example illustrates that although velocity is not a function of time but still there is an acceleration, because of the convective component (originating out of spatial gradients of flow velocities).

Example 3.17

Fluid flows through a convergent nozzle as shown in Fig. 3.20. Cross-sectional area of the nozzle is given as $A = A_0(1 - bx)$, where x is measured from the entrance, A_0 is the cross-sectional area of the nozzle at the entrance, and b is a constant. The free stream velocity is given as $u_\infty = c(1 + at)$, where c and a are dimensional constants. Find out an expression for acceleration assuming inviscid flow.

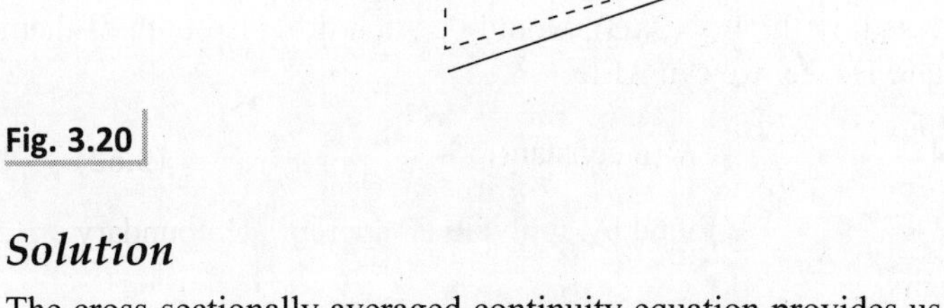

Fig. 3.20

Solution

The cross-sectionally averaged continuity equation provides us with

$$u_\infty A_0 = u(x, t)A(x)$$

or

$$u(x, t) = \frac{c(1+at)A_0}{A_0(1-bx)} = \frac{c(1+at)}{(1-bx)}$$

The acceleration field can be expressed as

$$a_x = \frac{\partial u}{\partial t} + u\frac{\partial u}{\partial x} = \frac{ca}{(1-bx)} + \frac{c(1+at)}{(1-bx)}\frac{cb(1+at)}{(1-bx)^2} = \frac{c}{(1-bx)}\left[a + \frac{cb(1+at)^2}{(1-bx)^2}\right]$$

Example 3.18

The two plates separated by a distance b form a channel (Fig. 3.21). One of the plates is porus and the other one is impermeable. A flow takes place within the channel so that the x component velocity u is a function of x only and its value at the inlet is u_0. There is a uniform inflow v_0 through the porus wall to the channel so that the velocity component v in the y direction within the channel is a function of y only. Considering the flow to be incompressible, find the expression of u as a function of x, and v as a function of y.

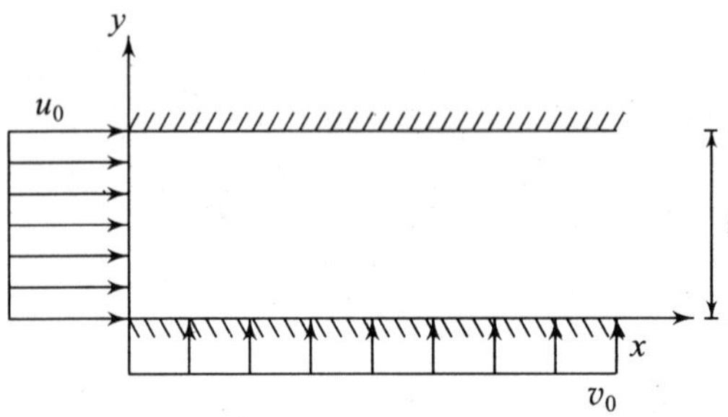

Fig. 3.21 Flow through a channel formed by two plates

Solution

The equation of continuity at any point within the channel can be written as

$$\frac{\partial u}{\partial x} + \frac{\partial v}{\partial y} = 0$$

or
$$\frac{\partial u}{\partial x} = -\frac{\partial v}{\partial y} \tag{3.63}$$

Since u is a function of x only and v is a function of y only, the equality of their derivatives, as expressed by the Eq. (3.63), would be valid provided both of them are equal to a constant. Hence, we can write

$$\frac{du}{dx} = -\frac{dv}{dy} = K \text{ (a constant)}$$

which give
$$\frac{du}{dx} = K \tag{3.64a}$$

$$\frac{dv}{dy} = -K \tag{3.64b}$$

Integration of Eqs (3.64a) and (3.64b) gives

$$u = Kx + C_1 \tag{3.65a}$$

$$v = -Ky + C_2 \tag{3.65b}$$

Using the boundary conditions

$$\text{at } x = 0 \qquad u = u_0$$

$$\text{at } y = 0 \qquad v = v_0$$

and at $y = b$ $v = 0$

in Eqs (3.65a) and (3.65b), we have

$$C_1 = u_0, C_2 = v_0, K = v_0/b$$

Substituting the values of K, C_1 and C_2 in Eqs (3.65a) and (3.65b), the final expressions for u and v are obtained as

$$u = u_0 + \frac{V_0}{b}x \quad \text{and} \quad v = v_0 - \frac{V_0}{b}y$$

3.5 STREAM FUNCTION

The concept of stream function is a direct consequence of the principle of continuity. Let us consider a two-dimensional incompressible flow parallel to the *xy* plane in a rectangular Cartesian coordinate system. The flow field in this case is defined by

$$u = u(x, y, t)$$
$$v = v(x, y, t)$$
$$w = 0$$

The equation of continuity is

$$\frac{\partial u}{\partial x} + \frac{\partial v}{\partial y} = 0 \tag{3.66}$$

If a function $\psi(x, y, t)$ is defined in the manner

$$u = \frac{\partial \psi}{\partial y} \tag{3.67a}$$

and $$v = -\frac{\partial \psi}{\partial x} \tag{3.67b}$$

so that it automatically satisfies the equation of continuity (Eq. (3.66)), then the *function* ψ is known as *stream function.* For a steady flow, ψ is a function of two variables x and y only. In case of a two-dimensional irrotational flow,

$$\frac{\partial v}{\partial x} - \frac{\partial u}{\partial y} = 0$$

so that

$$\frac{\partial}{\partial x}\left(-\frac{\partial \psi}{\partial x}\right) - \frac{\partial}{\partial y}\left(\frac{\partial \psi}{\partial y}\right) = 0$$

or $$\nabla^2 \psi = \frac{\partial^2 \psi}{\partial x^2} + \frac{\partial^2 \psi}{\partial y^2} = 0 \tag{3.68}$$

Thus, for an irrotational flow, stream function satisfies the *Laplace's equation.*

3.5.1 Constancy of ψ on a Streamline

Since ψ is a point function, it has a value at every point in the flow field. Hence, a change in the stream function ψ can be written as

$$d\psi = \frac{\partial \psi}{\partial x} dx + \frac{\partial \psi}{\partial y} dy = -v\, dx + u\, dy$$

Further, the equation of a streamline is given by

$$\frac{u}{dx} = \frac{v}{dy} \quad \text{or} \quad u\, dy - v\, dx = 0$$

It follows that $d\psi = 0$ on a streamline, i.e., the value of ψ is constant along a streamline. Therefore, the equation of a streamline can be expressed in terms of the stream function as

$$\psi(x, y) = \text{constant} \tag{3.69}$$

Once the function ψ is known, streamline can be drawn by joining the same values of ψ in the flow field.

3.5.2 Physical Significance of Stream Function

Figure 3.22(a) illustrates a two-dimensional flow. Let *A* be a fixed point, but *P* be any point in the plane of the flow. The points *A* and *P* are joined by the arbitrary lines *ABP* and *ACP*. For an incompressible steady flow, the volume flow rate across *ABP* into the space *ABPCA* (considering a unit width in a direction perpendicular to the plane of the flow) must be equal to that across *ACP*. A number of different paths connecting *A* and *P* (*ADP*, *AEP*,...) may be imagined but the volume flow rate across all the paths would be the same. This implies that the rate of flow across any curve between *A* and *P* depends only on the end points *A* and *P*.

Since *A* is fixed, the rate of flow across *ABP*, *ACP*, *ADP*, *AEP* (any path connecting *A* and *P*) is a function only of the position *P*. This function is known as the *stream function* ψ. The value of ψ at *P* therefore represents the volume flow rate across any line joining *P* to *A*. The value of ψ at *A* is made arbitrarily zero. The fixed point *A* may be the origin of coordinates, but this is not necessary. If a point P' is considered (Fig. 3.22(b)), PP' being along a streamline, then the rate of flow across the curve joining *A* to P' must be the same as across *AP*, since, by the definition of a streamline, there is no flow across PP'. The value of ψ thus remains same at P' and *P*. Since P' was taken as any point on the streamline through *P*, it follows that ψ is constant along a streamline. Thus the flow may be represented by a series of streamlines at equal increments of ψ. If another point P'' is considered (Fig. 3.22(b)) in the plane, such that PP'' is a small distance δn perpendicular to the streamline through *P* with $AP'' > AP$, then the volume flow rate across the curve AP'' is greater than that across *AP* by the increment $\delta\psi$ of the stream function from point *P* to P''. Let the average velocity perpendicular to PP'' (i.e., in the direction of streamline at *P*) be *V*, then

$$\delta\psi = V \cdot \delta n$$

or

$$V = \delta\psi / \delta n$$

Therefore, the velocity at a point can be expressed in terms of the stream function ψ as

$$V = \lim_{\delta n \to 0} \frac{\delta\psi}{\delta n} = \frac{\partial \psi}{\partial n}$$

This gives the mathematical definition of the stream function at the point *P*. The above concept can be visualised more easily by considering the flow between two adjacent streamlines in a rectangular Cartesian coordinate system (Fig. 3.22(c)). Let the values of the stream functions for the two streamlines be denoted by ψ and $\psi +$ $d\psi$. The volume flow rate dQ for an incompressible flow across any line, say *AB*, of unit width, joining any two points *A* and *B* on two streamlines, can be written as

$$dQ = d\psi$$

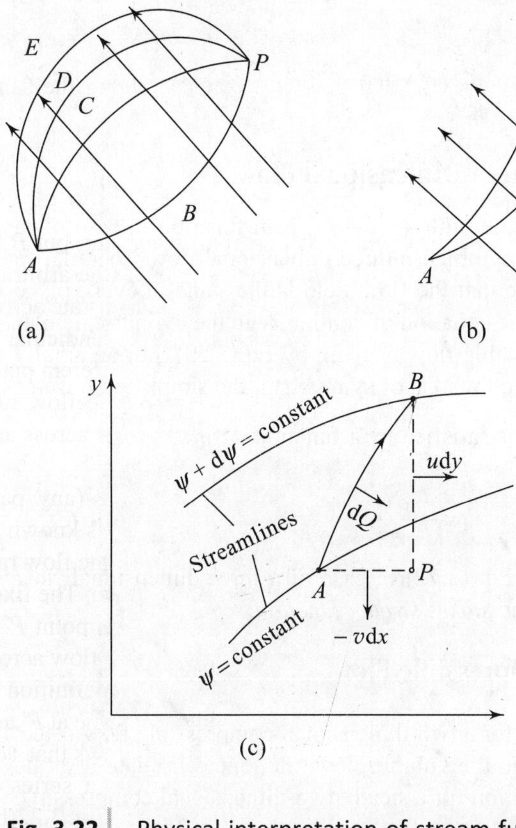

Fig. 3.22 Physical interpretation of stream function

As ψ is a function of space coordinates, *x* and *y*,

$$d\psi = \frac{\partial \psi}{\partial x} dx + \frac{\partial \psi}{\partial y} dy$$

Hence,
$$dQ = \frac{\partial \psi}{\partial x} dx + \frac{\partial \psi}{\partial y} dy \tag{3.70}$$

Again, the volume of fluid crossing the surface *AB* must be flowing out from surfaces *AP* and *BP* of unit width. Hence,

$$dQ = u\, dy - v\, dx \tag{3.71}$$

Comparing the Eqs (3.70) and (3.71), we get

$$u = \frac{\partial \psi}{\partial y} \quad \text{and} \quad v = -\frac{\partial \psi}{\partial x}$$

The stream function, in a polar coordinate system is defined as

$$v_r = \frac{1}{r}\frac{\partial \psi}{\partial \theta} \quad \text{and} \quad v_\theta = -\frac{\partial \psi}{\partial r}$$

The expressions for v_r and v_θ in terms of the stream function automatically satisfy the equation of continuity given by

$$\frac{\partial}{\partial r}(v_r\, r) + \frac{\partial}{\partial \theta}(v_\theta) = 0$$

3.5.3 Stream Function in a Three-Dimensional Flow

It is not possible to draw a streamline with a single stream function in case of a three-dimensional flow. An axially symmetric three-dimensional flow is similar to the two-dimensional case in a sense that the flow field is the same in every plane containing the axis of symmetry. The equation of continuity in the cylindrical polar coordinate system for an incompressible flow is given by Eq (3.59). For an axially symmetric flow (the axis $r = 0$ being the axis of symmetry), the simplified form of Eq. (3.59) without the term $\frac{1}{r}\frac{\partial v_\theta}{\partial \theta}$ is satisfied by a function defined as

$$r\, v_r = -\frac{\partial \psi}{\partial z}, \qquad r\, v_z = \frac{\partial \psi}{\partial r} \tag{3.72}$$

The function ψ, defined by the Eq. (3.72) in case of a three-dimensional flow with an axial symmetry, is called the *stokes stream function.*

3.5.4 Stream Function in Compressible Flow

Definition of the stream function ψ for a two-dimensional compressible flow offers no difficulty. Instead of relating it to the volume flow rate, one can relate it to the mass flow rate. The continuity equation for a steady two-dimensional compressible flow is given by

$$\frac{\partial}{\partial x}(\rho u) + \frac{\partial}{\partial y}(\rho v) = 0$$

Hence a stream function ψ can be defined which will satisfy the above equation of continuity as

$$\rho\, u = \rho_0 \frac{\partial \psi}{\partial y}$$

$$\rho\, v = -\rho_0 \frac{\partial \psi}{\partial x} \tag{3.73}$$

where ρ_0 is a reference density and is used in the manner as shown in Eq. (3.73), to retain the unit of ψ same as that in the case of an incompressible flow. Therefore, from a physical point of view, the difference in stream function between any two streamlines multiplied by the reference density ρ_0 will give the mass flow rate through the passage of unit width formed by the streamlines.

Example 3.19

A stream function is given by

$$\psi = 2x^2y + (3 + t)y^2$$

Find the flow rates across the faces of the triangular prism *OAB*, shown in Fig. 3.23, having a thickness of 1 unit in the *z* direction at time $t = 1$.

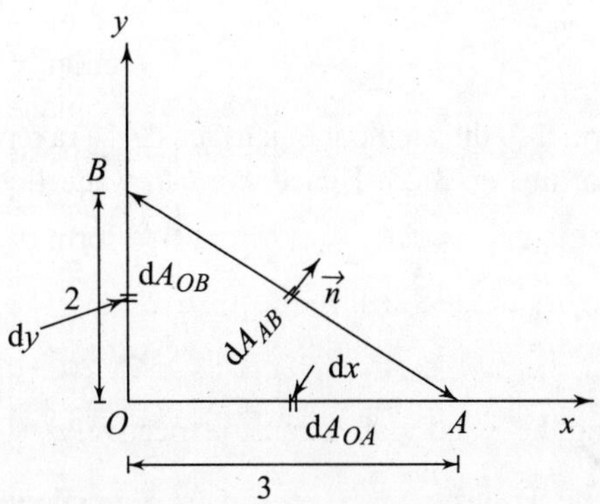

Fig. 3.23 The faces of a triangular prism

Solution

The velocity field corresponding to the given stream function can be written according to Eqs (3.67a) and (3.67b) as

$$u = \frac{\partial \psi}{\partial y} = 2x^2 + 2(3+t)y$$

$$v = -\frac{\partial \psi}{\partial x} = -4xy$$

where u and v are the velocity components along the x and y directions, respectively.

At $t = 1$

$$(u)_{t=1} = 2x^2 + 8y$$

$$(v)_{t=1} = -4xy$$

The volume flow rate across the face perpendicular to the x direction and with the edge *OB* as seen in the *xy* plane is found as

$$Q_{OB} = \int_{A_{OB}} (u)_{\substack{t=1\\x=0}} \, dA_{OB} = \int_0^2 8y \cdot dy = 16 \text{ units}$$

Similarly, the flow rate across the face with edge *OA* (as seen in the *xy* plane) and perpendicular to the *y* direction becomes

$$Q_{OA} = \int_{A_{OA}} (v)_{\substack{t=1\\y=0}} \, dA_{OA}$$

$$= 0$$

Since the *z* component of velocity is zero, the volume flow rates across the faces perpendicular to the *z* direction (i.e., face *OAB* and the face parallel to it and separated by a unit distance) become zero.

Volume flow rate across the inclined face with *AB* as the edge seen on the *xy* plane can be written as

$$Q_{AB} = \int \hat{n} \, dA_{AB} \cdot \vec{V} \tag{3.74}$$

where $\vec{n}$ is the unit vector along the normal to the element of surface dA_{AB}, taken positive when directed outwards as shown in Fig. 3.23. Hence, we can write Eq. (3.74) as

$$Q_{AB} = \int_{A_{AB}} [\hat{i} \, dy + \hat{j} \, dx] \cdot [\hat{i}(2x^2 + 8y) + \hat{j}(-4xy)]$$

(where, $\hat{i}$ and $\hat{j}$ are the unit vectors along the *x* and *y* directions, respectively)

$$= \int_{A_{AB}} [2x^2 + 8y] dy + \int_{A_{AB}} (-4xy) dx \tag{3.75}$$

Again from the geometry,

$$\frac{y}{3 - x} = \frac{2}{3}$$

along the surface *AB*.

Using the relation in Eq. (3.75), we get

$$Q_{AB} = \int_0^2 \left[2\left(3 - \frac{3}{2}y\right)^2 + 8y \right] dy - \int_0^3 4\left(2 - \frac{2}{3}x\right) x \, dx = 16 \text{ units}$$

Example 3.20

An incompressible flow around a circular cylinder of radius *a*, is represented by the stream function

$$\psi = -U r \sin\theta + \frac{U a^2 \sin\theta}{r}$$

where U represents the free stream velocity. Show that v_r (the radial component of velocity) = 0 along the circle, $r = a$. Find the values of θ at $r = a$, where $|\vec{V}| = U$.

Solution

In a polar coordinate system,

$$v_r = \frac{1}{r}\frac{\partial \psi}{\partial \theta}, \qquad v_\theta = -\frac{\partial \psi}{\partial r}$$

So
$$v_r = \frac{1}{r}\left[-U r\cos\theta + \frac{Ua^2}{r}\cos\theta\right] = -U\cos\theta\left(1 - \frac{a^2}{r^2}\right)$$

and
$$v_\theta = \left[U\sin\theta + \frac{Ua^2}{r^2}\sin\theta\right] = U\sin\theta\left(1 + \frac{a^2}{r^2}\right)$$

at $r = a$, $v_r = 0$ for all values of θ and $v_\theta = 2U\sin\theta$

Therefore, along the circle $r = a$, $\quad |\vec{V}| = |v_\theta| = |2U\sin\theta|$

Putting $|\vec{V}| = U$, we get $\sin\theta = \pm\dfrac{1}{2}$, i.e., when $\theta = +\,30°$, $150°$, $210°$ and $330°$.

Example 3.21

The velocity components for a steady flow are given as

$u = 0$, $v = -y^3 - 4z$, $w = 3y^2 z$. Determine: (i) whether the flow field is one-,two-, or three-dimensional, (ii) whether the flow is incompressible or compressible, and (iii) the stream function for the flow.

Solution

(i) Since the velocity field is a function of y and z only, the flow field is two-dimensional.

(ii) For an incompressible flow the continuity equation can be written in differential form as

$$\nabla.\vec{V} = \frac{\partial u}{\partial x} + \frac{\partial v}{\partial y} + \frac{\partial w}{\partial z} = 0$$

Here, $\quad u = 0$, $v = -y^3 - 4z$ and $w = 3y^2 z$

Hence, $\quad \dfrac{\partial u}{\partial x} = 0 \quad \dfrac{\partial v}{\partial y} = -3y^2$ and $\dfrac{\partial w}{\partial z} = 3y^2$

which, on substitution in the continuity equation satisfies it for all x, y, z and t values. This shows that the above velocity field represents a physically possible incompressible flow.

(iii) From the definition of stream function ψ,

$$v = \frac{\partial \psi}{\partial z}$$

$$\psi = \int v dz = -y^3 z - 2z^2 + f(y) + C_1$$

$$w = -\frac{\partial \psi}{\partial y}$$

$$\psi = -\int w dy = -y^3 z + g(z) + C_2$$

Comparing the two expressions for ψ, we find

$$f(y) = 0, g(z) = -2z^2$$

Hence, $$\psi = -y^3 z - 2z^2 + C$$

where, C is a constant.

It is important to note that solid wall itself is a stream function, because there is no flow across it. By physical sense, solid boundary of any shape is a streamline. We can give a particular value of stream function if the flow is two-dimensional and incompressible. Although classically we give $\psi = 0$ at the wall but we can give any value. It is just some sort of reference with respect to which one can find the other stream functions.

3.6 VELOCITY POTENTIAL

If $\nabla \times \vec{V} = 0$, then the velocity field is called the *conservative field* and $\vec{V}$ may be expressed in the form of gradient of a scalar function known as *velocity potential.* Thus, in that case,

$$\vec{V} = \nabla \phi \tag{3.76}$$

which implies

$$u = \frac{\partial \phi}{\partial x} \tag{3.76a}$$

$$v = \frac{\partial \phi}{\partial y} \tag{3.76b}$$

$$w = \frac{\partial \phi}{\partial z} \tag{3.76c}$$

In case of a two-dimensional incompressible flow, the continuity equation becomes

$$\frac{\partial u}{\partial x} + \frac{\partial v}{\partial y} + \frac{\partial w}{\partial z} = 0$$

Substituting u, v in terms of ϕ, we obtain

$$\frac{\partial^2 \phi}{\partial x^2}+\frac{\partial^2 \phi}{\partial y^2}+\frac{\partial^2 \phi}{\partial z^2}=0 \tag{3.77}$$

Thus, velocity potential automatically satisfies the Laplace's equation.

3.6.1 Relationship between Velocity Potential and Stream Function

Consider a two-dimensional, incompressible and irrotational flow so that both the stream function and velocity potential exist.

Since the flow is two-dimensional, $\phi = \phi\,(x, y)$

Thus,
$$d\phi = \frac{\partial \phi}{\partial x}dx + \frac{\partial \phi}{\partial y}dy$$
$$= udx + vdy$$

For constant ϕ, $d\phi$, i.e.,

$$\left.\frac{dy}{dx}\right|_{\phi=\text{cont}} = -\frac{u}{v} \tag{3.78}$$

Again for two-dimensional, incompressible flow, the stream function is expressed as

$$\psi = \psi(x, y)$$
$$d\psi = \frac{\partial \psi}{\partial x}dx + \frac{\partial \psi}{\partial y}dy$$
$$= -\,vdx + udy$$

For constant ψ, $d\psi$, = 0 i.e.,

$$\left.\frac{dy}{dx}\right|_{\psi=\text{cont}} = \frac{v}{u} \tag{3.79}$$

$$\left(\left.\frac{dy}{dx}\right|_{\phi=\text{const}}\right)\times\left(\left.\frac{dy}{dx}\right|_{\psi=\text{const}}\right) = -1\,(u,\, v \neq 0) \tag{3.80}$$

Equation. (3.80) implies that the lines of constant ϕ (equipotential line) and lines of constant ψ (streamline) are orthogonal to each other everywhere in the flow field except at certain points where the velocities are zero (stagnation points).

Example 3.22

The velocity potential function for a flow is given by $\phi = x^2 - y^2$. Verify that the flow is incompressible and then determine the stream function for the flow.

Solution

$$\phi = x^2 - y^2$$

From the definition of velocity potential

$$\vec{V} = \nabla\phi = \frac{\partial \phi}{\partial x}\hat{i} + \frac{\partial \phi}{\partial y}\hat{j}$$

Therefore, $\vec{V} = 2x\hat{i} - 2y\hat{j}, \qquad u = 2x, v = -2y$

Check for incompressible flow, $\dfrac{\partial u}{\partial x} + \dfrac{\partial v}{\partial y} = 2 - 2 = 0$

From the definition of stream function ψ,

$$u = \frac{\partial \psi}{\partial y}, \quad \psi = \int u\mathrm{d}y = 2xy + f(x) + C_1$$

$$v = -\frac{\partial \psi}{\partial x}, \quad \psi = -\int v\,\mathrm{d}x = 2xy + g(y) + C_2$$

comparing the two expressions for ψ, we find

$$f(x) = g(y) = 0$$

Hence, $\psi = 2xy + C$

where, C is a constant.

SUMMARY

- Kinematics of fluid characterises the different types of motion and associated deformation rates of fluid element, without any reference to the forcing parameters.
- Fluid motion is described by two methods, namely, the Lagrangian method and the Eulerian method. In the Lagrangian view, the velocity and other hydrodynamic parameters are specified for particles or elements of given identities, while, in the Eulerian view, these parameters are expressed as functions of location and time. The Lagrangian version of a flow field can be obtained from the integration of the set of equations describing the flow in the Eulerian version.
- A flow is defined to be steady when the flow velocities and fluid properties at any point do not change with time. Flow in which any of these parameters changes with time is termed as unsteady. A flow may appear steady or unsteady depending upon the choice of coordinate axes. A flow is said to be uniform when flow velocities and fluid properties do not change from point to point at any instant of time, or else the flow is non-uniform.
- The total derivative of velocity with respect to time is known as material or substantial acceleration, while the partial derivative of velocity with respect to time for a fixed location is known as temporal acceleration. Material acceleration = Temporal acceleration + Convective acceleration.

- A streamline at any instant of time is an imaginary curve or line in the flow field so that the tangent to the curve at any point represents the direction of the instantaneous velocity at that point. A path line is the trajectory of a fluid particle of a given identity. A streak line at any instant of time is the locus of temporary locations of all particles that have passed through a fixed point in the flow. In a steady flow, the streamlines, path lines, and streak lines are identical.
- Flow parameters, in general, become functions of time and space coordinates. A one-dimensional flow is that in which the flow parameters are functions of time and one space coordinate only.
- A fluid motion consists of translation, rotation, and continuous deformation. In a uniform flow, the fluid elements are simply translated without any deformation or rotation. The deformation and rotation of fluid elements are caused by the variations in velocity components with the space coordinates. The linear deformation or strain rate is defined as the rate of change of length of a linear fluid element per unit original length. The rate of angular deformation at a point is defined as the rate of change of angle between two linear elements at that point which were initially perpendicular to each other. The rotation at a point is defined as the arithmetic mean of the angular velocities of two perpendicular linear meeting at that point. The rotation of a fluid element in the absence of any deformation is known as pure or rigid body rotation.
- Continuity equation mathematically represents conservation of mass for fluid flow. In an incompressible flow field, the velocity vector is divergence-free. Further, for some cases, total derivative of density becomes zero, though that does not necessarily imply that density is a constant. However, a constant density flow is a special case of incompressible flow. The existence of a physically possible flow field is verified from the principle of conservation of mass.
- Vorticity of flow is defined as curl of the velocity vector, and is twice the angular velocity of flow. Circulation of flow is defined as the contour integral of the dot product of velocity with a oriented line element taken along the contour. When the vorticity (or circulation) at all points in the flow field is zero, the flow is said to be irrotational.
- The concept of stream function is a consequence of continuity. In a two-dimensional incompressible flow, the difference in stream functions between two points gives the volume flow rate (per unit width in a direction perpendicular to the plane of flow) across any line joining the points. The value of stream function is constant along a streamline.
- Irrotationality leads to the condition $\nabla \times \vec{V} = 0$ which demands $\vec{V} = \nabla \phi$, where ϕ is known as a potential function. Equipotential lines and streamlines are orthogonal to each other everywhere in the flow field except at the stagnation points.

EXERCISES

3.1 Choose the correct answer:

(i) A flow is said to be steady when
  (a) conditions change steadily with time
  (b) conditions do not change with time at any point
  (c) conditions do not change steadily with time at any point
  (d) the velocity does not change at all with time at any point
  (e) only when the velocity vector at any point remains constant with space and time

(ii) A streamline is a line
  (a) drawn normal to the velocity vector at any point
  (b) such that the streamlines divide the passage into equal number of parts
  (c) which is along the path of a particle
  (d) tangent to which is in the direction of velocity vector at every point

(iii) Streamline, pathline and streakline are identical when
  (a) the flow is uniform
  (b) the flow is steady
  (c) the flow velocities do not change steadily with time
  (e) the flow is neither steady nor uniform.

(iv) The material acceleration is zero for a
  (a) steady flow
  (b) steady and uniform flow
  (c) unsteady and uniform flow
  (d) unsteady and non-uniform flow

(v) A flow field satisfying $\nabla.\vec{V}=0$ as the continuity equation represents always a
  (a) steady and unifor flow
  (b) unsteady and non-uniform flow
  (c) unsteady and incompressible flow
  (d) unsteady and incompressible flow
  (d) incompressible flow

3.2 A two-dimensional incompressible flow field is given by $u = 2y$, $v = x$. Find the velocity and acceleration in the flow field at point A(1,1)

Ans. $(\vec{V} = 2\hat{i} + \hat{j}, \vec{a} = 2\hat{i} + 2\hat{j})$

3.3 Given the velocity field

$$\vec{V} = 10x^2 y\hat{i} + 15xy\hat{j} + (25t - 3xy)\hat{k}$$

Find the acceleration of a fluid particle at a point (1, 2, – 1) at time, $t = 0.5$.

Ans. (1531.90 units)

3.4 Given an unsteady temperature field $T = (xy + z + 3t)$K and unsteady velocity field $\vec{V} = xy\,\hat{i} + z\hat{j} + 5t\,\hat{k}$, what will be the rate of change of temperature of a particle at a point (2, – 2, 1) at time $t = 2$s?

Ans. (23 K/s)

3.5 A two-dimensional pressure field $p = 4x^3 - 2y^2$ is associated with a velocity field given by $\vec{V} = (x^2 - y^2 + x)\hat{i} - (2xy + y)\hat{j}$. Determine the rate of change of pressure at a point (2, 1).

Ans. (260 units)

3.6 The velocity field in a steady flow is given in a rectangular Cartesian coordinate system as $\vec{V} = 6x\hat{i} + (4y + 10)\hat{j} + 2t\hat{k}$. What is the path line of a particle which is at (2, 6, 4) at time $t = 2$s?

Ans. $[\{\ln x + \ln(4y + 10) + 15.77\}^2 - 100z = 0]$

3.7 The velocity field in the neighbourhood of a stagnation point is given by

$$u = U_0 x/L, v = -U_0 y/L, w = 0$$

(i) show that the acceleration vector is purely radial
(ii) if $L = 0.5$ m, what is the magnitude of U_0 if the total acceleration at $(x, y) = (L, L)$ is 10 m/s^2

Ans. (1.88 m/s)

3.8 For a steady two-dimensional incompressible flow through a nozzle, the velocity field is given by $\vec{V} = u_0(1 + 2x/L)\hat{i}$, where x is the distance along the axis of the nozzle from its inlet plane and L is the length of the nozzle. Find
(i) an expression of the acceleration of a particle flowing through the nozzle and
(ii) the time required for a fluid particle to travel from the inlet to the exit of the nozzle.

Ans. $\left(\dfrac{L}{2u_0}\ln 3\right)$

3.9 For a steady flow through a conical nozzle the axial velocity is approximately given by $u = U_0(1 - x/L)^{-2}$, where U_0 is the entry velocity and L is the distance from inlet plane to the apparent vertex of the cone. (i) Derive a general expression for the axial acceleration, and (ii) determine the acceleration at $x = 0$ and $x = 1.0$ m if $U_0 = 5$ m/s and $L = 2$m.

Ans. (25 m/s^2, 800 m/s^2)

3.10 Two large circular plates contain an incompressible fluid in between. The bottom plate is fixed and the top plate is moved downwards with a velocity V_0 causing the fluid to flow out in radial direction and azimuthal symmetry. Derive an expression of radial velocity and acceleration at a radial location r when the height between the plates is h. Consider the radial velocity across the plates to be uniform.

Ans. $\left(\dfrac{V_0 r}{2h}, \dfrac{V_0^2 r}{4h^2}\right)$

3.11 A fluid flows through a horizontal conical pipe having an inlet diameter of 200 mm and an outlet diameter of 400 mm and a length of 2 m. The velocity over any cross section may be considered to be uniform. Determine the convective and local acceleration at a section where the diameter is 300 mm for the following cases:

(i) Constant inlet discharge of 0.3 m^3/s.
(ii) Inlet discharge varying linearly from 0.3 m^3/s to 0.6 m^3/s over two seconds. The time of interest is when $t = 1$ second.

Ans. ((a) 0, – 12.01 m/s^2; (b) 2.12 m/s^2, – 27.02 m/s^2)

3.12 The velocity field for a steady two-dimensional flow in a cartesian coordinate system is given by $\vec{V} = by\hat{i} - ax\hat{j}$ where a and b are constants. Find the equation of stream line passing through the point $\left(1/\sqrt{2a}\right), \left(1/\sqrt{2b}\right)$

Ans. ($ax^2 + by^2 = 1$)

3.13 The velocity components in a two-dimensional flow field for an incompressible fluid are given by

$$u = e^x \cos h(y) \text{ and } v = -e^x \sin h(x)$$

Determine the equation of streamline for this flow.

Ans. ($\cos hx + \sin hy$ = constant)

3.14 A three-dimensional velocity field is given by $u = -x$, $v = 2y$, and $w = 5 - z$. Find the equation of streamline through (2, 2, 1).

Ans. ($x^2 y = 8, y(5 - z)^2 = 32$)

3.15 A three-dimensional velocity field is given by

$$u(x, y, z) = cx + 2w_0 y + u_0$$
$$v(x, y, z) = cy + v_0$$
$$w(x, y, z) = -2cz + w_0$$

where c, w_0, u_0, and v_0 are constants. Find the components of (i) rotational velocity, (ii) vorticity and (iii) the strain rates for the above flow field.

$$Ans. \begin{Bmatrix} \dot{\omega}_x = \dot{\omega}_y = 0, \dot{\omega}_z = -w_0 ; \Omega_x = \Omega_y = 0, \Omega_z = -2w_0 \\ \dot{\varepsilon}_{xx} = c, \dot{\varepsilon}_{yy} = c, \dot{\varepsilon}_{zz} = -2c ; \dot{\gamma}_{xy} = 2w_0, \dot{\gamma}_{yz} = \dot{\gamma}_{xz} = 0 \end{Bmatrix}$$

3.16 Verify whether the following flow fields are rotational. If so, determine the component of rotation about various axes.

(i)	(ii)	(iii)	(iv)
$u = xyz$	$u = xy$	$V_r = A/r$	$V_r = A/r$
$v = zx$	$v = \frac{1}{2}(x^2 - y^2)$	$V_0 = Br$	$V_0 = B/r$
$w = yz - xy^2$	$w = 0$	$V_z = 0$	$V_z = 0$

$$Ans. \begin{bmatrix} \text{(i) Rotational, } \omega_x = \frac{1}{2}(z - 2xy - x), \omega_y = \frac{1}{2} y(x - y), \omega_z = \frac{1}{2} z(1 - x); \\ \text{(ii) irrotational, (iii) rotational, } \omega_r = \omega_\theta = 0, \omega_z = B, \text{ (iv) irrotational} \end{bmatrix}$$

3.17 Show that the velocity field given by $\vec{V} = (a + by - cz)\hat{i} + (d - bx + ez)\hat{j} + (f + cx - ey)\hat{k}$ of a fluid represents a rigid body motion.

3.18 Which of the following velocity fields are kinematically possible for an incompressible flow?

(i) $\vec{V} = 5x\hat{i} + \left(3y + y^2\right)\hat{j}$

(ii) $\vec{V} = \left(x^2 + y^2\right)\hat{i} + \left(y^2 + z^2\right)\hat{j} - 2(x + y)z\hat{k}$

(iii) $\vec{V} = x^2 y\hat{i} + (x + y + z)\hat{j} + (z^2 + x^2)\hat{k}$

(iv) $u = -\dfrac{kx}{y},\ v = k\log(xy)$

(v) $u = \dfrac{k(x^2 - y)^2}{(x^2 + y^2)^2},\ v = \dfrac{2kxy}{(x^2 + y^2)^2}$

(vi) $v_r = \dfrac{m}{4\pi r},\ v_\theta = 0,\ v_z = 0$

where k and m are constants.

Ans. ((i) No, (ii) Yes, (iii) No, (iv) Yes, (v) No, (vi)Yes)

3.19 The x component of velocity in a two-dimensional incompressible flow field is prescribed as $u = By^3 - Ax^4$, where A and B are constants. Find out the y component of velocity. Assume that for all values of x, $v = 0$ at $y = 0$. Check whether the flow is irrotational or not.

Ans. ($v = 4Ax^3y$, No)

3.20 For the flows represented by the following stream functions, determine the velocity components and check for the irrotationality,

(i) $\psi = xy$

(ii) $\psi = \text{In}\,(x^2 + y^2)$

Ans. ((i) $u = x$, $v = -y$; irrotational flow

(ii) $u = \dfrac{2y}{x^2 + y^2},\ v = \dfrac{-2x}{x^2 + y^2}$; irrotational flow)

3.21 The x component of velocity in a two-dimensional flow field is given by $u = e^{-x}\cosh y + 1$. (a) Find the y component of velocity, if it vanishes at $y = 0$, (b) Also find the stream function that will give rise to these velocity components.

Ans. ((a) $e^{-x}\sinh y$ (b) $\psi = e^{-x}\sin y + y + \text{const}$)

3.22 In a two-dimensional incompressible flow over a solid plate, the velocity component perpendicular to the plate is $v = 2x^2y^2 + 3y^3x$, where x is the coordinate along the plate and y is perpendicular to the plate. Hence, find out (i) the velocity component along the plate, and (ii) an expression for stream function and verify whether the flow is irrotational or not.

Ans. ($u = -\dfrac{4}{3}x^3y - \dfrac{9}{2}x^2y^2$, $\psi = -\dfrac{2}{3}x^3y^2 - \dfrac{3}{2}x^2y^3$; rotational)

3.23 Prandtl has suggested that the velocity distribution for turbulent flow in conduits may be approximated by the equation $v = v_{\max}(y/r_0)^{1/7}$, where r_0 is the pipe radius and y is the distance from the pipe wall. Determine the expression of average velocity in terms of centre line velocity in the conduits.

Ans. ($v_{av} = 0.817\ v_{\max}$)

4

DYNAMICS OF INVISCID FLOWS: FUNDAMENTALS AND APPLICATIONS

4.1 INTRODUCTION

In Chapter 3, we dealt with the kinematics of fluid flow. Discussions in Chapter 3, accordingly, considered fluid motion disregarding the forcing parameters influencing the flow. In this chapter, we will discuss the role of some of the forcing parameters that influence the motion as well as their relation with the motion of the fluid. In particular, in this chapter, we will be neglecting the role of viscous effects, and analysing the fluid flow based on that consideration. Such idealised flows without viscous effects are known as *inviscid flows*. Although this might appear to be a hypothetical paradigm, it has interesting implications in several practical situations. For example, one may refer to flow outside the boundary layer (see Chapter 1) in which viscous effects are not relevant, despite the fluid possessing non-zero viscosity.

4.2 EQUATION OF MOTION FOR INVISCID FLOW IN CARTESIAN COORDINATES

Let us consider an elementary parallelopiped of fluid element as a control mass system in a frame of rectangular Cartesian coordinate system as shown in Fig. 4.1. Let b_x, b_y, b_z be the components of body forces acting per unit mass of the fluid element along the coordinate axes x, y and z, respectively. It is important to mention in this context that in the absence of viscous effects, the fluid is subjected to no shear stress and only normal stress acts on the system (expressed in terms of negative fluid pressure).

We can write Newton's second law of motion for the fluid element along the x direction, as

$$\sum F_x = (dm)\, a_x \quad \bullet$$

$$\underbrace{pdydz - \left(p + \frac{\partial p}{\partial x}dx\right)dydz}_{\text{surface force}} + \underbrace{\rho b_x dxdydz}_{\text{body force}} = \underbrace{\rho dxdydz}_{\text{mass}}\underbrace{\left[\frac{\partial u}{\partial t} + u\frac{\partial u}{\partial x} + v\frac{\partial u}{\partial y} + w\frac{\partial u}{\partial z}\right]}_{\text{acceleration}}$$

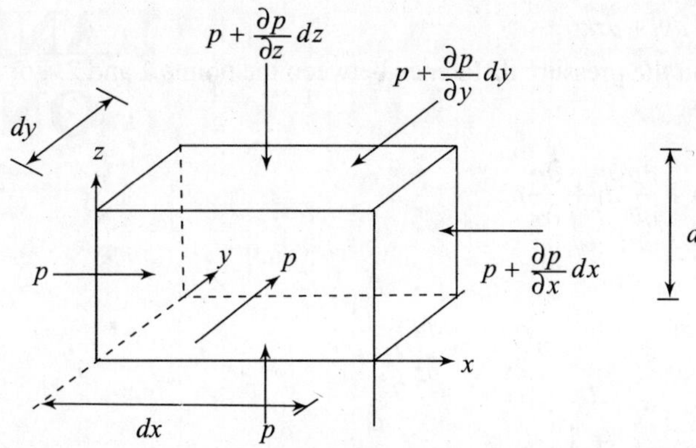

Fig. 4.1 A rectangular fluid element with surface forces in terms of pressure

After simplification, the above equation becomes

$$-\frac{\partial p}{\partial x} + \rho b_x = \rho\left[\frac{\partial u}{\partial t} + u\frac{\partial u}{\partial x} + v\frac{\partial u}{\partial y} + w\frac{\partial u}{\partial z}\right]$$

or
$$\rho\left[\frac{\partial u}{\partial t} + u\frac{\partial u}{\partial x} + v\frac{\partial u}{\partial y} + w\frac{\partial u}{\partial z}\right] = \rho b_x - \frac{\partial p}{\partial x} \tag{4.1}$$

Equation (4.1) is known as *Euler's equation* of motion along the x direction. Similarly, Euler's equation of motion along the y and z directions can be written as

$$\rho\left[\frac{\partial v}{\partial t} + u\frac{\partial v}{\partial x} + v\frac{\partial v}{\partial y} + w\frac{\partial v}{\partial z}\right] = \rho b_y - \frac{\partial p}{\partial y} \tag{4.2}$$

$$\rho\left[\frac{\partial w}{\partial t} + u\frac{\partial w}{\partial x} + v\frac{\partial w}{\partial y} + w\frac{\partial w}{\partial z}\right] = \rho b_z - \frac{\partial p}{\partial z} \tag{4.3}$$

Equations (4.1) – (4.3) can be cast into a single vector form as

$$\rho\frac{D\vec{V}}{Dt} = \rho\vec{b} - \nabla p \tag{4.4}$$

or
$$\rho\left[\frac{\partial V}{\partial t} + \left(\vec{V}.\nabla\right)\vec{V}\right] = \rho\vec{b} - \nabla p \tag{4.4a}$$

4.3 PRESSURE DIFFERENTIAL BETWEEN TWO POINTS IN AN INVISCID FLOW FIELD

Let us consider two points 1 and 2 very close to each other and the position vector $d\vec{l}$ directed from point 1 to 2 be such that

$$d\vec{l} = dx\hat{i} + dy\hat{j} + dz\hat{k}$$

Our interest is to find the pressure difference between the points 1 and 2. For that purpose, one can write

$$dp = \frac{\partial p}{\partial x}dx + \frac{\partial p}{\partial y}dy + \frac{\partial p}{\partial z}dz \tag{4.5}$$

From Eq. (4.1), we get

$$\frac{\partial p}{\partial x} = -\rho\left\{\frac{\partial u}{\partial t} + u\frac{\partial u}{\partial x} + v\frac{\partial u}{\partial y} + w\frac{\partial u}{\partial z}\right\} + \rho b_x$$

$$= -\rho\left\{\frac{\partial u}{\partial t} + \vec{V}.\nabla u\right\} + \rho b_x$$

Similarly, from Eqs (4.2), and (4.3), we get

$$\frac{\partial p}{\partial y} = -\rho\left\{\frac{\partial v}{\partial t} + \vec{V}.\nabla v\right\} + \rho b_y$$

$$\frac{\partial p}{\partial z} = -\rho\left\{\frac{\partial w}{\partial t} + \vec{V}.\nabla w\right\} + \rho b_z$$

Putting the expressions of $\frac{\partial p}{\partial x}$, $\frac{\partial p}{\partial y}$ and $\frac{\partial p}{\partial z}$ in Eq. (4.5), we get

$$dp = \underbrace{-\rho\left[\frac{\partial u}{\partial t}dx + \frac{\partial v}{\partial t}dy + \frac{\partial w}{\partial t}dz\right]}_{\text{Term 1}} \underbrace{-\rho\left[\left(\vec{V}.\nabla u\right)dx + \left(\vec{V}.\nabla v\right)dy + \left(\vec{V}.\nabla w\right)dz\right]}_{\text{Term 2}}$$

$$+\underbrace{\rho\left[b_x dx + b_y dy + b_z dz\right]}_{\text{Term 3}} \tag{4.5a}$$

We simplify various terms appearing in Eq. (4.5a), as follows:

Term 1 $= -\rho\left[\frac{\partial u}{\partial t}dx + \frac{\partial v}{\partial t}dy + \frac{\partial w}{\partial t}dz\right] = -\rho\left[\frac{\partial \vec{V}}{\partial t}.d\vec{l}\right]$

For simplifying Term 2, we first note the vector identity

$$\left(\vec{V}.\vec{\nabla}\right)\vec{V} = \frac{1}{2}\nabla\left(\vec{V}.\vec{V}\right) - \vec{V}\times\left(\nabla\times\vec{V}\right)$$

$$= \frac{1}{2}\nabla(\vec{V}.\vec{V}) - \vec{V}\times\vec{\xi} \text{ (noting that the vorticity, } \vec{\xi} = \nabla\times\vec{V}\text{)}$$

Term 2 $$= -\rho\left[(\vec{V}.\nabla u)dx + (\vec{V}.\nabla v)dy + (\vec{V}.\nabla w)dz\right] = -\rho\left[(\vec{V}.\vec{\nabla})\vec{V}.d\vec{l}\right]$$

$$= -\rho\left[\frac{1}{2}\nabla(\vec{V}.\vec{V}) - \vec{V}\times\vec{\xi}\right].d\vec{l} \text{ (using the above vector identity)}$$

$$= -\rho\left[\frac{1}{2}\left\{\hat{i}\frac{\partial V^2}{\partial x} + \hat{j}\frac{\partial V^2}{\partial y} + \hat{k}\frac{\partial V^2}{\partial z}\right\}.d\vec{l} - \left\{\vec{V}\times\vec{\xi}\right\}.d\vec{l}\right]$$

$$= -\rho\left[\frac{1}{2}\left\{\hat{i}\frac{\partial V^2}{\partial x} + \hat{j}\frac{\partial V^2}{\partial y} + \hat{k}\frac{\partial V^2}{\partial z}\right\}.(\hat{i}dx + \hat{j}dy + \hat{k}dz) - \left\{\vec{V}\times\vec{\xi}\right\}.d\vec{l}\right]$$

$$= -\rho\left[\frac{1}{2}\left\{\frac{\partial}{\partial x}(V^2)dx + \frac{\partial}{\partial y}(V^2)dy + \frac{\partial}{\partial z}(V^2)dz\right\}\right] + \rho\left[(\vec{V}\times\vec{\xi}).d\vec{l}\right]$$

$$= -\frac{1}{2}\rho(dV^2) + \rho\left[(\vec{V}\times\vec{\xi}).d\vec{l}\right]$$

Considering gravity as the only body force which is acting along negative z direction

(i.e., $b_x = 0, b_y = 0, b_z = -g$), Term 3 becomes

Term 3 $= \rho\left[b_x dx + b_y dy + b_z dz\right] = -\rho g dz$

Substituting Terms 1 through 3 in Eq. (4.5a), we obtain

$$dp = -\rho\left[\frac{\partial\vec{V}}{\partial t}.d\vec{l}\right] - \frac{1}{2}\rho d(V^2) + \rho\left\{(\vec{V}\times\vec{\xi}).d\vec{l}\right\} - \rho g dz$$

or $$dp + \frac{1}{2}\rho d(V^2) + \rho g dz + \rho\left[\frac{\partial\vec{V}}{\partial t}.d\vec{l}\right] - \underbrace{\rho\left[(\vec{V}\times\vec{\xi}).d\vec{l}\right]}_{A} = 0 \qquad (4.6)$$

Next, we will identify the cases for which the last term (A) in Eq. (4.6) becomes zero.

Case (1) When $d\vec{l}$ is along a streamline (since, in that Case $\vec{V}$ and $d\vec{l}$ are oriented in the same direction, by definition of streamline, $(\vec{V}\times\vec{\xi}).d\vec{l}$ becomes zero),

Or, Case (2), when $\vec{\xi} = 0$ (irrotational flow),

Or, Case (3), when $\vec{V}\times\vec{\xi}$ is perpendicular to $d\vec{l}$.

Cases (1) and (2) are more common in practical situations than Case (3). Although Case (3) is not commonly encountered, we cannot rule it out mathematically.

Considering any of the above cases to be valid, we will simplify Eq. (4.6) further for steady and unsteady flows.

4.3.1 Steady Flow

Under the situation, in consideration of any of the above three cases, Eq. (4.6) simplifies to

$$dp + \frac{1}{2}\rho d\left(V^2\right) + \rho g dz = 0$$

or

$$\frac{dp}{\rho} + \frac{1}{2}d\left(V^2\right) + g dz = 0 \tag{4.7}$$

If we consider $d\vec{l}$ to be oriented along a streamline [Case (1) above], Eq. (4.7) is known as Euler's equation of motion along a streamline. Equation (4.7) is valid for both compressible and incompressible flow.

Integrating Eq. (4.7) between any two points 1 and 2 along the same streamline (or between any points 1 and 2 in the flow field no matter whether the points are located on the same streamline or not, provided that the flow field is irrotational; see Case (2) above), we get

$$\int_1^2 \frac{dp}{\rho} + \int_1^2 \frac{1}{2} d\left(V^2\right) + \int_1^2 g dz = 0$$

If we further assume constant density flow, the above integration yields

$$\frac{p_2 - p_1}{\rho} + \frac{1}{2}\left(V_2^2 - V_1^2\right) + g\left(z_2 - z_1\right) = 0 \tag{4.8}$$

or

$$\frac{p_1}{\rho} + \frac{V_1^2}{2} + g z_1 = \frac{p_2}{\rho} + \frac{V_2^2}{2} + g z_2 \tag{4.8a}$$

Equations (4.8) or (4.8a) is known as *Bernoulli's equation.* To summarise, Eq. (4.8a) (Bernoulli's equation) is valid only when the flow is inviscid, steady, and incompressible. In addition, either of the following conditions needs to be satisfied: (i) points 1 and 2 are located on the same streamline, (ii) the flow field is irrotational, (iii) $\vec{V} \times \xi$ is perpendicular to $d\vec{l}$. For cases (ii) and (iii), the points 1 and 2 can be located anywhere in the flow field, and not necessarily on the same streamline.

It is important to note that that though we have considered here gravity as the only body force component, the form of Bernoulli's equation given by Eq. (4.8a) is valid under the situation of any conservative body force field so that the last terms of both LHS and RHS of Eq. (4.8a) are to be replaced by the corresponding potential energies per unit mass.

Equation (4.8a) may also be written in the following form:

$$\frac{p_1}{\rho g} + \frac{V_1^2}{2g} + z_1 = \frac{p_2}{\rho g} + \frac{V_2^2}{2g} + z_2 \tag{4.8b}$$

This particular form of Bernoulli's equation has each term of dimension of length, we shall see soon that these represent energy per unit weight in some form. In fluid engineering, energy per unit weight is technically known as *head*.

The physical consequences of each and every term in the Bernoulli's equation are as follows:

Second term of Eq. (4.8b) can be expressed as:

$\frac{1}{2}\frac{V_1^2}{g} = \frac{1}{2}\frac{mV_1^2}{mg}$; which is nothing but the kinetic energy per unit weight.

Third term of Eq. (4.8b) can be expressed as

$z_1 = \frac{mgz_1}{mg}$; which is nothing but the potential energy per unit weight.

To understand the physical meaning of the first term, let us consider fluid flows through a pipe, as shown in Fig. 4.2. For a flowing stream, a layer of fluid at any cross section has to push the adjacent neighbouring layer at its downstream in the direction of flow to make its way through and thus does work on it. The amount of work done can be calculated by considering a small amount of fluid of cross-sectional area A undergoes a small displacement Δx . Pressure remains constant over the small displacement as p. Work done to maintain the flow in the presence of pressure is $pA\Delta x$. The work done to maintain the flow in presence of pressure per unit weight is

$$= \frac{pA\Delta x}{\rho A\Delta x g} = \frac{p}{\rho g}$$

Therefore, the first term of Eq. (4.8b) represents the work done to maintain the flow in presence of pressure per unit weight. This is known as *flow energy* or *flow work.*

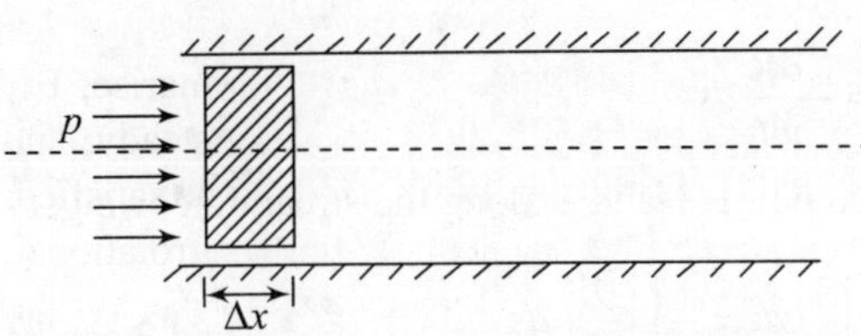

Fig. 4.2 Work done by a fluid to flow against a pressure

Bernoulli's equation along a streamline essentially states that the sum total of the flow energy, kinetic energy and potential energy per unit weight remains conserved as it is transmitted from one point to another in the flow field along a streamline under the assumptions of inviscid, steady, constant density flow. This is a sort of conservation of mechanical energy. The equation was developed first by Daniel Bernoulli in 1738 and is therefore referred to as Bernoulli's equation.

Interestingly, Bernoulli's equation can also be perceived as a mathematical depiction of mechanical energy balance between two sections of a conduit, provided that velocity profiles are uniform over each section (consistent with inviscid flow analysis), pressure is uniform over each section, and the differences in elevations of different points in the same section are negligible. This consideration has been invoked for the analysis presented in Section 4.7 and ahead, with an assumption that the other relevant restrictions of the Bernoulli's equation also apply.

4.3.2 Unsteady Flow Along a Streamline

When the term (A) becomes zero, Eq. (4.6) simplifies to

$$dp + \frac{1}{2}\rho d\left(V^2\right) + \rho g dz + \rho \frac{\partial \vec{V}}{\partial t}.d\vec{l} = 0$$

or
$$dp + \frac{1}{2}\rho d\left(V^2\right) + \rho g dz = -\rho \frac{\partial \vec{V}}{\partial t}.d\vec{l} \tag{4.9}$$

We further consider that the flow is along a streamline. Therefore, from the definition of streamline, the velocity $\vec{V}$ is oriented along the streamline, i.e.,

$$\vec{V} = \hat{\varepsilon}_s V$$

where $\hat{\varepsilon}_s$ is the unit vector in the streamwise direction. Similarly, we can write

$$d\vec{l} = \hat{\varepsilon}_s ds$$

Now, right-hand side of Eq. (4.9) can be written as

$$\rho \frac{\partial \vec{V}}{\partial t}.d\vec{l} = \rho \frac{\partial V}{\partial t} ds$$

Thus, Eq. (4.9) simplifies to

$$dp + \frac{1}{2}\rho d\left(V^2\right) + \rho g dz = -\rho \frac{\partial V}{\partial t} ds$$

or
$$\frac{dp}{\rho} + \frac{1}{2} d\left(V^2\right) + g dz = -\frac{\partial V}{\partial t} ds \tag{4.10}$$

Integrating Eq. (4.9) between any two points 1 and 2 along the streamline, we get

$$\int_1^2 \frac{dp}{\rho} + \int_1^2 \frac{1}{2} d\left(V^2\right) + \int_1^2 g dz = -\int_1^2 \frac{\partial V}{\partial t} ds$$

For constant density flow, the above equation yields

$$\frac{p_1}{\rho} + \frac{V_1^2}{2} + g z_1 = \frac{p_2}{\rho} + \frac{V_2^2}{2} + g z_2 + \int_1^2 \frac{\partial V}{\partial t} ds \tag{4.11}$$

Equation (4.11) is known as the unsteady form of Bernoulli's equation.

4.3.3 Unsteady Irrotational Flow

As discussed in Chapter 3, the velocity vector for an irrotational flow can be expressed as the gradient of a scalar function, the velocity potential (ϕ).

$$\vec{V} = \vec{\nabla}\phi = \frac{\partial \phi}{\partial x}\hat{i} + \frac{\partial \phi}{\partial y}\hat{j} + \frac{\partial \phi}{\partial z}\hat{k}$$

Since, $d\vec{l} = dx\hat{i} + dy\hat{j} + dz\hat{k}$, we have from Eq. (4.9)

$$\frac{\partial \vec{V}}{\partial t}.d\vec{l} = \frac{\partial}{\partial t}\left[\frac{\partial \phi}{\partial x}dx + \frac{\partial \phi}{\partial y}dy + \frac{\partial \phi}{\partial z}dz\right] = d\left(\frac{\partial \phi}{\partial t}\right)$$

Thus, under the assumption of inviscid and irrotational flow, Eq. (4.9) simplifies to

$$\frac{dp}{\rho} + \frac{1}{2}d\left(\nabla\phi^2\right) + gdz = -d\left(\frac{\partial \phi}{\partial t}\right) \tag{4.12}$$

Equation (4.12) is the Euler equation (unsteady form) in terms of the velocity potential. It is important to note that Eq. (4.12) is derived under the assumptions of inviscid and irrotational flow.

* The relationship between irrotational flow and inviscid flow is rather intriguing. In fact, viscous effect happens to be one of the important mechanisms that can convert an originally irrotational flow to a rotational one. For the sake of completeness, here we discuss some of the mechanisms that may be responsible for transferring an irrotational flow into a rotational one.

(a) **Presence of a solid boundary and viscous effects**: In fluid mechanics, solid boundaries are common for wall-bounded flows. Because of no-slip boundary condition (a combined consequence of viscous effects and the existence of a solid boundary), the fluid in contact with the solid boundary remains stationary. Consider the situation depicted in Fig. 4.3, in which *ABCD* is a rectangular fluid element in an originally irrotational flow. However coming in contact with the wall, the motion of the line element *AB* get 'frozen' whereas motion of the line element *DC* is relatively more unhindered. This imparts a rotational effect to the fluid element in its deformed configuration *ABC′D′*. This situation is very much analogous to the toppling effect induced on a person (analogous to the fluid element!) who jumps out of a running bus. The ground attempts to freeze the motion of the person, whereas his/her inertia attempts to carry him/her further.

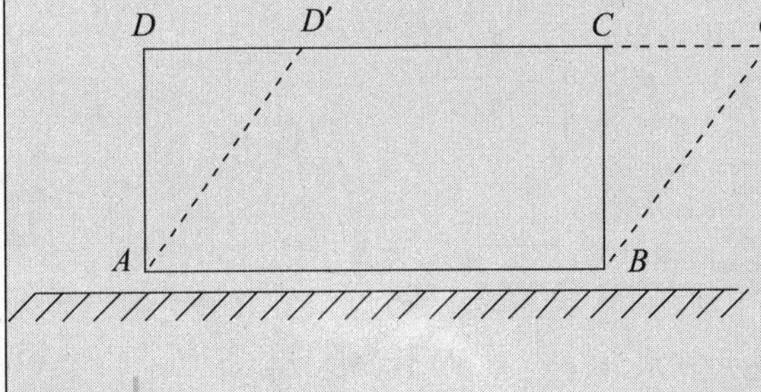

Fig. 4.3

(b) **Presence of shock waves**: Shock waves may be presented in highly compressible (supersonic, i.e., Mach number>1) flows in which there is an

* This portion may be omitted without loss of continuity

abrupt discontinuity in the fluid properties. The shock wave manifests itself through a wave front across which there is a jump in all flow properties. Across the shock front, there is a state that changes from supersonic to subsonic (i.e., Mach number<1). There may be situations in which a flow that was originally irrotational becomes rotational because of the presence of the shock wave, even if viscous effects are not important.

(c) **Thermal stratification**: Thermal stratification refers to the layering of fluid elements that occur due to the density gradient created by changes in temperature. In thermal stratification, a hotter, less dense fluid layer overlies a colder, denser fluid layer. Thermally stratified layers can make the flow rotational from an irrotational one.

(d) **Coriolis forces**: An originally irrotational flow may become rotational due to the presence of Coriolis forces. For example, there are rotational effects in the ocean currents which are predominantly created by the Coriolis effects.

Because of the fact that the above four factors are commonly present in nature, one cannot ensure the eternal irrotationality of an originally irrotational flow. However, in the absence of these factors, an originally irrotational flow will remain irrotational forever.

Example 4.1

In a frictionless, incompressible flow, the velocity field (in m/s) and the body force are given by $\vec{V} = Ax\hat{i} - Ay\hat{j}$ and $\vec{g} = -g\hat{k}$; the coordinates are measured in m. The pressure is p_0 at point (0, 0, 0). Obtain an expression for the pressure field, $p(x,y,z)$, in terms of the velocity field.

Solution

Given: $u = Ax$, $v = -Ay$, $w = 0$, $b_x = b_y = 0$, and $b_z = -g$
Putting the values of u, v, w, and b_x in Eq. (4.1), we get

$$\rho[(Ax)\times A + (-Ay)\times 0 + 0\times 0] = -\frac{\partial p}{\partial x} + 0$$

or

$$\frac{\partial p}{\partial x} = -\rho A^2 x$$

Integrating with respect to x, we obtain

$$p = -\rho A^2 \frac{x^2}{2} + f_1(y,z) \tag{4.13}$$

Putting the values of u, v, w, and b_y in Eq. (4.2), we have

$$\rho[A\times 0 + (-Ay)\times(-A) + 0\times 0] = -\frac{\partial p}{\partial y} + 0$$

or $$\frac{\partial p}{\partial y} = \rho A^2 y$$

Integrating with respect to y, we get

$$p = -\rho A^2 \frac{y^2}{2} + f_2(x, z) \tag{4.14}$$

Putting the values of u, v, w, and b_z in Eq. (4.3), it yields

$$0 = -\frac{\partial p}{\partial z} + \rho g$$

or $$\frac{\partial p}{\partial z} = -\rho g$$

Integrating with respect to z, we get

$$p = -\rho g z + f_3(x, y) \tag{4.15}$$

Equations (4.13), (4.14) and (4.15) are all essentially representing the same pressure field. Comparing the above three equations, thus, we obtain

$$f_1(y, z) = -\rho A^2 \frac{y^2}{2} - \rho g z + C,$$

$$f_1(x, z) = -\rho A^2 \frac{x^2}{2} - \rho g z + C \text{ and}$$

$$f_1(x, y) = -\rho A^2 \frac{x^2}{2} - \rho A^2 \frac{y^2}{2} + C \qquad \text{(where } C \text{ is a constant)}$$

Thus, $$p = -\rho A^2 \frac{x^2}{2} - \rho A^2 \frac{y^2}{2} - \rho g z + C$$

$$= -\frac{1}{2}\rho V^2 - \rho g z + C \text{ (since, } |V| = \sqrt{(Ax)^2 + (Ay)^2} \text{ and}$$
$$V.V = V^2 = A^2x^2 + A^2y^2)$$

or $$p + \frac{1}{2}\rho V^2 + \rho g z = C \tag{4.16}$$

Equation (4.16) essentially relates the pressure field with the velocity field. It is interesting to note here that for this flow,

the rate of angular deformation $\dot{\varepsilon}_{xy} = \frac{\partial u}{\partial y} + \frac{\partial v}{\partial x} = 0$

and rotation $$\omega_{xy} = \frac{1}{2}\left(\frac{\partial v}{\partial x} - \frac{\partial u}{\partial y}\right) = 0$$

This, in turn implies that it is effectively an inviscid (shear stress = 0) and irrotational (angular velocity = 0) flow. Since the fluid density is assumed to be constant in addition, Bernoulli's equation can be applied between any two points in the flow field, no matter whether those points are located on the same streamline or not.

Example 4.2

A rectangular chip floats on the top of a thin layer of air, above a bottom plate. Air is blown at a uniform velocity v_0 through holes in the bottom plate. For steady, inviscid, constant density flow, find out the weight of the chip that can be held in equilibrium by the air injected. Width of the chip perpendicular to the plane of the figure is L.

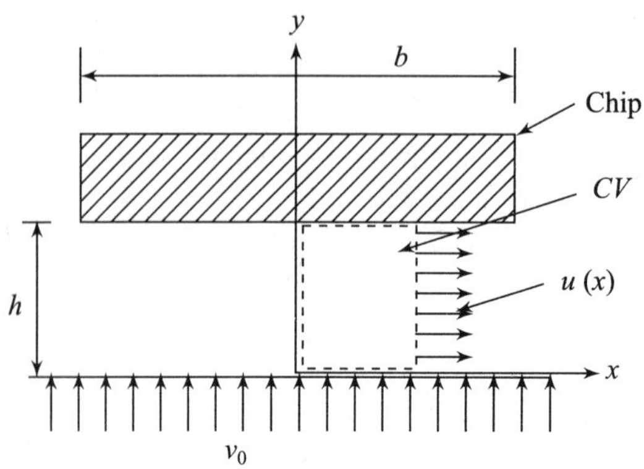

Fig. 4.4

Solution

Choose a fixed control volume as shown by the dashed line in Fig. 4.4

The rate at which air enters this control volume through the bottom is the same rate at which it leaves the control volume across the gap, to maintain continuity. Thus, one may write

$$v_0 x L = u(x) h L$$

or
$$u(x) = \frac{v_0 x}{h}$$

Note that here we considered a uniform velocity profile along y (u as function of x only), since the flow is assumed to be inviscid.

Acceleration of the fluid along the x direction is

$$a_x = u\frac{\partial u}{\partial x} = \frac{v_0^2}{h^2}x \tag{4.17}$$

Applying Euler's equation of motion along the x direction (Eq. (4.1)), one can get

$$\frac{\partial p}{\partial x} = -\rho a_x$$

$$\frac{\partial p}{\partial x} = -\rho \frac{v_0^2}{h^2} x$$

$$\therefore p = -\rho \frac{v_0^2}{h^2} \frac{x^2}{2} + C_1 \tag{4.18}$$

The integration constant C_1 can be found by applying an appropriate boundary condition as follows:

At $x = b/2,\ p = p_{atm}$

$$\therefore C_1 = p_{atm} + \rho \frac{v_0^2}{h^2} \frac{(b/2)^2}{2}$$

$$\therefore p = p_{atm} + \rho \frac{v_0^2}{h^2} \frac{(b/2)^2}{2} - \rho \frac{v_0^2}{h^2} \frac{x^2}{2} \tag{4.18a}$$

The net pressure force on the plate is upward. The net pressure force acting on the plate is determined to be

$$F_{net} = \int_A (p - p_{atm})\, dA$$

$$= 2\int_0^{b/2} (p - p_{atm}) L dx$$

$$= 2\int_0^{b/2} \left(\rho \frac{v_0^2}{h^2} \frac{(b/2)^2}{2} - \rho \frac{v_0^2}{h^2} \frac{x^2}{2} \right) L dx$$

or

$$F_{net} = \frac{\rho v_0^2 b^3 L}{12 h^2} \tag{4.19}$$

For equilibrium, the weight of the plate is balanced by the net pressure force. Hence,

$$W = \frac{\rho v_0^2 b^3 L}{12 h^2} \tag{4.19a}$$

4.4 EULER'S EQUATION OF MOTION IN STREAMLINE COORDINATE SYSTEM

Consider a fluid element in an inviscid flow field, as shown in Fig. 4.5(a). The coordinate s is the streamwise coordinate oriented along the streamline and the coordinate n is oriented normal to the streamline. The different forces acting on the fluid element along s are shown in Fig. 4.5(b).

If we apply Newton's second law in the streamwise (the s) direction to the fluid element, we get

$$\sum F_s = (\Delta m)\, a_s$$

$$p\Delta n - \left(p + \frac{\partial p}{\partial s}\Delta s \right)\Delta n - \rho g \Delta s \Delta n \cos\theta = \rho \Delta s \Delta n a_s$$

$$-\frac{\partial p}{\partial s} - \rho g \frac{\Delta z}{\Delta s} = \rho a_s \tag{4.20}$$

The advantage of using streamline coordinate is that flow is one-dimensional along the streamline, since the velocity vector is always tangential to the streamline. Since the velocity is a function of s and t only ($V=V(s, t)$), along a streamline, the total acceleration of a fluid particle in the streamwise direction is given by

$$a_s = \frac{DV}{Dt} = \frac{\partial V}{\partial t} + V\frac{\partial V}{\partial s}$$

Putting $a_s = \dfrac{\partial V}{\partial t} + V\dfrac{\partial V}{\partial s}$ into the Eq. (4.20), we obtain

$$-\frac{\partial p}{\partial s} - \rho g \frac{\Delta z}{\Delta s} = \rho\left(\frac{\partial V}{\partial t} + V\frac{\partial V}{\partial s} \right) \tag{4.21}$$

In the limit $\Delta s \to 0$, $\dfrac{\Delta z}{\Delta s}$ becomes $\dfrac{\partial z}{\partial s}$. Eq. (4.21) becomes

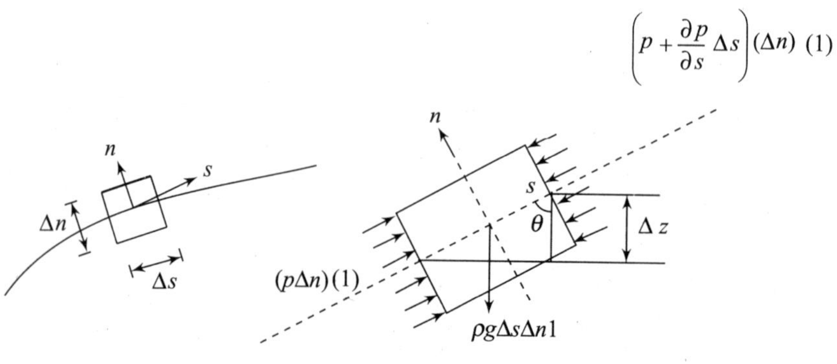

(a) Flow along a streamline (b) Forces acting on the fluid element along the streamline

Fig. 4.5

$$\rho\left(\frac{\partial V}{\partial t}+V\frac{\partial V}{\partial s}\right)=-\frac{\partial p}{\partial s}-\rho g\frac{\partial z}{\partial s} \tag{4.22}$$

The Eq. (4.22) is the Euler's equation of motion along a streamline.

The different forces acting on the fluid element along the direction normal to streamline are shown in Fig. 4.6. To obtain the equation of motion along a direction normal to the streamline (n direction), we apply Newton's second law in the normal (the n) direction to the fluid element.

$$\sum F_n=(\Delta m)\,a_n$$

$$p\Delta s-\left(p+\frac{\partial p}{\partial n}\Delta n\right)\Delta s-\rho g\Delta s\Delta n\cos\alpha=\rho\Delta s\Delta n a_n$$

$$-\frac{\partial p}{\partial n}-\rho g\frac{\Delta z}{\Delta n}=\rho a_n \tag{4.23}$$

In the limit $\Delta s, \Delta n\rightarrow 0$, the fluid element shrinks to a fluid particle so that one can consider that a fluid particle is moving along a curve. Equation (4.23) becomes

$$-\frac{\partial p}{\partial n}-\rho g\frac{\partial z}{\partial n}=\rho a_n \tag{4.24}$$

Further, the acceleration normal to the streamline is given by

$$a_n=-\frac{V^2}{r}$$

where r is the local radius of curvature of the streamline at that point. The acceleration a_n acts inward along the streamline normal and is usually known as centripetal acceleration. The equation of motion normal to the streamline, accordingly, can be written for a steady flow as

$$\frac{\partial p}{\partial n}+\rho g\frac{\partial z}{\partial n}=\rho\frac{V^2}{r} \tag{4.25}$$

For steady flow in a horizontal plane, Euler's equation normal to a streamline becomes

$$\frac{1}{\rho}\frac{\partial p}{\partial n}=\frac{V^2}{r} \tag{4.26}$$

Equation (4.26) indicates that there may occur a normal pressure gradient across the streamline due to streamline curvature. Thus, in regions where the streamlines are straight, the radius of curvature, r, is infinite and there is no pressure variation normal to the streamlines.

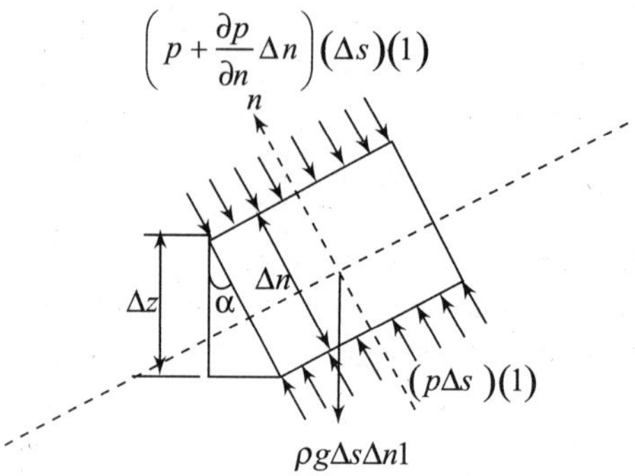

Fig. 4.6 Forces acting on a fluid element normal to the streamline

Example 4.3

Consider the velocity field $\vec{V} = Ax^2\hat{i} - 2Axy\hat{j}$, where A is a dimensional constant. Show that this is a possible incompressible flow. Derive an algebraic expression for the acceleration of a fluid particle. Estimate the radius of curvature of the streamline at $(x, y) = (1, 3)$.

Solution

For two-dimensional incompressible flow, the continuity equation becomes

$$\frac{\partial u}{\partial x} + \frac{\partial v}{\partial y} = 0$$

For the given flow,

$$\frac{\partial u}{\partial x} = 2Ax$$

$$\frac{\partial v}{\partial y} = -2Ax$$

Thus, $$\frac{\partial u}{\partial x} + \frac{\partial v}{\partial y} = 0 \tag{4.27}$$

Since the given velocity field satisfies the continuity equation for incompressible flow, it is a possible incompressible flow field.

The acceleration of a fluid particle can be obtained as follows:

$$\vec{a} = a_x\hat{i} + a_y\hat{j}$$

$$= a_s\hat{\varepsilon}_s + a_n\hat{\varepsilon}_n$$ (where s and n are streamline and cross-stream line coordinates)

$$a_x = u\frac{\partial u}{\partial x} + v\frac{\partial u}{\partial y} = \left(Ax^2\right)(2Ax) - (2Axy)(0) = 2A^2x^3$$

$$a_y = u\frac{\partial v}{\partial x} + v\frac{\partial v}{\partial y} = Ax^2(-2Ay) + (-2Axy)(-2Ax) = 2A^2x^2y$$

The acceleration of the fluid particle at (x, y) is

$$\vec{a} = a_x\hat{i} + a_y\hat{j} = 2A^2x^3\hat{i} + 2A^2x^2y\hat{j} \tag{4.28}$$

The acceleration of the fluid particle at (1,3) is

$$\vec{a} = 2A^2\left(1^3\right)\hat{i} + 2A^2\left(1^2\right)(3)\hat{j} = 2A^2\hat{i} + 6A^2\hat{j}$$

The velocity of the fluid particle at (1,3) is

$$\vec{V} = A\left(1^2\right)\hat{i} - 2A(1)(3)\hat{j} = A\hat{i} - 6A\hat{j}$$

The unit vector tangent to the streamline is

$$\hat{\varepsilon}_s = \frac{\vec{V}}{\left|\vec{V}\right|} = \frac{A\hat{i} - 6A\hat{j}}{\sqrt{(A)^2 + (6A)^2}} = 0.164A\hat{i} - 0.986A\hat{j} \tag{4.29}$$

The unit vector normal to the streamline is

$$\hat{\varepsilon}_n = \hat{\varepsilon}_s \times \hat{k} = \left(0.164A\hat{i} - 0.986A\hat{j}\right) \times \hat{k} = -0.986A\hat{i} - 0.164A\hat{j}$$

The normal component of acceleration is

$$a_n = \vec{a}.\hat{\varepsilon}_n = \left(2A^2\hat{i} + 6A^2\hat{j}\right).\left(-0.986A\hat{i} - 0.164A\hat{j}\right) = -2.956A^3 \tag{4.30}$$

The radius of curvature can be found as follows:

$$a_n = -\frac{V^2}{r}$$

i.e,

$$r = -\frac{V^2}{a_n} = \frac{37A^2}{2.956A^3} = \frac{0.013}{A}$$

Example 4.4

The radial variation of velocity at the midsection of the 180° bend shown is given by rV_θ = constant as shown in Fig 4.7. The cross section of the bend is square. Assume that the velocity is not a function of z. Derive an expression for the pressure difference between the outside and the inside of the bend. Express your answer in terms of the mass flow rate, the fluid density, the geometric parameters, R_1 and R_2, and the depth of the bend, $b = R_2 - R_1$.

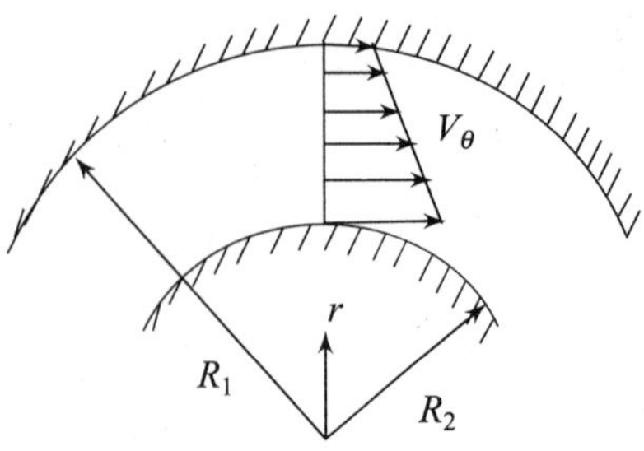

Fig. 4.7

Solution

Applying Euler's equation along r, we obtain

$$\frac{1}{\rho}\frac{\partial p}{\partial r} = \frac{V^2}{r}$$

$$\frac{dp}{dr} = \rho\frac{V^2}{r} = \rho\frac{V_\theta{}^2}{r} \text{ where } V_\theta = \frac{c}{r}$$

$$\frac{dp}{dr} = \rho\frac{c^2}{r^3} \tag{4.31}$$

Integrating, we have

$$p_2 - p_1 = \rho c^2 \int_{R_1}^{R_2} r^{-3} dr = \rho c^2 \left[\frac{r^{-2}}{-2}\right]_{R_1}^{R_2}$$

$$= \frac{\rho c^2}{2}\left[\frac{R_2^2 - R_1^2}{R_1^2 R_2^2}\right] \tag{4.32}$$

The constant, c, can be written in terms of the mass flow rate, m as given by

$$\dot{m} = \int_{R_1}^{R_2} \rho V_\theta drb = \rho b \int_{R_1}^{R_2} \frac{c}{r} dr \tag{4.33}$$

$$= \rho bc \ln\frac{R_2}{R_1}$$

or $$c = \frac{\dot{m}}{\rho b \ln\frac{R_2}{R_1}} \tag{4.33a}$$

Substituting the value of c in Eq. (4.32), one can write

$$p_2 - p_1 = \frac{\dot{m}^2}{2\rho b^2 (\ln \frac{R_2}{R_1})^2} \left[\frac{R_2^2 - R_1^2}{R_1^2 R_2^2} \right] \tag{4.33b}$$

It is interesting to note that $p_2 - p_1 = \rho \frac{\left(V_1^2 - V_2^2\right)}{2}$, although 1 and 2 are not located on the same streamline. This is not surprising, as the flow field given by $rV_\theta =$ constant is irrotational.

Example 4.5

A cylindrical tank of diameter D contains liquid to an initial height of h_0, as shown in Fig. 4.8. At time $t = 0$, a small stopper of diameter d is removed from the bottom. For constant density, inviscid flow with no losses, derive (i) a differential equation for the free-surface height $h(t)$ during draining, and (ii) an expression for the time t_0 to drain the entire tank.

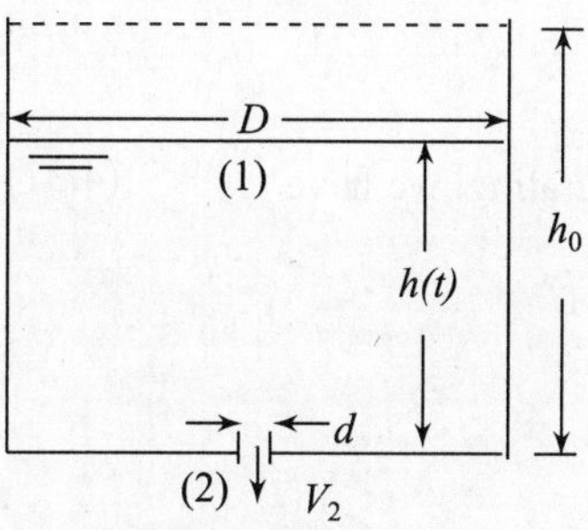

Fig. 4.8

Solution

(i) Applying continuity equation between the sections 1 and 2 (Fig. 4.8), one can write

$$V_2 = V_1 \frac{A_1}{A_2} = V_1 \left(\frac{D}{d} \right)^2 \tag{4.34}$$

Applying unsteady Bernoulli's equation between along a streamline connecting the centrepoints of the sections 1 and 2, we get

$$\frac{p_1}{\rho} + \frac{V_1^2}{2} + gz_1 = \frac{p_2}{\rho} + \frac{V_2^2}{2} + gz_2 + \int_1^2 \frac{\partial V}{\partial t} ds$$

Noting that the integral term $\int_1^2 \frac{\partial V}{\partial t} ds \approx \frac{dV_1}{dt} h$, and $p_1 = p_2 = p_{atm}$, we get

$$\frac{V_2^2 - V_1^2}{2} - g(z_1 - z_2) = -\frac{dV_1}{dt}h$$

Now, $\dfrac{\left|\dfrac{dV_1}{dt}h\right|}{|gh|} \langle\langle 1$, so that the term $\dfrac{dV_1}{dt}h$ can be neglected. Thus, one can write

$$\frac{V_2^2 - V_1^2}{2} = gh \tag{4.35}$$

Using Eq. (4.34) in Eq. (4.35), we get

$$\frac{V_1^2}{2}\left[\left(\frac{D}{d}\right)^4 - 1\right] = gh$$

or $$V_1 = C\sqrt{2gh} \tag{4.36}$$

where $C = \dfrac{1}{\sqrt{\left(\dfrac{D}{d}\right)^4 - 1}}$

Again, substituting $V_1 = -\dfrac{dh}{dt}$, and then integrating, we have

$$\int_{h_o}^{h(t)} \frac{dh}{h^{1/2}} = -C\sqrt{2g}\int_o^t dt,$$

or $$h(t) = \left[h_o^{1/2} - C\left(\frac{g}{2}\right)^{\frac{1}{2}} t\right]^2 \tag{4.37}$$

(ii) The tank is empty when $h(t) = 0$, then from Eq. (4.37), we get

$$0 = \left[h_o^{1/2} - C\left(\frac{g}{2}\right)^{\frac{1}{2}} t_e\right]^2$$

where t_e is the tank emptying time. Simplifying, we get

$$t_e = \left[\frac{2h_0}{Cg}\right]^{\frac{1}{2}} = \left[\frac{2h_0\left\{\left(\dfrac{D}{d}\right)^4 - 1\right\}}{g}\right]^{\frac{1}{2}} \tag{4.37a}$$

Example 4.6

Two circular plates are separated by a liquid layer as shown in Fig. 4.9. The lower plate is stationary; the upper plate moves downward at constant speed, V. The radius of the top plate is R. The liquid is squeezed out in the transverse direction between the plates. Find an expression for pressure distribution, $p(r)$, assuming inviscid flow.

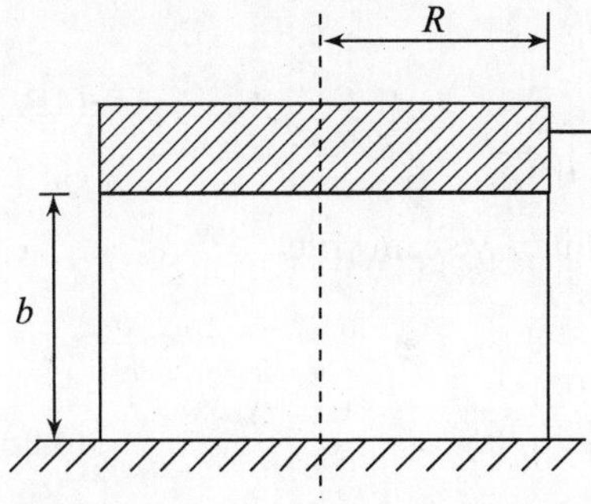

Fig. 4.9

Solution

Consider a control volume of radius r, coaxial with the plate and of height b. By continuity, the fluid squeezes out of the gap between the plates cross the control surface at the same rate at which it is forced to move by the downward motion of the upper plate. Thus, one may write (considering uniform cross-radial velocity profile due to inviscid flow)

$$\pi r^2 V = 2\pi r b V_r$$

$$V_r = \frac{Vr}{2b} \tag{4.38}$$

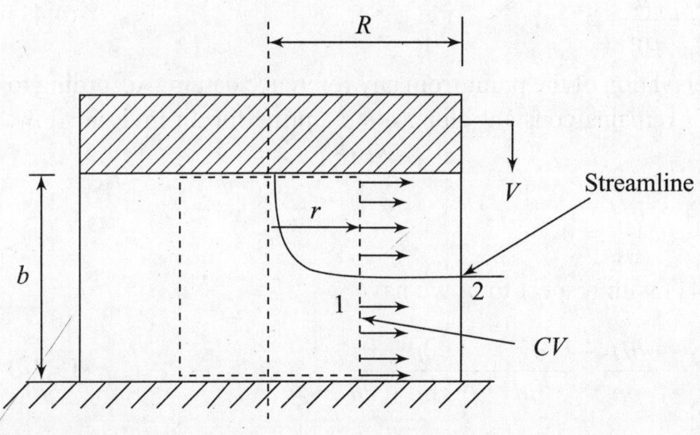

Fig. 4.9(a)

The temporal variation of the flow velocity is, thus,

$$\frac{\partial V_r}{\partial t} = -\frac{Vr}{2b^2}\frac{db}{dt}$$

$$= \left(-\frac{Vr}{2b^2}\right)(-V) = \frac{V^2 r}{2b^2} \tag{4.39}$$

Applying unsteady Bernoulli's equation between points 1 and 2 located on the same radial streamline, we get

$$\frac{p_1}{\rho} + \frac{V_1^2}{2} + gz_1 = \frac{p_2}{\rho} + \frac{V_2^2}{2} + gz_2 + \int_1^2 \frac{\partial V_r}{\partial t} ds$$

Since the points 1 and 2 are on same horizontal line, we can write

$$\frac{p_1}{\rho} + \frac{\left(\frac{Vr}{2b}\right)^2}{2} = \frac{p_{atm}}{\rho} + \frac{\left(\frac{VR}{2b}\right)^2}{2} + \int_1^2 \frac{\partial V_r}{\partial t} dr$$

or,

$$\frac{p_1 - p_{atm}}{\rho} = \frac{V^2}{8b^2}\left(R^2 - r^2\right) + \int_r^R \frac{V^2 r}{2b^2} dr \qquad \text{(using Eq. (4.39))}$$

i.e.,

$$\frac{p_1 - p_{atm}}{\rho} = \frac{V^2}{8b^2}\left(R^2 - r^2\right) + \frac{V^2}{4b^2}\left[R^2 - r^2\right] \tag{4.40}$$

4.5 PRINCIPLE OF MECHANICAL ENERGY CONSERVATION AND ITS APPLICATIONS TO VORTEX FLOWS

The total mechanical energy per unit weight (also known as total head, *H)* can be expressed in terms of velocity, pressure and vertical elevation as

$$H = \frac{V^2}{2g} + \frac{p}{\rho g} + z \tag{4.41}$$

where z is the vertical elevation of the point from any reference datum. According to Bernoulli's equation, H remains constant along a streamline for an inviscid flow. Hence,

$$\frac{\partial H}{\partial s} = 0$$

Differentiating Eq. (4.41) with respect to n, we have

$$\frac{\partial H}{\partial n} = \frac{V}{g}\frac{\partial V}{\partial n} + \frac{1}{\rho g}\frac{\partial p}{\partial n} + \frac{\partial z}{\partial n} \tag{4.42}$$

Substituting $\frac{\partial p}{\partial n}$ from Eq. (4.25) into Eq. (4.42) we have

$$\frac{\partial H}{\partial n} = \frac{V}{g}\frac{\partial V}{\partial n} + \frac{V^2}{gR} - \frac{\partial z}{\partial n} + \frac{\partial z}{\partial n}$$

or,

$$\frac{\partial H}{\partial n} = \frac{V}{g}\left(\frac{\partial V}{\partial n} + \frac{V}{R}\right) \tag{4.43}$$

Equation (4.43) physically implies the variation of total mechanical energy in a direction normal to the streamline for an inviscid flow.

4.5.1 Plane Circular Vortex Flows

For plane circular vortex flows for which there exists only tangential component of velocity (v_θ) (radial component of velocity $v_r = 0$), Eq. (4.43) can be written for the variation of total mechanical energy with radius with respect to a polar coordinate system as

$$\frac{dH}{dr} = \frac{v_\theta}{g}\left(\frac{dv_\theta}{dr} + \frac{v_\theta}{r}\right) \tag{4.44}$$

4.5.1.1 *Free Vortex Flows*

Free vortex flows are the plane circular vortex flows where the total mechanical energy remains constant in the entire flow field. There is neither any addition nor any destruction of mechanical energy in the flow field. Therefore, the total mechanical energy does not vary from streamline to streamline. Hence from Eq. (4.44), we have,

$$\frac{\mathrm{d}H}{\mathrm{d}r} = \frac{v_\theta}{g}\left(\frac{\mathrm{d}v_\theta}{\mathrm{d}r} + \frac{v_\theta}{r}\right) = 0$$

or

$$\frac{1}{r}\left[\frac{\mathrm{d}}{\mathrm{d}r}(v_\theta r)\right] = 0 \tag{4.45}$$

Integration of Eq. (4.45) gives

$$v_\theta = \frac{C}{r} \tag{4.46}$$

The Eq. (4.46) describes the velocity field in a free vortex flow, where C is a constant in the entire flow field. The vorticity in a polar coordinate system is defined by Eq. (3.27a) as

$$\Omega = \frac{\partial v_\theta}{\partial r} - \frac{1}{r}\frac{\partial v_r}{\partial \theta} + \frac{v_\theta}{r}$$

In case of vortex flows, it can be written as

$$\Omega = \frac{dv_\theta}{dr} + \frac{v_\theta}{r}$$

For a free vortex flow, described by Eq. (4.46), Ω becomes zero. Therefore we conclude that a free vortex flow is irrotational, and hence, it is also referred to as irrotational vortex. It has been shown in Section. 4.3 that the total mechanical energy remains same throughout in an irrotational flow field. Therefore, irrotationality is a direct consequence of the constancy of total mechanical energy in the entire flow field and *vice versa*. The interesting feature in a free vortex flow is that as $r \to 0$, $v_\theta \to \infty$, [Eq. (4.46)]. It mathematically signifies a point of singularity at $r = 0$ which, in practice, is impossible. In fact, the definition of a free vortex flow cannot be extended as $r = 0$ is approached. In a real fluid, friction becomes dominant as $r \to 0$ and so a fluid in this central region tends to rotate as a solid body. Therefore, the singularity at $r = 0$ does not render the theory of irrotational vortex useless, since, in practical problems, our concern is with conditions away from the central core.

4.5.1.2 *Pressure Distribution in a Free Vortex Flow*

Pressure distribution in a vortex flow is usually found out by integrating the equation of motion in the r direction. The equation of motion in the radial direction for a vortex flow can be written with the help of Eq. (4.25) as

$$\frac{1}{\rho}\frac{dp}{dr} = \frac{v_\theta^2}{r} - g\frac{dz}{dr} \tag{4.47}$$

Integrating Eq. (4.47) with respect to dr, and considering the flow to be incompressible we have,

$$\frac{p}{\rho} = \int \frac{v_\theta^2}{r}\, dr - gz + A \tag{4.48}$$

where A is a constant to be found out from a suitable boundary condition.

For a free vortex flow,

$$v_\theta = \frac{C}{r}$$

Hence, Eq. (4.48.) becomes

$$\frac{p}{\rho} = -\frac{C^2}{2r^2} - gz + A \tag{4.49}$$

If the pressure at some radius $r = r_a$, is known to be the atmospheric pressure p_{atm}, then Eq. (4.49) can be written as

$$\frac{p - p_{atm}}{\rho} = \frac{C^2}{2}\left(\frac{1}{r_a^2} - \frac{1}{r^2}\right) - g(z - z_a)$$

$$= \frac{(v_\theta)_{r=r_a}^2}{2} - \frac{v_\theta^2}{2} - g(z - z_a) \tag{4.50}$$

where z and z_a are the vertical elevations (measured from any arbitrary datum) at r and r_a. Equation (4.50) can also be derived by a straightforward application of Bernoulli's equation between any two points at $r = r_a$ and $r = r$. In a free vortex flow, the total mechanical energy remains constant. There is neither any energy interaction between an outside source and the flow, nor is there any dissipation of mechanical energy within the flow. The fluid rotates by virtue of some rotation previously imparted to it or because of some internal action. Some examples are a whirlpool in a river, the rotatory flow that often arises in a shallow vessel when liquid flows out through a hole in the bottom (as is often seen when water flows out from a bath tub or a wash basin), and flow in a centrifugal pump case just outside the impeller.

4.5.1.3 *Cylindrical Free Vortex*

A cylindrical free vortex motion is conceived in a cylindrical coordinate system with axis z directing vertically upwards (Fig. 4.10), where at each horizontal cross section, there exists a planar free vortex motion with tangenital velocity given by Eq. (4.46). The total energy at any point remains constant and can be written as

$$\frac{p}{\rho} + \frac{C^2}{2r^2} + gz = H \text{ (constant)} \tag{4.51}$$

The pressure distribution along the radius can be found from Eq. (4.51) by considering z as constant; again, for any constant pressure p, values of z, determining a surface of equal pressure, can also be found from Eq. (4.51). If p is measured in gauge pressure, then the value of z, where $p = 0$ determines the free surface (Fig. 4.10), if one exists.

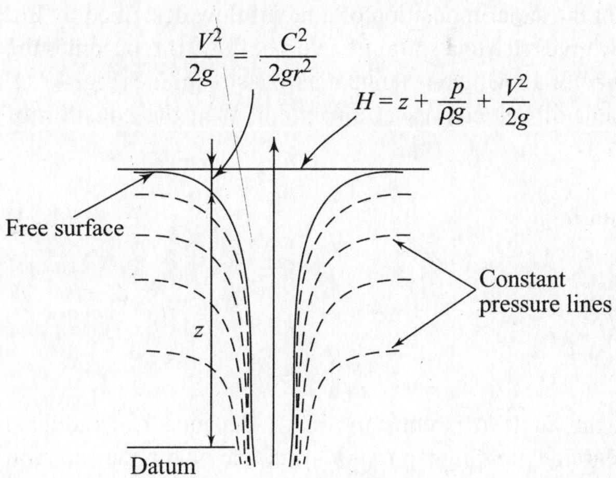

Fig. 4.10 Cylindrical free vortex

A plane spiral free vortex flow in a two-dimensional frame of reference is described in a sense that the tangential and radial velocity components at any point with respect to a polar coordinate system are inversely proportional to the radial coordinate of the point. Therefore, the flow field (Fig. 4.11) can be mathematically defined as

$$v_\theta = \frac{c_1}{r} \tag{4.52}$$

and

$$v_r = \frac{c_2}{r} \tag{4.53}$$

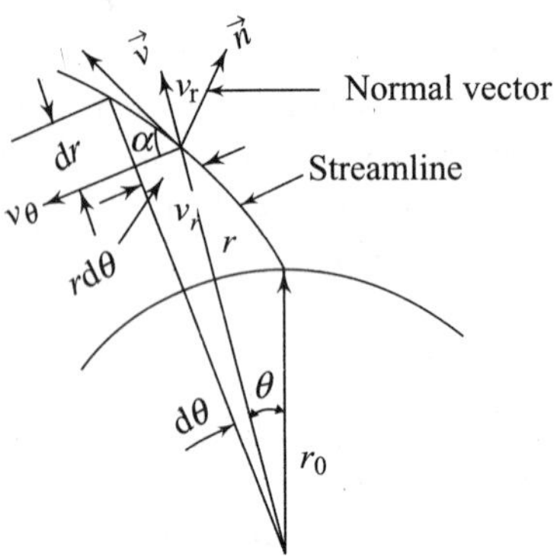

Fig. 4.11 Geometry of spiral flow

Therefore, we can say that the superimposition of a radial flow described by Eq. (4.53) with a free vortex flow gives rise to a spiral free vortex flow. If α becomes the angle between the velocity vector $\vec{V}$, which is tangential to a streamline (Fig. 4.11), and the tangential component of velocity v_θ at any point, then the equation of streamline can be expressed as

$$\frac{1}{r}\frac{dr}{d\theta} = \tan\alpha \tag{4.54}$$

Again, we can write

$$\tan\alpha = \frac{v_r}{v_\theta} = \frac{c_2}{c_1}$$

It follows therefore that the angle α is constant, i.e., independent of radius r. Hence Eq. (4.54) can be integrated, treating $\tan\alpha$ as constant, to obtain the equation of streamlines as

$$r = r_0\, e^{\theta\tan\alpha} = r_0\, e^{\left(\frac{c_2}{c_1}\right)\theta} \tag{4.55}$$

where r_0 is the radius at $\theta = 0$ (Fig. 4.11). Equation (4.55) shows that the pattern of streamlines are logarithmic spirals. Vorticity $\vec{\Omega}$ as defined by Eq. (4.52) becomes zero for the flow field described by Eqs (4.52) and (4.53). Therefore, the spiral free vortex flow is also irrotational like a circular free vortex flow and hence the total energy remains constant in the entire flow field. The outflow through a circular hole in the bottom of a shallow vessel resembles closely to a spiral free vortex flow.

4.5.1.5 *Forced Vortex Flows*

Flows where streamlines are concentric circles and the tangential velocity is directly proportional to the radius of curvature are known as *plane circular forced vortex flows*. The flow field is described in a polar coordinate system as

$$v_\theta = \omega r \tag{4.56 a}$$

and

$$v_r = 0 \tag{4.56 b}$$

All fluid particles rotate with the same angular velocity ω like a solid body. Hence, a forced vortex flow is termed as a *solid body rotation*. The vorticity Ω for the flow field can be calculated as

$$\Omega = \frac{\partial v_\theta}{\partial r} - \frac{1}{r}\frac{\partial v_r}{\partial \theta} + \frac{v_\theta}{r}$$

$$= \omega - 0 + \omega = 2\omega$$

Therefore, a forced vortex motion is not irrotational; rather it is a rotational flow with a constant vorticity 2ω. Equation (4.43) is used to determine the distribution of mechanical energy across the radius as

$$\frac{dH}{dr} = \frac{v_\theta}{g}\left(\frac{dv_\theta}{dr} + \frac{v_\theta}{r}\right) = \frac{2\omega^2 r}{g}$$

Integrating the equation between the two radii on the same horizontal plane, we have

$$H_2 - H_1 = \frac{\omega^2}{g}\,(r_2^2 - r_1^2) \tag{4.57}$$

Thus, we see from Eq. (4.57) that the total head (total energy per unit weight) increases with an increase in radius. The total mechanical energy at any point is the sum of the kinetic energy, the flow work or pressure energy, and the potential energy. Therefore, the difference in total head between any two points in the same horizontal plane can be written as

$$H_2 - H_1 = \left[\frac{p_2}{\rho g} - \frac{p_1}{\rho g}\right] + \left[\frac{v_2^2}{2g} - \frac{v_1^2}{2g}\right]$$

$$= \left[\frac{p_2}{\rho g} - \frac{p_1}{\rho g}\right] + \frac{\omega^2}{2g}\,[r_2^2 - r_1^2]$$

Substituting this expression of $H_2 - H_1$ in Eq. (4.57), we get

$$\frac{p_2 - p_1}{\rho} = \frac{\omega^2}{2}\,[r_2^2 - r_1^2] \tag{4.58}$$

The same equation can also be obtained by integrating the equation of motion in a radial direction as

$$\int_1^2 \frac{1}{\rho}\frac{\mathrm{d}p}{\mathrm{d}r}\,\mathrm{d}r = \int_1^2 \frac{v_\theta^2}{r}\,\mathrm{d}r = \omega^2 \int_1^2 r\,\mathrm{d}r$$

or
$$\frac{p_2 - p_1}{\rho} = \frac{\omega^2}{2}\,[r_2^2 - r_1^2]$$

To maintain a forced vortex flow, mechanical energy has to be spent from outside and thus an external torque is always necessary to be applied continuously. Forced vortex can be generated by rotating a vessel containing a fluid so that the angular velocity is the same at all points. A paddle rotating in a large mass of fluid creates a forced vortex flow near its diameter. Another common example is the motion of liquid within a centrifugal pump or of gas in a centrifugal compressor.

4.5.1.6 *Cylindrical Forced Vortex*

A cylindrical forced vortex motion is realised in a three-dimensional space. It can be generated by rotating a cylindrical vessel containing a fluid (Fig. 4.12a). At any horizontal plane, the tangential velocity satisfies the Eq. (4.56a). The pressure head $p/\rho g$ at any point in the fluid is equal to z, the depth of the point below the free surface, if one exists (Fig. 4.12a). By writing the Eq. (4.58) between the points a and o at the same horizontal plane (Fig. 4.12a) we have,

$$z - z_0 = \frac{\omega^2 r^2}{2g} \tag{4.59}$$

Equation (4.59) represents the equation of free surface which, if r is perpendicular to z (i.e., the axis of rotation is vertical), is a paraboloid of revolution. If the liquid is confined within a vessel (Fig. 4.12b), the free surface may not exist, but the pressure along any radius will vary in the same way as if there were a free surface. Hence, the two are equivalent.

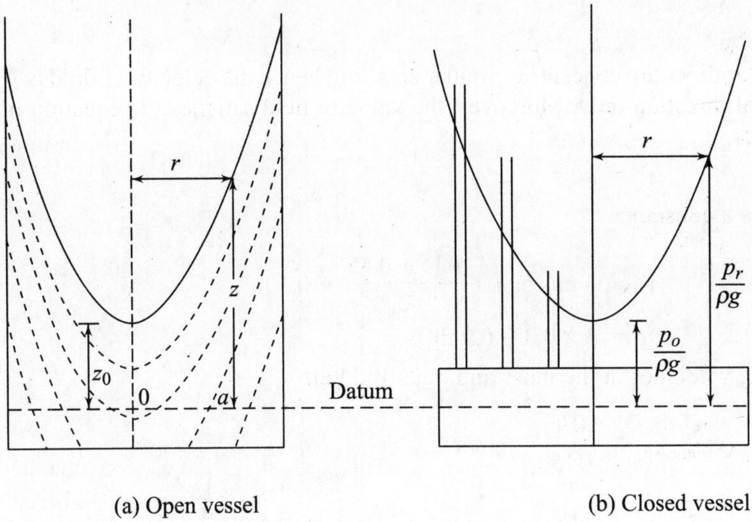

Fig. 4.12 Cylindrical forced vortex

4.5.1.7 *Spiral Forced Vortex*

Superimposition of purely radial flow (inwards or outwards) with a plane circular forced vortex results in a spiral forced vortex flow.

Example 4.7

Water flows through a right-angled bend (Fig. 4.13) formed by two concentric circular arcs in a horizontal plane with the inner and outer radii of 0.15 m and 0.45 m, respectively. The centre-line velocity is 3 m/s. Assuming a two-dimensional free vortex flow, determine (i) the tangential and normal accelerations at the inner and outer walls of the bend, (ii) the pressure gradients normal to the streamline at the inner and outer walls of the bend, and (iii) the pressure difference between the inner and outer walls of the bend.

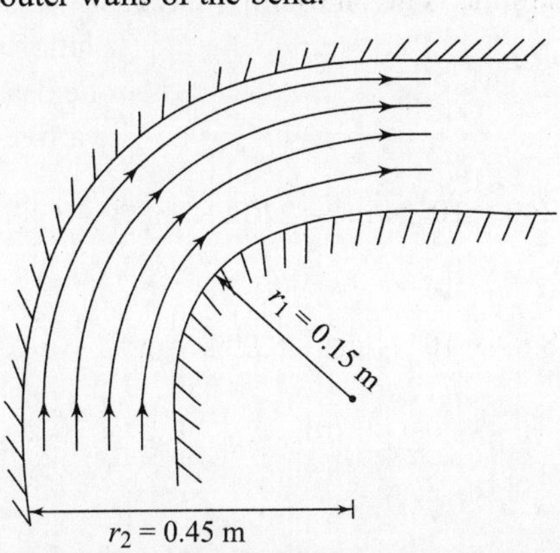

Fig. 4.13 Flow of water through a right-angled bend

Solution

Here the streamlines are concentric circular arcs, and hence the velocity of fluid is in the tangential direction only. Moreover, the velocity field satisfies the equation of free vortex as

$$v_\theta = c/r$$

where c is a constant.

We have $\quad v_\theta = 3 \text{ m/s at } r = \dfrac{0.15 + 0.45}{2} = 0.3 \text{ m}$

Hence, $\quad c = 3 \times 0.3 = 0.9 \text{ m}^2\text{/s}$

Therefore, velocities at the inner and outer radii are

$$(v_\theta)_{\text{at } r = r_1} = \frac{0.9}{0.15} = 6 \text{ m/s}$$

$$(v_\theta)_{\text{at } r = r_2} = \frac{0.9}{0.45} = 2 \text{ m/s}$$

(i) The accelerations along the streamline and normal to it can be written as a_s

$$\text{(acceleration along the streamline)} = v_\theta \frac{\partial v_\theta}{r\partial \theta} = 0$$

$$a_n \text{ (acceleration normal to streamline)} = \frac{-v_\theta^2}{r}$$

Therefore, $\quad (a_n)_{r = r_1} = \dfrac{-6 \times 6}{0.15} = -240 \text{ m/s}^2$

$$(a_n)_{r = r_2} = \frac{-2 \times 2}{0.45} = -8.89 \text{ m/s}^2$$

Minus sign indicates that the accelerations are radially inwards.

(ii) The pressure gradient normal to the streamline is given by Eq. (4.47) as

$$\frac{\mathrm{d}p}{\mathrm{d}r} = \rho \frac{v_\theta^2}{r}$$

Therefore, $\quad \left(\dfrac{\mathrm{d}p}{\mathrm{d}r}\right)_{\text{at } r = r_1} = 1000 \times \dfrac{6 \times 6}{0{\cdot}15} = 240 \times 10^3 \text{ N/m}^2 = 240 \text{ kN/m}^2$

and $\quad \left(\dfrac{\mathrm{d}p}{\mathrm{d}r}\right)_{\text{at } r = r_2} = 1000 \times \dfrac{2 \times 2}{0.45} = 8.89 \times 10^3 \text{ N/m}^2 = 8.89 \text{ kN/m}^2$

(iii) At any radius r,

$$\frac{\mathrm{d}p}{\mathrm{d}r} = \rho \frac{v_\theta^2}{r}$$

Hence,
$$\int_{r_1}^{r_2} \frac{dp}{dr}\, dr = \int_{r_1}^{r_2} \rho \frac{c^2}{r^3}\, dr$$

or
$$p_2 - p_1 = \frac{\rho}{2}\,(c^2/r_1^2 - c^2/r_2^2) = \frac{\rho}{2}\,(v_{\theta_1}^2 - v_{\theta_2}^2)$$

$$= \frac{1000}{2} \times (36 - 4)\ \text{N/m}^2 = 16\ \text{kN/m}^2$$

where p_1 and p_2 are the pressures at inner and outer walls of the bend respectively.

Example 4.8

A hollow cylinder of 0.6 m diameter, open at the top, contains a liquid and spins about its vertical axis, producing a forced vortex motion. (i) Calculate the height of the vessel so that the liquid just reaches the top of the vessel and begins to uncover the base at 100 rpm. (ii) If the speed is now increased to 130 rpm, what area of the base will be uncovered?

Solution

(i) The situation when the liquid just reaches the top of the vessel and begins to uncover the base is shown in Fig. 4.14a. Let H be the height of the cylinder. The difference in pressure between the point 1 (at the centre), and 2 (at the outer wall), at the bottom surface of the vessel can be found from Eq. (4.47) as,

$$\int_1^2 \frac{dp}{dr}\, dr = \int_1^R \rho \frac{v_\theta^2}{r}\, dr$$

$$= \frac{\rho}{2}\,(\omega^2 r^2)_0^R$$

or
$$p_2 - p_1 = \rho \frac{\omega^2 R^2}{2} \tag{4.60}$$

where R is the radius and ω is the angular velocity of the vessel.

Since point 1 is on the free surface (Fig. 4.14a) and point 2 is at a depth H below the free surface,

$$p_1 = p_{atm}\ \text{(atmospheric pressure)}$$
$$p_2 = p_{atm} + \rho g H$$

Substituting the values of p_1 and p_2 in Eq. (4.60), we get

$$H = \frac{\omega^2 R^2}{2g}$$

$$= \left(\frac{2\pi \times 100}{60}\right)^2 \frac{(0.3)^2}{2 \times 9.81} = 0.503\ \text{m}$$

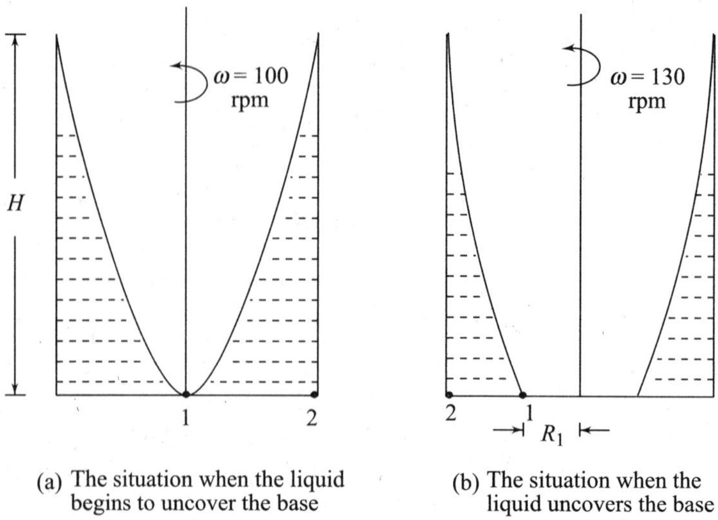

Fig. 4.14 Liquid under uniform rotation in an open vessel

(ii) When the speed is increased beyond 100 rpm, the maximum centrifugal head $\omega^2 R^2/2g$ will be more than the maximum static pressure head $\rho g H$, and hence the liquid being detached from the centre will uncover the base as shown in Fig. 4.14b. Let R_1 be the radius where the free surface meets the base; then the pressure difference between points 1 and 2 (Fig. 4.14b) can be written as

$$p_2 - p_1 = \frac{\rho\omega^2}{2}\,(R^2 - R_1^2)$$

but

$$p_2 - p_1 = \rho g H$$

Hence,

$$\rho g H = \frac{\rho\omega^2}{2}\,(R^2 - R_1^2)$$

or

$$R_1^2 = R^2 - \frac{2gH}{\omega^2}$$

$$= (0.3)^2 - \frac{2 \times 9.81 \times 0.503}{(2\pi \times 130/60)^2} = 0.037 \text{ m}^2$$

Therefore, uncovered area at the base $= \pi R_1^2 = \pi(0.037) = 0.116 \text{ m}^2$.

Example 4.9

A closed cylinder 0.4 m in diameter and 0.4 m in height is filled with oil of specific gravity 0.80. If the cylinder is rotated about its vertical axis at a speed of 200 rpm, calculate the thrust of oil on top and bottom covers of the cylinder.

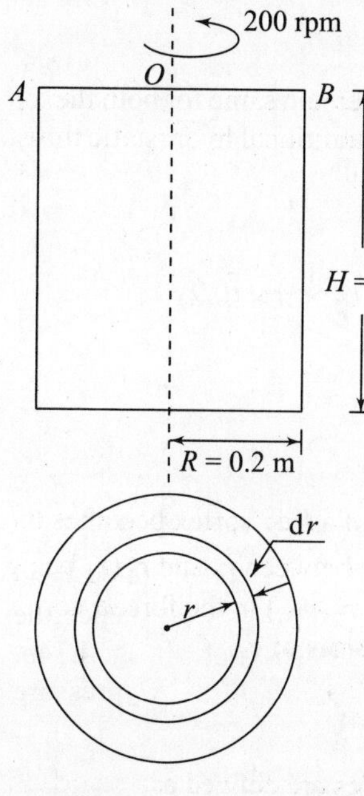

Fig. 4.15 A rotating closed cylinder filled with oil

Solution

In the top plane *AB* of the cylinder (Fig. 4.15), pressure head at any radial distance *r* is given by

$$p/\rho g = \omega^2 r^2/2g$$

where, ω is the angular velocity of the cylinder.

Considering a thin annular ring of radius r and thickness dr (Fig. 4.15), and summing up the forces on all such elemental rings, we have

The thrust on top plane,

$$F_T = \int_0^R p 2\pi r \, \mathrm{d}r$$

$$= \pi\rho\,\omega^2 \int_0^R r^3 \, \mathrm{d}r$$

$$= \frac{\pi\rho\omega^2 R^4}{4}$$

Here,

$$\omega = \frac{2\pi \times 200}{60} = 20.94 \text{ rad/sec}$$

R (the radius of the cylinder) = 0.2 m

Therefore,
$$F_T = \frac{\pi \times 0.8 \times 10^3 \times (20.94)^2\,(0.2)^4}{4}$$
$$= 440.81 \text{ N}$$

The radial distribution of pressure due to rotation will remain same for both the top and bottom covers. But the bottom cover experiences an additional hydrostatic thrust due to the weight of liquid above it.

Hence, the thrust at the bottom cover

$$\begin{aligned} F &= F_T + \rho g\, H \times \pi R^2 \\ &= 440.81 + 0.8 \times 10^3 \times 9.81 \times 0.4 \times \pi \times (0.2)^2 \\ &= 835.29 \text{ N} \end{aligned}$$

Example 4.10

At a radial location r_1 in a horizontal plane, the velocity of a free vortex becomes the same as that of a forced vortex. If the pressure difference between r_1 and r_2 (r_2 being another radial location in the same horizontal plane with $r_2 > r_1$) in the forced vortex becomes twice that in the free vortex, determine r_2 in terms of r_1.

Solution

At any radius r, the tangential velocities for the two vortices are defined as

$$v_{\theta \text{ free vortex}} = c/r \text{ (where } c \text{ is a constant throughout the flow)}$$
$$v_{\theta \text{ forced vortex}} = \omega r \text{ (where } \omega \text{ is the angular velocity)}$$

From the equality of two velocities at $r = r_1$,

$$c/r_1 = \omega\, r_1$$

or
$$c/\omega = r_1^2 \tag{4.61}$$

The pressure difference between the points $r = r_1$ and $r = r_2$ for the two vortices can be written as

$$(p_2 - p_1)_{\text{free vortex}} = \rho \frac{c^2}{2}\,(1/r_1^2 - 1/r_2^2)$$

$$(p_2 - p_1)_{\text{forced vortex}} = \rho \frac{\omega^2}{2}\,(r_2^2 - r_1^2)$$

From the condition given in the problem,

$$\rho \frac{\omega^2}{2}\,(r_2^2 - r_1^2) = 2\rho \frac{c^2}{2}\,(1/r_1^2 - 1/r_2^2)$$

or $$r_1^2 r_2^2 = 2c^2/\omega^2 \tag{4.62}$$

Finally, from Eq. (4.61) and (4.62), we have

$$r_2^2/r_1^2 = 2$$

or $$r_2 = \sqrt{2}\, r_1$$

Example 4.11

The velocity of air at the outer edge of a tornado, where the pressure is 750 mm of Hg and diameter 30, is 12 m/s. Calculate the velocity and pressure of air at a radius of 2, from its axis. Consider the density of air to be constant and equals to 1.2 kg/m^3 (specific gravity of mercury = 13.6).

Solution

The flow field in a tornado (except near the centre) is simulated by a free vortex motion. Therefore, the velocity at a radius of 2 m is given by

$$(v_\theta)_{\text{at } r = 2\text{m}} = (v_\theta)_{\text{at } r = 15 \text{ m}} \times 15/2 = 12 \times 15/2 = 90 \text{ m/s}$$

Let p_0 and p be the absolute pressures at the outer edge of the tornado and at a radius of 2 m from its axis respectively. Then, for a free vortex,

$$\frac{p_0}{\rho g} - \frac{p}{\rho g} = \frac{(90)^2 - (12)^2}{2g}$$

$$= 405.50 \text{ m of air}$$

where ρ is the density of air

It is given that

$$\frac{p_0}{\rho g} = \frac{750 \times 10^{-3} \times 13.6 \times 10^3}{1.2}$$

$$= 8500 \text{ m of air}$$

Hence, $$p/\rho g = 8500 - 405.50$$

$$= 8094.5 \text{ m of air}$$

$$= 714.22 \text{ mm of Hg}$$

4.6 PRINCIPLES OF A HYDRAULIC SIPHON: APPLICATION OF BERNOULLI'S EQUATION

Fluids always flow from a higher energy level to a lower energy level. Here, by energy, we mean the total mechanical energy. Consider a container T containing some liquid (Fig. 4.16). If one end of a pipe S, completely filled in with the same liquid, is dipped into the container as shown in Fig. 4.16 with other end being open and vertically below the free surface of the liquid in the container T, then the liquid will continuously flow from the container T through the pipe S and will get discharged at the end B. This is known as the siphonic action by which the tank T

containing the liquid can be made empty. The pipe *S*, under the situation, is known as a hydraulic siphon or simply a siphon. The justification of flow through the pipe *S* can be made in the following way.

If we write the Bernoulli's equation, neglecting the frictional effects, at the two points *A* and *B* as shown in Fig. 4.16, we have

$$\frac{p_A}{\rho g} + 0 + z_A = \frac{p_B}{\rho g} + \frac{V_B^2}{2g} + z_B \tag{4.63a}$$

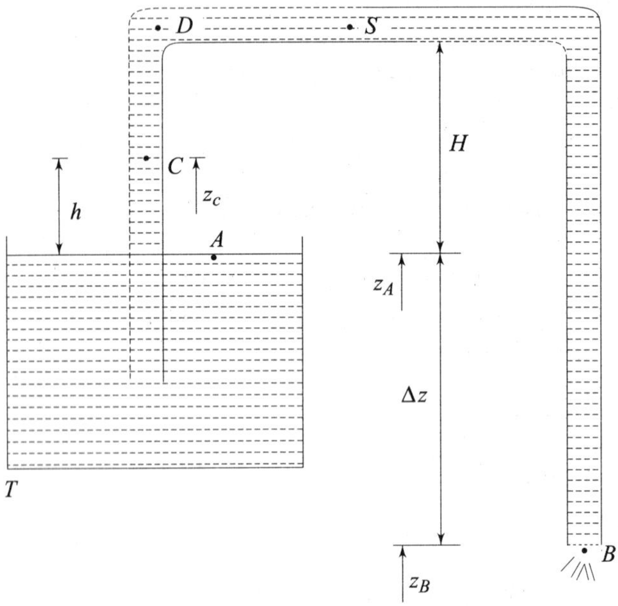

Fig. 4.16 Hydraulic siphon

The pressures at *A* and *B* are same and equal to the atmospheric pressure. Velocity at *A* is negligible compared to that at *B*, since the area of the tank *T* is very large compared to that of the tube *S*. Hence we get from Eq. (4.63a)

$$V_B = \sqrt{2g(z_A - z_B)} = \sqrt{2g\,\Delta z} \tag{4.63b}$$

Equation (4.63b) shows that a velocity head at *B* is created at the expense of the difference in potential head between *A* and *B* and thus justifies the flow from tank *T* through the pipe *S*.

The frictional effect due to viscosity of the fluid is taken care of by writing the Eq. (4.63a) in a modified from as

$$\frac{p_A}{\rho g} + 0 + z_A = \frac{p_B}{\rho g} + \frac{V_B^2}{2g} + z_B + h_L$$

which gives $$V_B = \sqrt{2g(\Delta z - h_L)} \tag{4.63c}$$

Since $$p_A = p_B = p_{atm} \text{ (atmospheric pressure)}$$

Here h_L is the loss of total head due to fluid friction in the flow from A to B. Hence, the velocity V_B expressed by the Eq. (4.63c) becomes less than that predicted by the Eq. (4.63b) in the absence of friction. Let us consider a point C in the pipe (Fig. 4.16), and apply the Bernoulli's equation between A and C. Then, neglecting frictional losses, we have

$$\frac{p_A}{\rho g} + 0 + z_A = \frac{p_C}{\rho g} + \frac{V_C^2}{2g} + z_C \tag{4.64a}$$

Considering the cross-sectional area of the pipe to be uniform, we have, from continuity, $V_B = V_C$, and the Eq. (4.64a) can be written as

$$\frac{p_C}{\rho g} = \frac{p_{atm}}{\rho g} - \frac{V_B^2}{2g} - h \tag{4.64b}$$

(Since, $p_A = p_{atm}$, the atmospheric pressure and $z_C - z_A = h$)

With the consideration of frictional losses, Eq. (4.64b) becomes

$$\frac{p_C}{\rho g} = \frac{p_{atm}}{\rho g} - \frac{V_B^2}{2g} - h - h'_L \tag{4.64c}$$

where h'_L is the loss of head due to friction in the flow from A to C. Therefore, it is found that the pressure at C is below the atmospheric pressure by the amount ($V_B^2/2g + h + h'_L$). This implies physically that a part of the pressure head at A is responsible for the gain in the velocity and potential head of the fluid at C plus the head which is utilised to overcome the friction in the path of flow. Usually the frictional loss h'_L is small due to the low velocity of flow and one can neglect it with respect to the change in potential head. Now it is obvious that the minimum pressure in the flow would be attained at the topmost part of the siphon, for example, point D, where the potential head is maximum. From the application of Bernoulli's equation between A and D, neglecting losses, we have

$$\frac{p_D}{\rho g} = \frac{p_{atm}}{\rho g} - \frac{V_B^2}{2g} - H$$

(Since the pipe is uniform, velocity at D equals to that at B)

From the Eq. (4.63b) $$\frac{V_B^2}{2g} = \Delta z$$

Therefore, it becomes, $$\frac{p_D}{\rho g} = \frac{p_{atm}}{\rho g} - (\Delta z + H) \tag{4.65}$$

The Eq. (4.65) can also be obtained by the application of Bernoulli's equation between D and B.

If the pressure of a liquid becomes equal to its vapour pressure at the existing temperature, then the liquid starts boiling and pockets of vapour are formed which create vapour locks to the flow and the flow is stopped. The vapour pockets are formed where the pressure is sufficiently low. These pockets are suddenly collapsed—either because they are carried along by the liquid until they arrive at a region of higher pressure or because the pressure increases again at the point in question. It results in cavities and the surrounding liquid rushes in to fill it, creating a very high pressure which can seriously damage the solid surface. This phenomenon is known as *cavitation*. In ordinary circumstances, liquids contain some dissolved air. This air is released as the pressure is reduced, and it too may form pockets in the liquid as air locks. Therefore to avoid this, the absolute pressure in a flow of liquid should never be allowed to fall to a pressure below which the air locking problem starts in practice. For water, this minimum pressure is about 20 Kpa (2 m of water). Therefore, the phenomenon of cavitation puts a constraint in the design of any hydraulic circuit where there is a chance for the liquid to attain a pressure below that of the atmosphere. For a siphon, this condition has to be checked at point D, so that $p_D > p_{\min},$ where $p_{\min}$ is the pressure for air locking or vapour locking to start.

Example 4.12

A tube is used as a siphon to discharge an oil of specific gravity 0.8 from a large open vessel into a drain at atmospheric pressure as shown in Fig. 4.17. Calculate (i) the velocity of oil through the siphon, (ii) the pressure at points A and B,(iii) the pressure at the highest point C,(iv) the maximum height of C that can be accommodated above the level in the vessel, and (v) the maximum vertical depth of the right limb of the siphon. (Take the vapour pressure of liquid at the working temperature to be 29.5 kPa and the atmospheric pressure as 101 kPa. Neglect friction)

Solution

(i) Applying Bernoullis equation between points 1 and D, we get

$$\frac{p_{\text{atm}}}{\rho g} + \frac{V_1^2}{2g} + 5.5 = \frac{p_D}{\rho g} + \frac{V_D^2}{2g} + 0 \qquad (4.66)$$

The horizontal plane through D is taken as the datum. Since the siphon discharges into the atmosphere, the pressure at the exit is atmospheric.

Hence, $\qquad p_D = p_{\text{atm}}$

Again, $V_1 << V_D$, since the area of tank is much larger compared to that of the tube. Therefore, Eq. (4.66) can be written as

$$\frac{p_{\text{atm}}}{\rho g} + 0 + 5.5 = \frac{p_{\text{atm}}}{\rho g} + \frac{V_D^2}{2g}$$

or $$V_D = \sqrt{2 \times 9.81 \times 5.5} = 10.39 \text{ m/s}$$

If the cross-sectional area of the siphon tube is uniform, the velocity of oil through the siphon will be uniform and equals to 10.39 m/s.

(ii) The points A and B are on the same horizontal plane, while point A is outside the tube, B is inside it.

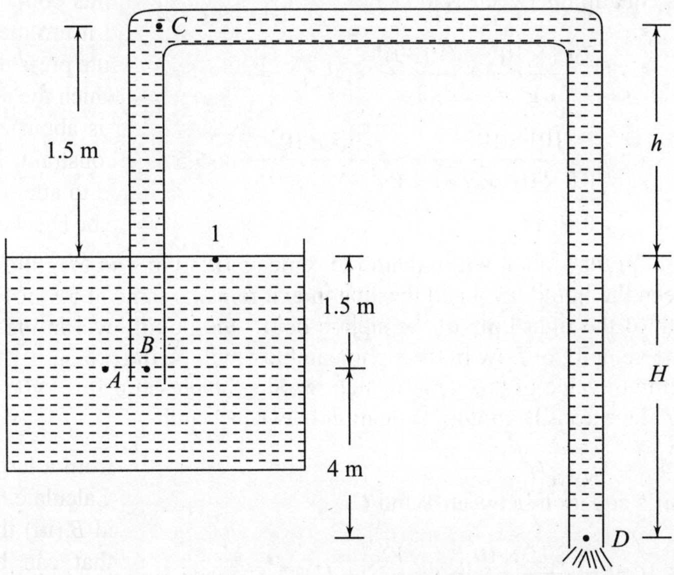

Fig. 4.17 A siphon discharging oil from a vessel to the atmosphere

The pressure at A is $p_A = p_{\text{atm}} + 1.5 \times 0.8 \times 10^3 \times 9.81$ Pa

$$= (101 + 11.77) \times 10^3 \text{ Pa}$$

$$= 112.77 \text{ kPa}$$

Applying Bernoulli's equation between A and B,

$$\frac{p_A}{\rho g} + 0 + 4.0 = \frac{p_B}{\rho g} + \frac{V_B^2}{2g} + 4.0 \tag{4.67}$$

or $$p_B = p_A - \rho \frac{V_B^2}{2}$$

We have $V_B = V_D = 10.39$ m/s and $p_A = 112.77$ kPa

Hence, $$p_B = 112.77 - 0.8 \times 10^3 \times \frac{(10.39)^2}{2 \times 10^3} = 69.59 \text{ kPa}$$

(iii) Applying Bernoulli's equation between 1 and C,

$$\frac{p_{\text{atm}}}{\rho g} + 0 + 5.5 = \frac{p_c}{\rho g} + \frac{(10.39)^2}{2g} + 7 \tag{4.68}$$

or $$\frac{p_C}{\rho g} = \frac{p_{\text{atm}}}{\rho g} - 7$$

or $$p_C = 101 - \frac{7 \times 0.8 \times 10^3 \times 9.81}{10^3} = 46.06 \text{ kPa}$$

(iv) For a maximum height of *c* above the liquid level, the pressure at *C* will be the vapour pressure of the liquid at working temperature. Let *h* be this height. Then applying Bernoulli's equation between 1 and *c*.

$$\frac{p_{\text{atm}}}{\rho g} + 0 + 5.5 = \frac{29.5 \times 10^3}{\rho g} + \frac{(10.39)^2}{2g} + 5.5 + h$$

or $$h = \frac{101 \times 10^3}{0.8 \times 10^3 \times 9.81} - \frac{29.5 \times 10^3}{0.8 \times 10^3 \times 9.81} - 5.5$$

$$= 3.61 \text{ m}$$

(The velocity of oil in the siphon will remain the same at 10.39 m/s so long the vertical height between the liquid level and the siphon exit remains same at 5.5 m.)

(v) More is the depth of the right limb of the siphon below the liquid level in the tank, more will be the velocity of flow in the siphon and less will be the pressure at *C*. Let *H* be the maximum value of this depth which renders the pressure at *C* to be the vapour pressure. Then from Bernoulli's equation between 1 and *D*.

$$V_D = \sqrt{2gH}$$

Again from Bernoulli's equation between 1 and *C*.

$$\frac{p_{\text{atm}}}{\rho g} + 0 + H = \frac{29.5 \times 10^3}{0.8 \times 10^3 \times 9.81} + \frac{V_D^2}{2g} + H + 1.5 \text{ [Since } V_C = V_D] \quad (4.69)$$

or $$H = \frac{p_{\text{atm}}}{\rho g} - \frac{29.5 \times 10^3}{0.8 \times 10^3 \times 9.81} - 1.5$$

$$= \frac{101 \times 10^3}{0.8 \times 10^3 \times 9.81} - \frac{29.5 \times 10^3}{0.8 \times 10^3 \times 9.81} - 1.5$$

$$= 7.61 \text{ m}$$

4.7 APPLICATION OF BERNOULLI'S EQUATION FOR MEASUREMENT OF FLOW RATE THROUGH PIPES

Flow rate through a pipe is usually measured by providing a coaxial area contraction within the pipe and by recording the pressure drop across the contraction. Determination of the flow rate from the measurement of the concerned pressure drop depends on the straightforward application of Bernoulli's equation. Three different flow meters primarily operate on this principle. These are (i) Venturimeter, (ii) Orificemeter, and (iii) Flow nozzle.

4.7.1 Venturimeter

A venturimeter is essentially a short pipe (Fig. 4.18) consisting of two conical parts with a short portion of uniform cross section in between. This short portion has the minimum area and is known as the throat. The two conical portions have the same base diameter, but one has a shorter length with a larger cone angle while the other has a larger length with a smaller cone angle.

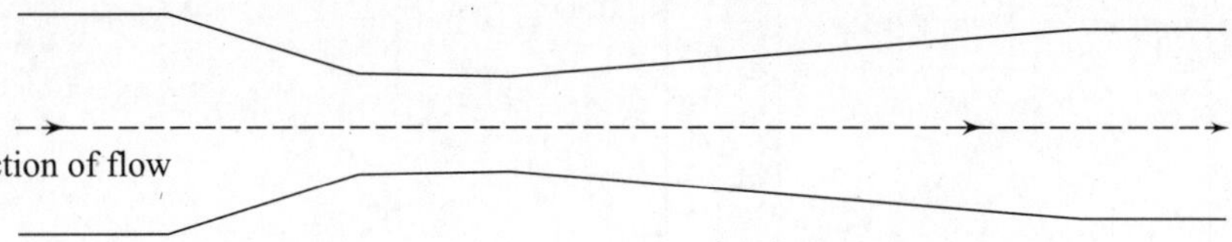

Fig. 4.18 A venturimeter

The venturimeter is always used in a way that the upstream part of the flow takes place through the short conical portion while the downstream part of the flow through the long one. This ensures a rapid converging passage and a gradual diverging passage in the direction of flow to avoid loss of energy due to separation. In course of a flow through the converging part, the velocity increases in the direction of flow according to the principle of continuity, while the pressure decreases according to Bernoulli's theorem. The velocity reaches its maximum value and pressure reaches its minimum value at the throat. Subsequently, a decrease in the velocity and an increase in the pressure take place in course of flow through the divergent part. This typical variation of fluid velocity and pressure by allowing it to flow through such a constricted convergent-divergent passage was first demonstrated by an Italian scientist, Giovanni Battista Venturi, in 1797.

Figure 4.19 shows that a venturimeter is inserted in an inclined pipe line in a vertical plane to measure the flow rate through the pipe. Let us consider a steady, ideal and one-dimensional (along the axis of the venturimeter) flow of fluid. Under this situation, the velocity and pressure at any section will be uniform. Let the velocity and pressure at the inlet (Section 1) are V_1 and p_1, respectively, while those at the throat (Section 2) are V_2 and p_2. Now, applying mechanical energy balance equation between sections 1 and 2, we get

$$\frac{p_1}{\rho g} + \frac{V_1^2}{2g} + z_1 = \frac{p_2}{\rho g} + \frac{V_2^2}{2g} + z_2 \tag{4.70}$$

or
$$\frac{V_2^2 - V_1^2}{2g} = \frac{p_1 - p_2}{\rho g} + z_1 - z_2 \tag{4.71}$$

where ρ is the density of fluid flowing through the venturimeter. From continuity,

$$V_2 A_2 = V_1 A_1 \tag{4.72}$$

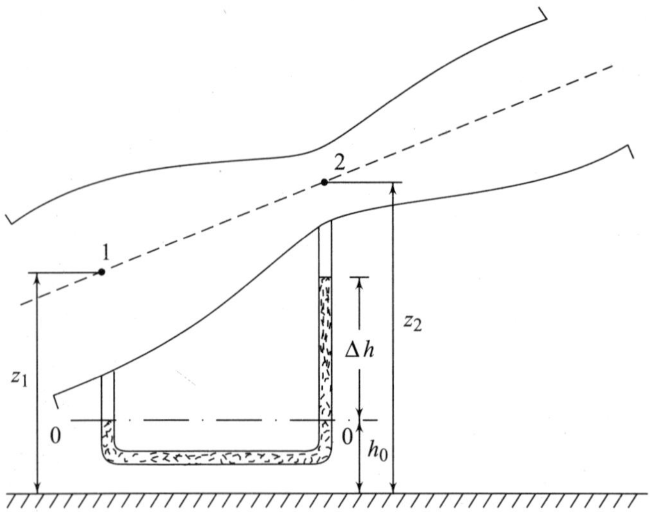

Fig. 4.19 Measurement of flow by a venturimeter

where A_2 and A_1 are the cross-sectional areas of the venturimeter at its throat and inlet, respectively. With the help of Eq. (4.72), Eq. (4.71) can be written as

$$\frac{V_2^2}{2g}\left(1-\frac{A_2^2}{A_1^2}\right) = \left(\frac{p_1}{\rho g}+z_1\right)-\left(\frac{p_2}{\rho g}+z_2\right)$$

or

$$V_2 = \frac{1}{\sqrt{1-\frac{A_2^2}{A_1^2}}}\sqrt{2g\,(h_1^* - h_2^*)} \tag{4.73}$$

where h_1^* and h_2^* are the piezometric pressure heads at 1 and 2 respectively, and are defined as

$$h_1^* = \frac{p_1}{\rho g} + z_1 \tag{4.74a}$$

$$h_2^* = \frac{p_2}{\rho g} + z_2 \tag{4.74b}$$

Hence, the volume flow rate through the pipe is given by

$$Q = A_2 V_2 = \frac{A_2}{\sqrt{1-\frac{A_2^2}{A_1^2}}}\sqrt{2g\,(h_1^* - h_2^*)} \tag{4.75}$$

If the pressure difference between 1 and 2 is measured by a manometer, as shown in Fig. 4.19, we can write

$$p_1 + \rho g\,(z_1 - h_0) = p_2 + \rho g\,(z_2 - h_0 - \Delta h) + \Delta h\,\rho_m g$$

or $$(p_1 + \rho g z_1) - (p_2 + \rho g z_2) = (\rho_m - \rho) g \, \Delta h$$

or $$\left(\frac{p_1}{\rho g} + z_1\right) - \left(\frac{p_2}{\rho g} + z_2\right) = \left(\frac{\rho_m}{\rho} - 1\right) \Delta h$$

or $$h_1^* - h_2^* = \left(\frac{\rho_m}{\rho} - 1\right) \Delta h \tag{4.76}$$

where ρ_m is the density of the manometric liquid. Equation (4.76) shows that a manometer always registers a direct reading of the difference in piezometric pressures. Now, substitution of $h_1^* - h_2^*$ from Eq. (4.76) in Eq. (4.75) gives

$$Q = \frac{A_1 A_2}{\sqrt{A_1^2 - A_2^2}} \sqrt{2g (\rho_m/\rho - 1) \Delta h} \tag{4.77}$$

If the pipe along with the venturimeter is horizontal, then $z_1 = z_2$; and hence $h_1^* - h_2^*$ becomes $h_1 - h_2$, where h_1 and h_2 are the static pressure heads $\left(h_1 = \frac{p_1}{\rho g}, h_2 = \frac{p_2}{\rho g}\right)$. The manometric equation [Eq. (5.59)] then becomes

$$h_1 - h_2 = \left[\frac{\rho_m}{\rho} - 1\right] \Delta h$$

Therefore, it is interesting to note that the final expression of flow rate, given by Eq. (4.77), in terms of manometer deflection Δh, remains the same irrespective of whether the pipeline along with the venturimeter connection is horizontal or not. Measured values of Δh, the difference in piezometric pressures between Sections 1 and 2, for a real fluid will always be greater than that assumed in case of an ideal fluid because of frictional losses in addition to the change in momentum. Therefore, Eq. (4.77) always overestimates the actual flow rate since we substitute the value of Δh from actual measurement in practice. In order to take this into account, a multiplying factor C_d, called the coefficient of discharge, is incorporated in the Eq. (4.77) as

$$Q_{\text{actual}} = C_d \frac{A_1 A_2}{\sqrt{A_1^2 - A_2^2}} \sqrt{2g (\rho_m/\rho - 1) \Delta h} \tag{4.78}$$

The coefficient of discharge C_d is always less than unity and is defined as

$$C_d = \frac{\text{Actual rate of discharge}}{\text{Theoretical rate of discharge}}$$

where, the theoretical discharge rate is predicted by the Eq. (4.77) with the measured value of Δh, and the actual rate of discharge is the discharge rate measured in practice. Value of C_d for a venturimeter usually lies between 0.95 and 0.98.

Example 4.13

Water flows through a 300 mm × 150 mm venturimeter at the rate of 0.037 m^3/s and the differential gauge is deflected 1 m, as shown in Fig. 4.20. Specific gravity of the gauge liquid is 1.25. Determine the coefficient of discharge of the meter.

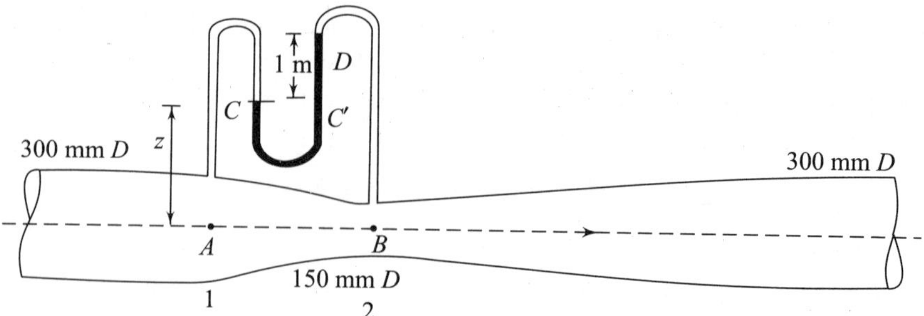

Fig. 4.20 A venturimeter measuring the flow of water through a pipe

Solution

Applying Bernoulli's equation between A and B, and considering the fluid to be inviscid, we get

$$\frac{p_A}{\rho g} + \frac{V_A^2}{2g} + 0 = \frac{p_B}{\rho g} + \frac{V_B^2}{2g} + 0 \tag{4.79}$$

(the axis of the venturimeter is considered to be horizontal)
Again from continuity,

$$V_A^2 = (A_B/A_A)^2 V_B^2 \tag{4.80}$$

Solving for V_B from Eq. (4.79) with the help of Eq. (4.80), we have

$$V_B = \sqrt{\frac{2(p_A - p_B)/\rho}{1-(A_B/A_A)^2}}$$

The actual rate of discharge Q can be written as

$$Q = C_D A_B V_B$$

$$= C_D A_B \sqrt{\frac{2(p_A - p_B)/\rho}{1-(A_B/A_A)^2}} \tag{4.81}$$

where C_D is the coefficient of discharge.
From the principle of hydrostatics applied to the differential gauge, we get

$$(p_A/\rho g - z) = p_B/\rho g - (z+1) + 1.25 \times 1$$

or $$\frac{p_A - p_B}{\rho g} = 0.25\,\text{m}$$

Hence, from Eq. (4.81), we can write

$$0.037 = C_D \frac{\pi}{4} (0.15)^2 \sqrt{2 \times 9.81(0.25)/(1 - 1/16)}$$

which gives $$C_D = 0.976$$

Example 4.14

A necked-down or venturi section of a pipe flow develops a low pressure which can be used to aspirate liquid upward from a reservoir as shown in Fig. 4.21. Derive an expression for the exit velocity V_2 which is just sufficient to cause Sthe reservoir liquid to rise in the tube up to Section 1 (Fig. 4.21).

Consider the liquids originally flowing through the pipe and that to be pumped from the reservoir are same (neglect frictional losses).

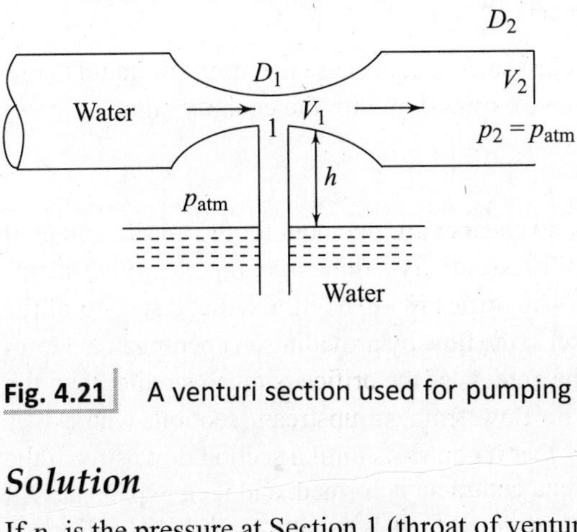

Fig. 4.21 A venturi section used for pumping water from a reservoir

Solution

If p_1 is the pressure at Section 1 (throat of venturi tube), then for the liquid from the reservoir to rise through the tube,

$$p_{atm} - p_1 \geq \rho g h$$

where p_{atm} is the atmospheric pressure acting on the free surface of the liquid in the reservoir.

From continuity, $$V_1 = V_2 (D_2/D_1)^2$$

Applying the Bernoulli's equation in consideration of the pipe axis to be horizontal, and without any loss of head in the flow, we have

$$p_1 + \frac{1}{2}\rho V_1^2 = p_2 + \frac{1}{2}\rho V_2^2$$

Here, $p_2 = p_{atm}$ as given in the problem.

Hence, $$p_{atm} - p_1 = \frac{1}{2}\rho[V_1^2 - V_2^2]$$

$$= \frac{1}{2}\rho\left[\left(\frac{D_2}{D_1}\right)^4 - 1\right]V_2^2$$

For the liquid to rise through the tube,

$$p_{\text{atm}} - p_1 \geq \rho g h$$

or

$$(1/2)\ \rho\left[\left(\frac{D_2}{D_1}\right)^4 - 1\right] V_2^2 \geq \rho g h$$

Therefore,

$$V_2 \geq \frac{\sqrt{2gh}}{\left[\left(\frac{D_2}{D_1}\right)^4 - 1\right]^{1/2}}$$

The existing velocity V_2, which is just sufficient to cause the reservoir liquid to rise through the tube, is given by the above expression with the equality sign.

4.7.2 Orificemeter

An orificemeter provides a simpler and cheaper arrangement for the measurement of flow through a pipe. An orificemeter is essentially a thin circular plate with a sharp-edged concentric circular hole in it. The orifice plate, being fixed at a section of the pipe, (Fig. 4.22) creates an obstruction to the flow by providing an opening in the form of an orifice to the flow passage. The area A_0 of the orifice is much smaller than the cross-sectional area of the pipe. The flow from an upstream section, where it is uniform, adjusts itself in such a way that it contracts until a section downstream the orifice plate is reached, where the vena contracta is formed, and then expands to fill the passage of the pipe. One of the pressure tapings is usually provided at a distance of one diameter upstream the orifice plate where the flow is almost uniform (Section 1–1) and the other at a distance of half a diameter downstream the orifice plate. Considering the fluid to be ideal and the downstream pressure taping to be at the vena contracta (Section *c–c*), we can write, by applying mechanical energy conservation between Section 1–1 and Section *c–c*,

$$\frac{p_1^*}{\rho g} + \frac{V_1^2}{2g} = \frac{p_c^*}{\rho g} + \frac{V_c^2}{2g} \tag{4.82}$$

where p_1^* and p_c^* are the piezometric pressures at Section 1–1 and *c–c*, respectively. From the equation of continuity,

$$V_1 A_1 = V_c A_c \tag{4.83}$$

where A_c is the area of the vena contracta. With the help of Eq. (4.83), Eq. (4.82) can be written as

$$V_c = \sqrt{\frac{2(p_1^* - p_c^*)}{\rho(1 - A_c^2/A_1^2)}} \tag{4.84}$$

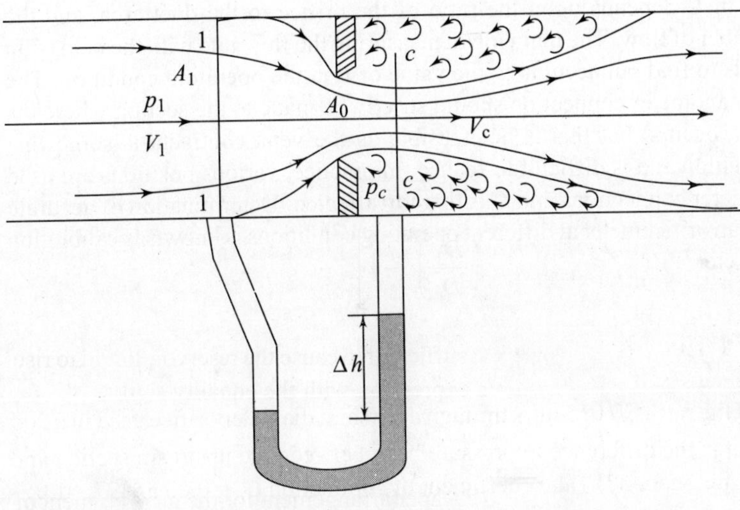

Fig. 4.22 Flow through an orificemeter

Recalling the fact that the measured value of the piezometric pressure drop for a real fluid is always more due to friction than that assumed in case of an inviscid flow, a coefficient of velocity C_v (always less than 1) has to be introduced to determine the actual velocity V_c when the pressure drop $p_1^* - p_c^*$ in Eq. (4.84) is substituted by its measured value in terms of the manometer deflection Δh

Hence,
$$V_c = C_v \sqrt{\frac{2g(\rho_m/\rho - 1)\Delta h}{(1 - A_c^2/A_1^2)}} \tag{4.85}$$

where Δh is the difference in liquid levels in the manometer and ρ_m is the density of the manometric liquid.

Volumetric flow rate $\quad Q = A_c V_c$ (4.86)

If a coefficient of contraction C_c is defined as, $C_c = A_c/A_0$, where A_0 is the area of the orifice, then Eq.(4.86) can be written, with the help of Eq. (4.85) as

$$Q = C_c A_0 C_v \sqrt{\frac{2g(\rho_m/\rho - 1)}{(1 - C_c^2 A_0^2/A_1^2)}}\, \Delta h$$

$$= C_v C_c A_0 \sqrt{\frac{2g}{(1 - C_c^2 A_0^2/A_1^2}} \sqrt{(\rho_m/\rho - 1)\Delta h}$$

$$= C\sqrt{(\rho_m/\rho - 1)\Delta h} \tag{4.87}$$

with, $$C = C_d A_0 \sqrt{\frac{2g}{(1 - C_c^2 A_0^2/A_1^2)}}, \text{ where } (C_d = C_v C_c)$$

The value of C depends upon the ratio of the orifice to the duct area, and the Reynolds number of flow. The main job in measuring the flow rate with the help of an orificemeter, is to find out accurately the value of C at the operating condition. The downstream manometer connection should strictly be made to the section where the vena contracta occurs, but this is not feasible as the vena contracta is somewhat variable in position and is difficult to realise. In practice, various positions are used for the manometer connections and C is thereby affected. Determination of accurate values of C of an orificemeter at different operating conditions is known as calibration of the orificemeter.

Example 4.15

Water flows at the rate of 0.015 m³/s through a 100 mm diameter orifice used in a 200 mm pipe. What is the difference in pressure head between the upstream section and the vena contracta section? (Take coefficient of contraction $C_c = 0.60$ and $C_v = 1.0$).

Solution

We know from Eq. (4.87)

$$Q = C\sqrt{\frac{\Delta p}{\rho g}}$$

where, $C = C_v C_c A_0 \sqrt{\frac{2g}{(1 - C_c^2 A_0^2/A_1^2}}$

For the present problem,

$$C = 1.0 \times 0.60 \times \frac{\pi}{4}(0.1)^2 \sqrt{\frac{2 \times 9.81}{[(1 - (0.60)^2 (1/2)^4)]}}$$

$$= 0.0211$$

Hence, $$0.015 = 0.0211\sqrt{\Delta p/\rho g}$$

or $$\Delta p/\rho g = 0.505 \text{ m of water}$$

4.7.3 Flow Nozzle

The flow nozzle as shown in Fig. 4.23 is essentially a venturimeter with the divergent part omitted. Therefore the basic equations for calculation of flow rate are the same as those for a venturimeter. The dissipation of energy downstream of the throat due to flow separation is greater than that for a venturimeter. But this disadvantage is often offset by the lower cost of the nozzle. The downstream connection of the manometer may not necessarily be at the throat of the nozzle or at a point sufficiently for from the nozzle. The deviations however are taken care of in the values of C_d. The coefficient C_d depends on the shape of the nozzle, the ratio of pipe to nozzle diameter and the Reynolds number of flow.

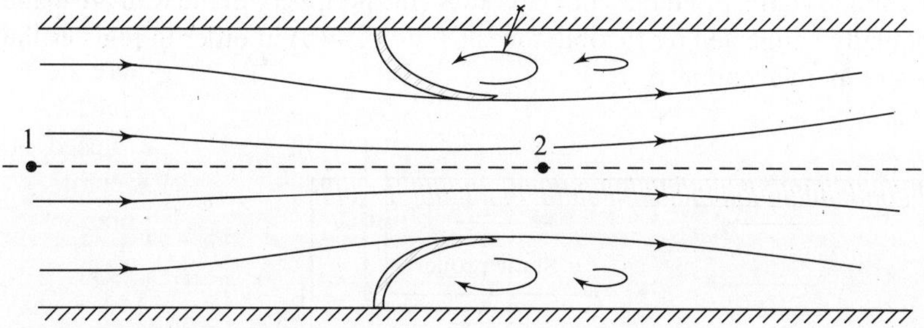

Fig. 4.23 A flow nozzle

A comparative picture of the typical values of C_d, accuracy, and the cost of three flowmeters (venturimeter, orificemeter and flow nozzle) is given below:

Type of Flowmeter	*Accuracy*	*Cost*	*Loss of Total Head*	*TypicalValues of C_d*
Venturimeter	High	High	Low	0.95 to 0.98
Orificemeter	Low	Low	High	0.60 to 0.65
Flow nozzle	Intermediate between a venturimeter and an orificemeter			0.70 to 0.80

4.7.4 Concept of Static and Stagnation Pressures and Application of Pitot Tube in Flow Measurements

4.7.4.1 *Static Pressure*

The thermodynamic or hydrostatic pressure caused by molecular collisions is known as *static pressure in a fluid flow* and is usually referred to as the pressure p. When the fluid is at rest, this pressure p is the same in all directions and is categorically known as the *hydrostatic pressure*. For the flow of a real and *Stoksian fluid* (the fluid which obeys Stoke's law as explained in Section. 8.2) the static or thermodynamic pressure becomes equal to the negative of the arithmetic average of the normal stresses at a point (mechanical pressure). The static pressure is a parameter to describe the state of a flowing fluid. Physically it is the pressure of fluid that one can measure while moving with the flow. Let us consider the flow of a fluid through a closed passage as shown in Fig. 4.24(a). If a hole is made at the wall and is connected to any pressure measuring device, it will then sense the static pressure at the wall. This type of hole at the wall is known as a wall tap. The fact that a wall tap actually senses the static pressure can be appreciated by noticing that there is no component of velocity along the axis of the hole. In most circumstances, for example, in case of parallel flows, the static pressure at the cross section remains the same. The wall tap under this situation registers the static pressure at that cross section. In practice, instead of a single wall tap, a number of taps along the periphery of the wall are made and are mutually connected by flexible tubes (Fig. 4.24 (b) in order to register the static pressure more accurately.

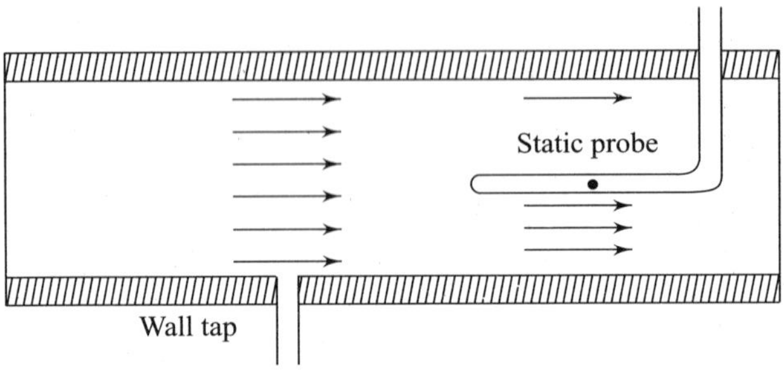

(a) Single wall tap

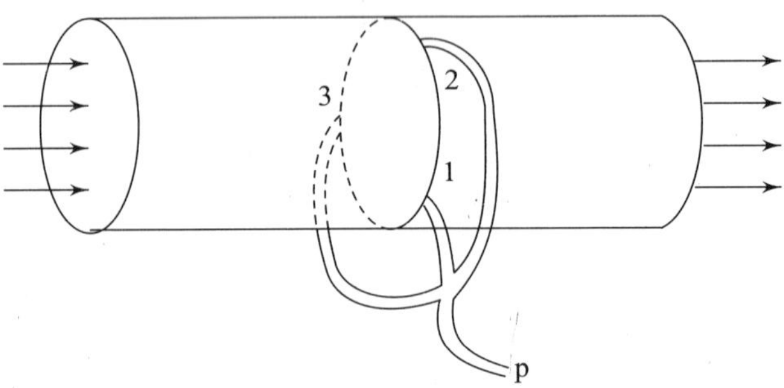

(b) Multiple wall tap

Fig. 4.24 Measurement of static pressure

4.7.4.2 *Stagnation Pressure*

The stagnation pressure at a point in a fluid flow is the pressure which could result if the fluid were brought to rest isentropically. The word 'isentropically' implies the sense that the entire kinetic energy of a fluid particle is utilised to increase its pressure only. This is possible only in a *reversible adiabatic process* known as *isentropic process*. Let us consider the flow of fluid through a closed passage (Fig. 4.25). At Section 1–1 let the velocity and static pressure of the fluid be uniform. Consider a point *A* on that section just in front of which a right-angled tube with one end facing the flow and the other end closed is placed. When equilibrium is attained, the fluid in the tube will be at rest, and the pressure at any point in the tube including the point *B* will be more than that at *A* where the flow velocity exists. By the application of

Bernoulli's equation between the points B and A, in consideration of the flow to be inviscid and incompressible, we have

$$p_0 = p + \frac{\rho V^2}{2} \tag{4.88}$$

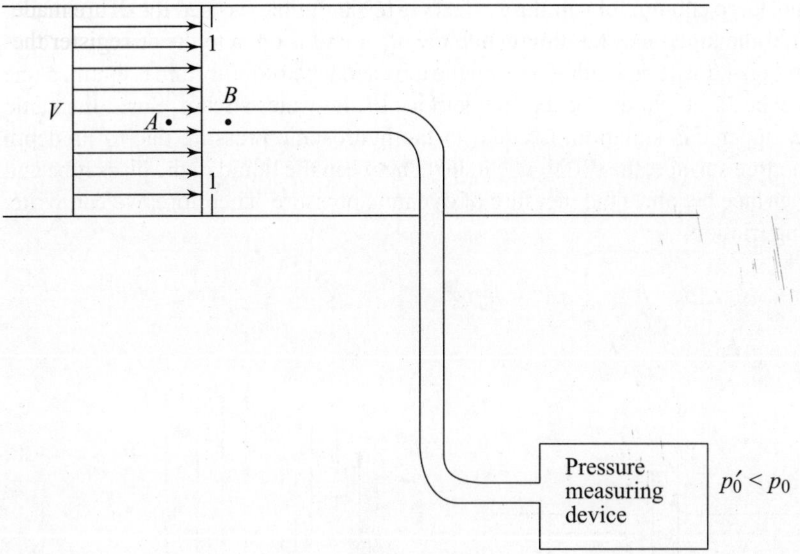

Fig. 4.25 Measurement of stagnation pressure

where p and V are the pressure and velocity, respectively, at the point A at Section 1–1, and p_0 is the pressure at B, which, according to the definition, refers to the stagnation pressure at point A. It is found from Eq. (4.88) that the stagnation pressure p_0 consists of two terms, the static pressure p and the term $\rho V^2/2$, which is known as *dynamic pressure*. Therefore Eq. (4.88) can be written for a better understanding as

$$\underset{\text{Stagnation pressure}}{p_0} = \underset{\text{Static pressure}}{p} + \underset{\text{Dynamic pressure}}{\frac{1}{2}\rho V^2} \tag{4.89}$$

or

$$V = \sqrt{2(p_0 - p)/\rho} \tag{4.90}$$

Therefore, it appears from Eq. (4.90), that from a measurement of both static and stagnation pressure in a flowing fluid, the velocity of flow can be determined. But it is difficult to measure the stagnation pressure in practice for a real fluid due to friction. The pressure p_0' in the stagnation tube indicated by any pressure measuring device (Fig. 4.25) will always be less than p_0, since a part of the kinetic energy will be converted into intermolecular energy due to fluid friction. This is taken care of by an empirical factor C in determining the velocity from Eq. (4.90) as

$$V = C\sqrt{2(p_0 - p)/\rho} \tag{4.91}$$

4.7.4.3 *Pitot Tube for Flow Measurement*

The principle of flow measurement by the Pitot tube was adopted first by a French Scientist, Henri Pitot, in 1732 for measuring velocities in the river. A right-angled glass tube, large enough for capillary effects to be negligible, is used for this purpose. One end of the tube faces the flow while the other end is open to the atmosphere, as shown in Fig. 4.26a. The liquid flows up the tube and when equilibrium is attained, the liquid reaches a height above the free surface of the water stream. Since the static pressure, under this situation, is equal to the hydrostatic pressure due to its depth below the free surface, the difference in level between the liquid in the glass tube and the free surface becomes the measure of dynamic pressure. Therefore, we can write, neglecting friction,

$$p_0 - p = \frac{1}{2}\rho V^2 = h\rho g$$

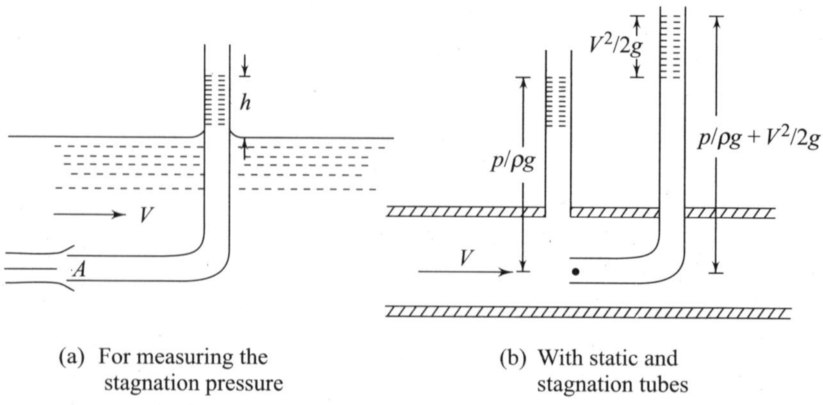

Fig. 4.26 Simple pitot tube

where p_0, p and V are the stagnation pressure, static pressure and velocity respectively at point A (Fig. 4.26a).

or $$V = \sqrt{2gh}$$

Such a tube is known as a Pitot tube and provides one of the most accurate means of measuring fluid velocity. For an open stream of liquid with a free surface, this single tube is sufficient to determine the velocity. But for a fluid flowing through a closed duct, the Pitot tube measures only the stagnation pressure and so the static pressure must be measured separately. Measurement of static pressure in this case is made at the boundary of the wall (Fig. 4.26b). The axis of the tube measuring the static pressure must be perpendicular to the boundary and free from burrs, so that the boundary is smooth and hence the streamlines adjacent to it are not curved. This is done to sense the static pressure only without any part of the dynamic pressure. A Pitot tube is also inserted as shown (Fig. 4.26b) to sense the stagnation pressure. The ends of the Pitot tube, measuring the stagnation pressure, and the piezometric tube, measuring the static pressure, may be connected to a suitable differential manometer for the determination of flow velocity and hence the flow rate.

4.7.4.4 *Pitot Static Tube*

The tubes recording static pressure and the stagnation pressure (Fig. 4.26b) are usually combined into one instrument known as *the Pitot static tube* (Fig. 4.27). The tube for sensing the static pressure is known as the static tube which surrounds the Pitot tube that measures the stagnation pressure. Two or more holes are drilled radially through the outer wall of the static tube into the annular space. The position of these static holes is important. Downstream of the nose N, the flow is accelerated somewhat with consequent reduction in static pressure. But in front of the supporting stem, there is a reduction in velocity and increase in pressure. The static holes should therefore be at the position where the two opposing effects are counterbalanced and the reading corresponds to the undisturbed static pressure. Finally, the flow velocity is given by

$$V = C\sqrt{2\Delta p/\rho} \tag{4.92}$$

where Δp is the difference between stagnation and static pressures. The factor C takes care of the non-idealities, due to friction, in converting the dynamic head into pressure head and depends, to a large extent, on the geometry of the pitot tube. The value of C is usually determined from calibration test of the Pitot tube.

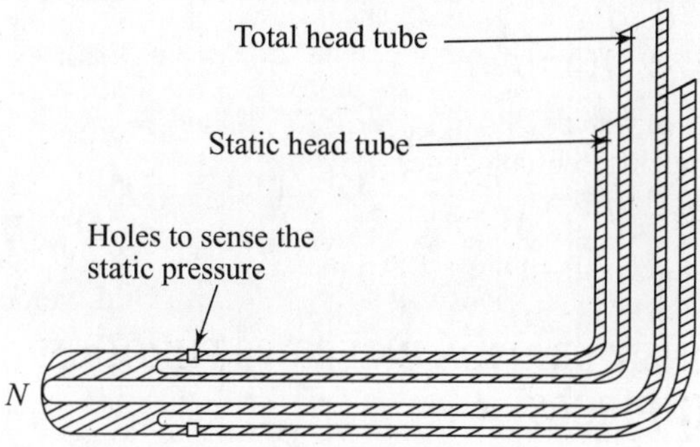

Fig. 4.27 Pitat static tube

Example 4.16

Air flows through a duct, and the Pitotstatic tube measuring the velocity is attached to a differential manometer containing water. If the deflection of the manometer is 100 mm, calculate the air velocity, assuming the density of air is constant and equals to 1.22 kg/m^3, and that the coefficient of the tube is 0.98.

Solution

From the differential manometer,

$$\frac{\Delta p}{\rho g} = \frac{(0.1)\times(9.81)\times 10^3}{1.22\times 9.81}$$

$$= 81.97 \text{ m of air}$$

where, Δp is the difference in stagnation and static pressures as measured by the differential manometer. Velocity of air is calculated using Eq. (4.92) as

$$V = 0.98\sqrt{2\times 9.81\times(81.97)}$$

$$= 39.3 \text{ m/s}$$

Example 4.17

Water flows at a velocity of 1.417 m/s. A differential gauge which contains a liquid of specific gravity 1.25 is attached to a Pitot static tube. What is the deflection of the gauge fluid? (Assume the coefficient of the tube to be 1).

Solution

We know from Eq. (4.92)

$$V = C\sqrt{2g(\Delta p/\rho g)}$$

Therefore, for the present case,

$$1.417 = 1.00\sqrt{2\times 9.81(\Delta p/\rho g)}$$

Hence, $\Delta p/\rho g = 0.1023$ m of water

From the manometric equation of the differential gauge,

$$0.1023 = (1.25-1)h$$

which gives the deflection of the gauge $h = 0.409$ m $= 409$ mm

4.8 SOME PRACTICAL CONSIDERATION ON FLOWS THROUGH ORIFICES AND MOUTHPIECES

An orifice is a small aperture through which a fluid passes. The thickness of an orifice in the direction of flow is very small in comparison to its other dimensions. If a tank containing a liquid has a hole made on the side or base through which a liquid flows, then such a hole may be termed as an orifice. The rate of flow of the liquid through such an orifice at a given time will depend partly on the shape, size and form of the orifice. An orifice usually has a sharp edge so that there is minimum contact with the fluid and consequently minimum frictional resistance at the sides of the orifice. If a sharp edge is not provided, the flow depends on the thickness of the orifice and the roughness of its boundary surface too.

4.8.1 Flow from an Orifice at the Side of a Tank under a Constant Head

Let us consider a tank containing a liquid and with an orifice at its side wall as shown in Fig. 4.28. The orifice has a sharp edge with the bevelled side facing downstream. Let the height of the free surface of liquid above the centre line of the orifice be kept fixed by some adjustable arrangements of inflow to the tank. The liquid issues from the orifice as a free jet under the influence of gravity only. The streamlines approaching the orifice converge towards it. Since an instantaneous change of direction is not possible, the streamlines continue to converge beyond the orifice until they become parallel at the Section *c–c* (Fig. 4.28). For an ideal fluid, streamlines will strictly be parallel at an infinite distance, but however fluid friction in practice produce parallel flow at only a short distance from the orifice. The area of the jet at the Section *c–c* is lower than the area of the orifice. Section *c–c* is known as the vena contracta. The contraction of the jet can be attributed to the action of a lateral force on the jet due to a change in the direction of flow velocity when the fluid approaches the orifice. Since the streamlines become parallel at vena contracta, the pressure at this section is assumed to be uniform. If the pressure difference due to surface tension is neglected, the pressure in the jet at vena contracta becomes equal to that of the ambience surrounding the jet. Considering the flow to be steady and frictional effects to be negligible, we can write by the application of Bernoulli's equation between two points 1 and 2 on a particular stream-line with point 2 being at vena contracta (Fig 4.28).

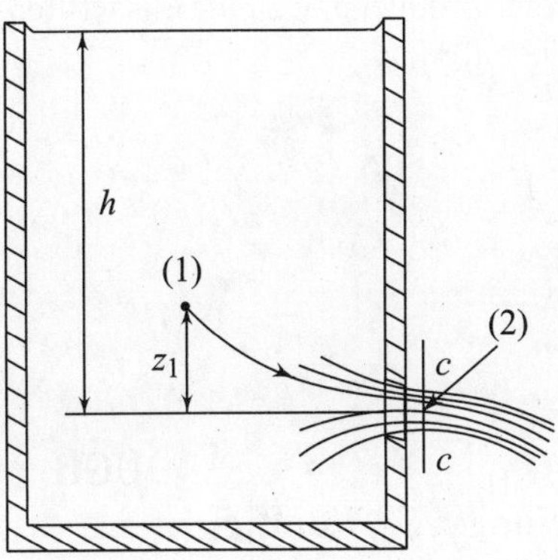

Fig. 4.28 Flow from a sharp edged orifice

$$p_1/\rho g + V_1^2/2g + z_1 = p_{atm}/\rho g + V_2^2/2g + 0 \tag{4.93}$$

The horizontal plane through the centre of the orifice has been taken as the datum level for determining the potential head. If the area of the tank is large enough as compared to that of the orifice, the velocity at Point 1 becomes negligibly small and pressure p_1 equals to the hydrostatic pressure at that point as

$$p_1 = p_{atm} + rg(h - z_1).$$

Therefore, Eq. (4.93) becomes

$$p_{atm}/\rho g + (h - z_1) + 0 + z_1 = p_{atm}/\rho g + V_2^2/2g \tag{4.94}$$

or

$$V_2 = \sqrt{2gh} \tag{4.95}$$

If the orifice is small in comparison to h, the velocity of the jet is constant across the vena contracta. The Eq. (4.95) states that the velocity with which a jet of liquid escapes from a small orifice is proportional to the square root of the head above the orifice, and is known as *Torricelli's formula*. The velocity V_2 in Eq. (4.95) represents the ideal velocity since the frictional effects were neglected in the derivation. Therefore, a multiplying factor C_v, known as *coefficient of velocity* is introduced to determine the actual velocity as

$$V_{2\,\text{actual}} = C_v \sqrt{2gh}$$

Since the role of friction is to reduce the velocity, C_v is always less than unity. The rate of discharge through the orifice can then be written as

$$Q = a_c C_v \sqrt{2gh} \tag{4.96}$$

where a_c is the cross-sectional area of the jet at vena contracta. Defining a *coefficient of contraction* C_c as the ratio of the area of vena contracta to the area of orifice, Eq. (4.96) can be written as

$$Q = C_c C_v a_0 \sqrt{2gh} \tag{4.97}$$

where, a_0 is the cross-sectional area of the orifice. The product of C_c and C_v is written as C_d and is termed as *coefficient of discharge*. Therefore,

$$Q = C_d a_0 \sqrt{2gh}$$

or

$$C_d = \frac{Q}{a_0 \sqrt{2gh}}$$

$$= \frac{\text{Actual discharge}}{\text{Ideal discharge}}$$

4.8.2 Determination of Coefficient of Velocity C_v, Coefficient of Contraction C_c and Coefficient of Discharge C_d

All the coefficients C_v, C_c and C_d of an orifice depend on the shape and size of the orifice. The values of C_v, C_c and C_d are determined experimentally as described below.

Consider the tank in Fig. 4.29. Let H be the height of the liquid, maintained constant, above the centre line of the orifice. The Section c–c is at vena contracta. The jet of liquid coming out of the orifice is acted upon by gravity only with a downward acceleration of g. Therefore, the horizontal component of velocity u of the jet remains constant. Let P be a point on the jet such that x and z are the horizontal and vertical coordinates respectively of P from the vena contracta c–c as shown in Fig. 4.29. Considering the flow of a fluid particle from c–c to P along the jet, we can write

$$x = ut$$

and $$z = gt^2/2$$

(where t is the time taken by the fluid particle to move from c–c to P).

Eliminating t from the two equations, we have

$$\frac{x^2}{u^2} = \frac{2z}{g}$$

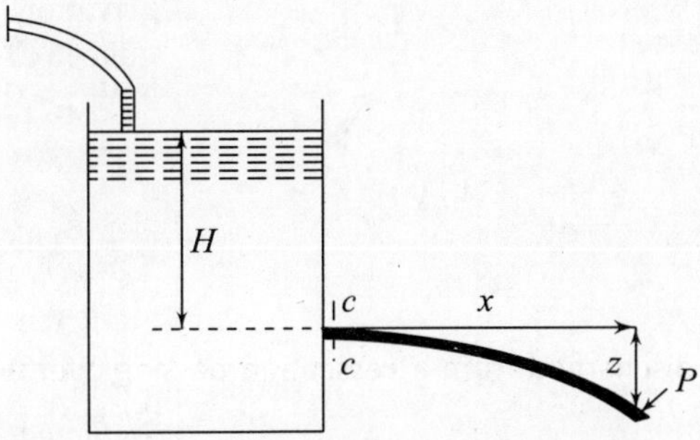

Fig. 4.29 Trajectory of a liquid jet discharged from a sharp- edged orifice

or $$u^2 = \frac{gx^2}{2z} \tag{4.98}$$

But $$C_v = \frac{u}{\sqrt{2gH}}$$

Substituting for u from Eq. (4.98)

$$C_v = \frac{x}{\sqrt{4zH}} \tag{4.99}$$

Therefore, the coefficient of velocity C_v of an orifice under a given value of H can be found from Eq. (4.99) with the measured values of x and z. The coefficient of discharge is determined by measuring the actual quantity of liquid discharged through the orifice in a given time under a constant head, and then dividing this quantity by the theoretical discharge. The theoretical discharge rate is calculated from known values of liquid head H and orifice area a_0, as $Q_{\text{theo}} = a_0\sqrt{2gH}$. The coefficient of contraction C_c is usually found out by dividing the value of C_d by the measured value of C_v. The coefficient of discharge varies with the head 'H' and the type of orifice. For a sharp-edged orifice, typical values of C_d lie between 0.60 and 0.65, while the values of C_v vary between 0.97 and 0.99.

Example 4.18

For the configuration shown, (Fig. 4.30) calculate the minimum or just sufficient head H in the vessel and the corresponding discharge which can pass over the plate. (Take $C_v = 1, C_d = 0.8$)

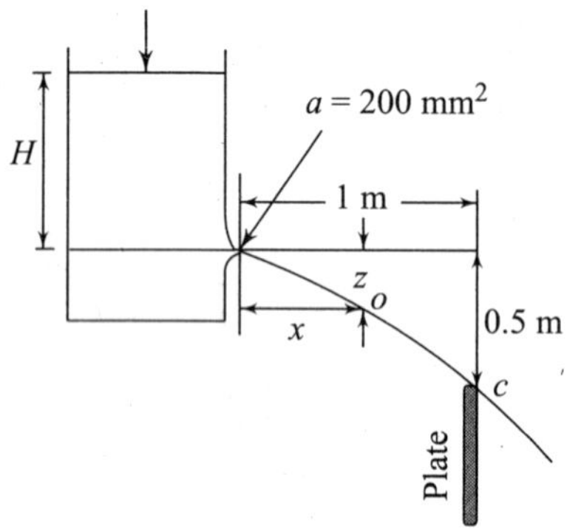

Fig. 4.30 Trajectory of a liquid jet discharged from a vessel and passing over a plate

Solution

For a point O on the trajectory,

$$x = u_1 t$$

and

$$z = \frac{1}{2} g t^2$$

where t is the time taken for any liquid particle to reach the point O after being ejected from the orifice with a velocity u_1. Eliminating t from the two equations, we get,

$$z = \frac{g}{2} \frac{x^2}{u_1^2}$$

which shows that the trajectory must be parabolic.

Again, for $C_v = 1$,

$$u_1 = \sqrt{2gH}$$

Therefore,

$$z = \frac{x^2}{4H}$$

or $$H = \frac{x^2}{4z}$$

Hence, the minimum head H to pass over c

$$= (1)^2/4 \times (0.5) = 0.5 \text{ m}$$

The corresponding minimum discharge

$$Q = \sqrt{2 \times 9.81 \times 0.5} \times (2 \times 10^{-4}) \times 0.8 \text{ m}^3/\text{s}$$
$$= 0.0005 \text{ m}^3/\text{s}$$

Example 4.19

Two identical orifices are mounted on one side of a vertical tank (Fig. 4.31). The height of water above the upper orifice is 3 m. If the jets of water from the two orifices intersect at a horizontal distance of 8 m from the tank, estimate the vertical distance between the two orifices. Calculate the vertical distance of the point of intersection of the jets from the water level in the tank. Assume $C_v = 1$ for the orifices.

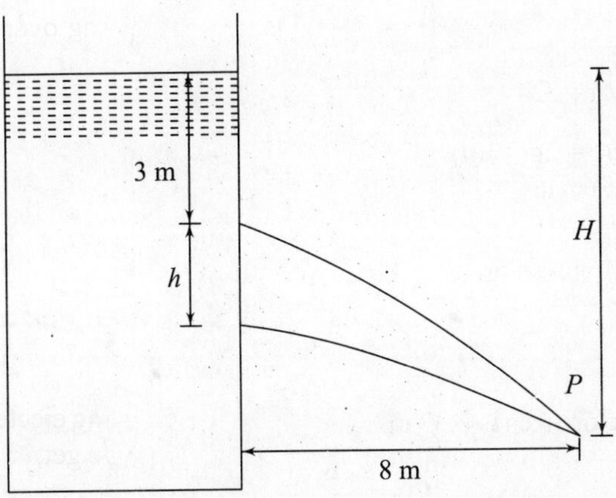

Fig. 4.31 Trajectory of water jets discharged from two orifices at the side of a tank

Solution

Let, P be the point of intersection of two jets as shown in Fig. 4.31. If t is the time taken for any liquid particle flowing in the jet from the upper orifice to reach the point P from the plane of the orifice, then,

$$8 = u_1 t \tag{4.100a}$$

and $$(H - 3) = \frac{1}{2} g t^2 \tag{4.100b}$$

where, u_1 is the velocity of discharge at the plane of the upper orifice and H is the vertical distance of P from the water level in the tank. Eliminating t from Eqs (4.100a) and (4.100b),

$$(H-3) = \frac{1}{2} g \frac{64}{u_1^2}$$

or
$$\frac{u_1^2}{g}(H-3) = 32 \tag{4.101}$$

again, applying the Bernoulli's equation between the top water level and the discharge plane of the upper orifice,

$$u_1^2 = 2g \times 3 = 6g$$

Substituting this value of u_1^2 in Eq. (4.101), we have

$$3(H-3) = 16$$

or
$$H = 8.33 \text{ m}$$

Similarly, for the jet from the lower orifice,

$$8 = u_2 t$$

and,
$$(H-3-h) = \frac{1}{2} g t^2$$

Eliminating t from the above two,

$$\frac{u_2^2}{g}(H-3-h) = 32$$

again,
$$u_2^2 = 2g(3+h)$$

Hence,
$$(3+h)(H-3-h) = 16$$

or
$$3(H-3) - 3h + h(H-3) - h^2 = 16$$

Substituting $H = 8.33$ in the above expression we get

$$h^2 - 2.33h = 0$$

or
$$h(h-2.33) = 0$$

which gives $h = 0$ and $h = 2.33$ m.

Therefore, the distance between the orifices is 2.33 m.

Example 4.20

A fireman must reach a window 40 m above the ground (Fig. 4.32) with a water jet, from a nozzle of 30 mm diameter discharging 30 kg/s. Assuming the nozzle discharge to be at a height of 2 m from the ground, determine the greatest distance from the building where the fireman can stand so that the jet can reach the window.

Solution

Let u_1 be the velocity of discharge from the nozzle,

then,
$$u_1 = \frac{\dot{m}}{\rho A} = \frac{30}{1000 \times (\pi/4) \times (0.03)^2} = 42.44 \text{ m/s}$$

(Note that $u_1^2/2g$ should be more than the height of the window for the jet to reach at all. In this case $u_1^2/2g = 91.80$ m which is greater than 40 m.)

Let α be the angle of the nozzle with horizontal. Considering the time taken by a fluid particle in the jet to reach the window at point 2 from the discharge point 1 as t, we can write

$$x = 42.44 \cos \alpha t \tag{4.102a}$$

and

$$38 = 42.44 \sin \alpha t - 1/2\, gt^2 \tag{4.102b}$$

where, x is the horizontal distance between the nozzle and window. Eliminating t from Eqs (4.102a)and (4.102b), we have

$$38 = x \tan \alpha - \frac{9.81}{2 \times (42.44)^2} \frac{x^2}{\cos^2 \alpha} \tag{4.103}$$

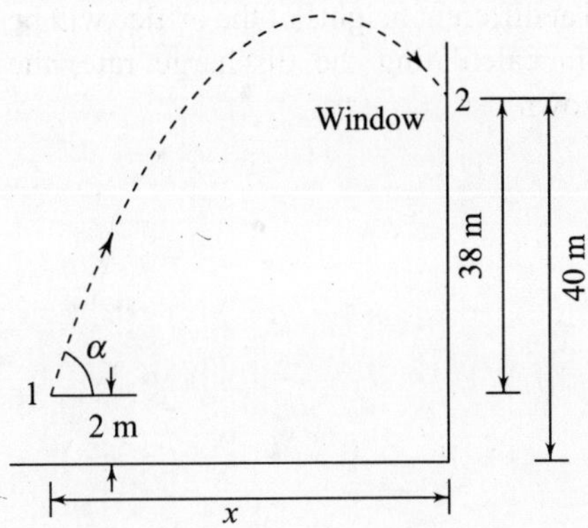

Fig. 4.32 Trajectory of a water jet issued from a nozzle to reach a window above the ground

The value of x depends upon the value of α. For maximum value of x we require an optimization with α.

Differentiating each term of Eq. (4.103) with respect to α, we get

$$0 = x \sec^2 \alpha + \tan \alpha \frac{dx}{d\alpha} - 0.0027\, x^2 (2 \sec^2 \alpha \tan \alpha)$$

$$- \frac{0.0027}{\cos^2 \alpha}\left(2x \frac{dx}{d\alpha}\right)$$

For maximum x, $\frac{dx}{d\alpha} = 0$

Hence, $x \sec^2 \alpha - 2 \times 0.0027\, x^2 \sec^2 \alpha \tan \alpha = 0$

$$x \tan \alpha = \frac{1}{2 \times 0.0027} = 185.2 \text{ m} \tag{4.104}$$

Solving for x and α from Eq. (4.103) and (4.104), we get

$$38 = 185.2 - 0.0027\, x^2/\cos^2 \alpha$$

which gives $x/\cos \alpha = 233.5$ m

Again, $$x \tan \alpha = 185.2 \text{ m}$$

$$\sin \alpha = \frac{185.2}{233.5} = 0.793$$

or $$\alpha = 52.5°$$

and $$x = 142 \text{ m}$$

4.8.3 Large Vertical Orifices

If a vertical orifice is so large that its height is comparable to the height of the liquid in the tank, then the variation in the liquid head at different heights of the orifice will be considerable. To take this into account in calculating the discharge rate, the geometrical shape of the orifice has to be known.

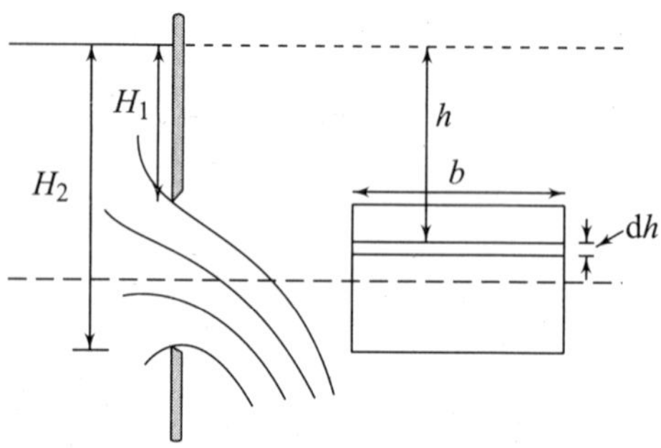

Fig. 4.33 Large Vertical orifice

Consider a large orifice with a rectangular cross section, as shown in Fig. 4.33. Let the heights of the liquid level be H_1 and H_2 above the top and the lower edges of the orifice respectively. Let b be the breadth of the orifice. Consider, at a depth h from the liquid level, a horizontal strip of the orifice of thickness dh. The velocity of liquid coming out from this strip will be equal to $\sqrt{2gH}$. Hence, the rate of discharge through the elemental strip

$$= C_d \times \text{area} \times \text{velocity} = C_d \times b\, \text{d}h \times \sqrt{2gh}$$

Therefore, the rate of discharge through the entire orifice

$$= \int_{H_1}^{H_2} C_d \times b\, \text{d}h \times \sqrt{2gh}$$

$$= C_d\, b\, \sqrt{2g} \int_{H_1}^{H_2} h^{1/2}\, dh$$

$$= \frac{2}{3} C_d b \sqrt{2g}\, (H_2^{3/2} - H_1^{3/2})$$

Here C_d is assumed to be constant throughout the orifice.

4.8.4 Drowned or Submerged Orifice

A drowned or submerged orifice is one which does not discharge into the open atmosphere, but discharges into a liquid of the same kind. The orifice illustrated in Fig. 4.34 is an example of a submerged orifice. It discharges liquid from one side of the tank to the other side, where, the heights of the liquid are maintained constant on both the sides. The formation of the vena contracta takes place but the pressure there corresponds to the head h_2 (Fig. 4.34). Application of Bernoulli's equation between points 1 and 2 on a streamline (Fig. 4.34) gives

$$p_1/\rho g + z_1 + V_1^2/2g = p_2/\rho g + 0 + V_2^2/2g$$

or

$$(h_1 - z_1) + z_1 + 0 = h_2 + V_2^2/2g$$

(since $V_1 << V_2$)

or

$$V_2 = \sqrt{2g(h_1 - h_2)} \tag{4.105}$$

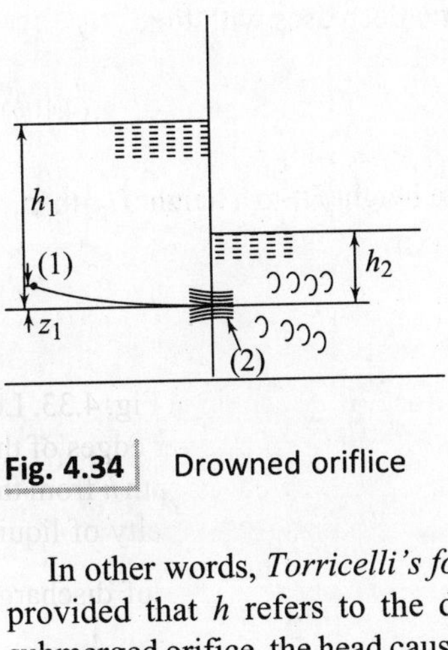

Fig. 4.34 Drowned oriflice

In other words, *Torricelli's formula* as expressed by Eq. (4.95) is still applicable provided that h refers to the difference of head across the orifice. For a large submerged orifice, the head causing flow through it at any height remains same as $(h_1 - h_2)$, and hence the Eq. (4.105) is valid. This is because the variations in head with the orifice height from both the sides cancel each other.

4.8.5 Time of Emptying Tank

Let us consider a tank of uniform cross-sectional area A (Fig. 4.35) to contain a liquid of height H_1 from the base. Let the liquid be discharged through an orifice of area a in the base of the tank and the height of the liquid in the tank fall accordingly. If at any instant, h is the height of the liquid level which falls by an amount dh during an infinitesimal time interval dt, then from continuity, (the volume displaced by liquid level in the tank equals to the volumetric flow through the orifice), we can write

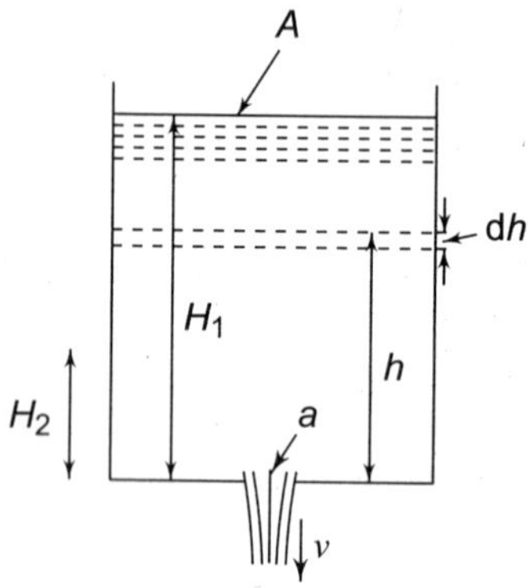

Fig. 4.35 Discharge from a tank of uniform area

$$-A\,\mathrm{d}h = C_d a\,\sqrt{2gh}\,\mathrm{d}t$$

The minus sign is introduced because the height h decreases with time.

Therefore, $$\mathrm{d}t = \frac{-A\,\mathrm{d}h}{C_d a\sqrt{2gh}} \qquad (4.106)$$

If T is the time taken for the liquid level to fall from a height H_1 to a height H_2, then,

$$\int_0^T \mathrm{d}t = \frac{-A}{C_d a\sqrt{2g}} \int_{H_1}^{H_2} h^{-1/2}\,\mathrm{d}h$$

From which,

$$T = \frac{2A}{C_d a\sqrt{2g}}\left(H_1^{1/2} - H_2^{1/2}\right) \qquad (4.107a)$$

If the tank is completely emptied, $H_2 = 0$.
Then,

$$T = \frac{2A}{C_d a\sqrt{2g}} H_1^{1/2} \qquad (4.107b)$$

4.8.5.1 *Time of Emptying Tank with Non-uniform Cross Section*

In the foregoing problem, we have considered the cross-sectional area A of the tank to be uniform and therefore while integrating the right-hand side of Eq. (4.106) with respect to h, the area A was considered to be constant. For a tank where A varies with height, a functional relationship between A and h has to be found out from the geometry of the tank so that the relationship can be introduced in the Eq. (4.106) for its integration to determine the time of emptying. An example to determine the time of emptying a hemispherical vessel is given below.

Let R be the radius of a hemispherical vessel as shown in Fig. 4.36, and let the liquid level fall from H_1 to H_2 in time T due to the discharge from an orifice at the bottom of the vessel. Consider the instant when the liquid level is at a height h from the bottom of the vessel, and the radius of the vessel's cross section at this level be x. If dh is the decrease in the liquid level in time dt, then from continuity,

$$-A_h\,\mathrm{d}h = C_d a\sqrt{2gh}\,\mathrm{d}t$$

or

$$\mathrm{d}t = \frac{-A_h\,\mathrm{d}h}{C_d a\sqrt{2gh}} \tag{4.108}$$

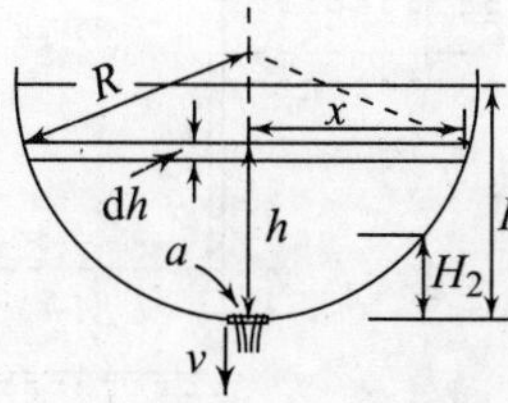

Fig. 4.36 Discharge from a hemispherical vessel

where, A_h is the cross-sectional area of the vessel at height h

Here, $A_h = \pi x^2$

From the geometry of the vessel,

$$x^2 = R^2 - (R-h)^2$$
$$= 2Rh - h^2$$

Therefore, Eq. (4.108) becomes

$$\mathrm{d}t = \frac{-\pi(2Rh - h^2)}{C_d a\sqrt{2gh}}\,\mathrm{d}h$$

or

$$\int_0^T \mathrm{d}t = \frac{-\pi}{C_d a\sqrt{2g}}\int_{H_1}^{H_2}(2Rh^{1/2} - h^{3/2})\,\mathrm{d}h$$

or

$$T = \frac{-2\pi}{C_d a\sqrt{2g}}\left[\frac{2}{3}R\left(H_1^{3/2} - H_2^{3/2}\right) - 1/5\left(H_1^{5/2} - H_2^{5/2}\right)\right] \tag{4.109a}$$

If the vessel is initially full and is completely emptied afterwards, then, $H_1 = R$ and $H_2 = 0$. Equation (4.109a) then becomes

$$T = \frac{2\pi}{C_d a\sqrt{2g}}\left(\frac{2}{3}R^{5/2} - \frac{1}{5}R^{5/2}\right)$$

$$= \frac{14\pi R^{5/2}}{15C_d a\sqrt{2g}} \tag{4.109b}$$

Here, T is the time of emptying the vessel.

4.8.6 Time of Flow from One Tank to Another

Let us consider a liquid flowing from a tank of area A_1 to another tank of area A_2 through an orifice between the tanks as shown in the Fig. 4.37. Under this situation, the liquid level falls in one tank while it rises in the other one. The orifice will be drowned. The head causing flow at any instant of time will be equal to the difference between the instantaneous liquid levels in the tanks. At a certain instant, let this difference in liquid levels between the tanks be h, and in time dt, the small quantity of fluid that passes through the orifice causes the liquid level in tank A_1 to fall by an amount dH. The liquid level in tank A_2 will then rise by an amount dH (A_1/A_2).

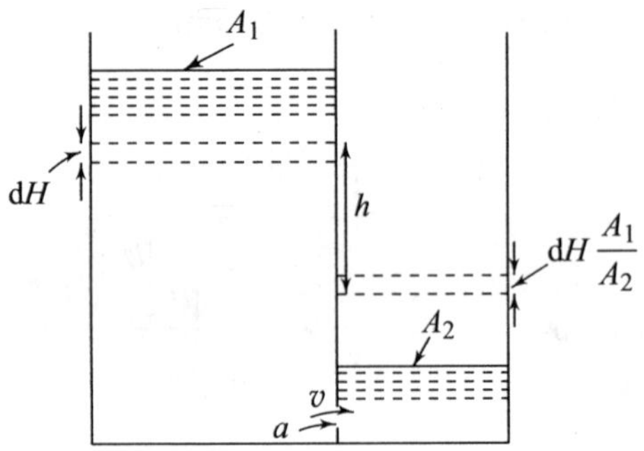

Fig. 4.37 Flow of liquid from one tank to another

Hence, the difference in levels after a time dt becomes

$$h - \mathrm{d}H\,(1 + A_1/A_2)$$

Therefore, the change in head causing flow

$$\mathrm{d}h = h - [h - \mathrm{d}H\,(1 + A_1/A_2)]$$
$$= \mathrm{d}H\,(1 + A_1/A_2) \tag{4.110}$$

From principle of continuity in tank A_1, we can write

$$-A_1 \mathrm{d}H = C_d\, a\,\sqrt{2gh}\;\mathrm{d}t$$

Therefore,

$$\mathrm{d}t = \frac{-A_1\,\mathrm{d}H}{C_d a\sqrt{2gh}} \tag{4.111}$$

Substituting for dH from Eq. (4.110) in Eq. (4.111), we have

$$\mathrm{d}t = \frac{-A_1\,\mathrm{d}h}{C_d a\,(1 + A_1/A_2)\sqrt{2gh}} \tag{4.112}$$

If T is the time taken to bring the difference in levels between the tanks from H_1 to H_2, then,

$$\int_0^T \mathrm{d}t = \frac{-A_1}{C_d a\,(1 + A_1/A_2)\sqrt{2g}} \int_{H_1}^{H_2} h^{-1/2}\,\mathrm{d}h$$

or

$$T = \frac{2A_1\,(H_1^{1/2} - H_2^{1/2})}{C_d a\,(1 + A_1/A_2)\sqrt{2g}} \tag{4.113}$$

The flow of liquid from one tank to the other will stop automatically when the head causing the flow, i.e., the difference in liquid levels between the tanks will become zero. If T_1 is the time taken to make this equalisation of the liquid levels, then from Eq. (4.114),

$$T_1 = \frac{2A_1 H_1^{1/2}}{C_d a(1 + A_1/A_2)\sqrt{2g}} \tag{4.114}$$

where H_1 represents the initial value of the difference in the liquid levels.

Example 4.21

A tank has the form of a frustum of a cone, with a diameter of 2.44 m at the top and 1.22 m at the bottom as shown in Fig. 4.38. The bottom contains a circular orifice whose coefficient of discharge is 0.60. What diameter of the orifice will empty the tank in 6 minutes if the full depth is 3.05 m?

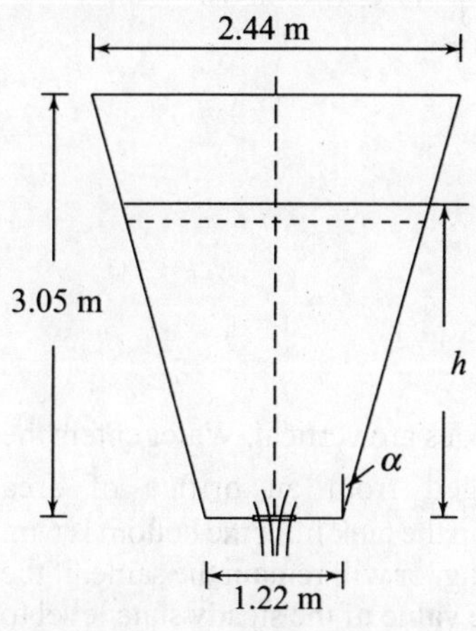

Fig. 4.38 A tank in the form of a frustum of a cone

Solution

Let the diameter of the orifice be d_0, and at any instant t, the height of the liquid level above the orifice be h. Then during an infinitesimal time dt, discharge through the orifice is

$$q = C_d \frac{\pi d_0^2}{4}\sqrt{2gh}\ \mathrm{d}t$$

$$= 0.60 \times \frac{1}{4}\pi d_0^2 \sqrt{2gh}\ \mathrm{d}t$$

If the liquid level in the tank falls by an amount dh during this time, then from continuity,

$$-A_h \, dh = 0.60 \frac{\pi d_0^2}{4} \sqrt{2gh} \, dt \tag{4.115}$$

where A_h is the area of the tank at height h.

From the geometry of the tank (Fig. 4.38),

$$\tan \alpha = \frac{(2.44 - 1.22)}{2 \times 3.05} = 0.2$$

Therefore the diameter of the tank at height $h = 1.22 + 2 \times 0.2\, h$

Hence, $$A_h = (\pi/4)(1.22 + 0.4h)^2$$

Substituting the value of A_h in Eq. (4.115), we have

$$0.60 \times (1/4)\, \pi d_0^2 \sqrt{2 \times 9.81 h}\, dt = -\pi/4\, (1.22 + 0.4h)^2\, dh$$

or $$d_0^2 \int dt = \frac{1}{0.60 \times \sqrt{2 \times 9.81}} \int_{3.05}^{0} (1.22 + 0.4\, h)^2\, h^{-1/2} dh$$

Since, the time of emptying $\int dt = 360$ seconds

$$d_0^2 = \frac{1}{0.60 \times \sqrt{2 \times 9.81} \times 360} \int_{3.05}^{0} (1.22 + 0.4h^2)\, h^{-1/2}\, dh$$

Integrating and solving for d_0, we get

$$d_0^2 = 0.010 \text{ m}^2$$

or $$d_0 = 0.1 \text{ m} = 100 \text{ mm}$$

Example 4.22

A concrete tank is 10 m long and 6 m wide, and its sides are vertical. Water enters the tank at the rate of 0.1 m^3/s and is discharged from an orifice of area 0.05 m^2 at its bottom (Fig. 4.39). Initial level of water in the tank from the bottom is 5 m. Find whether the liquid level will start rising, or falling, or will remain the same, if the liquid level changes (either rise or fall), then find the value of the steady state level to which the liquid will reach. Find also the time taken for the change in the liquid level to be 60% of its total change (Take $C_d = 0.60$).

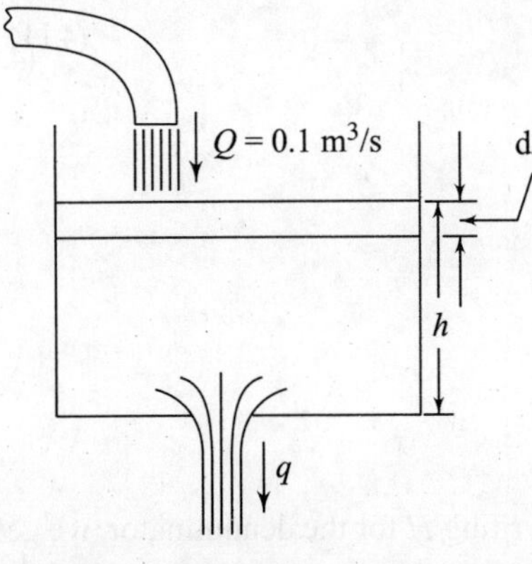

Fig. 4.39 A tank, with an inflow, discharging water from an orifice at the bottom

Solution

The rise or fall of liquid level at any instant will depend upon the relative magnitudes of instantaneous rate of inflow to and outflow from the tank. Here, the rate of inflow Q is constant and equals to 0.1 m^3/s. Initially, the discharge rate from the orifice

$$q = 0.6 \times 0.05\,\sqrt{2 \times 9.81 \times 5}$$
$$= 0.297\ \text{m}^3/\text{s}$$

Since $q > Q$, the liquid level will start falling. As the liquid level falls, the discharge rate through the orifice decreases, and when it equals to the rate of inflow, the liquid level will neither rise nor fall further. Let H_s be this height of steady liquid level from the bottom of the tank.

Then $$0.6 \times 0.05 \times \sqrt{2 \times 9.81 \times H_s} = 0.1$$

or $$H_s = \frac{(0.1)^2}{2 \times 9.81 \times (0.6)^2 \times (0.05)^2} = 0.57\ \text{m}$$

Total change in the liquid level = (5 – 0.57) = 4.43 m. The liquid will attain a level of 2.34 m (= 5 – 0.6 × 4.43) when the change in the level will be 60% of its final value 4.43.

Consider at any instant t, the height of liquid level in the tank to be h, and let this height fall by dh in a small interval of time dt.

The amount of inflow during this time = $Q\,dt = 0.1\,dt$

and the amount of discharge $$= 0.6 \times 0.05\,\sqrt{2gh}\;dt$$

$$= 0.133\,\sqrt{h}\;dt$$

From continuity,

$$-A\,dh = 0.133\,\sqrt{h}\;dt - 0.1\,dt$$

where A is the area of the tank $= 6 \times 10 = 60\ \text{m}^2$

Hence,
$$dt = \frac{60\,dh}{(0.1 - 0.133\sqrt{h})} \tag{4.116}$$

Let
$$H = 0.1 - 0.133\sqrt{h}$$

Then,
$$h = \frac{(0.1 - H)^2}{0.0177}$$

$$dh = -\frac{2(0.1 - H)}{0.0177}\,dH$$

Substituting the value of dh in Eq. (4.116) and writing H for the denominator, we get

$$dt = \frac{-120(0.1 - H)\,dH}{0.0177\,H}$$

$$= -6780\left(\frac{0.1}{H} - 1\right)dH$$

If T is the time taken for the liquid level to fall from 5 m to 2.34 m, then

$$\int_0^T dt = -6780\,[0.1 \ln(0.1 - 0.133\sqrt{h}) - (0.1 - 0.133\sqrt{h})]_5^{2.34}$$

$$= -6780\left[0.1 \ln \frac{0.1 - 0.133\sqrt{2.34}}{0.1 - 0.133\sqrt{5}} + 0.133(\sqrt{2.34} - \sqrt{5})\right]$$

$$= 5080 \text{ s}$$
$$= 1.41 \text{ h}$$

4.8.7 External Mouthpieces

The discharge through an orifice may be increased by fitting a short length of pipe to the outside. This is because, the vena contracta gets the opportunity to expand and fill the pipe. Therefore, the coefficient of contraction becomes unity. The increase in the value of C_c thus increases the discharge rate despite a little decrease in the value of C_v due to frictional losses in the pipe.

Consider the tank in Fig. 4.40. A cylindrical piece of pipe is attached to the orifice towards the outside of the tank. This pipe is known as cylindrical mouthpiece.

Let a = Cross-sectional area of the mouthpiece

a_c = Cross-sectional area of flow at vena contracta

V = Velocity at outlet of pipe

V_c = Velocity at vena contracta

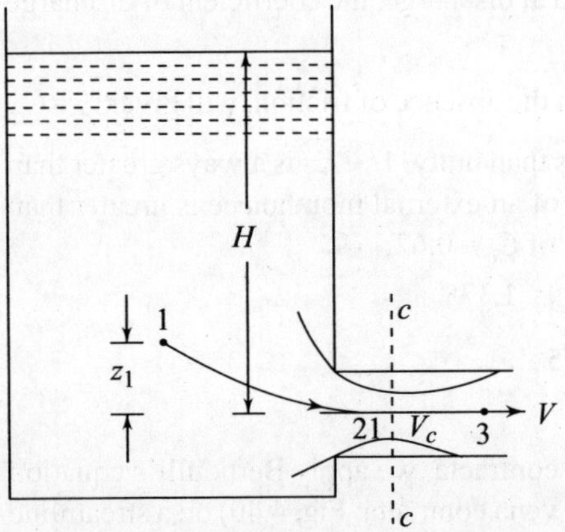

Fig. 4.40 Flow through an external mouthpiece

Applying Bernoulli's equation between the points 1 and 3 (Point 3 being at the plane of discharge) on a streamline, we get

$$p_{atm}/\rho g + (H - z_1) + 0 + z_1 = p_{atm}/\rho g + V^2/2g + 0 + h_L \tag{4.117}$$

(where p_{atm} is the atmospheric pressure. Velocity in the tank is considered to be negligible as compared to that in the pipe) h_L is the loss of head. If friction is neglected because of the short length of the pipe, h_L represents only the loss of head due to contraction. Hence,

$$h_L = \frac{(V_c - V)^2}{2g}$$

Again, from continuity,

$$a_c V_c = aV$$

or

$$V = C_c V_c$$

where, C_c (the coefficient of contraction) = a_c/a

Therefore, $$h_L = \frac{V^2}{2g}\,(1/C_c - 1)^2$$

Hence Eq. (4.117) becomes,

$$H = \frac{V^2}{2g}\,[1 + (1/C_c - 1)^2] = K\frac{V^2}{2g}$$

where $$K = [1 + (1/C_c - 1)^2]$$

The coefficient of velocity C_v can be written as

$$C_v = \frac{V}{\sqrt{2gH}} = \frac{\sqrt{2gH/K}}{\sqrt{2gH}}$$

or

$$C_v = \sqrt{1/K}$$

Since there is no contraction of flow area at discharge, the coefficient of discharge $C_d = C_v = 1/\sqrt{K}$.

On the other hand, an orifice of area a, in the absence of friction, will give $C_d = C_c$. It can be proved that for all values of C_c less than unity, $1/\sqrt{K}$ is always greater than C_c, and hence, the coefficient of discharge of an external mouthpiece is greater than that of an orifice. Assuming a typical value of $C_c = 0.62$.

$$K = [1 + (1/0.62 - 1)^2] = 1.375$$

Hence $\quad (C_d)_{\text{mouthpiece}} = 1/\sqrt{1.375} = 0.855$

While $\quad (C_d)_{\text{orifice}} = C_c = 0.62$

In order to find the pressure at the vena contracta, we apply Bernoulli's equation between points 1 and 2 (Point 2 being at the vena contracta; Fig. 4.40) on a streamline as

$$p_{\text{atm}}/\rho g + (H - z_1) + 0 + z_1 = p_c/\rho g + V_c^2/2g + 0 \tag{4.118}$$

or
$$p_c/\rho g = p_{\text{atm}}/\rho g + H - V_c^2/2g$$

but
$$H = KV^2/2g$$

and
$$V_c = V/C_c$$

Therefore,
$$p_c/\rho g = p_{\text{atm}}/\rho g + KV^2/2g - (1/C_c^2)V^2/2g$$
$$= p_{\text{atm}}/\rho g - (1/C_c^2 - K)\, V^2/2g$$

It was stated earlier that $1/C_c^2$ is always greater than K for all values of $C_c < 1$. Hence, the pressure at vena contracta is always lower than the atmospheric pressure. When $C_c = 0.62$,

$$K = (1/0.62 - 1)^2 + 1 = 1.375$$

Then,
$$p_c/\rho g = p_{\text{atm}}/\rho g - [1/(0.62)^2 - 1.375]\, V^2/2g$$
$$= p_{\text{atm}}/\rho g - 1.225\, V^2/2g$$

Therefore, the influence of the mouthpiece on the rate of discharge can also be looked at from an angle of decrease in pressure at the vena contracta in increasing the effective head causing flow.

4.8.7.1 *Convergent Divergent Mouthpiece*

The losses due to contraction in a mouthpiece may be considerably reduced if the mouthpiece is convergent up to the vena contracta and becomes divergent afterwards. In this case, the geometry of the mouthpiece is made almost to the shape of the jet. If frictional losses are neglected, the coefficient of discharge for this type of mouthpiece becomes unity. Such a mouthpiece is shown in Fig. 4.41. Applying Bernoulli's equation between points 1 and 2 (Point 2 being at the vena contracta cc; Fig. 4.41), we have

$$p_{\text{atm}}/\rho g + (H - z_1) + 0 + z_1$$
$$= p_c/\rho g + V_c^2/2g + 0$$

Hence,
$$V_c^2/2g = p_{\text{atm}}/\rho g + H - p_c/\rho g \tag{4.119}$$

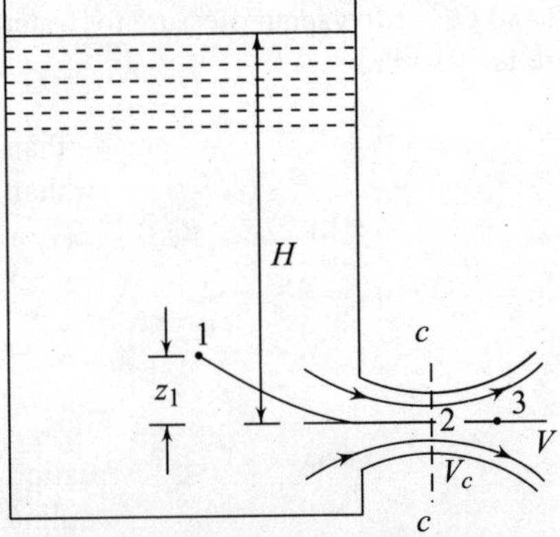

Fig. 4.41 Flow through a convergent divergent mouthpiece

Again, application of Bernoulli's equation between points 1 and 3 (Point 3 being on the plane of discharge, Fig. 4.41) gives,

$$p_{atm}/\rho g + (H - z_1) + 0 + z_1 = p_{atm}/\rho g + V^2/2g + 0$$

or
$$V^2/2g = H \tag{4.120}$$

The loss of head due to contraction does not take place under this situation. From Eqs (4.119) and (4.120),

$$\frac{V_c}{V} = \sqrt{1 + \frac{H_a - H_c}{H}} \tag{4.121}$$

where, $H_a = p_{atm}/\rho g$ (the atmospheric pressure head)

$H_c = p_c/\rho g$ (the pressure head at vena contracta cc)

From continuity, $V_c a_c = Va$

Therefore, Eq. (4.121) becomes

$$\frac{a}{a_c} = \sqrt{1 + \frac{H_a - H_c}{H}} \tag{4.122}$$

The maximum ratio of a and a_c to avoid separation is given by

$$(a/a_c)_{max} = \sqrt{1 + \frac{H_a - H_{c\ minimum}}{H}} \tag{4.123}$$

where, $H_{c\ minimum}$ is the minimum head at cc to avoid cavitation.

Example 4.23

For the 100 mm diameter short tube acting as a mouthpiece in a tanks as shown in Fig. 4.42, (i) what flow of water at 24 °C will occur under a head of 9.2 m? (ii) What is the pressure head at the vena contracta section *c*? (iii) What maximum head can be used if the tube is to flow full at exit? (Take $C_d = 0.82$ and $C_v = 1.0$; vapour pressure for water at 24 °C is 3 kPa absolute, atmospheric pressure is 101 kPa).

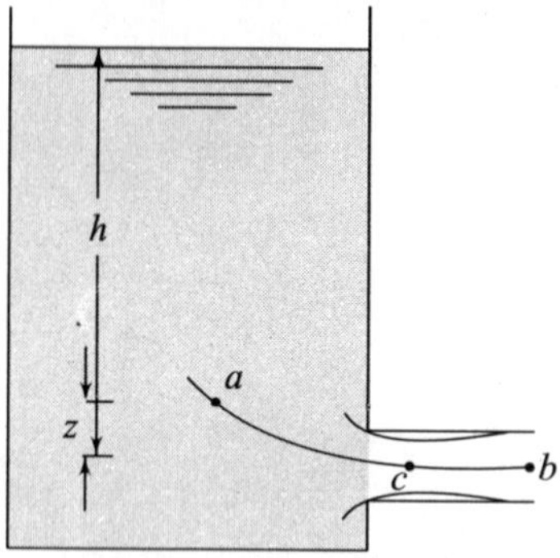

Fig. 4.42 A tank with a short tube as a mouthpiece

Solution

(i) Applying Bernoulli's equation between the points *a* and *b* (the point *b* being at the exit plane) on a streamline (Fig. 4.42), with the horizontal plane through *b* as datum, we can write

$$\frac{p_{\text{atm}}}{\rho g} + (h - z) + 0 + z = \frac{p_{\text{atm}}}{\rho g} + \frac{V_b^2}{2g} + 0 + \frac{V_b^2}{2g}\left(\frac{1}{C_c} - 1\right)^2 \quad (4.124)$$

where C_c is the coefficient of contraction ($= C_d/C_v = 0.82/1 = 0.82$). With the values given,

$$9.2 = \frac{V_b^2}{2g}\left[1 + \left(\frac{1}{0.82} - 1\right)^2\right]$$

which gives $V_b =$ 13.12 m/s

Then $Q = C_d \times A \times V_b = 0.82 \times (\pi/4)\,(0.1)^2 \times (13.12) = 0.084\ \text{m}^3/\text{s}$

(ii) Applying Bernoulli's equation between the points *a* and *c* on a streamline (the point *c* being at the vena contracta section), we get

$$\frac{p_{\text{atm}}}{\rho g} + (h - z) + 0 + z = \frac{p_c}{\rho g} + \frac{V_c^2}{2g} + 0 \quad (4.125)$$

Again from continuity,

$$A_b \times V_b = A_c \times V_c$$

where A_b and A_c are the areas at the exit and the vena contracta, respectively

Hence, $$V_c = \frac{V_b}{A_c / A_b} = \frac{V_b}{C_c} = \frac{13.12}{0.82} = 16 \text{ m/s}$$

Substituting V_c in Eq. (4.125), we have

$$\frac{p_{\text{atm}}}{\rho g} + 9.2 = \frac{p_c}{\rho g} + \frac{(16)^2}{2g}$$

which gives $$P_c/rg = \left(\frac{p_{\text{atm}}}{\rho g} - 3.85\right) \text{ m of water}$$

Therefore the pressure at the vena contracta = 3.85 m of water vacuum.

(iii) As the head causing flow through the short tube is increased, the velocity of flow at any section will increase and the pressure head at c will be reduced. For a steady flow with the tube full at exit, the pressure at c must not be less than the vapour pressure of the liquid at the working temperature. For any head h (the height of the liquid level in the tank above the centre line of the tube), we get from Eq. (4.125)

$$\frac{p_{\text{atm}}}{\rho g} + h = \frac{p_c}{\rho g} + \frac{V_c^2}{2g} \tag{4.126}$$

Again $$V_c = A_b V_b / A_c = V_b / C_c$$

Again, from Eq. (4.124),

$$V_b^2 = \frac{2gh}{\left(\frac{1}{C_c} - 1\right) + 1}$$

Therefore, $$\frac{V_c^2}{2g} = \frac{h}{C_c^2\left[\left(\frac{1}{C_c} - 1\right)^2 + 1\right]}$$

$$= \frac{h}{(0.82)^2\left[\left(\frac{1}{0.82} - 1\right)^2 + 1\right]} = 1.42\,h$$

Substituting this value of V_c in Eq. (4.126), we get

$$\frac{p_{\text{atm}}}{\rho g} + h = \frac{p_c}{\rho g} + 1.42\,h$$

or $$0.42\,h = \frac{p_{atm}}{\rho g} - \frac{p_c}{\rho g}$$

For the maximum head h, $p_c = p_v$, the vapour pressure of water at the working temperature.

For the present case, $$\frac{p_v}{\rho g} = \frac{3 \times 10^3}{10^3 \times 9.81} = 0.306 \text{ m}$$

While $$\frac{p_{atm}}{\rho g} = \frac{101 \times 10^3}{10^3 \times 9.81} = 10.296 \text{ m}$$

Hence, $$h_{max} = \frac{10.296 - 0.306}{0.42} = 23.78 \text{ m}$$

Example 4.24

An external mouthpiece converges from the inlet up to the vena contracta to the shape of the jet and then it diverges gradually. The diameter at the vena contracta is 20 mm and the total head over the centre of the mouthpiece is 1.44 m of water above the atmospheric pressure. The head loss in flow through the converging passage and through the diverging passage may be taken as one per cent and five per cent, respectively, of the total head at the inlet to the mouthpiece. What is the maximum discharge that can be drawn through the outlet and what should be the corresponding diameter at the outlet. Assume that the pressure in the system may be permitted to fall to 8 m of water below the atmospheric pressure head, and the liquid conveyed is water.

Solution

In terms of meters of water, total head available at the inlet to the mouthpiece $h_1 = 1.44$ m above the atmospheric pressure.

Loss of head in the converging passage $= 0.01 \times 1.44$

$$= 0.0144 \text{ m}$$

Loss of head in the divergent part $= 0.05 \times 1.44$

$$= 0.0720 \text{ m}$$

Total head available at the vena contracta

$= 1.44 - 0.0144 = 1.4256$ m above the atmospheric pressure

At the vena contracta, we can write

$$\frac{p_c}{\rho g} + \frac{V_c^2}{2g} = 1.4256 + \frac{p_{atm}}{\rho g}$$

For a maximum velocity V_c, the pressure p_c will attain its lower limit which is 8 m below the atmospheric pressure. Therefore,

$$\frac{p_{atm}}{\rho g} - 8 + \frac{V_c^2}{2g} = 1.4256 + \frac{p_{atm}}{\rho g}$$

which gives $V_c = 13.6$ m/s

Therefore, the maximum possible discharge becomes

$$Q_{max} = 13.6 \,' p(0.02)^2/4 = 0.0043 \text{ m}^3\text{/s}$$

Pressure at the exit is atmospheric. Application of Bernoulli's equation between the vena contracta section and the exit section gives

$$\frac{p_{atm}}{\rho g} + 1.4256 = \frac{p_{atm}}{\rho g} + \frac{V_2^2}{2g} + 0.0720$$

Hence, V_2, the exit velocity = 5.15 m/s

Therefore, the diameter d_2 at the exit is given by

$$(\pi d_2^2/4) \times 5.15 = 0.0043$$

or

$$d_2 = 0.0326 \text{ m} = 32.60 \text{ mm}$$

SUMMARY

- Euler's equation of motion describes the dynamics of inviscid flows. Bernouilli's equation explicates the fact that the sum total of flow energy, kinetic energy and potential energy transmitted in a steady, constant density, inviscid flow field remains conserved between points 1 and 2. In addition, either of the following conditions needs to be satisfied for the Bernoulli's equation to remain applicable (i) points 1 and 2 are located on the same streamline, **or** (ii) points 1 and 2 may be located anywhere in the flow field (not necessarily on the same streamline) provided the flow is irrotational ($\vec{\xi} = \nabla \times \vec{V} = 0$), **or** ($\vec{V} \times \vec{\xi}$) is perpendicular to the line element joining points 1 and 2.
- Flows having only tangential velocities with streamlines as concentric circles are known as plane circular vortex flows. A free vortex flow is an irrotational vortex flow where the total mechanical energy of the fluid elements remains same in the entire flow field and the tangential velocity is inversely proportional to the radius of curvature. A forced vortex flow is a rotational vortex flow where the tangential velocity is directly proportional to the radius of curvature. Pressure in vortex flows increases with an increase in radius of curvature. Spiral vortex flows are obtained as a result of superimposition of a plane circular vortex flow with a purely radial flow.
- The flow through a siphon takes place because of a difference in potential head between the entrance and exit of the tube. The maximum height of a siphon tube above the liquid level at atmospheric pressure is limited by the minimum pressure inside the tube which is never allowed to fall below the

vapour pressure of the working liquid at the existing temperature, to avoid vapour locking in the flow.

- Venturimeter, orificemeter and flow nozzle are typical flowmeters which measure the rate of flow of a fluid through a pipe by providing a coaxial area contraction within the pipe, thus creating a pressure drop across the contraction. The flow rate is measured by determining the velocity of flow at the constricted section in terms of the pressure drop by the application of Bernoulli's equation. The pressure drop is recorded experimentally. A venturimeter is a short pipe consisting of two conical parts with a short uniform cross section, in between, known as throat. An orificementer is a thin circular plate with a sharp-edged concentric circular hole in it. A flow nozzle is a short conical tube providing only a convergent passage to the flow. In a comparison between the three flowmeters, a venturimeter is the most accurate but the most expensive, while the orificemeter is the least expensive but the least accurate. The flow nozzle falls in between these two.
- The static pressure in a fluid is the thermodynamic pressure defining the state of fluid and becomes equal to the negative of the arithmetic average of the normal stresses at a point (mechanical pressure) in case of a real and Stokesian fluid. The stagnation pressure at a point in a fluid flow is the pressure which could result if the fluid were brought to rest isentropically. The difference between the stagnation and static pressure is the pressure equivalence of the velocity head ($\frac{1}{2}\rho V^2$) and is known as *dynamic pressure.* An instrument which contains tubes to record the stagnation and static pressures in a flow to finally determine the flow velocity and flow rate is known as a *Pitot static tube*.
- An orifice is a small aperture through which a fluid passes. Liquid contained in a tank is usually discharged through a small orifice at its side. A drowned or submerged orifice is one which does not discharge into the open atmosphere, but discharges into a liquid of the same kind. The discharge through an orifice is increased by fitting a short length of pipe to the outside known as the external mouthpiece. The discharge rate is increased due to a decrease in the pressure at the vena contracta within the mouthpiece, resulting in an increase in the effective head causing the flow.

EXERCISES

4.1 Choose the correct answer:

(i) Euler's equation of motion

(a) is a statement of energy balance
(b) is a preliminary step to derive Bernoulli's equation
(c) statement of conservation of momentum for a real fluid
(d) statement of conservation of momentum for an incompressible flow
(e) statement of conservation of momentum for the flow of an inviscid

(ii) From the following statements, choose the correct one related to Bernoulli's equation

$$V^2/2 + p/\rho + gz = \text{constant}$$

(a) The equation is valid for the steady flow of an incompressible ideal or real fluid along a stream tube.
(b) The energy equation for the flow of a frictionless fluid of constant density along a streamline with gravity as the only body force.
(c) The equation is derived from dynamic consideration involving gravity, viscous and inertia forces.
(d) The constant in the equation varies across streamlines if the flow is irrotational.

(iii) When is Bernoulli's equation applicable between any two points in a flow field?
(a) The flow is steady, constant density and rotational
(b) The flow is steady, variable density and irrotational
(c) The flow is unsteady, constant density and irrotational
(d) The flow is steady, constant density and irrotational

(iii) Stagnation pressure at a point in the fluid flow is the pressure
(a) which could result if the fluid were brought to rest isentropically
(b) which could result if the fluid were brought to rest isothermally
(c) at stagnation point
(d) None of these

4.2 The velocity components in an inviscid, constant density ($=1000$ kg/m^3), steady flow field are given as follows: $u = \frac{A}{2}(x+y+z)$, $v = \frac{A}{2}(x+y+z)$, $w = -A(x+y+z)$, where A is a dimensional constant, with a numerical value of 1 unit. Consider a directed line segment in the flow field, connecting the points $P_1(0,0,0)$ and $P_2(-3,3,0)$. The pressure is given as zero gauge at the origin. (i) Can Bernoulli's equation be applied to find the change in pressure experienced on moving from the point P_1 to the point P_2 along the direction P_1P_2? (ii) Starting from Euler's equation of motion in differential form, find the pressure at P_2. What is the stagnation pressure at the same point?

Ans.((i) No (ii) Zero (gauge), Zero (gauge)

4.3 Water comes out of a tank through a nozzle, as an open jet. As a result, the level of water in the tank continuously falls. A streamline in the tank conceptually identifies in the Fig. 4.43 (Spanning from Point 1 to Point 2), the curvilinear length of which (Spanning from Point 1 to the Point 2) is approximately kh, where $k = 1.5$. For mathematical analysis, the following assumptions can be made: (i) velocity of flow along the streamline is approximately V_1, and (ii) viscous effects are negligible. The ratio of area of cross section of the tank to that of the nozzle is 2:1. At a given instant of time, $h = 5$m and $V_2 = 1$m/s, what is the local component of acceleration of flow at Point 2, at that instant?

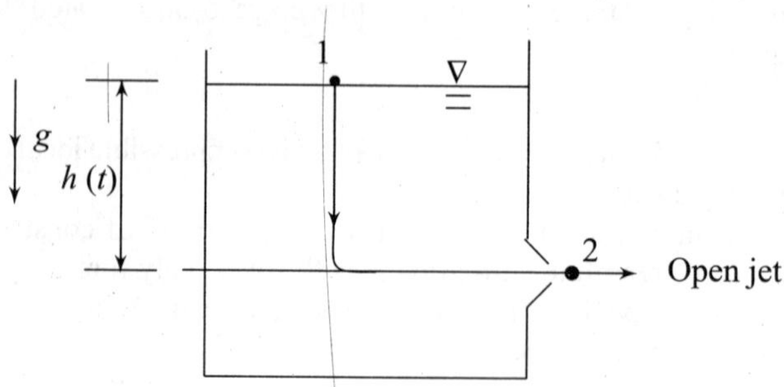

Fig. 4.43

Ans. (6.47 m/s^2)

4.4 A 0.3 m diameter pipe contains a short section in which the diameter is gradually reduced to 0.15 m and then enlarged again to 0.3 m. The 0.15 m section is 0.6 m below section A in the 0.3 m pipe where the pressure is 517 kN/m^2. If a differential manometer containing mercury is attached to the 0.3 m and 0.15 m section, what is the deflection of the manometer when the flow of water is 0.12 m^3/s downward? Assume the flow to be inviscid.

Ans. (175 mm)

4.5 At point A in a pipeline carrying water, the diameter is 1 m, the pressure is 98 kPa and the velocity is 1 m/s. At point B, 2 m higher than A, the diameter is 0.5 m and the pressure is 20 kPa. Determine the direction of flow.

Ans. (From A to B)

4.6 A tornado may be modeled as a combination of vortices with $v_r = v_z = 0$ and $v_\theta = v_\theta(r)$, such that $v_\theta = \omega r \qquad r \le R$

$$v_\theta = \frac{\omega R^2}{r} \qquad r \ge R$$

Determine whether the flow pattern is irrotational in either the inner or outer region. Using the r-momentum equation, determine the pressure distribution $p(r)$ in the tornado. Assume $p = p_\infty$ at $r \to \infty$. Find the location and magnitude of the lowest pressure.

Ans. (Rotational for $r \le R$, irrotational for $r \ge R$

$$p = p_\infty - \rho\omega^2 R^2\left(1 - \frac{r^2}{2R^2}\right) \qquad r \le R$$

$$p = p_\infty - \rho\omega^2 R^2\left(\frac{R}{r}\right)^2 \qquad r \ge R$$

$p_{\min}$ is at $r = 0$, $p = p_\infty - \rho\omega^2 R^2$)

4.7 A horizontal cylinder of internal diameter 100 mm is filled with water and rotated about its axis with an angular velocity of 3000 rpm. Calculate the pressure at the ends of the horizontal and vertical diameter.

Ans. (Ends of horizontal diameter: 124.1 kN/m^2; Vertical diameter: Top end: 123.6 kN/m^2; Bottom end: 124.6 kN/m^2)

4.8 An open cylindrical tank 2m high and 1 m in diameter, is filled with water to a depth of 1m. If the cylinder rotates about its vertical axis. Determine (a) the maximum angular velocity of the cylinder without spilling any water (b) the depth at the center and gauge pressure at the bottom 0·2 m from the center at the condition of maximum angular velocity

Ans. ((a) 6.26 rad/s, (b) zero, 783.75 Pa)

4.9 A hollow cone filled with a liquid, with its apex downwards, has a base diameter d and a vertical height h. At what speed should it spin about its vertical axis so that the kinetic energy of the rotating liquid is maximum? What per cent of the total volume of the cone is then occupied by the liquid?

Ans. ($\omega = \sqrt{12gh/5d^2}$, 55%)

4.10 A vessel with a fluid moves vertically upward with an acceleration of $g/2$, and simultaneously rotates about the vertical axis of symmetry with an angular velocity of ω. Derive an equation for the free surface of the liquid in a Cartesian coordinate system.

Ans. (($z = \omega^2(x^2 + y^2)/3g$))

4.11 A paddle wheel of 100 mm diameter rotates at 150 rpm inside a closed concentric vertical cylinder of 300 mm diameter completely filled with water. (i) Assuming a two-dimensional flow in a horizontal plane, find the difference in pressure between the cylinder surface and the centre of the wheel. (ii) If provision is made for an outward radial flow which has a velocity of 1m/s at the periphery of the wheel, what is the resultant velocity at a radius of 100 mm and its inclination to the radial direction?

Ans. ((i) 0.582 kN/m^2 (ii) 0.636 m/s, 38.13°)

4.12 The velocity of water at the outer edge of a whirlpool where the water level is horizontal and in the same plane as the bulk of the liquid, is 2 m/s and the diameter is 500 mm. Calculate the depth of free surface at a diameter of 100 mm from the eye of the whirlpool.

Ans. (4.89 m)

4.13 In a flapper valve, air enters at the centre of the lower disk through a 10 mm pipe with a velocity of 10 m/s. It then moves radially to the outer circumference. The two disks forming the valve are of 150 mm diameter and 5 mm apart. The air pressure at inlet is 1.5 kN/m^2 gauge. Assuming the air density to be constant at 1.2 kg/m^3, estimate the net force acting on the upper plate.

Ans. (27.443 N)

4.14 Water flows upward through a vertical 300 mm × 150 mm venturimeter whose coefficient is 0.98. The deflection of a differential gauge is 1.18 m of liquid of specific gravity 1.25, as shown in Fig. 4.44. Determine the flow rate in m^3/s.

Ans. (0.044 m^3/s)

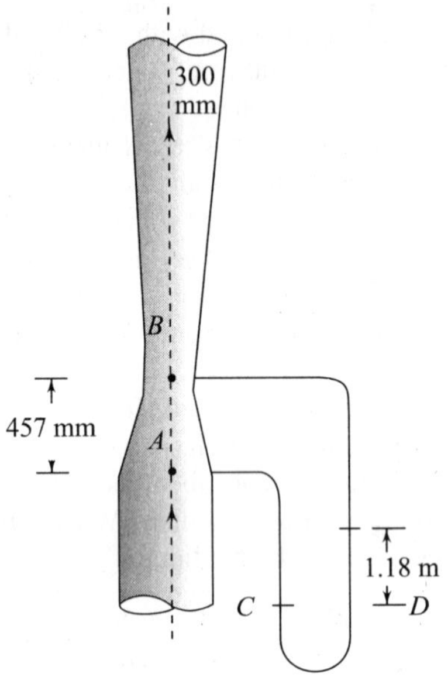

Fig. 4.44 A vertical venturimeter

4.15 A vertical venturimeter carries a liquid of specific gravity 0.8 and has an inlet and throat diameter of 150 mm and 75 mm, respectively. The pressure connection at the throat is 150 mm above that at the inlet. If the actual rate of flow is 40 liters/s and the coefficient of discharge is 0.96, calculate (i) the pressure difference between inlet and throat, and (ii) the difference in levels of mercury in a vertical U-tube manometer connected between these points.

Ans. ((i) 34.53 kN/m^2, (ii) 0.275 m)

4.16 A venturimeter is placed at 30° to the horizontal (sloping upwards in the direction of flow) to a pipe line carrying an oil of specific gravity 0·8. A differential manometer with mercury as the manometric fluid is attached to the inlet and throat of theventurimeter. The manometer shows a deflection (the difference in height between the menisci of mercury at the two limbs) of 100 mm. The pipe diameter is 200 mm, while the diameter of venturi throat is 100 mm. (a) Find the volume flow rate of oil if the coefficient of discharge of the venturimeter is 0·96. (b) What will be the reading of differential manometer if the venturimeter is turned horizontal? The length of venturimeter between the inlet and the throat is 320 mm.

Ans. ((a) 0.044 m^3 /s, (b) 110 mm)

4.17 Figure 4.45 shows the weight W supported by the pressure difference created in the venturi and applied across the piston. At the inlet the air velocity is 10 m/s and the area is 100 cm^2. The throat area is 4 cm^2. Given that the air density is equal to 1.2 kg/m^3, calculate the weight W that will be in equilibrium. Assume ideal flow in the duct, no friction between the piston and cylinder, and the piston has no mass.

Ans. (18.72 N)

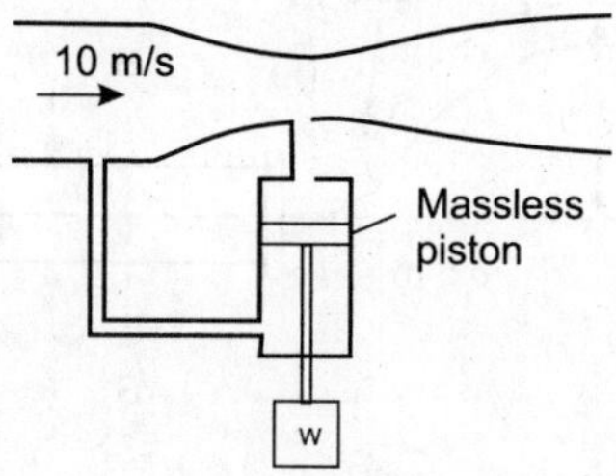

Fig. 4.45

4.18 The loss of head from the entrance to the throat of a 254 mm × 127 mm venturimeter is 1/6 times the throat velocity head. If the mercury in the differential gauge attached to the meter deflects 101.6 mm, what is the flow of water through the venturimeter?

Ans. (0.06 m^3/s)

4.19 The air supply to an oil-engine is measured by inducting air directly from the atmosphere into a large reservoir through a sharp-edged orifice of 50 mm diameter. The pressure difference across the orifice is measured by an alcohol manometer set at a slope of $\sin^{-1} 0.1$ to the horizontal. Calculate the volume flow rate of air if the manometer reading is 271 mm. Specific gravity of alcohol is 0.80, the coefficient of discharge for the orifice is 0.62 and atmospheric pressure and temperature are 775 mm of Hg and 15.8 °C, respectively. (Take C_c, the coefficient of contraction = 0.6.)

Ans. (0.022 m^3/s)

4.20 Flow of air at 49°C is measured by a Pitot static tube. If the velocity of air is 18.29 m/s and the coefficient of the tube is 0.95, what differential reading will be shown in a water manometer? Assume the density of air to be constant at 1.2 kg/m^3.

Ans. (22.70 mm)

4.21 A fire-fighting hose with an exit nozzle diameter = 50 mm discharges water at the rateof 3000 litres per minute. The height of the nozzle exit from the pump is 4 m andwater level in the sump is 1 m below the pump (see Fig. 4.46). Calculate the head and power developed by the pump. Assume there are no losses in the pipeline and pump.

Ans. (38.17 m, 18.72 kW)

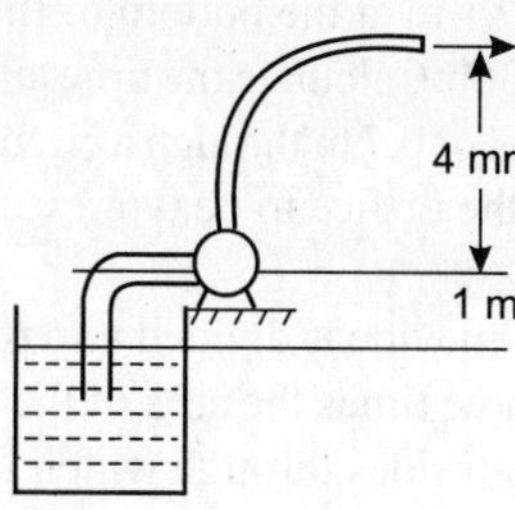

Fig. 4.46

4.22 A free jet of water is produced using a 75 mm diameter nozzle attached to a 200 mm diameter pipe, as shown in Fig. 4.47. If the average velocity of water at plane B is 3.8 m/s, calculate the velocity of water at point A in the free jet. Neglect friction losses in the nozzle and pipe. *Ans.* (19.4 m/s)

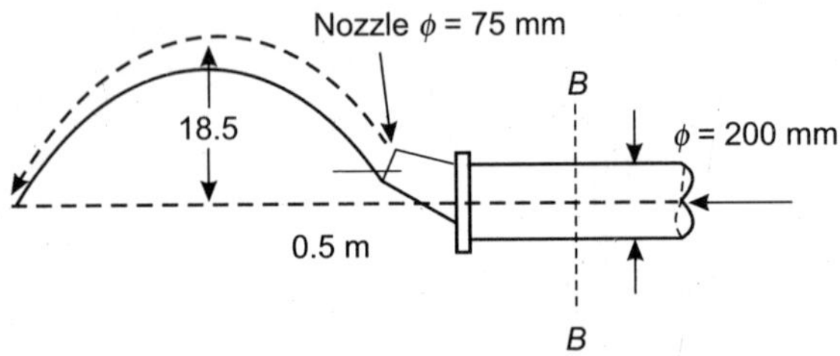

Fig. 4.47

4.23 What is the size of an orifice required to discharge 0.016 m^3/s of water under a head of 8.69 m? (Consider the coefficient of discharge to be unity.)

Ans. (area: 1225 mm^2)

4.24 A sharpe-edged orifice has a diameter of 25.4 mm and coefficients of velocity and contraction of 0.98 and 0.62, respectively. If the jet drops 939 mm in a horizontal distance of 2496 mm, determine the flow in m^3/s and the head on the orifice.

Ans. (0.0018 m^3/s, 1727 mm)

4.25 A vertical triangular orifice in the wall of a reservoir has a base 0.9 m long which is 0.6 m below its vertex and 1.2 m below the water surface. Determine the rate of theoretical discharge.

Ans. (1.19 m^3/s)

4.26 An orifice in the side of a large tank is rectangular in shape, 1.2 m broad and 0.6 m deep. The water level on one side of the orifice is 1.2 m above the top edge; the water level on the other side of the orifice is 0.3 m below the top edge. Find the discharge per second if the coefficient of discharge of the orifice is 0.62.

Ans. (2.36 m^3/s)

4.27 Two orifices in the side of a tank are one above the other and are vertically 1.829 m apart. The total depth of water in the tank is 4.267 m and the height of the water surface from the upper orifice is 1.219 m. For the same values of C_v, show that the jets will strike the horizontal plane, on which the tank rests, at the same point.

4.28 Water issues out of a conical tank whose radius of cross section varies linearly with height from 0.1 m at the bottom of the tank. The slope of the tank wall with the vertical is 30°. Calculate the time taken for the tank to be emptied from an initial water level of 0.7 m through a circular orifice of 20 mm diameter at the base. Take C_d of the orifice to be 0.6.

Ans. (437.97s)

4.29 A tank 3 m long and 1.5 m wide is divided into two parts by a partition so that the area of one part is three times the area of the other. The partition contains a square orifice of 75 mm sides through which the water may flow from one part to the other. If water level in the smaller division is 3 m above that of the larger, find the time taken to reduce the difference of water level to 0.6 m. C_d of the orifice is 0.6.

Ans. (108s)

4.30 A cylindrical tank is placed with its axis vertical and is provided with a circular orifice, 80 mm in diameter, at the bottom. Water flows into the tank at a uniform rate, and is discharged through the orifice. It is found that it takes 107 s for the water height in the tank to rise from 0.6 m to 0.75 m and 120 s for it to rise from 1.2 m to 1.28 m. Find the rate of inflow and the cross-sectional area of the tank. Assume a coefficient of discharge of 0.62 for the orifice.

Ans. (0.019 m^3/s, 5.48 m^2)

4.31 Calculate the coefficient of discharge from a projecting mouthpiece in the side of a water tank assuming that the only loss is that due to the sudden enlargement of water stream in the mouthpiece. Take a coefficient of contraction 0.64.

Ans. (0.871)

5

INTEGRAL FORMS OF CONSERVATION EQUATIONS

5.1 REYNOLDS TRANSPORT THEOREM (RTT)

It is important to note that the laws of Newtonian mechanics were originally stated for particles or particle systems of invariant mass. As such, the classical statements of the basic laws of mechanics and thermodynamics are all postulated on a system-based approach (Lagrangian approach). However, because of its continuously deformable nature, fluid flow may be more conveniently studied from a control volume (CV)-based approach (Eulerian approach), rather than a system-based approach. To apply the classical laws from a control volume perspective, one may, therefore need to apply some kind of mathematical transformation that can express the rate of change of a physical parameter with respect to a system in terms of that with respect to a control volume. This transformation is achieved by a general theorem, known as *Reynolds transport theorem*.

5.1.1 Derivation of Reynolds Transport Theorem

To obtain a relation between system-based and control-volume-based formulation, we refer to the situation as depicted in Fig. 5.1. As shown in the figure, at any time instant t, a control mass system occupies a region in space that is given by the combinations of zones I and II. At time $t + \Delta t$ the control mass system has moved somewhat, as the collection of particles constituting the control mass system now occupies a new configuration, given by the combinations of zones II and III. Let N be an extensive property (for example, total mass, momentum, energy) associated with the system at time t, and let η be the same property but expressed per unit mass at any location. The quantity N may be any scalar or vector property of the fluid. As $\Delta t \to 0$, we may identify a common region between the system configurations at times t and $t + \Delta t$. This common region (Zone II) here signifies the control volume; essentially an identified region in space across which the transported entity may flow. Our objective is to express the time rate of change in N for the system in terms of the time rate of change in N for the control volume.

In order to determine the rate of change of N within the system, we first note that the property of the system at time t is given by

$$(N)_t = (N_I)_t + (N_{II})_t$$

The property of the system at time $t + \Delta t$ is given by

$$(N)_{t+\Delta t} = (N_{II})_{t+\Delta t} + (N_{III})_{t+\Delta t}$$

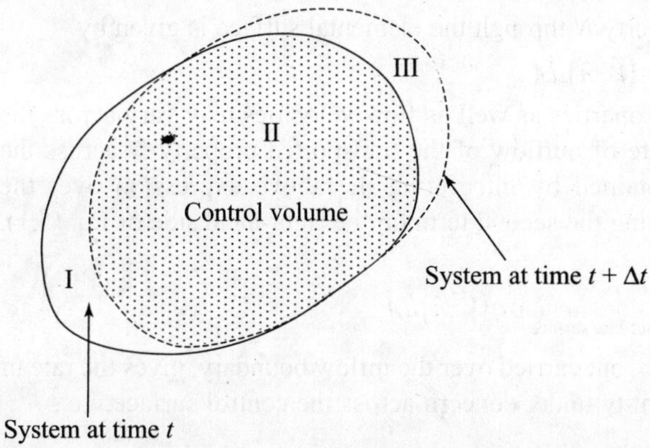

Fig. 5.1 System and control volume configuration

From the definition of a derivative, the rate of change of N, as observed from the system perspective is given by

$$\left.\frac{dN}{dt}\right|_{\text{system}} \equiv \lim_{\Delta t\to 0}\frac{N_{t+\Delta t}-N_t}{\Delta t}$$

$$= \underbrace{\lim_{\Delta t\to 0}\frac{(N_{II})_{t+\Delta t}-(N_{II})_t}{\Delta t}}_{\text{Term 1}}+\underbrace{\lim_{\Delta t\to 0}\frac{(N_{III})_{t+\Delta t}}{\Delta t}}_{\text{Term 2}}-\underbrace{\lim_{\Delta t\to 0}\frac{(N_I)_t}{\Delta t}}_{\text{Term 3}} \qquad (5.1)$$

Since $\lim \Delta t \to 0$, system approaches the control volume, the first term on the right-hand side of Eq. (5.1) simplifies to (considering ρ as the fluid density)

$$\left.\frac{\partial N}{\partial t}\right|_{CV} = \frac{\partial}{\partial t}\int_{CV}\eta\rho d\forall$$

where $d\forall$ is a volume element arbitrarily chosen within the CV. Note the use of the partial derivative with respect to time here, as we refer to time variations at invariant spatial locations (Zone II) considered in this term.

To understand the physical meaning of the last two terms on the right-hand side of Eq. (5.1), we consider an elemental area dA on the surface of the CV across which fluid is leaving with a velocity $\vec{V}$, as shown in Fig. 5.2. Let $\hat{n}$ be a unit vector oriented along outward normal to dA. Let θ be the angle between $\vec{V}$ and $\hat{n}$. Then the mass of fluid swept through dA in time Δt is given by

$$\Delta m = \rho\vec{V}dA\cos\theta\Delta t = \left(\rho\vec{V}\cdot\hat{n}\right)dA\Delta t$$

The rate of mass flow of fluid through the elemental surface dA is given by

$$\Delta\dot{m} = \left(\rho\vec{V}\cdot\hat{n}\right)dA$$

The rate of flow of the property N through the elemental surface is given by

$$\Delta \dot{N} = \rho \eta \left(\vec{V} \cdot \hat{n} \right) dA$$

Considering that the fluid properties as well as flow velocities may vary across the control surface (CS), the rate of outflow of the transported property N across the control surface may be obtained by integrating the above expression over the outflow boundary, representing the second term on the right-hand side of Eq. (5.1), to yield

$$\dot{N}_{\text{outflow}} = \int_{\text{outflow surface}} \rho \eta \left(\vec{V} \cdot \hat{n} \right) dA$$

The same integral expression, but carried over the inflow boundary, gives the rate of inward transport of the quantity under concern across the control surface, i.e.,

$$\dot{N}_{\text{inflow}} = \int_{\text{inflow surface}} \rho \eta \left(\vec{V} \cdot \hat{n} \right) dA$$

It can be mentioned in this context that $\hat{n}$ is the outward normal unit vector to dA. Therefore, $\vec{V} \cdot \hat{n}$ is positive for outflow from the CV and negative for inflow into the CV and its integral over the entire control surface, i.e., $\int_{CS} \eta \left(\rho \vec{V} \cdot \hat{n} \right) dA$ gives the net rate of outflow, i.e., outflow minus inflow, of the property N from the CV (sum of the second and third terms on the right-hand side of Eq. 5.1).

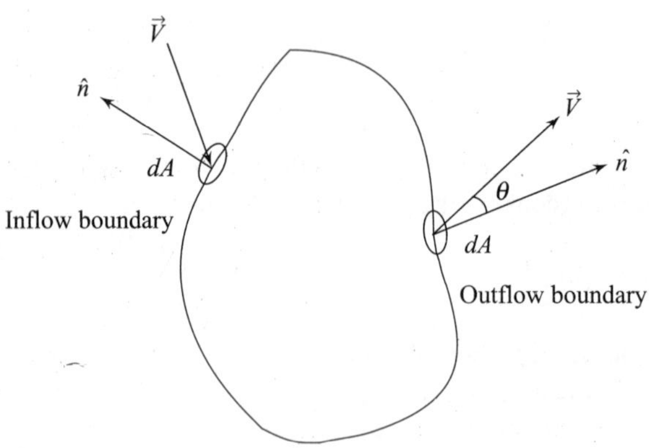

Fig. 5.2

Thus, from Eq. (5.1), one may write

$$\left. \frac{dN}{dt} \right|_{\text{system}} = \frac{\partial}{\partial t} \int_{CV} \eta \rho \, d\forall + \int_{CS} \eta \left(\rho \vec{V} \cdot \hat{n} \right) dA \tag{5.2}$$

Equation (5.2) is the mathematical statement of the Reynolds *Transport Theorem (RTT).* The left-hand side of Eq. (5.2) represents rate of change of property N with respect to the system. First term on the right-hand side represents the same but with respect to CV.The second term on the right-hand side represents the net rate of outflow (i.e., outflow-inflow) of the same property across the CS. Equation. (5.2), therefore, relates the rate of change with respect to the CV to that with respect to the system. In words, this relationship may be rephrased as 'the time rate of change of property N within a control mass system is equal to the time rate of change of property N within the control volume plus the net rate of efflux of the property N across the control surface.'

It is important to mention in this context that in the derivation of RTT the control volume is considered to be fixed in a frame of reference in which the velocity field is described. Moreover, for evaluation of rates of outflow and inflow of properties, the velocities must be expressed relative to the CV, so that Eq. (5.2) is more precisely described as

$$\left.\frac{dN}{dt}\right|_{\text{system}} = \frac{\partial}{\partial t}\int_{CV} \eta\rho d\forall + \int_{CS} \eta\left(\rho\vec{V}_r\cdot\hat{n}\right)dA \tag{5.3}$$

where $\vec{V}_r$ is the velocity of fluid relative to the control volume,

$$\vec{V}_r = \vec{V} - \vec{V}_c \tag{5.4}$$

where $\vec{V}$ and $\vec{V}_c$ are the velocities of fluid and the control volume respectively, as observed in a fixed reference frame.

5.1.2 Application of RTT to Conservation of Mass

For applying Eq. (5.3) to express mass conservation, we set $N = m$ i.e., $\eta = 1$. Then Eq. (5.3) becomes

$$\left.\frac{dm}{dt}\right|_{\text{system}} = \frac{\partial}{\partial t}\int_{CV} \rho d\forall + \int_{CS}\left(\rho\vec{V}_r\cdot\hat{n}\right)dA$$

For a control mass system (defined as a system with fixed mass and identity),

$$\left.\frac{dm}{dt}\right|_{\text{system}} = 0$$

This gives

$$0 = \frac{\partial}{\partial t}\int_{CV} \rho d\forall + \int_{CS}\left(\rho\vec{V}_r\cdot\hat{n}\right)dA \tag{5.5}$$

Next, we may consider a non-deformable CV as a specific example, so that $\forall$ is not a function of t. In that case, we can take $\frac{\partial}{\partial t}$ inside and outside the first integral appearing on the right-hand side of Eq. (5.5) interchangeably. Further, we also assume a stationary CV, so that $\vec{V}_r = \vec{V}$. Considering these two assumptions, Eq. (5.5) simplifies to

$$0 = \int_{CV}\frac{\partial\rho}{\partial t}d\forall + \int_{CS}\rho\left(\vec{V}\cdot\hat{n}\right)dA \tag{5.6}$$

Applying Gauss divergence theorem to the second term on the right-hand side, Eq. (5.6) becomes

$$0 = \int_{CV}\frac{\partial\rho}{\partial t}d\forall + \int_{CV}\nabla\cdot\left(\rho\vec{V}\right)d\forall$$

i.e,

$$0 = \int_{CV}\left[\frac{\partial\rho}{\partial t} + \nabla\cdot\left(\rho\vec{V}\right)\right]d\forall$$

Since the choice of the elemental CV is arbitrary, we have

$$\frac{\partial \rho}{\partial t} + \nabla \cdot (\rho \vec{V}) = 0 \tag{5.7}$$

Equation (5.7) is the well-known continuity equation in its differential form.

Example 5.1

A two-dimensional wedge-shaped fluid element as shown in Fig. 5.3 below is subjected to the following velocity field: $u = V_0 \frac{x}{L}$, $v = -V_0 \frac{y}{L}$ and $w = 0$, where u, v and w are the velocity components along the x, y and z directions respectively, and V_0 is constant. For constant density flow, find the volume flow rate through the surface AC, per unit width.

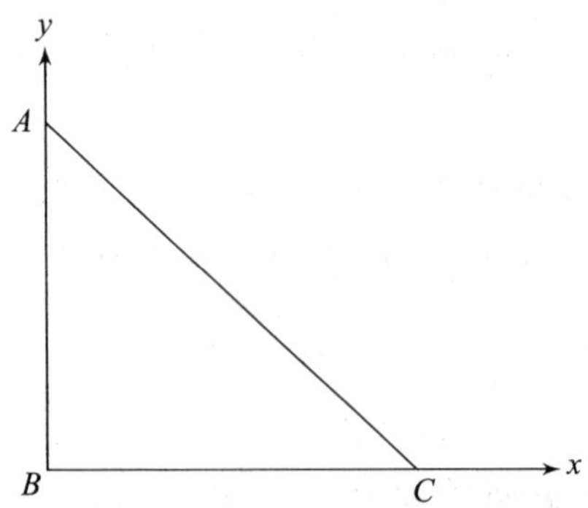

Fig. 5.3

Solution

Choose a fixed control volume as shown by the dotted line in Fig. 5.3 (a).

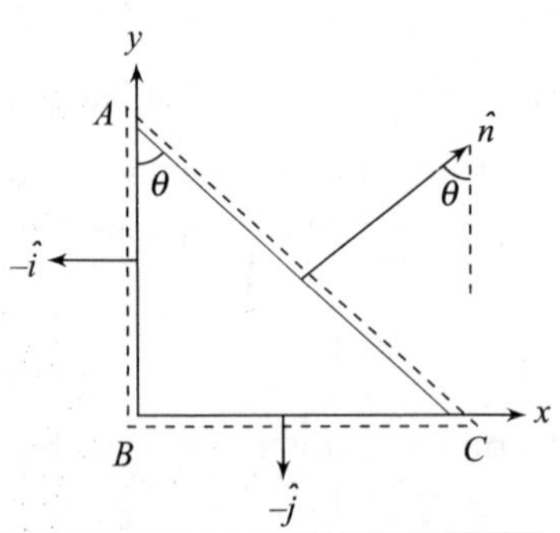

Fig. 5.3(a)

From the conservation of mass for the control volume, we get

$$0 = \frac{\partial}{\partial t}\int_{CV} \rho d\forall + \int_{CS} (\rho\vec{V}\cdot\hat{n})dA$$

Since density is constant, we can write

$$\frac{\partial}{\partial t}\left[\int_{CV} d\forall\right] + \int_{CS} (\vec{V}\cdot\hat{n})dA = 0$$

Now, $\int_{CV} d\forall = \forall$, and hence

$$\frac{\partial(\forall)_{CV}}{\partial t} + \int_{CS} (\vec{V}.\hat{n})dA = 0 \tag{5.8}$$

For non-deformable control volume, first term of Eq. (5.8) is zero. Thus,

$$\int_{CS} (\vec{V}\cdot\hat{n})dA = 0$$

The velocity field everywhere has the form $V = \hat{i}u + \hat{j}v$. This must be evaluated along each surface. Surface AB is the plane $x = 0$. The unit outward normal is $\hat{n} = -\hat{i}$, as shown in Fig. 5.3 (a). The normal velocity is

$$\left(\vec{V}.\hat{n}\right)_{AB} = \left(\hat{i}u + \hat{j}v\right)\cdot\hat{i} = u|_{x=0} = 0$$

The volume flow through surface AB is thus,

$$Q_{AB} = \int_{AB} (\vec{V}\cdot\hat{n})dA = 0$$

Section BC is the plane $y = 0$. The unit normal is $\hat{n} = -\hat{j}$, as shown in Fig. 5.3 (a). The normal velocity is

$$\left(\vec{V}\cdot\hat{n}\right)_{BC} = \left(\hat{i}u + \hat{j}v\right)\cdot\left(-\hat{j}\right) = -v|_{y=0} = 0$$

Thus, normal velocity is zero all along surface BC. Hence $Q_{BC} = \int_{BC} (\vec{V}.\hat{n})dA = 0$.

Mass is conserved in this constant-density flow, and there are no net sources or sinks within the control volume. Therefore, the sum of the volume flow through surfaces AB, BC and AC will be zero, i.e.,

$$Q_{AB} + Q_{BC} + Q_{AC} = 0$$

or

$$Q_{BC} = 0$$

The volume flow rate through the surface AC is zero.

Example 5.2

A fluid flows through a constant head tank as shown in Fig 5.4. fluid is entering into the tank with a uniform velocity U_∞ and is leaving with a velocity u such that

$u = U_0(1 - \frac{y_1^2}{H^2})$, where y_1 is the local transverse coordinate, as shown in Fig. 5.4.

Width perpendicular to the plane of paper is b. For constant density flow, express U_0 in terms of U_∞.

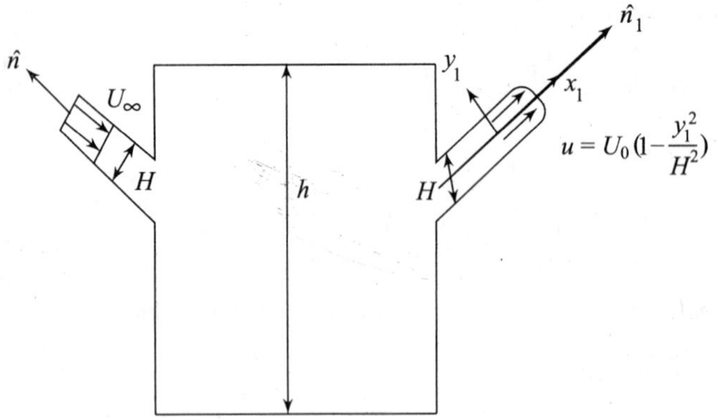

Fig. 5.4

Solution

Choose a fixed control volume as shown by the dotted line in Fig. 5.4(a).

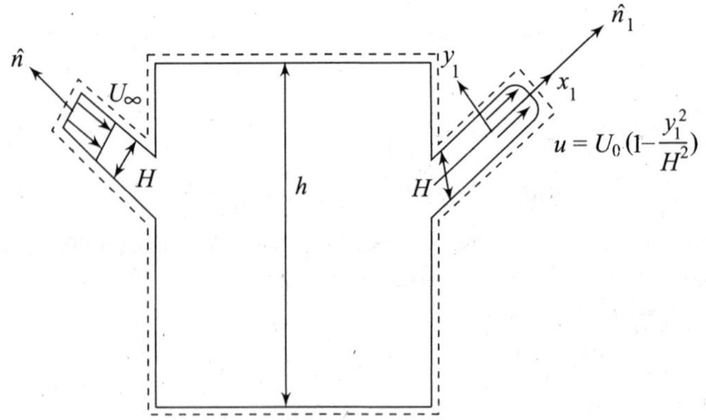

Fig. 5.4(a)

From the conservation of mass for the control volume, we get

$$0 = \frac{\partial}{\partial t}\int_{CV} \rho d\forall + \int_{CS} (\rho \vec{V}\cdot\hat{n})dA$$

Since density is constant, we can write

$$\frac{\partial}{\partial t}\left[\int_{CV} d\forall\right] + \int_{CS}(\vec{V}\cdot\hat{n})dA = 0$$

Now, $\int_{CV} d\forall = \forall$, and hence

$$\frac{\partial(\forall)_{CV}}{\partial t} + \int_{CS} (\vec{V}.\hat{n}) dA = 0 \qquad (5.9)$$

For a non-deformable control volume, the first term on the left-hand side of Eq. (5.9) is zero. Thus,

$$\int_{CS} (\vec{V} \cdot \hat{n}) dA = 0$$

$$-U_{\infty} Hb + \int_{-\frac{H}{2}}^{\frac{H}{2}} U_0 (1 - \frac{y_1^2}{H^2}) dy_1 b = 0$$

After simplification,

$$U_0 = \frac{12}{11} U_{\infty}$$

Example 5.3

Fluid flows through a variable head tank as shown in Fig 5.5. Fluid is entering the tank with a uniform velocity V_1 through a pipe of cross-sectional area A_1 as well as with a uniform velocity V_2 through a pipe of cross-sectional area A_2. Height of the tank from the bottom h. For constant density flow, find an expression for the change in fluid height.

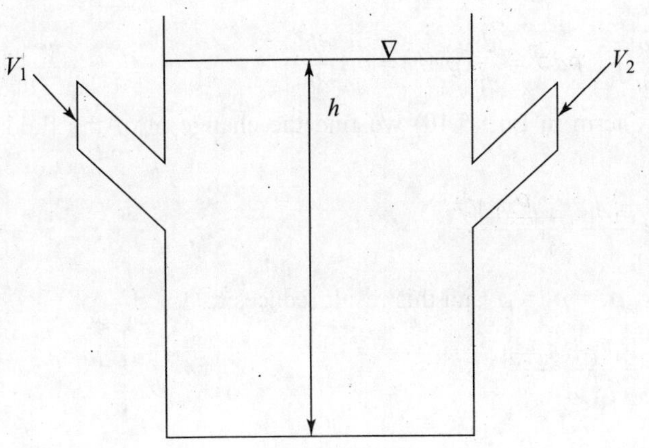

Fig. 5.5

Solution

Choose a fixed control volume as shown by the dotted line in Fig. 5.5 (a).

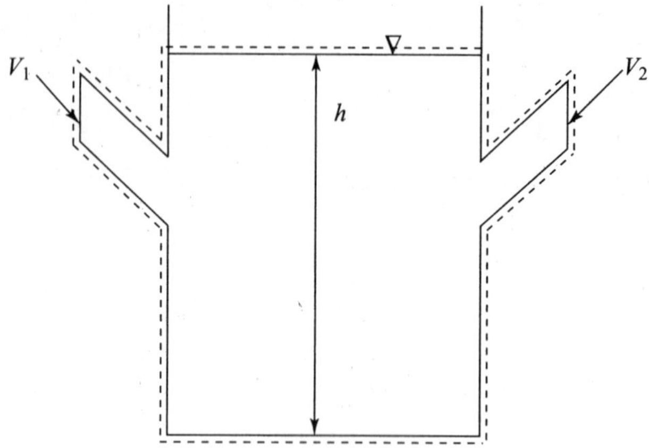

Fig. 5.5(a)

Flow within the control volume is unsteady. Writing the conservation of mass for the control volume results in

$$0 = \frac{\partial}{\partial t}\int_{CV} \rho d\forall + \int_{CS} (\rho \vec{V} \cdot \hat{n}) dA$$

$$0 = \frac{\partial}{\partial t}\int_{CV} \rho d\forall - \rho_1 A_1 V_1 - \rho_2 A_2 V_2 \tag{5.10}$$

Now if A is the tank cross-sectional area, the unsteady term can be evaluated as follows:

$$\frac{\partial}{\partial t}\int_{CV} \rho d\forall = \frac{d}{dt}(\rho A h) = \rho A \frac{dh}{dt}$$

Substituting the unsteady term in Eq. (5.10) we find the change in of the fluid, height

$$\frac{dh}{dt} = \frac{\rho_1 A_1 V_1 + \rho_2 A_2 V_2}{\rho A}$$

For constant density flow, $\rho_1 = \rho_2 = \rho$, and this result reduces to

$$\frac{dh}{dt} = \frac{A_1 V_1 + A_2 V_2}{A}$$

Example 5.4

A conical tank of half-angle θ, with radius R and height H, drains though a hole of radius r_e in its bottom, as shown in Fig. 5.6. The speed of the liquid leaving the tank is approximately $V_e \approx \sqrt{2gh}$, where h is the instantaneous height of the liquid free surface above the hole at a time t. Find the rate of change of surface level in the tank at the instant when $h = H/2$.

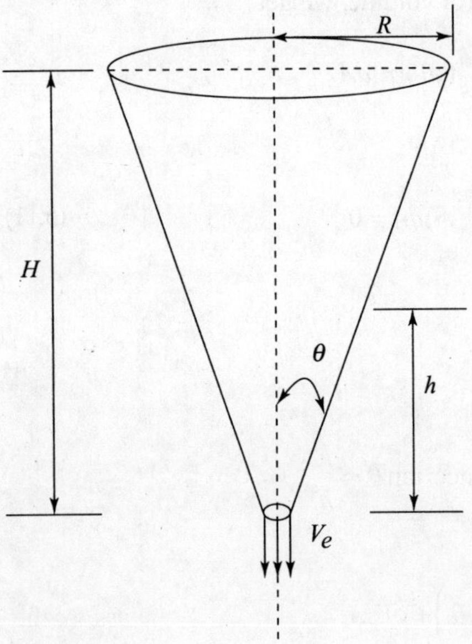

Fig. 5.6

Solution

We choose a CV fixed with the free surface of the liquid, as shown by the dotted line in the Fig. 5.6(a).

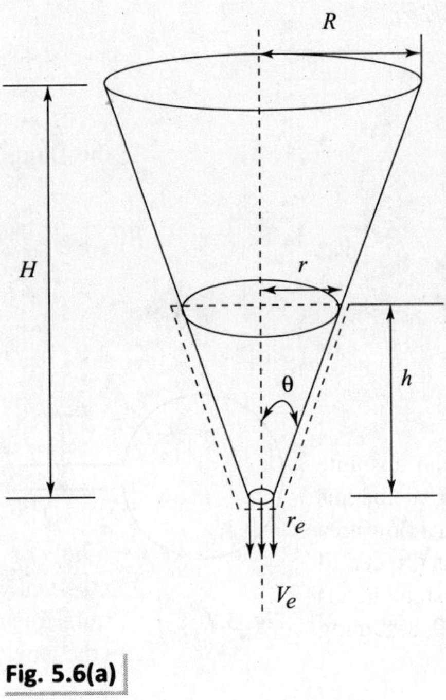

Fig. 5.6(a)

From the conservation of mass for the control volume, we get

$$0= \frac{\partial}{\partial t}\int_{CV} \rho d\forall + \int_{CS} (\rho\vec{V}\cdot\hat{n})dA$$

Since density is constant, we can write

$$\frac{\partial}{\partial t}\left[\int_{CV} d\forall\right] + \int_{CS} (\vec{V}.\hat{n})dA = 0 \tag{5.11}$$

Volume of the CV at that instant,

$$\forall = \frac{1}{3}\pi r^2 h$$

$$= \frac{1}{3}\pi h^3 \tan^2\theta \text{ (since } \tan\theta = \frac{r}{h}\text{)}$$

From Eq. (5.11), we obtain

$$\therefore \quad 0= \rho\frac{d}{dt}\left(\frac{1}{3}\pi h^3 \tan^2\theta\right) + \rho V_e A_e$$

$$\rho\frac{d}{dt}\left(\frac{1}{3}\pi h^3 \tan^2\theta\right) + \rho V_e \pi r_e^2 = 0$$

$$\frac{1}{3}3\pi h^2 \tan^2\theta\frac{dh}{dt} + \sqrt{2gh}\pi r_e^2 = 0$$

$$h^{3/2} \tan^2\theta\frac{dh}{dt} = -\sqrt{2g}r_e^2$$

$$\frac{dh}{dt} = -\frac{\sqrt{2g}r_e^2}{h^{3/2} \tan^2\theta}$$

When, $h = H/2$

$$\frac{dh}{dt} = -\frac{\sqrt{2g}r_e^2}{(H/2)^{3/2} \tan^2\theta}$$

Example 5.5

A rigid tank of volume $\forall$ contains air at an absolute pressure of P and temperature T. At $t = 0$, air begins escaping from the tank through a valve with a flow area of A_1. The air passing through the valve has a speed of V_1 and a density of ρ_1. Determine the instantaneous rate of change of density in the tank at $t = 0$, assuming it to be uniform within the tank.

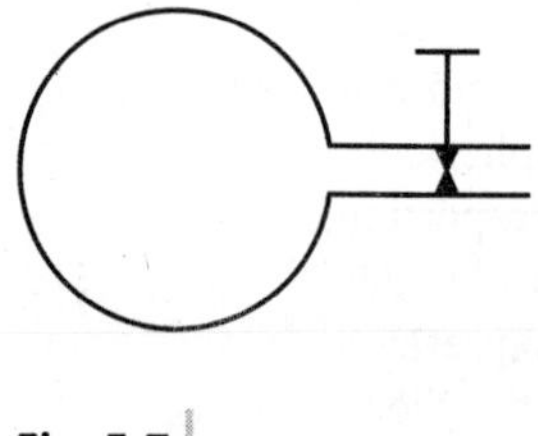

Fig. 5.7

Solution

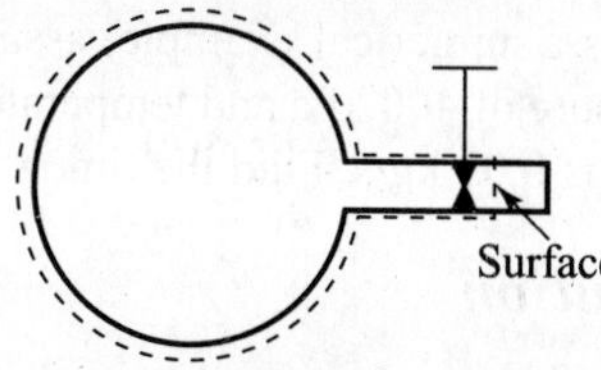

Fig. 5.7(a)

Choose a fixed control volume as shown by the dashed line in Fig. 5.7(a).

From conservation of mass for the control volume, we get

$$0 = \frac{\partial}{\partial t}\int_{CV} \rho d\forall + \int_{CS} (\rho \vec{V}.\hat{n})dA \quad (5.12)$$

Assuming that the properties in the tank are uniform, but time–dependent, the above equation can be written in the form

$$\frac{\partial}{\partial t}\left[\rho \int_{CV} d\forall\right] + \int_{CS} \rho(\vec{V}.\hat{n})dA = 0 \qquad (5.12a)$$

Now, $\int_{CV} d\forall = \forall$. Hence,

$$\frac{\partial(\rho\forall)_{CV}}{\partial t} + \int_{CS} \rho(\vec{V}\cdot\hat{n})dA = 0$$

The only place where mass crosses the boundary of the control volume is at surface (1). Hence,

$$\int_{CS} \rho(\vec{V}\cdot\hat{n})dA = \int_{A_1} \rho(\vec{V}\cdot\hat{n})dA$$

The flow is assumed uniform over surface (1), so that

$$\frac{\partial(\rho\forall)}{\partial t} + \rho_1 V_1 A_1 = 0$$

Since the volume, $\forall$, of the tank is not a function of time,

$$\forall\frac{\partial\rho}{\partial t} + \rho_1 V_1 A_1 = 0$$

$$\frac{\partial\rho}{\partial t} = -\frac{\rho_1 V_1 A_1}{\forall}$$

Example 5.6

Compressed air exhausts from a small hole in a rigid spherical tank at the mass flow rate of $\dot{m}_e$, which is proportional to the density (ρ) of the tank. If ρ_0 is the initial density in the tank of volume, $\forall$, derive an expression for the density change as a function of time after the hole is opened. Assume uniform density within the tank.

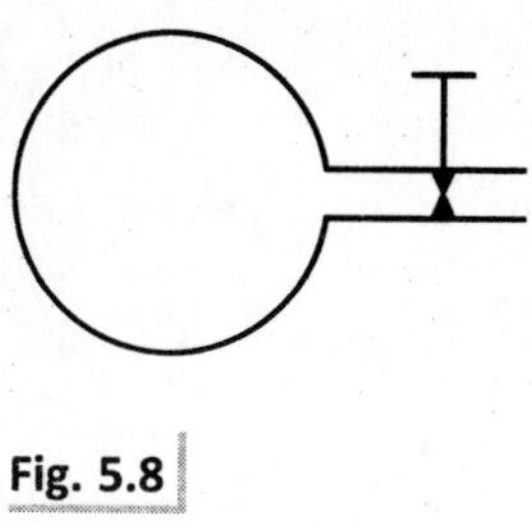

Fig. 5.8

As a numerical example, assume diameter of the tank as 60 cm with an initial pressure of 400 kPa and temperature of 400 K. Initial exhaust rate of air through the hole is 0.02 kg/s. Find the time required for the tank density to drop by 40 %.

Solution

Choose a fixed control volume as shown by the dotted line in Fig. 5.8 (a).
From the conservation of mass for the fixed control volume, we get

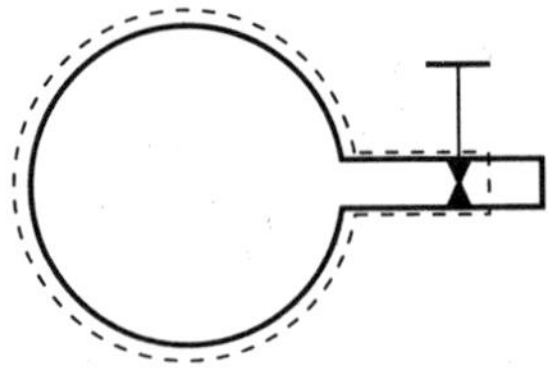

Fig. 5.8(a)

$$0=\frac{\partial}{\partial t}\int_{CV}\rho d\forall+\int_{CS}(\rho\vec{V}.\hat{n})dA \qquad (5.13)$$

Assuming that the properties in the tank are uniform, but time-dependent, the above equation can be written in the form,

$$\frac{\partial}{\partial t}\left[\rho\int_{CV}d\forall\right]+\int_{CS}\rho(\vec{V}.\hat{n})dA=0 \qquad (5.13a)$$

Now, $\int_{CV} d\forall=\forall_{\text{tank}}$, is not a function of time. Hence,

$$\forall_{\text{tank}}\frac{d\rho}{dt}+\dot{m}_e=0 \qquad (5.14)$$

Since $\dot{m}_e \propto \rho$, we assume $\dot{m}_e = k\rho$, where k is a proportionality constant. Thus,

$$\forall_{\text{tank}}\frac{d\rho}{dt}+k\rho=0 \qquad (5.14a)$$

Integrating,

$$\forall_{\text{tank}}\frac{d\rho}{dt}+k\rho=0$$

$$\frac{\rho}{\rho_0}=\exp\left[-\frac{k}{\forall_{\text{tank}}}t\right] \qquad (5.15)$$

Now, $p_0 = 400$ kPa, and $T_0 = 400$ K, then $\rho_0=\frac{p_0}{RT_0}=\frac{400}{0.287\times 400}$ =3.48 kg/m^3

$$\dot{m}_{e0}=k\rho_0=0.02$$

or $$k=\frac{0.02}{3.48}=0.005747\text{ m}^3/\text{s}$$

The tank volume is $\forall_{\text{tank}}=\frac{\pi}{6}D^3=\frac{\pi}{6}(0.6\text{ m})^3=0.113\text{ m}^3$

For the given final conditions,

$$\frac{\rho}{\rho_0} = 0.4 = \exp\left[-\frac{0.005747}{0.113}t\right]$$

$$t = 18\text{ s}$$

5.1.3 Conservation of Momentum or Momentum Theorem

The principle of conservation of momentum as applied to a control volume is usually referred to as the momentum theorem.

5.1.3.1 *Conservation of Linear Momentum*

Let N be the linear momentum, $\int_{\text{system}} dm\vec{V}$, of the system and η be the linear momentum per unit mass, i.e., the velocity $\vec{V}$. Then Eq. (5.3) becomes

$$\frac{d}{dt}\left(\int_{\text{system}} dm\vec{V}\right) = \frac{\partial}{\partial t}\int_{CV} \rho\vec{V}d\forall + \int_{CS} \rho\vec{V}\left(\vec{V}_r \cdot \hat{n}\right) dA \tag{5.16}$$

The velocity $\vec{V}$ defined in Eq. (5.16) is the fluid velocity relative to an inertial (non-accelerating) frame of reference.

Now, $\frac{d}{dt}\left(\int_{\text{system}} dm\vec{V}\right) = \int_{\text{system}} \frac{d\vec{V}}{dt} dm$ (since mass of a system is invariant with time)

$$= \int_{\text{system}} \vec{a}dm$$

$$= \sum \vec{F}_{\text{system}} \quad \text{(by Newton's second law of motion)}$$

We have derived the RTT with a limit as $\Delta t \to 0$. Hence the left-hand side of Eq. (5.16) essentially becomes the resultant force acting on the control volume ($\sum \vec{F}_{CV}$). Thus, Eq. (5.16) may be written as

$$\sum \vec{F}_{CV} = \frac{\partial}{\partial t}\int_{CV} \rho\vec{V}d\forall + \int_{CS} \rho\vec{V}\left(\vec{V}_r . \hat{n}\right) dA \tag{5.17}$$

In other words, the momentum theorem may be stated as follows: the resultant force acting on a control volume is equal to the time rate of increase of linear momentum within the control volume plus the net efflux of linear momentum from the control surface.

Example 5.7

A plate with frictionless surfaces is oriented inclined to an incipient water jet, as shown in Fig. 5.9. Velocity is uniform over each flow section. The fluid flow may be approximated as incompressible and inviscid. The angle made by the plate with the vertical is θ. Express Q_1 and Q_2 as functions of Q_0 and θ. Assume density of the water is constant. Neglect the change in height between the various points.

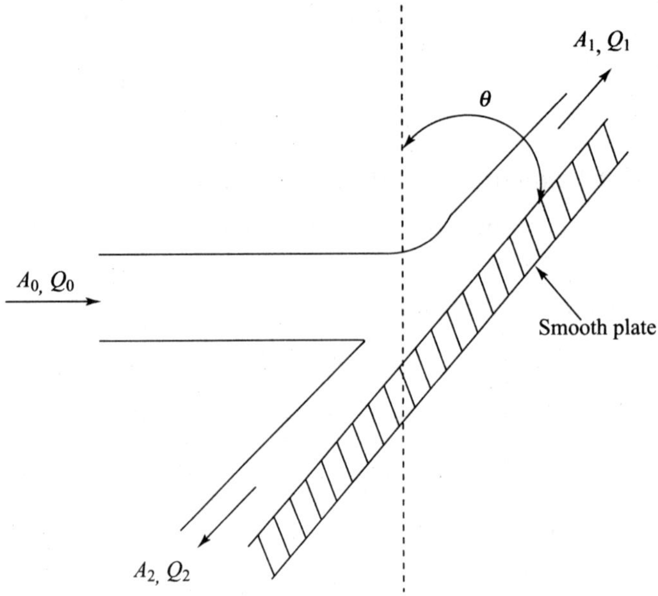

Fig. 5.9

Solution

Choose a fixed CV as shown by the dotted line in Fig. 5.9(a).

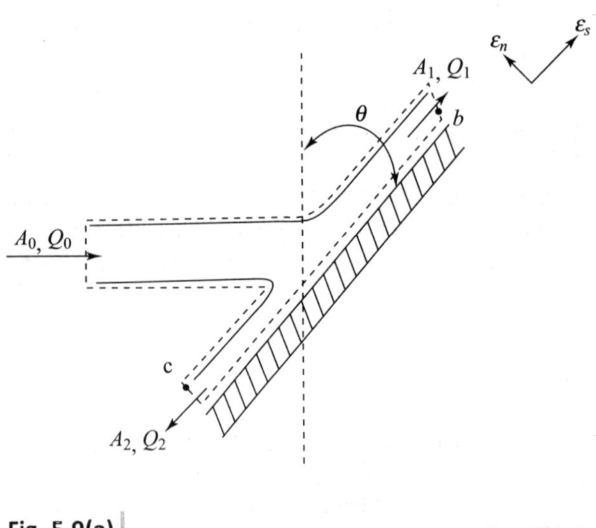

Fig. 5.9(a)

From the conservation of mass for the control volume, we get

$$0 = 0 + \rho(-Q_0) + \rho Q_1 + \rho Q_2$$

or $$Q_1 + Q_2 = Q_0 \tag{5.18}$$

or $$A_1 V_1 + A_2 V_2 = A_0 V_0$$

where V_0, V_1 and V_2 are the uniform velocities at section 0, 1 and 2, respectively, and A_0, A_1 and A_2 are the corresponding cross-sectional areas.

From the conservation of linear momentum of the control volume following Eq. (5.17), one can write

$$F_S\hat{\varepsilon}_S + F_n\hat{\varepsilon}_n = 0 + \rho(V_0 \sin\theta\,\hat{\varepsilon}_s - V_0 \cos\theta\,\hat{\varepsilon}_n)(-Q_0) + \rho V_1 \hat{\varepsilon}_S Q_1 + (-\rho V_2 \hat{\varepsilon}_S)(Q_2) \tag{5.19}$$

where $\hat{\varepsilon}_S$ and $\hat{\varepsilon}_n$ are unit vectors tangential and normal to the inclined plates, respectively. Further, F_s and F_n are the respective force components exerted by the plate on the water. Equating the force components along the surface of the plate we get

$$F_s = -\rho V_0 \sin\theta + \rho V_1 Q_1 - \rho V_2 Q_2 \tag{5.20}$$

Since the plate is frictionless, the net tangential force acting on the plate is zero. Therefore, Eq. (5.20) becomes

$$0 = -\rho V_0 \sin\theta + \rho V_1 Q_1 - \rho V_2 Q_2$$

or $$Q_2 V_2 - Q_1 V_1 = -V_0 Q_0 \sin\theta \tag{5.21}$$

Since the flow is steady, inviscid, and of constant density, Bernoulli's equation may be applied along a streamline. Accordingly, applying Bernoulli's equation along a streamline connecting the points and b, we get

$$\frac{p_a}{\rho} + \frac{V_a^{\ 2}}{2} + gz_a = \frac{p_b}{\rho} + \frac{V_b^{\ 2}}{2} + gz_b$$

Since $p_a = p_b$ and $z_a = z_b$, $V_a = V_b$, i.e, $V_0 = V_1$

Similarly, $V_0 = V_2$

Thus, from Eqs (5.18) and (5.21), we get

$$Q_1 = \frac{Q_0}{2}(1 + \sin\theta) \text{ and } \frac{Q_0}{2}(1 - \sin\theta)$$

Note: From Eq. (5.19), equating the force components normal to the plate, we get

$$F_n = \rho V_0 \cos\theta Q_0 = \rho A_0 V_0^2 \cos\theta\,. \tag{5.22}$$

This is the force exerted by the plate on the water present in the control volume. Therefore, the force exerted by the water on the plate is $-\rho A_0 V_0^2 \cos\theta$.

Example 5.8

A fluid of constant density ρ flows over a stationary flat plate. The relative velocity between the solid boundary and the fluid in contact with that is zero (no-slip boundary condition). The flow domain may be conceptually divided into two parts, viz., a region adhering to the plate in which viscous effects are important and an outer region where viscous effects are negligible. The imaginary line that demarcates these two regions is the edge of the so called boundary layer, as shown in the Fig. 5.10. Incipient free stream velocity on the plate is U_∞. The velocity distribution within the boundary layer is approximated by $\frac{u}{U_\infty} = 2\frac{y}{\delta} - \left(\frac{y}{\delta}\right)^2$. The boundary layer thickness at location B is δ_0. The plate width perpendicular to the plane of the figure is w. Find the net drag force exerted by the plate on the fluid over the length L.

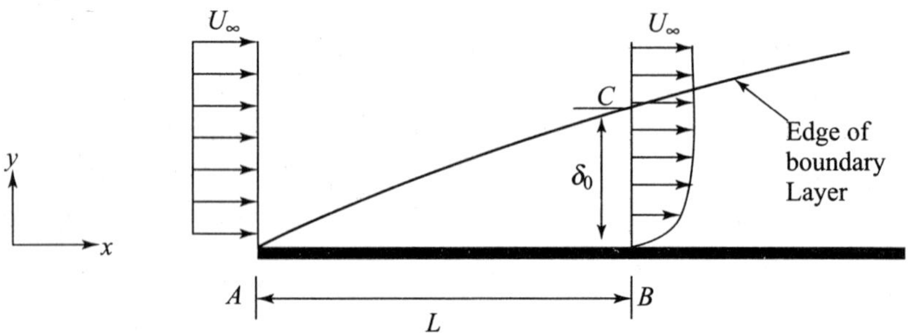

Fig. 5.10

Solution

Choose a fixed CV as shown by the dashed line in Fig. 5.10 (a).

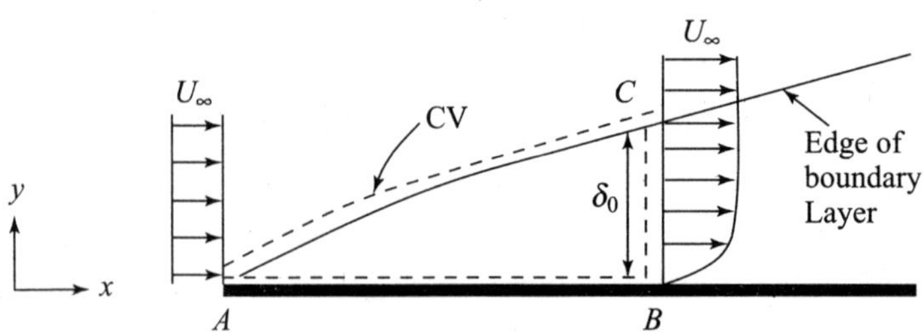

Fig. 5.10(a)

From conservation of mass for the CV, we get

$$0 = 0 + \rho \int_0^{\delta_0} u\,dy\,w + \dot{m}_{AC}$$

$$\dot{m}_{AC} = -\rho \int_0^{\delta_0} u\,dy\,w$$

Applying the x component of the linear momentum conservation equation, we have

$$F_x = 0 + \int_0^{\delta_0} \rho u\, uwdy + u_{AC} \cdot \dot{m}_{AC}$$

$$F_x = \rho w\left[\int_0^{\delta_0} u^2 dy - U_\infty \int_0^{\delta_0} udy\right] = \rho w \int_0^{\delta_0}\left(u^2 - uU_\infty\right)dy \qquad (5.23)$$

Substituting $\dfrac{u}{U_\infty} = 2\dfrac{y}{\delta} - \left(\dfrac{y}{\delta}\right)^2$ into Eq. (5.23), and letting $\eta = \dfrac{y}{\delta}$, we obtain

$$F_x = 2\rho w\left[U_\infty^2 \delta_0 \int_0^1 (\eta - \frac{1}{2}\eta^2)^2 d\eta - U_\infty^2 \delta_0 \int_0^1 (\eta - \frac{1}{2}\eta^2) d\eta\right]$$

$$= 2\rho w U_\infty^2 \delta_0 \left[\int_0^1 (\eta^2 + \frac{1}{4}\eta^4 - \eta^3) d\eta - \int_0^1 (\eta - \frac{1}{2}\eta^2) d\eta\right]$$

$$= 2\rho w U_\infty^2 \delta_0 \left[\frac{3}{2}\frac{\eta^3}{3} - \frac{\eta^2}{2} - \frac{\eta^4}{4} + \frac{1}{4}\frac{\eta^5}{5}\right]_0^1$$

$$= 2\rho w U_\infty^2 \delta_0 \left[\frac{3}{2}\frac{1}{3} - \frac{1}{2} - \frac{1}{4} + \frac{1}{4}\frac{1}{5}\right]$$

$$= -\frac{2}{5}\rho w U_\infty^2 \delta_0$$

This is the force exerted by the plate on the fluid. Note that the minus sign indicates a force along the negative x direction, i.e, the force tends to slow the fluid down (hence the name drag force). Also note that by Newton's third law, the drag force exerted by the fluid on the plate will be $\dfrac{2}{5}\rho w U_\infty^2 \delta_0$, which is a force along the positive x direction.

Alternative Choice of the Control Volume (CV)

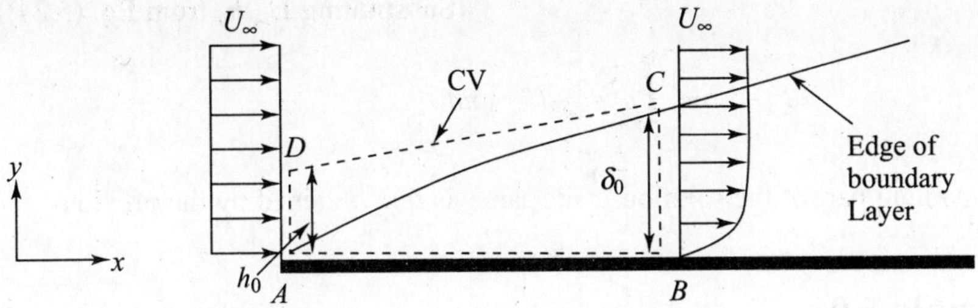

Fig. 5.10(b)

The very fact that there is a flow across AC in Fig. 5.10(a) renders our calculations somewhat tedious, which may be simplified to some extent if we choose another reference line instead of AC such that the net flow across that new reference line is zero. Such a line may be chosen by considering the streamline passing through C (i.e., DC in Fig. 5.10(b)), since by definition there is no flow across a streamline. This motivates us to choose the CV as $ABCD$, as indicated in Fig. 5.10(b). Further, because of no penetration boundary condition at the plate (i.e., zero normal component of flow velocity at the wall), there is no flux across the surfaces AB. Thus fluxes flow out only across the surfaces AD and BC.

From conservation of mass for the CV, we get

$$0 = 0 + \int_{AD} \rho(\vec{V}\cdot\hat{n})dA + \int_{BC} \rho(\vec{V}\cdot\hat{n})dA$$

i.e.,

$$0 = 0 + \rho\int_0^{h_0} (-U_\infty)\, wdy + \int_0^{\delta_o} \rho uwdy$$

i.e.,

$$U_\infty h_0 = \int_0^{\delta_0} udy \tag{5.24}$$

Applying the x component of momentum conservation equation for the CV, we have

$$F_x = \int_{AD} \rho u(\vec{V}.\hat{n})dA + \int_{BC} \rho u(\vec{V}.\hat{n})dA$$

$$= \rho\int_0^{h_0} U_\infty(-U_\infty)\, wdy + \rho\int_0^{\delta_o} u\,(u)\, wdy$$

$$= -\rho U_\infty^2 wh_0 + \rho w\int_0^{\delta_0} u^2 dy$$

$$= -\rho U_\infty w\int_0^{\delta_0} udy + \rho w\int_0^{\delta_0} u^2 dy$$

(Substituting $U_\infty\, h_0$ from Eq. (5.24))

Thus,

$$F_x = \rho w\left[\int_0^{\delta_0} u^2 dy - U_\infty \int_0^{\delta_0} udy\right]$$

The remaining part of the solution is the same as that obtained by the previous method.

Example 5.9

A fluid of constant density ρ flows over a solid body of arbitrary shape with a uniform free stream velocity U_∞, as shown in Fig. 5.11. The velocity distribution at the downstream of the body is given by $\dfrac{u}{U_\infty} = \sin\left(\dfrac{\pi}{2}\dfrac{y}{a}\right)$. The width of flow perpendicular to the plane of the figure is w. Assuming that the flow is symmetrical with respect to the centre line, find the total drag force on the solid body exerted by the fluid.

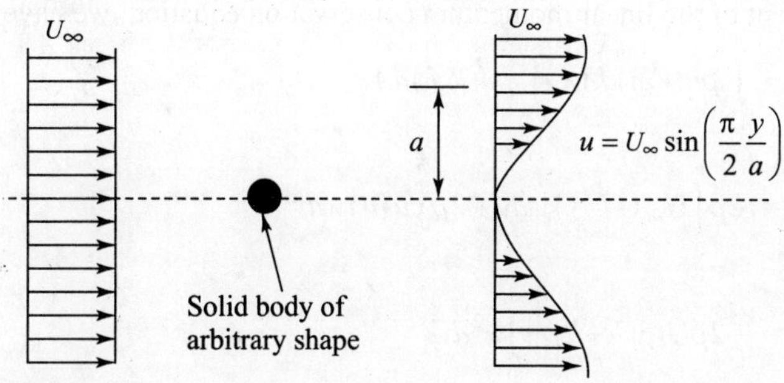

Fig. 5.11

Solution

Choose a CV *ABCD* as shown in Fig. 5.11(a), such that *AD* and *BC* are streamlines, i.e., there are no fluxes across these two surfaces.

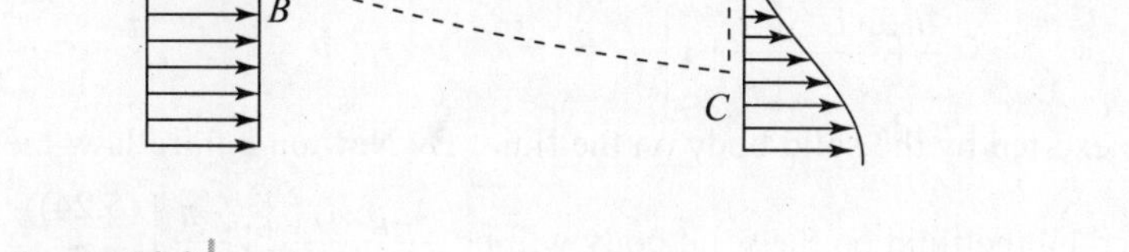

Fig. 5.11(a)

Writing the conservation of mass for the CV results in

$$0 = 0 + \int_{CS} \rho\left(\vec{V}\cdot\hat{n}\right)dA$$

$$0 = 0 + \int_{AB} \rho(\vec{V}\cdot\hat{n})dA + \int_{CD} \rho(\vec{V}\cdot\hat{n})\, dA$$

$$0 = 0 + 2\rho\int_0^h (-U_\infty) w dy + \int_0^a uw dy$$

$$U_\infty h = \int_0^a u dy \tag{5.25}$$

Applying the x component of the linear momentum conservation equation, we have

$$\sum F_x = \int_{AB} \rho u(\vec{V}.\hat{n})dA + \int_{CD} \rho u(\vec{V}.\hat{n})dA$$

$$= 2\rho\int_0^h U_\infty(-U_\infty) w dy + 2\rho\int_0^a u(u) w dy$$

$$= -2\rho U_\infty^2 wh + 2\rho w\int_0^a u^2 dy$$

$$= 2\rho w\left[\int_0^a u^2 dy - U_\infty \int_0^a u dy\right] \tag{5.26}$$

Substituting, $\dfrac{u}{U_\infty} = \sin\left(\dfrac{\pi}{2}\dfrac{y}{a}\right)$ in Eq. (5.26), and letting $\theta = \dfrac{\pi}{2}\dfrac{y}{a}$, we obtain

$$\sum F_X = 2\rho w U_\infty^2 . \frac{2a}{\pi}\left[\int_0^{\frac{\pi}{2}} \sin^2\theta d\theta - \int_0^{\frac{\pi}{2}} \sin\theta d\theta\right]$$

$$= \frac{4a\rho w U_\infty^2}{\pi}\left[\frac{\theta}{2} - \frac{\sin 2\theta}{4} + \cos\theta\right]_0^{\frac{\pi}{2}}$$

$$= \frac{4a\rho w U_\infty^2}{\pi}\left[\frac{\pi}{4} - 1\right]$$

This is the force exerted by the solid body on the fluid. By Newton's third law, the drag force exerted by the fluid on the solid body will be $\dfrac{4a\rho w U_\infty^2}{\pi}\left[1 - \dfrac{\pi}{4}\right]$.

Example 5.10

Water flows steadily through the 90° reducing elbow, as shown in the Fig. 5.12. At the inlet to the elbow, the pressure, velocity and cross-sectional area are p_1, V_1 and A_1, respectively. At the outlet, the corresponding values are p_2, V_2 and A_2, respectively. The weight of the elbow is W. The elbow discharges to the atmosphere. Determine the force required to hold the elbow in place. Neglect the non-uniformity in the velocity profile.

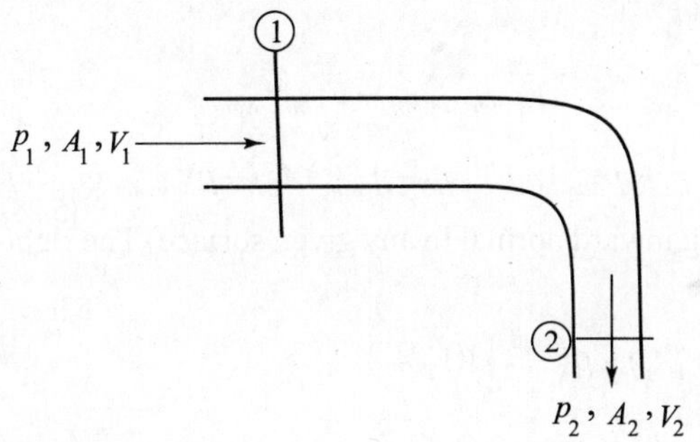

Fig. 5.12

Solution

Choose a fixed CV (water + elbow) as shown by the dashed line in Fig. 5.12(a).

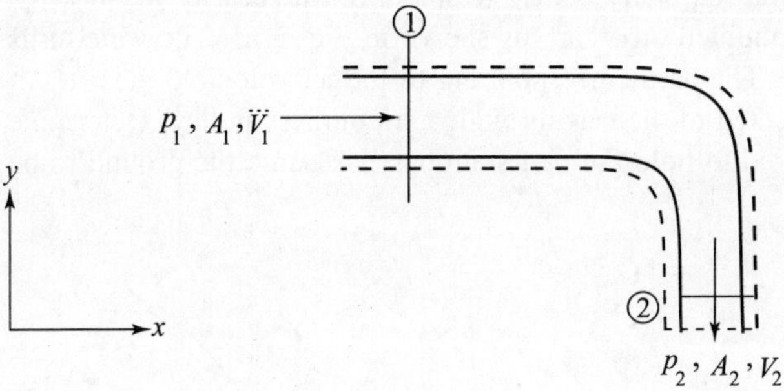

Fig. 5.12(a)

The forces acting on the control volume include those due to

- Pressure p_1 acting on area A_1
- Pressure p_2 acting on area A_2
- Weight of the elbow
- Weight of the water in the control volume
- Reacting force acting on control volume (force exerted by the pipe supports on the elbow).

It is extremely important to mention here that the resultant force due to uniform pressure distribution acting over a closed contour is zero. Therefore, a uniform atmospheric pressure around the CV exerts no resultant force acting on the same. Thus, any pressure difference relative to the atmospheric pressure (i.e., gauge pressure) is only capable enough of exerting any net force on the CV due to pressure. Here, such deviation occurs only over sections 1 and 2, with the corresponding gauge pressures as $(p_1 - p_{atm})$ and $(p_2 - p_{atm})$, respectively.

Now, writing the linear momentum conservation equation for the CV results in

$$\sum \vec{F} = 0 + \int_{CS} \rho \vec{V} \left(\vec{V} \cdot \hat{n} \right) dA \tag{5.27}$$

The left-hand side of Eq. (5.27)

$$\sum \vec{F} = \vec{F}_{reaction} + \vec{F}_{pressure} + \vec{F}_{water\ weight} + \vec{F}_{elbow\ weight}$$

or

$$\sum \vec{F} = \vec{F}_{reaction} + (p_1 - p_{atm})A_1\hat{i} + (p_2 - p_{atm})A_2\hat{j} - \rho\forall_{water}g\hat{j} - W\hat{j}$$

(Noting that pressure is always acting inward normal to any given surface) The right-hand side of Eq. (5.27) becomes

$$\int_{CS} \rho\vec{V}(\vec{V}\cdot\hat{n})dA = \rho(V_1\hat{i})(-V_1A_1) + \rho(-V_2\hat{j})(V_2A_2)$$

Thus, Eq. (5.27) simplifies to

$$\vec{F}_{reaction} + (p_1 - p_{atm})A_1\hat{i} + (p_2 - p_{atm})A_2\hat{j} - \rho\forall_{water}g\hat{j} - W\hat{j} = \rho(V_1\hat{i})(-V_1A_1) + \rho(-V_2\hat{j})(V_2A_2)$$

$$\vec{F}_{reaction} = -(p_1 - p_{atm})A_1\hat{i} - (p_2 - p_{atm})A_2\hat{j} + \rho\forall_{water}g\hat{j} + W\hat{j} + \rho(V_1\hat{i})(-V_1A_1) + \rho(-V_2\hat{j})(V_2A_2)$$

Example 5.11

A jet of water issuing from a stationary nozzle with a uniform velocity, V, strikes a frictionless turning vane mounted on a cart, as shown in Fig. 5.13. The vane turns the jet through an angle θ. The area corresponding to the jet velocity, V, is A. An external mass, M, is connected to the cart through a frictionless pulley. Determine the magnitude of M required to hold the cart stationary. Assume the ground to be frictionless.

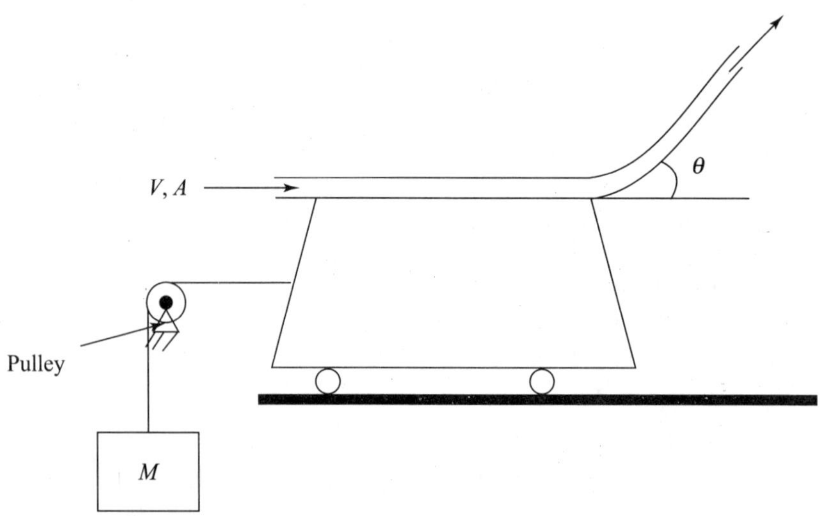

Fig. 5.13

Solution

Choose a fixed CV, as shown by the dashed line in Fig. 5.13(a).

Applying the x component of the linear momentum equation to the inertial CV, we have

$$F_x = \int_A \rho \vec{V}\left(\vec{V}.\hat{n}\right)dA = \rho V\cos\theta(AV) + \rho(V)(-AV) = -T \quad (5.28)$$

where T is the tension the string (see Fig. 5.13(b), which shows forces acting on the CV).

$$T = \rho AV^2(1 - \cos\theta)$$

Since, for equilibrium, tension in the string equals to Mg, because of frictionless nature of the pulley, we have

$$T = \rho AV^2(1 - \cos\theta) = Mg$$

or

$$M = \frac{\rho AV^2(1-\cos\theta)}{g} \quad (5.29)$$

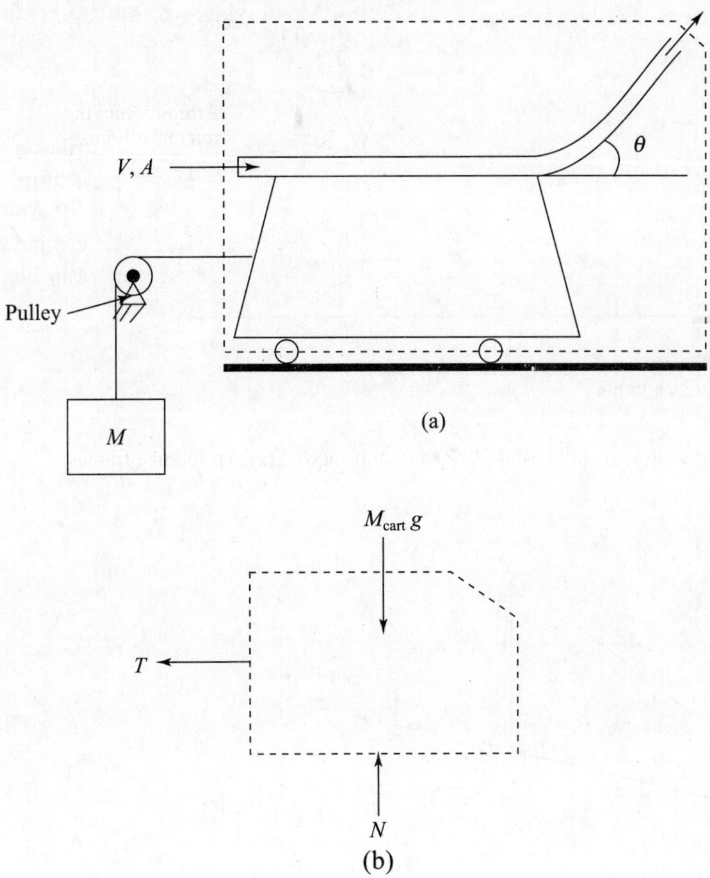

Fig. 5.13

5.1.3.2 *Analysis for Accelerating Reference Frame*

Till this stage, we have considered CV analysis with respect to stationary reference frames. However, for situations involving arbitrarily moving CVs (linearly accelerating and/or rotating), one may require to apply suitable transformations for extending laws originally derived for inertial reference frames to arbitrarily moving reference frames. To have a basic understanding of the underlying principles, we first consider a fixed vector, $\vec{A}$, relative to an arbitrarily moving reference frame *xyz*; whereas *XYZ* is a stationary reference frame. We consider *xyz* to be rotating at an angular velocity, $\vec{\omega}$, relative to *XYZ*. We further consider, the vector $\vec{A}$ to be rotating in the plane of figure, for the sake of illustration, without loss of generality. Position of the vector $\vec{A}$ at time $t + \Delta t$ and is shown in Fig. 5.14 (b).

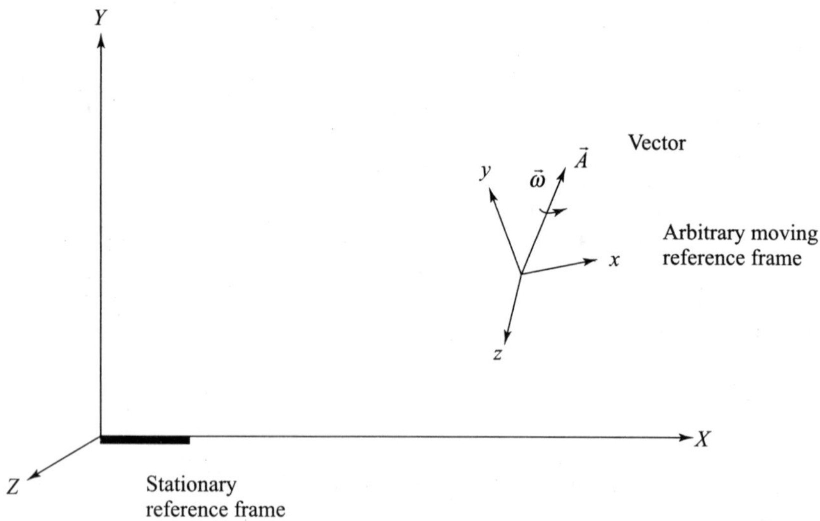

(a) Location of a vector $\vec{A}$ in inertial (*XYZ*) and non-inertial (*xyz*) reference frames

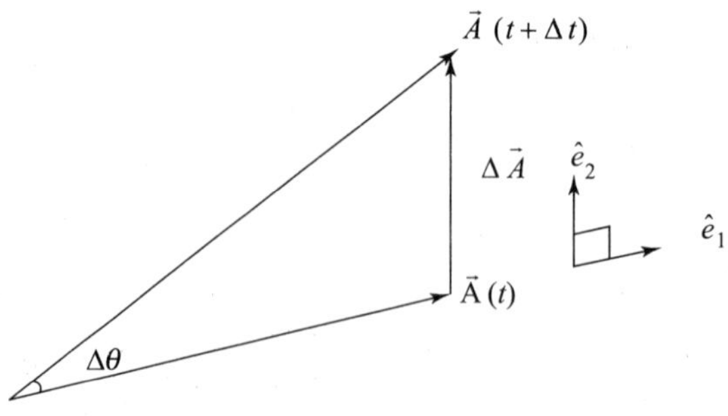

(b) Vector triangle depicting the change in a fixed vector $\vec{A}$ in reference frame *xyz* over a time interval Δt; $\hat{e}_1$ and $\hat{e}_2$ being the unit vectors along radial and cross-radial directions, such that $\hat{e}_1 \times \hat{e}_2 = \hat{e}_3$. Here $\hat{e}_3$ is the direction along which the angular velocity vector is oriented.

Fig. 5.14

Let us consider that the vector $\vec{A}$ sweeps an angle $\Delta\theta$ over the time interval Δt. Therefore, the change in the vector is

$$\Delta\vec{A} = A\Delta\theta\hat{e}_2$$

$$\left.\frac{d\vec{A}}{dt}\right|_{XYZ} = A\hat{e}_2 \lim_{\Delta t\to 0}\frac{\Delta\theta}{\Delta t}$$

$$= A\hat{e}_2\omega$$

Also, $\vec{\omega}\times\vec{A} = \omega\hat{e}_3 \times A\hat{e}_1 = \omega A\hat{e}_2$

Hence, $\left.\frac{d\vec{A}}{dt}\right|_{XYZ} = \vec{\omega}\times\vec{A}$

For an arbitrary vector $\vec{A}$ in reference frame *xyz* (which may translate relative to XYZ, in addition to its rotation), we may generalise the above as

$$\left.\frac{d\vec{A}}{dt}\right|_{XYZ} = \left.\frac{d\vec{A}}{dt}\right|_{xyz} + \vec{\omega}\times\vec{A} \tag{5.30}$$

Equation (5.18) is known as *Chasles' theorem.*

For illustration, we may now apply this theorem for estimating velocity and acceleration of a particle. We consider the particle *P* having a position vector $\vec{r}$ relative to the arbitrarily moving reference frame *xyz*, and a position vector $\vec{R}$ relative to the stationary reference frame *XYZ*, as depicted in Fig. 5.15.

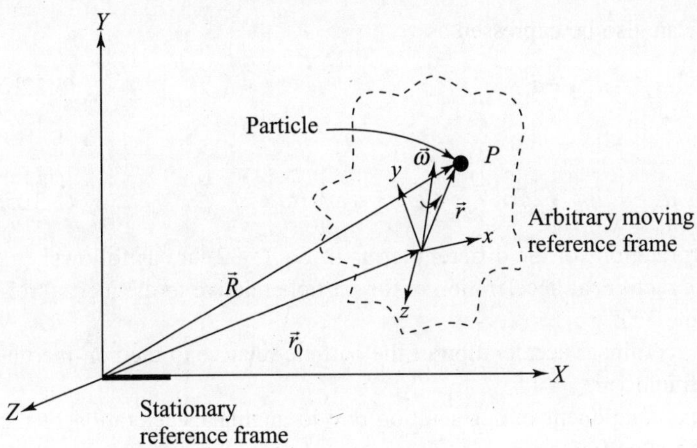

Fig. 5.15 Location of a particle in inertial (*XYZ*) and non-inertial (*xyz*) reference frames

From the geometry of the figure,

$$\vec{R} = \vec{r}_0 + \vec{r}$$

The velocity of the particle at the point *P* with respect to inertial reference frame, *XYZ*, is

$$\vec{V}_{XYZ} = \left.\frac{d\vec{R}}{dt}\right|_{XYZ} = \left.\frac{d\vec{r}_0}{dt}\right|_{XYZ} + \left.\frac{d\vec{r}}{dt}\right|_{XYZ} = \dot{\vec{r}}_0 + \dot{\vec{r}}_{xyz} + \vec{\omega}\times\vec{r} = \vec{V}_{CV} + \vec{V}_{xyz} + \vec{\omega}\times\vec{r}$$

(by Chasles' theorem)

or
$$\vec{V}_{XYZ} = \vec{V}_{CV} + \vec{V}_{xyz} + \vec{\omega}\times\vec{r} \tag{5.31}$$

where $\vec{V}_{CV}$ is the velocity of the control volume and $\vec{V}_{xyz}$ is the velocity of the particle relative to non-inertial reference frame *xyz*. On right-hand side of Eq (5.31), all quantities are expressed relative to *xyz*.

Acceleration of the particle with respect to inertial reference frame, *XYZ*, is given by

$$\vec{a}_{XYZ} = \left.\frac{d^2\vec{R}}{dt^2}\right|_{XYZ} = \ddot{\vec{r}}_0 + \left\{\frac{d}{dt}\left[\dot{\vec{r}}\right] + \vec{\omega}\times\dot{\vec{r}}\right\} + \left\{\frac{d}{dt}[\vec{\omega}\times\vec{r}]_{xyz} + \vec{\omega}\times(\vec{\omega}\times\vec{r})\right\}$$

(Differentiating Eq. (5.31) over *XYZ* and applying Chasles' theorem to transfer the result relative to *xyz*)

$$\vec{a}_{XYZ} = \ddot{\vec{r}}_0 + \ddot{\vec{r}} + 2\vec{\omega}\times\dot{\vec{r}} + \dot{\vec{\omega}}\times\vec{r} + \vec{\omega}\times(\vec{\omega}\times\vec{r}) \tag{5.32}$$

or
$$\vec{a}_{XYZ} = \vec{a}_{CV} + \vec{a}_{xyz} + 2\vec{\omega}\times\vec{V}_{xyz} + \dot{\vec{\omega}}\times\vec{r} + \vec{\omega}\times(\vec{\omega}\times\vec{r}) \tag{5.32a}$$

Equation. (5.32a) can also be expressed as

$$\vec{a}_{XYZ} = \vec{a}_{xyz} + \vec{a}_{rel} \tag{5.32b}$$

where

$$\vec{a}_{rel} = \vec{a}_{CV} + 2\vec{\omega}\times\vec{V}_{xyz} + \dot{\vec{\omega}}\times\vec{r} + \vec{\omega}\times(\vec{\omega}\times\vec{r}) \tag{5.32c}$$

The physical interpretations of the different terms in Eq. (5.32) are as follows:

1. $\vec{a}_{XYZ}$ is the rectilinear acceleration of the particle relative to the inertial reference frame *XYZ*.
2. $\vec{a}_{xyz}$ is the rectilinear acceleration of the particle relative to the non-inertial reference frame *xyz*.
3. $\dot{\vec{\omega}}\times\vec{r}$ is the component of acceleration due to angular acceleration of the moving reference frame.
4. $2\vec{\omega}\times\dot{\vec{r}}$ is the Coriolis component of acceleration due to translation of the particle relative to the rotating reference frame *xyz*.
5. $\vec{\omega}\times(\vec{\omega}\times\vec{r})$ is the centripetal acceleration, due to angular velocity of the moving reference frame *xyz*.

5.1.3.3 *Reynolds Transport Theorem Applied for Linear Momentum Conservation for Control Volume with Arbitrary Acceleration*

Let N be the linear momentum, $\int_{\text{system}} dm\vec{V}_{xyz}$, of the system in an arbitrary moving reference frame (xyz) and η be the linear momentum per unit mass, i.e., $\vec{V}_{xyz}$. Then Eq. (5.3) becomes

$$\frac{d}{dt}\left[\int_{\text{system}} dm\vec{V}_{xyz}\right] = \frac{\partial}{\partial t}\int_{CV} \rho\vec{V}_{xyz} d\forall + \int_{CS} \rho\vec{V}_{xyz}\left(\vec{V}_{xyz}.\hat{n}\right) dA \tag{5.33}$$

Left-hand side of Eq. (5.33) can be written as

$$\frac{d}{dt}\int_{\text{system}} dm\vec{V}_{xyz} = \int_{\text{system}} dm\frac{d\vec{V}_{xyz}}{dt} \quad \text{(Since system mass is time invariant)}$$

$$= \int_{\text{system}} dm\vec{a}_{xyz}$$

$$= \int_{\text{system}} dm[\vec{a}_{XYZ} - \vec{a}_{rel}]$$

$$= \sum F_{CV} - \int_{CV} \rho\vec{a}_{rel} d\forall \tag{5.34}$$

$$\left(\text{since} \int_{\text{system}} dm\vec{a}_{XYZ} = \sum \vec{F}_{\text{system}}\right)$$

The first term on the right-hand side of Eq. (5.34) represents mass times acceleration as viewed from the inertial reference frame, XYZ, and according to Newton's 2nd law of motion, the same is equal to the resultant force acting on the system. In the limiting case of $\Delta t \to 0$, the same is equal to the resultant force acting on the control volume. Therefore, the right-hand side of Eq. (5.33) becomes

$$\sum \vec{F}_{CV} - \int_{CV} \vec{a}_{rel} dm$$

Thus, Eq. (5.35) becomes

$$\sum \vec{F}_{C\forall} - \int_{CV} \vec{a}_{rel} dm = \frac{\partial}{\partial t}\int_{CV} \rho\vec{V}_{xyz} d\forall + \int_{CS} \rho\vec{V}_{xyz}\left(\vec{V}_{xyz}.\hat{n}\right) dA \tag{5.35}$$

From Eq. (5.35), it is evident that there is a correction term on the left-hand side because of the acceleration of the control volume (non-inertial reference frame).

Example 5.12

A vane, with a turning angle θ, is attached to a cart which is moving with uniform velocity, V_c, on a frictionless track. The vane receives a jet of water, which leaves a stationary nozzle horizontally with a velocity V. Determine the resultant force exerted by the water jet on the cart. Assume the water flow to be inviscid.

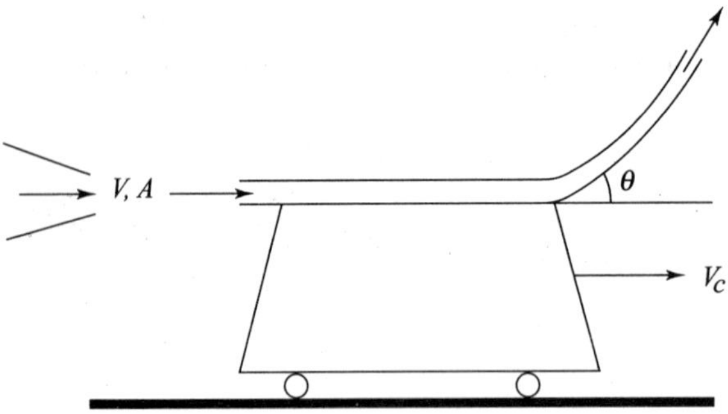

Fig. 5.16

Solution

Choose the CV and coordinate systems as shown in Fig. 5.16(a). *XY* is the inertial reference frame, while reference frame *xy* moves with the cart.

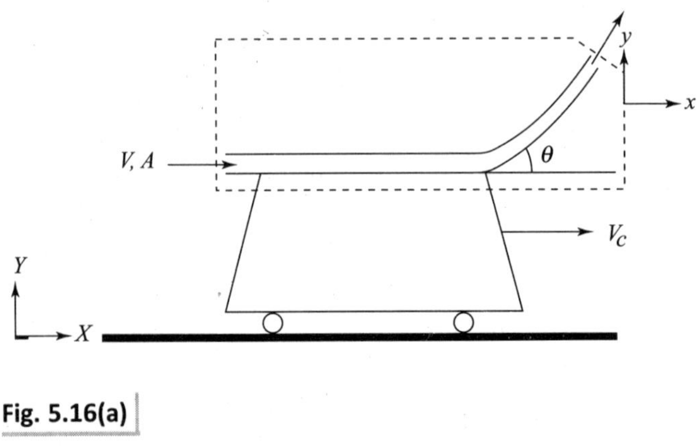

Fig. 5.16(a)

Applying the linear momentum conservation equation to the moving CV, we have

$$\sum \vec{F}_{CV} - 0 = 0 + \rho \int_{CS} \vec{V}_{xyz} \left(\vec{V}_{xyz} \cdot \hat{n} \right) dA \tag{5.36}$$

Note that the correction term on the left-hand side is zero, since the reference frame *xyz* is non-accelerating. The right-hand side of Eq. (5.36) may be further expanded by noting that because of the frictionless nature of the flow, the speed of water relative to the cart remains unaltered as it enters and leaves the CV (this may be ascertained by applying Bernoulli's equation along a streamline connecting the inlet and the exit), although the flow direction gets altered, thereby giving rise to a rate of change of linear momentum. Mathematically, the scenario may be represented as

$$\sum \vec{F}_{CV} = \rho\left[-(V - V_c)\hat{i}\,(V - V_c)A\right] + \rho\left[(V - V_c)\left(\cos\theta\,\hat{i} + \sin\theta\,\hat{j}\right)(V - V_c)A\right] \tag{5.37}$$

The *x* component of the force is

$$F_x = \rho[-(V-V_c)(V-V_c)A] + \rho[(V-V_c)\cos\theta(V-V_c)A]$$

$$F_x = \rho A(V-V_c)^2(\cos\theta - 1) \qquad (5.37a)$$

The y component of the force is

$$F_y = \rho[(V-V_c)\sin\theta(V-V_c)A] = \rho A(V-V_c)^2 \sin\theta \qquad (5.37b)$$

These are the x and y components of force exerted by the cart on the water jet. By Newton's third law, the x and y components of force exerted by the water on the cart are $\rho A(V - V_c)^2 (1 - \cos\theta)$ and $-\rho A(V - V_c)^2 \sin\theta$, respectively.

Example 5.13

A vane is attached to a block as shown in Fig. 5.17. The block moves under the influence of a liquid jet. The mass of the block is M_b. The coefficient of kinetic friction for motion of the block along the surface is μ_k. Obtain a differential equation for the speed of the block as it accelerates from rest. Also find the terminal speed of the block.

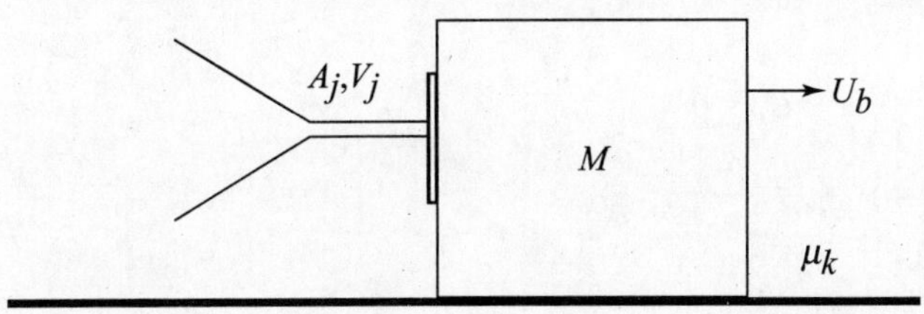

Fig. 5.17

Solution

Choose the CV and coordinate systems as shown in Fig. 5.17(a). XY is the inertial reference frame, while reference frame xy moves with the cart.

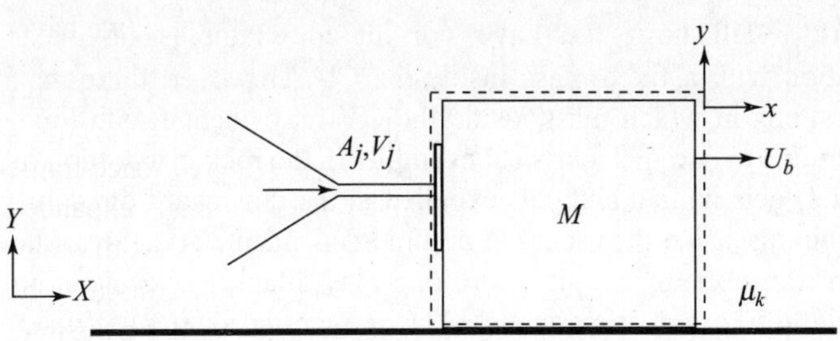

Fig. 5.17(a)

Applying x component of linear momentum conservation equation to the linearly accelerating control volume, we get

$$-\mu_k M_b g - M_b \frac{dU_b}{dt} = 0 + \rho A(V_j - U_b)[-A(V_j - U_b)] \tag{5.38}$$

It is important to note here that the CV constitutes of a solid portion (block) and a fluid portion (liquid). The fractional occupancy of the liquid in the CV is small enough so that its contribution to the velocity relative to the *xy* reference frame (i.e., V_{xy}) may be practically neglected. Thus, the first term on the right-hand side of the momentum equation turns out to be approximately zero. In addition, the exit of water relative to CV has orientations only along the vertical (one stream upward and another stream downward), leaving no outward flux of momentum along the *x* direction. Accordingly, one may obtain the linear momentum equation (Eq. (5.38)) in the following simplified form:

$$M_b \frac{dU_b}{dt} - \rho A(V_j - U_b)^2 + \mu_k M_b g = 0 \tag{5.39}$$

Equation (5.39) may be rearranged as

$$\frac{dU_b}{dt} = \frac{\rho A(V_j - U_b)^2}{M_b} - \mu_k g$$

With an initial condition:

$$t = 0, \qquad U_b = 0$$

At terminal speed, $\frac{dU_b}{dt} = 0$ and $U_b = U_{bt}$, so

$$0 = \frac{\rho A(V_j - U_{bt})^2}{M_b} - \mu_k g$$

or

$$V_j - U_{bt} = \sqrt{\frac{\mu_k M_b g}{\rho A}}$$

or

$$U_{bt} = V_j - \sqrt{\frac{\mu_k M_b g}{\rho A}} \tag{5.40}$$

Motion of a Rocket Till now, we have considered examples of linear momentum conservation with a fixed mass inside the CV. However, there may occur interesting situations in which mass inside the CV may change with time. Motion of a rocket pertains to one such classical example. In the rocket, a portion of the mass included (fuel) gets ignited and gets expelled in the downward direction, leaving the mass remaining inside the rocket to be time-dependent. To analyse the mass and momentum transport in that situation, one may attach an accelerating reference frame *xy* with the rocket. We assume *XY* as the inertial reference frame. The CV is shown in the Fig. 5.18 by a dotted line, which encloses the rocket, cuts through the exit jet, and accelerates upward at the rocket speed.

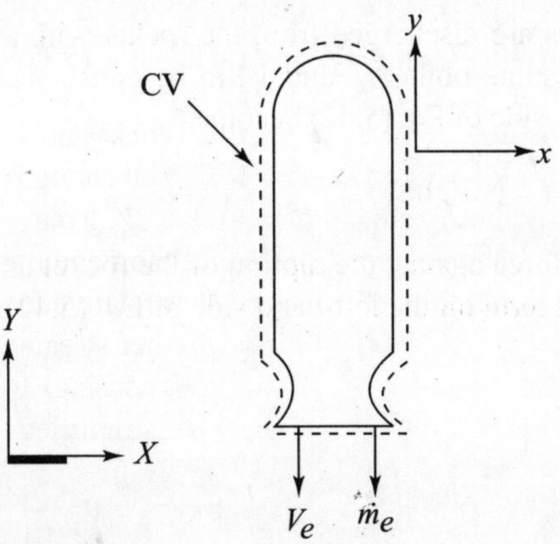

Fig. 5.18

Further we may use conservation of mass for the CV to get

$$0 = \frac{\partial}{\partial t}\int_{CV} dm + \int_{CS}\left(\rho \vec{V}_{xyz}\cdot\hat{n}\right) dA$$

$$= \frac{dM_{CV}}{dt} + \dot{m}_e$$

$$\int_{M_0}^{M} dM_{CV} = -\int_{0}^{t} \dot{m}_e dt$$

Noting that at $t = 0$, $M_{CV} = M_0$ and at $t = t$, $M_{CV} = M$ one may integrate the above to get

Then, $\quad M - M_0 = -\dot{m}_e t$

or $\quad M - M_0 = -\dot{m}_e t \qquad (5.41)$

The linear momentum conservation equation becomes

$$\sum \vec{F} - \int_{CV} \vec{a}_{rel} dm = \frac{\partial}{\partial t}\int_{CV} \rho \vec{V}_{xyz} d\forall + \int_{CS} \rho \vec{V}_{xyz}\left(\vec{V}_{xyz}.\hat{n}\right) dA \qquad (5.42)$$

where $\vec{a}_{rel} = \vec{a}_{CV} + 2\vec{\omega}\times\vec{V}_{xyz} + \dot{\vec{\omega}}\times\vec{r} + \vec{\omega}\times(\vec{\omega}\times\vec{r})$

Since the reference frame has no angular motion, $\vec{a}_{rel} = \vec{a}_{CV}$. Thus, the second term on the left-hand side of Eq. (5.42) becomes

$$-\int_{CV} \vec{a}_{rel} dm = -M\frac{dV_{C\forall}}{dt}\hat{j}$$

where M is the instantaneous mass of the rocket (including unburnt fuel).

Let $\dot{m}_e$ be the rate at which spent gases are discharged from the rocket with a velocity, V_e, relative to the rocket. We assume both $\dot{m}_e$ and V_e to be constants. Accordingly, the last term on the right-hand side of Eq. (5.42) becomes

$$\int_{CS} \rho \vec{V}_{xyz} \left(\vec{V}_{xyz} . \hat{n}\right) dA = (-V_e)\dot{m}_e \hat{j}$$

Considering gravity and the resistive drag force against the motion of the rocket as the sole external forces on the CV, the first term on the left-hand side of Eq. (5.42) becomes

$$\sum \vec{F} = (-D - Mg)\hat{j}$$

Thus, Eq. (5.42) becomes

$$-D - Mg - M\frac{dV_{CV}}{dt} = -\dot{m}_e V_e \tag{5.43}$$

It is important to mention here that in arriving at Eq. (5.43), we make a simplistic assumption that the term $\frac{\partial}{\partial t}\int_{CV} \rho \vec{V}_{xyz} d\forall$ is negligible. This, in turn, implicates that we neglect time dependences of fluid velocities relative to the CV (rocket) within the rocket itself as well as time dependence of density within the rocket (which is strictly not true).

Substituting Eq (5.41) into (5.43), we obtain

$$-D - (M_0 - \dot{m}_e t)g - (M_0 - \dot{m}_e t)\frac{dV_{CV}}{dt} = -\dot{m}_e V_e \tag{5.43a}$$

Neglecting the drag force for simplicity, one may write

$$\frac{dV_{CV}}{dt} = \frac{\dot{m}_e V_e}{(M_0 - \dot{m}_e t)} - g \tag{5.44}$$

$$\int_0^{V_{CV}} dV_{CV} = \int_0^t \frac{\dot{m}_e V_e}{(M_0 - \dot{m}_e t)} dt - g\int_0^t dt$$

Assuming that the velocity of the control volume at time $t = 0$, we have

$$V_{CV} = V_e \ln \frac{M_0}{M_0 - \dot{m}_e t} - gt \tag{5.45}$$

where V_{CV} is the velocity of the control volume at time t.

Example 5.14

A cart is propelled by a liquid jet issuing horizontally from a tank as shown in Fig. 5.19. The initial mass of the block is M_0. The tank is pressurised so that the jet speed may be considered to be a constant. Obtain an expression for the speed of the block as it accelerates from rest. Assume that the jet issues with a velocity V in the leftward direction, relative to the cart. Also assume the track to be frictionless.

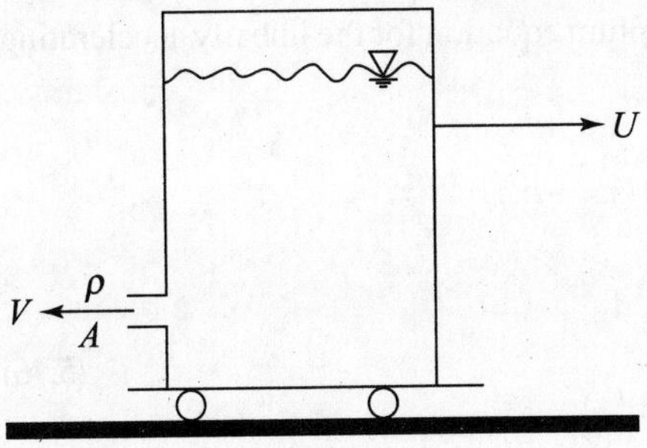

Fig. 5.19

Solution

We attach the reference frame *xy* to the control volume moving with cart (Fig. 5.19 (a)), whereas the reference frame *XY* is inertial.

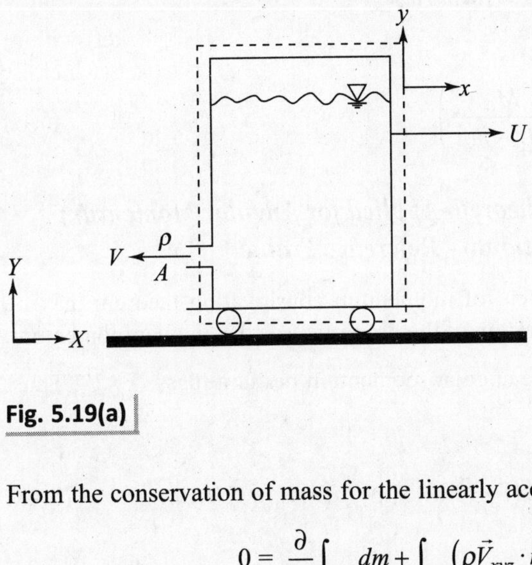

Fig. 5.19(a)

From the conservation of mass for the linearly accelerating control volume, we get

$$0 = \frac{\partial}{\partial t}\int_{CV} dm + \int_{CS}\left(\rho\vec{V}_{xyz}\cdot\hat{n}\right)dA$$

$$= \frac{dM_{CV}}{dt} + \dot{m}_e$$

$$\int_{M_0}^{M} dM_{CV} = -\int_0^t \dot{m}_e dt$$

or $$M = M_0 - \dot{m}_e t$$

Where $M = M_{CV}$.

Writing the x component of the linear momentum equation for the linearly accelerating CV, we get

$$-M\frac{dU}{dt} = -\dot{m}_e V = -(\rho A V)V = -\rho A V^2$$

or

$$\frac{dU}{dt} = \frac{\rho A V^2}{M} = \frac{\rho A V^2}{M_0 - \dot{m}_e t} \tag{5.46}$$

Separating the variables and integrating,

$$\int_0^U dU = \int_0^t \frac{\rho A V^2}{M_0 - \dot{m}_e t} dt$$

$$U = -\frac{\rho A V^2}{\dot{m}_e} \ln[M_0 - \dot{m}_e t]_0^t$$

or

$$U = V \ln\left(\frac{M_0}{M_0 - \dot{m}_e t}\right) \tag{5.47}$$

5.1.3.4 *Reynolds Transport Theorem Applied for Angular Momentum Conservation Relative to Stationary Reference Frame*

The angular momentum or moment of momentum conservation theorem may be derived from Eq. (5.3) in consideration of the property N as the angular momentum, $\vec{r} \times m\vec{V}$, and accordingly η as the angular momentum per unit mass, $\vec{r} \times \vec{V}$. Thus, one can write

$$\frac{d}{dt}\int_{\text{system}} \vec{r} \times dm\vec{V} = \frac{\partial}{\partial t}\int_{CV} \rho(\vec{r} \times \vec{V})\, d\forall + \int_{CS} \rho(\vec{r} \times \vec{V})(\vec{V}_r \cdot \hat{n})\, dA \tag{5.48}$$

where $\vec{r}$ is the position vector of any arbitrary point having flow velocity as $\vec{V}$. The left-hand side of Eq. (5.48) may be simplified as

$$\frac{d}{dt}\int_{\text{system}} \vec{r} \times dm\vec{V} = \int_{\text{system}} \frac{d}{dt}(\vec{r} \times \vec{V})dm$$

(Since system mass is invariant with time)

$$= \int_{\text{system}} \vec{r} \times \frac{d\vec{V}}{dt} dm \quad \left(\text{since } \frac{d\vec{r}}{dt} \times \vec{V} = \vec{V} \times \vec{V} = \bar{0}\right)$$

The term on the left-hand side of Eq. (5.48) is the time rate of change of angular momentum of a system, while the first and second terms on the right-hand side of the equation are the time rate of increase of angular momentum within a control volume and the rate of net efflux of angular momentum across the control surface.

The velocity $\vec{V}$ defining the angular momentum in Eq. (5.48) is described in an inertial frame of reference. Therefore, the term $\frac{d}{dt}\left(\vec{r}\times m\vec{V}\right)_{\text{system}}$ can be substituted by the net moment $\sum M$ applied to the system or to the coinciding control volume. Hence, one can write Eq. (5.48) as

$$\sum M = \frac{\partial}{\partial t}\int_{CV}\rho\left(\vec{r}\times\vec{V}\right)d\forall + \int_{CS}\rho\left(\vec{r}\times\vec{V}\right)\left(\vec{V}_r.\hat{n}\right)dA \tag{5.49}$$

The Eq. (5.49) is the analytical statement of angular momentum theorem applied to a control volume.

5.1.3.5 *Reynolds Transport Theorem Applied for Angular Momentum Conservation for Control Volume with Arbitrary Acceleration*

Let N be the angular momentum, $\vec{r}\times m\vec{V}_{xyz}$, of the system in an arbitrary moving reference frame and η be the linear momentum per unit mass, i.e. $\vec{r}\times\vec{V}_{xyz}$. Then Eq. (5.3) becomes

$$\frac{d}{dt}\int_{\text{system}}\vec{r}\times dm\vec{V}_{xyz} = \frac{\partial}{\partial t}\int_{CV}\rho\left(\vec{r}\times\vec{V}_{xyz}\right)d\forall + \int_{CS}\rho\left(\vec{r}\times\vec{V}_{xyz}\right)\left(\vec{V}_{xyz}.\hat{n}\right)dA \tag{5.50}$$

The left-hand side of Eq. (5.50) may be simplified, following the simplification of Eq. (5.48), as

$$\frac{d}{dt}\int_{\text{system}}\vec{r}\times dm\vec{V}_{xyz} = \int_{\text{system}}\frac{d}{dt}\left(\vec{r}\times\vec{V}\right)dm$$

(Since system mass is invariant with time)

$$= \int_{\text{system}}\vec{r}\times\frac{d\vec{V}}{dt}dm \quad \left(\text{since } \frac{d\vec{r}}{dt}\times\vec{V} = \vec{V}\times\vec{V} = \vec{0}\right)$$

After simplification, Eq. (5.50) becomes

$$\sum\vec{M} - \int_{CV}\left(\vec{r}\times\vec{a}_{rel}\right)dm = \frac{\partial}{\partial t}\int_{CV}\rho\left(\vec{r}\times\vec{V}_{xyz}\right)d\forall + \int_{CS}\rho\left(\vec{r}\times\vec{V}_{xyz}\right)\left(\vec{V}_{xyz}.\hat{n}\right)dA \tag{5.51}$$

Where $\vec{a}_{rel} = \vec{a}_{CV} + 2\vec{\omega}\times\vec{V}_{xyz} + \dot{\vec{\omega}}\times\vec{r} + \vec{\omega}\times\left(\vec{\omega}\times\vec{r}\right)$

Example 5.15

A lawn sprinkler discharges water upward and outward from the horizontal plane, as shown in Fig. 5.20. Initially, the angular speed of the sprinkler is zero. The total volumetric rate of discharges through the sprinkler is Q. If the area of cross section of the sprinkler arms and the discharging nozzles is A, derive an expression for the angular velocity, ω, of the sprinkler as a function of time.

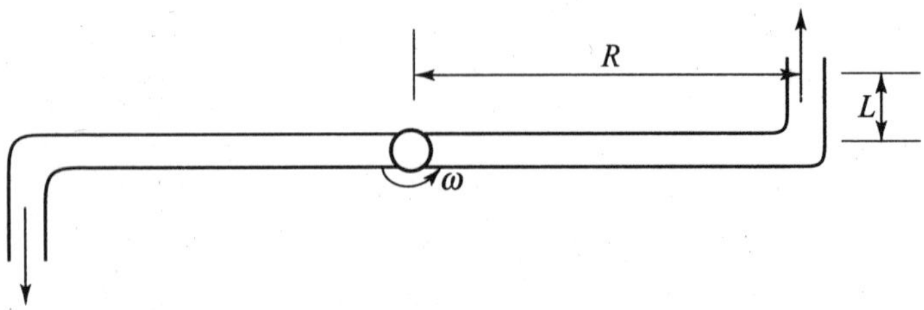

Fig. 5.20

Solution

Approach 1: Moving Reference Frame

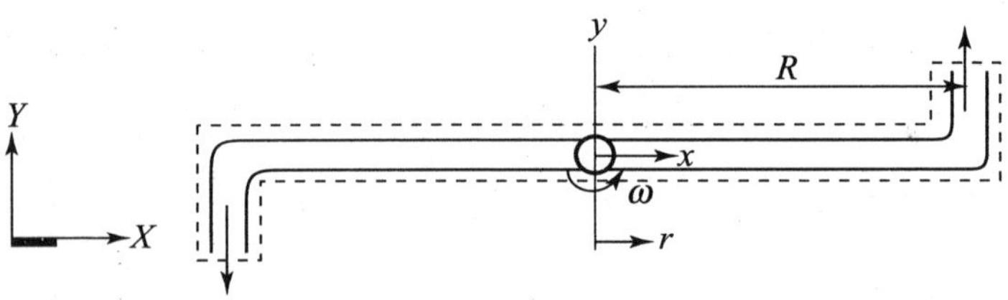

Fig. 5.20(a)

Angular momentum conservation equation for a moving reference frame attached with the rotating sprinkler, as shown in Fig. 5.20(a), results in

$$\underbrace{\sum \vec{M}}_{\text{Term 1}} - \underbrace{\int_{CV} (\vec{r} \times \vec{a}_{rel})\, dm}_{\text{Term 2}} = \frac{\partial}{\partial t} \underbrace{\int_{CV} \rho \left(\vec{r} \times \vec{V}_{xyz}\right) d\forall}_{\text{Term 3}} + \underbrace{\int_{CS} \rho \left(\vec{r} \times \vec{V}_{xyz}\right)\left(\vec{V}_{xyz} . \hat{n}\right) dA}_{\text{Term 4}} \quad (5.52)$$

Term 1 = $\sum \vec{M} = 0$ (since the sprinkler is freely rotating about its pivot)

For the term 2, we note that

$$\vec{a}_{rel} = \vec{a}_{CV} + 2\vec{\omega} \times \vec{V}_{xyz} + \dot{\vec{\omega}} \times \vec{r} + \vec{\omega} \times (\vec{\omega} \times \vec{r})$$

or $$\vec{a}_{rel} = 0 + 2\omega\hat{k}\times V_{xyz}\hat{i} + \dot{\omega}\hat{k}\times r\hat{i} + \omega\hat{k}\times\left(\omega\hat{k}\times r\hat{i}\right)$$

or $$\vec{a}_{rel} = 2\omega V_{xyz}\hat{j} + \dot{\omega} r\hat{j} - \omega^2 r\hat{i}$$

$$\vec{r}\times\vec{a}_{rel} = r\hat{i}\times\left[2\omega V_{xyz}\hat{j} + \dot{\omega} r\hat{j} - \omega^2\hat{i}\right] = \left(2\omega V_{xyz} + \dot{\omega} r\right)\hat{k}$$

Thus, $$\text{Term 2} = \int_{CV}(\vec{r}\times\vec{a}_{rel})\,dm = 2\int_0^R\left(2\omega V_{xyz} + \dot{\omega} r\right)\rho A dr\hat{k}$$

$$= 2\rho A\left(\omega V_{xyz}R^2 + \dot{\omega}\frac{R^3}{3}\right)\hat{k}$$

$$\text{Term 3} = \frac{\partial}{\partial t}\int_{CV}\rho\left(\vec{r}\times\vec{V}_{xyz}\right)d\forall = \bar{0} \quad \text{(Since } \vec{V}_{xyz} \text{ is time invariant)}$$

$$\text{Term 4} = \int_{CS}\rho\left(\vec{r}\times\vec{V}_{xyz}\right)\left(\vec{V}_{xyz}.\hat{n}\right)dA$$

$$= 2\times\rho R\hat{i}\times V_{xyz}\hat{j}\frac{Q}{2}$$

(Neglecting small length of the bent portion)

$$= \rho R V_{xyz} Q\hat{k}$$

Note here that multiplier 2 in the equation preceding the above is indicative of the fact that there are two outflow boundaries with a flow rate of $\frac{Q}{2}$ through each boundary. Substituting Terms 1 through 4 into Eq. (5.52), we obtain

$$-2\rho A\left(\omega V_{xyz}R^2 + \dot{\omega}\frac{R^3}{3}\right) = \rho R V_{xyz} Q$$

where $$V_{xyz} = \frac{Q}{2A} \quad \text{(Relative speed of flow through each limb)}$$

$$-2\rho A\left(\omega\frac{Q}{2A}R^2 + \frac{d\omega}{dt}\frac{R^3}{3}\right) = \rho R\frac{Q}{2A}Q \tag{5.53}$$

Equation (5.53) may be cast in the form,

$$\frac{d\omega}{dt} = a - b\omega \tag{5.53a}$$

where $$a = -\frac{3Q^2}{4AR^2},\ b = \frac{3Q}{2AR}$$

$$\int_0^{\omega} \frac{d\omega}{a - b\omega} = \int_0^t dt$$

Integrating and rearranging the terms, we get

$$\omega = \frac{a}{b}\left[1 - \frac{1}{\exp(bt)}\right] \tag{5.54}$$

Approach 2: Stationary Reference Frame

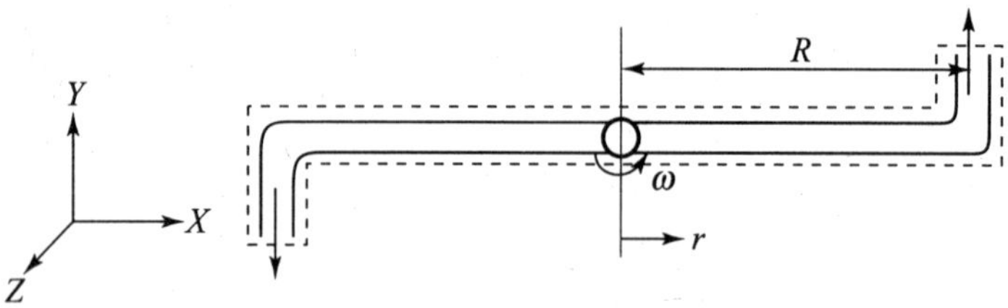

Fig. 5.20(b)

Writing the angular momentum conservation equation relative to the stationary reference frame *XYZ*, as shown in Fig. 5.20(b), results in

$$\underbrace{\sum \vec{M} = \frac{\partial}{\partial t}\int_{CV} \rho\left(\vec{r} \times \vec{V}\right) d\forall}_{\text{Term 1}} + \underbrace{\int_{CS} \rho\left(\vec{r} \times \vec{V}\right)\left(\vec{V}_r . \hat{n}\right) dA}_{\text{Term 2}} \tag{5.55}$$

Term 1:

$$\sum \vec{M} = 0$$

For Term 2, we note that:

$$\vec{V} = \vec{V}_{\text{sprinkler}} + \vec{V}_r$$

where $\vec{V}_r$ is the velocity of water relative to sprinkler.

Thus, considering the right arm (horizontal portion of the sprinkler),

$$\vec{V} = \left(\omega\hat{k} \times r\hat{i}\right) + V_r\hat{i} = \omega r\hat{j} + V_r\hat{i}$$

$$\vec{r} \times \vec{V} = r\hat{i} \times \left(\omega r\hat{j} + V_r\hat{i}\right) = \omega r^2 \hat{k}$$

$$\text{Term 2} = \frac{\partial}{\partial t}\int_{CV} \rho\left(\vec{r} \times \vec{V}\right) d\forall = 2\frac{\partial}{\partial t}\int_0^R \rho\left(\vec{r} \times \vec{V}\right) d\forall$$

$$= 2\frac{\partial}{\partial t}\int_0^R \rho\omega r^2 \hat{k} A dr = 2\rho A \frac{R^3}{3}\frac{d\omega}{dt}\hat{k}$$

$$\text{Term }3 = \int_{CS} \rho\left(\vec{r}\times\vec{V}\right)\left(\vec{V}_r.\hat{n}\right)dA$$

$$= 2\left[\rho R\hat{i}\times\left\{\omega R\hat{j} + V_r\hat{j}\right\}\frac{Q}{2}\right]$$

$$= \left(\rho\omega R^2 Q + \rho R V_r Q\right)\hat{k}$$

Substituting terms 1 through 3 into Eq. (5.55), we obtain

$$0 = 2\rho A\frac{R^3}{3}\frac{d\omega}{dt} + 2\rho\omega R^2 Q + \rho R V_r Q$$

The solution of this differential equation may be obtained in a manner similar to the approach adapted in the context of moving-reference-frame-based method adapted for this problem.

SUMMARY

- *Reynolds transport theorem* relates the rate of change of a physical parameter with respect to a system in terms of that with respect to a control volume. In other words, this relationship may be stated as 'the time rate of change of property *N* within a system, is equal to the time rate of change of property *N* within the control volume, plus the net rate of efflux of the property *N* across the control surface'.
- Differential form of the continuity equation can be derived from Reynolds transport theorem by applying the law of conservation of mass to a control volume.
- The statement of the law of conservation of linear momentum as applied to a control volume is known as the momentum theorem. This theorem states that the resultant force acting on a control volume is equal to the time rate of increase of linear momentum within the control volume plus the net efflux of linear momentum from the control surface.
- The law of conservation of angular momentum as applied to a control volume states that the resultant moment acting on a control volume is equal to the time rate of increase of the angular momentum within the control volume, plus the net efflux of the angular momentum from the control surface.
- Reynolds transport theorem applied for linear momentum conservation can be extended to a non-inertial reference frame using *Chasles' theorem*. There is a correction term in force acting on the control volume because of the acceleration of the control volume (non-inertial reference frame).
- Similar to linear momentum conservation, Reynolds transport theorem can also be extended to non-inertial reference frame for angular momentum conservation and there is a correction term in the moment acting on the control volume.

EXERCISES

5.1 A fluid flows steadily through a parallel-plate channel of height $2h$. The fluid is entering the channel with a uniform velocity U_{av}. The velocity distribution at a section downstream is $u = u_{\max}\left[1-\left(\frac{y}{h}\right)^2\right]$, where u_{max} is the maximum velocity. For constant density flow, express u_{max} in term of U_{av}.

Ans. $\left(u_{\max} = \frac{3}{2}U_{av}\right)$

52 Thin liquid film of fluid falling slowly down an inclined wall as shown in Fig. 5.21. The velocity profile at a section downstream is $u = U\left[2\frac{y}{h}-\left(\frac{y}{h}\right)^2\right]$, where U is the surface velocity. Find the volume flow rate per unit width.

Ans. $\left(\frac{5}{3}Uh\right)$

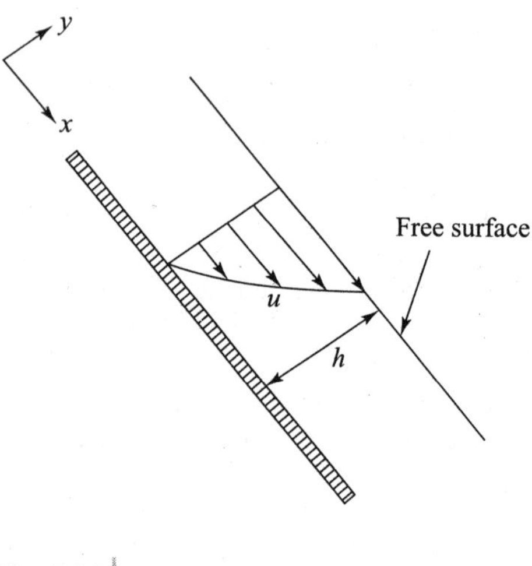

Fig. 5.21

5.3 A fluid flows through a variable head cylindrical tank as shown in Fig 5.22. The fluid is entering the tank with a uniform velocity V_2, through a pipe of diameter d_2, as well as with a volume flow rate of Q_3, through a pipe on the top of the tank. The fluid is leaving with a uniform velocity V_1, through a pipe of diameter d_1. The height of the tank from the bottom is h. For constant density flow, find an expression for the change in height of the fluid.

Ans. $\left(\frac{dh}{dt} = \frac{4Q}{\pi D^2} + \frac{d_2^2V_2 - d_1^2V_1}{D^2}\right)$

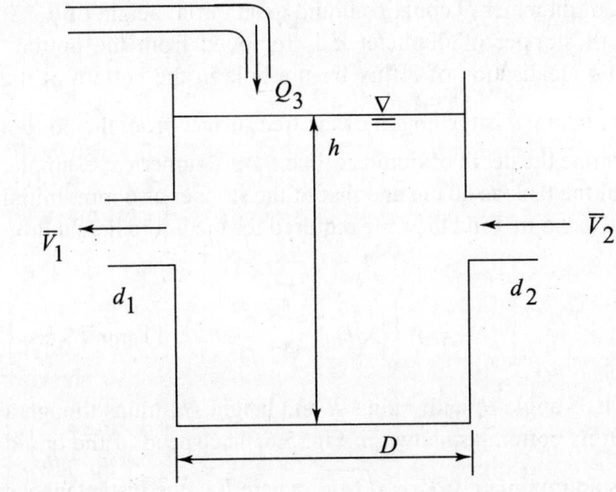

Fig. 5.22

5.4 Two circular plates are separated by a liquid layer as shown in Fig. 5.23. The lower plate is stationary and the upper plate moves downward at constant speed, V_0. The radius of the top plate is R. The liquid is squeezed out in the transverse direction between the plates. Find an expression for $V(r)$, assuming inviscid flow.

Ans. $\left(V(r) = V_0 \dfrac{r}{2h}\right)$

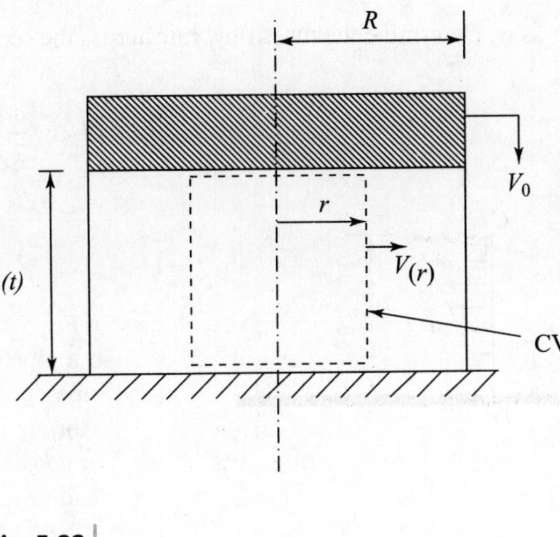

Fig. 5.23

5.5 A cylindrical tank of diameter D contains liquid to an initial height of h_0. At time $t = 0$, a small stopper of diameter d is removed from the bottom. Assume Torricelli's idealisation of efflux from a hole in the bottom of the tank as $V = \sqrt{2gh}$, where h is the height of the free surface from the bottom of the tank. Determine the depth of liquid at time t. As a numerical example, assume diameter of the tank as 60 cm and that of the stopper as 6 mm. Initial height of the liquid is 0.5 m. Find the time required for the liquid height drop by 40 %.

$$Ans. \left(\left(\sqrt{h_0} - \sqrt{\frac{g}{2}} \frac{d^2}{D^2} t \right)^2, \text{ 11 min 59 sec} \right)$$

5.6 A conical tank of half-angle θ, with radius R and height H, drains though a hole of radius r_e in its bottom as shown in Fig. 5.6. The speed of the liquid leaving the tank is approximately $V_e \approx \sqrt{2gh}$, where h is the instantaneous height of the liquid free surface above the hole at a time t. Obtain an expression for the time required to drain the tank.

$$Ans. \left(\frac{\frac{2}{5} H^{5/2} \tan^2 \theta}{\sqrt{2g} r_e^2} \right)$$

5.7 A fluid of constant density ρ flows over a stationary, smooth flat plate with an incipient free stream velocity U_∞, as shown in Fig. 5.24. The thickness of the boundary layer at section CD is δ. The velocity distribution within the boundary layer is approximated by $\frac{u}{U_\infty} = \frac{y}{\delta}$. The plate width perpendicular to the plane of the Figure. is w. Determine the mass flow rate across the section BC.

$$Ans. \left(\rho U_\infty w \frac{\delta}{2} \right)$$

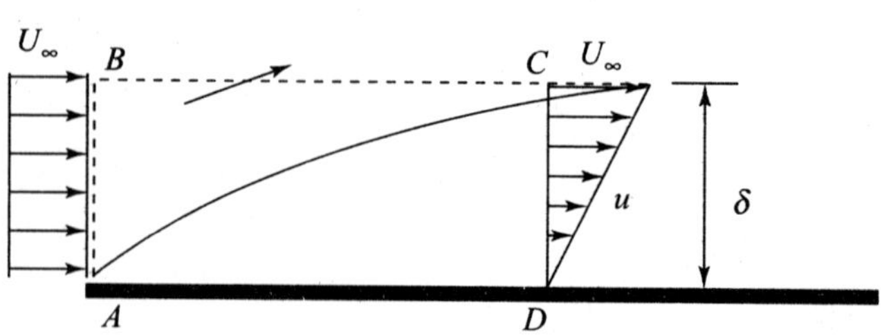

Fig. 5.24

5.8 A fluid of constant density ρ flows steadily past a porous plate with a uniform free stream velocity U_∞, as shown in Fig 5.25. Constant suction is applied along the porous section. The velocity distribution at section *CD* is given by $\dfrac{u}{U_\infty} = \dfrac{3}{2}\dfrac{y}{\delta} - \dfrac{1}{2}\left(\dfrac{y}{\delta}\right)^3$. Determine the mass flow rate across the section *BC*.

Ans. $\left(\rho U_\infty \left(\dfrac{3}{8}\delta - 0.1L\right)\right)$

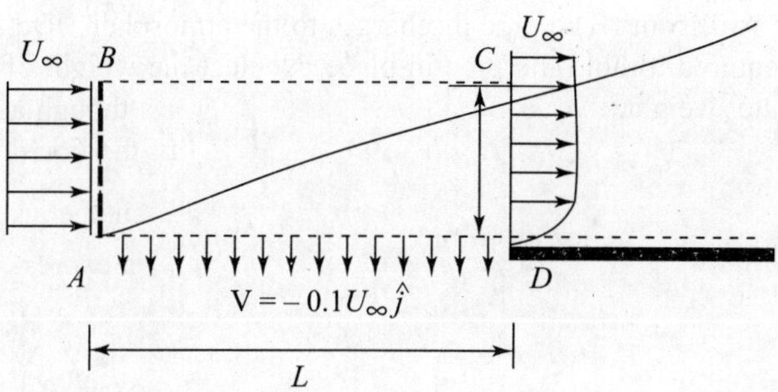

Fig. 5.25

5.9 A spherical balloon of initial radius R_0 is being filled through section 1 (Fig. 5.26), where the area is A_1, velocity (uniform) is V_1, and fluid density is ρ_1. The spatial average density within the balloon is $\rho_b(t)$, which is given by $\rho_b = \rho_1 = ke^{\frac{t}{\tau}}$, where τ and k are constants, and t is the time. Find an expression for the rate of change of the radius of the balloon when its radius is same as R_0.

Ans. $\left(\dfrac{A_1V_1}{4\pi R_0^2} - \dfrac{R_0}{3\tau}\right)$

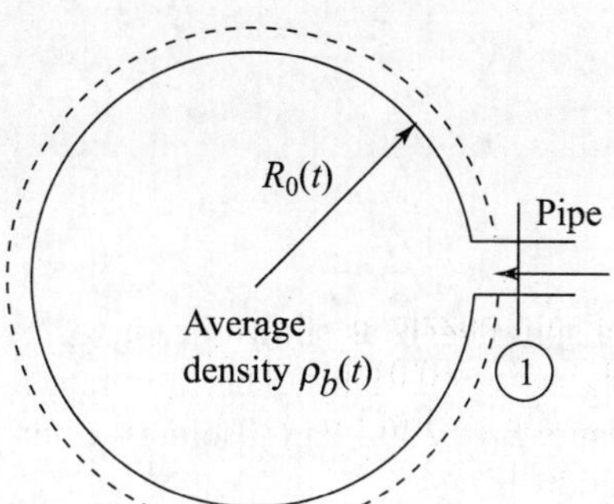

Fig. 5.26

5.10 A jet of water of cross-sectional area A, issuing from a stationary nozzle with a velocity V strikes a vertical plate concentrically. Develop an expression for the force required to hold the plate in place. Water flowing at the rate of 0.034 m^3/s strikes a flat plate held normal to its path. If the force exerted on the plate in the direction of incoming water jet is 720 N, calculate the diameter of the stream of water.

Ans. (ρAV, 45 mm)

5.11 Water flows steadily through the 180° reducing pipe bend as shown in the Fig. 5.27. At the inlet to the pipe, the pressure, velocity and cross-sectional area are 80 kPa (gauge), 5 m/s and 0.3 m^2, respectively. At the outlet, the cross-sectional area is 0.15 m^2. The pipe discharges to the atmosphere. Determine the force required to hold the pipe in place. Neglect the weight of the bend and the water weight.

Ans. (46.5 kN)

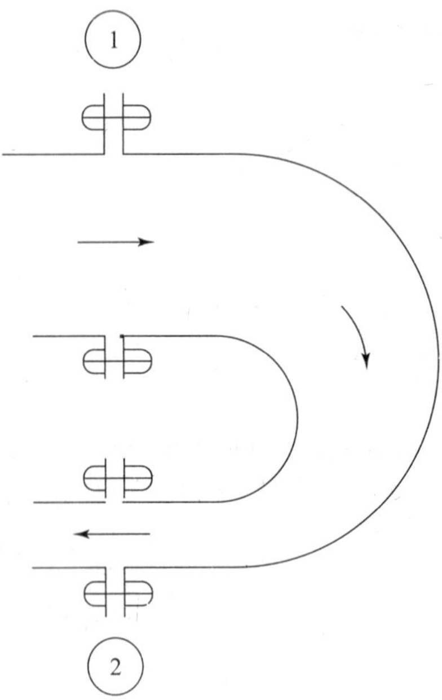

Fig. 5.27

5.12 Water exits to the atmosphere through a split nozzle as shown in Fig. 5.28. The duct areas are A_1 = 0.02 m^2 and $A_2 = A_3$ = 0.01 m^2. The flow rate is $Q_2 = Q_3$ = 0.04 m^2/s, and the inlet pressure p_1 = 150 kPa (absolute). Compute the force on the flange bolts at Section 1.

Ans. (1577.13 N)

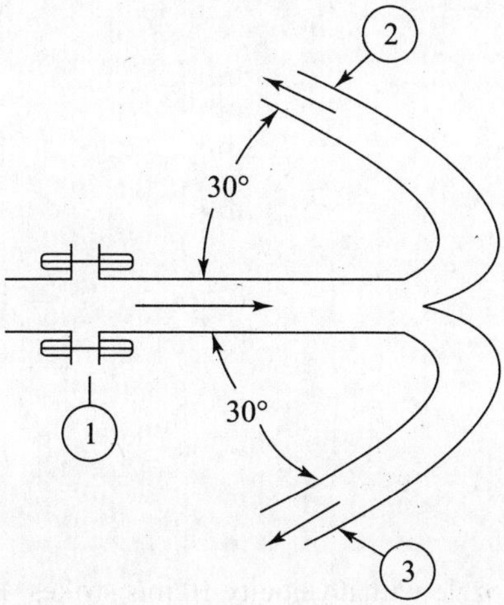

Fig. 5.28

5.13 A fluid of constant density ρ, flows over a stationary flat plate with a uniform free stream velocity U_∞, as shown in Fig 5.29. The velocity distribution at the downstream of the body is given by $\dfrac{u}{U_\infty} = \dfrac{3}{2}\dfrac{y}{\delta} - \dfrac{1}{2}\left(\dfrac{y}{\delta}\right)^3$. The width of flow perpendicular to the plane of the figure is w. Find the net drag force exerted by the plate on the fluid over the length L.

Ans. $\left(\dfrac{39}{140}\rho w U_\infty^2 \delta_0\right)$

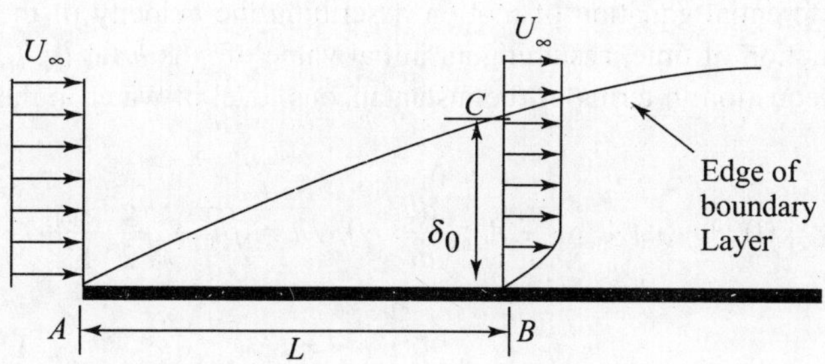

Fig. 5.29

5.14 A vertical jet issuing from a nozzle with a velocity of 10 m/s impinges on a horizontal plate bearing a total load of 200 N (Fig. 5.30). The fluid is an oil of density 800 kg/m^3 and the nozzle exit diameter is 60 mm. Obtain a general expression for the speed of the fluid jet as a function of height, h. Find the height to which the plate will rise and remain stationary.

Ans. (1.12 m)

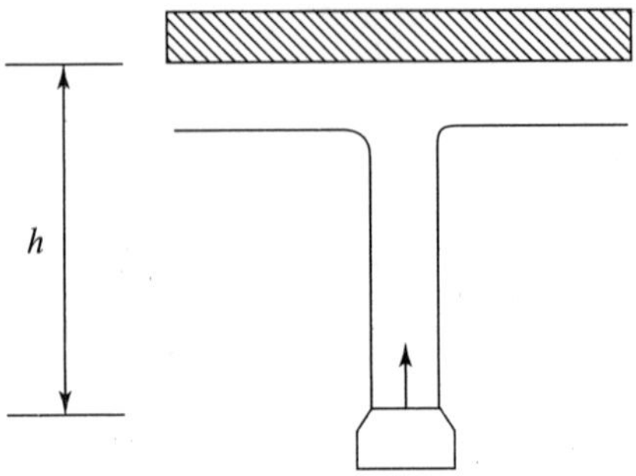

Fig. 5.30

5.15 A jet of water issuing from a stationary nozzle with a velocity 10 m/s strikes a turning vane mounted on a cart, as shown in Fig. 5.13. The vane turns the jet through an angle 30°. The area corresponding to jet velocity 10 m/s A. An external mass M is connected to the cart through a frictionless pulley. Determine the magnitude of M required to hold the cart stationary.

Ans. 4.4 kg

5.16 A tank, weighing 150 N when empty, contains water with a density of 1000 kg/m^3. The tank is of cylindrical shape with a diameter of 1 m, and is kept on a smooth slab of ice (Fig. 5.31). Coefficient of static friction between the tank and the ice slab is 0.01 and the coefficient of kinetic friction between the two is 0.001. Assume Torricelli's idealisation of efflux from a hole on the side of the tank as $V = \sqrt{2gh}$. The hole's diameter is 9 cm.

(i) Does the tank move from its initial state of rest towards the right, if $h = 0.6$ m?

(ii) Derive a differential equation of motion describing the velocity of the tank as a function of time, assuming an initial value of h as h_0 at $t = 0$. Express the equation in terms of the instantaneous level of water in the tank.

$$Ans. \left(\text{(i) No, (ii) } A_{\text{hole}}V + A_{\text{tank}}\frac{dh}{dt} = 0, \sqrt{h} = \sqrt{h_0} + \sqrt{\frac{g}{2}}\frac{A_{\text{hole}}}{A_{\text{tank}}}t \right)$$

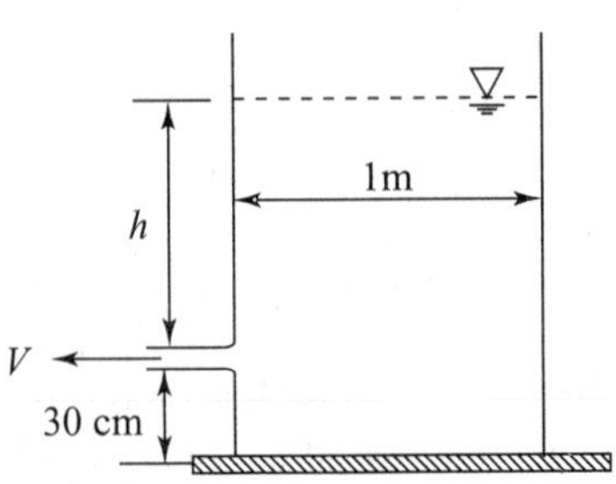

Fig. 5.31

5.17 A container with vertical walls, filled with water, rolls without resistance along a smooth horizontal track. A water jet issues horizontally through a small nozzle located on the wall of the container (Fig. 5.32). The jet has a velocity V_f relative to the container and has a cross-sectional area A_f. The initial mass of the filled container is M_0 and its velocity is u_0 at $t = 0$. One face of the container is struck by a horizontal water jet of cross-sectional area A_j and velocity V_j, so as to slow it down. Obtain an expression for the variation of the velocity of the container as a function of time.

$$Ans. \left(-\left(M_0 - \dot{m}_f t\right)\frac{dV_{CV}}{dt} = -\rho_f A_f V_f^2 + \rho A\left(V + V_{CV}\right)^2\right)$$

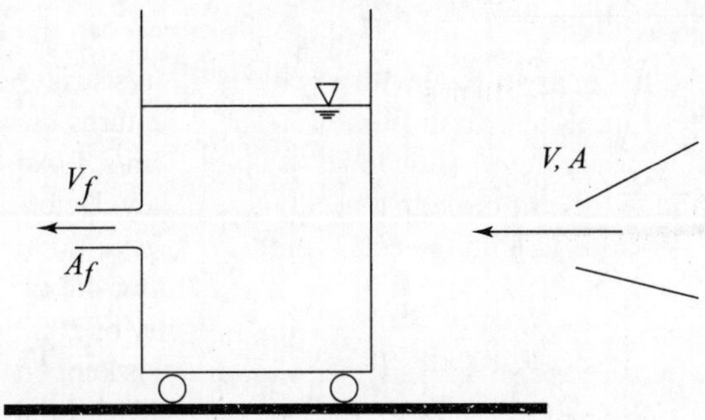

Fig. 5.32

5.18 A vertical jet issuing from a nozzle with a velocity of 10 m/s impinges on a horizontal plate bearing a total load of 200 N (Fig. 5.33). The fluid is an oil of density 800 kg/m^3 and the nozzle exit diameter is 60 mm. When the plate is 1.12 m above the nozzle exit, it is moving upward at 5 m/s. Find the vertical acceleration of the plate at this instant.

Ans. (1.24 m/s^2)

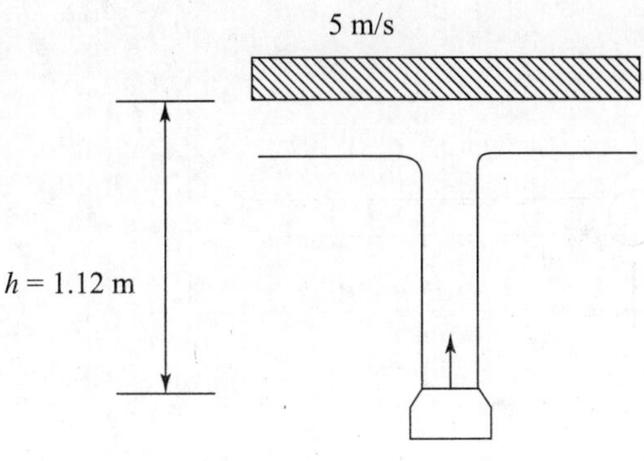

Fig. 5.33

5.19 A vane, with turning angle $\theta = 120°$, is attached to a cart which is moving with uniform velocity 10 m/s on a frictionless track (Fig. 5.34). The vane receives a jet of water of 10 cm diameter which leaves a stationary nozzle horizontally with a velocity Eq. 30 m/s. Determine the resultant force exerted by the water jet on the cart. Assume the water flow to be inviscid.

Ans. (5.44 kN)

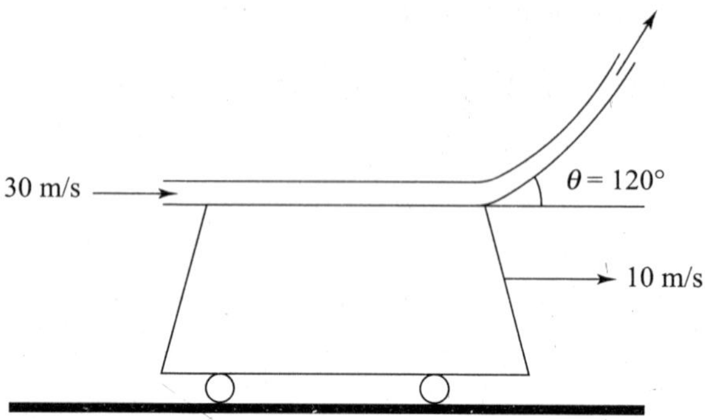

Fig. 5.34

5.20 A rocket, with an initial mass of 250 kg, is to be launched vertically. The spent gases are discharged from the rocket at a constant rate of 9 kg/s and with a constant velocity of 3000 m/s relative to the rocket. Neglecting the air drag, find the rocket speed after 10 s.

Ans. (1240.76 m/s)

5.21 A lawn sprinkler is supplied with water at a volume flow rate of Q. The water runs along two equal arms of radius R and is discharged through nozzles of cross-sectional area A, at an angle of θ as shown in Fig. 5.35. Find the steady-state angular velocity ω assuming no friction at the pivot. Neglect small length of the bent portion.

Ans. $\left(-\dfrac{Q}{2AR}\right)$

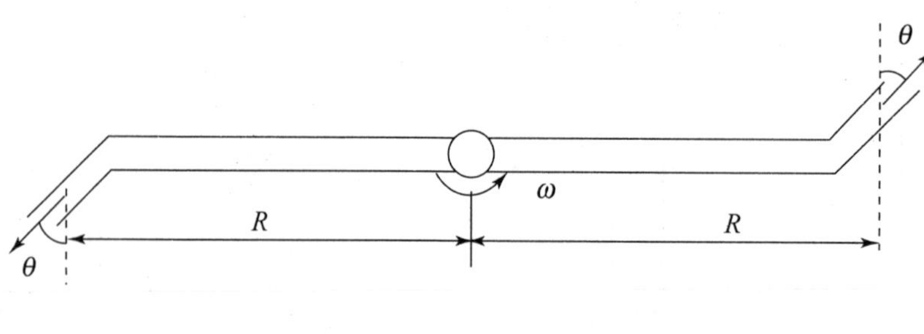

Fig. 5.35

6

PRINCIPLES OF PHYSICAL SIMILARITY AND DIMENSIONAL ANALYSIS

6.1 INTRODUCTION

Solutions to engineering problems, due to their complex nature, are determined mostly from experiments. Due to economic advantages, saving of time and ease of investigations, it is not possible in a number of instances, to perform experiments in the laboratory under identical conditions, in relation to the operating parameters prevailing in practice. Therefore, laboratory tests are usually carried out under altered conditions of the operating variables from the actual ones in practice. These variables in case of problems relating to fluid flow are pressure, velocity, geometrical dimensions of the working systems, and the physical properties of the working fluid. The pertinent questions arising out of this situation are the following:

(i) How can we apply the test results from laboratory experiments to the actual problems for another set of conditions in practice?

(ii) When the performance of a system is governed by a large number of operating parameters as the input variables, a large number of experiments are required to be carried out accordingly to determine the influences of each and every operating parameter on the performance of the system. Is it possible, by any way, to reduce the large number of experiments, involving huge labour, time, and cost, to a lesser one in achieving the same objective?

A positive clue in answering the above two questions lies in the principle of physical similarity. This principle makes it possible and justifiable (i) to apply the results taken from tests under one set of conditions to another set of conditions and (ii) to predict the influences of a large number of independent operating variables on the performance of a system from an experiment with a limited number of operating variables. Therefore, a large part of the progress made in the study of mechanics of fluids and in the engineering applications of the subject has come from experiments conducted on scale models. No aircraft is now built before exhaustive tests are carried out on small models in a wind tunnel. The behaviour and power requirements of a ship are calculated in advance from the results of tests in which a small model of the ship is towed through water. Flood control of rivers, spillways of dams, harbour works, performances of fluid machines like turbines, pumps and propellers, and similar large-scale projects are studied in detail with models in the laboratory. In a number of situations, tests are conducted with one fluid and the results are applied to situations in which another fluid is used.

6.2 CONCEPT AND TYPES OF PHYSICAL SIMILARITY

The primary and fundamental requirement for the physical similarity between two problems is that the physics of the problems must be the same. For example, a fully developed flow through a closed conduit can never be made, (under any situation) physically similar to a flow in an open channel, since the flow in the earlier case is governed by viscous and pressure forces while the gravity force is dominant in the latter case to maintain the flow. Therefore, the laws of similarity have to be sought between problems described by the same physics. We shall first define physical similarity as a general proposition. Two systems, described by the same physics, but operating under different sets of conditions are said to be physically similar in respect of certain specified physical quantities when the ratio of corresponding magnitudes of these quantities between the two systems is the same everywhere. If the specified physical quantities are geometrical dimensions, the similarity is called *geometric similarity*, if the quantities are related to motion, the similarity is called *kinematic similarity* and if the quantities refer to force, then the similarity is termed as *dynamic similarity*. In the field of mechanics, these three similarities together constitute the complete similarity between problems of the same kind.

6.2.1 Geometric Similarity

Geometric similarity is the similarity of shape. This is probably the type of similarity most commonly encountered and most easily understood. In geometrically similar systems, the ratio of any length in one system to the corresponding length in the other system is the same everywhere. This ratio is usually known as the scale factor. Therefore, geometrically similar objects are similar in their shapes, i.e., proportionate in their physical dimensions, but differ in size. In investigations of physical similarity, the full size or actual scale systems are known as *prototypes* while the laboratory scale systems are referred to as *models*. As already indicated, use of the same fluid with both the prototype and the model is not necessary, nor is the model necessarily smaller than the prototype. The flow of fluid through an injection nozzle or a curburettor, for example, would be more easily studied by using a model much larger than the prototype. The model and prototype may be of identical size, although the two may then differ in regard to other factors such as velocity, and properties of the fluid. If l_1 and l_2 are the two characteristic physical dimensions of any object, then the requirement of geometrical similarity is

$$\frac{l_{1m}}{l_{1p}} = \frac{l_{2m}}{l_{2p}} = l_r$$

(The second suffices m and p refer to model and prototype, respectively) where l_r is the scale factor or sometimes known as the *model ratio*. Figure 6.1 shows three pairs of geometrically similar objects, namely, a right circular cylinder, a parallolepiped, and a triangular prism.

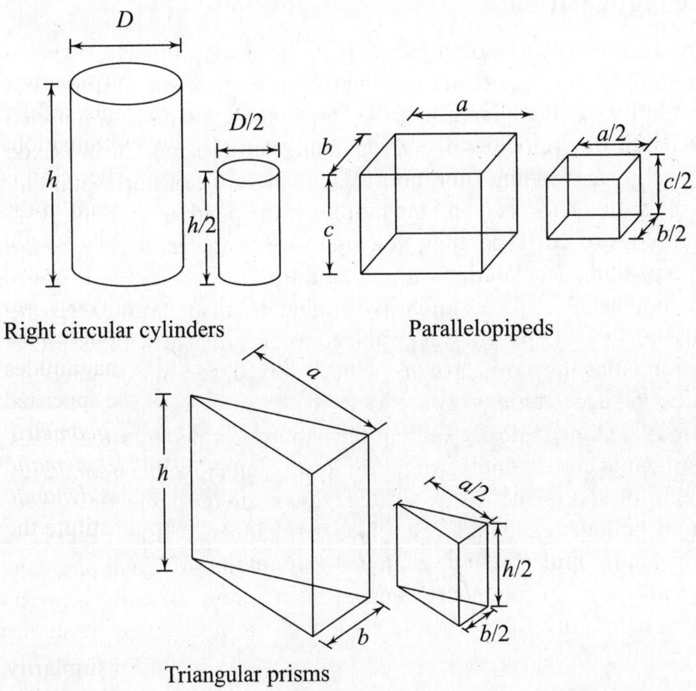

Fig. 6.1 Geometrically similar objects

It can be mentioned in this context that roughness of the surface should also be geometrically similar. Geometric similarity is perhaps the most obvious requirement in a model system designed to correspond to a given prototype system. A perfect geometric similarity is not always easy to attain. For a small model, the surface roughness might not be reduced according to the scale factor unless the model surfaces can be made very much smoother than those of the prototype. If for any reason the scale factor is not the same throughout, a distorted model results.

Sometimes it may so happen that to have a perfect geometric similarity within the available laboratory space, physics of the problem changes. For example, in case of large prototypes, such as rivers, the size of the model is limited by the available floor space of the laboratory; but if a very low scale factor is used in reducing both the horizontal and vertical lengths, this may result in a stream so shallow that surface tension has a considerable effect and moreover, the flow may be laminar instead of turbulent. In this situation, a distorted model may be unavoidable (a lower scale factor for horizontal lengths while a relatively higher scale factor for vertical lengths). The extent to which perfect geometric similarity should be sought therefore depends on the problem being investigated, and the accuracy required from the solution.

6.2.2 Kinematic Similarity

Kinematic similarity is similarity of motion. Since motions are described by distance and time, kinematic similarity implies similarity of lengths (i.e., geometrical similarity) and in addition, similarity of time intervals. If the corresponding lengths in the two systems are in a fixed ratio, the velocities of corresponding particles must be in a fixed ratio of magnitude of corresponding time intervals. If the ratio of corresponding lengths, known as the scale factor, is l_r and the ratio of corresponding time intervals is t_r, then the magnitudes of corresponding velocities are in the ratio l_r/t_r and the magnitudes of corresponding accelerations are in the ratio l_r/t_r^2.

A well-known example of kinematic similarity is found in a planetarium. Here the galaxies of stars and planets in space are reproduced in accordance with a certain length scale and in simulating the motions of the planets, a fixed ratio of time intervals (and hence velocities and accelerations) is used.

When fluid motions are kinematically similar, the patterns formed by streamlines are geometrically similar at corresponding times. Since the impermeable boundaries also represent streamlines, kinematically similar flows are possible only past geometrically similar boundaries. Therefore, geometric similarity is a necessary condition for the kinematic similarity to be achieved, but not the sufficient one. For example, geometrically similar boundaries may ensure geometrically similar streamlines in the near vicinity of the boundary but not at a distance from the boundary.

6.2.3 Dynamic Similarity

Dynamic similarity is the similarity of forces. In dynamically similar systems, the magnitudes of forces at similar points in each system are in a fixed ratio. In other words, the ratio of magnitudes of any two forces in one system must be the same as the magnitude ratio of the corresponding forces in other systems. In a system involving flow of fluid, different forces due to different causes may act on a fluid element. These forces are as follows:

Viscous force (due to viscosity)	$\vec{F}_v$
Pressure force (due to difference in pressure)	$\vec{F}_p$
Gravity force (due to gravitational attraction)	$\vec{F}_g$
Capillary force (due to surface tension)	$\vec{F}_c$
Compressibility force (due to elasticity)	$\vec{F}_e$

According to Newton's law, the resultant F_R of all these forces, will cause the acceleration of a fluid element. Hence,

$$\vec{F}_R = \vec{F}_V + \vec{F}_p + \vec{F}_g + \vec{F}_c + \vec{F}_e \tag{6.1}$$

Moreover, the inertia force $\vec{F}_i$ is defined as equal and opposite to the resultant accelerating force $\vec{F}_R$. Therefore, Eq. (6.1) may also be expressed as

$$\vec{F}_v + \vec{F}_p + \vec{F}_g + \vec{F}_c + \vec{F}_e + \vec{F}_i = 0$$

For dynamic similarity, the magnitude ratios of these forces have to be same for both the prototype and the model. The inertia force $\vec{F}_i$ is usually taken as the common one to describe the ratios as

$$\frac{|\vec{F}_v|}{|\vec{F}_i|}, \frac{|\vec{F}_p|}{|\vec{F}_i|}, \frac{|\vec{F}_g|}{|\vec{F}_i|}, \frac{|\vec{F}_c|}{|\vec{F}_i|}, \frac{|\vec{F}_e|}{|\vec{F}_i|}$$

A fluid motion, under all such forces is characterised by (i) hydrodynamic parameters like pressure, velocity and acceleration due to gravity, (ii) rheological and other physical properties of the fluid involved, and (iii) geometrical dimensions of the system. Now it becomes important to express the magnitudes of different forces in terms of these parameters, so as to know the extent of their influences on the different forces acting on a fluid element in the course of its flow.

Inertia Force $\vec{F}_i$ The inertia force acting on a fluid element is equal in magnitude to the mass of the element multiplied by its acceleration. The mass of a fluid element is proportional to ρl^3, where ρ is the density of fluid and l is the characteristic geometrical dimension of the system. The acceleration of a fluid element in any direction is the rate at which its velocity in that direction changes with time and is therefore proportional in magnitude to some characteristic velocity V, divided by some specified interval of time t. The time interval t, is proportional to the characteristic length l, divided by the characteristic velocity V, so that the acceleration becomes proportional to V^2/l. The magnitude of inertia force is thus proportional to $\rho l^3 V^2/l = \rho l^2 V^2$. This can be written as

$$|\vec{F}_i| \propto \rho l^2 V^2 \tag{6.2a}$$

Viscous Force $\vec{F}_v$ The viscous force arises from shear stress in a flow of fluid. Therefore, we can write

Magnitude of viscous force $\vec{F}_v$ = Shear stress × Surface area over which the shear stress acts

Again, shear stress = μ (Viscosity) × Rate of shear strain

where, rate of shear strain $\propto$ velocity gradient $\propto \frac{V}{l}$ and surface are $\propto l^2$

Hence,
$$|\vec{F}_v| \propto \mu \frac{V}{l} l^2$$
$$\propto \mu V l \tag{6.2b}$$

Pressure Force $\vec{F}_p$ The pressure force arises due to the difference of pressure in a flow field. Hence, it can be written as

$$|\vec{F}_p| \propto \Delta p \cdot l^2 \tag{6.2c}$$

where Δp is some characteristic pressure difference in the flow.

Gravity Force $\vec{F}_g$ The gravity force on a fluid element is its weight.

Hence,
$$|\vec{F}_g| \propto \rho l^3 g \tag{6.2d}$$

where g is the acceleration due to gravity or weight per unit mass.

Capillary or Surface Tension Force $\vec{F}_c$ The capillary force arises due to the existence of an interface between two fluids. The surface tension force acts tangential to a surface and is equal to the coefficient of surface tension σ multiplied by the length of a linear element on the surface perpendicular to which the force acts. Therefore,

$$\left|\vec{F}_c\right| \propto \sigma l \tag{6.2e}$$

Compressibility or Elastic Force $\vec{F}_e$ Elastic force comes into consideration due to the compressibility of the fluid in course of its flow. It has been shown in Eq. (1.3) that for a given compression (a decrease in volume), the increase in pressure is proportional to the bulk modulus of elasticity E and gives rise to a force known as the *elastic force*.

Hence,
$$\left|\vec{F}_e\right| \propto El^2 \tag{6.2f}$$

The flow of a fluid in practice does not involve all the forces simultaneously. Therefore, the pertinent dimensionless parameters for dynamic similarity are derived from the ratios of dominant forces causing the flow.

6.2.4 Dynamic Similarity of Flows governed by Viscous, Pressure and Inertia Forces

The criteria of dynamic similarity for the flows controlled by viscous, pressure and inertia forces are derived from the ratios of the representative magnitudes of these forces with the help of Eqs (6.2a) to (6.2c) as follows:

$$\frac{\text{Viscous force}}{\text{Inertia force}} = \frac{\left|\vec{F}_v\right|}{\left|\vec{F}_i\right|} \propto \frac{\mu V l}{\rho V^2 l^2} = \frac{\mu}{\rho l V} \tag{6.3a}$$

$$\frac{\text{Pressure force}}{\text{Inertia force}} = \frac{\left|\vec{F}_p\right|}{\left|\vec{F}_i\right|} \propto \frac{\Delta p l^2}{\rho l^2 V^2} = \frac{\Delta p}{\rho V^2} \tag{6.3b}$$

The term $\rho l V/\mu$ is known as *Reynolds number*, Re, after the name of the scientist who first developed it and is thus proportional to the magnitude ratio of inertia force to viscous force. The term $\Delta p/\rho V^2$ on the RHS of Eq. (6.3b) is known as *Euler number*, Eu, after the name of the scientist who first derived it. Therefore, the dimensionless terms Re and Eu represent the criteria of dynamic similarity for the flows which are affected only by viscous, pressure and inertia forces. Such instances, for example, are (i) the full flow of fluid in a completely closed conduit, (ii) flow of air past a low-speed aircraft and (iii) the flow of water past a submarine deeply submerged to produce no waves on the surface. Hence, for a complete dynamic similarity to exist between the prototype and the model for this class of flows, the Reynolds number, Re, and the Euler number, Eu, have to be same for the two (prototype and model). Thus,

$$\frac{\rho_p l_p V_p}{\mu_p} = \frac{\rho_m l_m V_m}{\mu_m} \tag{6.3c}$$

$$\frac{\Delta p_p}{\rho_p V_p^2} = \frac{\Delta p_m}{\rho_m V_m^2} \tag{6.3d}$$

where the suffix p and suffix m refer to the parameters for prototype and model respectively. In practice, the pressure drop is the dependent variable, and hence it is compared to the two systems with the help of Eq. (6.3d), while the equality of Reynolds number (Eq. (6.3c)) along with the equalities of other parameters in relation to kinematic and geometric similarities are maintained.

The characteristic geometrical dimension l, and the reference velocity V, in the expression of the Reynolds number may be any geometrical dimension and any velocity which are significant in determining the pattern of flow. For internal flows through a closed duct, the hydraulic diameter of the duct D_h, and the average flow velocity at a section are invariably used for l and V, respectively. The hydraulic diameter D_h, is defined as $D_h = 4A/P$, where A and P are the cross-sectional area and wetted perimeter, respectively.

6.2.5 Dynamic Similarity of Flows with Gravity, Pressure and Inertia Forces

A flow of the type, where significant forces are gravity force, pressure force and inertia force, is found when a free surface is present. One example is the flow of a liquid in an open channel; another is the wave motion caused by the passage of a ship through water. Other instances are the flows over weirs and spillways. The condition for dynamic similarity of such flows requires the equality of the Euler number Eu (the magnitude ratio of pressure to inertia force), and the equality in the magnitude ratio of gravity to inertia force at corresponding points in the systems being compared.

From Eqs (6.2a) and (6.2d),

$$\frac{\text{Gravity force}}{\text{Inertia force}} = \frac{|\vec{F}_g|}{|\vec{F}_i|} \propto \frac{\rho l^3 g}{\rho l^2 V^2} = \frac{lg}{V^2} \tag{6.3e}$$

In practice, it is often convenient to use the square root of this ratio to have the first power of the velocity. From a physical point of view, equality of $(lg)^{1/2}/V$ implies equality of lg/V^2 as regard to the concept of dynamic similarity. The reciprocal of the term $(lg)^{1/2}/V$ is known as *Froude number* after William Froude who first suggested the use of this number in the study of naval architecture. Hence Froude number, $\text{Fr} = V/(lg)^{1/2}$. Therefore, the primary requirement for dynamic similarity between the prototype and the model involving flow of fluid with gravity as the dominant force, is the equality of Froude number, Fr, i.e.,

$$\frac{(l_p g_p)^{1/2}}{V_p} = \frac{(l_m g_m)^{1/2}}{V_m} \tag{6.3f}$$

6.2.6 Dynamic Similarity of Flows with Surface Tension as the Dominant Force

Surface tension forces are important in certain classes of practical problems such as (i) flows in which capillary waves appear, (ii) flows of small jets and thin sheets of liquid injected by a nozzle in air (iii) flow of a thin sheet of liquid over a solid surface. Here the significant parameter for dynamic similarity is the magnitude ratio of the surface tension force to the inertia force, and can be written with the help of Eqs (6.2a) and (6.2e) as

$$\frac{|\vec{F}_c|}{|\vec{F}_i|} \propto \frac{\sigma l}{\rho l^2 V^2} = \frac{\sigma}{\rho V^2 l} \tag{6.3g}$$

The term $\sigma/\rho V^2 l$ is usually known as *Weber number*, Wb, after the German naval architect Moritz Weber who first suggested the use of this term as a relevant parameter.

6.2.7 Dynamic Similarity of Flows with Elastic Force

When the compressibility of fluid in the course of its flow becomes important, the elastic force along with the pressure and inertia forces has to be considered. Therefore, the magnitude ratio of inertia to elastic force becomes a relevant parameter for dynamic similarity under this situation. With the help of Eqs (6.2a) and (6.2f)

$$\frac{\text{Inertia force}}{\text{Elastic force}} = \frac{|\vec{F}_i|}{|\vec{F}_e|} \propto \frac{\rho l^2 V^2}{E l^2} = \frac{\rho V^2}{E} \tag{6.3h}$$

The parameter $\rho V^2/E$ is known as the *Cauchy number*, after the French mathematician A.L. Cauchy. If we consider the flow to be isentropic, then it can be written

$$\frac{|\vec{F}_i|}{|\vec{F}_e|} \propto \frac{\rho V^2}{E_s} \tag{6.3i}$$

where E_s is the *isentropic bulk modulus* of elasticity. It is shown in Chapter 14 that the velocity with which a sound wave propagates through a fluid medium equals to $\sqrt{E_s/\rho}$. Hence, the term $\rho V^2/E_s$ can be written as V^2/a^2, where a is the acoustic velocity in the fluid medium. The ratio V/a is known as the *Mach number,* Ma, after an Austrian physicist Earnst Mach. It has been shown in Chapter 1, Eq. (1.29) that the effects of compressibility become important when the Mach number exceeds 0.33. The situation arises in the flow of air past high-speed aircraft, missiles, propellers and rotory compressors. In these cases equality of the Mach number is a condition for dynamic similarity.

Therefore, $\quad V_p/a_p = V_m/a_m \qquad (6.3j)$

It is appropriate at this point to summarise the ratios of forces arising in the context of dynamic similarity for different situations of flow as discussed above. This is shown in Table 6.1.

Table 6.1

Pertinent dimensionless term as the criterion of dynamic similarity in different situations of fluid flow	*(Representative) magnitude ratio of the forces*	*Name*	*Recommended symbol*
$\rho l V/m$	$\dfrac{\text{Inertia force}}{\text{Viscous force}}$	Reynolds number	Re
$\Delta p/\rho V^2$	$\dfrac{\text{Pressure force}}{\text{Inertia force}}$	Euler number	Eu
$V/(lg)^{1/2}$	$\dfrac{\text{Inertia force}}{\text{Gravity force}}$	Froude number	Fr
$\sigma/\rho V^2 l$	$\dfrac{\text{Surface tension force}}{\text{Inertia force}}$	Weber number	Wb
$V/\sqrt{E_s/\rho}$	$\dfrac{\text{Inertia force}}{\text{Elastic force}}$	Mach number	Ma

6.3 THE APPLICATION OF DYNAMIC SIMILARITY—DIMENSIONAL ANALYSIS

We have already seen that a number of dimensionless parameters, representing the magnitude ratios of certain physical variables, namely, geometrical dimension, velocity and force become the criteria of complete physical similarity between systems governed by the same physical phenomenon. Therefore, a physical problem may be characterised by a group of dimensionless similarity parameters or variables rather than by the original dimensional variables. This gives a clue to the reduction in the number of parameters requiring separate consideration in an experimental investigation. For example, if the Reynolds number Re = $\rho V D_h/\mu$ is considered as the independent variable, in case of a flow of fluid through a closed duct of hydraulic diameter D_h, then a change in Re may be caused through a change in flow velocity V, only. Thus a range of Re can be covered simply by the variation in V, without varying other independent dimensional variables ρ, D_h and μ. In fact, the variation in the Reynolds number physically implies the variation in any of the dimensional parameters defining it, though the change in Re, may be obtained through the variation in any one parameter, say the velocity V. A number of such dimensionless parameters in relation to dynamic similarity are shown in Table 6.1. Sometimes it becomes difficult to derive these parameters straightforward from an estimation of the representative order of magnitudes of the forces involved. An alternative method of determining these dimensionless parameters by a mathematical technique is known as *dimensional analysis*. The requirement of dimensional homogeneity imposes conditions on the quantities involved in a physical problem, and these restrictions, placed in the form of an algebraic function by the requirement of dimensional homogeneity, play the central role in dimensional analysis. There are two existing approaches; one due to *Buckingham* and the other due to *Rayleigh*. Before discussing the description of these two methods, a few examples of the dimensions of physical quantities are given as follows.

6.3.1 Dimensions of Physical Quantities

All physical quantities are expressed by magnitudes and units. For example, the velocity and acceleration of a fluid particle are 8 m/s and 10 m/s^2 respectively. Here the dimensions of velocity and acceleration are ms^{-1} and ms^{-2} respectively. In SI (System International) units, the primary physical quantities which are assigned base dimensions are mass, length, time, temperature, current, and luminous intensity. Of these, the first four are used in fluid mechanics and they are symbolised as M (mass), L (length), T (time), and θ (temperature).

Any physical quantity can be expressed in terms of these primary quantities by using the basic mathematical definition of the quantity. The resulting expression is known as the dimension of the quantity. For example, shear stress τ is defined as force/area. Again, force = mass × acceleration

$$\text{Dimensions of acceleration} = \text{Dimensions of velocity/Dimension of time}$$

$$= \frac{\text{Dimension of distance}}{(\text{Dimension of time})^2}$$

$$= \frac{\mathrm{L}}{\mathrm{T}^2}$$

$$\text{Dimension of area} = (\text{Length})^2 = \mathrm{L}^2$$

Hence, dimensions of shear stress $\tau = \mathrm{ML/T^2L^2} = \mathrm{ML^{-1}T^{-2}}$ (6.4)

To find out the dimension of viscosity, as another example, one has to consider Newton's law (Eq. 1.1) for the definition of viscosity as

$$\tau = \mu\, du/dy$$

or

$$\mu = \frac{\tau}{(du/dy)}$$

The dimension of velocity gradient du/dy can be written as

$$\text{dimension of } du/dy = \text{dimension of } u/\text{dimension of } y = \mathrm{L/TL} = \mathrm{T^{-1}}$$

The dimenison of shear stress τ is given in Eq. (6.4).

$$\text{Hence, dimension of } \mu = \frac{\text{Dimension of } \tau}{\text{Dimension of } du/dy} = \frac{ML^{-1}T^{-2}}{T^{-1}}$$

$$= ML^{-1}T^{-1}$$

Dimensions of various physical quantities commonly encountered in problems on fluid flow are given in Table 6.2.

Table 6.2 Dimensions of physical quantities

Physical quantity	*Dimension*
Mass	M
Length	L
Time	T
Temperature	θ
Velocity	LT^{-1}
Angular velocity	T^{-1}
Acceleration	LT^{-2}
Angular acceleration	T^{-2}
Force, Thrust, Weight	MLT^{-2}
Stress, Pressure	$ML^{-1}T^{-2}$
Momentum	MLT^{-1}
Angular momentum	ML^2T^{-1}
Moment, Torque	ML^2T^{-2}
Work, Energy	ML^2T^{-2}
Power	ML^2T^{-3}
Stream function	L^2T^{-1}
Vorticity, Shear rate	T^{-1}
Velocity potential	L^2T^{-1}
Density	ML^{-3}
Coefficient of dynamic viscosity	$ML^{-1}T^{-1}$
Coefficient of kinematic viscosity	L^2T^{-1}
Surface tension	MT^{-2}
Bulk modulus of elasticity	$ML^{-1}T^{-2}$

6.3.2 Buckingham's Pi Theorem

When a physical phenomenon is described by m number of variables like $x_1, x_2, x_3, \ldots x_m$, we may express the phenomenon analytically by an implicit functional relationship of the controlling variables as

$$f(x_1, x_2, x_3, .., x_m) = 0 \tag{6.5}$$

Let n be the number of fundamental dimensions like mass, length, time, temperature, etc., involved in these m variables, then according to Buckingham's π theorem, and following the conditions of dimensional homogeneity of the variables, the phenomenon can be described in terms of $(m - n)$ dimensionless groups like π_1, π_2, ... π_{m-n}. The π terms, which represent the dimensionless groups or parameters consist of different combinations of a number of dimensional variables out of the m variables defining the problem. Therefore, the analytical version of the phenomenon given by Eq. (6.5) can be reduced to

$$F(\pi_1, \pi_2, ... \pi_{m-n}) = 0 \tag{6.6}$$

This physically implies that the phenomenon which is basically described by m dimensional variables, is ultimately controlled by $(m-n)$ dimensionless parameters known as π terms. One out of these π terms includes the dependent physical variable and thus represents the dependent π term which is a function of the remaining independent π terms

6.3.2.1 *Determination of π terms*

The number of π terms is fixed by the Pi theorem. The next step is the determination of π terms as follows:

Any group of n (n = number of fundamental dimensions) variables out of m (m = total number of variables defining the problem) is first chosen. These n variables are referred to as repeating variables. Then the π terms are formed by the product of these repeating variables raised to arbitrary unknown integer exponents and any one of the excluded $(m - n)$ variables. For example, $x_1\, x_2 \ldots x_n$ are taken as the repeating variables. Then,

$$\pi_1 = x_1^{a_1}\, x_2^{a_2} \ldots x_n^{a_n}\, x_{n+1}$$

$$\pi_2 = x_1^{a_1}\, x_2^{a2} \ldots x_n^{a_n}\, x_{n+2}$$

....

$$\pi_{m-n} = x_1^{a_1}\, x_2^{a_2} \ldots x_n^{a_n}\, x_m$$

The sets of integer exponents $a_1\, a_2 \ldots a_n$ are different for each π term. Since π terms are dimensionless, it requires that if all the variables in any π term are expressed in terms of their fundamental dimensions, the exponent of all the basic dimensions must be zero. This leads to a system of n linear equations in $a_1\, a_2 \ldots a_n$ which gives a unique solution for the exponents. Thus the values of $a_1\, a_2 \ldots a_n$ for each π term are known and hence the π terms are uniquely defined. In selecting the repeating variables, the following points have to be considered:

(i) The repeating variables must include among them all the n fundamental dimensions, not necessarily in each one but collectively.

(ii) The dependent variable or the output parameter of the physical phenomenon should not be included in the repeating variables.

It can be mentioned in this context that when $m < n$, there is no solution which means no physical phenomenon is described under this situation. Moreover, when $m = n$, there is a unique solution of the variables involved and hence all the parameters have fixed values. This situation also does not represent a physical phenomenon or process. Therefore all feasible phenomena in paractice are defined with $m > n$. When $m = n + 1$, then, according to the Pi theorem, the number of π term is one and the phenomenon can be expressed as

$$f(\pi_1) = 0$$

where, the non-dimensional term π_1 is some specific combination of $n + 1$ variables involved in the problem.

When $m > n + 1$, the number of π terms are more than one. The most important point to discuss is that a number of choices regarding the repeating variables arise in this case. Again, it is true that if one of the repeating variables is changed, it results in a different set of π terms. Therefore the interesting question is which set of repeating variables is to be chosen, to arrive at the correct set of π terms to describe the problem. The answer to this question lies in the fact that different sets of π terms resulting from the use of different sets of repeating variables are not independent. Thus, any one of such interdependent sets is meaningful in describing the same physical phenomenon.

From any set of such π terms, one can obtain the other meaningful sets from some combination of the π terms of the existing set without altering their total numbers $(m–n)$ as fixed by the Pi theorem. The following two examples will make the understanding of Buckingham's Pi theorem clear.

Example 6.1

The vertical displacement h of a freely falling body from its point of projection at any time t, is determined by the acceleration due to gravity g. Find the relationship of h with t and g using Buckingham's Pi theorem.

Solution

The above phenomenon can be described by the functional relation as

$$F(h, t, g) = 0 \tag{6.7}$$

Here , the number of variables $m = 3$ (h, t, and g) and they can be expressed in terms of two fundamental dimensions L and T. Hence, the number of π terms $= m - n = 3 - 2 = 1$. In determining this π term, the number of repeating variables to be taken is 2. Since h is the dependent variable, the only choice left for the repeating variables is with t and g.

Therefore,

$$\pi_1 = t^a g^b h \tag{6.8}$$

By substituting the fundamental dimensions of the variables on the left- and- right-hand sides of Eq. (6.8), we get

$$L^0T^0 = T^a (LT^{-2})^b L$$

Equating the exponents of T and L on both the sides of the above equation we have

$$a - 2b = 0$$

and $$b+1=0$$

which give,

$$a=-2$$

$$b=-1$$

Hence, $$\pi_1 = h/gt^2$$

Therefore the functional relationship (Eq. (6.7)) of the variables describing the phenomenon of free fall of a body under gravity can be written in terms of the dimensionless parameter (π_1) as

$$f\left(\frac{h}{gt^2}\right)=0 \tag{6.9}$$

From elementary classical mechanics we know that $\frac{h}{gt^2}=\frac{1}{2}$. One should know, in this context, that the Pi theorem can only determine the pertinent dimensionless groups describing the problem but not the exact functional relationship between them.

Example 6.2

For a steady, fully developed laminar flow through a duct, the pressure drop per unit length of the duct $\Delta\, p/l$ is constant in the direction of flow and depends on the average flow velocity V, the hydraulic diameter of the duct D_h, the density ρ, and the viscosity μ, of the fluid. Find out the pertinent dimensionless groups governing the problem by the use of Buckingham's π theorem.

Solution

The variables involved in the problem are

$$\frac{\Delta p}{l}, V, D_h, \rho, \ \mu$$

Hence, $$m=5.$$

The fundamental dimensions in which these five variables can be expressed are M (mass), L (length) and T time. Therefore, $n=3$. According to Pi theorem, the number of independent π terms is (5–3)=2, and the problem can be expressed as

$$f(\pi_1\ \pi_2)=0 \tag{6.10}$$

In determining π_1 and π_2, the number of repeating variables that can be taken is 3. The term $\Delta\, p/l$ being the dependent variable should not be taken as the repeating one. Therefore, choices are left with V, D_h, ρ and μ. Incidentally, any combination of three out of these four quantities involves all the fundamental dimensions M, L and T. Hence any one of the following four possible sets of repeating variables can be used:

$$V, D_h, \rho$$
$$V, D_h, \mu$$
$$D_h, \rho, \mu$$
$$V, \rho, \mu$$

Let us first use the set V, D_h and ρ. Then the π terms can be written as

$$\pi_1 = V^a D_h^b \rho^c \Delta p/l \qquad (6.11)$$

$$\pi_2 = V^a D_h^b \rho^c \mu \qquad (6.12)$$

Expressing the Eqs (6.11) and (6.12) in terms of the fundamental dimensions of the variables, we get

$$\mathrm{M^0L^0T^0} = (\mathrm{LT^{-1}})^a (\mathrm{L})^b (\mathrm{ML^{-3}})^c \mathrm{ML^{-2}T^{-2}} \qquad (6.13)$$

$$\mathrm{M^0L^0T^0} = (\mathrm{LT^{-1}})^a (\mathrm{L})^b (\mathrm{ML^{-3}})^c \mathrm{ML^{-1}T^{-1}} \qquad (6.14)$$

Equating the exponents of M, L and T on both sides of Eq. (6.13) we have,

$$c + 1 = 0$$

$$a + b - 3c - 2 = 0$$

$$-a - 2 = 0$$

which give $a = -2, b = 1$ and $c = -1$

$$\therefore \qquad p_1 = \frac{\Delta p D_h}{l \rho V^2}$$

Similarly from Eq. (6.14)

$$c + 1 = 0$$

$$a + b - 3c - 1 = 0$$

$$-a - 1 = 0$$

which give $a = -1, b = -1$, and $c = -1$

Therefore, $\pi_2 = \dfrac{\mu}{V D_h \rho}$

Hence, Eq. (6.10) can be written as

$$F\left(\frac{\Delta p D_h}{l\rho V^2}, \frac{\mu}{V D_h \rho}\right) = 0 \qquad (6.15)$$

The term π_2 is the reciprocal of Reynolds number, Re, as defined earlier.Equation (6.15) can also be expressed as

$$f\left(\frac{\Delta p D_h}{l\rho V^2}, \frac{V D_h \rho}{\mu}\right) = 0 \qquad (6.16)$$

or

$$\frac{\Delta p D_h}{l\rho V^2} = \phi(\mathrm{Re}) \qquad (6.17)$$

The term π_1, i.e., $\dfrac{\Delta p D_h}{l\rho V^2}$ is known as the *friction factor* in relation to a fully developed flow through a closed duct.

Let us now choose V, D_h and μ as the repeating variables.

Then

$$\pi_1 = V^a D_h^b \mu^c (\Delta p/l) \qquad (6.18)$$

$$\pi_2 = V^a D_h^b \mu^c \rho \tag{6.19}$$

Expressing the right-hand side of Eq. (6.18) in terms of fundamental dimensions, we have

$$M^0L^0T^0 = (LT^{-1})^a L^b (ML^{-1}T^{-1})^c ML^{-2}T^{-2}$$

Equating the exponents of M, L and T from above,

$$c + 1 = 0$$
$$a + b - c - 2 = 0$$
$$-a - c - 2 = 0$$

Finally, $a = -1, b = 2, c = -1$

Therefore, $$\pi_1 = \frac{\Delta p}{l} \frac{D_h^2}{V\mu}$$

Similarly, equating the exponents of fundamental dimensions of the variables on both sides of Eq. (6.19) we get

$$a = 1, b = 1, c = -1$$

Therefore, $$\pi_2 = \frac{\rho V D_h}{\mu}$$

Hence, the same problem which was defined by Eq. (6.15) can also be defined by the equation

$$f\left(\frac{\Delta p}{l} \frac{D_h^2}{V\mu}, \frac{\rho V D_h}{\mu}\right) = 0 \tag{6.20}$$

Though the Eqs (6.15) and (6.20) are not identical, but they are interdependent. Now if we write the two sets of π terms obtained straightforward from the application of π theorem as

	π_1	π_2
Set 1:	$\dfrac{\Delta p D_h}{l \rho V^2}$,	$\dfrac{\mu}{\rho V D_h}$
Set 2:	$\dfrac{\Delta p D_h^2}{l V \mu}$,	$\dfrac{\rho V D_h}{\mu}$

We observe that

$$(1/\pi_2) \text{ of Set 2} = (\pi_2) \text{ of Set 1}$$

and $$(\pi_1/\pi_2) \text{ of Set 2} = (\pi_1) \text{ of Set 1}$$

Therefore, it can be concluded that, from one set of π terms, one can obtain the other set by some combination of the π terms of the existing set. It is justified both mathematically and physically that the functional relationship of π terms representing a problem in the form

$$f(\pi_1, \pi_2, \ldots \pi_r) = 0$$

is equivalent to any implicit functional relationship between other π terms obtained from any arbitrary mathematical combination of π terms of the existing set, provided the total number of independent π terms remains the same. For example, Eq. (6.20) and Eq. (6.15) can be defined in terms of π parameters of the Set 2 as $f(\pi_1, \pi_2) = 0$ and $F(\pi_1/\pi_2, 1/\pi_2) = 0$, respectively.

Table 6.3 shows different mutually interdependent sets of π terms obtained from all possible combinations of the repeating variables of Example 6.2. Though the different sets of π terms as shown in Column 2 of Table 6.3 are mathematically meaningful, many of them lack physical significance. The physically meaningful parameters of the problem are $\Delta p\ D_h/l\rho V^2$ are $\rho V D_h/\mu$ and are known as *friction factor* and *Reynolds number*, respectively. Therefore while selecting the repeating variables, for a fluid flow problem, it is desirable to choose one variable with geometric characteristics, another variable with flow characteristics and yet another variable with fluid properties. This ensures that the dimensionless parameters obtained will be the meaningful ones with respect to their physical interpretations.

Table 6.3 Different sets of π terms resulting from different combinations of repeating variables of a pipe flow problem

Repeating variables	*Set of π terms* π_1	π_2	*Functional relation*
V, D_h, ρ	$\dfrac{\Delta p D_h}{l\rho V^2}$	$\dfrac{\mu}{\rho V D_h}$	$F\left(\dfrac{\Delta p D_h}{l\rho V^2}, \dfrac{\mu}{\rho V D_h}\right) = 0$
V, D_h, μ	$\dfrac{\Delta p D_h^2}{l V \mu}$	$\dfrac{\rho V D_h}{\mu}$	$f\left(\dfrac{\Delta p D_h^2}{l V \mu}, \dfrac{\rho V D_h}{\mu}\right) = 0$
D_h, ρ, μ	$\dfrac{\Delta p D_h^3 \rho}{l \mu^2}$	$\dfrac{\rho V D_h}{\mu}$	$\phi\left(\dfrac{\Delta p D_h^3 \rho}{l \mu^2}, \dfrac{\rho V D_h}{\mu}\right) = 0$
V, ρ, m	$\dfrac{\Delta p \mu}{l V^3 \rho^2}$	$\dfrac{\rho V D_h}{\mu}$	$\psi\left(\dfrac{\Delta p \mu}{l V^3 \rho^2}, \dfrac{\rho V D_h}{\mu}\right) = 0$

The above discussion on Buckingham's π theorem can be summarised as follows:

(i) List the m physical quantities involved in a particular problem. Note the number n, of the fundamental dimensions to express the m quantities. There will be $(m–n)$ π terms.

(ii) Select n of the m quantities, excluding any dependent variable, none dimensionless and no two having the same dimensions. All fundamental dimensions must be included collectively in the quantities chosen.

(iii) The first π term can be expressed as the product of the chosen quantities each raised to an unknown exponent and one other quantity.

(iv) Retain the quantities chosen in (ii) as repeating variables and then choose one of the remaining variables to establish the next π term in a similar manner as described in (iii). Repeat this procedure for the successive π terms.

(v) For each π term, solve for the unknown exponents by dimensional analysis.
(vi) If a quantity out of m physical variables is dimensionless, it is a π term.
(vii) If any two physical quantities have the same dimensions, their ratio will be one of the π terms.
(viii) Any π term may be replaced by the term, raised to an exponent. For example, π_3 may be replaced by $\pi_3^{\,2}$ or π_2 by $\sqrt{\pi_2}$.
(ix) Any π term may be replaced by multiplying it by a numerical constant. For example, π_1 may be replaced by $3\,\pi_1$.

Example 6.3

Drag force F, on a high speed air craft depends on the velocity of flight V, the characteristic geometrical dimension of the air craft l, the density ρ, viscosity μ, and isentropic bulk modulus of elasticity E_s, of ambient air. Using Buckingham's π theorem, find out the independent dimensionless quantities which describe the phenomenon of drag on the aircraft.

Solution

The physical variables involved in the problem are F, V, l, ρ, μ and E_s; and they are 6 in number. The fundamental dimensions involved with these variables are 3 in number and they are, namely, M, L, T. Therefore, according to the π theorem, the number of independent π terms are $(6 - 3) = 3$.

Now to determine these π terms, V, l and ρ are chosen as the repeating variables. Then the π terms can be written as

$$\pi_1 = V^a\, l^b\, \rho^c\, F$$
$$\pi_2 = V^a\, l^b\, \rho^c\, \mu$$
$$\pi_3 = V^a\, l^b\, \rho^c\, E_s$$

The variables of the above equations can be expressed in terms of their fundamental dimensions as

$$M^0L^0T^0 = (LT^{-1})^a\, L^b\, (ML^{-3})^c\, MLT^{-2} \tag{6.21}$$
$$M^0L^0T^0 = (LT^{-1})^a\, L^b\, (ML^{-3})^c\, ML^{-1}T^{-1} \tag{6.22}$$
$$M^0L^0T^0 = (LT^{-1})^a L^b\, (ML^{-3})^c\, ML^{-1}T^{-2} \tag{6.23}$$

Equating the exponents of M, L and T on both sides of the equations we have, from Eq. (6.21),

$$c + 1 = 0$$
$$a + b - 3c + 1 = 0$$
$$-a - 2 = 0$$

which, give, $\quad a = -2, b = -2$, and $c = -1$

Therefore, $\quad \pi_1 = \dfrac{F}{\rho V^2 l^2}$

From Eq. (6.22), $\quad c + 1 = 0$

$$a + b - 3c - 1 = 0$$
$$-a - 1 = 0$$

which give, $a = -1, b = -1$ and $c = -1$

Therefore, $\pi_2 = \dfrac{\mu}{\rho V l}$

From Eq. (6.23), $c + 1 = 0$

$$a + b - 3c - 1 = 0$$
$$-a - 2 = 0$$

which give $a = -2, b = 0$ and $c = -1$

Therefore, $\pi_3 = \dfrac{E_s}{V^2 \rho}$

$$= \frac{E_s / \rho}{V^2}$$

Hence, the independent dimensionless parameters describing the problem are

$$\pi_1 = \frac{F}{\rho V^2 l^2} \qquad \pi_2 = \frac{\mu}{\rho V l} \qquad \text{and} \qquad \pi_3 = \frac{E_s / \rho}{V^2}$$

Now we see that $\dfrac{1}{\pi_2} = \dfrac{\rho V l}{\mu} = \text{Re (Reynolds number)}$

and $\dfrac{1}{\sqrt{\pi_3}} = \dfrac{V}{\sqrt{E_s / \rho}} = \dfrac{V}{a} = \text{Ma (Mach number)}$

where a is the local speed of sound.

Therefore, the problem of drag on an aircraft can be expressed by an implicit functional relationship of the pertinent dimensionless parameters as

$$f\left(\frac{F}{\rho V^2 l^2}, \frac{\rho V l}{\mu}, \frac{V}{a}\right) = 0$$

or
$$\frac{F}{\rho V^2 l^2} = \phi\left(\frac{\rho V l}{\mu}, \frac{V}{a}\right) \tag{6.24}$$

The term $F/\rho V^2 l^2$ is known as drag coefficient C_D. Hence Eq. (6.24) can be written as

$$C_D = f(\text{Re}, \text{Ma})$$

Example 6.4

An aircraft is to fly at a height of 9 km (where the temperature and pressure are – 45°C and 30.2 kPa, respectively) at 400 m/s. A $1/20^{th}$ scale model is tested in a pressurised wind-tunnel in which the air is at 15 °C. For complete dynamic similarity what pressure and velocity should be used in the wind tunnel? (For air, $\mu \propto T^{3/2}/(T + 117)$, $E_s = \gamma p$, $p = \rho RT$ where the temperature T is in kelvin, γ is the ratio of specific heats).

Solution

We find from Eq. (6.24) that for complete dynamic similarity the Reynolds number, Re and Mach number, Ma for the model must be the same with those of the prototype. From the equality of Mach number Ma,

$$\frac{V_m}{a_m} = \frac{V_p}{a_p}$$

or

$$V_m = V_p \frac{a_m}{a_p}$$

$$= V_p \sqrt{\frac{E_{s_m}/\rho_m}{E_{s_p}/\rho_p}}$$

(Subscripts m and p refer to the model and prototype, respectively.)

$$= V_p \sqrt{\frac{\gamma\, p_m}{\gamma\, p_p} \cdot \frac{\rho_p}{\rho_m}}$$

(Since $a = \sqrt{E_s/\rho}$, and $E = \gamma p$)

Again from the equation of state,

$$\frac{p_m}{\rho_m}\frac{\rho_p}{p_p} = \frac{T_m}{T_p}$$

Hence,

$$V_m = V_p \sqrt{\frac{T_m}{T_p}}$$

$$= 400 \text{ m/s} \sqrt{\frac{(273.15 + 15)}{(273.15 - 45)}}$$

$$= 450 \text{ m/s}$$

From the equality of Reynolds number,

$$\frac{\rho_m V_m l_m}{\mu_m} = \frac{\rho_p V_p l_p}{\mu_p}$$

or

$$\frac{\rho_m}{\rho_p} = \frac{V_p}{V_m} \cdot \frac{l_p}{l_m} \cdot \frac{\mu_m}{\mu_p}$$

or

$$\frac{p_m}{p_p} = \frac{V_p}{V_m} \cdot \frac{l_p}{l_m} \cdot \frac{T_m}{T_p}, \frac{\mu_m}{\mu_p}$$

$$= \frac{V_p}{V_m} \cdot \frac{l_p}{l_m} \left[\frac{T_m}{T_p} \right]^{5/2} \left[\frac{T_p + 117}{T_m + 117} \right]$$

[Since $\mu \propto T^{3/2} / (T + 117)$]

$$\text{Therefore } p_m = 30.2 \text{ kPa} \left(\frac{400}{450} \right) (20) \left(\frac{273.15 + 15}{273.15 - 45} \right)^{5/2} \left(\frac{273.15 - 45 + 117}{273.15 + 15 + 117} \right)$$

$$= 821 \text{ kPa}$$

Example 6.5

An agitator of diameter D, requires power P, to rotate at a constant speed N, in a liquid of density ρ, and viscosity μ. Show (i) with the help of the Pi theorem that $P = \rho N^3 D^5 F(\rho N D^2/\mu)$ and (ii) An agitator of 225 mm diameter rotating at 23 rev/s in water requires a driving torque of 1.1 Nm. Calculate the corresponding speed and the torque required to drive a similar agitator of 675 mm diameter rotating in air (Viscosities: air 1.86×10^5 Pas, water 1.01×10^{-3} Pas. Densities: air 1.20 kg/m^3, water 1000 kg/m^3).

Solution

(i) The problem is described by 5 variables as

$$F(P, N, D, \rho, \mu) = 0$$

These variables are expressed by 3 fundamental dimensions M, L, and T. Therefore, the number of π terms = (5 – 3) = 2. N, D, and ρ are taken as the repeating variables in determining the π terms.

Then,

$$\pi_1 = N^a D^b \rho^c P \tag{6.25}$$

$$\pi_2 = N^a D^b \rho^c \mu \tag{6.26}$$

Substituting the variables of Eq. (6.25) and (6.26) in terms of their fundamental dimensions M, L and T we get,

$$\text{M}^0 \text{L}^0 \text{T}^0 = (\text{T}^{-1})^a (\text{L})^b (\text{ML}^{-3})^c \text{ML}^2 \text{T}^{-3} \tag{6.27}$$

$$\text{M}^0 \text{L}^0 \text{T}^0 = (\text{T}^{-1})^a (\text{L})^b (\text{ML}^{-3})^c \text{ML}^{-1} \text{T}^{-1} \tag{6.28}$$

Equating the exponents of M, L and T from Eq. (6.27), we get

$$c + 1 = 0$$

$$b - 3c + 2 = 0$$

$$-a - 3 = 0$$

which give $a = -3, b = -5, c = -1$

and hence

$$\pi_1 = \frac{P}{\rho N^3 D^5}$$

Similarly, from Eq. (6.28)

$$c + 1 = 0$$

$$b - 3c - 1 = 0$$

$$-a - 1 = 0$$

which give $a = -1, b = -2, c = -1$

Hence $$\pi_2 = \frac{\mu}{\rho N D^2}$$

Therefore, the problem can be expressed in terms of independent dimensionless parameters as

$$f\left(\frac{P}{\rho N^3 D^5}, \frac{\mu}{\rho N D^2}\right) = 0$$

which is equivalent to

$$\psi\left(\frac{P}{\rho N^3 D^5}, \frac{\rho N D^2}{\mu}\right) = 0$$

or $$\frac{P}{\rho N^3 D^5} = F\left(\frac{\rho N D^2}{\mu}\right)$$

or $$P = \rho N^3 D^5 \, F\left(\frac{\rho N D^2}{\mu}\right)$$

(ii)

$D_1 = 225\,\text{mm}$	$D_2 = 675\,\text{mm}$
$N_1 = 23\ \text{rev/s}$	$N_2 = ?$
$\rho_1 = 1000\ \text{kg/m}^3$	$\rho_2 = 1.20\ \text{kg/m}^3$
$\mu_1 = 1.01 \times 10^{-3}\ \text{Pas}$	$\mu_2 = 1.86 \times 10^{-5}\ \text{Pas}$
$P_1 = 2\pi \times 23 \times 1.1\ \text{W}$	$P_2 = ?$

From the condition of similarity as established above,

$$\frac{\rho_2 N_2 D_2^2}{\mu_2} = \frac{\rho_1 N_1 D_1^2}{\mu_1}$$

$$N_2 = N_1 \left(\frac{D_1}{D_2}\right)^2 \frac{\rho_1}{\rho_2} \frac{\mu_2}{\mu_1}$$

$$= 23\ \text{rev/s} \left(\frac{225}{675}\right)^2 \frac{1000}{1.20} \frac{1.86 \times 10^{-5}}{1.01 \times 10^{-3}}$$

$$= 39.22\ \text{rev/s}$$

again, $$\frac{P_2}{\rho_2 N_2^3 D_2^5} = \frac{P_1}{\rho_1 N_1^3 D_1^5}$$

or
$$\frac{P_2}{P_1} = \left(\frac{D_2}{D_1}\right)^5 \left(\frac{N_2}{N_1}\right)^3 \frac{\rho_2}{\rho_1}$$

or
$$\frac{T_2}{T_1} = \left(\frac{D_2}{D_1}\right)^5 \left(\frac{N_2}{N_1}\right)^2 \frac{\rho_2}{\rho_1}$$

where, T represents the torque and satisfies the relation $P = 2\,\pi NT$

Hence,
$$T_2 = T_1 \left(\frac{D_2}{D_1}\right)^5 \left(\frac{N_2}{N_1}\right)^2 \frac{\rho_2}{\rho_1}$$

$$= 1.1\,\text{Nm} \left(\frac{675}{225}\right)^5 \left(\frac{39.22}{23}\right)^2 \frac{1.20}{1000}$$

$$= 0.933\,\text{Nm}$$

Example 6.6

A torpedo-shaped object 900 mm diameter is to move in air at 60 m/s and its drag is to be estimated from tests in water on a half scale model. Determine the necessary speed of the model and the drag of the full scale object if that of the model is 1140 N. (Fluid properties are same as in Example 6.5 (ii)).

Solution

The dimensionless parameters representing the criteria of similarity, have to be determined first. The drag force F on the object depends upon its velocity V, diameter D, the density ρ, and viscosity μ of air. Now we use Buckhingham's π theorem to find the dimensionless parameters. The five variables F, V, D, ρ and μ are expressed by three fundamental dimensions M, L and T. Therefore the number of π terms is $(5-3) = 2$.

We choose V, D and ρ as the repeating variables

Then,
$$\pi_1 = V^a D^b \rho^c F$$
$$\pi_2 = V^a D^b \rho^c \mu$$

Expressing the variables in the equations above in terms of their fundamental dimensions, we have

$$M^0 L^0 T^0 = (LT^{-1})^a L^b (ML^{-3})^c MLT^{-2} \quad (6.29)$$
$$M^0 L^0 T^0 = (LT^{-1})^a L^b (ML^{-3})^c ML^{-1} T^{-1} \quad (6.30)$$

Equating the exponents of M, L and T on both the sides of the above equations, we get $a = -2$, $b = -2$, $c = -1$ from Eq. (6.29), and $a = -1$, $b = -1$, $c = -1$ from Eq. (6.30)

Hence,
$$\pi_1 = \frac{F}{\rho V^2 D^2} \quad \text{and} \quad \pi_2 = \frac{\mu}{\rho V D}$$

The problem can now be expressed mathematically as

$$f\left(\frac{F}{\rho V^2 D^2}, \frac{\rho V D}{\mu}\right) = 0$$

or

$$\frac{F}{\rho V^2 D^2} = \phi\left(\frac{\rho V D}{\mu}\right) \tag{6.31}$$

For dynamic similarity, the Reynolds numbers ($\rho VD/\mu$) of both the model and prototype have to be same so that drag force F, of the model and prototype can be compared from the π_1 term which is known as the *drag coefficient*. Therefore, we can write

$$\frac{\rho_m V_m D_m}{\mu_m} = \frac{\rho_p V_p D_p}{\mu_p}$$

(Subscripts m and p refer to the model and prototype, respectively)

or

$$V_m = V_p\left(\frac{D_p}{D_m}\right)\left(\frac{\rho_p}{\rho_m}\right)\left(\frac{\mu_m}{\mu_p}\right)$$

Here,

$$\frac{D_m}{D_p} = \frac{1}{2}$$

Hence,

$$V_m = 60 \times (2) \times \left(\frac{1.20}{1000}\right)\left(\frac{1.01 \times 10^{-3}}{1.86 \times 10^{-5}}\right)$$

$$= 7.82 \text{ m/s}$$

At the same value of Re, we can write from Eq. (6.31)

$$\frac{F_p}{\rho_p V_p^2 D_p^2} = \frac{F_m}{\rho_m V_m^2 D_m^2}$$

or

$$F_p = F_m\left(\frac{D_p}{D_m}\right)^2\left(\frac{\rho_p}{\rho_m}\right)\left(\frac{V_p}{V_m}\right)^2$$

$$= 1140\,(4)\left(\frac{60}{7.82}\right)^2\left(\frac{1.20}{1000}\right) \text{N}$$

$$= 322 \text{ N}$$

Example 6.7

A fully developed laminar incompressible flow between two flat plates with one plate moving with a uniform velocity U, with respect to other is known as *Couette flow*. In a Couette flow, the velocity u, at a point depends on its location y (measured perpendicularly from one of the plates), the distance of separation h, between the plates, the relative velocity U, between the plates, the pressures gradient dp/dx imposed on the flow, and the viscosity μ of the fluid. Find a relation in dimensionless form to express u in terms of the independent variables as described above.

Solution

The Buckingham's π theorem is used for this purpose. The variables describing a Couette flow are u, U, y h, dp/dx and μ. Therefore, m (the total no. of variables) = 6.

n (the number of fundamental dimensions in which the six variables are expressed) = 3 (M, L and T)

Hence no. of independent π terms is 6 – 3 = 3

To determine these π terms, U, h and μ are taken as repeating variables. Then,

$$\pi_1 = U^a h^b \mu^c u$$
$$\pi_2 = U^a h^b \mu^c y$$
$$\pi_3 = U^a h^b \mu^c \, \mathrm{d}p/\mathrm{d}x$$

The above three equations can be expressed in terms of the fundamental dimensions of each variable as

$$\mathrm{M^0L^0T^0} = (\mathrm{LT^{-1}})^a (\mathrm{L})^b (\mathrm{ML^{-1}T^{-1}})^c \, \mathrm{LT^{-1}} \tag{6.32}$$
$$\mathrm{M^0L^0T^0} = (\mathrm{LT^{-1}})^a (\mathrm{L})^b (\mathrm{ML^{-1}T^{-1}})^c \mathrm{L} \tag{6.33}$$
$$\mathrm{M^0L^0T^0} = (\mathrm{LT^{-1}})^a (\mathrm{L})^b (\mathrm{ML^{-1}T^{-1}})^c \, \mathrm{ML^{-2}T^{-2}} \tag{6.34}$$

Equating the exponents of M, L and T on both sides of the above equations we get the following:

From Eq. (6.32): $c = 0$

$$a + b - c + 1 = 0$$
$$-a - c - 1 = 0$$

which give $a = -1, \quad b = 0 \quad$ and $\quad c = 0$

Therefore, $\pi_1 = \dfrac{u}{U}$

From equation (6.33): $c = 0$

$$a + b - c + 1 = 0$$
$$-c - a = 0$$

which give $a = 0, b = -1 \quad$ and $\quad c = 0$

Therefore, $\pi_2 = \dfrac{y}{h}$

It is known from one of the corolaries of the π theorem, as discussed earlier, that if any two physical quantities defining a problem have the same dimensions, the ratio of the quantities is a π term. Therefore, there is no need of evaluating the terms π_1 and π_2 through a routine application of π theorem as done here; instead they can be written straight forward as $\pi_1 = u/U$ and $\pi_2 = y/h$.

From Eq. (6.34)

$$c + 1 = 0$$
$$a + b - c - 2 = 0$$
$$-a - c - 2 = 0$$

which give $a = -1, b = 2$ and $c = -1$

Therefore, $\pi_3 = \dfrac{h^2}{\mu U} \dfrac{\mathrm{d}p}{\mathrm{d}x}$

Hence, the governing relation amongst the different variables of a couette flow in dimensionless form is

$$f\left(\frac{u}{U}, \frac{y}{h}, \frac{h^2}{\mu U}\frac{dp}{dx}\right) = 0$$

or
$$\frac{u}{U} = F\left(\frac{y}{h}, \frac{h^2}{\mu U}\frac{dp}{dx}\right) \tag{6.35}$$

It is interesting to note, in this context, that from the exact solution of *Navier Stokes* equation, the expression of velocity profile in case of a couette flow has been derived in Chapter 8 (Sec. 8.3.2) and is given by Eq. (6.34) as

$$\frac{u}{U} = y/h - \left(\frac{h^2}{2\mu U}\frac{dp}{dx}\right)\frac{y}{h}\left(1 - \frac{y}{h}\right)$$

However, π theorem can never determine this explicit functional form of the relation between the variables.

Example 6.8

A 1/30 model of a ship with 900 m^2 wetted area, towed in water at 2 m/s, experiences a resistance of 20 N. Calculate, (i) the corresponding speed of the ship, (ii) the wave making drag on the ship, (iii) the skin-friction drag if the skin-drag coefficient for the model is 0.004 and for the prototype 0.015, (iv) the total drag on the ship, and (v) the power to propel the ship.

Solution

First of all we should identify the pertinent dimensionless parameters that describe the ship resistance problem. For this we have to physically define the problem as follows:

The total drag force F, on a ship depends on ship velocity V, its characteristic geometrical length l, acceleration due to gravity g, density ρ, and viscosity μ, of the fluid. Therefore, the total number of variables which describe the problem = 6 and the number of fundamental dimensions involved with the variables = 3.

Hence, according to the π theorem, number of independent

$$\pi \text{ terms} = 6 - 3 = 3$$

V, l and ρ are chosen as the repeating variables.

Then
$$\pi_1 = V^a l^b \rho^c F$$
$$\pi_2 = V^a l^b \rho^c g$$
$$\pi_3 = V^a l^b \rho^c \mu$$

Expressing the π terms by the dimensional formula of the variables involved we can write

$$M^0 L^0 T^0 = (LT^{-1})^a (L)^b (ML^{-3})^c (MLT^{-2}) \tag{6.36}$$
$$M^0 L^0 T^0 = (LT^{-1})^a (L)^b (ML^{-3})^c (LT^{-2}) \tag{6.37}$$
$$M^0 L^0 T^0 = (LT^{-1})^a (L)^b (ML^{-3})^c (ML^{-1}T^{-1}) \tag{6.38}$$

Equating the exponents of the fundamental dimensions on both sides of the above equations we have

From Eq. (6.36):

$$c + 1 = 0$$
$$a + b - 3c + 1 = 0$$
$$-a - 2 = 0$$

which give $a = -2, b = -2$ and $c = -1$

Therefore, $\pi_1 = \dfrac{F}{\rho V^2 l^2}$

From Eq. (6.37)

$$c = 0$$
$$a + b - 3c + 1 = 0$$
$$-a - 2 = 0$$

which give $a = -2, b = 1$ and $c = 0$

Therefore, $\pi_2 = \dfrac{\text{lg}}{V^2}$

π_2 is the reciprocal of the square of the *Froude number, Fr.*

From Eq. (6.38)

$$c + 1 = 0$$
$$a + b - 3c\ 1 = 0$$
$$-a - 1 = 0$$

which give $a = -1, b = -1$ and $c = -1$.

Hence,

$$\pi_3 = \frac{\mu}{V l \rho}$$

which is the reciprocal of the Reynolds number, Re. Hence the problem of ship resistance can be expressed as

$$f\left(\frac{F}{\rho V^2 l^2}, \frac{V^2}{lg}, \frac{V l \rho}{\mu}\right) = 0$$

or

$$F = \rho V^2 l^2 \phi\left(\frac{V^2}{lg}, \frac{V l \rho}{\mu}\right) \tag{6.39}$$

Therefore, it is found from Eq. (6.39) that the total resistance depends on both the Reynolds number and the Froude number. For complete similarity between a prototype and its model, the Reynolds number must be the same, i.e.,

$$\frac{V_p l_p \rho_p}{\mu_p} = \frac{V_m l_m \rho_m}{\mu_m} \tag{6.40}$$

and also the Froude number must be the same, that i.e.,

$$\frac{V_p}{(l_p g_p)^{1/2}} = \frac{V_m}{(l_m g_m)^{1/2}} \tag{6.41}$$

Equation (6.40) gives $V_m/V_p = (l_p/l_m)(\nu_m/\nu_p)$ where, ν (kinematic viscosity) $= \mu/\rho$. On the other hand, Eq. (6.41) gives $V_m/V_p = (l_m/l_p)^{1/2}$ since, in practice, g_m cannot be different from g_p. For testing small models these conditions are incompatible. The two conditions together require $(l_m/l_p)^{3/2} = \nu_m/\nu_p$ and since both the model and prototype usually operate in water, this condition for the scale factor cannot be satisfied. There is, in fact, no practicable liquid which would enable ν_m to be less than ν_p. Therefore, it concludes that the similarity of viscous forces (represented by the Reynolds number) and similarity of gravity forces (represented by the Froude number) cannot be achieved simultaneously between the model and the prototype.

The way out of the difficulty was suggested by Froude. The assumption is made that the total resistance is the sum of three distinct parts: (a) the wave-making resistance; (b) skin friction; and (c) the eddy-making resistance. The part (a) is usually uninfluenced by viscosity but depends on gravity and is therefore independent of the Reynolds number, Re. Part (c), in most cases, is a small portion of the total resistance and varies little with the Reynolds number. Part (b) depends only on the Reynolds number. Therefore, it is usual to lump (c) together with (a). These assumptions allow us to express the function of Re and Fr in Eq. (6.39) as the sum of two separate functions, ϕ_1 (Re) + ϕ_2 (Fr). Now the skin friction part may be estimated by assuming that it has the same value as that for a flat plate, with the same length and wetted surface area, which moves end on through the water at same velocity. Hence, the function ϕ_1 (Re) is provided by the empirical information of drag resistance on such surfaces. Since the part of the resistance which depends on the Reynolds number is separately determined, the test on the model is conducted at the corresponding velocity which gives equality of the Froude number between the model and the prototype; thus dynamic similarity for the wave-making resistance is obtained. Therefore, the solution of present problem (Example 6.8) is made as follows:

From the equality of the Froude number,

$$\frac{V_m}{\sqrt{l_m g}} = \frac{V_p}{\sqrt{l_p g}}$$

(i) The corresponding speed of the ship $V_p = \sqrt{\dfrac{l_p}{l_m}} \cdot V_m$

$$= \sqrt{30} \times 2 \text{ m/s}$$

$$= 10.95 \text{ m/s}$$

Area ratio, $= \dfrac{A_m}{A_p} = \left(\dfrac{1}{30}\right)^2 = \dfrac{1}{900}$

Therefore A_m (area of the model) $= \dfrac{900 \text{ m}^2}{900} = 1 \text{ m}^2$

(ii) If F_w and F_s represent the wave-making resistance and skin friction resistance of the ship respectively, then from the definition of the drag coefficient C_D, we can write

$$F_{s_m} = \frac{1}{2} \rho_m \times V_m^2 \times A_m \times C_D$$

$$= \frac{1}{2} \times 1000 \times 2^2 \times .004$$

$$= 8\text{N}$$

Now the total resistance on the model $F_m = F_{s_m} + F_{w_m}$

Hence, $$F_{w_m} = F_m - F_{s_m} = 20 - 8 = 12 \text{ N}$$

Now from dynamic similarity for wave making resistance

$$\frac{F_{w_p}}{\rho_p V_p^2 l_p^2} = \frac{F_{w_m}}{\rho_m V_m^2 l_m^2}$$

or $$F_{w_p} = F_{wm} \frac{\rho_p}{\rho_m} \left(\frac{V_p}{V_m}\right)^2 \left(\frac{l_p}{l_m}\right)^2$$

$$= 12 \times 1 \times 30 \times 900 \text{ N} = 324 \text{ kN}$$

(iii) F_{s_p} (skin friction of the prototype)

$$= \frac{1}{2} ' 1000 \times (10.95)^2 ' 900 ' 0.015 \text{ N}$$

$$= 809.34 \text{ kN}$$

(iv) Therefore, F_p (total drag resistance of the prototype)

$$= 324 + 809.34 = 1133.34 \text{ kN}$$

(v) Propulsive power required $= 1133.34 \times 10.95 = 12410$ kW

$$= 12.41 \text{ MW}$$

Example 6.9

In the study of vortex shedding phenomenon due to the presence of a bluff body in a flow through a closed duct, the following parameters are found to be important: velocity of flow V, density of liquid ρ, coefficient of dynamic viscosity of liquid μ, hydraulic diameter of the duct D_h, the width of the body B, and the frequency of vortex shedding n. Obtain the dimensionless parameters governing the phenomenon.

Solution

The problem is described by 6 variables V, ρ, μ, D_h, B, and n. The number of fundamental dimensions in which the variables can be expressed = 3. Therefore, the number of independent π terms is (6–3) = 3. We use the Buckingham's π theorem to find the π terms and choose ρ, V, D_h as the repeating variables.

Hence, $$\pi_1 = \rho^a V^b D_h^c \mu$$
$$\pi_2 = \rho^a V^b D_h^c B$$
$$\pi_3 = \rho^a V^b D_h^c n$$

Expressing the equations in terms of the fundamental dimensions of the variables we have

$$M^0L^0T^0 = (ML^{-3})^a (LT^{-1})^b (L)^c (ML^{-1}T^{-1}) \tag{6.42}$$

$$M^0L^0T^0 = (ML^{-3})^a (LT^{-1})^b (L)^c L \tag{6.43}$$

$$M^0L^0T^0 = (ML^{-3})^a (LT^{-1})^b (L)^c T^{-1} \tag{6.44}$$

Equating the exponents of M, L and T in the above equations we get, From Eq. (6.42),

$$a + 1 = 0$$

$$-3a + b + c - 1 = 0, -b - 1 = 0$$

which give $a = -1, b = -1 \text{ and } c = -1$

Hence, $p_1 = m/rVD_h$

From Eq. (6.43),

$$a = 0$$

$$-3a + b + c + 1 = 0$$

$$-b = 0$$

which give $a = b = 0, c = -1$

Hence, $p_2 = B/D_h$

From Eq. (6.44),

$$a = 0$$

$$-3a + b + c = 0$$

$$-b - 1 = 0$$

which give $a = 0, b = -1, c = 1$

Hence, $p_3 = (n D_h)/V$

Therefore, the governing dimensionless parameters are

$\left(\dfrac{\rho V D_h}{\mu}\right)$ $(= 1/p_1)$ the Reynolds number

$\dfrac{B}{D_h}$ $(= p_2)$ ratio of the width of the body to hydraulic diameter of the duct

$\dfrac{n D_h}{V}$ $(= p_3)$ the Strouhal number

6.3.3 Rayleigh's Indicial Method

This alternative method is also based on the fundamental principle ofdimensional homogeneity of physical variables involved in a problem. Here the dependent variable is expressed as a product of all the independent variables raised to an unknown integer exponent. Equating the indices of *n* fundamental dimensions of the variables involved, *n* independent equations are obtained which are solved for the purpose of obtaining the dimensionless groups. Let us illustrate this method by solving the pipe flow problem in Example 2. Here, the dependent variable $\Delta p/l$ can be written as

$$\frac{\Delta p}{l} = A\, V^a\, D_h^b\, \rho^c\, \mu^d \tag{6.45}$$

where, A is a dimensionless constant.

Inserting the dimensions of each variable in the above equation, we obtain,

$$\mathrm{M\,L^{-2}\,T^{-2} = A\,(LT^{-1})^a\,(L)^b\,(ML^{-3})^c\,(ML^{-1}T^{-1})^d}$$

Equating the indices of M, L, and T on both sides, we get

$$\begin{aligned} c + d &= 1 \\ a + b - 3c - d &= -2 \\ -a - d &= -2 \end{aligned} \tag{6.46}$$

There are three equations and four unknowns. Solving these equations in terms of the unknown d, we have

$$\begin{aligned} a &= 2 - d \\ b &= -d - 1 \\ c &= 1 - d \end{aligned}$$

Hence, Eq. (6.45) can be written as

$$\frac{\Delta p}{l} = A\, V^{2-d}\, D_h^{-d-1}\, \rho^{1-d}\, \mu^d$$

or

$$\frac{\Delta p}{l} = \frac{AV^2\rho}{D_h}\left(\frac{\mu}{VD_h\rho}\right)^d$$

or

$$\frac{\Delta p D_h}{l\rho V^2} = A\left(\frac{\mu}{VD_h\rho}\right)^d \tag{6.47}$$

Therefore we see that there are two independent dimensionless terms of the problem, namely,

$$\frac{\Delta p D_h}{l\rho V^2} \quad \text{and} \quad \frac{\mu}{VD_h\rho}$$

It should be mentioned in this context that both Buckingham's method and Rayleigh's method of dimensional analysis determine only the relevant independent dimensionless parameters of a problem, but not the exact relationship between them. For example, the numerical values of A and d in the Eq. (6.47) can never be known from dimensional analysis. They are found out from experiments. If the system of Eq. (6.46) is solved for the unknown c, it results,

$$\frac{\Delta p}{l}\frac{D_h^2}{V\mu} = A\left(\frac{\rho V D_h}{\mu}\right)^c$$

Therefore, different interdependent sets of dimensionless terms are obtained with the change of unknown indices in terms of which the set of indicial equations are solved. This is similar to the situations arising with different possible choices of repeating variables in Buckingham's Pi theorem.

Example 6.10

The time period τ of a simple pendulum depends on its effective length l and the local acceleration due to gravity g. Using both Buckingham's Pi theorem and Rayleigh's indicial method, find the functional relationship between the variables involved.

Solution

Application of Buckingham's Pi theorem:
The variables of the problem are τ, l and g and the fundamental dimensions involved in these variables are L (length) and T (time). Therefore the no. of independent π term $= (3 - 2) = 1$, since τ is the dependent variable, the only choice left for the repeating variables to be l and g.

Hence, $$\pi_1 = l^a g^b \tau$$

Expressing the equation in terms of the fundamental dimensions of the variables we get $L^0T^0 = L^a (LT^{-2})^b T$. Equating the exponents of L and T on both sides of the equation, we have,

$$a + b = 0, \quad \text{and} \quad -2b + 1 = 0$$

which give $$a = -\frac{1}{2}, \quad b = \frac{1}{2}; \quad \text{and hence } \pi_1 = \tau\sqrt{\frac{g}{l}}$$

Therefore, the required functional relationship between the variables of the problem is

$$f\left(\tau\sqrt{\frac{g}{l}}\right) = 0 \tag{6.48}$$

Application of Rayleigh's indicial method:
Since τ is the dependent variable, it can be expressed as

$$\tau = A\, l^a g^b \tag{6.49}$$

where A is a non-dimensional constant. The Eq. (6.49) can be written in terms of the fundamental dimensions of the variables as

$$T = A L^a (LT^{-2})^b$$

Equating the exponent of L and T on both sides of the equation, we get, $a + b = 0$ and

$-2b = 1$ which give $a = \dfrac{1}{2}$ and $b = -\dfrac{1}{2}$.

Hence, Eq. (6.49) becomes $\tau = A\sqrt{\dfrac{l}{g}}$

or $$\tau\sqrt{\frac{g}{l}} = A$$

Therefore, it is concluded that the dimensionless governing parameter of the problem is $\tau\sqrt{\dfrac{g}{l}}$. From elementary physics, we know that $A = 2\pi$.

Example 6.11

The capillary rise h, of a fluid of density ρ, and surface tension σ, in a tube of diameter D, depends upon the contact angle ϕ, and acceleration due to gravity g. Find an expression for h in terms of dimensionless variables by Rayleigh's indicial method.

Solution

Capillary rise h, is the dependent variable of the problem and can be expressed in terms of the independent variables as

$$h = A\,\rho^a\,\sigma^b\,D^c\,g^d\,\phi \tag{6.50}$$

where A is a dimensionless constant.

(ϕ is not raised to any exponent, since it is a dimensionless variable and hence an independent π term).

Expressing the variables in terms of their fundamental dimensions in the above equation, we get

$$\text{L} = \text{A}\,(\text{ML}^{-3})^a\,(\text{MT}^{-2})^b\,\text{L}^c\,(\text{LT}^{-2})^d$$

Equating the exponents of M, L and T in LHS and RHS of the equation, we have

$$a + b = 0$$
$$-3a + c + d = 1$$
$$-2b - 2d = 0$$

Solving these three equations in terms of a, we get

$$b = -a$$
$$c = 1 + 2a$$
$$d = a$$

Substituting these values in Eq. (6.52), we get

$$h = A\,D\left(\frac{\rho\, g\, D^2}{\sigma}\right)^a \phi$$

or

$$\frac{h}{D} = A\left(\frac{\rho\, g\, D^2}{\sigma}\right)^a \phi$$

This is the required expression.

SUMMARY

- Physical similarities are always sought between the problems of same physics. The complete physical similarity requires geometric similarity, kinematic similarity and dynamic similarity to exist simultaneously.
- In geometric similarity, the ratios of the corresponding geometrical dimensions between the systems remain the same. In kinematic similarity, the ratios of corresponding motions and in dynamic similarity, the ratios of corresponding forces between the systems remain the same.

- For prediction of the performance characteristics of actual systems in practice from the results of model scale experiments in laboratories, complete physical similarity has to be achieved between the prototype and the model.
- Dimensional homogeneity of physical quantities implies that the number of dimensionless independent variables are smaller as compared to the number of their dimensional counterparts to describe a physical phenomenon. The dimensionless variables represent the criteria of similarity. Buckingham's P_i theorem states that if a physical problem is described by m dimensional variables which can be expressed by n fundamental dimensions, then the number of independent dimensionless variables defining the problem will be $m - n$. These dimensionless variables are known as π terms. The independent π terms of a physical problem are determined either by Buckingham's P_i theorem or by Rayleigh's indicial method.

REFERENCE

1. Feynaman, R. P. et al., *Lectures on Physics, Volume II, Addison Wesley*, USA. 1964

EXERCISES

6.1 Choose the correct answer:

(i) The repeating variables in a dimensional analysis should
  (a) be equal in number to that of the fundamental dimensions involved in the problem variables
  (b) include the dependent variable
  (c) have at least one variable containing all the fundamental dimensions
  (d) collectively contain all the fundamental dimensions

(ii) A dimensionless group formed with the variables ρ (density), ω (angular velocity), μ (dynamic viscosity), and D (characteristic diameter) is
  (a) $\rho\,\omega\,\mu\,/D^2$
  (b) $\rho\,\omega\,D^2/\mu$
  (c) $\mu D^2\,\rho\,\omega$
  (d) $\rho\,\omega\,\mu\,D$

(iii) In similitude with gravity force, where equality of Froude number exists, the acceleration ratio a_r becomes
  (a) L_r^2
  (b) 1.0
  (c) $1/L_r$
  (d) $L_r^{5/2}$
  (where L_r is the geometrical scale factor)

6.2 Show that for a flow governed by gravity, inertia and *pressure forces*, the ratio of volume flow rates in two dynamically similar systems equals to the 5/2 power of the length ratio.

6.3 For a flow governed by gravity and surface tension force, evaluate the length scale ratio which will ensure complete similarity between the model and the prototype.

Ans. ($l_m / l_p = \sqrt{(\sigma_m/\sigma_p)/(\rho_m/\rho_p)}$)

6.4 The tensile force F inside a pendulum depends on the mass m, length l, period t, and angular amplitude a of the pendulum. Using Buckingham Pi theorem, obtain an expression for the tensile force.

Ans. ($\frac{Ft^2}{ml}$ = constant)

6.5 Liquid flow across an orifice loses useful power which is dependent on the liquid density ρ and viscosity μ, the volume flow rate Q, and the orifice diameter D. Using Buckingham's Pi theorem, establish a functional relationship amongst the above variables.

Ans. ($\frac{\rho Q}{\mu D}$ = constant))

6.6 Using the Buckingham's π theorem, show that the velocity U through a circular orifice is given by

$$U = (2gH)^{0.5}\ \phi(D/H, \rho UH/\mu)$$

where H is the head causing flow, D is the diameter of the orifice, μ is the coefficient of dynamic viscosity, ρ is the density of fluid flowing through the orifice and g is the acceleration due to gravity.

6.7 The speed of propagation U of a capillary wave in deep water depends upon density ρ of water, wavelength λ, and surface tension coefficient σ. (a) Using Buckinghham Pi theorem, establish a relationship of U with ρ, λ and σ. (b) For a given surface tension and wavelength, how does the propagation speed changes if the density is halved?

Ans. ((a) $\frac{\rho U^2 \lambda}{\sigma}$ = constant , (b) increased by a factor of $\sqrt{2}$)

6.8 For rotodynamic fluid machines of a given shape, and handling an incompressible fluid, the relevant variables involved are D (the rotor diameter), Q (the volume flow rate through the machine), N (the rotational speed of the machine), gH (the difference of head across the machine, i.e., energy per unit mass), ρ (the density of fluid), μ (the dynamic viscosity of the fluid) and P (the power transferred between fluid and rotor). Show with the help of Buckingham's p_i theorem that the relationship between the variables can be expressed by a functional form of the pertinent dimensionless parameters as

$$\phi(Q/ND^3, gH/N^2D^2, \rho ND^2/\mu, P/\rho N^3 D^5) = 0$$

6.9 The pressure drop Δp in pipe flows of liquids depends on the time required t to pass a volume of a given liquid through, on this volume $\forall$, and on the density ρ as well as the viscosity μ of the liquid. Using Buckingham Pi theorem, obtain an expression for the pressure drop in a functional form.

$$Ans. \left(\Delta p = \frac{\mu}{t}\psi\left(\frac{\rho \forall^{2/3}}{\mu t}\right)\right)$$

6.10 The shape of a drop of liquid pulsates as it falls. The period of oscillation t is observed to depend on the surface tension σ, the mean radius of the drop r, and the liquid density ρ. Show with the help of Buckingham Pi theorem that

$$t \propto \sqrt{\frac{\sigma}{\rho r^3}}$$

6.11 Using Buckingham Pi theorem, reduce the relationship $n = G(g, A, \rho, m)$ for the frequency n of the wing beat of a flying insect, where g stands for the gravitational acceleration; A, the wing area; ρ, the air density; and m, the mass of the insect.

$$Ans. \left(n = \frac{g^{1/2}}{A^{1/4}}\psi\left(\frac{\rho A^{3/2}}{m}\right)\right)$$

6.12 In a two-dimensional motion of a projectile, the range R depends upon the x component of velocity V_x, the y component of velocity V_y, and the acceleration due to gravity g. Show with the help of Rayleigh's indicial method of dimensional analysis that

$$\mathrm{R} = \frac{V_x^2}{g} f\left(\frac{V_y}{V_x}\right)$$

6.13 The boundary layer thickness δ at any section for a flow past a flat plate depends upon the distance x measured along the plate from the leading edge to the section, free stream velocity U and the kinematic viscosity ν of the fluid. Show with the help of Rayleigh's indicial method of dimensional analysis that

$$\frac{\delta}{x} \propto (Ux/\nu) \text{ or } \frac{\delta}{x} = f\left(\frac{Ux}{\nu}\right)$$

6.14 A high speed liquid sheet in ambient air is disintegrated into drops of liquid due to hydrodynamic instability. The drop diameter d depends upon the velocity V of liquid sheet, the thickness h of the liquid sheet, the surface tension coefficient σ of the liquid and density ρ of ambient air. Show, with the help of both (i) Buckingham's Pi theorem and (ii) Rayleigh's indicial method, that the functio-nal relationship amongst the above variables can be expressed as $d/h = \phi\,(\sigma/\rho V^2 h)$.

6.15 Assume that nothing is known about the particle motion under gravity beyond $\chi = G(v_o, g, t)$ where χ, v_0, g, and t are respectively the displacement, initial velocity, gravitational acceleration, and time. Perform a dimensional analysis to explain the situation.

6.16 During the flow through a pipe, it is observed that there exists a critical average flow velocity $\vec{V}_{cr}$ beyond which the flow becomes turbulent. It is also known that $\vec{V}_{cr}$ is influenced by the diameter of the pipe D, the density ρ and the viscosity μ of the fluid. Perform a dimensional analysis to explain the situation.

6.17 A model of a reservoir is drained in 4 minutes by opening a sluice gate. The model scale is 1:225. How long should it take to empty the prototype?

Ans. (60 minutes)

6.18 Evaluate the model scale when both viscous and gravity forces are necessary to secure similitude. What should be the model scale if oil of kinematic viscosity 92.9×10^{-6} m^2/s is used in the model tests and if the prototype liquid has a kinematic viscosity of 743.2×10^{-6} m^2/s?

Ans. ($l_m/l_p = (\nu_m/\nu_p)^{2/3}$; $l_m = 0.25\, l_p$)

6.19 A 1/6 model automobile is tested in a wind tunnel with same air properties as the prototype. The prototype automobile runs on the roads at a velocity of 60 km/hr. For dynamically similar conditions, the drag measured on the model is 500 N. Determine the drag of the prototype and the power required to overcome this drag.

Ans. (500N, 8.33 kN)

6.20 A hydraulic jump occurring in a stilling basin is to be studied in a 1:36 scale model. The prototype jump has an initial velocity of 10 m/s, an entrance Froude number of 6.0 and a power loss of 2 kW per meter width of basin. Determine (a) thecorresponding model velocity, (b) model Froude number and (c) power loss per meter width of the model.

Ans. ((a) 1·67 m/s, (b) 6·0, (c) 0·26 W)

6.21 A sphere advancing at 1.5 m/s in a stationary mass of water experiences a drag of 4.5 N. Find the flow velocity required for dynamic similarity of another sphere twice the diameter, placed in a wind tunnel. Calculate the drag at this speed if the kinematic viscosity of air is 13 times that of water and its density is 1.25 kg/m^3.

Ans. (9.75 m/s, 0.951 N)

6.22 Calculate the thrust required to run a motor boat 5 m long at 100 m/s in a lake if the force required to tow a 1:30 model in a reservoir is 5 N. Neglect the viscous resistance due to water in comparison to the wave-making resistance.

Ans. (135 kN)

6.23 The flow rate over a spillway is 120 m^3/s. What is the length scale for a dynamically similar model if a flow rate of 0.75 m^3/s is available in the laboratory? On part of such model, a force of 2.8 N is measured. What is the corresponding force on the prototype spillway? (viscosity and surface tension effects are negligible.)

Ans. ($l_m = 0.13\, l_p$, 1.27 kN)

6.24 The flow through a closed, circular-sectioned pipe may be metered by measuring the speed of rotation of a propeller having its axis along the pipe centre line. Derive a relation between the volume flow rate and the rotational speed of the propeller, in terms of the diameters of the pipe and the propeller and the density and viscosity of the fluid. A propeller of 75 mm diameter, installed in a 150 mm pipe carrying water at 42.5 litres/s, was found to rotate at 20.7 rev/s. In a similar physical situation, a propeller rotates in air flow through a pipe of 750 mm diameter. Estimate the diameter and rotational speed of the propeller and the volume flow rate of air. The density of air is 1.25 kg/m^3 and its viscosity 1.93×10^{-5} Pas. The viscosity of water is 1.145×10^{-3} Pas.

Ans. (375 mm, 11.16 rev/s, 2.86 m^3/s)

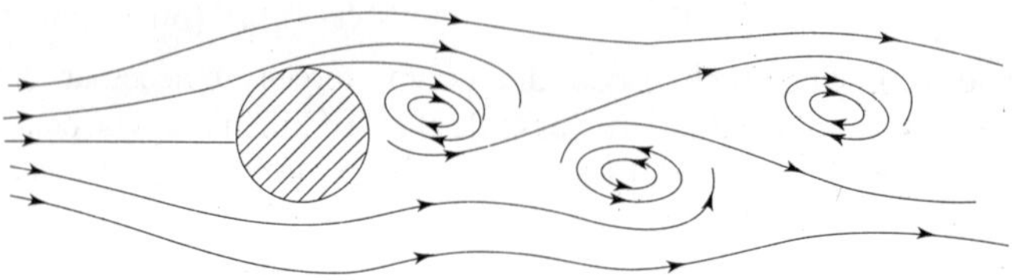

Fig. 6.2 Vortex shedding past a cylinder [1]

6.25 The vortices are shed from the rear of a cylinder placed in a cross flow. The vortices alternately leave the top and bottom of the cylinder, as shown in Fig. 6.2. The vortex shedding frequency, f, is thought to depend on ρ, V, D and μ.

(i) Use dimensional analysis to develop a functional relationship for f.

(ii) The vortex shedding occurs in standard air on two different cylinders with a diameter ratio of 2. Determine the velocity ratio for dynamic similarity, and ratio of the vortex shedding frequencies.

Ans. $\left(\frac{fD}{V} = \phi\left(\frac{\rho V D}{\mu}\right)\right)$

7

FLOW OF IDEAL FLUIDS

7.1 INTRODUCTION

Flows at high Reynolds number reveal that the viscous effects are confined within the boundary layers. Far away from the solid surface, the flow is nearly inviscid and in many cases it is incompressible. We now aim at developing techniques for analyses of inviscid incompressible flows.

Incompressible flow is a constant density flow, and we assume ρ to be constant. We visualise a fluid element of defined mass moving along a streamline in an incompressible flow. Because the density is constant, we can write

$$\nabla \cdot \vec{\mathbf{V}} = 0 \tag{7.1}$$

Over and above, if the fluid element does not rotate as it moves along the streamline, or to be precise, if its motion is translational (and deformation with no rotation) only, the flow is termed as *irrotational flow*. It has already been shown in Sec. 3.3.5 that the motion of a fluid element can in general have translation, deformation and rotation. The rate of rotation of the fluid element can be measured by the average rate of rotation of two perpendicular line segments. The average rate of rotation ω_z about the z axis is expressed in terms of the gradients of velocity components (refer to Chapter 3) as

$$\omega_z = \frac{1}{2}\left[\frac{\partial v}{\partial x} - \frac{\partial u}{\partial y}\right]$$

Similarly, the other two components of rotation are

$$\omega_x = \frac{1}{2}\left[\frac{\partial w}{\partial y} - \frac{\partial v}{\partial z}\right] \quad \text{and} \quad \omega_y = \frac{1}{2}\left[\frac{\partial u}{\partial z} - \frac{\partial w}{\partial x}\right]$$

As such, they are components of $\vec{\omega}$, which is given by

$$\vec{\omega} = \frac{1}{2}\left(\nabla \times \vec{\mathbf{V}}\right)$$

In a two-dimensional flow, ω_z is the only non-trivial component of the rate of rotation. Imagine a pathline of a fluid particle shown in Fig. 7.1. Rate of spin of the particle is ω_z. The flow in which this spin is zero throughout is known as irrotational flow. A generalised statement is more appropriate: For irrotational flows, $\nabla \times \vec{\mathbf{V}} = 0$ in the flow field.

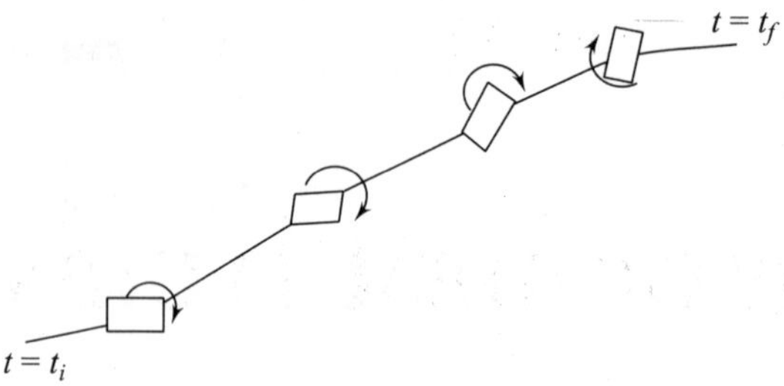

Fig. 7.1 Pathline of a fluid particle

Therefore for an irrotational flow, the velocity $\vec{\mathbf{V}}$ can be expressed as the gradient of a scalar function called the velocity potential, denoted by ϕ

$$\vec{\mathbf{V}} = \nabla \phi \tag{7.2}$$

Combination of Eqs (7.1) and (7.2) yields,

$$\nabla^2 \phi = 0 \tag{7.3}$$

From Eq. (7.3) we see that an inviscid, incompressible, irrotational flow is governed by Laplace's equation.

Laplace's equation is linear, hence any number of particular solutions of Eq. (7.3) added together will yield another solution. This concept forms the building-block of the solution of inviscid, incompressible, irrotational flows. A complicated flow pattern for an inviscid, incompressible, irrotational flow can be synthesised by adding together a number of elementary flows which are also inviscid, incompressible and irrotational.

The analysis of Laplace's Eq. (7.3) and finding out the potential functions are known as *potential flow theory* and the inviscid, incompressible, irrotational flow is often called as potential flow. However, the following elementary flows can constitute several complex potential-flow problems:

1. Uniform flow
2. Source or sink
3. Vortex

7.2 ELEMENTARY FLOWS IN A TWO-DIMENSIONAL PLANE

7.2.1 Uniform Flow

In this flow, velocity is uniform along the y axis and there exists only one component of velocity which is in the x direction. Magnitude of the velocity is U_0.

From Eq. (7.2) ,we can write

$$\hat{\mathbf{i}} U_0 + \hat{\mathbf{j}} 0 = \hat{\mathbf{i}} \frac{\partial \phi}{\partial x} + \hat{\mathbf{j}} \frac{\partial \phi}{\partial y}$$

or
$$\frac{\partial \phi}{\partial x} = U_0 \qquad \frac{\partial \phi}{\partial y} = 0$$

Hence,
$$\phi = U_0 x + C_1 \tag{7.4}$$

Recall from Sec. 3.5 that in a two-dimensional flow field, the flow can also be described by stream function ψ. In the case of uniform flow,

$$\frac{\partial \psi}{\partial y} = U_0 \quad \text{and} \quad -\frac{\partial \psi}{\partial x} = 0$$

so that
$$\Psi = U_0 y + K_1 \tag{7.5}$$

The constants of integration C_1 and K_1 in Eqs (7.4) and (7.5) are arbitrary. The values of ψ and ϕ for different streamlines and velocity potential lines may change but flow pattern is unaltered. The constants of integration may be omitted and it is possible to write

$$\psi = U_0 y, \qquad \phi = U_0 x \tag{7.6}$$

These are plotted in Fig. 7.2(a) and consist of a rectangular mesh of straight streamlines and orthogonal straight potential lines. It is conventional to put arrows on the streamlines showing the direction of flow.

In terms of polar $(r - \theta)$ coordinate, Eq. (7.6) becomes

$$\psi = U_0 r \sin \theta, \qquad \phi = U_0 r \cos \theta \tag{7.7}$$

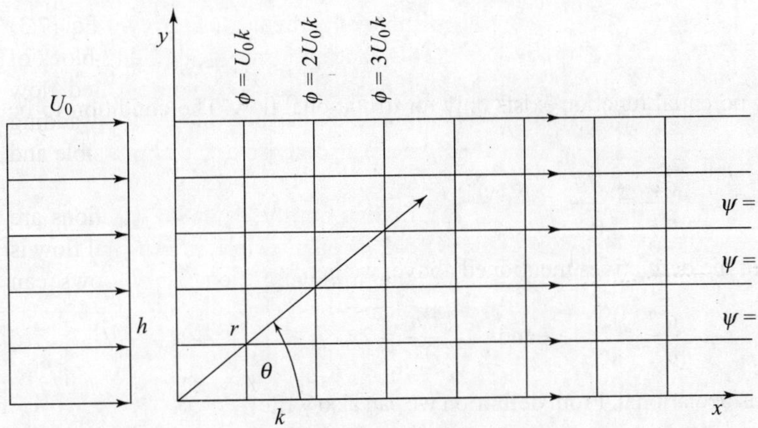

(a) Flownet for a uniform stream

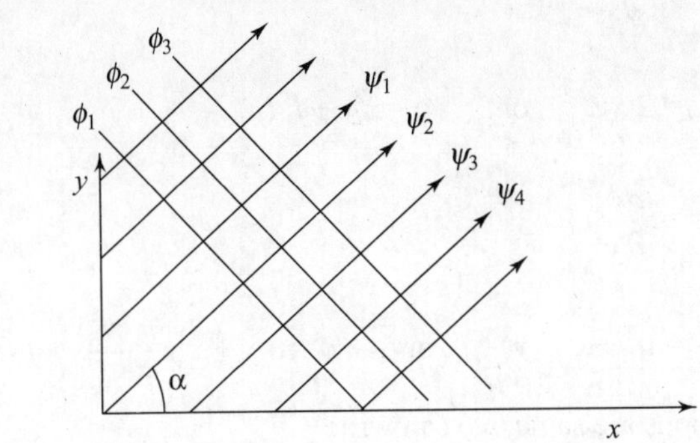

(b) Flownet for a uniform stream with an angle α with the x axis

Fig. 7.2 Streamlines and velocity potential lines

If we consider a uniform stream at an angle α to the x axis as shown in Fig. 7.2(b), we require that

$$u = U_0 \cos\alpha = \frac{\partial \psi}{\partial y} = \frac{\partial \phi}{\partial x}$$

and

$$v = U_0 \sin\alpha = -\frac{\partial \psi}{\partial x} = \frac{\partial \phi}{\partial y} \tag{7.8}$$

Integrating, we obtain for a uniform velocity U_0 at an angle α, the stream function and velocity potential respectively as

$$\psi = U_0(y\cos\alpha - x\sin\alpha), \quad \phi = U_0(x\cos\alpha + y\sin\alpha) \tag{7.9}$$

Example 7.1

The velocity components of two-dimensional incompressible flow are $u = 2xy$ and $v = a^2 + x^2 - y^2$. Show that a velocity potential function exists and find out the velocity potential.

Solution

The velocity potential function exists only for irrotational flow. The condition to be satisfied is

$$\frac{\partial v}{\partial x} = \frac{\partial u}{\partial y}$$

Evaluating the derivatives mentioned above, we get

$$\frac{\partial v}{\partial x} = 2x \quad \text{and} \quad \frac{\partial u}{\partial y} = 2x$$

The flow is irrotational. From definition we can also write

$$u = \frac{\partial \phi}{\partial x}$$

$$\frac{\partial \phi}{\partial x} = 2\,xy \quad \text{or} \quad \phi = x^2 y + f_1(y)$$

Also

$$v = \frac{\partial \phi}{\partial y}$$

or

$$\frac{\partial \phi}{\partial y} = a^2 + x^2 - y^2 \quad \text{or} \quad \phi = a^2 y + x^2 y - \frac{y^3}{3} + f_2(x)$$

Since both the solutions are same, we can write

$$x^2 y + f_1(y) = a^2 y + x^2 y - \frac{y^3}{3} + f_2(x)$$

or $$f_1(y) = a^2 y - \frac{y^3}{3} + f_2(x)$$

In order to keep the above expression valid for all the values of y, $f_2(x)$ has to be a constant.

Thus, $$\phi = a^2 y + x^2 y - \frac{y^3}{3} + \text{constant}$$

Since ϕ = constant and represents a family of lines, ϕ may be written without a constant as

$$\phi = a^2 y + x^2 y - \frac{y^3}{3}$$

Example 7.2

The flow of an incompressible fluid is defined by $u = 2, v = 8x$. Does a stream function exist? If so, find its expression.

Solution

Compliance of continuity describes the existence of a stream function

$$\frac{\partial u}{\partial x} + \frac{\partial v}{\partial y} = \frac{\partial (2)}{\partial x} + \frac{\partial}{\partial y}(8x) = 0$$

Therefore, the stream function exists.
Now we can write

$$d\psi = \frac{\partial \psi}{\partial x} dx + \frac{\partial \psi}{\partial y} dy$$

or $$d\psi = -v dx + u dy$$

or $$d\psi = -8x dx + 2dy$$

or $$\psi = -4x^2 + 2y + C$$

Dropping the constant C, $\psi = -4x^2 + 2y$.

Example 7.3

Does a velocity potential function $\phi = 2(x^2 + 2y - y^2)$ describe the possible flow of an incompressible fluid? If so, find out the equation for the velocity vector $\vec{V}$. Also determine the equation for streamlines.

Solution

For the given ϕ, in order to describe an incompressible flow, we check with the Laplace's equation,

$$\frac{\partial^2 \phi}{\partial x^2} + \frac{\partial^2 \phi}{\partial y^2} = 2(2) + 2(-2) = 0$$

So, a flow field exists.

The velocity components are

$$u = \frac{\partial \phi}{\partial x} = 2(2x) = 4x$$

$$v = \frac{\partial \phi}{\partial y} = 2\,(2 - 2y) = 4 - 4y$$

Velocity vector $\vec{\mathbf{V}} = 4x\,\hat{\mathbf{i}} + (4 - 4y)\,\hat{\mathbf{j}}$

Stream function ψ can be expressed as

$$d\psi = \frac{\partial \psi}{\partial x}\,dx + \frac{\partial \psi}{\partial y}\,dy$$

or $$d\psi = -\,v dx + u dy$$

or $$d\psi = -\,(4 - 4y)\,dx + 4x\,dy$$

$$\psi = -\int (4 - 4y)dx + \int 4x\,dy + C$$

$$\psi = -\,4x + 4xy + 4xy + C$$

Dropping the constant C, stream function becomes

$$\psi = 4\,(2xy - x)$$

7.2.2 Source or Sink

Consider a flow with straight streamlines emerging from a point, where the velocity along each streamline varies inversely with distance from the point, as shown in Fig. 7.3. Only the radial component of velocity is non-trivial ($v_\theta = 0$, $v_z = 0$).

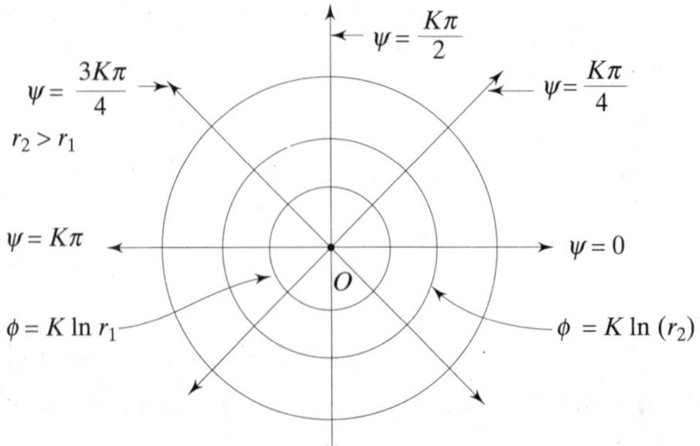

Fig. 7.3 Flownet for a source flow

Such a flow is called *source flow*. In a steady source flow the amount of fluid crossing any given cylindrical surface of radius r, and unit length is constant ($\dot{m}$)

$$\dot{m} = 2\pi\, r\, v_r\, \rho$$

or $$v_r = \frac{\dot{m}}{2\pi\rho}\cdot\frac{1}{r} = \frac{\Lambda}{2\pi}\cdot\frac{1}{r} = \frac{K}{r} \tag{7.10a}$$

where, K is the source strength

$$K = \frac{\dot{m}}{2\pi\rho} = \frac{\Lambda}{2\pi} \tag{7.10b}$$

and Λ is the volume flow rate:

Again recall from Sec. 4.2.2 that the definition of stream function in cylindrical polar coordinate states that

$$v_r = \frac{1}{r}\frac{\partial\psi}{\partial\theta} \quad \text{and} \quad v_\theta = -\frac{\partial\psi}{\partial r} \tag{7.11}$$

Now, for the source flow, it can be said that

$$\frac{1}{r}\frac{\partial\psi}{\partial\theta} = \frac{K}{r} \tag{7.12}$$

and $$-\frac{\partial\psi}{\partial r} = 0 \tag{7.13}$$

Combining Eqs (7.12) and (7.13), we get

$$\psi = K\theta + C_1 \tag{7.14}$$

However, this flow is also irrotational and we can write

$$\hat{\mathbf{i}}\, v_r + \hat{\mathbf{j}}\, v_\theta = \hat{\mathbf{i}}\frac{\partial\phi}{\partial r} + \hat{\mathbf{j}}\frac{1}{r}\frac{\partial\phi}{\partial\theta}$$

or $$v_r = \frac{\partial\phi}{\partial r} \quad \text{and} \quad v_\theta = 0 = \frac{1}{r}\frac{\partial\phi}{\partial\theta}$$

or $$\frac{\partial\phi}{\partial r} = v_r = \frac{K}{r} \quad \text{or} \quad \phi = K\ln r + C_2 \tag{7.15}$$

Likewise in uniform flow, the integration constants C_1 and C_2 in Eqs (7.14) and (7.15) have no effect on the basic structure of velocity and pressure in the flow. The equations for streamlines and velocity potential lines for source flow become

$$\psi = K\theta \quad \text{and} \quad \phi = K\ln r \tag{7.16}$$

where K is defined as the source strength and is proportional to Λ, which is the rate of volume flow from the source per unit depth perpendicular to the page as shown in Fig. 7.3. If Λ is negative, we have sink flow, where the flow is in the opposite direction of the source flow. In Fig. 7.3, the point O is the origin of the radial streamlines. We visualise that point O is a point source or sink that induces radial flow in the neighbourhood. The point source or sink is a point of singularity in the flow field (because v_r becomes infinite). It can also be visualised that point O in Fig. 7.3 is simply a point formed by the intersection of plane of the paper and a line perpendicular to the paper. The line perpendicular to the paper is a line source, with volume flow rate (Λ) per unit length. However, for sink, the stream function and velocity potential function are

$$\psi = -K\theta \quad \text{and} \quad \phi = -K\ln r \tag{7.17}$$

Example 7.4

The radial velocity of a flow is described by $v_r = \dfrac{k}{\sqrt{r}} \cos\theta$.

If $v_\theta = 0$ at $\theta = 0$, find out v_θ and the stream function for the flow.

Solution

$$v_r = \frac{1}{r}\cdot\frac{\partial\psi}{\partial\theta} = \frac{k}{\sqrt{r}}\cos\theta$$

or
$$\frac{\partial\psi}{\partial\theta} = k\sqrt{r}\cos\theta$$

or
$$\psi = k\sqrt{r}\sin\theta = f(r)$$

Now,
$$v_\theta = -\frac{\partial\psi}{\partial r} = \frac{k}{2\sqrt{r}}\sin\theta + f'(r)$$

We know,
$$v_\theta = 0 \text{ at } \theta = 0, \text{ which depicts}$$
$$f'(r) = 0 \quad \text{and} \quad f(r) = \text{constant}$$

Therefore,
$$v_\theta = \frac{k}{2\sqrt{r}}\sin\theta \text{ and } \psi = k\sqrt{r}\sin\theta$$

7.2.3 Vortex Flow

In this flow all the streamlines are concentric circles about a given point where the velocity along each streamline is inversely proportional to the distance from the centre, as shown in Fig. 7.4. Such a flow is called *vortex (free vortex) flow*. This flow is necessarily irrotational.

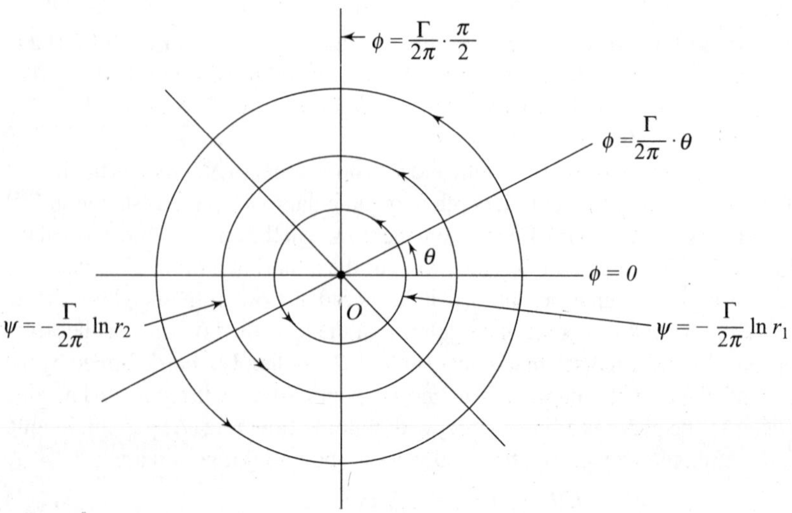

Fig. 7.4 Flownet for a vortex (free vortex)

In *purely circulatory (free vortex flow) motion*, we can write the tangential velocity as

$$v_\theta = \frac{\text{Circulation constant}}{r}$$

$$v_\theta = \frac{\Gamma/2\pi}{r} \tag{7.18}$$

where Γ is circulation,

Also, for purely circulatory motion one can write

$$v_r = 0 \tag{7.19}$$

With the definition of stream function, it is evident that

$$v_\theta = -\frac{\partial \psi}{\partial r} \quad \text{and} \quad v_r = \frac{1}{r}\frac{\partial \psi}{\partial \theta}$$

Combining Eqs (7.18) and (7.19) with the above said relations for stream function, it is possible to write

$$\psi = -\frac{\Gamma}{2\pi}\ln r + C_1 \tag{7.20}$$

Because of irrotationality, it should satisfy

$$\hat{\mathbf{i}}\,v_r + \hat{\mathbf{j}}\,v_\theta = \hat{\mathbf{i}}\frac{\partial \phi}{\partial r} + \hat{\mathbf{j}}\frac{1}{r}\frac{\partial \phi}{\partial \theta}$$

Eqs (7.18) and (7.19) and the above solution of Laplace's equation yields,

$$\phi = \frac{\Gamma}{2\pi}\theta + C_2 \tag{7.21}$$

The integration constants C_1 and C_2 have no effect whatsoever on the structure of velocities or pressures in the flow. Therefore like other elementary flows, we shall consistently ignore such constants. It is clear that the *streamlines* for vortex flow are *circles* while the *potential* lines are *radial*. These are given by

$$\psi = -\frac{\Gamma}{2\pi}\ln r \quad \text{and} \quad \phi = \frac{\Gamma}{2\pi}\theta \tag{7.22}$$

In Fig. 7.4, point O can be imagined as a point vortex that induces the circulatory flow around it. The point vortex is a singularity in the flow field (v_θ becomes infinite). It is also discerned that the point O in Fig. 7.4 is simply a point formed by the intersection of the plane of a paper and a line perpendicular to the plane. This line is called *vortex filament* of strength Γ, where Γ is the circulation around the *vortex filament* and the circulation is defined as

$$\Gamma = \oint \vec{\mathbf{V}} \cdot \mathbf{d}\vec{\mathbf{s}} \tag{7.23}$$

In Eq. (7.23), the line integral of the velocity component tangent to a curve of elemental length ds, is taken around a closed curve. It may be stated that the circulation for a closed path in an irrotational flow field is zero. However, the circulation for a given path in an irrotational flow containing a finite number of singular points is constant. In general this circulation constant Γ denotes the algebraic strength of the vortex filament contained within the closed curve.

From Eq. (7.23) we can write

$$\Gamma = \oint \vec{\mathbf{V}} \cdot \mathbf{d}\vec{\mathbf{s}} = \oint (u\mathrm{d}x + v\mathrm{d}y + w\mathrm{d}z)$$

For a two-dimensional flow

$$\Gamma = \oint (u\mathrm{d}x + v\,\mathrm{d}y)$$

or

$$\Gamma = \oint V \cos\alpha \,\mathrm{d}s \tag{7.24}$$

Consider a fluid element as shown in Fig. 7.5. Circulation is positive in the anti-clockwise direction (not a mandatory but general convention).

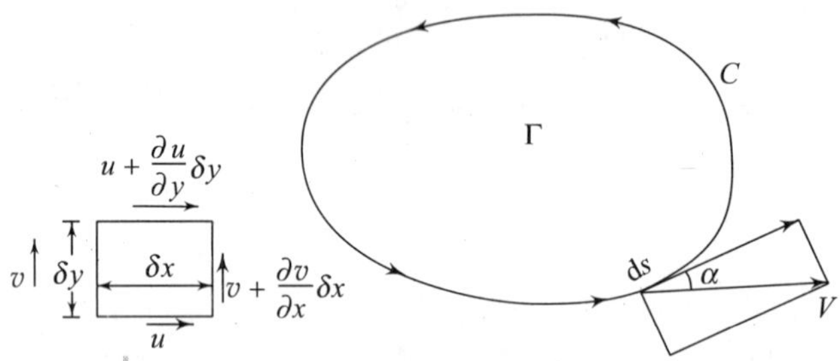

Fig. 7.5 Circulation in a flow field

$$\delta\Gamma = u\,\delta x + \left(v + \frac{\partial v}{\partial x}\delta x\right)\delta y - \left(u + \frac{\partial u}{\partial y}\delta y\right)\delta x - v\delta y$$

or

$$\delta\Gamma = \left(\frac{\partial v}{\partial x} - \frac{\partial u}{\partial y}\right)\delta x\,\delta y$$

or

$$\delta\Gamma = \Omega_z\,\delta A$$

or

$$\delta\Gamma/\delta A = \Omega_z \tag{7.25}$$

Physically, circulation per unit area is the vorticity of the flow.

Now, for a free vortex flow, the tangential velocity is given by Eq. (7.18) as

$$v_\theta = \frac{\Gamma/2\pi}{r} = \frac{C}{r}$$

For a circular path (refer Fig. 7.5)

$$\alpha = 0, \qquad V = v_\theta = \frac{C}{r}$$

Thus,

$$\Gamma = \oint_0^{2\pi} \frac{C}{r}\,r\,\mathrm{d}\theta = 2\pi C \tag{7.26}$$

It may be noted that although free vortex is basically an irrotational motion, the circulation for a given path containing a singular point (including the origin) is constant ($2\pi C$) and independent of the radius of a circular streamline. However, if the circulation is calculated in a free vortex flow along any closed contour excluding the singular point (the origin), it should be zero. Let us look at Fig. 7.6(a) and take a closed contour *ABCD* in order to find out circulation about the point, *P* around *ABCD*

$$\Gamma_{ABCD} = -v_{\theta_{AB}} r_1 \, d\theta - v_{r_{BC}} (r_2 - r_1) + v_{\theta_{CD}} r_2 \, d\theta + v_{r_{DA}} (r_2 - r_1)$$

There is no radial flow

$$v_{r_{BC}} = v_{r_{DA}} = 0, \; v_{\theta_{AB}} = \frac{C}{r_1} \text{ and } v_{\theta_{CD}} = \frac{C}{r_2}$$

$$\Gamma_{ABCD} = \frac{-C}{r_1} \cdot r_1 d\theta + \frac{C}{r_2} \cdot r_2 d\theta = 0 \tag{7.27}$$

If there exists a solid body rotation at constant ω induced by some external mechanism, the flow should be called a *forced vortex motion* (Fig. 7.6(b)) and we can write

$$v_\theta = \omega r \qquad \text{and}$$

$$\Gamma = \oint v_\theta \, ds = \oint_0^{2\pi} \omega r \cdot r d\theta = 2\pi r^2 \omega \tag{7.28}$$

Equation (7.28) predicts that the circulation is zero at the origin and it increases with increasing radius. The variation is parabolic.

It may be mentioned that the free vortex (irrotational) flow at the origin (Fig. 7.6(a)) is impossible because of mathematical singularity. However, physically there should exist a rotational (forced vortex) core which is shown by the dotted line. Below are given two statements which are related to Kelvin's circulation theorem (1869) and Cauchy's theorem on irrotational motion (1815), respectively:

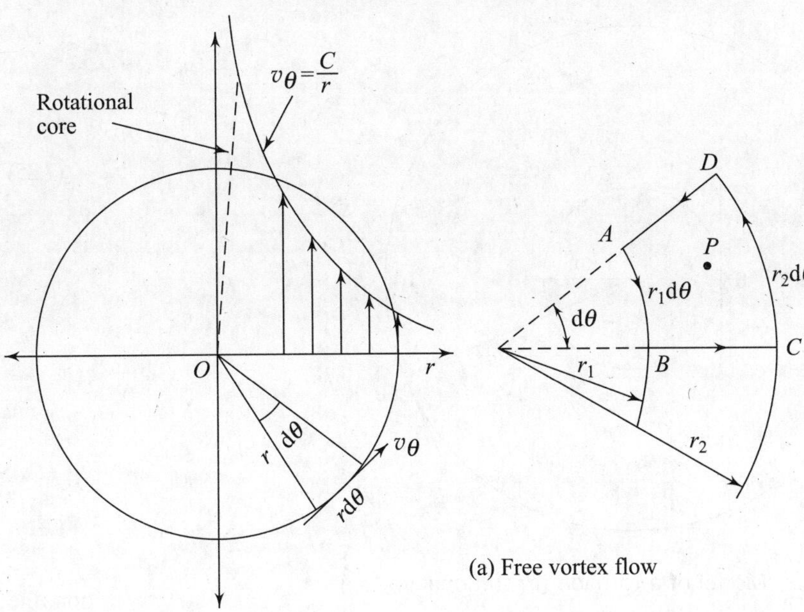

(a) Free vortex flow

Fig. 7.6 (*Contd.*)

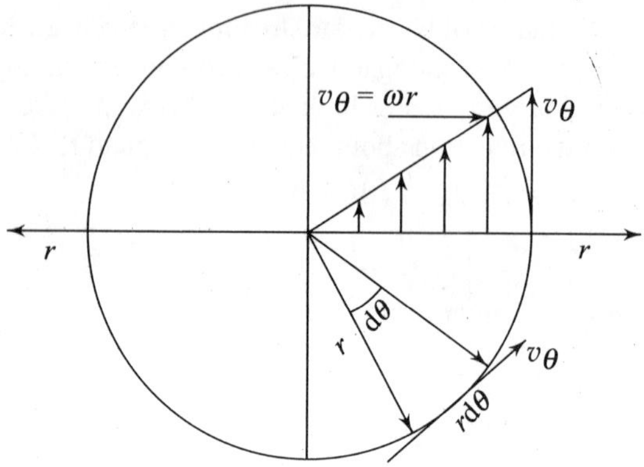

(b) Forced vortex flow

Fig. 7.6 Vortex Flow

(i) The circulation around any closed contour is invariant with time in an inviscid fluid.
(ii) A body of inviscid fluid in irrotational motion continues to move irrotationally.

Example 7.5

The wind velocity at a location 5 km away from the centre of a tornado (consider inviscid, irrotational vortex motion shown in Fig 7.7) was measured as 30 km/hr and

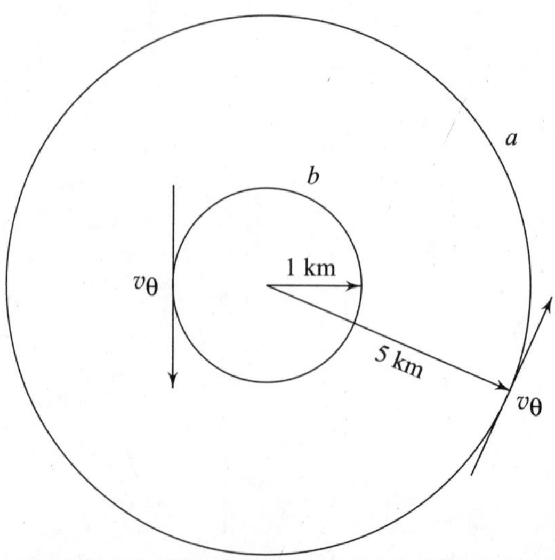

Fig. 7.7 Model of a tornado (irrotational vortex)

the barometric pressure was 750 mm of Hg. Calculate the wind velocity 1 km from the tornado centre and its barometric pressure. (Density of air = 1.2 kg/m^3, density of mercury = 13.6×10^3 kg/m^3).

Solution

$$p_a = 750 \text{ mm of Hg} = 0.75 \text{ m of Hg}$$

$$= 0.75 \times 13.6 \times 10^3 \times 9.81 = 100.062 \text{ kN/m}^2$$

From free vortex consideration, we can write

$$r_a \, v_{\theta a} = r_b \, v_{\theta b} = C$$

$$r_a = 5000 \text{ m}, v_{\theta a} = \frac{30 \times 1000}{60 \times 60} = 8.33 \text{ m/s}$$

$$C = \text{circulation constant} = 41650 \text{ m}^2/\text{s}$$

at $$r_b = 1000 \text{ m},$$

$$v_{\theta b} = \frac{41650}{1000} = 41.65 \text{ m/s}$$

Bernoulli's equation between points *a* and *b*,

$$\frac{p_a}{\rho g} + \frac{V_{\theta a}^2}{2g} = \frac{p_b}{\rho g} + \frac{V_{\theta b}^2}{2g}$$

$$\frac{100062}{1.2 \times 9.81} + \frac{(8.33)^2}{2 \times 9.81} = \frac{p_b}{\rho g} + \frac{(41.65)^2}{2 \times 9.81}$$

or $$\frac{p_b}{\rho g} = 8500 + 3.536 - 88.416$$

or $$\frac{p_b}{\rho g} = 8415.12$$

or $$p_b = 99062 \text{ N/m}^2 = 99.062 \text{ kN/m}^2$$

7.3 SUPERPOSITION OF ELEMENTARY FLOWS

We can now form different flow patterns by superimposing the velocity potential and stream functions of the elementary flows stated above.

7.3.1 Doublet

In order to develop a doublet, imagine a source and a sink of equal strength K at equal distance s from the origin along the x axis, as shown in Fig. 7.8

From any point $P(x, y)$ in the field, r_1 and r_2 are drawn to the source and the sink. The polar coordinates of this point (r, θ) have been shown.

The potential functions of the two flows may be superimposed to describe the potential for the combined flow at P as

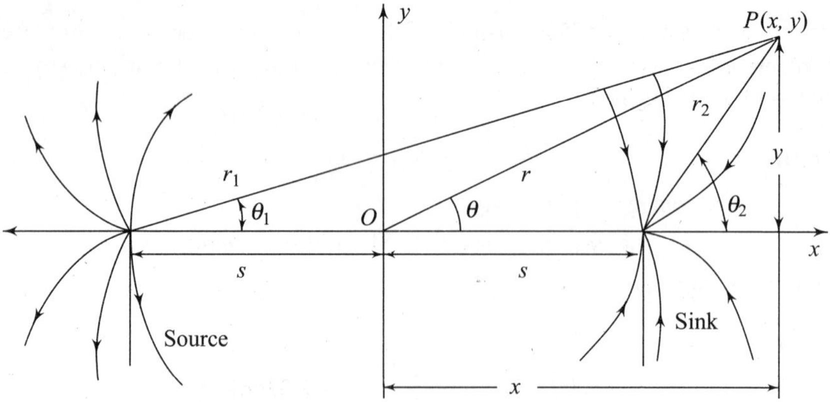

Fig. 7.8 Superposition of a source and a sink

$$\phi = K \ln r_1 - K \ln r_2 \tag{7.29}$$

Similarly,

$$\psi = K(\theta_1 - \theta_2) = -K\alpha \tag{7.30}$$

where $\alpha = (\theta_2 - \theta_1)$

We can also write

$$\tan\theta_1 = \frac{y}{x+s} \quad \text{and} \quad \tan\theta_2 = \frac{y}{x-s} \tag{7.31}$$

$$r_1 = \sqrt{r^2 + s^2 + 2rs\cos\theta} \quad \text{and} \quad r_2 = \sqrt{r^2 + s^2 - 2rs\cos\theta} \tag{7.32}$$

Now using the above mentioned relations we find

$$\tan(\theta_2 - \theta_1) = \frac{\tan\theta_2 - \tan\theta_1}{1 + \tan\theta_2\tan\theta_1}$$

or

$$\tan\alpha = \left[\frac{yx + ys - yx + ys}{x^2 - s^2}\right] \Big/ \left(1 + \frac{y^2}{x^2 - s^2}\right)$$

or

$$\tan\alpha = \frac{2ys}{x^2 + y^2 - s^2} \tag{7.33}$$

Hence the stream function and the velocity potential function are formed by combining Eqs (7.30) and (7.33), as well as Eqs (7.29) and (7.32) respectively,

$$\psi = -K\tan^{-1}\left(\frac{2ys}{x^2 + y^2 - s^2}\right) \tag{7.34}$$

$$\phi = \frac{K}{2}\ln\left(\frac{r^2 + s^2 + 2rs\cos\theta}{r^2 + s^2 - 2rs\cos\theta}\right) \tag{7.35}$$

Doublet is a special case when a source as well as a sink are brought together in such a way that $s \to 0$ and at the same time the strength K ($\Lambda/2\pi$) is increased to an infinite value. These are assumed to be accomplished in a manner which makes the product of s and Λ/π (in limiting case) a finite value χ. Under the aforesaid circumstances,

$$\psi = -\frac{\Lambda}{2\pi} \cdot \frac{2ys}{x^2 + y^2 - s^2}$$

[Since in the limiting case $\tan^{-1} \alpha = \alpha$]

$$\psi = -\chi \cdot \frac{y}{x^2 + y^2} = \frac{-\chi \sin\theta}{r} \tag{7.36}$$

From Eq. (7.35), we get

$$\phi = \frac{\Lambda}{4\pi} \left[\ln (r^2 + s^2 + 2rs \cos \theta) - \ln (r^2 + s^2 - 2rs \cos \theta)\right]$$

or $$\phi = \frac{\Lambda}{4\pi}\left[\ln\left\{\left(r^2+s^2\right)\left(1+\frac{2rs\cos\theta}{r^2+s^2}\right)\right\} - \ln\left\{\left(r^2+s^2\right)\left(1-\frac{2rs\cos\theta}{r^2+s^2}\right)\right\}\right]$$

or $$\phi = \frac{\Lambda}{4\pi}\left[\left\{\frac{2rs\cos\theta}{r^2+s^2} - \frac{1}{2}\left(\frac{2rs\cos\theta}{r^2+s^2}\right)^2 + \frac{1}{3}\left[\frac{2rs\cos\theta}{r^2+s^2}\right]^3 + \cdots\right\}\right.$$

$$\left. - \left\{-\frac{2rs\cos\theta}{r^2+s^2} - \frac{1}{2}\left(\frac{2rs\cos\theta}{r^2+s^2}\right)^2 - \frac{1}{3}\left[\frac{2rs\cos\theta}{r^2+s^2}\right]^3 + \cdots\right\}\right]$$

or $$\phi = \frac{\Lambda}{4\pi}\left[\frac{4rs\cos\theta}{r^2+s^2} + \frac{2}{3}\left(\frac{2rs\cos\theta}{r^2+s^2}\right)^3 + \cdots\right]$$

In the limiting condition the above expression can be written as

$$\phi \approx \frac{\chi r\cos\theta}{r^2+s^2}$$

or $$\phi \approx \frac{\chi\cos\theta}{r} \tag{7.37}$$

We can see that the streamlines associated with the doublet are

$$-\frac{\chi\sin\theta}{r} = C_1$$

If we replace $\sin\theta$ by y/r, and the minus sign be absorbed in C_1, we get

$$\chi\frac{y}{r^2} = C_1 \tag{7.38a}$$

In terms of Cartesian coordinate, it is possible to write

$$x^2 + y^2 - \frac{\chi}{C_1} y = 0 \tag{7.38b}$$

Equation (7.38b) represents a family of circles. For $x = 0$, there are two values of y, one of which is zero. The centres of the circles fall on the y axis. On the circle, where $y = 0$, x has to be zero for all the values of the constant. It is obvious that the family of circles formed due to different values of C_1 must be tangent to the x axis at the origin. These streamlines are illustrated in Fig. 7.9. Due to the initial positions of the source and the sink in the development of the doublet, it is certain that the flow will emerge in the negative x direction from the origin and it will converge via the positive x direction of the origin.

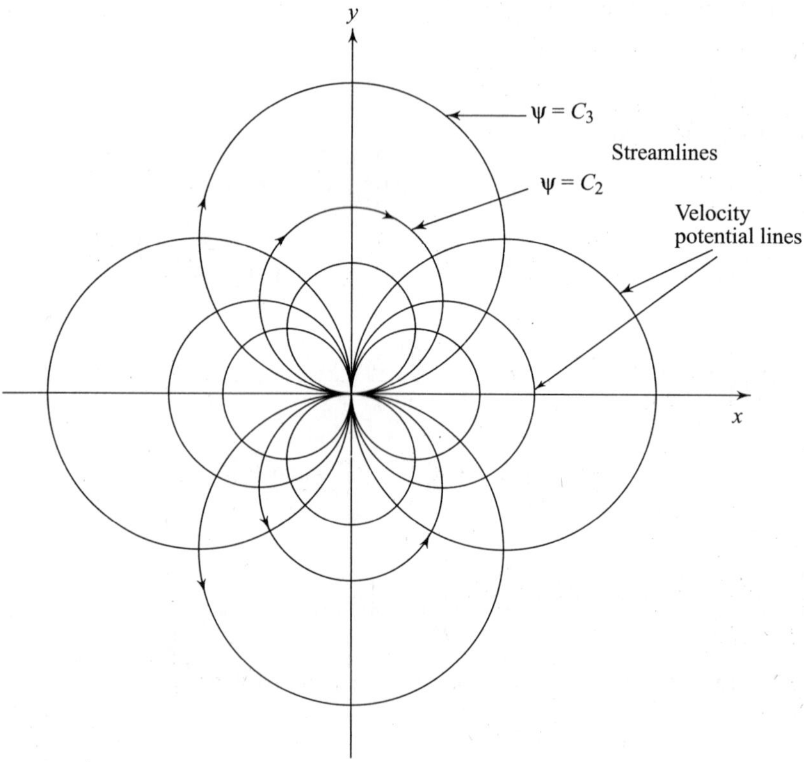

Fig. 7.9 Streamlines and velocity potential lines for a doublet

However, the velocity potential lines are

$$\frac{\chi \cos\theta}{r} = K_1$$

In Cartesian coordinate this equation becomes

$$x^2 + y^2 - \frac{\chi}{K_1} x = 0 \tag{7.39}$$

Once again we shall obtain a family of circles. The centres will fall on the x axis. For $y = 0$ there are two values of x, one of which is zero. When $x = 0$, y has to be zero for all values of the constant. Therefore these circles are tangent to the y axis at the origin.

The orthogonality of constant ψ and constant ϕ lines are maintained as we iron out the procedure of drawing constant value lines (Fig. 7.9). In addition to the determination of the stream function and velocity potential, it is observed from Eq. (7.37) that for a doublet

$$v_r = \frac{\partial \phi}{\partial r} = \frac{-\chi \cos \theta}{r^2} \tag{7.40}$$

As the centre of the doublet is approached, the radial velocity tends to be infinite. It shows that the doublet flow has a singularity. Since the circulation about a singular point of a source or a sink is zero for any strength, it is obvious that the circulation about the singular point in a doublet flow must be zero. It follows that for all paths in a doublet flow $\Gamma = 0$

$$\Gamma = \oint \vec{\mathbf{V}} \cdot \mathbf{d}\vec{\mathbf{s}} = 0 \tag{7.41}$$

Applying Stokes theorem between the line integral and the area integral,

$$\Gamma = \iint (\nabla \times \vec{\mathbf{V}})\, d\vec{A} = 0 \tag{7.42}$$

From Eq. (7.42), the obvious conclusion is $\nabla \times \vec{\mathbf{V}} = 0$, i.e., doublet flow is an irrotational flow.

At large distances from a doublet, the flow approximates the disturbances of a two-dimensional airfoil. The influence of an airfoil as felt at distant walls may be approximated mathematically by a combination of doublets with varying strengths. Thus the cruise conditions of a two-dimensional airfoil can be simulated by the superposition of a uniform flow and a doublet sheet of varying strengths.

7.3.2 Flow about a Cylinder Without Circulation

Inviscid-incompressible flow about a cylinder in uniform flow is equivalent to the superposition of a uniform flow and a doublet. The doublet has its axis of development parallel to the direction of the uniform flow. The combined potential of this flow is given by

$$\phi = U_0 x + \frac{\chi \cos \theta}{r} \tag{7.43}$$

and consequently the stream function becomes

$$\psi = U_0 y - \frac{\chi \sin \theta}{r} \tag{7.44}$$

In our analysis, we shall draw streamlines in the flow field. In two-dimensional flow, a streamline may be interpreted as the edge of a surface on which the velocity vector should always be tangent and there is no flow in the direction normal to it. The latter is identically the characteristics of a solid impervious boundary. Hence, a streamline may also be considered as the contour of an *impervious two-dimensional body*. Figure 7.10 shows a set of streamlines. The streamline *C-D* may be considered as the edge of a two-dimensional body while the remaining streamlines form the flow about the boundary.

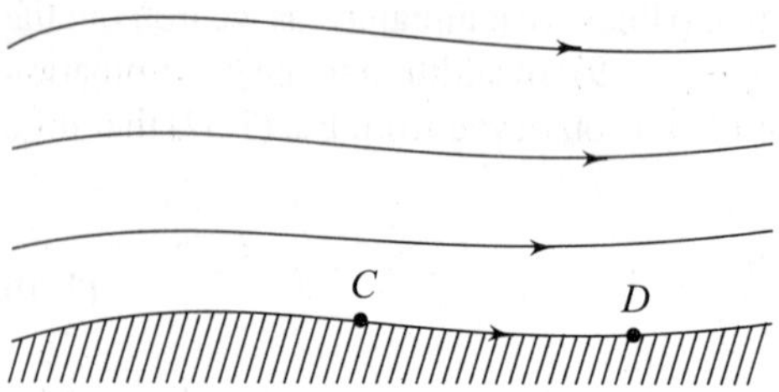

Fig. 7.10 Surface streamline

Now we follow the essential steps involving the superposition of elementary flows in order to form a flow about the body of interest. A streamline has to be determined which encloses an area whose shape is of practical importance in fluid flow. This streamline will describe the boundary of a two-dimensional solid body. The remaining streamlines outside this solid region will constitute the flow about this body.

Let us look for the streamline whose value is zero. Thus, we obtain

$$U_0 y - \frac{\chi \sin\theta}{r} = 0 \tag{7.45}$$

replacing y by $r \sin\theta$, we have

$$\sin\theta\left(U_0 r - \frac{\chi}{r}\right) = 0 \tag{7.46}$$

If $\theta = 0$ or $\theta = \pi$, the equation is satisfied. This indicates that the x axis is a part of the streamline $\psi = 0$. When the quantity in the parentheses is zero, the equation is identically satisfied. Hence it follows that

$$r = \left(\frac{\chi}{U_0}\right)^{1/2} \tag{7.47}$$

It can be said that there is a circle of radius $\left(\frac{\chi}{U_0}\right)^{1/2}$ which is an intrinsic part of the streamline $\psi = 0$. This is shown in Fig. 7.11. Let us look at the points of intersection of the circle and the x axis, i.e., the points A and B. The polar coordinates of these points are

$$r = \left(\frac{\chi}{U_0}\right)^{1/2}, \qquad \theta = \pi, \text{ for point A}$$

$$r = \left(\frac{\chi}{U_0}\right)^{1/2}, \qquad \theta = 0, \text{ for point B}$$

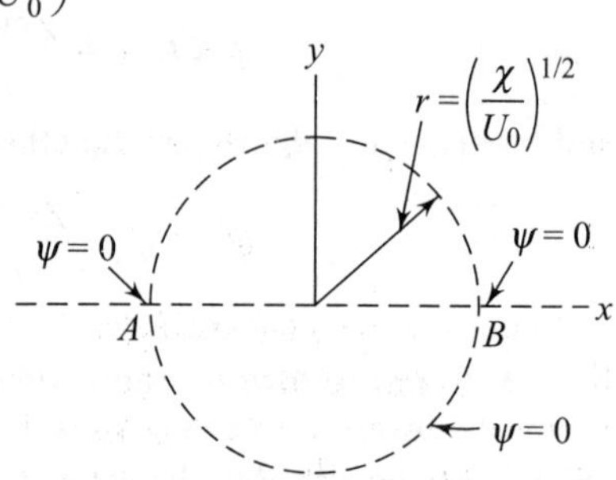

Fig. 7.11 Streamline $\psi = 0$ in a superimposed flow of doublet and uniform stream

The velocity at these points are found out by taking partial derivatives of the velocity potential in two orthogonal directions and then substituting the proper values of the coordinates. Thus

$$v_r = \frac{\partial \phi}{\partial r} = U_0 \cos\theta - \frac{\chi \cos\theta}{r^2} \tag{7.48a}$$

$$v_\theta = \frac{1}{r}\frac{\partial \phi}{\partial \theta} = -U_0 \sin\theta - \frac{\chi \sin\theta}{r^2} \tag{7.48b}$$

At point A $\left[\theta = \pi,\ r = \left(\frac{\chi}{U_0}\right)^{1/2}\right]$

$$v_r = 0, v_\theta = 0$$

At point B $\left[\theta = 0,\ r = \left(\frac{\chi}{U_0}\right)^{1/2}\right]$

$$v_r = 0, v_\theta = 0$$

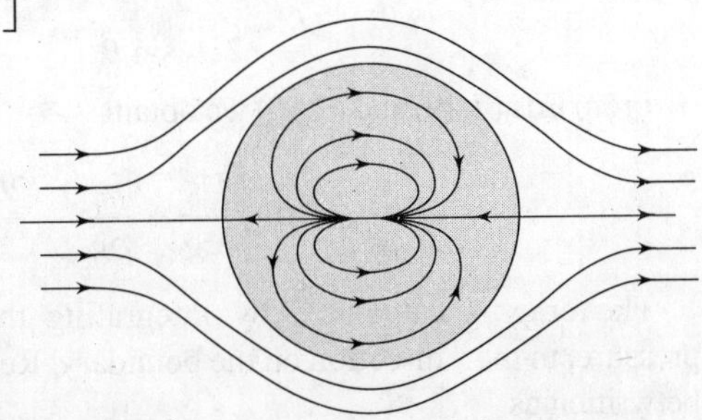

Fig. 7.12 Inviscid flow past a cylinder

The points A and B are clearly the stagnation points through which the flow divides and subsequently reunites forming a zone of circular bluff body.

The circular region, enclosed by part of the streamline $\psi = 0$ could be imagined as a solid cylinder in an inviscid flow. At a large distance from the cylinder the flow is moving uniformly in a cross-flow configuration.

Figure 7.12 shows the streamlines of the flow. The streamlines outside the circle describe the flow pattern of the inviscid irrotational flow across a cylinder. However, the streamlines inside the circle may be disregarded since this region is considered as a solid obstacle.

7.3.3 Lift and Drag for Flow Past a Cylinder Without Circulation

Lift and drag are the forces per unit length on the cylinder in the directions normal and parallel respectively, to the direction of uniform flow.

Pressure for the combined doublet and uniform flow becomes uniform at large distances from the cylinder where the influence of doublet is indeed small. Let us imagine the pressure p_0 is known as well as uniform velocity U_0. Now we can apply Bernoulli's equation between infinity and the points on the boundary of the cylinder. Neglecting the variation of potential energy between the aforesaid point at infinity and any point on the surface of the cylinder, we can write

$$\frac{p_0}{\rho g} + \frac{U_0^2}{2g} = \frac{p_b}{\rho g} + \frac{U_b^2}{2g} \tag{7.49}$$

where, the subscript b indicates the surface of the cylinder. As we know, since fluids cannot penetrate a solid boundary, the velocity U_b should be in the transverse direction only or in other words, only v_θ component of velocity be present on the streamline $\psi = 0$.

Thus, at $\quad r = \left(\dfrac{\chi}{U_0}\right)^{1/2}$

$$U_b = v_\theta \Big|_{\text{at } r = (\chi/U_0)^{1/2}} = -\frac{1}{r}\frac{\partial \phi}{\partial \theta}\Big|_{\text{at } r = (\chi/U_0)^{1/2}} \tag{7.50}$$

$$= -2U_0 \sin\theta$$

From Eqs (7.49) and (7.50) we obtain

$$p_b = \rho g\left[\frac{U_0^2}{2g} + \frac{p_0}{\rho g} - \frac{(2U_0 \sin\theta)^2}{2g}\right] \tag{7.51}$$

The drag is calculated by integrating the force components arising out of pressure, in the x direction on the boundary. Referring to Fig. 7.13, the drag force can be written as

$$D = -\int_0^{2\pi} p_b \cos\theta \left(\frac{\chi}{U_0}\right)^{1/2} d\theta$$

or

$$D = -\int_0^{2\pi} \rho g \left(\frac{\chi}{U_0}\right)^{1/2} \left[\frac{U_0^2}{2g} + \frac{p_0}{\rho g} - \frac{(2U_0 \sin\theta)^2}{2g}\right] \cos\theta\, d\theta$$

$$D = -\int_0^{2\pi} \left[p_0 + \frac{\rho U_0^2}{2}(1 - 4\sin^2\theta)\right]\left(\frac{\chi}{U_0}\right)^{1/2} \cos\theta\, d\theta \tag{7.52}$$

Similarly, the lift force

$$L = -\int_0^{2\pi} p_b \sin\theta \left(\frac{\chi}{U_0}\right)^{1/2} d\theta \tag{7.53}$$

The Eqs (7.52) and (7.53) produce $D = 0$ and $L = 0$ after the integration is carried out.

However, in reality, the cylinder will always experience some drag force. This contradiction between the inviscid flow result and the experiment is usually known as *D'Almbert paradox*. The reason for the discrepancy lies in completely ignoring the viscous effects throughout the flow field. Effect of the thin region adjacent to the solid boundary is of paramount importance in determining drag force. However, the lift may often be predicted by the present technique. We shall appreciate this fact in a subsequent section.

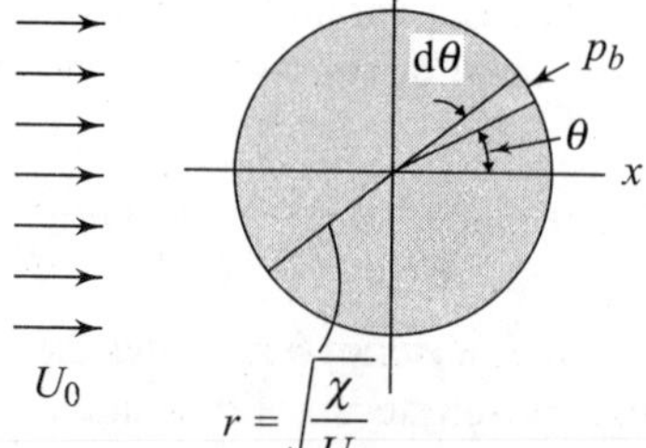

Fig 7.13 Calculation of drag on a cylinder

Bernoulli' s equation can be used to calculate the pressure distribution on the cylinder surface

$$\frac{p_b(\theta)}{\rho g} + \frac{U_b^2(\theta)}{2g} = \frac{p_0}{\rho g} + \frac{U_0^2}{2g}$$

$$\frac{p_b(\theta) - p_0}{\rho} = \frac{U_0^2}{2}\left[1 - 4\sin^2\theta\right]$$

The pressure coefficient, c_p is therefore

$$C_p = \frac{p_b(\theta) - p_0}{\frac{1}{2}\rho U_0^2} = [1 - 4\sin^2\theta] \tag{7.54}$$

The pressure distribution on a cylinder is shown (Fig.7.14) below.

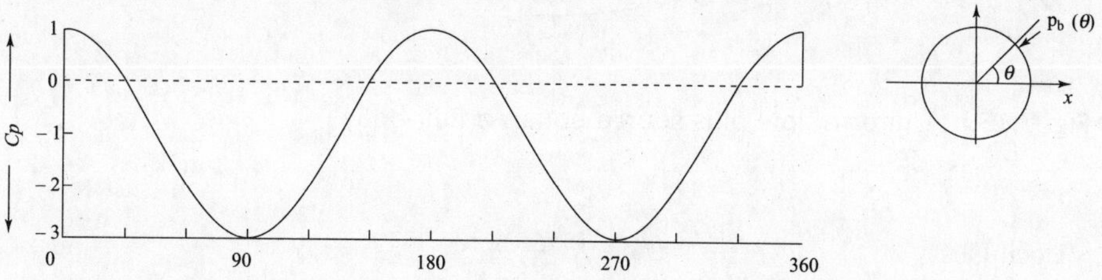

Fig. 7.14 Variation of coefficient of pressure with angle

Example 7.6

A two-dimensional source of volume flow rate $\Lambda = 2.5\ \text{m}^2/\text{s}$ is located in a uniform flow (U_0) of 2 m/s. Determine the stagnation point and the maximum thickness of the resulting half body.

Solution

We have already constructed different flow patterns by superimposing elementary flows. An interesting body shape appears if we superimpose a uniform flow over an isolated source or sink which is known as Rankine half-body (refer Fig. 7.15). Let the source be located at the origin.

(i) Then the stream function of combination is

$$\psi = U_0 y + \frac{\Lambda}{2\pi}\tan^{-1}\left(\frac{y}{x}\right)$$

or

$$\psi = U_0 y + K\tan^{-1}\left(\frac{y}{x}\right)$$

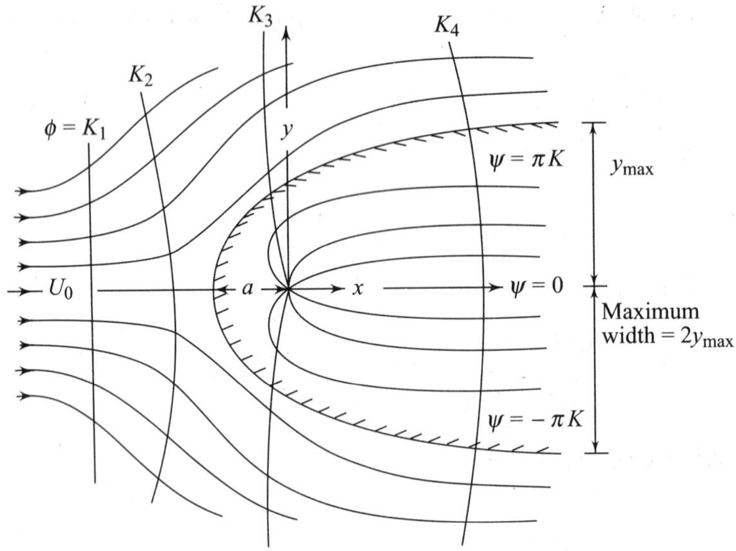

Fig. 7.15 Uniform flow plus source equals a half-body

Velocity, $$u = \frac{\partial \psi}{\partial y} = U_0 + K\frac{x}{x^2 + y^2}$$

Similarly, $$v = -\frac{\partial \psi}{\partial x} = +K\frac{y}{x^2 + y^2}$$

At the stagnation point $u = 0$, $v = 0$, demand from the above equation $y = 0$ and

$$x = \frac{\Lambda}{2\pi U_0} = -\frac{2.5}{2\pi \times 2} = -0.2\text{ m}$$

The coordinates of stagnation points are $(-a, 0)$ or $(-0.2, 0)$
The value of stream function at the stagnation point is

$$\psi(-0.2, 0) = 0 + \frac{\Lambda}{2\pi}\tan^{-1} 0$$

or $$\psi_{stag} = 0 + \frac{2.5}{2\pi}\pi = 1.25\text{ m}^2/\text{s}$$

The half-body is described by dividing streamline,

$$\psi = \frac{\Lambda}{2} = \pi \cdot \frac{\Lambda}{2\pi} = \pi K$$

or $$U_0 y + \frac{\Lambda}{2\pi}\tan^{-1}\frac{y}{x} = \frac{\Lambda}{2}$$

or $$U_0 y + \frac{\Lambda\theta}{2\pi} = \frac{\Lambda}{2} \quad \text{or} \quad y = \frac{\Lambda\left(1 - \frac{\theta}{\pi}\right)}{2U_0}$$

at $$\theta = 0 \quad y_{max} = \frac{\Lambda}{2U_0}, \text{ the maximum ordinate}$$

at $$\theta = \frac{\pi}{2}, \quad y = \frac{\Lambda}{4U_0}, \text{ the upper ordinate at the origin}$$

at $$\theta = \pi, \quad y = 0, \text{ the stagnation point}$$

at $$\theta = \frac{3\pi}{2}, \quad y = -\frac{\Lambda}{4U_0}, \text{ the lower ordinate at the origin}$$

(ii) However, the equation of the half-body becomes

$$U_0 y + \frac{\Lambda}{2\pi}\tan^{-1}\left(\frac{y}{x}\right) = \frac{\Lambda}{2}$$

The maximum thickness occurs as $x \to \infty$

$$2y + \frac{1.25}{\pi}\tan^{-1} 0 = 1.25$$

$$y = \frac{1.25}{2} = 0.625 \text{ m}$$

The maximum thickness $= 2y_{max} = 1.25$

Example 7.7

A source at the origin and a uniform flow at 5m/s are superimposed. The half-body which is formed has a maximum width of 2 m. Calculate (i)The location of stagnation point, (ii) width of the body at the origin, and (iii) velocity at a point $\left(0.7, \frac{\pi}{2}\right)$.

Solution

(i) We have seen in Example 7.5, that

at $$\theta = 0, \quad y_{max} = \frac{\Lambda}{2U_0} = \frac{2}{2} = 1 \text{ m}$$

or $$\Lambda = 2 \times 5 \times 1 = 10 \text{ m}^2/\text{s}$$

for stagnation point,

$$x = -\frac{K}{U_0} = -\frac{\Lambda}{2\pi U_0} = \frac{-10}{2\pi \times 5} = -0.32 \text{ m}$$

and $$y = 0$$

(ii) at $\theta = \dfrac{\pi}{2}, \quad y = \dfrac{\Lambda}{4U_0}$ [from Example 7.5]

at $\theta = \dfrac{\pi}{2}, \quad y = \dfrac{10}{4 \times 5} = 0.5 \text{ m}$

The width of the body at the origin is $2 \times 0.5 = 1$ m

(iii) In polar coordinate,

$$\psi = U_0 r \sin\theta + K\theta, \qquad \text{where} \qquad K = \frac{\Lambda}{2\pi}$$

$$v_r = \frac{1}{r}\frac{\partial \psi}{\partial \theta} = U_0 \cos\theta + \frac{\Lambda}{2\pi r}$$

$$v_\theta = -\frac{\partial \psi}{\partial r} = -U_0 \sin\theta$$

at the point $(0.7, \pi/2)$,

$$v_r = \frac{\Lambda}{2\pi r} = \frac{10}{2\pi \times 0.7} = 2.27 \text{ m/s}$$

$$v_\theta = -U_0 \sin\theta = -5 \sin\frac{\pi}{2} = -5 \text{ m/s}$$

$$V_{\text{resultant}} = \sqrt{(2.27)^2 + (5)^2} = 5.49 \text{ m/s}$$

Example 7.8

A line source discharging a flow at 0.6 m^2/s per unit length is located at (–1,0) and a sink of volume flow rate 1.2 m^2/s is located at (2,0). For a dynamic pressure of 10 N/m^2 at the origin, determine the velocity and dynamic pressure at (1,1).

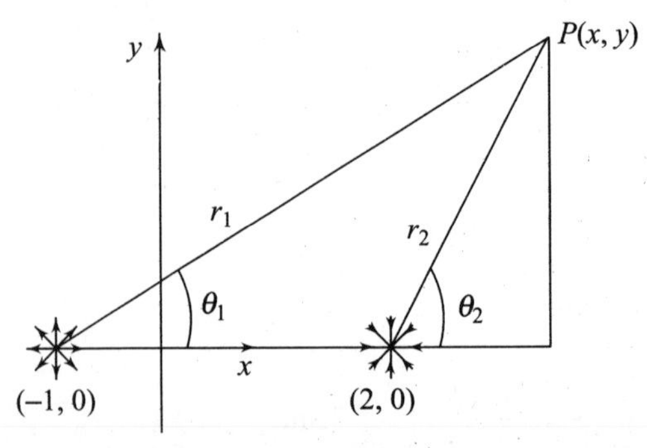

Fig. 7.16 Source and sink pair

Solution

ψ at P may be expressed as

$$\psi = K_1\theta_1 - K_2\theta_2$$

or
$$\psi = \frac{\Lambda_1}{2\pi}\theta_1 - \frac{\Lambda_2}{2\pi}\theta_2$$

or
$$\psi = \frac{0.6}{2\pi}\tan^{-1}\left(\frac{y}{x+1}\right) - \frac{1.2}{2\pi}\tan^{-1}\left(\frac{y}{x-2}\right)$$

$$u = \frac{\partial \psi}{\partial y} = \frac{0.6}{2\pi}\left[\frac{(x+1)}{(x+1)^2+y^2}\right] - \frac{1.2}{2\pi}\left[\frac{(x-2)}{(x-2)^2+y^2}\right]$$

$$v = -\frac{\partial \psi}{\partial x} = \frac{0.6}{2\pi}\left[\frac{y}{(x+1)^2+y^2}\right] - \frac{1.2}{2\pi}\left[\frac{y}{(x-2)^2+y^2}\right]$$

at the origin (0,0)

$$u = \frac{0.6}{2\pi} - \frac{1.2}{2\pi}\left(\frac{-2}{4}\right) = \frac{0.6}{2\pi} + \frac{0.6}{2\pi} = \frac{0.6}{\pi} = 0.1909 \text{ m/s}$$

$$v = 0$$

Dynamic pressure $\frac{1}{2}\rho V^2 = 10 \text{ N/m}^2$

or
$$\rho = \frac{20}{V^2} = 548.3 \text{ kg/m}^3$$

At point (1,1)

$$u = \frac{0.6}{2\pi}\cdot\frac{2}{5} + \frac{1.2}{2\pi}\cdot\frac{1}{2} = \frac{0.6}{5\pi} + \frac{0.6}{2\pi}$$

$$= 0.0381 + 0.0954 = 0.1335 \text{ m/s}$$

$$v = \frac{0.6}{2\pi}\cdot\frac{1}{5} - \frac{1.2}{2\pi}\cdot\frac{1}{2} = \frac{0.6}{10\pi} + \frac{0.6}{2\pi}$$

$$= 0.019 - 0.0954 = -0.0764 \text{ m/s}$$

$$V_{\text{resultant}} = 0.1538 \text{ m/s}$$

Dynamic pressure at (1,1) $= \frac{1}{2} \times 548.3 \times (0.1538)^2$

$$= 6.48 \text{ N/m}^2$$

Example 7.9

A source with volume flow rate 0.2 m^2/s and a vortex with strength1 m^2/s are located at the origin. Determine the equations for velocity potential and stream function. What should be the resultant velocity at $x = 0.9$ m and $y = 0.8$ m?

Solution

For the source, $\psi = K_1\theta, \qquad \phi = K_1 \ln r$

For the vortex, $\psi = -K_2 \ln r, \qquad \phi = K_2\theta$

Combined, $$\psi = \frac{0.2}{2\pi}\theta - \frac{1}{2\pi}\ln r = \frac{1}{\pi}\left[0.1\,\theta - \frac{1}{2}\ln r\right]$$

Combined, $$\phi = \frac{0.2}{2\pi}\ln r + \frac{1}{2\pi}\theta = \frac{1}{\pi}\left[0.1\ln r + \frac{1}{2}\theta\right]$$

Now, $$v_r = \frac{\partial\phi}{\partial r} = \frac{1}{10\pi r}$$

$$v_\theta = \frac{1}{r}\frac{\partial\phi}{\partial\theta} = \frac{1}{2\pi r}$$

at $x = 0.9$ m and $y = 0.8$ m

$$r = \sqrt{(0.9)^2 + (0.8)^2} = 1.204\text{ m}$$

$$v_r(0.9, 0.8) = \frac{1}{10\times\pi\times 1.204} = 0.026\text{ m/s}$$

$$v_\theta(0.9, 0.8) = \frac{1}{2\pi r} = \frac{1}{2\times\pi\times 1.204}$$

$$= 0.132\text{ m/s}$$

$$V_{\text{resultant}} = \sqrt{(0.026)^2 + (0.132)^2} = 0.134\text{ m/s}$$

7.3.4 Flow about a Rotating Cylinder

In addition to the superimposed uniform flow and a doublet, a vortex is thrown at the doublet centre. This will simulate a rotating cylinder in uniform stream. We shall see that the pressure distribution will result in a force, a component of which will culminate in lift force. The phenomenon of generation of lift by a rotating object placed in a stream is known as *Magnus effect*. The velocity potential and stream functions for the combination of doublet, vortex and uniform flow are

$$\phi = U_0 x + \frac{\chi\cos\theta}{r} - \frac{\Gamma}{2\pi}\theta \text{ (clockwise rotation)} \tag{7.55 a}$$

$$\psi = U_0 y - \frac{\chi\sin\theta}{r} + \frac{\Gamma}{2\pi}\ln r \text{ (clockwise rotation)} \tag{7.55 b}$$

By making use of either the stream function or velocity potential function, the velocity components are

$$v_r = \frac{1}{r}\frac{\partial\psi}{\partial\theta} = \left(U_0 - \frac{\chi}{r^2}\right)\cos\theta \tag{7.56}$$

$$v_\theta = \frac{1}{r}\frac{\partial\phi}{\partial\theta} = -\left(U_0 + \frac{\chi}{r^2}\right)\sin\theta - \frac{\Gamma}{2\pi r} \tag{7.57}$$

Implicit in the above derivation are $x = r\cos\theta$ and $y = r\sin\theta$. At the stagnation points the velocity components must vanish. From Eq. (7.56), we get

$$\cos\theta\left(U_0 - \frac{\chi}{r^2}\right) = 0 \tag{7.58}$$

From Eq. (7.58) it is evident that a zero radial velocity component may occur at $\theta = \pm\frac{\pi}{2}$ and along the circle, $r = \left(\frac{\chi}{U_0}\right)^{1/2}$. Eq. (7.57) depicts that a zero transverse velocity requires

$$\sin\theta = \frac{-\Gamma/2\pi r}{U_0 + (\chi/r^2)} \quad \text{or} \quad \theta = \sin^{-1}\left[\frac{-\Gamma/2\pi r}{U_0 + \frac{\chi}{r^2}}\right] \tag{7.59}$$

However, at the stagnation point, both radial and transverse velocity components must be zero.

So, the location of stragnation point occurs at

$$r = \left(\frac{\chi}{U_0}\right)^{1/2}$$

and
$$\theta = \sin^{-1}\frac{\left\{-\Gamma\Big/\left(2\pi\left(\frac{\chi}{U_0}\right)^{1/2}\right)\right\}}{\left[U_0 + \chi\Big/\left(\frac{\chi}{U_0}\right)\right]}$$

or
$$\theta = \sin^{-1}\left[\frac{-\Gamma}{2\pi\left(\frac{\chi}{U_0}\right)^{1/2}}\cdot\frac{1}{2U_0}\right]$$

or
$$\theta = \sin^{-1}\left[\frac{-\Gamma}{4\pi(\chi U_0)^{1/2}}\right] \tag{7.60}$$

There will be two stagnation points since there are two angles for a given sine except for $\sin^{-1}(\pm 1)$.

The streamline passing through these points may be determined by evaluating ψ at these points. Substitution of the stagnation coordinate (r, θ) into the stream function (Eq. 7.55 b) yields,

$$\psi = \left[U_0\left(\frac{\chi}{U_0}\right)^{1/2} - \frac{\chi}{\left(\frac{\chi}{U_0}\right)^{1/2}}\right]\sin\sin^{-1}\left[\frac{-\Gamma}{4\pi(\chi U_0)^{1/2}}\right] + \frac{\Gamma}{2\pi}\ln\left(\frac{\chi}{U_0}\right)^{1/2}$$

$$\psi = \left[(U_0\chi)^{1/2} - (U_0\chi)^{1/2}\right]\left[\frac{-\Gamma}{4\pi(\chi U_0)^{1/2}}\right] + \frac{\Gamma}{2\pi}\ln\left(\frac{\chi}{U_0}\right)^{1/2}$$

or
$$\psi_{stag} = \frac{\Gamma}{2\pi}\ln\left(\frac{\chi}{U_0}\right)^{1/2} \tag{7.61}$$

Equating the general expression for stream function to the above constant, we get

$$U_0 r \sin\theta - \frac{\chi\sin\theta}{r} + \frac{\Gamma}{2\pi}\ln r = \frac{\Gamma}{2\pi}\ln\left(\frac{\chi}{U_0}\right)^{1/2}$$

By rearranging we can write

$$\sin\theta\left[U_0 r - \frac{\chi}{r}\right] + \frac{\Gamma}{2\pi}\left[\ln r - \ln\left(\frac{\chi}{U_0}\right)^{1/2}\right] = 0 \tag{7.62}$$

All points along the circle $r = \left(\frac{\chi}{U_0}\right)^{1/2}$ satisfy Eq. (7.62), since for this value of r, each quantity within parentheses in the equation is zero. Considering the interior of the circle (on which $\psi = 0$) to be a solid cylinder, the outer streamline pattern is shown in Fig. 7.17.

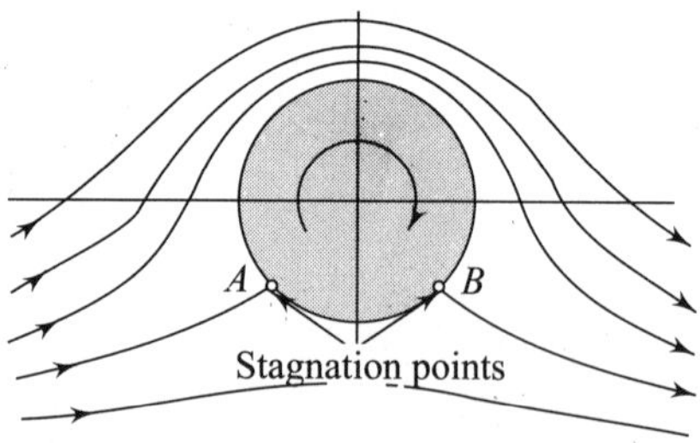

Fig. 7.17 Flow past a cylinder with circulation

A further look into Eq. (7.60) explains that at the stagnation point

$$\theta = \sin^{-1}\left[\frac{-(\Gamma/2\pi)}{2(\chi U_0)^{1/2}}\right]$$

or
$$\theta = \sin^{-1}\left[\frac{-(\Gamma/2\pi)}{2U_0 r}\right] \tag{7.63}$$

The limiting case arises for $\frac{(\Gamma/2\pi)}{U_0 r} = 2$, where $\theta = \sin^{-1}(-1) = -90°$ and two stagnation points meet at the bottom as shown in Fig. 7.18.

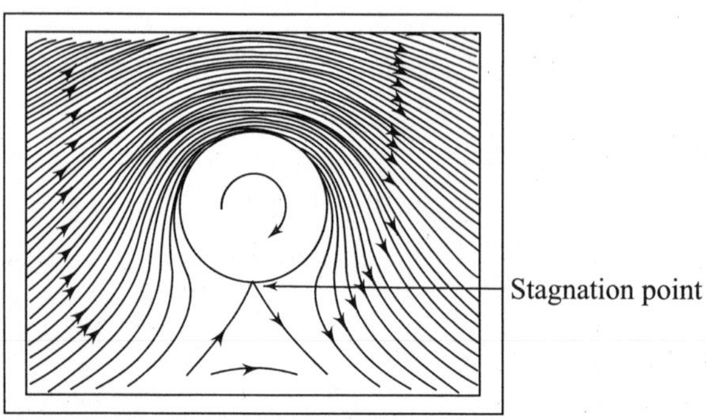

Fig. 7.18 Flow past a circular cylinder with circulation value $\frac{\Gamma/2\pi}{U_0 r} = 2$

However, in all these cases the effects of the vortex and doublet become negligibly small as one moves a large distance from the cylinder. The flow is assumed to be uniform at infinity. We have already seen that the change in strength Γ of the vortex changes the flow pattern, particularly the position of the stagnation points but the radius of the cylinder remains unchanged.

7.3.5 Lift and Drag for Flow about a Rotating Cylinder

The pressure at large distances from the cylinder is uniform and given by p_0. Deploying Bernoulli's equation between the points at infinity and on the boundary of the cylinder,

$$p_b = \rho g\left[\frac{U_0^2}{2g}+\frac{p_0}{\rho g}-\frac{U_b^2}{2g}\right] \tag{7.64}$$

The velocity U_b is as such $v_\theta \Big|_{r=\left(\frac{\chi}{U_0}\right)^{1/2}}$

Hence, $$U_b = \frac{1}{r}\frac{\partial\phi}{\partial\theta} = -2U_0\sin\theta - \frac{\Gamma}{2\pi}\left[\frac{U_0}{\chi}\right]^{1/2} \tag{7.65}$$

From Eqs (7.64) and (7.65) we can write

$$p_b = \rho g\left[\frac{U_0^2}{2g}+\frac{p_0}{\rho g}-\frac{\left[-2U_0\sin\theta - \frac{\Gamma}{2\pi}\left(\frac{U_0}{\chi}\right)^{\frac{1}{2}}\right]^2}{2g}\right] \tag{7.66}$$

The lift may calculated as (refer Fig. 7.12)

$$L = \int_0^{2\pi} p_b \sin\theta \left[\frac{\chi}{U_0}\right]^{\frac{1}{2}} d\theta$$

or $$L = -\int_0^{2\pi}\left\{\frac{\rho U_0^2}{2}+p_0-\frac{\rho\left[-2U_0\sin\theta-\frac{\Gamma}{2\pi}\left(\frac{U_0}{\chi}\right)^{\frac{1}{2}}\right]^2}{2}\right\}$$

$$\left[\frac{\chi}{U_0}\right]^{\frac{1}{2}}(\sin\theta)\, d\theta$$

or $$L = -\int_0^{2\pi}\left[\frac{\rho U_0^2}{2}\left(\frac{\chi}{U_0}\right)^{\frac{1}{2}}\sin\theta + p_0\left(\frac{\chi}{U_0}\right)^{\frac{1}{2}}\sin\theta - \frac{\rho}{2}\left\{4U_0^2\sin^2\theta\right.\right.$$

$$+\frac{4U_0\Gamma\sin\theta}{2\pi}\left(\frac{U_0}{\chi}\right)^{\frac{1}{2}}+\frac{\Gamma^2}{4\pi^2}\left[\frac{U_0}{\chi}\right]\Bigg\}\left(\frac{\chi}{U_0}\right)^{\frac{1}{2}}\sin\theta\Bigg]d\theta$$

or $$L=-\int_0^{2\pi}\Bigg[\frac{\rho U_0^2}{2}\left(\frac{\chi}{U_0}\right)^{\frac{1}{2}}\sin\theta+p_0\left(\frac{\chi}{U_0}\right)^{\frac{1}{2}}\sin\theta-2\rho U_0^2\sin^3\theta$$

$$\left(\frac{\chi}{U_0}\right)^{\frac{1}{2}}-\frac{\rho U_0\Gamma}{\pi}\sin^2\theta-\frac{\rho\Gamma^2}{8\pi^2}\left(\frac{U_0}{\chi}\right)^{\frac{1}{2}}\sin\theta\Bigg]d\theta$$

or $$L=\rho U_0\Gamma \tag{7.67}$$

The drag force, which includes the multiplication by cos θ (and integration over 2π) is zero.

Thus, the inviscid flow also demonstrates lift. It can be seen that the lift becomes a simple formula involving only the density of the medium, free stream velocity and circulation. In addition, it can also be shown that in two-dimensional incompressible steady flow about a boundary of any shape, the lift is always a product of these three quantities.

The validity of Eq. (7.67) for any two-dimensional incompressible steady potential flow around a body of any shape, not necessarily a circular cylinder, is known as the *Kutta-Joukowski theorem,* named after the German fluid dynamist Wilhelm Kutta (1867–1944) and Russian mathematician Nikolai J. Joukowski (1847–1921). A very popular example of the lift force acting on a rotating body is observed in the game of soccer. If a player imparts rotation on the ball while shooting it, instead of following the usual trajectory, the ball will swerve in the air and puzzle the goalkeeper. The swerve in the air can be controlled by varying the strength of circulation, i.e., the amount of rotation. In 1924, a man named *Flettner* had a ship built in Germany which possessed two rotating cylinders to generate thrust normal to wind blowing past the ship. The Flettner design did not gain any popularity but it is of considerable scientific interest (shown in Fig. 7.19).

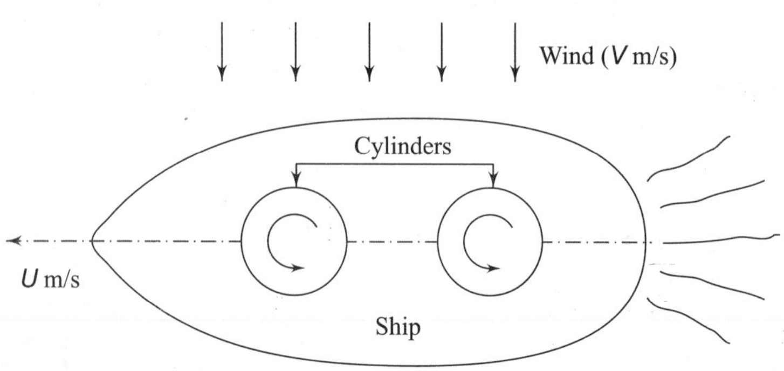

Fig. 7.19 Schematic diagram of the plan view of Flettner's ship

Example 7.10

A 300 mm diameter circular cylinder is rotated about its axis in a stream of water having a uniform velocity of 5 m/s. Estimate the rotational speed when both the stagnation points coincide. Estimate the lift force experienced by the cylinder under such condition. ρ of water may be assumed to be 1000 kg/m^3.

Solution

Stagnation point is given by

$$\theta = \sin^{-1}\left(\frac{-\Gamma}{4\pi r U_0}\right)$$

When both the stagnation points coincide, the two angles are equal and $\theta = -90°$. The stagnation point is at the lower surface [Fig. 7.18].

Thus, $$\frac{\Gamma}{4\pi r U_0} = 1$$

or $$\Gamma = 4\pi r U_0$$

If the cylinder is rotating at an angular speed ω, the circulation due to the equivalent forced vortex is

$$\Gamma = \oint (\omega r)\, r\, d\theta = 2\pi r^2 \omega$$

$$2\pi \omega r^2 = 4\pi r U_0$$

$$\omega = \frac{2U_0}{r}$$

or $$\omega = \frac{2 \times 5}{0.15} = 66.67 \text{ rad/s}$$

and $$\Gamma = 4\pi \times 0.15 \times 5$$
$$= 9.42 \text{ m}^2/\text{s}$$

Lift force $$= L = \rho U \Gamma$$

or $$L = 1000 \times 5 \times 9.42 \left[\frac{\text{kg}}{\text{m}^3} \cdot \frac{\text{m}}{\text{s}} \cdot \frac{\text{m}^2}{\text{s}}\right]$$

or $$L = 47100 \text{ N/m}$$

As such, the lift is calculated per metre length of the cylinder

So, $$\text{Lift} = 47.1 \text{ kN/m}^2$$

7.4 AEROFOIL THEORY

Aerofoils are streamline-shaped wings which are used in airplanes and turbo-machinery. These shapes are such that the drag force is a very small fraction of the lift. The following nomenclatures are used for defining an aerofoil (refer to Fig. 7.20).

The chord (c) is the distance between the leading edge and trailing edge. The length of an aerofoil, normal to the cross section (i.e., normal to the plane of a paper) is called the span of aerofoil. The camber line represents the mean profile of the aerofoil. Some important geometrical parameters for an aerofoil are the ratio of maximum thickness to chord (t/c) and the ratio of maximum camber to chord (h/c). When these ratios are small, an aerofoil can be considered to be thin. For the analysis of flow, a thin aerofoil is represented by its camber.

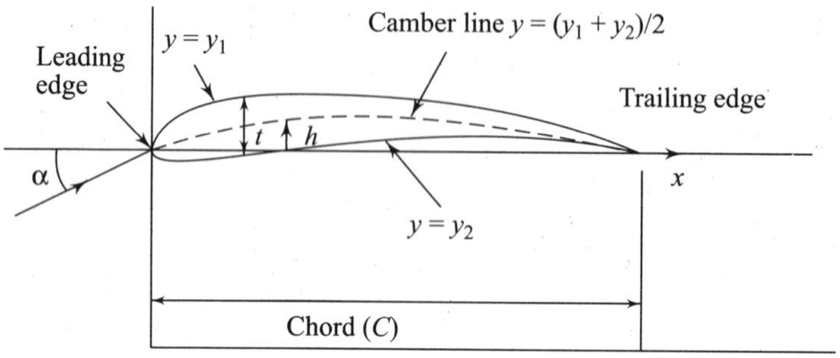

Fig. 7.20 Aerofoil section

The theory of thick cambered aerofoils is an advanced topic. Basically it uses a complex-variable mapping which transforms the inviscid flow across a rotating cylinder into the flow about an aerofoil shape with circulation.

7.4.1 Flow Around a Thin Aerofoil

Thin aerofoil theory is based upon the superposition of uniform flow at infinity and a continuous distribution of clockwise free vortex on the camber line having circulation density $\gamma(s)$ per unit length. The circulation density $\gamma(s)$ should be such that the resultant flow is tangent to the camber line at every point.

Since the slope of the camber line is assumed to be small, $\gamma(s)\mathrm{d}s = \gamma(\eta)\mathrm{d}\eta$ (refer Fig. 7.21). The total circulation around the profile is given by

$$\Gamma = \int_0^C \gamma(\eta)\, \mathrm{d}\eta \tag{7.68}$$

A vortical motion of strength $\gamma \mathrm{d}\eta$ at $x = \eta$ develops a velocity at the point P which may be expressed as

$$\mathrm{d}v = \frac{\gamma(\eta)\, \mathrm{d}\eta}{2\pi(\eta - x)} \quad \text{acting upwards}$$

The total induced velocity in the upward direction at P due to the entire vortex distribution along the camber line is

$$v(x) = \frac{1}{2\pi} \int_0^C \frac{\gamma(\eta)\, \mathrm{d}\eta}{(\eta - x)} \tag{7.69}$$

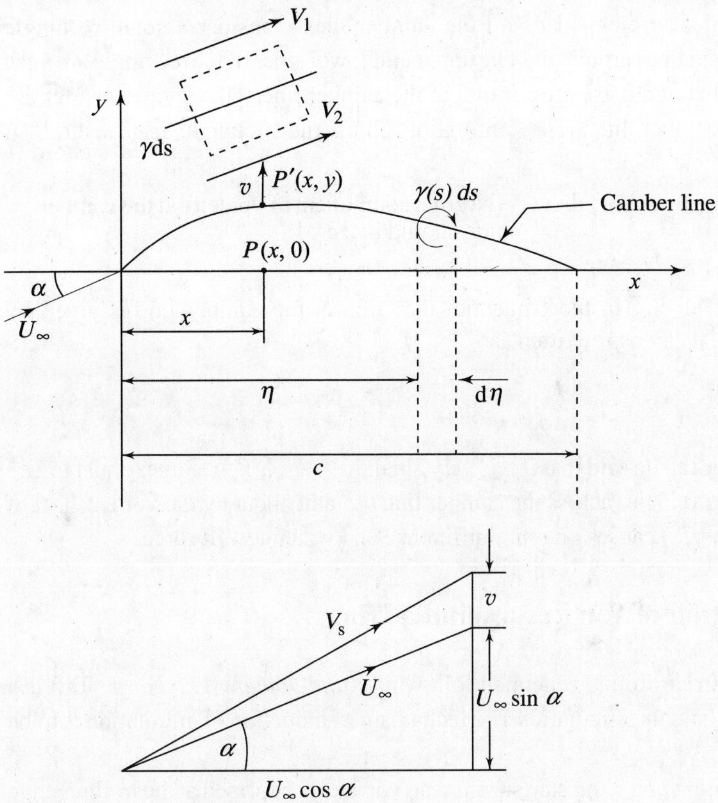

Fig. 7.21 Flow around thin aerofoil

For a small camber (having small α), this expression is identically valid for the induced velocity at P' due to the vortex sheet of variable strength $\gamma(s)$ on the camber line. The resultant velocity due to U_∞ and $v(x)$ must be tangential to the camber line so that the slope of a camber line may be expressed as

$$\frac{dy}{dx} = \frac{U_\infty \sin\alpha + v}{U_\infty \cos\alpha} = \tan\alpha + \frac{v}{U_\infty \cos\alpha}$$

or

$$\frac{dy}{dx} = \alpha + \frac{v}{U_\infty} \quad \text{[since } \alpha \text{ is very small]} \tag{7.70}$$

From Eqs (7.69) and (7.70) we can write

$$\frac{dy}{dx} = \alpha + \frac{1}{2\pi U_\infty} \int_0^C \frac{\gamma(\eta)\, d\eta}{(\eta - x)} \tag{7.71)*$$

* For a given aerofoil, the left-hand side term of the integral Eq. (7.71) is a known function. Finding out $\gamma(\eta)$ from it is a formidable task. This exercise is not being discussed in this text. Interested readers may refer to the books by Glauert [1] and Batchelor [2]. If $\gamma(\eta)$ is determined, the circulation Γ and consequently the lift $L = \rho U_\infty \Gamma$ can easily be calculated.

Let us consider an element ds on the camber line. Consider a small rectangle (drawn with dotted line) around ds. The upper and lower sides of the rectangle are very close to each other and these are parallel to the camber line. The other two sides are normal to the camber line. The circulation along the rectangle is measured in clockwise direction as

$$V_1\,ds - V_2\,ds = \gamma ds \qquad \text{[Normal component of velocity at the camber line should be zero]}$$

or
$$V_1 - V_2 = \gamma \tag{7.72}$$

If the mean velocity in the tangential direction at the camber line is given by $V_s = (V_1 + V_2)/2$, it can be rewritten as

$$V_1 = V_s + \frac{\gamma}{2} \qquad \text{and} \qquad V_2 = V_s - \frac{\gamma}{2}$$

In the event, it can be said that if v is very small [$v << U_\infty$], V_s becomes equal to U_∞. The difference in velocity across the camber line brought about by the vortex sheet of variable strength $\gamma(s)$ causes pressure difference and generates lift force.

7.4.2 Generation of Vortices around a Wing

The lift around an aerofoil is generated following *Kutta-Joukowski theorem*. Lift is a product of ρ, U_∞ and the circulation Γ. Mechanism of induction of circulation is to be understood clearly.

When the motion of a wing starts from rest, vortices are formed at the trailing edge (refer Fig. 7.22).

At the start, there is a velocity discontinuity at the trailing edge. This is eventual because near the trailing edge, the velocity at the bottom surface is higher than that at the top surface. This discrepancy in velocity culminates in the formation of vortices at the trailing edge. Figure 7.22 (a) depicts the formation of starting vortex by impulsively moving aerofoil. However, the starting vortices induce a counter circulation as shown in Figure 7.22 (b). The circulation around a path ($ABCD$) enclosing the wing and just shed (starting) vortex must be zero. Here we refer to Kelvin's theorem once again.

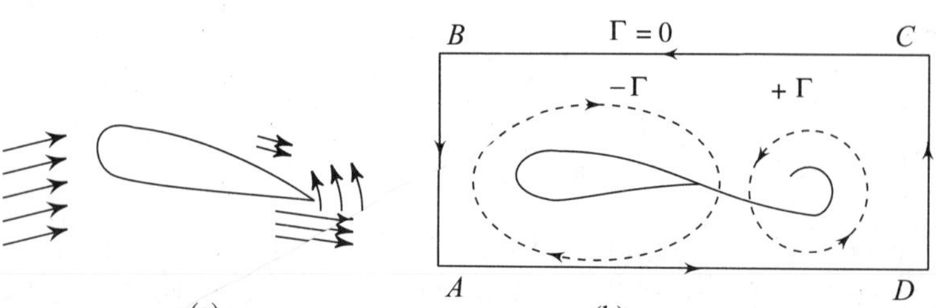

Fig. 7.22 Vortices generated when an aerofoil just begins to move

Initially the flow starts with the zero circulation around the closed path. Thereafter, due to the change in angle of attack or flow velocity, if a fresh starting vortex is shed, the circulation around the wing will adjust itself so that a net zero vorticity is set around the closed path.

The discussions in the previous section were for two-dimensional, infinite span wings. But real wings have finite span or finite aspect ratio λ, defined as

$$\lambda = \frac{b^2}{A_s} \tag{7.73}$$

where b is the span length and A_s is the plan form area as seen from the top. For a wing of finite span, the end conditions affect both the lift and the drag. In the leading edge region, pressure at the bottom surface of a wing is higher than that at the top surface. The longitudinal vortices are generated at the edges of finite wing owing to pressure differences between the bottom surface directly facing the flow and the top surface (refer Fig. 7.23). This is very prominent for small aspect ratio delta wings which are used in high-performance aircrafts as shown in Fig. 7.24.

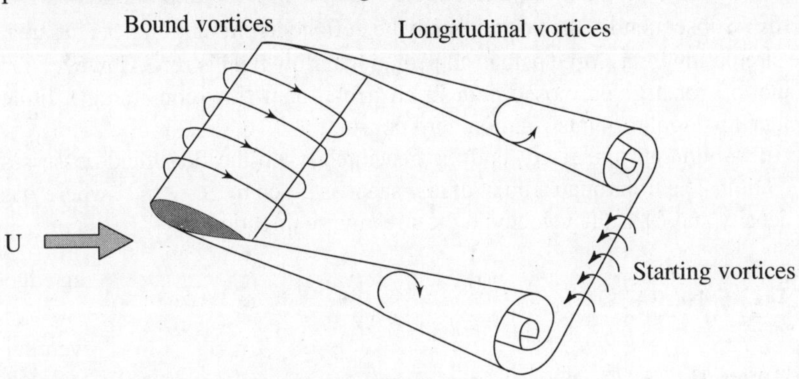

Fig. 7.23 Vortices around a finite wing

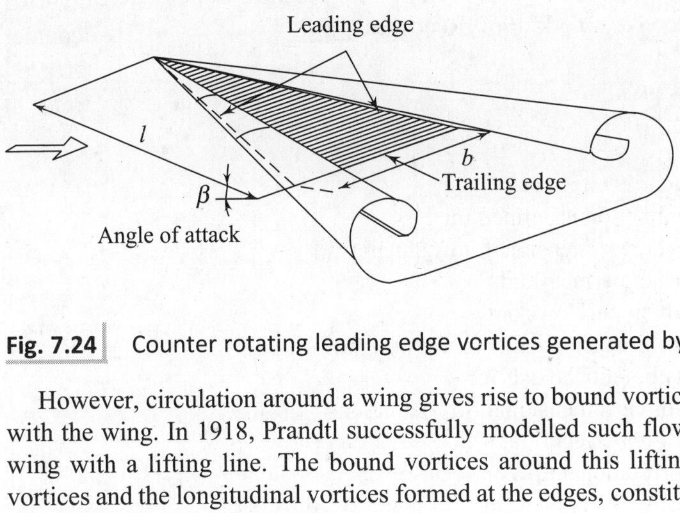

Fig. 7.24 Counter rotating leading edge vortices generated by a delta-wing

However, circulation around a wing gives rise to bound vortices that move along with the wing. In 1918, Prandtl successfully modelled such flows by replacing the wing with a lifting line. The bound vortices around this lifting line, the starting vortices and the longitudinal vortices formed at the edges, constitute a closed vortex ring as shown in Fig. 7.23.

SUMMARY

This chapter has given a brief description of inviscid, incompressible, irrotational flows.

- Irrotationality leads to the condition $\nabla \times \vec{V} = 0$ which demands $\vec{V} = \nabla\phi$, where ϕ is known as a potential function. For a potential flow $\nabla^2\phi = 0$.
- The stream function ψ also obeys the Laplace's equation $\nabla^2 \psi = 0$ for the potential flows. Laplace's equation is linear, hence any number of particular solutions of Laplace's equation added together will yield another solution. So a complicated flow for an inviscid, incompressible, irrotational condition can be synthesised by adding together a number of elementary flows which are also inviscid, incompressible and irrotational. This is called the method of superposition.
- Some inviscid flow configurations of practical importance are solved by using the method of superposition. The circulation in a flow field is defined as $\Gamma = \int \vec{V}\cdot d\vec{s}$. Subsequently, the vorticity may be defined as circulation per unit area. The circulation for a closed path in an irrotational flow field is zero. However, the circulation for a given closed path in an irrotational flow containing a finite number of singular points is a non-zero constant.
- The lift around an immersed body is generated when the flow field possesses circulation. The lift around a body of any shape is given by $L = \rho U_0 \Gamma$, where ρ is the density and U_0 is the velocity in the streamwise direction.

REFERENCES

1. Glauert, H., *The Elements of Aerofoil and Airscrew Theory,* Cambridge University Press, London , 1926, 1948.
2. Batchelor, G.K., *An Introduction to Fluid Dynamics,* Cambridge University Press, London/New York, 1967, Reprinted 1970.

EXERCISES

7.1 Choose the correct answer :

(i) Stream function is defined for
  (a) all three-dimensional flow situations
  (b) flow of perfect fluid
  (c) irrotational flows only
  (d) two-dimensional incompressible flows

(ii) Velocity potential exists for
  (a) all three-dimensional flow situations
  (b) flow of perfect fluid
  (c) all irrotational flows
  (d) steady irrotational flow

(iii) The continuity equation for a fluid states that
(a) mass flow rate through a stream tube is constant
(b) the derivatives of velocity components exist at every point
(c) the velocity is tangential to stream lines
(d) the stream function exists for steady flows

(iv) $\nabla . \vec{V} = 0$ implies that
(a) dilatation rate for a fluid particle is zero
(b) net mass flux from a control volume in any flow situation is zero
(c) the fluid is compressible
(d) density is a function of time only

(v) Momentum theorem is valid only if the fluid is
(a) incompressible (b) in irrotational motion
(c) inviscid (d) irrespective of the above restrictions.

(vi) Circulation is defined as
(a) line integral of velocity about any path
(b) integral of tangential component of velocity about a path
(c) line integral of velocity about a closed path
(d) line integral of tangential component of velocity about a closed path.

(vii) In an irrotational flow, Stokes theorem implies that circulation is zero
(a) around two-dimensional infinite bodies
(b) in simply connected regions
(c) in multiply connected regions
(d) without any restriction

(viii) The curl of a given velocity field indicates
(a) the rate of increase or decrease of flow at a point
(b) the rate of twisting of the lines of flow
(c) the deformation rate
(d) None of the above

7.2 Prove that the streamlines $\psi(r, \theta)$ in a polar coordinate system are orthogonal to the velocity potential lines $\phi(r, \theta)$.

7.3 The x and y components of velocity in a two-dimensional incompressible flow are given by

$$u = 3x + 3y \quad \text{and} \quad v = 2x - 3y$$

Derive an expression for the stream function. Show that the flow is rotational. Calculate the vorticity in the flow field.

Ans. $(-x^2 + 3xy + (3/2)y^2, \nabla \times \vec{V} = -1)$

7.4 Determine whether the function $A(r + a^2/r) \cos \theta - (B/2\pi)\theta$ (A, B and a are constants) is a possible velocity potential for inviscid incompressible flow.

Ans. (possible)

7.5 A source of volume flow rate 2 m^2/s is located at origin and another source of volume flow rate 4 m^2/s is located at (3,0). Find out the velocity components at (2, 2).

Ans. ($u = -0.048$ m/s, $v = 0.334$ m/s)

7.6 A source of volume flow rate 5 m^2/s at the origin and a uniform flow of velocity 4 m/s combine to form a two-dimensional half-body. Find out the maximum width of the half-body.

Ans. (1.25 m)

7.7 A source and a sink of equal volume flow rate 10 m^2/s are located 2 m apart. If a uniform flow of 5 m/s is superimposed, find out the location of the stagnation points.

Ans. (1.28, 0) and (– 1.28, 0)

7.8 Find the stream function and the velocity potential for flow around a closed contour formed by superposition of a source, sink and a rectilinear flow. The source and sink are of equal strength and they are located at equal distances on either side of the origin along the x-axis.

$$Ans.\ (\psi = -\frac{\Lambda}{2\pi}\tan^{-1}\left(\frac{2ay}{x^2+y^2-a^2}\right) + U_0 y + \text{constant}$$

$$\phi = \frac{\Lambda}{2\pi}\ln\left(\frac{x^2+y^2+a^2+2ax}{x^2+y^2+a^2-2ax}\right) + U_0 x + \text{constant}\ )$$

7.9 The discharge of 30 m^2/s pollutants from a chemical plant into 10 m deep river, flowing at 0.3 m/s, can be modelled as a two-dimensional source across the river depth. It is found that the fishes in a certain zone die whereas those outside the zone are unaffected. Find out the extent of this critical zone, if the point of discharge is in the midplane of a wide river.

Ans. (Rankine half-body with stagnation point (15.91, 0) and $2y_{max} = 100$ m)

7.10 The Flettner rotor ship (Fig. 7.15) makes use of two rotating vertical cylinders. Each has a diameter of 3 m and length of 15 m. If they rotate at a speed of 720 rpm, calculate the magnitude of magnus force developed by the rotors in a breeze of 10 m/s. Assume air density as 1.22 kg/m^3.

Ans. (390.083 kN)

7.11 Find out the strength of a doublet which simulates a physical situation of 2 m diameter cylinder in a uniform flow of 15 m/s.

Ans. ($\Lambda = 47.124$ m^3/s per metre)

7.12 A kite having cross-sectional area of $0.5\ m \times 0.5\ m$ and weight of 4.9 N, flies in the air at an angle of 10° with the horizontal plane. The string of the kite makes an angle of 45° with the horizontal plane. The tension applied by the person flying the kite is N. If the velocity of the wind at the layer where the kite is flying is 15 m/s, find out the lift and drag coefficients of the kite. Take ρ of air as 1.2 kg/m^3, $F_D = C_D \rho A \frac{U_0^2}{2}$ and $F_L = C_L \rho A \frac{U_0^2}{2}$.

Ans. ($C_D = 0.419$, $C_L = 0.564$)

7.13 Consider a forced vortex rotated at an angular speed ω. Evaluate the circulation around any closed path in a forced vortex flow.

Derive the expression of hydrodynamic pressure as a function of radius for (i) a free vortex and (ii) a forced vortex.

Ans. (Forced vortex $\Gamma = 2\pi r^2 \omega$; (i) $p = -\rho C^2/r^2 + C_2$ (ii) $p = \rho\omega^2 r^2/2 + C_1$)

7.14 In a tornado (free vortex motion), at a radius of 20 cm the velocity (tangential component is the only non-trivial component) and pressure are 22.5 m/s and 117.72 kPa absolute, respectively. Find out the pressure at a radius of 4.5 m. The fluid is air and assumed to have density equal to 1.2 kg/m^3.

Ans. (196.705 kPa)

7.15 A tornado may be modelled as a circulating flow shown in Fig. 7.25 with $v_r = v_z = 0$

$$v_\theta = \omega r \text{ for } r \le R$$

$$= \frac{\omega R^2}{r} \text{ for } r \ge R$$

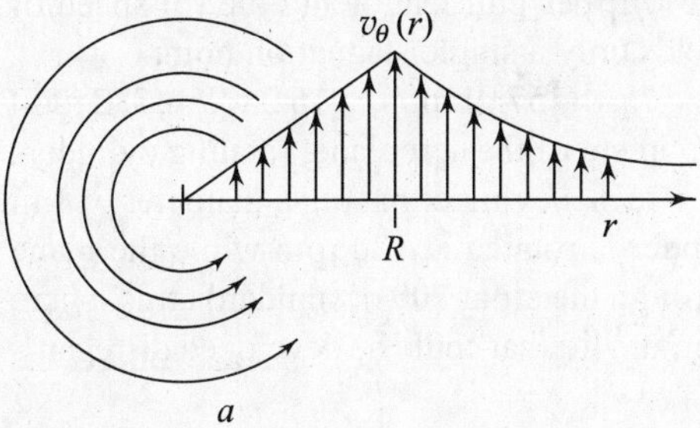

Fig. 7.25 Model of a tornado (Combination of free and forced vortex)

Determine whether the flow pattern is irrotational in either the inner ($r < R$) or the outer ($r > R$) region. Using r momentum equation, determine the pressure distribution $p\,(r)$ in the tornado, assuming $p = p_\infty$ at $r \to \infty$. Find out the location of the minimum pressure.

Ans. ($p = p_\infty - \rho\omega^2 R^2 (1 - r^2/2R^2)$ for $r < R$;
$p = p_\infty - \rho\omega^2 R^4 / 2r^2$ for $r > R$)

7.16 Verify that the stream function and the velocity potential for flow around a circular cylinder with circulation satisfies their respective Laplace equations.

7.17 A 2 m diameter cylinder is rotating at 1400 rpm in an air stream flowing at 20 m/s. Calculate the lift and drag forces per unit depth of the cylinder. Assume air density as 1.22 kg/m^3.

Ans. ($L = 22.476$ kN, $D = 0$)

7.18 Flow past a rotating cylinder can be simulated by superposition of a doublet, a uniform flow and a vortex. The peripheral velocity of the rotating cylinder alone is given by v_θ at $r = R$ (R is the radius of the cylinder). Use the expression for the combined velocity potential for the superimposed uniform flow, doublet and vortex flow (clockwise rotation) and show that the resultant velocity at any point on the cylinder is given by $-2U_o \sin\theta - v_\theta$ (at $r = R$). The angle θ is the angular position of the point of interest. A cylinder rotates at 360 rpm around its own axis which is perpendicular to the uniform air stream (density 1.24 kg/m^3) having a velocity of 25 m/s. The cylinder is 2 m in diameter. Find out (a) circulation, Γ (b) lift per unit length and the (c) position of the stagnation points.

Ans. (236.87 m^2/s, 7343 N/m, – 48.93° and 228.93°)

7.19 Flow past a rotating cylinder can be simulated by superposition of a doublet, a uniform flow and a vortex. The peripheral velocity of the rotating cylinder alone is given by v_θ at $r = R$ (R is the radius of the cylinder). Use the expression for the combined velocity potential for the superimposed uniform flow, doublet and vortex flow (clockwise rotation) and show that the resultant velocity at any point on the cylinder is given by $-2U_o \sin\theta - v_\theta$ (at $r = R$). The angle θ is the angular position of the point of interest. A cylinder rotates at 240 rpm around its own axis which is perpendicular to the uniform air stream (density 1.24 kg/m^3) having a velocity of 20 m/s. The cylinder is 2 m in diameter. Find out (a) circulation, Γ (b) lift per unit length and the (c) speed of rotation of the cylinder, which yields only a single stagnation point.

Ans. (157.91 m^2/s, 3916.25 N/m, 382 rpm)

7.20 In the early attempts for the development of the aeroplane, rotating cylinders were used as the airfoils. Consider such a cylinder having a diameter of 1 m and a length of 10 m. If this cylinder is rotated at 120 rpm while the plane moves at a speed of 120 km/hr through the air at 700 m standard atmosphere ($\rho = 0.59$ kg/m^3), estimate the minimum lift that could be developed disregarding end effects.

Ans. (3880 N)

7.21 If a cylinder (placed with its axis normal to the free stream) is to be used as a Pitot-static tube for the measurement of the free-stream velocity, find the locations of piezometric holes on the cylinder.

Assume ideal fluid flow. The velocity components in the flow field are

$$v_r = U_0\left(1 - \frac{R^2}{r^2}\right)\cos\theta$$

$$v_\theta = -U_0\left(1 + \frac{R^2}{r^2}\right)\sin\theta$$

Ans. (180°, 210°)

Note: A Pitot-static tube measures the velocity head, i.e., the difference between the stagnation pressure and static pressure in the flow.

8

DYNAMICS OF VISCOUS FLOWS

8.1 INTRODUCTION

In Chapter 4, we dealt with fluid flow analysis using integral forms of the conservation equations that do not explicitly take into account point-to-point variations in the flow domain, but rather express the sense of overall conservation within the control volume in an integral form. However, in many cases, information on point-to-point variations in the flow field may be necessary for analysing fluid flow problems. In this chapter, we will be developing a differential equation-based approach to address this issue, considering that viscous effects are important in dictating the flow characteristics. It is important to mention here that the derivation of the governing equations, as presented in this chapter, is valid for both incompressible as well as compressible viscous flows. However, in the subsequent portion of the chapter, solutions to these equations are presented only for certain simple cases of steady, viscous, incompressible flows for which elementary analytical solutions exist. Solutions of conservation equations for compressible flows are beyond the scope of this chapter.

8.2 CONSERVATION OF LINEAR MOMENTUM IN DIFFERENTIAL FORM

Our first attempt in this chapter will be to derive a differential equation for linear momentum conservation, starting from the corresponding integral form. Considering a stationary reference frame and non-deformable control volume, the integral form of conservation of linear momentum may be written as (see Chapter 5),

$$\sum \vec{F}_{CV} = \frac{\partial}{\partial t}\int_{CV} \rho \vec{V} d\forall + \int_{CS} \rho \vec{V}\left(\vec{V}\cdot\hat{n}\right) dA$$

or

$$\sum \vec{F}_{CV} = \int_{CV} \frac{\partial}{\partial t} \rho \vec{V} d\forall + \int_{CS} \rho \vec{V}\left(\vec{V}\cdot\hat{n}\right) dA \tag{8.1}$$

where $\sum \vec{F}_{CV}$ is the resultant force on the control volume (CV). The net force acting on the control volume along the ith direction is given by

$$F_{CV,i} = \text{Net surface force along } i + \text{Net body force along } i$$

i.e., $$F_{CV,i} = F_{\text{surface},i} + F_{\text{body},i} \quad (8.2)$$

where $F_{\text{surface},i}$ is the net surface force acting on the CV along i^{th} direction and $F_{\text{body},i}$ is the net body force acting on the CV along the same direction. In the subsequent discussions, we will delineate the manner in which these forces may be expressed in a structured mathematical form.

Referring to Fig 8.1, let T^n be the traction vector acting on an elemental plane of a control surface having unit normal vector $\hat{n}$. Let T_i^n be the ith component of the traction vector T^n (note that $i = 1$ means component along x, $i = 2$ means component along y, and $i = 3$ means component along z in a Cartesian indexing system). In special cases when the direction $\hat{n}$ coincides with any of the coordinate directions (say, j), the notations T_i^j and τ_{ji} may be used equivalently. The later ones are also known to constitute the stress tensor components (see Chapter 1). Note that each of the indices i and j may vary between 1 and 3.

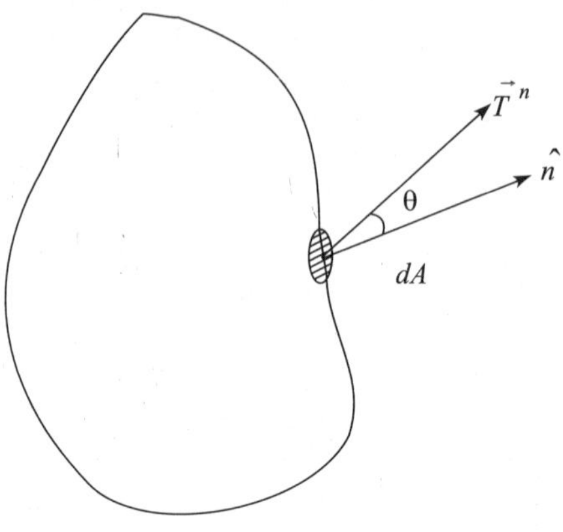

Fig. 8.1

For any arbitrary orientation $\hat{n}$, the traction vector at a point may be expressed in terms of the stress tensor components, by utilising Cauchy's theorem (see Chapter 1), as follows:

$$T_i^n = \sum_{j=1}^{3} \tau_{ji} n_j = \sum_{j=1}^{3} \tau_{ij} n_j \quad \text{(since, from angular momentum conservation, } \tau_{ji} = \tau_{ij}\text{)}$$

where, $\hat{n} = n_1\hat{i} + n_2\hat{j} + n_3\hat{k}$ and $\tau_{ij} n_j = \tau_{i1}n_1 + \tau_{i2}n_2 + \tau_{i3}n_3 = \vec{\tau}_i \cdot \hat{n}$ (defining $\vec{\tau}_i$ as $\vec{\tau}_i = \tau_{i1}\hat{i} + \tau_{i2}\hat{j} + \tau_{i3}\hat{k}$).

Thus, $$F_{\text{surface},i} = \int_{CS} T_i^n \, dA = \int (\vec{\tau}_i \cdot \hat{n}) \, dA$$

Let b_i be the ith component of the body force per unit mass. Therefore, the ith component of the body force acting on an elemental volume $d\forall$ within the CV is $\rho b_i d\forall$. Thus, the net body force along i is given by

$$F_{\text{body},i} = \int_{CV} b_i \rho d\forall$$

Hence, Eq. (8.2) may be rewritten as

$$\sum F_{CV,i} = \int_{CS} (\vec{\tau}_i \cdot \hat{n}) dA + \int_{CV} \rho b_i d\forall$$

Accordingly, the ith component of the vector equation (8.1) may be written as

$$\int_{CS} (\vec{\tau}_i \cdot \hat{n}) dA + \int_{CV} \rho b_i d\forall = \int_{CV} \frac{\partial}{\partial t}(\rho u_i) d\forall + \int_{CS} (\rho u_i \vec{V}) \cdot \hat{n}\, dA \tag{8.3}$$

where u_i is the ith component of the velocity vector $\vec{V}$. Now, using the divergence theorem ($\int_{CS} \vec{F} \cdot \hat{n}\, dA = \int_{CV} \nabla \cdot \vec{F} d\forall$), the area integrals appearing in Eq. (8.3) may be converted into volume integrals, to yield

$$\int_{CV} (\nabla \cdot \vec{\tau}_i) d\forall + \int_{CV} \rho b_i d\forall = \int_{CV} \frac{\partial}{\partial t}(\rho u_i) d\forall + \int_{CV} \nabla \cdot (\rho u_i \vec{V}) d\forall \tag{8.4}$$

or,

$$\int_{CV} \left[(\nabla \cdot \vec{\tau}_i) + \rho b_i - \frac{\partial}{\partial t}(\rho u_i) - \nabla \cdot (\rho u_i \vec{V}) \right] d\forall = 0 \tag{8.4a}$$

Since the choice of the elemental CV is arbitrary, we have

$$(\nabla \cdot \vec{\tau}_i) + \rho b_i - \frac{\partial}{\partial t}(\rho u_i) - \nabla \cdot (\rho u_i \vec{V}) = 0,$$

or equivalently,

$$\frac{\partial}{\partial t}(\rho u_i) + \nabla \cdot (\rho u_i \vec{V}) = \nabla \cdot \vec{\tau}_i + \rho b_i \tag{8.5}$$

Equation (8.5) may be alternatively expressed in Cartesian index notation as

$$\frac{\partial}{\partial t}(\rho u_i) + \frac{\partial}{\partial x_j}(\rho u_i u_j) = \frac{\partial}{\partial x_j}(\tau_{ij}) + \rho b_i \tag{8.5a}$$

Equation (8.5) or (8.5a) is known as the Navier's equation of equilibrium along the ith direction.

*Eq. (8.5a) has been derived starting with the application of linear momentum conservation to a finite control volume. It can be mentioned in this context that the same equation can be derived by straightforward application of Newton's second law of motion to a fluid element as a control mass system. In order to illustrate the same, one may consider a fluid element of volume $d\forall = dx_1 dx_2 dx_3$ and mass dm in a Cartesian reference frame with coordinate axes as shown in Fig. 8.2 and write a force balance on the same. It is important to reiterate in this context that in the representation of stress through τ_{ij} notation, the first subscript (i) denotes the direction of the normal to the plane on which the stress acts, while the second subscript (j) denotes the direction of action of the force which causes the stress. As mentioned earlier, $i = 1$ means component along x, $i = 2$ means component along y, and $i = 3$ means component along z. If the stress components at the centre of the differential element are taken to be τ_{11}, τ_{21}, and τ_{31}, then the stress components on different planes caused by the forces in x_1 direction (obtained by a Taylor series expansion about the centre of the element) are only shown in Fig. 8.2.

*This portion may be omitted without loss of continuity

The net surface force in the x_1 direction is given by

$$dF_{Sx_1} = \left(\tau_{11} + \frac{\partial \tau_{11}}{\partial x_1}\frac{dx_1}{2}\right)dx_2\,dx_3 - \left(\tau_{11} - \frac{\partial \tau_{11}}{\partial x_1}\frac{dx_1}{2}\right)dx_2\,dx_3 + \left(\tau_{21} + \frac{\partial \tau_{21}}{\partial x_2}\frac{dx_2}{2}\right)dx_1\,dx_3$$

$$-\left(\tau_{21} - \frac{\partial \tau_{21}}{\partial x_2}\frac{dx_2}{2}\right)dx_1\,dx_3 + \left(\tau_{31} + \frac{\partial \tau_{31}}{\partial x_3}\frac{dx_3}{2}\right)dx_1\,dx_2 - \left(\tau_{31} - \frac{\partial \tau_{31}}{\partial x_3}\frac{dx_3}{2}\right)dx_1\,dx_2$$

or, $$dF_{Sx_1} = \left(\frac{\partial \tau_{11}}{\partial x_1} + \frac{\partial \tau_{21}}{\partial x_2} + \frac{\partial \tau_{31}}{\partial x_3}\right)dx_1\,dx_2\,dx_3$$

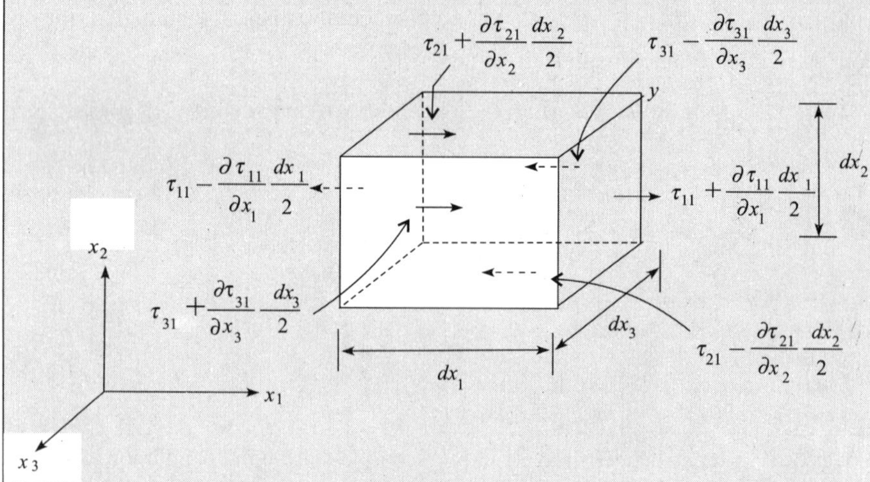

Fig. 8.2

Let b_1 be the component of the body force per unit mass in the x_1 direction. Therefore, component of the body force acting in the x_1 direction is $\rho b_1 dx_1 dx_2 dx_3$. Thus, the net force in the x_1 direction, dF_{x_1}, is given by

$$dF_{x_1} = \left(\frac{\partial \tau_{11}}{\partial x_1} + \frac{\partial \tau_{21}}{\partial x_2} + \frac{\partial \tau_{31}}{\partial x_3} + \rho b_1\right)dx_1\,dx_2\,dx_3$$

Writing Newton's second law of motion to the fluid element along x_1 direction, we obtain

$$dm\frac{Du_1}{Dt} = \left(\frac{\partial \tau_{11}}{\partial x_1} + \frac{\partial \tau_{21}}{\partial x_2} + \frac{\partial \tau_{31}}{\partial x_3} + \rho b_1\right)dx_1 dx_2 dx_3$$

or $$\rho\frac{Du_1}{Dt} = \frac{\partial \tau_{11}}{\partial x_1} + \frac{\partial \tau_{21}}{\partial x_2} + \frac{\partial \tau_{31}}{\partial x_3} + \rho b_1 \qquad (8.5b)$$

One can derive similar expressions for the force components in the x_2 and x_3 directions. These can be generalised, so that the equation of motion in ith direction may be expressed in Cartesian index notation as

$$\rho \frac{Du_i}{Dt} = \frac{\partial}{\partial x_j}(\tau_{ji}) + \rho b_i$$

Since from angular momentum conservation, $\tau_{ji} = \tau_{ij}$, the above equation can be expressed as

$$\rho \frac{Du_i}{Dt} = \frac{\partial}{\partial x_j}(\tau_{ij}) + \rho b_i \tag{8.5c}$$

While the left-hand side of Eq. (8.5a) implies the net rate of momentum change in control volume plus the net rate of momentum efflux from the control volume, the left-hand side of Eq. (8.5c) represents the rate of change in momentum per unit volume of a fluid element as control mass system. The equality of the two can be proved with the aid of continuity equation as follows:

$$\frac{\partial}{\partial t}(\rho u_i) + \frac{\partial}{\partial x_j}(\rho u_j u_i) = u_i \frac{\partial \rho}{\partial t} + \rho \frac{\partial u_i}{\partial t} + u_i \frac{\partial}{\partial x_j}(\rho u_j) + \rho u_j \frac{\partial u_i}{\partial x_j}$$

$$= u_i \left[\underbrace{\frac{\partial \rho}{\partial t} + \frac{\partial}{\partial x_j}(\rho u_j)}_{\substack{0 \\ \text{From continuity equation}}} \right] + \rho \left[\underbrace{\frac{\partial u_i}{\partial t} + u_j \frac{\partial u_i}{\partial x_j}}_{\frac{Du_i}{Dt}} \right]$$

$$= \rho \frac{Du_i}{Dt}$$

Equation (8.5a) is fairly generic in nature, but not mathematically closed. This is because of the fact that in addition to flow velocities, it contains τ_{ij} s also as additional unknowns, for which additional governing equations need to be prescribed. This task can be accomplished by expressing τ_{ij} s in terms of the primary variables (velocity and pressure), which is by no means a trivial act. The non-triviality in such mathematical descriptions stems from the fact that different types of fluids are constitutively different. In other words, the relationship between stress and the rate of deformation (also known as constitutive relationship for fluids) for different types of fluids may intrinsically differ in their mathematical forms.

In an effort to obtain more fruitful insight on the mathematical description of fluid stresses, one may note that it is possible to split the net stress into two parts; one being the function of deformation (called deviatoric component) and another being independent of deformation (called hydrostatic component). The hydrostatic component of stress, independent of fluid deformation, would exist even if the fluid is at rest. The deviatoric component of stress, on the other hand, is directly linked to the fluid deformation. Accordingly, one may express τ_{ij} as

$$\tau_{ij} = \tau_{ij}^{\text{hydrostatic}} + \tau_{ij}^{\text{deviatoric}} \tag{8.6}$$

The philosophy of writing the stress tensor as above (Eq. (8.6)) is that the expression of stress field must have continuity to that under the situation of hydrostatics where the deformation rates are zero.

The deviatoric component of the stress tensor is related to the velocity gradients. The velocity gradient tensor can be decomposed into symmetric and antisymmetric parts, as

$$\frac{\partial u_i}{\partial x_j} = \underbrace{\frac{1}{2}\left(\frac{\partial u_i}{\partial x_j} + \frac{\partial u_j}{\partial x_i}\right)}_{e_{ij}} + \underbrace{\frac{1}{2}\left(\frac{\partial u_i}{\partial x_j} - \frac{\partial u_j}{\partial x_i}\right)}_{\omega_{ij}}$$

The symmetric part represents fluid deformation and the antisymmetric part represents fluid rotation. The antisymmetric part cannot by itself generate stress. Therefore, deviatoric stress tensor should be naturally related to symmetric part of the velocity gradient tensor. Mathematically, we can write

$$\tau_{ij}^{\text{deviatoric}} = f(\text{rate of deformation}) \tag{8.7}$$

In deriving the explicit form of the relation given by Eq.(8.7), the following assumptions are made:

- The relationship between the stress tensor and the deformation rate tensor is linear. The fluids which obey the linear relationship are known as Newtonian fluids.
- The fluid is homogeneous and isotropic so that the relationship given by Eq. (8.8) is invariant to coordinate transformation comprising rotation, translation and reflection. Thus, the relation implies a physical law.

Based on the above assumptions, and following a rigorous mathematical derivation, Eq. (8.7) can be written as

$$\tau_{ij}^{\text{deviatoric}} = \lambda \frac{\partial u_k}{\partial x_k}\delta_{ij} + \mu\left(\frac{\partial u_i}{\partial x_j} + \frac{\partial u_j}{\partial x_i}\right) \tag{8.8}$$

Here μ is the viscosity of the fluid which relates to the deviatoric component of the stress linearly with the rate of shear deformation $\left(\frac{\partial u_j}{\partial x_i} + \frac{\partial u_i}{\partial x_j}\right)$ for homogeneous and isotropic Newtonian fluids. λ is known as the 2nd coefficient of viscosity, which relates the deviatoric component of the stress linearly with the rate of volumetric deformation $\left(\frac{\partial u_k}{\partial x_k}\right)$.

* Equation (8.8) may be arrived at by starting from Eq. (8.7). In order to achieve that, it may first be noted that for Newtonian fluids, the functional relationship between the deviatoric stress tensor and the velocity gradient tensor is linear. It is also important to recognise in this context that both deviatoric stress and velocity gradient tensors are second order tensors. To linearly map one of them into the other (as necessary for Newtonian fluids), one requires a fourth-order tensor as a mediator, so that one may write

$$\tau_{ij}^{\text{deviatoric}} = C_{ijkl}\, e_{kl} \tag{8.9}$$

where $e_{kl} = \frac{1}{2}\left[\frac{\partial u_k}{\partial x_l} + \frac{\partial u_l}{\partial x_k}\right]$ and C_{ijkl} is a fourth-order tensor (having 81 components that depend on the thermodynamic state of the medium).

---

*This portion may be omitted without loss of continuity

For a homogeneous and isotropic fluid, the C_{ijkl} components are position independent and are also invariant to rotation of the coordinate axes. To move deeper into the underlying implications, let us attempt to form a scalar starting from C_{ijkl}. To do that, clearly we can employ four vectors $\vec{A}, \vec{B}, \vec{C}, \vec{D}$, each of which may offer with one index, so that the product $C_{ijkl}A_iB_jC_kD_l$ becomes a scalar (note that repeated index implies an invisible summation over the same from 1 to 3). Further, if the scalar is isotropic, then that is independent of the absolute orientations of the individual vectors but just dependent on the angle between those taken two at a time. Since the angle between the two vectors is also given by their dot product, one may write, for an isotropic scalar C_{ijkl},

$$C_{ijkl}A_iB_jC_kD_l = \alpha(\vec{A}\cdot\vec{B})(\vec{C}\cdot\vec{D}) + \beta(\vec{A}\cdot\vec{C})(\vec{B}\cdot\vec{D}) + \gamma(\vec{A}\cdot\vec{D})(\vec{B}\cdot\vec{C}) \quad (8.10)$$

where α, β, and γ are constant scalars. In terms of index notations:

$$C_{ijkl}\ A_iB_jC_kD_l = \alpha A_iB_iC_kD_k + \beta A_iC_iB_jD_j + \gamma A_iD_iB_jC_j$$

or $$C_{ijkl}\ A_iB_jC_kD_l = A_iB_jC_kD_l\left[\alpha\delta_{ij}\delta_{kl} + \beta\,\delta_{ik}\delta_{jl} + \gamma\delta_{il}\delta_{jk}\right] \quad (8.11)$$

where δ_{ij} is the Kronecker delta and is given by

$$\delta_{ij} = 1 \quad \text{if} \quad i = j$$
$$= 0 \quad \text{if} \quad i \neq j$$

In Eq. (8.11), we have used the generic vector transformation: $B_i = B_j\delta_{ij}$. From Eq. (8.11), we get

$$C_{ijkl} = \alpha\delta_{ij}\delta_{kl} + \beta\,\delta_{ik}\delta_{jl} + \gamma\delta_{il}\delta_{jk} \quad (8.12)$$

Equation (8.12) expresses the homogeneous and isotropic tensor C_{ijkl} in terms of 3 homogeneous and isotropic scalars α, β, and γ. Further, since $\tau_{ij}^{\text{deviatoric}}$ is a symmetric tensor

$$\tau_{ij}^{\text{deviatoric}} = \tau_{ji}^{\text{deviatoric}}$$

so that

$$C_{ijkl}\, e_{kl} = C_{jikl}\, e_{kl}$$

From Eq. (8.12) we get

$$C_{jikl} = \alpha\delta_{ji}\delta_{kl} + \beta\,\delta_{jk}\delta_{il} + \gamma\delta_{jl}\delta_{ik} \quad (8.13)$$

Comparing Eqs (8.12) and (8.13), and noting that δ_{ij} is a symmetric tensor ($\delta_{ij} = \delta_{ji}$), we have

$$(\beta - \gamma)\left(\delta_{ik}\delta_{jl} - \delta_{il}\delta_{jk}\right) = 0$$

Since $\left(\delta_{ik}\delta_{jl} - \delta_{il}\delta_{jk}\right)$ is not equal to zero in general, one must have

$$(\beta - \gamma) = 0 \quad (8.14)$$

which implies

$$C_{ijkl} = \alpha\delta_{ij}\delta_{kl} + \beta\,\delta_{ik}\delta_{jl} + \beta\delta_{il}\delta_{jk}$$

Thus, $$\tau_{ij}^{\text{deviatoric}} = C_{ijkl}e_{kl} = \alpha\delta_{ij}\delta_{kl}e_{kl} + \beta\,\delta_{ik}\delta_{jl}e_{kl} + \beta\delta_{il}\delta_{jk}e_{kl}$$

or $$\tau_{ij}^{\text{deviatoric}} = \alpha\delta_{ij}e_{kk} + \beta\, e_{ij} + \beta e_{ji} \quad (\text{since } \delta_{mn} \neq 0 \text{ only if } n = m)$$

or, $$\tau_{ij}^{\text{deviatoric}} = \alpha\delta_{ij}e_{kk} + 2\beta\, e_{ij} \quad (\text{ since } e_{ij} = e_{ji})$$

Further, since, $2\, e_{ij} = \left(\frac{\partial u_i}{\partial x_j} + \frac{\partial u_j}{\partial x_i}\right)$, which is the rate of deformation, β must be physically the viscosity (μ) of the fluid that relates the deviatoric stress in a Newtonian fluid with the rate of deformation. The coefficient α is given a general notation λ in standard texts, which is known as the second coefficient of viscosity and is related to the volumetric rate of dilation $\left(e_{kk} = \frac{\partial u_1}{\partial x_1} + \frac{\partial u_2}{\partial x_2} + \frac{\partial u_3}{\partial x_3} = \frac{\partial u_k}{\partial x_k}\right)$. Therefore, the deviatoric stress tensor component for homogeneous and isotropic Newtonian fluids can be written as

$$\tau_{ij}^{\text{deviatoric}} = \lambda \frac{\partial u_k}{\partial x_k}\delta_{ij} + \mu\left(\frac{\partial u_i}{\partial x_j} + \frac{\partial u_j}{\partial x_i}\right)$$

Regarding the description of the hydrostatic components of stress tensor, it is important to reiterate here that in a fluid at rest there are only normal components of stress on a surface that are independent of the orientation of the surface. In other words, the hydrostatic stress tensor is isotropic or spherically symmetric. Because the stress in a static fluid is isotropic, it must be of the form

$$\tau_{ij}^{\text{hydrostatic}} = -p\delta_{ij} \tag{8.15}$$

where p is the thermodynamic pressure, which may be related to density and temperature by an equation of state for compressible fluids. It is important to recognise here that the hydrostatic component of stress is present not only for fluid at rest, but is also equally active for fluid under motion. Only, for fluid under motion, an additional stress component in terms of deviatoric stress acts, which is absent for fluid at rest. A negative sign is introduced in Eq. (8.15) because of the fact that the normal components of stress are regarded as positive if they indicate tension, whereas pressure by definition is compressive in nature. Thus, one may write

$$\tau_{ij} = \tau_{ij}^{\text{hydrostatic}} + \tau_{ij}^{\text{deviatoric}}$$

or, $$\tau_{ij} = -p\delta_{ij} + \lambda\frac{\partial u_k}{\partial x_k}\delta_{ij} + \mu\left(\frac{\partial u_j}{\partial x_i} + \frac{\partial u_i}{\partial x_j}\right) \tag{8.16}$$

Equation. (8.16) is a representation of the general constitutive behaviour of a homogeneous and isotropic Newtonian fluid.

To proceed further, we first consider the normal components of the stress tensor,

$$\left.\begin{aligned} \tau_{11} &= -p + \lambda\frac{\partial u_k}{\partial x_k} + 2\mu\frac{\partial u_1}{\partial x_1} \\ \tau_{22} &= -p + \lambda\frac{\partial u_k}{\partial x_k} + 2\mu\frac{\partial u_2}{\partial x_2} \\ \tau_{33} &= -p + \lambda\frac{\partial u_k}{\partial x_k} + 2\mu\frac{\partial u_3}{\partial x_3} \end{aligned}\right\} \tag{8.17}$$

From Eq. (8.17), one gets

$$\frac{\tau_{11}+\tau_{22}+\tau_{33}}{3}=-p+\left(\lambda+\frac{2}{3}\mu\right)\frac{\partial u_k}{\partial x_k}$$

or

$$-p_m=-p+\left(\lambda+\frac{2}{3}\mu\right)\frac{\partial u_k}{\partial x_k} \tag{8.18}$$

Here, $\frac{\tau_{11}+\tau_{22}+\tau_{33}}{3}=-p_m$, is the mechanical pressure, which is defined as the arithmetic average of the normal components of the stress tensor. A negative sign is introduced because the normal components of stress are regarded as positive if they indicate tension, whereas pressure by definition is compressive in nature. It is important to mention in this context that, in general, mechanical pressure is not the same as thermodynamic pressure. Mechanical pressure represents the effects of translational mode of energy of the system of molecules, whereas thermodynamic pressure represents combined effects of translational, rotational and vibrational modes of energy of the system of molecules. Distinction between them, therefore, is because of rotational and vibrational modes of energy of the system of molecules. There are some substances for which the rotational and vibrational modes of energy are zero (for example, dilute monatomic gases), so that the mechanical pressure is identically equal to thermodynamic pressure. For such substances, Eq. (8.18) yields

$$\lambda+\frac{2}{3}\mu=0$$

or

$$\lambda=-\frac{2}{3}\mu$$

For a fluid in general, therefore $p_m \neq p$. However, for fluids in local thermodynamic equilibrium, the various modes of energy corresponding to the thermodynamic pressure almost instantaneously get converted into the translational modes of energy manifested in terms of mechanical pressure. Thus, for most cases, we have

$$\lambda+\frac{2}{3}\mu=0 \tag{8.19}$$

Equation (8.19) is known as the Stokes hypothesis, which follows from physical arguments. Fluids obeying this hypothesis are called Stokesian fluids. Interestingly, note that for incompressible flows, $\frac{\partial u_k}{\partial x_k}=0$, so that $p_m=p$, is trivially independent of invoking the Stokes hypothesis.

To have a deeper understanding on the possible applicability of the Stokes hypothesis, let us consider a bubble, inside which there is a gas that changes its thermodynamic states. With a change in state, the effect may be perceived as an equivalent change in mechanical pressure as well. This change in mechanical pressure may be the same as equating the corresponding change in thermodynamic pressure only if sufficient time is allowed over which the system is made to locally equilibrate. This time required is called *relaxation time*. If the change that is taking place is so fast that the time scale of the change is faster than the relaxation time (for example, consider the case of a bubble that is alternatively expanding and contracting at a very high frequency), then the fluid will not be able to come to a state of equilibrium before a new change of state has taken place. Under such circumstances, $p_m \neq p$, and Stokes hypothesis will not work. However, for most practical applications, Stokes hypothesis works, since usually the relaxation time-scales for fluids are very fast (at least, significantly faster than the time-scale of imposition of the thermodynamic change for most processes).

It is also interesting to note that following Stokes hypothesis, λ is negative for positive μ. Considering the expression for τ_{11}, we may observe that the contributor of volumetric deformation to τ_{11} is negative if the fluid is expanding (i.e., if $\frac{\partial u_k}{\partial x_k}$ is positive). Thus, higher the value of $\frac{\partial u_k}{\partial x_k}$ (considering it to be positive), the corresponding proportionate change in τ_{11} is actually less, because it is multiplied by a negative term. It means if a fluid element is already expanding, the proportionate enhancement in stress to expand it further is less.

To derive the final forms of the differential equation of motion, we will now apply Eq. (8.16) to Eq. (8.5a). This, in turn, implies

$$\frac{\partial}{\partial t}(\rho u_i)+\frac{\partial}{\partial x_j}(\rho u_j u_i)=\frac{\partial}{\partial x_j}\left[-p\delta_{ij}+\lambda\frac{\partial u_k}{\partial x_k}\delta_{ij}+\mu\left(\frac{\partial u_j}{\partial x_i}+\frac{\partial u_i}{\partial x_j}\right)\right]+\rho b_i$$

$$=-\frac{\partial p}{\partial x_i}+\frac{\partial}{\partial x_i}\left(\lambda\frac{\partial u_k}{\partial x_k}\right)+\frac{\partial}{\partial x_j}\left[\mu\left(\frac{\partial u_j}{\partial x_i}+\frac{\partial u_i}{\partial x_j}\right)\right]+\rho b_i \qquad (8.20)$$

Since $\frac{\partial}{\partial x_j}\left(\mu\frac{\partial u_j}{\partial x_i}\right)=\frac{\partial}{\partial x_i}\left(\mu\frac{\partial u_j}{\partial x_j}\right)=\frac{\partial}{\partial x_i}\left(\mu\frac{\partial u_k}{\partial x_k}\right)$, Eq. (8.20) may be rewritten as

$$\frac{\partial}{\partial t}(\rho u_i)+\frac{\partial}{\partial x_j}(\rho u_j u_i)=-\frac{\partial p}{\partial x_i}+\frac{\partial}{\partial x_j}\left(\mu\frac{\partial u_i}{\partial x_j}\right)+\frac{\partial}{\partial x_i}\left[(\lambda+\mu)\frac{\partial u_k}{\partial x_k}\right]+\rho b_i \qquad (8.21)$$

Since, using continuity equation, left-hand side of Eq. (8.21) may also be written as $\rho\left[\frac{\partial u_i}{\partial t}+u_j\frac{\partial u_i}{\partial x_j}\right]=\rho\frac{Du_i}{Dt}$, it also follows

$$\rho\frac{Du_i}{Dt}=-\frac{\partial p}{\partial x_i}+\frac{\partial}{\partial x_j}\left(\mu\frac{\partial u_i}{\partial x_j}\right)+\frac{\partial}{\partial x_i}\left[(\lambda+\mu)\frac{\partial u_k}{\partial x_k}\right]+\rho b_i \qquad (8.22)$$

While Eq. (8.21) is known as the **conservative form** of the momentum equation (since it is directly derived from control volume conservation considerations), Eq. (8.22) is the **non-conservative form** of the same. Equations (8.21) and (8.22) are valid for both incompressible as well as compressible flows.

Using Stokes hypothesis ($\lambda = -\frac{2}{3}\mu$), Eqs. (8.21) and (8.22) can be written as

$$\frac{\partial}{\partial t}(\rho u_i) + \frac{\partial}{\partial x_j}(\rho u_j u_i) = -\frac{\partial p}{\partial x_i} + \frac{\partial}{\partial x_j}\left(\mu\frac{\partial u_i}{\partial x_j}\right) + \frac{\partial}{\partial x_i}\left[\frac{\mu}{3}\frac{\partial u_k}{\partial x_k}\right] + \rho b_i \tag{8.23a}$$

$$\rho\frac{Du_i}{Dt} = -\frac{\partial p}{\partial x_i} + \frac{\partial}{\partial x_j}\left(\mu\frac{\partial u_i}{\partial x_j}\right) + \frac{\partial}{\partial x_i}\left[\frac{\mu}{3}\frac{\partial u_k}{\partial x_k}\right] + \rho b_i \tag{8.23b}$$

Equation (8.23a), or equivalently (8.23b), is known as the Navier Stokes equation. It is important to mention here that these equations are non-linear second-order partial differential equations. Unknowns in these equations are the three velocity components, density, and pressure. Hence, in addition to the three components of the momentum conservation equation (Navier Stokes equation), one also needs to invoke the continuity equation, the equation of state (to relate pressure with density and temperature) and energy equation (governing differential equation for temperature) to close the unknowns with the requisite set of independent equations.

It is important to mention here that we often deal with incompressible flows, for which the Navier Stokes equation becomes

$$\rho\frac{Du_i}{Dt} = -\frac{\partial p}{\partial x_i} + \frac{\partial}{\partial x_j}\left(\mu\frac{\partial u_i}{\partial x_j}\right) + \rho b_i \tag{8.24}$$

Equation. (8.24) is essentially a mathematical representation of Newtons's second law of motion from an Eulerian perspective. The left-hand side is nothing but mass times acceleration per unit volume, whereas the right-hand side is the resultant force per unit volume. The first term on the right-hand side is force due to pressure gradient, the second term on the right-hand side is force due to viscous effects, and the third term on the right-hand side is the body forces (all forces expressed per unit volume).

The corresponding expression in vector notation, considering fluids with position independent viscosity, may be written as

$$\underbrace{\rho\frac{D\vec{u}}{Dt}}_{\substack{\text{mass}\times\text{acceleration}\\ \text{per unit volume}}} = \underbrace{-\nabla p}_{\substack{\text{force due to}\\ \text{pressure gradient}\\ \text{per unit volume}}} + \underbrace{\mu\nabla^2\vec{u}}_{\substack{\text{viscous shear force}\\ \text{per unit volume}}} + \underbrace{\rho\vec{b}}_{\substack{\text{body force per}\\ \text{unit volume}}} \tag{8.25}$$

If viscous effects are negligible, we obtain the Euler equation for inviscid flow from Eq. (8.25) as

$$\rho\frac{D\vec{u}}{Dt} = -\nabla p + \rho\vec{b} \tag{8.26}$$

It is important to mention here that many physical situations involve the action of a constant body force, for example, the gravity force. In that case, considering $\vec{g}$ as acceleration due to gravity vector, one may replace $\vec{b}$ in Eq. (8.25) with $\vec{g}$, and write the same in a compact form as

$$\rho \frac{D\vec{u}}{Dt} = -\nabla p^* + \mu \nabla^2 \vec{u} \tag{8.27}$$

where $\nabla p^* = \nabla p - \rho \vec{g}$

Denoting z_v as a coordinate axis aligned with the vertically upward direction (opposite in sense to the action of $\vec{g}$), the above implies

$$\frac{\partial p^*}{\partial z_v} = \frac{\partial p}{\partial z_v} + \rho g$$

For constant density flow,

$$p^* = p + \rho g z_v \text{ (with suitable choice of reference datum for } z_v\text{)}$$

Here p^* is known as piezometric pressure. From Eq. (8.27), it eventually follows that one can couple the effect of a constant body force vector with the pressure gradient term in the Navier Stokes equation, by defining a modified pressure. For the special case of gravity force as the particular body force, this modified pressure term is nothing but the piezometric pressure. Thus, when the pressure gradient term denotes the piezometric pressure gradient, no separate accounting of gravity force becomes necessary in the Navier Stokes equation, since it is already incorporated in the definition of piezometric pressure.

Scalar components of Eq. (8.25) in a Cartesian coordinate system may be written as (where $\vec{u} = u\hat{i} + v\hat{j} + w\hat{k}$, and u, v and w are the velocity components along x, y, and z directions, respectively).

$$\rho\left[\frac{\partial u}{\partial t} + u\frac{\partial u}{\partial x} + v\frac{\partial u}{\partial y} + w\frac{\partial u}{\partial z}\right] = -\frac{\partial p}{\partial x} + \mu\left(\frac{\partial^2 u}{\partial x^2} + \frac{\partial^2 u}{\partial y^2} + \frac{\partial^2 u}{\partial z^2}\right) + \rho b_x \tag{8.28a}$$

$$\rho\left[\frac{\partial v}{\partial t} + u\frac{\partial v}{\partial x} + v\frac{\partial v}{\partial y} + w\frac{\partial v}{\partial z}\right] = -\frac{\partial p}{\partial y} + \mu\left(\frac{\partial^2 v}{\partial x^2} + \frac{\partial^2 v}{\partial y^2} + \frac{\partial^2 v}{\partial z^2}\right) + \rho b_y \tag{8.28b}$$

$$\rho\left[\frac{\partial w}{\partial t} + u\frac{\partial w}{\partial x} + v\frac{\partial w}{\partial y} + w\frac{\partial w}{\partial z}\right] = -\frac{\partial p}{\partial z} + \mu\left(\frac{\partial^2 w}{\partial x^2} + \frac{\partial^2 w}{\partial y^2} + \frac{\partial^2 w}{\partial z^2}\right) + \rho b_z \tag{8.28c}$$

Equations 8.28 (a), (b), and (c) are mathematically closed with the aid of the continuity equation.

Navier Stokes equations in cylindrical coordinates (Fig. 8.3) are useful in solving many problems. If v_r, v_θ and v_z denote the velocity components along the radial, azimuthal and axial directions respectively, then for incompressible flow Eq. (8.25) leads to the following system of momentum equations in r, θ, and z directions as follows:

$$\begin{aligned}\rho\left(\frac{\partial v_r}{\partial t} + v_r\frac{\partial v_r}{\partial r} + \frac{v_\theta}{r}\frac{\partial v_r}{\partial \theta} - \frac{v_\theta^2}{r} + v_z\frac{\partial v_r}{\partial z}\right) &= -\frac{\partial p}{\partial r} \\ +\mu\left[\frac{\partial}{\partial r}\left(\frac{1}{r}\frac{\partial}{\partial r}(r v_r)\right) + \frac{1}{r^2}\frac{\partial^2 v_r}{\partial \theta^2} + \frac{\partial^2 v_r}{\partial z^2} - \frac{2}{r^2}\frac{\partial v_\theta}{\partial \theta}\right] &+ \rho b_r\end{aligned} \tag{8.28d}$$

$$\rho\left(\frac{\partial v_\theta}{\partial t}+v_r\frac{\partial v_\theta}{\partial r}+\frac{v_\theta}{r}\frac{\partial v_\theta}{\partial \theta}+\frac{v_r v_\theta}{r}+v_z\frac{\partial v_\theta}{\partial z}\right)=-\frac{1}{r}\frac{\partial p}{\partial \theta}$$
$$+\mu\left[\frac{\partial}{\partial r}\left(\frac{1}{r}\frac{\partial}{\partial r}(rv_\theta)\right)+\frac{1}{r^2}\frac{\partial^2 v_\theta}{\partial \theta^2}+\frac{\partial^2 v_\theta}{\partial z^2}+\frac{2}{r^2}\frac{\partial v_r}{\partial \theta}\right]+\rho b_\theta \tag{8.28e}$$

$$\rho\left(\frac{\partial v_z}{\partial t}+v_r\frac{\partial v_z}{\partial r}+v_\theta\frac{\partial v_z}{\partial \theta}+v_z\frac{\partial v_z}{\partial z}\right)=-\frac{\partial p}{\partial z}$$
$$+\mu\left[\frac{1}{r}\frac{\partial}{\partial r}\left(r\frac{\partial v_z}{\partial r}\right)+\frac{1}{r^2}\frac{\partial^2 v_z}{\partial \theta^2}+\frac{\partial^2 v_z}{\partial z^2}\right]+\rho b_z \tag{8.28f}$$

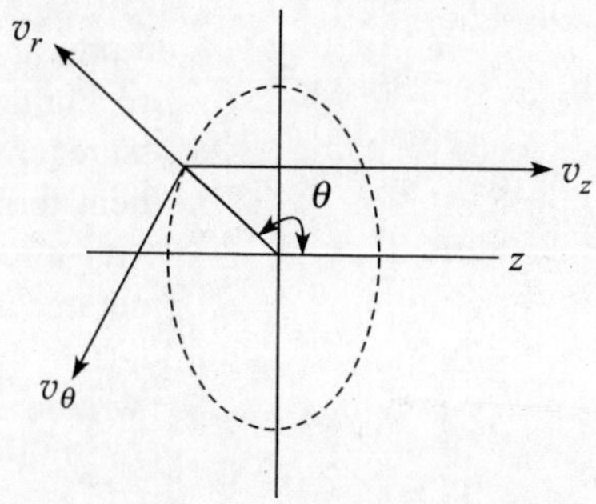

Fig. 8.3 Cylindrical polar coordinate and the velocity components

In spherical coordinates (Fig. 8.4), Eq. (8.25) leads to the system of momentum equations in the r, θ, and ϕ directions as follows:

$$\rho\left(\frac{\partial v_r}{\partial t}+v_r\frac{\partial v_r}{\partial r}+\frac{v_\theta}{r}\frac{\partial v_r}{\partial \theta}+\frac{v_\phi}{r\sin\theta}\frac{\partial v_r}{\partial \phi}-\frac{v_\theta^2+v_\phi^2}{r}\right)=-\frac{\partial p}{\partial r}$$
$$+\mu\left[\frac{1}{r^2}\frac{\partial^2}{\partial r^2}(r^2 v_r)+\frac{1}{r^2\sin\theta}\frac{\partial}{\partial \theta}\left(\sin\theta\frac{\partial v_r}{\partial \theta}\right)+\frac{1}{r^2\sin\theta}\frac{\partial^2 v_r}{\partial \phi^2}\right]+\rho b_r \tag{8.28g}$$

$$\rho\left(\frac{\partial v_\theta}{\partial t}+v_r\frac{\partial v_\theta}{\partial r}+\frac{v_\theta}{r}\frac{\partial v_\theta}{\partial \theta}+\frac{v_\phi}{r\sin\theta}\frac{\partial v_\theta}{\partial \phi}+\frac{v_r v_\theta-v_\phi^2\cot\theta}{r}\right)=-\frac{1}{r}\frac{\partial p}{\partial \theta}$$
$$+\mu\left[\frac{1}{r^2}\frac{\partial^2}{\partial r^2}(r^2 v_\theta)+\frac{1}{r^2}\frac{\partial}{\partial \theta}\left(\frac{1}{\sin\theta}\frac{\partial}{\partial \theta}(v_\theta\sin\theta)\right)\right.$$
$$\left.+\frac{1}{r^2\sin\theta}\frac{\partial^2 v_\theta}{\partial \phi^2}+\frac{2}{r^2}\frac{\partial v_r}{\partial \theta}-\frac{2}{r^2}\frac{\cot\theta}{\sin\theta}\frac{\partial v_\theta}{\partial \phi}\right]+\rho b_\theta \tag{8.28h}$$

$$\rho\left(\frac{\partial v_\phi}{\partial t}+v_r\frac{\partial v_\phi}{\partial r}+\frac{v_\theta}{r}\frac{\partial v_\phi}{\partial \theta}+\frac{v_\phi}{r\sin\theta}\frac{\partial v_\phi}{\partial \phi}+\frac{v_r v_\phi+v_\theta v_\phi\cot\theta}{r}\right)=-\frac{1}{r\sin\theta}\frac{\partial p}{\partial \phi}$$

$$+\mu\left[\frac{1}{r^2}\frac{\partial}{\partial r}\left(r^2\frac{\partial v_\phi}{\partial r}\right)+\frac{1}{r^2}\frac{\partial}{\partial \theta}\left(\frac{1}{\sin\theta}\frac{\partial}{\partial\theta}\left(v_\phi \sin\theta\right)\right)\right.$$

$$\left.+\frac{1}{r^2\sin\theta}\frac{\partial^2 v_\phi}{\partial\phi^2}+\frac{2}{r^2\sin\theta}\frac{\partial v_r}{\partial\phi}+\frac{2}{r^2}\frac{\cot\theta}{\sin\theta}\frac{\partial v_\theta}{\partial\phi}\right]+\rho b_\phi \quad (8.28i)$$

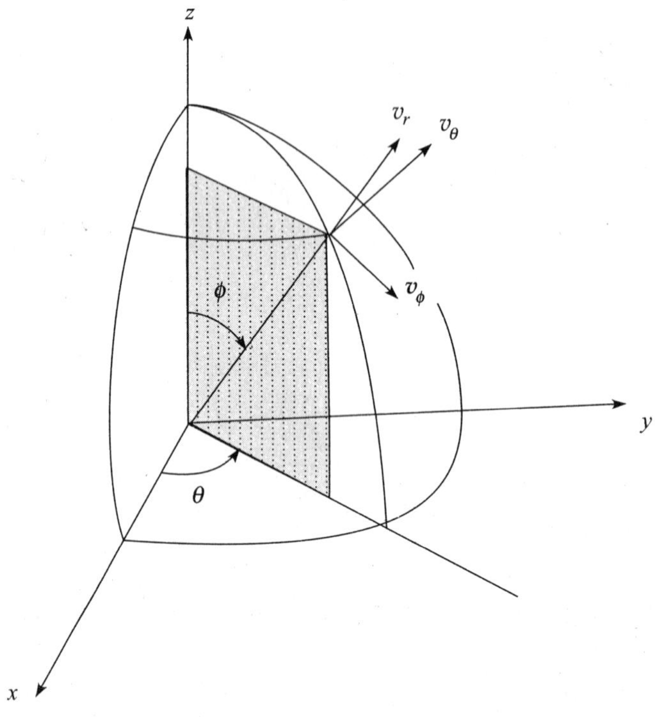

Fig. 8.4 Spherical coordinate and the velocity components

8.3 SOME EXACT SOLUTIONS OF NAVIER STOKES EQUATION FOR STEADY INCOMPRESSIBLE FLOWS

Navier Stokes equation represents a set of partial differential equations that are coupled and non-linear. Therefore, in most cases, analytical solutions to these equations are intractable, and one needs to resort to numerical techniques by employing suitable computational fluid dynamics (CFD) algorithms. However, there are a few special cases in which the Navier Stokes equations may be analytically solved. Some such cases are discussed below.

8.3.1 Fully Developed Laminar Flow Between Two Infinite Parallel Plates or Plane Poiseuille Flow

Let us consider that an incompressible free stream with uniform velocity (u_∞) enters the gap between two parallel plates (see Fig. 8.5). Each plate offers a frictional resistance to the fluid so that adjacent to the plates the fluid is effectively slowed down. As discussed in Chapter 1, boundary layers tend to grow past the respective plates so that the velocity profile continuously changes as we move along the axial direction. Velocity in the core region (outside the boundary layer) increases progressively along that direction to compensate for a slowing down effect in the progressively thickening boundary layer (the net mass flow rate must be conserved at each section). In the core region, viscous effects are not felt. The boundary layers from the two plates meet at section AA, as shown in Fig. 8.5, so that beyond this section the entire fluid (including the centerline fluid) feels the effect of the walls. Beyond section AA, the flow is said to be fully developed in a sense that the velocity profile does not vary with the axial coordinate (x) any further, i.e., $\frac{\partial u}{\partial x} = 0$ for all y. The region from the channel entrance to section AA is known as the developing region and the corresponding length is known as the entrance length. Obtaining the velocity profile in the developing region is not trivial and is beyond the scope of this introductory text. Rather, our objective here will be to obtain an expression for the cross-sectional velocity variation in this fully developed region.

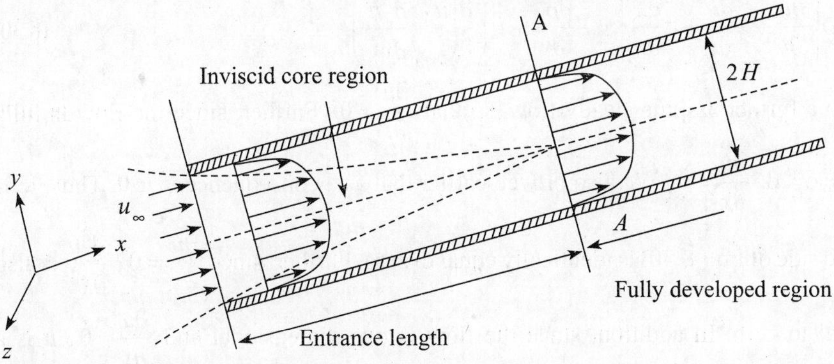

Fig. 8.5 Parallel flow in a straight inclined channel

First we consider an important simplification by considering the width of the plates along the z direction to be infinitely large as compared to its height ($2H$), so that there are no gradients of flow variables along the z direction. This renders the basic flow consideration to be two-dimensional, for which the continuity equation under incompressible flow conditions reads:

$$\frac{\partial u}{\partial x}+\frac{\partial v}{\partial y}=0$$

For fully developed flow, $\frac{\partial u}{\partial x}=0$

Hence, $$\frac{\partial v}{\partial y}=0$$

$$\Rightarrow \quad v \neq v(y)$$

Since $v = 0$ at $y = \pm H$ as a result of the no-penetration at the walls, v is identically equal to zero for all y.

Now, considering the two-dimensional form of Eq (8.28b), one may write

$$\rho\left[\frac{\partial v}{\partial t}+u\frac{\partial v}{\partial x}+v\frac{\partial v}{\partial y}\right]=-\frac{\partial p^*}{\partial y}+\mu\left(\frac{\partial^2 v}{\partial x^2}+\frac{\partial^2 v}{\partial y^2}\right)$$

Since v is identically equal to zero for fully developed flow, the above leads to

$$\frac{\partial p^*}{\partial y}=0 \tag{8.29}$$

It implies that p^* is a function of x only.

Next, we consider the two-dimenisonal form of Eq. (8.28a),

$$\rho\left[\frac{\partial u}{\partial t}+u\frac{\partial u}{\partial x}+v\frac{\partial u}{\partial y}\right]=-\frac{\partial p^*}{\partial x}+\mu\left(\frac{\partial^2 u}{\partial x^2}+\frac{\partial^2 u}{\partial y^2}\right) \tag{8.30}$$

We further assume steady flow, so that $\frac{\partial u}{\partial t}=0$. Further, since the flow is fully developed, $\frac{\partial u}{\partial x}=0$. We also proved earlier that as a consequence, $v = 0$. Thus, left-hand side of Eq.(8.30) is identically equal to zero. Further, since $\frac{\partial u}{\partial x}=0$, $\frac{\partial^2 u}{\partial x^2}$ is also equal to zero. In addition, since the flow is two-dimensional and $\frac{\partial u}{\partial x}=0$, u is a function of y only, so that $\frac{\partial^2 u}{\partial y^2}$ can be written as $\frac{d^2 u}{dy^2}$. Hence, Eq. (8.30) simplifies to

$$\frac{\partial p^*}{\partial x}=\mu\frac{d^2 u}{dy^2} \tag{8.30a}$$

Further, from Eq. (8.29) it follows that p^* is a function of x only. Therefore, Eq. (8.30a) can be written as

$$\frac{dp^*}{dx} = \mu \frac{d^2u}{dy^2} \tag{8.30b}$$

Since p^* is a function of x only and u is a function of y only, it becomes

$$\mu \frac{d^2u}{dy^2} = \frac{dp^*}{dx} = \text{ constant}=c(say) \tag{8.31}$$

It can be mentioned in this context that Eq. (8.31) implies p^* is a linear function of x, while u is a quadratic function of y.

Integrating Eq. (8.31), we have

$$\frac{du}{dy} = \frac{c}{\mu} y + c_1$$

$$u = \frac{c}{2\mu} y^2 + c_1 y + c_2 \tag{8.32}$$

where c_1 and c_2 are arbitrary, independent constants of integration.

The boundary conditions are at $y = H$, $u = 0$; and at $y = -H$, $u = 0$ (Note that one may use an alternative boundary condition: $\frac{du}{dy} = 0$ at $y = 0$; since because of symmetry, maximum of u occurs at $y = 0$). Applying the boundary conditions, the constants of integration are evaluated as $c_1 = 0$ and $c_2 = -\frac{cH^2}{2}$. Therefore, from Eq. (8.32) we get

$$u = \frac{-1}{2\mu}\frac{dp^*}{dx}\left(H^2 - y^2\right) = \frac{H^2}{2\mu}\left(-\frac{dp^*}{dx}\right)\left(1 - \frac{y^2}{H^2}\right) \tag{8.33}$$

Equation (8.33) implies that the velocity profile is parabolic. This typical flow is usually termed as *plane Poiseuille flow*. Eq. (8.33) implies that the flow is always associated with a negative piezometric pressure gradient.

The average velocity, (which is physically an equivalent uniform velocity field that could have given rise to the same volume flow rate as that induced by the variable velocity field under consideration) $\bar{u}$, is given by

$$\bar{u} = \frac{\int_A u dA}{A} = \frac{\int_{-H}^{H} u dy b}{b2H}$$

$$= \frac{2\int_0^H u dy}{2H} = \frac{1}{2\mu}\frac{dp^*}{dx}\frac{\int_0^H \left(y^2 - H^2\right) dy}{H}$$

$$= \frac{1}{2\mu H}\frac{dp^*}{dx}\left(\frac{H^3}{3} - H^2\right) = -\frac{1}{3\mu}\frac{dp^*}{dx}H^2$$

$$\bar{u} = -\frac{1}{3\mu}\frac{dp^*}{dx}H^2 \tag{8.34}$$

Using Eq. (8.34) in Eq. (8.33), it follows that

$$\frac{u}{\bar{u}} = \frac{3}{2}\left(1 - \frac{y^2}{H^2}\right) \tag{8.35}$$

The maximum velocity occurs at the centerline $y = 0$;

$$u_{\max} = -\frac{1}{2\mu}\frac{dp^*}{dx}H^2$$

So that
$$\frac{\bar{u}}{u_{\max}} = \frac{2}{3} \tag{8.36}$$

Shear stress at the wall is given by

$$\tau_w = \tau_{xy}\big|_{wall} = \mu\left(\frac{\partial u}{\partial y} + \frac{\partial v}{\partial x}\right)\bigg|_{wall} = \mu\left(\frac{\partial u}{\partial y}\right)\bigg|_{y=\pm H} = \left(-\frac{dp^*}{dx}\right)\frac{H}{\mu}$$

(using Eq. (8.33))

Using Eq. (8.34), it follows that

$$\tau_w = \frac{3\mu\bar{u}}{H} \tag{8.37}$$

The skin friction coefficient C_f, which is a dimensionless representation of the wall shear stress, is defined as

$$C_f = \frac{|\tau_w|}{\frac{1}{2}\rho\bar{u}^2}$$

$$= \frac{\frac{3\mu\bar{u}}{H}}{\frac{1}{2}\rho\bar{u}^2} = \frac{6\mu}{\mu H\bar{u}}$$

$$C_f = \frac{12}{\rho\frac{\bar{u}2H}{\mu}} = \frac{12}{\text{Re}_{2H}} \tag{8.38}$$

where $\text{Re}_{2H}\left(= \rho\frac{\bar{u}2H}{\mu}\right)$ is the Reynolds number of flow based on average flow velocity and the channel height $2H$.

Example 8.1

Water at 20°C is flowing between a two-dimensional channel in which the top and bottom walls are 1.5 mm apart. If the average velocity is 2 m/s, find out (i) the maximum velocity, (ii) the pressure drop, and (iii) the wall shearing stress [$\mu = 0.00101$ kg/m.s]

Solution

(i) The maximum velocity is given by

$$u_{max} = \frac{3}{2}\bar{u} = \frac{3}{2}[2] = 3 \text{ m/s}$$

(ii) The pressure drop in a two-dimensional straight channel is given by

$$\frac{dp}{dx} = \frac{-2\mu u_{max}}{H^2}, \text{ where } H = \text{half the channel height}$$

$$= \frac{-2(0.00101)3}{[1.5/(2\times 1000)]^2}$$

$$= -10773.33 \text{ N/m}^3$$

or

$$= -10773.33 \text{ N/m}^2 \text{ per m}$$

(iii) The wall shearing stress in a channel flow is given by

$$\tau_w = \mu \left.\frac{\partial u}{\partial y}\right|_{wall} = -H\frac{dp}{dx} = \frac{-1.5}{2\times 1000}(-10773.33)$$

or

$$\tau_w = 8.080 \text{ N/m}^2$$

Example 8.2

Two viscous, incompressible, immiscible fluids of same density (= ρ) but different viscosities (viscosity of the lower fluid layer = μ_1 and that of the upper fluid layer = $\mu_2 < \mu_1$) flow in separate layers between parallel boundaries located at $y = \pm H$, as shown in the Fig. 8.6. Thickness of each fluid layer is identical and their interface is flat. The flow is driven by a constant favourable pressure gradient of $\frac{dp}{dx}$. Derive expressions for the velocity profiles in the fluid layers. Also, make a sketch of the velocity profiles. Assume the flow to be steady and the plates to be of infinitely large width.

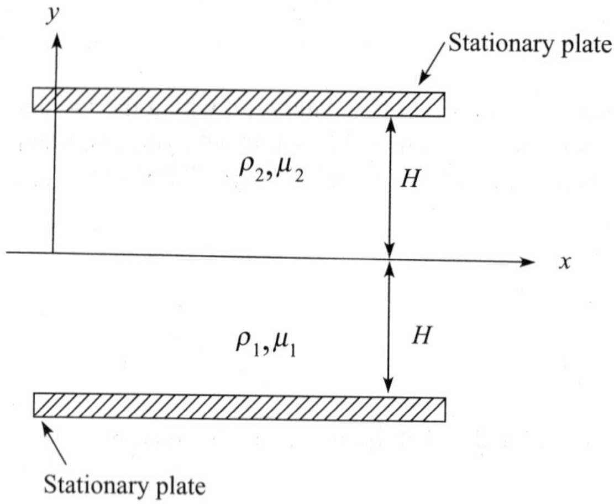

Fig. 8.6

Solution

Large lateral width of the plate renders the basic flow consideration to be two-dimensional, for which the continuity equation under incompressible flow conditions reads:

$$\frac{\partial u}{\partial x} + \frac{\partial v}{\partial y} = 0$$

For fully developed flow, $\frac{\partial u}{\partial x} = 0$

Hence, $$\frac{\partial v}{\partial y} = 0$$

$$\Rightarrow v \neq v(y)$$

Since $v = 0$ at $y = \pm H$ as a result of the no-penetration at the walls, v is identically equal to zero for all y, i.e.,

$$\therefore \quad v = 0 \text{ at } -H \leq y \leq H$$

Now, considering x momentum equation, we have, for steady flow,

$$\rho\left[u\frac{\partial u}{\partial x} + v\frac{\partial u}{\partial y}\right] = -\frac{dp}{dx} + \mu\left(\frac{\partial^2 u}{\partial x^2} + \frac{\partial^2 u}{\partial y^2}\right)$$

Using derivations identical to those presented in Sec. 8.3.1, it follows that

$$\underbrace{\frac{d^2u}{dy^2}}_{\text{Function of } y \text{ only}} = \underbrace{\frac{1}{\mu}\frac{dp}{dx} = \text{constant}}_{\text{Function of } x \text{ only}}$$

For $-H \le y \le 0$

$$\frac{d^2u_1}{dy^2} = \frac{1}{\mu_1}\frac{dp}{dx}$$

Integrating the above equation, we have

$$\frac{du_1}{dy} = \frac{1}{\mu_1}\frac{dp}{dx}y + C_1$$

$$u_1 = \frac{1}{\mu_1}\frac{dp}{dx}\frac{y^2}{2} + C_1 y + C_2 \tag{8.39a}$$

where C_1 and C_2 are constants of integration.

For $0 \le y \le H$

$$\frac{d^2u_2}{dy^2} = \frac{1}{\mu_2}\frac{dp}{dx}$$

Integrating the above equation, we have

$$\frac{du_2}{dy} = \frac{1}{\mu_2}\frac{dp}{dx}y + C_3$$

$$u_2 = \frac{1}{\mu_2}\frac{dp}{dx}\frac{y^2}{2} + C_3 y + C_4 \tag{8.39b}$$

where C_3 and C_4 are constants of integration.

Equations (8.39a) and (8.39b) are subjected to the following boundary conditions.

At $y = -H,\ u_1 = 0,$

At $y = H,\ u_2 = 0,$

At $y = 0,\ u_1 = u_2$ (continuity of flow velocity)

At $y = 0,\ \mu_1 \frac{du_1}{dy} = \mu_2 \frac{du_2}{dy}$ (continuity of shear stress)

From the boundary conditions at $y = -H$, $u_1 = 0$, and at $y = H$, $u_2 = 0$, we get

$$C_2 = C_1 H - \frac{1}{\mu_1} \frac{dp}{dx} \frac{H^2}{2}$$

$$C_4 = -C_3 H - \frac{1}{\mu_2} \frac{dp}{dx} \frac{H^2}{2}$$

From the boundary condition at $y = 0$, $\mu_1 \frac{du_1}{dy} = \mu_2 \frac{du_2}{dy}$, we have

$$\mu_1 C_1 = \mu_2 C_3$$

$$\Rightarrow \qquad C_1 = \frac{\mu_2}{\mu_1} C_3$$

From the boundary condition at $y = 0$, $u_1 = u_2$, one can write

$$C_2 = C_4$$

or $$C_1 H - \frac{1}{\mu_1} \frac{dp}{dx} \frac{H^2}{2} = -C_3 H - \frac{1}{\mu_2} \frac{dp}{dx} \frac{H^2}{2}$$

or $$(C_1 + C_3) H = \frac{dp}{dx} \frac{H^2}{2} \left[\frac{1}{\mu_1} - \frac{1}{\mu_2} \right]$$

or $$\left(\frac{\mu_2}{\mu_1} + 1 \right) C_3 = \frac{dp}{dx} \frac{H}{2} \left[\frac{1}{\mu_1} - \frac{1}{\mu_2} \right]$$

or $$C_3 = \frac{dp}{dx} \frac{H}{2} \frac{(\mu_2 - \mu_1)}{(\mu_2 + \mu_1) \mu_2}$$

$$C_1 = \frac{dp}{dx} \frac{H}{2} \frac{(\mu_2 - \mu_1)}{(\mu_2 + \mu_1) \mu_1}$$

Substituting the values of C_1, C_2, C_3 and C_4 in the Eqs (8.39a) and (8.39b), we have

$$u_1 = \frac{1}{\mu_1} \frac{dp}{dx} \frac{1}{2} \left[y^2 - H^2 + \frac{(\mu_2 - \mu_1)}{(\mu_2 + \mu_1)} H (y + H) \right]$$

$$u_2 = \frac{1}{\mu_2} \frac{dp}{dx} \frac{1}{2} \left[y^2 - H^2 + \frac{(\mu_2 - \mu_1)}{(\mu_2 + \mu_1)} H (y - H) \right]$$

The velocity profiles are shown in Fig. 8.7.

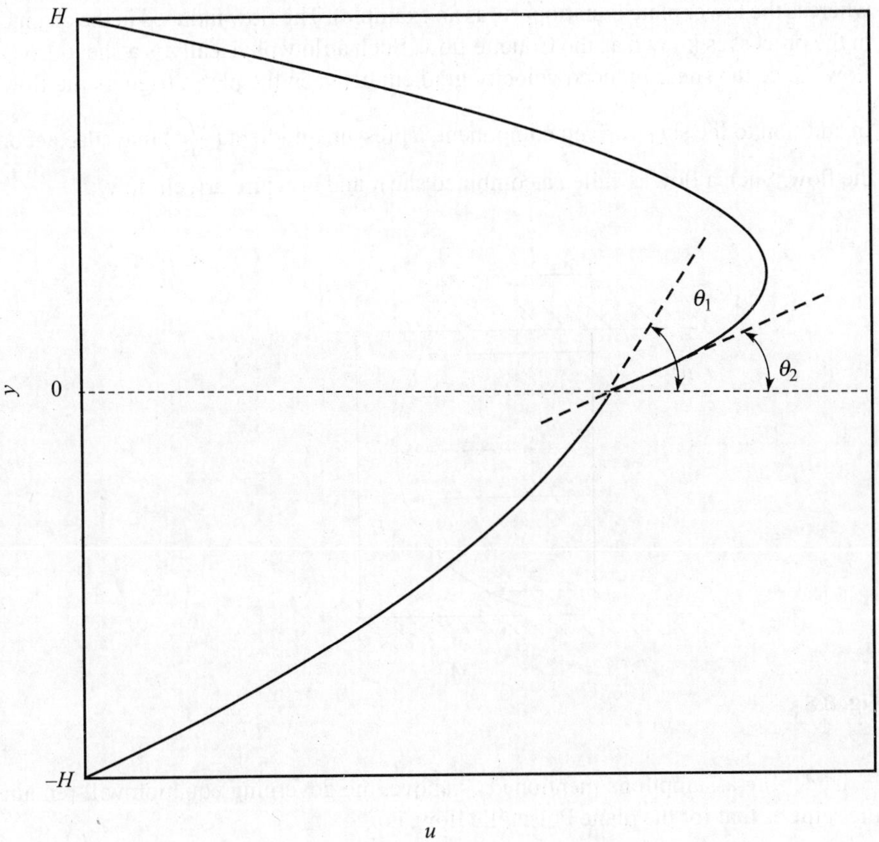

Fig. 8.7 Velocity profiles for two viscous, incompressible, immiscible fluids flowing between two stationary parallel plates

Note from the above figure that since $\mu_2 < \mu_1$, $\left.\dfrac{du_2}{dy}\right|_{y=0} > \left.\dfrac{du_1}{dy}\right|_{y=0}$

$$\left(\text{so as to satisfy } \mu_1 \left.\frac{du_1}{dy}\right|_{y=0} = \mu_2 \left.\frac{du_2}{dy}\right|_{y=0}\right)$$

or
$$\left.\frac{dy}{du_1}\right|_{y=0} > \left.\frac{dy}{du_2}\right|_{y=0},$$

or equivalently, $\tan\theta_1 > \tan\theta_2$.

8.3.2 Couette Flow

Consider steady, incompressible, viscous flow of a Newtonian fluid between two parallel plates located a distance H apart, as shown in Fig. 8.8, such that one of the plates moves with a velocity relative to the other (without loss of generality, we assume that the upper plate moves towards the right with a constant velocity U, whereas the lower plate is stationary, as an example). The flow induced in the channel in the process is known as the Couette flow. Such a flow physically is a shear driven flow, since the shear induced velocity gradient between the plates triggers the flow. In addition to the shear driven component, a pressure gradient ($\frac{dp}{dx}$) may also act on the flow. Such a flow is called a combined shear and pressure driven flow.

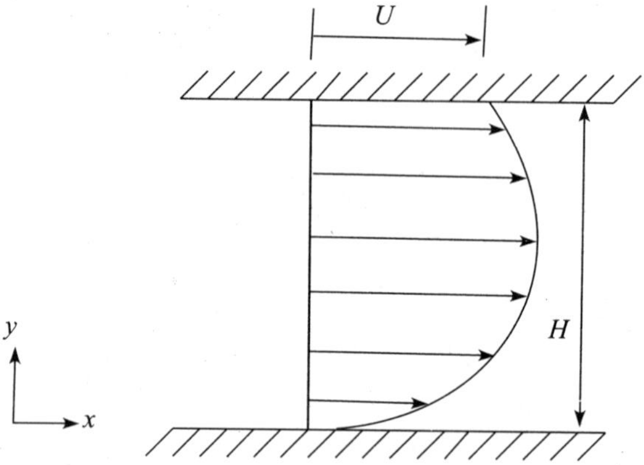

Fig. 8.8

Under the assumptions mentioned as above, the governing equation will remain the same as that for the plane Poiseuille flow, i.e.,

$$\mu \frac{d^2u}{dy^2} = \frac{dp^*}{dx} \tag{8.40}$$

Subjected to the boundary conditions,

at $y = 0, \quad u = 0$

and at $y = H, \quad u = U$

Integrating Eq. (8.40), we get

$$u = \frac{1}{2\mu}\frac{dp^*}{dx}y^2 + C_1 y + C_2$$

Applying the boundary conditions, the constants of integration are evaluated as

$$C_2 = 0$$

and $$C_1 = \frac{U}{H} - \frac{1}{2\mu}\frac{dp^*}{dx}H$$

Therefore,

$$u = U\frac{y}{H} + UP\frac{y}{H}\left(1 - \frac{y}{H}\right) \tag{8.41}$$

where $$P = \frac{H^2}{2\mu U}\left(-\frac{dp^*}{dx}\right)$$

when $P = 0$, the velocity field turns out to be

$$u = U\frac{y}{H} \tag{8.42}$$

This implies that when there is no piezometric pressure gradient, the velocity varies linearly from zero at the fixed plates to U at the moving plate and hence the shear stress across the section of the channel remains constant. This situation is known as simple Couette flow. For a simple Couètte flow, thus the shear rate (du/dy), turns out to be U/H, which remains constant within the channel.

Therefore it is observed from Eq. (8.41) that Couette flow is a superimposition of simple Couette flow (purely shear driven flow) and plane Poisseuille flow(purely pressure driven flow)

The quantitative description of non-dimensional velocity distribution across the channel, depicted by Eq. (8.42), is shown in Fig. 8.9(a).

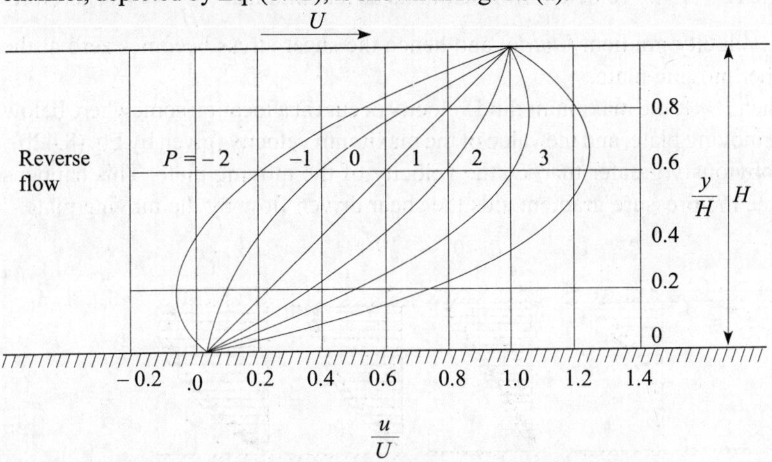

(a) Velocity distribution of the Couette flow

Fig. 8.9(a)

The location of maximum or minimum velocity in the channel is found by setting the derivative du/dy equal to zero. From Eq. (8.42), we can write

$$\frac{du}{dy} = \frac{U}{H} + \frac{PU}{H}\left(1 - 2\frac{y}{H}\right)$$

For maximum or minimum velocity,

$$\frac{du}{dy} = 0$$

which gives $$\frac{y_{max}}{H} = \frac{1}{2} + \frac{1}{2P} \tag{8.42a}$$

where y_{max} is the y corresponding to the location of maximum or minimum velocity. Substituting $y = y_{max}$ in Eq. (8.42b), one write

$$\left.\begin{aligned} u_{max} &= \frac{U(1+P)^2}{4P} \qquad \text{for } P \geq 1 \\ u_{min} &= \frac{U(1+P)^2}{4P} \qquad \text{for } P \leq 1 \end{aligned}\right\} \tag{8.42b}$$

Some interesting features in the flow field are observed for different values of P, which are shown in Fig. 8.9(b) and are described below.

Let us first consider the case of favourable pressure gradient in the direction of flow ($P > 0$). The following may be observed from Eq. (8.42a):

(i) When P lies between 0 and 1, the velocity profile (given by Eq. (8.41)) exhibits neither a maximum nor a minimum in the flow field. The velocity increases monotonically from zero at the stationary plate to U at the moving plate.

(ii) When $P = 1$, the velocity profile exhibits its maximum at $\frac{y_{max}}{H} = 1$. Therefore, the velocity gradient (du/dy) and hence the shear stress becomes zero at the upper moving plate.

(iii) When $P > 1$, the maximum fluid velocity occurs at a location somewhere below the moving plate, and the value of the maximum velocity (given by Eq. (8.42b)) is obviously greater than U, the velocity of the moving plate. This happens since the pressure gradient aids the shear driven flow by the moving plate.

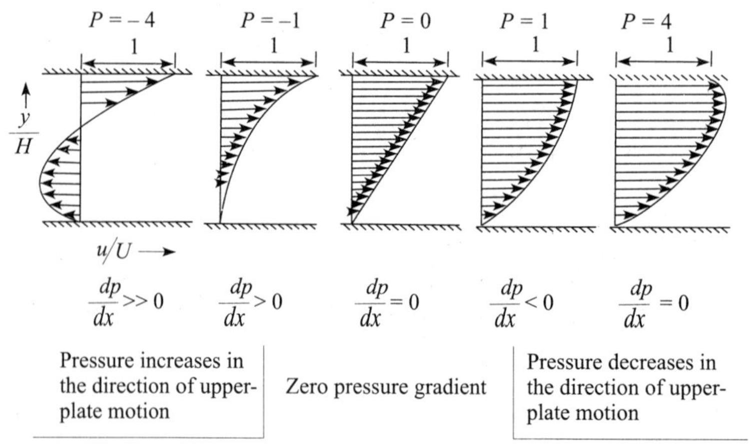

(b) Velocity profile for the Couette flow for varoius values of pressure gradient

Fig. 8.9(b)

Let us now consider the case of adverse pressure gradient (($P < 0$) in the direction of flow. The following may be observed from Eq. (8.42a).

(i) When P lies between 0 and – 1, the velocity profile exhibits neither a maximum nor a minimum within the flow field. The velocity decreases monotonically from U at the moving plate to zero at the stationary plate.

(ii) When $P = -1$, the minimum velocity is attained at the stationary plate. The velocity gradient (du/dy) and hence the shear stress at the stationary plate becomes zero. This implies the onset of flow reversal.

(iii) When $P < -1$, the minimum velocity is attained at locations given by $\frac{y_{max}}{H} > 0$, which means that there occurs a back flow near the fixed plate.

The volumetric flow rate per unit width of the plates can be written as

$$Q = \int_0^h u dy$$

With the help of Eq. (8.41), the above equation becomes

$$Q = \left(\frac{1}{2} + \frac{P}{6}\right) UH \tag{8.42c}$$

The average velocity $\bar{u}$ is given by

$$\bar{u} = \left(\frac{1}{2} + \frac{P}{6}\right) U \tag{8.42d}$$

The Eq. (8.42c) implies that when $P = -3$, there is no flow across the passage formed by the plates. This is due to the fact that the influence of adverse pressure gradient balances the dragging action by the moving plate. In other words, shear driven flow balances the opposing pressure driven flow. There is a pure recirculating flow within the passage under the situation.

Example 8.3

Two infinite plates are h distance apart as in Fig. 8.10. There is a fluid of viscosity μ between the plates and the pressure is constant. The upper plate is moving at speed $U = 4$ m/s. The height of the channel $h = 1.8$ cm. Calculate the shear stress at the upper and lower walls if $\mu = 0.44$ kg/m.s and $\rho = 888$ kg/m^3.

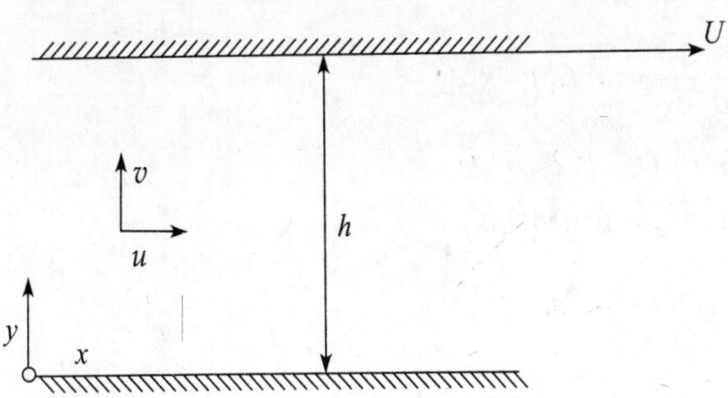

Fig. 8.10 Parallel flow between two plates with upper plate moving

Solution

Re $= \rho h U/\mu = (888)\,(1.8/100)\,(4)/0.44 = 145$. So, the flow is laminar and $\tau = \mu \dfrac{\partial u}{\partial y}$, u at any y is given by $\dfrac{U}{h}\,y$.

Shear stresses at the two walls are of equal magnitude, therefore,

$$\tau = \mu \frac{\partial u}{\partial y} = \mu \frac{(U-0)}{h} = (0.44)\,(4)/(1.8/100)$$
$$= 97.8 \text{ Pa}$$

Example 8.4

Water at 60°C flows between two large flat plates. The lower plate moves to the left at a speed of 0.3 m/s. The plate spacing is 3 mm and the flow is laminar. Determine the pressure gradient required to produce zero net flow at a cross section. ($\mu = 4.7 \times 10^{-4}$ Ns/m^2 at 60 °C)

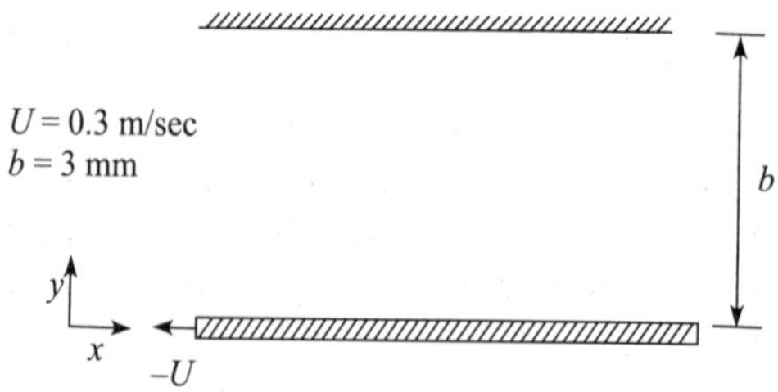

Fig. 8.11

Solution

Governing equation: $\mu \dfrac{d^2u}{dy^2} = \dfrac{dp}{dx}$

$$u = \frac{1}{2\mu}\frac{dp}{dx}y^2 + C_1 y + C_2$$

at $y = 0$, $u = -U$, $\quad C_2 = -U$

at $y = b$, $u = 0$, which yields

$$0 = \frac{1}{2\mu}\frac{dp}{dx}b^2 + C_1 b - U,$$

or

$$C_1 = \frac{U}{b} - \frac{1}{2\mu}\cdot\frac{dp}{dx}b$$

$$u = \frac{1}{2\mu}\frac{dp}{dx}(y^2 - by) + U\left(\frac{y}{b} - 1\right)$$

Now, $$Q = \int_0^b u\,\mathrm{d}y = \int_0^b \left[\frac{1}{2\mu}\frac{\mathrm{d}p}{\mathrm{d}x}(y^2 - by) + U\left(\frac{y}{b} - 1\right)\right]\mathrm{d}y$$

or $$Q = -\frac{1}{12\mu}\frac{\mathrm{d}p}{\mathrm{d}x}b^3 - \frac{Ub}{2}a$$

For $Q = 0$, with $\mu = 4.7 \times 10^{-4}\ \mathrm{Ns/m^2}$

$$\frac{\mathrm{d}p}{\mathrm{d}x} = -\frac{6U\mu}{b^2} = \frac{-6 \times 0.3 \times 4.7 \times 10^{-4}}{(0.003)^2} = -94.0\ \mathrm{N/m^2 \cdot m}$$

8.3.3 Thin Film Flows

Consider a steady fully developed flow of a thin liquid film falling slowly down an inclined wall, as shown in Fig. 8.12. The films thickness is h. The liquid constituting the film is assumed to have constant physical properties, for simplicity.

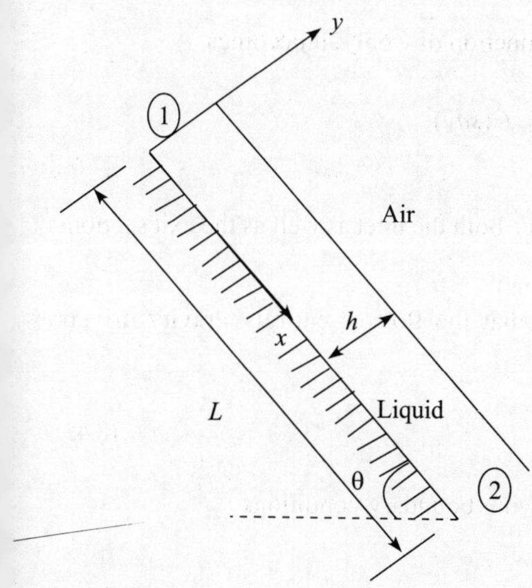

Fig. 8.12

With similar considerations as the flow between two parallel plates, we may first invoke the continuity equation to show that $v = 0$ for all values of y. Thus the y momentum equation reduces to

$$0 = -\frac{\partial p}{\partial y} - \rho g \cos\theta$$

or, $$p = -\rho g \cos\theta\, y + f(x)$$

Note here that we have used the static pressure itself instead of the piezometric pressure for the pressure gradient term, because of which an additional accounting of the gravity force in terms of a body force becomes necessary.
The x momentum equation can be written as

$$\rho\left[\frac{\partial u}{\partial t}+u\frac{\partial u}{\partial x}+v\frac{\partial u}{\partial y}\right]=-\frac{\partial p}{\partial x}+\mu\left(\frac{\partial^2 u}{\partial x^2}+\frac{\partial^2 u}{\partial y^2}\right)+\rho g\sin\theta$$

Following a procedure similar to the one adopted for analysing flow in parallel plate channels, one may obtain:

$$\mu\frac{d^2 u}{\partial y^2}=\frac{\partial p}{\partial x}-\rho g\sin\theta$$

If the films thickness is considered to be small enough then the variation of pressure across the film may be neglected and then p effectively becomes a function of x only, so that one may write

$$\mu\frac{d^2 u}{\partial y^2}=\frac{dp}{dx}-\rho g\sin\theta$$

Since p is a function of x only and u is a function of y only, it becomes

$$\mu\frac{d^2 u}{dy^2}+\rho g\sin\theta=\frac{dp}{dx}\Rightarrow \text{constant}=C(say)$$

Hence, $\frac{dp}{dx}=\frac{p_2-p_1}{L}=\frac{p_a-p_a}{L}=0$ (since both the inlet as well as the exit sections are exposed to the same ambient conditions)
Thus, $C = 0$, which physically implies that the flow is entirely gravity driven. Accordingly, one may write

$$\mu\frac{d^2 u}{dy^2}+\rho g\sin\theta=0 \tag{8.43}$$

The above equation is subjected to following boundary conditions:

(i) At $y=0, u=0$

(ii) At $y=h$, $\mu_{\text{liquid}}\left.\frac{du}{dy}\right|_{\text{liquid}}=\mu_{\text{air}}\left.\frac{du}{dy}\right|_{\text{air}}$ (continuity of shear stress)

Since $\mu_{\text{air}}\ll\mu_{\text{liquid}}$ the boundary condition (ii) may be approximated as $\mu_{\text{liquid}}\left.\frac{du}{dy}\right|_{\text{liquid}}\approx 0$ at $y=h$, for most practical conditions. Under these circumstances, solution to Eq. (8.43) may be obtained as

$$u=\frac{1}{2\mu}\rho g\sin\theta\left(2hy-y^2\right) \tag{8.44}$$

Example 8.5

A continuous belt (Fig. 8.13) passing upward through a chemical bath at velocity U_0, picks up a liquid film of thickness h, density ρ, and viscosity μ. Gravity tends to make the liquid drain down, but the movement of the belt keep the fluid from running off completely. Assume that the flow is fully developed and that the atmosphere produces no shear at the outer surface of the film. State clearly the boundary conditions to be satisfied by velocity at $y = 0$ and $y = h$. Obtain an expression for the velocity profile.

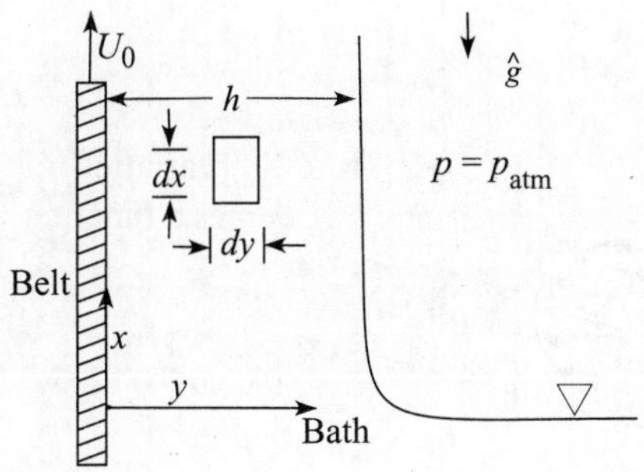

Fig. 8.13

Solution

The governing equation is

$$\mu \frac{d^2u}{dy^2} = \rho g$$

or

$$\mu \frac{du}{dy} = \rho g y + C_1$$

or

$$\frac{du}{dy} = \frac{\rho g y}{\mu} + \frac{C_1}{\mu}$$

$$u = \frac{\rho g y^2}{2\mu} + \frac{C_1}{\mu} y + C_2$$

$$\text{at } y = 0,\ u = U_0, \quad \text{so} \quad C_2 = U_0$$

$$\text{at } y = h,\ \tau = 0, \quad \text{so} \quad \frac{du}{dy} = 0 \text{ and } C_1 = -\rho g h$$

$$u = \frac{\rho g y^2}{2\mu} - \frac{\rho g h y}{\mu} + U_0 = \frac{\rho g}{\mu}\left(\frac{y^2}{2} - hy\right) + U_0$$

8.3.4 Fully Developed Flow Through Circular Tube / Pipe (Hagen Poiseuille Flow)

Consider a steady, fully developed laminar flow through a circular pipe/tube, as shown in Fig. 8.14. For flow in a circular tube with axial symmetry it is convenient to employ a cylindrical coordinate system. The continuity equation in cylindrical co-ordinates is given by

$$\frac{1}{r}\frac{\partial}{\partial r}(\rho r v_r)+\frac{1}{r}\frac{\partial}{\partial \theta}(\rho v_\theta)+\frac{\partial}{\partial z}(\rho v_z)=0 \tag{8.45}$$

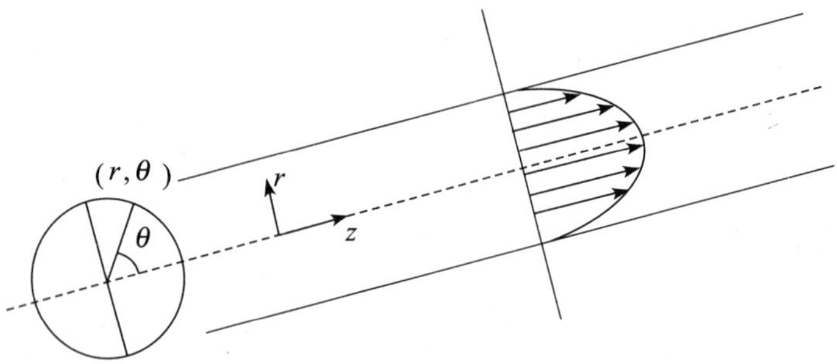

Fig. 8.14 Hagen Poiseuille flow through a circular tube/pipe

We assume that the flow is fully developed, i.e., $\frac{\partial v_z}{\partial z}=0$. We also assume that there is no rotational component in the flow ($v_\theta=0$) and the flow is axially symmetric ($\frac{\partial}{\partial \theta}$(any variable) $= 0$). Then, Eq. (8.45) reduces to

$$\frac{\partial}{\partial r}(r v_r)=0$$

The above implies that rv_r is not a function of r. This, in turn, by virtue of no-penetration boundary condition at the wall, implies that $v_r = 0$ everywhere in the flow field (except at $r = 0$, which is a singularity in this case). Hence, for fully developed flow, there is only one velocity component, $v_z = v_z(r)$.

Next, we invoke the r momentum equation, which reads

$$\rho\left(\frac{\partial v_r}{\partial t}+v_r\frac{\partial v_r}{\partial r}+\frac{v_\theta}{r}\frac{\partial v_r}{\partial \theta}-\frac{v_\theta^2}{r}+v_z\frac{\partial v_r}{\partial z}\right)=-\frac{\partial p^*}{\partial r}$$
$$+\mu\left[\frac{\partial}{\partial r}\left(\frac{1}{r}\frac{\partial}{\partial r}(r v_r)\right)+\frac{1}{r^2}\frac{\partial^2 v_r}{\partial \theta^2}+\frac{\partial^2 v_r}{\partial z^2}-\frac{2}{r^2}\frac{\partial v_\theta}{\partial \theta}\right] \tag{8.46}$$

With the simplifying assumptions mentioned earlier, Eq. (8.46) simplifies to

$$0 = -\frac{\partial p^*}{\partial r} \tag{8.47}$$

In other words, p^* is not a function of r. Therefore, p^* is a function of z only. Next, we invoke the z momentum equation in cylindrical coordinates as

$$\rho\left(\frac{\partial v_z}{\partial t} + v_r\frac{\partial v_z}{\partial r} + \frac{v_\theta}{r}\frac{\partial v_z}{\partial \theta} + v_z\frac{\partial v_z}{\partial z}\right) = -\frac{\partial p^*}{\partial z} + \mu\left[\frac{1}{r}\frac{\partial}{\partial r}\left(r\frac{\partial v_z}{\partial r}\right) + \frac{1}{r^2}\frac{\partial^2 v_z}{\partial \theta^2} + \frac{\partial^2 v_z}{\partial z^2}\right] \tag{8.48}$$

Considering the simplifying assumptions mentioned earlier, Eq. (8.48) becomes

$$0 = -\frac{\partial p^*}{\partial z} + \mu\frac{1}{r}\frac{\partial}{\partial r}\left(r\frac{\partial v_z}{\partial r}\right)$$

or

$$\frac{dp^*}{dz} = \mu\frac{1}{r}\frac{d}{dr}\left(r\frac{dv_z}{dr}\right) \tag{8.49}$$

Since p^* is a function of z only and u is a function of r only, Eq. (8.49) simplifies to

$$\frac{dp^*}{dz} = \mu\frac{1}{r}\frac{d}{dr}\left[r\frac{dv_z}{dr}\right] \Rightarrow \text{constant} = C \ \text{(say)} \tag{8.50}$$

Thus, $\mu\dfrac{1}{r}\dfrac{d}{dr}\left[r\dfrac{dv_z}{dr}\right] = C$

Multiplying both sides by $\dfrac{r}{\mu}$, we integrate once to get

$$r\frac{dv_z}{dr} = \frac{Cr^2}{2\mu} + C_1$$

where C_1 is a constant of integration. Dividing both sides by r, we integrate once more to obtain

$$v_z = \frac{Cr^2}{4\mu} + C_1 \ln r + C_2 \tag{8.51}$$

where C_2 is another constant of integration. The boundary conditions are no-slip at the wall and finite velocity at the centreline. The former (i.e., $v_z = 0$ at $r = R$) results in

$$0 = \frac{Cr^2}{4\mu} + C_1 \ln r + C_2$$

Further, to ensure finite velocity at the channel centreline (avoiding a logarithmic singularity) we must have, $C_1 = 0$ (this also follows from the consideration of maximum velocity at the centreline).

Thus,

$$C_2 = \frac{CR^2}{4\mu}$$

Therefore, $$v_z = -\frac{C}{4\mu}\left(R^2 - r^2\right)$$

or, equivalently, $$v_z = -\frac{R^2}{4\mu}\frac{dp^*}{dz}\left(1-\frac{r^2}{R^2}\right) \tag{8.52}$$

This shows that the axial velocity profile in a steady, fully developed laminar flow through a circular tube has a parabolic variation along r.
The maximum velocity occurs at the centreline, as given by

$$v_{z,\,\max} = -\frac{R^2}{4\mu}\frac{dp^*}{dz} \tag{8.53}$$

Therefore, we can write

$$\frac{v_z}{v_{z,\max}} = \left(1-\frac{r^2}{R^2}\right) \tag{8.53a}$$

The volume flow rate in a pipe is

$$Q = \int v_z dA$$

The average velocity is given by

$$\overline{v}_z = \frac{Q}{A} = \frac{\int_o^R v_z 2\pi r dr}{\pi R^2}$$

Using Eq. (8.52), we have

$$\overline{v}_z = -\frac{R^2}{8\mu}\frac{dp^*}{dz} \tag{8.54}$$

Comparing Eq. (8.53) and (8.54), one can write

$$\overline{v}_z = \frac{v_{z,\max}}{2} \tag{8.54a}$$

One can also write

$$\frac{v_z}{\overline{v}_z} = 2\left(1-\frac{r^2}{R^2}\right) \tag{8.55}$$

Shear stress at any location can be written as

$$\tau = \tau_{rz} = \mu\left(\frac{\partial v_r}{\partial z} + \frac{\partial v_z}{\partial r}\right)$$

$$= \mu\frac{dv_z}{dr}$$

(Since, $v_r = 0$ under the present situation and v_z is a function of r only).

With the help of Eq. (8.52), we can write

$$\tau = \frac{r}{2}\frac{dp^*}{dz} \tag{8.56}$$

Equation (8.56) indicates that shear stress varies linearly with the radial distance from the axis.

Therefore, the wall shear stress is

$$\tau_w = \frac{R}{2}\frac{dp^*}{dz} \tag{8.56a}$$

The skin friction coefficient C_f (also known as Fanning friction coefficient), is defined as

$$C_f = \frac{|\tau_w|}{\frac{1}{2}\rho \bar{v}_z^{\,2}}$$

Using Eqs (8.54) and (8.56a), it becomes

$$C_f = \frac{16}{\rho \bar{v}_z D/\mu}$$

The quantity in the denominator at the above equation is defined as Reynolds number for pipe flow as

$$\mathrm{Re}_D = \frac{\rho \bar{v}_z D}{\mu}$$

Hence,
$$C_f = \frac{16}{\mathrm{Re}_D} \tag{8.57}$$

We note from Eq (8.54) that

$$\frac{dp^*}{dz} = -\frac{8\mu \bar{v}_z}{R^2} \tag{8.58}$$

Sine $\frac{dp^*}{dz}$ is linear, we can write in consideration of p_1^* and p_2^* being the piezometric pressures at upstream and downstream, respectively, over a length L

$$\frac{dp^*}{dz} = \frac{p_2^* - p_1^*}{L} = -\frac{\Delta p^*}{L}$$

where $\Delta p^* = p_1^* - p_2^*$

Hence Eq. (8.58) can be written as

$$\Delta p^* = \frac{8\mu \bar{v}_z L}{R^2} = \frac{32\mu \bar{v}_z L}{D^2} \tag{8.59}$$

where D is the pipe diameter.

Since $\bar{v} = \dfrac{4Q}{\pi D^2}$, it becomes

$$\Delta p^* = \frac{128\mu QL}{\pi D^4} \tag{8.60}$$

Equation (8.60) is known as the *Hagen Poiseuille* equation. The equation is sometimes expressed in terms of head loss $h_f = \Delta p^* / \rho g$ as

$$h_f = \frac{128\mu QL}{\rho g \pi D^4} \tag{8.61}$$

It is important to mention here that frictional head losses in a pipe may alternatively be represented by a non-dimensional factor f, known as Darcy friction factor (for details, see the Chapter 11), defined such that the following equation (known as *Darcy-Weisbach* equation) holds:

$$h_f = f \frac{L}{D} \frac{\bar{v}_z^2}{2g} \tag{8.62}$$

Equations (8.59) and (8.62) yield,

$$f \frac{L}{D} \frac{\bar{v}_z^2}{2g} = \frac{32\mu \bar{v}_z L}{\rho g D^2}$$

or

$$f = \frac{64}{\dfrac{\rho \bar{v}_z D}{\mu}} = \frac{64}{\mathrm{Re}_D} \tag{8.63}$$

On comparison between Eqs (8.57) and (8.63), it appears

$$f = 4C_f \tag{8.63a}$$

i.e., the Darcy friction factor is 4 times the Fanning friction coefficient.

Example 8.6

The analysis of a fully developed laminar flow through a pipe can alternatively be derived from control volume approach. Derive the expression $v_z = \dfrac{R^2}{4\mu}\left(-\dfrac{\mathrm{dp}}{\mathrm{dz}}\right)\left(1 - \dfrac{r^2}{R^2}\right)$ accordingly.

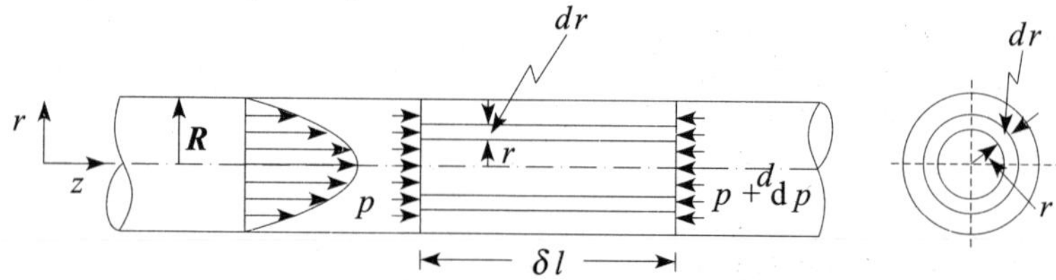

Fig. 8.15 Fully developed laminar flow through a pipe

Solution

Let us have a look at Fig. 8.15. The fluid moves due to the axial pressure gradient while the pressure across a section may be regarded as constant. Due to viscous friction, individual layers act on each other producing a shearing stress which is proportional to $\frac{\partial v_z}{\partial r}$.

In order to establish the condition of equilibrium, we consider a fluid cylinder of length δl and radius r. Now we can write

$$[p - (p + \mathrm{d}p)]\ \pi r^2 = -\ \tau\ 2\ \pi\ r\ \delta l$$

or
$$-\ \mathrm{d}p\ \pi r^2 = -\ \mu \frac{\partial v_z}{\partial r} 2\ \pi\ r\ \delta l$$

or
$$\frac{\partial v_z}{\partial r} = \frac{1}{2\mu}\frac{\mathrm{d}p}{\mathrm{d}l} r = \frac{1}{2\mu}\frac{\mathrm{d}p}{\mathrm{d}z} r$$

Upon integration,

$$v_z = \frac{1}{4\mu}\frac{\mathrm{d}p}{\mathrm{d}z} r^2 + K$$

at $r=R, v_z = 0$, hence $K = -\left(\frac{1}{4\mu}\frac{\mathrm{d}p}{\mathrm{d}z}\right) \cdot R^2$

So, $v_z = \frac{R^2}{4\mu}\left(-\frac{\mathrm{d}p}{\mathrm{d}z}\right)\left(1 - \frac{r^2}{R^2}\right)$

Example 8.7

In the laminar flow of a fluid in a circular pipe, the velocity profile is exactly a parabola. The rate of discharge is then represented by volume of a paraboloid. Prove that for this case the ratio of the maximum velocity to mean velocity is 2.

Solution

See Fig. 8.15. For a paraboloid,

$$v_z = v_{z_{\max}}\left[1 - \left(\frac{r}{R}\right)^2\right]$$

$$Q = \int v_z\ \mathrm{d}A = \int_0^R v_{z_{\max}}\left[1 - \left(\frac{r}{R}\right)^2\right](2\pi r\ \mathrm{d}r)$$

$$= 2\pi\, v_{z_{\max}}\left[\frac{r^2}{2} - \frac{r^4}{4R^2}\right]_0^R = 2\pi\, v_{z_{\max}}\left[\frac{R^2}{2} - \frac{R^2}{4}\right]$$

$$= v_{z_{\max}}\left(\frac{\pi R^2}{2}\right)$$

$$v_{z_{\text{mean}}} = \frac{Q}{A} = \frac{v_{z_{\max}}(\pi R^2/2)}{(\pi R^2)} = \frac{v_{z_{\max}}}{2}$$

Thus,

$$\frac{v_{z_{\max}}}{v_{z_{\text{mean}}}} = 2$$

Example 8.8

The velocity distribution for a fully developed laminar flow in a pipe is given by

$$v_z = -\frac{R^2}{4\mu}\cdot\frac{\partial p}{\partial z}[1-(r/R)^2]$$

Determine the radial distance from the pipe axis at which the velocity equals the average velocity.

Solution

For a fully developed laminar flow in a pipe, we can write

$$v_z = -\frac{R^2}{4\mu}\frac{\partial p}{\partial z}\left[1-\left(\frac{r}{R}\right)^2\right]$$

$$\bar{v}_z = \frac{Q}{A} = \frac{1}{\pi R^2}\int_0^R \left\{-\frac{R^2}{4\mu}\frac{\partial p}{\partial z}\left[1-\left(\frac{r}{R}\right)^2\right]\right\} 2\pi r\,dr$$

$$= -\frac{R^2}{8\mu}\frac{\partial p}{\partial z}$$

Now, for $v_z = \bar{v}_z$ we have,

$$\frac{R^2}{4\mu}\frac{\partial p}{\partial z}\left[1-\left(\frac{r}{R}\right)^2\right] = -\frac{R^2}{8\mu}\frac{\partial p}{\partial z}$$

or

$$1-\left(\frac{r}{R}\right)^2 = \frac{1}{2}$$

or

$$\left(\frac{r}{R}\right)^2 = \frac{1}{2} \quad \text{or} \quad r = \frac{R}{\sqrt{2}} = 0.707\text{ R}$$

8.3.5 Flow Between Two Concentric Large Rotating Cylinders

Consider steady, incompressible flow in the annulus of two cylinders, where R_1 and R_2 are the radii of inner and outer cylinders, respectively, and the cylinders move with different rotational speeds, ω_1 and ω_2, respectively (Fig. 8.16).

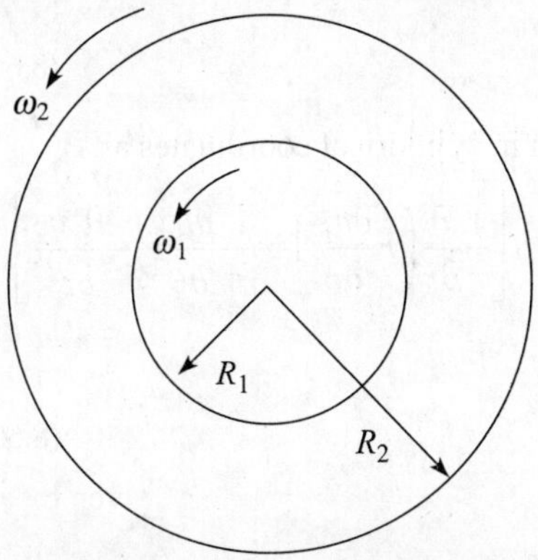

Fig. 8.16

We assume that the length of the cylinders are large enough so that the end effects can be neglected and $\frac{\partial}{\partial z}(\text{any property}) = 0$. Since both the cylinders rotate without any preference on θ, $\frac{\partial}{\partial z}(\text{any property}) = 0$ (axially symmetric flow). To analyse this problem, we first invoke the continuity equation.

The continuity equation for this problem becomes

$$\frac{\partial \rho}{\partial t} + \frac{1}{r}\frac{\partial}{\partial r}(\rho r v_r) + \frac{1}{r}\frac{\partial}{\partial \theta}(\rho v_\theta) + \frac{\partial}{\partial z}(\rho v_z) = 0$$

or
$$0 + \frac{1}{r}\frac{\partial}{\partial r}(\rho r v_r) + \frac{1}{r}0 + 0 = 0$$

or
$$\frac{1}{r}\frac{\partial}{\partial r}(r v_r) = 0$$

or
$$r v_r = \text{constant} \tag{8.64}$$

Since $v_r = 0$ at both the inner and outer cylinders, it follows that $v_r = 0$ everywhere and the motion can only be purely circumferential, $v_\theta = v_\theta(r)$.

Next, we invoke the r momentum equation, which reads

$$\rho\left(\frac{\partial v_r}{\partial t} + v_r\frac{\partial v_r}{\partial r} + \frac{v_\theta}{r}\frac{\partial v_r}{\partial \theta} - \frac{v_\theta^2}{r} + v_z\frac{\partial v_r}{\partial z}\right) = -\frac{\partial p}{\partial r}$$
$$+\mu\left[\frac{\partial}{\partial r}\left(\frac{1}{r}\frac{\partial}{\partial r}(r v_r)\right) + \frac{1}{r^2}\frac{\partial^2 v_r}{\partial \theta^2} + \frac{\partial^2 v_r}{\partial z^2} - \frac{2}{r^2}\frac{\partial v_\theta}{\partial \theta}\right]$$

With the simplifying considerations mentioned as above, the r momentum equation finally becomes

$$-\frac{\rho v_\theta^2}{r} = -\frac{\partial p}{\partial r} \tag{8.65}$$

Next, we consider the z momentum equation in cylindrical coordinates as

$$\rho\left(\frac{\partial v_z}{\partial t}+v_r\frac{\partial v_z}{\partial r}+v_\theta\frac{\partial v_z}{\partial \theta}+v_z\frac{\partial v_z}{\partial z}\right)=-\frac{\partial p}{\partial z}+\mu\left[\frac{1}{r}\frac{\partial}{\partial r}\left(r\frac{\partial v_z}{\partial r}\right)+\frac{1}{r^2}\frac{\partial^2 v_z}{\partial \theta^2}+\frac{\partial^2 v_z}{\partial z^2}\right]$$

Since $v_z = 0$, it follows from the above that

$$-\frac{\partial p}{\partial z} = 0 \tag{8.66}$$

The θ momentum equation may be written as

$$\rho\left(\frac{\partial v_\theta}{\partial t}+v_r\frac{\partial v_\theta}{\partial r}+\frac{v_\theta}{r}\frac{\partial v_\theta}{\partial \theta}+\frac{v_r v_\theta}{r}+v_z\frac{\partial v_\theta}{\partial z}\right)=-\frac{1}{r}\frac{\partial p}{\partial \theta}$$
$$+\mu\left[\frac{\partial}{\partial r}\left(\frac{1}{r}\frac{\partial}{\partial r}(r v_\theta)\right)+\frac{1}{r^2}\frac{\partial^2 v_\theta}{\partial \theta^2}+\frac{\partial^2 v_\theta}{\partial z^2}+\frac{2}{r^2}\frac{\partial v_r}{\partial \theta}\right]$$

Since $v_r = 0$, $\frac{\partial}{\partial \theta}(\text{any variable}) = 0$, $v_z = 0$ and $v_\theta = v_\theta\, r$ only, it follows from the above that

$$\mu\frac{\partial}{\partial r}\left[\frac{1}{r}\frac{\partial}{\partial r}(r v_\theta)\right] = 0 \tag{8.67}$$

Since v_θ is a function of r only, Eq. (8.67) can be rewritten as

$$\frac{d}{dr}\left[\frac{1}{r}\frac{d}{dr}(r v_\theta)\right] = 0$$

Integrating Eq. (8.67), it follows that

$$v_\theta = C_1\frac{r}{2}+\frac{C_2}{r} \tag{8.68}$$

The constants are readily found by imposing the boundary conditions at the inner and outer cylinders:

Outer at $r = R_2$, $\quad v_\theta = \omega_2 R_2 = C_1\frac{R_2}{2}+\frac{C_2}{R_2}$

Inner at $r = R_1$, $\quad v_\theta = \omega_1 R_1 = C_1\frac{R_1}{2}+\frac{C_2}{R_1}$

Thus, $\quad C_1 = 2\left[\omega_1-\frac{R_2^2}{(R_2^2-R_1^2)}(\omega_1-\omega_2)\right]$

and $\quad C_2 = \frac{R_1^2 R_2^2}{(R_2^2-R_1^2)}(\omega_1-\omega_2)$

Finally, the velocity distribution is given by

$$v_\theta = \frac{1}{\left(R_2^2 - R_1^2\right)}\left[\left(\omega_2 R_2^2 - \omega_1 R_1^2\right) r + \frac{R_1^2 R_2^2}{r}(\omega_1 - \omega_2)\right] \quad (8.69)$$

$$\frac{v_\theta}{r} = \frac{1}{\left(R_2^2 - R_1^2\right)}\left[\left(\omega_2 R_2^2 - \omega_1 R_1^2\right) + \frac{R_1^2 R_2^2}{r^2}(\omega_1 - \omega_2)\right]$$

Shear stress at any location is given by

$$\tau_{r\theta} = \mu\left[r\frac{\partial}{\partial r}\left(\frac{v_\theta}{r}\right) + \frac{1}{r}\frac{\partial v_r}{\partial \theta}\right]$$

or

$$\tau_{r\theta} = \mu\left[r.\frac{1}{\left(R_2^2 - R_1^2\right)}.(-2\frac{R_1^2 R_2^2}{r^3})(\omega_1 - \omega_2) + 0\right]$$

or

$$\tau_{r\theta} = -\frac{2\mu R_1^2 R_2^2}{\left(R_2^2 - R_1^2\right)}(\omega_1 - \omega_2)\frac{1}{r^2} \quad (8.70)$$

As an interesting limiting case, we may consider a single solid cylinder rotating in a fluid of infinite extent, so that $R_2 \to \infty$, and $\omega_2 = 0$. In that case, the constant C_1 appearing in Eq. (8.68) becomes identically zero (so as to predict a finite v_θ), so that $v_\theta = \frac{C_2}{r}$, which is essentially reminiscent of a free vortex flow. The corresponding shear stress distribution is given as

$$\tau_{r\theta} = -\frac{2\mu C_2}{r^2} \quad (8.71)$$

Next, it would be interesting to calculate the shear force per unit volume. For that, we utilise the vector form of the Navier Stokes equation and note that $\mu\nabla^2\vec{V}$ is nothing but the viscous shear force per unit volume. Next, we use the vector identity

$$\nabla^2\vec{V} = \nabla\left(\nabla\cdot\vec{V}\right) - \nabla\times\left(\nabla\times\vec{V}\right)$$

and utilise that $\nabla\cdot\vec{V} = 0$ for incompressible flow, to get

$$\mu\nabla^2\vec{V} = -\mu\nabla\times\left(\nabla\times\vec{V}\right)$$

$$= -\mu\nabla\times\xi \quad (8.72)$$

where $\vec{\xi} = \nabla\times\vec{V}$ is the vorticity vector. Since $v_\theta = \frac{C_2}{r}$, an irrotational flow field, the vorticity vector becomes null, and hence from Eq. (8.72) it follows that the viscous shear force per unit volume becomes identically zero. *Thus, this case confers to an interesting situation in which the shear force per unit volume is zero, despite the shear stress ($\tau_{r\theta}$) being non-zero.*

Example 8.9

Consider steady flow of an incompressible Newtonian fluid (density = ρ, viscosity = μ) between two infinitely long concentric circular cylinders of inner radius a and outer radius b (Fig. 8.17). Both the cylinders rotate with the same angular velocity ω. In addition, the inner cylinder moves along the z direction, with a constant velocity of U_i.

(i) Assuming steady flow with no axial pressure gradient, derive the velocity field between the two cylinders.

(ii) Assuming the pressure at the surface of the inner cylinder to be p_i, derive the pressure field between the cylinders.

Solution

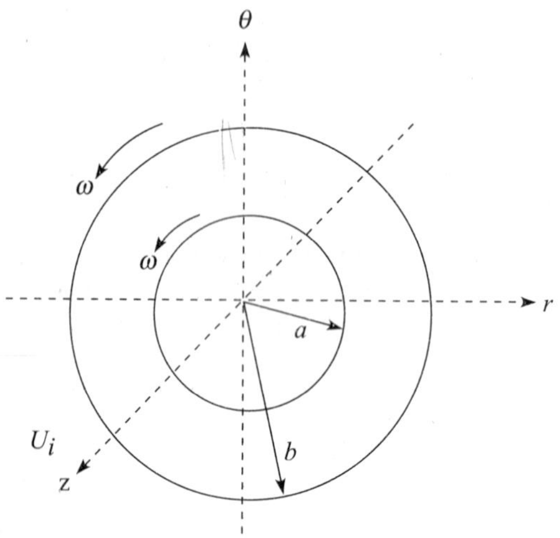

Fig. 8.17

Infinitely long cylinder implies $\frac{\partial}{\partial z}(\text{any variable}) = 0$.

Since, the flow is rotational symmetric, we have ($\frac{\partial}{\partial z}(\text{any variable}) = 0$).

No axial pressure gradient means $\frac{\partial p}{\partial z} = 0$ and for steady flow $\frac{\partial}{\partial z}(\text{any variable}) = 0$)

The continuity equation in cylindrical coordinates under the above conditions is given by

$$\frac{\partial}{\partial r}(\rho r v_r) = 0$$

which implies that rv_r is not a function of r. Now, by virtue of no penetration boundary condition $v_r = 0$ at $r = a, b$, which implies that $v_r = 0$ everywhere in the flow field (except at $r = 0$, which is a singularity in this case).

From the z momentum equation in cylindrical coordinates (Eq. 8.28f), we get, using the above simplifying considerations

$$0 = \frac{\partial}{\partial r}\left(r\frac{\partial v_z}{\partial r}\right)$$

Integrating the above equation (noting that v_z is a function of r only), we have

$$r\frac{dv_z}{dr} = C_1$$

$$v_z = C_1 \,\text{In}\, r + C_2 \tag{8.73a}$$

where C_1 and C_2 are constants of integration.

Equation (8.73a) is subjected to the following boundary conditions:

$$v_z = U_i \quad \text{at} \quad r = a \quad \text{and} \quad v_z = 0 \quad \text{at} \quad r = b.$$

The constants of integrations are accordingly found to be

$$C_1 = \frac{U_i}{\ln\frac{a}{b}} \text{ and } C_1 = -\frac{U_i}{\ln\frac{a}{b}}\ln b$$

Thus, Eq. (8.73a) becomes

$$v_z = \frac{U_i}{\ln\frac{a}{b}}\ln\frac{r}{b}$$

From the θ momentum equation in cylindrical coordinates (Eq. 8.28e) and using the simplifying considerations specified as before, we get

$$0 = \frac{\partial}{\partial r}\left(\frac{1}{r}\frac{\partial}{\partial r}(rv_\theta)\right)$$

Integrating the above equation (noting that v_θ is a function of r only), we have

$$\frac{1}{r}\frac{\partial}{\partial r}(rv_\theta) = C_3$$

$$v_\theta = C_3\frac{r}{2} + \frac{C_4}{r} \tag{8.73b}$$

Equation (8.73b) is subjected to the following boundary conditions:

At $r = a$, $v_\theta = a\omega$ and at $r = b$, $v_\theta = b\omega$.

The constant of integrations are found to be

$$C_3 = 2\omega \quad \text{and} \quad C_4 = 0$$

Finally, Eq. (8.73b) becomes

$$v_\theta = \omega r$$

(ii) The r momentum equation in cylindrical co-ordinates (Eq. 8.28d), using the simplifying considerations specified as before, reads

$$-\rho\frac{v_\theta^2}{r} = -\frac{\partial p}{\partial r}$$

or
$$\frac{dp}{dr} = \rho\frac{v_\theta^2}{r} \text{ (Since } p \text{ is a function of } r \text{ only)}$$

or
$$\frac{dp}{dr} = \rho\frac{\omega^2 r^2}{r}$$

Integrating the above equation, we have

$$\int_{p_i}^{p} dp = \int_a^r \rho\omega^2 r dr$$

$$p = p_i + \frac{\rho\omega^2}{2}\left(r^2 - a^2\right)$$

8.4 LOW REYNOLDS NUMBER FLOW

We have seen in Chapter 6 that Reynolds number is the ratio of inertia force to viscous force. For flow at low Reynolds number, the inertia terms in the Navier-Stokes equations become small as compared to viscous terms. As such, when the inertia terms are omitted from the equations of motion, the analyses are valid for only Re << 1. Consequently, this approximation, linearises the Navier–Stokes equations and for some problems, makes it amenable to analytical solutions. We shall discuss such flows in this section. Motions at very low Reynolds number are sometimes referred to as creeping motion.

8.4.1 Theory of Hydrodynamic Lubrication

A thin film of oil, confined between the interspace of moving parts, may acquire high pressures up to 100 MPa which is capable of supporting load and reducing friction. The salient features of this type of motion can be understood from a study of slipper bearing (Fig. 8.18). The slipper moves with a constant velocity U, past the bearing plate. This slipper face and the bearing plate are not parallel but slightly inclined at an angle of α. A typical bearing has a gap width of 0.025 mm or less, and the convergence between the walls may be of the order of 1/5000. It is assumed that the sliding surfaces are very large in transverse direction so that the problem can be considered two dimensional.

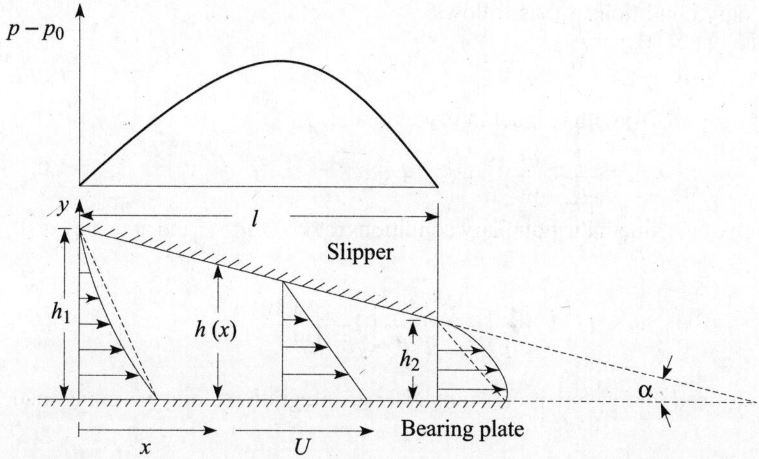

Fig. 8.18 Flow in a slipper bearing

For the analysis, we may assume that the slipper is at rest and the plate is forced to move with a constant velocity U. The height $h(x)$ of the wedge between the block and the guide is assumed to be very small as compared with the length l of the block. This motion is different from that we have considered while discussing Couette flow. The essential difference lies in the fact that here the two walls are inclined at an angle to each other. Due to the gradual reduction of narrowing passage, the convective acceleration $u\left(\frac{\partial u}{\partial x}\right)$ is distinctly not zero. However, a relative estimation of inertia term with respect to viscous term suggests that, for all practical purposes, inertia terms can be neglected. The estimate is done in the following way:

$$\frac{\text{Inertia force}}{\text{Viscous force}} = \frac{\rho u(\partial u/\partial x)}{\mu(\partial^2 u/\partial y^2)} = \frac{\rho U^2/l}{\mu U/h^2} = \frac{\rho U l}{\mu}\left(\frac{h}{l}\right)^2$$

The inertia force can be neglected with respect to viscous force if the modified Reynolds number,

$$R^* = \frac{U l}{\nu}\left(\frac{h}{l}\right)^2 << 1$$

The equation for motion in the y direction can be omitted since the v component of velocity is very small with respect to u. Besides, in the x momentum equation, $\partial^2 u/\partial x^2$ can be neglected as compared with $\partial^2 u/\partial y^2$ because the former is smaller than the latter by a factor of the order of $(h/l)^2$. With these simplifications the equations of motion reduce to

$$\mu \frac{\partial^2 u}{\partial y^2} = \frac{\mathrm{d}p}{\mathrm{d}x} \tag{8.74}$$

The equation of continuity can be written as

$$Q = \int_0^{h(x)} u \, \mathrm{d}y \tag{8.75}$$

The boundary conditions are as follows:

at $y = 0, u = U$ at $x = 0, p = p_0$

at $y = h, u = 0$ and at $x = l, p = p_0$ (8.76)

Integrating Eq. (8.74) with respect to y, we obtain

$$u = \frac{1}{2\mu}\cdot\frac{dp}{dx}\, y^2 + C_1 y + C_2$$

Application of the kinematic boundary conditions (at $y = 0, u = U$ and at $y = h, u = 0$), yields

$$u = U\left(1 - \frac{y}{h}\right) - \frac{h^2}{2\mu}\cdot\frac{dp}{dx}\left(1 - \frac{y}{h}\right)\frac{y}{h} \tag{8.77}$$

It is to be noticed that $\left(\frac{dp}{dx}\right)$ is constant as far as integration along y is concerned, but p and $\frac{dp}{dx}$ vary along the x axis. At the point of maximum pressure, $\frac{dp}{dx} = 0$, hence,

$$u = U\left(1 - \frac{y}{h}\right) \tag{8.78}$$

Equation (8.78) depicts that the velocity profile along y is linear at the location of maximum pressure. The gap at this location may be denoted as h^*.

Now, substituting Eq. (8.77) into Eq. (8.75) and integrating, we get

$$Q = \frac{Uh}{2} - \frac{p' h^3}{12\,\mu}$$

or

$$p' = 12\,\mu\left(\frac{U}{2h^2} - \frac{Q}{h^3}\right) \tag{8.79}$$

where $p' = dp/dx$.

Integrating Eq. (8.79) with respect to x, we obtain

$$\int \frac{dp}{dx}\,dx = 6\mu U \int \frac{dx}{(h_1 - \alpha x)^2} - 12\,\mu Q \int \frac{dx}{(h_1 - \alpha x)^3} + C_3 \tag{8.80a}$$

or

$$p = \frac{6\mu U}{\alpha(h_1 - \alpha x)} - \frac{6\mu Q}{\alpha(h_1 - \alpha x)^2} + C_3 \tag{8.80b}$$

where $\alpha = (h_1 - h_2)/l$ and C_3 is a constant.

Since the pressure must be the same ($p = p_0$), at the ends of the bearing, namely, $p = p_0$ at $x = 0$ and $p = p_0$ at $x = l$, the unknowns in the above equations can be determined by applying the pressure boundary conditions. We obtain

$$Q = \frac{U h_1 h_2}{h_1 + h_2} \quad \text{and} \quad C_3 = p_0 - \frac{6\,\mu U}{\alpha(h_1 + h_2)}$$

With these values inserted, the equation for pressure distribution (8.80) becomes

$$p = p_0 + \frac{6\,\mu U x\,(h - h_2)}{h^2\,(h_1 + h_2)}$$

or

$$p - p_0 = \frac{6\,\mu U x\,(h - h_2)}{h^2\,(h_1 + h_2)} \tag{8.81}$$

It may be seen from Eq. (8.81) that, if the gap is uniform, i.e., $h = h_1 = h_2$, the gauge pressure will be zero. Furthermore, it can be said that very high pressure can be developed by keeping the film thickness very small. Figure 8.18 shows the distribution of pressure throughout the bearing.

The total load bearing capacity per unit width is

$$P = \int_0^l (p - p_0)\,dx = \frac{6\mu U}{h_1 + h_2}\int_0^l \frac{x(h - h_2)}{h^2}\,dx$$

After substituting $h = h_1 - \alpha x$ with $\alpha = (h_1 - h_2)/l$ in the above equation and performing the integration,

$$P = \frac{6\mu U l^2}{(h_1 - h_2)^2}\left[\ln\frac{h_1}{h_2} - 2\left\{\frac{h_1 - h_2}{h_1 + h_2}\right\}\right] \qquad (8.82)$$

The shear stress at the bearing plate is

$$\tau_0 = -\mu\frac{\partial u}{\partial y}\bigg|_{y=0} = \left(p'\frac{h}{2} + \mu\frac{U}{h}\right) \qquad (8.83)$$

Substituting the value of p' from Eq. (8.79) and then invoking the value of Q in Eq. (8.83), the final expression for shear stress becomes

$$\tau_0 = 4\mu\frac{U}{h} - \frac{6\mu U h_1 h_2}{h^2(h_1 + h_2)}$$

The drag force required to move the lower surface at speed U is expressed by

$$D = \int_0^l \tau_0\,dx = \frac{\mu U l}{h_1 - h_2}\left[4\ln\frac{h_1}{h_2} - 6\frac{h_1 - h_2}{h_1 + h_2}\right] \qquad (8.84)$$

Michell thrust bearing, named after A.G.M. Michell, works on the principles based on the theory of hydrodynamic lubrication. The journal bearing (Fig. 8.19) develops its force by the same action, except that the surfaces are curved.

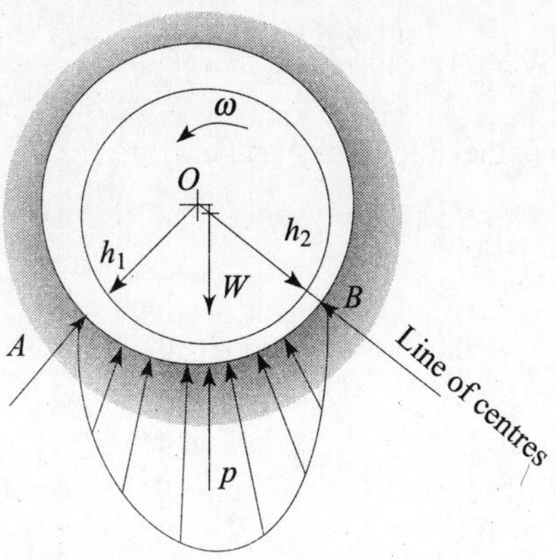

Fig. 8.19 Hydrodynamic action of a journal bearing

Example 8.10

A slipper (slider) and plate (guide), both 0.5 m wide, constitutes a bearing as shown in Fig. 8.20. Density of the fluid, $\rho = 9.00\ kg/m^3$ and viscosity, $\mu = 0.1\ Ns/m^2$.

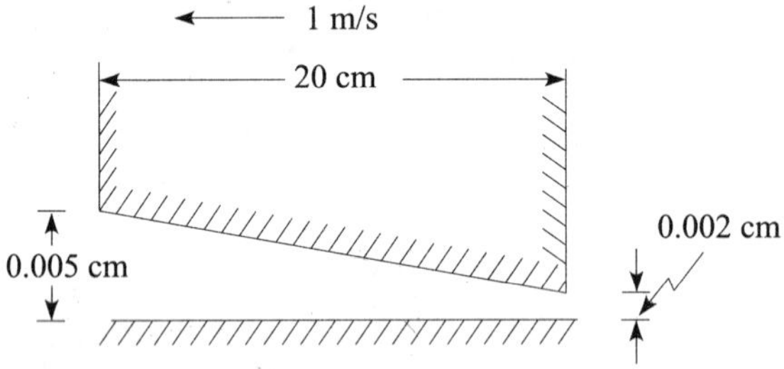

Fig. 8.20 Slipper and plate both 0.5 m wide

Find out the (i) load carrying capacity of the bearing, (ii) drag, and (iii) power lost in the bearing.

Solution

(i) Considering the width as b and using Eq. (8.82) for the load carrying capacity,

$$P = \frac{6\pi U l^2 b}{(h_1 - h_2)^2}\left[\ln\frac{h_1}{h_2} - 2\frac{h_1 - h_2}{h_1 + h_2}\right]$$

$$= \left[\frac{6 \times 0.1 \times 1 \times 0.2 \times 0.2 \times 0.5}{(0.005 - 0.002)^2}\right] \times \left[\ln\frac{0.005}{0.002} - 2\frac{0.005 - 0.002}{0.005 + 0.002}\right]$$

$$= \frac{0.012}{9.0 \times 10^{-6}}(0.9163 - 0.8571) = 78.93\ \text{N}$$

(ii) Making use of Eq. (8.84) for width b, the drag force may be written as

$$D = \frac{\mu U l b}{h_1 - h_2}\left[4\ln\frac{h_1}{h_2} - 6\frac{h_1 - h_2}{h_1 + h_2}\right]$$

$$= \frac{0.1 \times 1 \times 0.2 \times 0.5}{(0.005 - 0.002)}\left[4\ln\frac{0.005}{0.002} - 6\frac{0.005 - 0.002}{0.005 + 0.002}\right]$$

$$= \frac{0.01}{0.003}(3.6651 - 2.5714) = 3.645\ \text{N}$$

(iii) Power lost = Drag × Velocity

$$= 3.645 \times 1 = 3.645\ \text{W}$$

Example 8.11

A cylindrical journal bearing supports a load directed vertically upwards, with the shaft rotating clockwise. Sketch the position of the shaft centre with respect to that of the bearing (hole), if no cavitation is present. Give explanation. No equations are required.

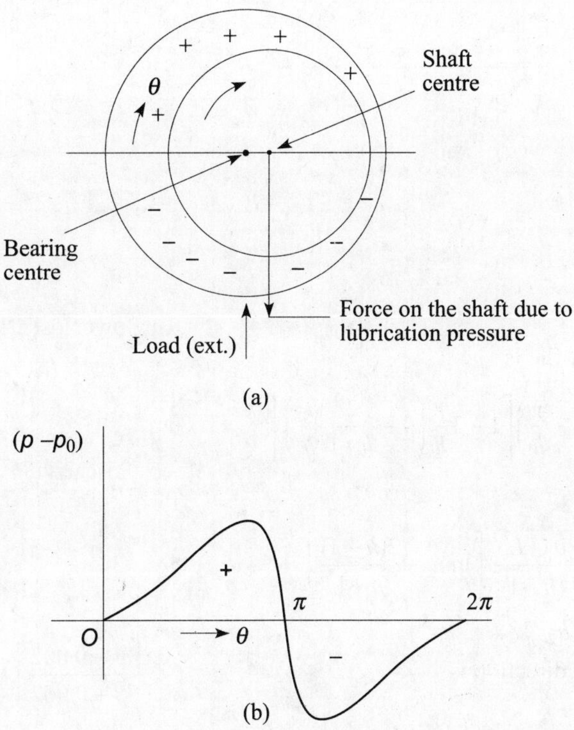

Fig. 8.21

Solution

In this case, the pressure distribution is symmetric, as shown in (Fig 8.21), where θ is measured in the direction of the rotation of the shaft from the position of maximum clearance. Thus, as shown, the shaft centre is to the *right* of the bearing centre, and the line of centres is *horizontal*.

Example 8.12

For the following thrust bearing (Fig. 8.22), show that the force on the straight slider in the x direction is the same as that on the guide.

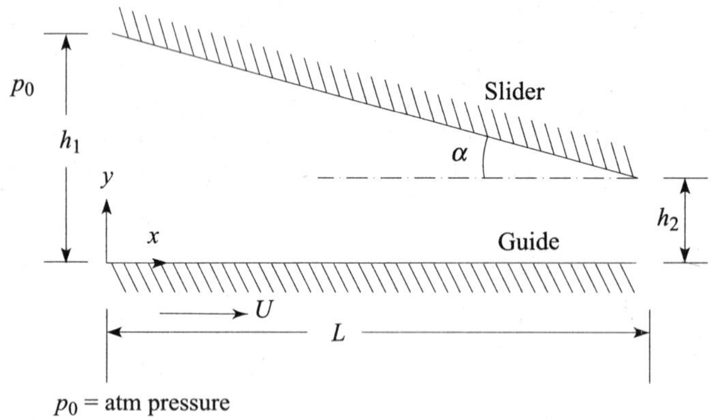

Fig. 8.22

Solution

It is given that the velocity profile is

$$\frac{u}{U} = \left(1 - \frac{y}{h}\right)\left[1 - 3\frac{y}{h}\left(1 - \frac{2}{n+1}\frac{h_1}{h}\right)\right]$$

and load

$$P = \frac{6\mu U L^2}{h_2^2 (n-1)^2}\left[\ln n - \frac{2(n-1)}{n+1}\right]$$

where $\qquad n = h_1/h_2$

Force on the slider in the x direction is

$$F_s = \int_0^L \tau_s \, \mathrm{d}x(1)\frac{\cos\alpha}{\cos\alpha} + \int_0^L (p - p_0)\frac{\mathrm{d}x}{\cos\alpha}(1)\sin\alpha$$

$$= \int_0^L \tau_s \, \mathrm{d}x + \tan\alpha \int_0^L (p - p_0)\,\mathrm{d}x$$

Now, $\qquad \tau_s = -\mu\left(\frac{\partial u}{\partial y}\right)_{y=h}$

and $\qquad u = U\left(1 - \frac{y}{h}\right)\left[1 - 3\frac{y}{h}\left(1 - \frac{2}{n+1}\cdot\frac{h_1}{h}\right)\right]$, so we get

$$\frac{\partial u}{\partial y} = U\left[-\frac{1}{h} - 3\left(\frac{1}{h} - \frac{2y}{h^2}\right)\left(1 - \frac{2}{n+1}\cdot\frac{h_1}{h}\right)\right]$$

$$\therefore \qquad \tau_s = -\mu U\left[-\frac{1}{h} + \frac{3}{h}\left(1 - \frac{2}{n+1}\cdot\frac{h_1}{h}\right)\right]$$

$$= \mu U\left[-\frac{2}{h} + \frac{6}{n+1}\cdot\frac{h_1}{h^2}\right]$$

Also, $$h = h_1 - (h_1 - h_2)\,\frac{x}{L}$$

Therefore, $$\mathrm{d}h = -\,\frac{h_1 - h_2}{L}\,\mathrm{d}x$$

Thus, $$\int_0^L \tau_s\,\mathrm{d}x = \frac{L}{h_1 - h_2}\int_{h_2}^{h_1} \tau_s\,\mathrm{d}h$$

$$= \frac{\mu L U}{h_1 - h_2}\int_{h_2}^{h_1}\left(-\frac{2}{h} + \frac{6}{n+1}\cdot\frac{h_1}{h^2}\right)\mathrm{d}h$$

$$= \frac{\mu U L}{h_1 - h_2}\left(-2\ln h - \frac{6}{n+1}\cdot\frac{h_1}{h}\right)_{h_2}^{h_1}$$

$$= \frac{\mu U L}{h_2(n-1)}\left[-2\ln n - \frac{6h_1}{n+1}\left(\frac{1}{h_1} - \frac{1}{h_2}\right)\right]$$

$$= \frac{\mu UL}{h_2(n-1)}\left[-2\ln n + \frac{6(n-1)}{n+1}\right]$$

Also, load $$P = \int_0^L (p - p_0)\,\frac{\mathrm{d}x\cos\alpha}{\cos\alpha}$$

(neglecting contribution of τ_s to load; α is small)

Therefore, $$F_s = \int_0^L \tau_s\,\mathrm{d}x + \tan\alpha\,(P)$$

or $$F_s = \frac{\mu UL}{h_2(n-1)}\left(-2\ln n + \frac{6(n-1)}{n+1}\right) + \left[\frac{h_1 - h_2}{L}\right]\times \frac{6\mu UL^2}{h_2^2(n-1)^2}\left(\ln n - \frac{2(n-1)}{n+1}\right)$$

$$= \frac{\mu UL}{h_2(n-1)}\left(4\ln n - \frac{6(n-1)}{n+1}\right)$$

Now the force on the guide is

$$F_G = \int_0^L \tau_G \, dx$$

But
$$\tau_G = \mu \left(\frac{\partial u}{\partial y} \right)_{y=0}$$

$$= \mu U \left[\frac{1}{h} + \frac{3}{h} \left(1 - \frac{2}{n+1} \cdot \frac{h_1}{h} \right) \right]$$

The expression is
$$\tau_G = \mu U \left[\frac{4}{h} - \frac{6}{n+1} \cdot \frac{h_1}{h^2} \right]$$

Therefore,
$$F_G = \int_0^L \tau_G dx$$

$$= \frac{L}{h_1 - h_2} \int_{h_2}^{h_1} \tau_G \, dh \qquad \text{(as before)}$$

$$= \frac{\mu UL}{h_2 (n-1)} \int_{h_2}^{h_1} \left(\frac{4}{h} - \frac{6}{n+1} \cdot \frac{h_1}{h^2} \right) dh$$

$$= \frac{\mu UL}{h_2 (n-1)} \left[4 \ln n + \frac{6h_1}{n+1} \left(\frac{1}{h_1} - \frac{1}{h_2} \right) \right]$$

$$= \frac{\mu UL}{h_2 (n-1)} \left[4 \ln n - \frac{6(n-1)}{n+1} \right]$$

which is the same as F_s.

8.4.2 Low Reynolds Number Flow around a Sphere

In 1851 Stokes obtained the solution for the slow motion of a viscous fluid past a sphere. In his analysis, Stokes neglected the inertia terms of Navier–Stokes equations. Such flows are called creeping flows. The flow patterns are identical in all planes parallel to the mainstream and passing through the centre of the sphere.

Consider the low Reynolds number flow around a sphere of radius R placed in a uniform stream, U_∞ (Fig. 8.23). For fluids with high viscosity or for fluids flowing at very slow speeds (Re < 1) the inertia terms on the left-hand side of Eq. (8.25) can be neglected when compared with the viscous terms. The body force is neglected and steady flow is considered. The Navier-Stokes equations, Eq. (8.25) and the continuity equation, reduce respectively, to

$$\nabla p = \mu \nabla^2 \vec{u} \tag{8.85}$$

$$\nabla \cdot \vec{u} = 0 \tag{8.86}$$

The divergence of Eq. (8.85) yields

$$\nabla^2 p = \mu \nabla \cdot \left(\nabla^2\right)\vec{u} = \mu \nabla^2 (\nabla \cdot \vec{u}) = 0 \tag{8.87}$$

Equation (8.87) depicts that the pressure p, satisfies the Laplace equation. Therefore for very slow motions or for the motion of highly viscous fluids pressure is a harmonic function.

Equations (8.85), (8.86) and (8.87) may by deployed to describe a steady uniform flow around a sphere at rest (Fig. 8.23). This problem was first solved by Stokes and is often referred to as Stokes' problem. Let the origin of the coordinate system be chosen at the centre of the sphere, the z axis be the direction of uniform flow far away from the sphere, and $r = \left(x^2 + y^2 + z^2\right)^{1/2}$ be the distance of an arbitrary point from the origin. The boundary conditions which must be satisfied by the flow are as follows:

$$\left.\begin{array}{llll} w = u = v = 0 & & & at\ r = R \\ w = U_{\infty}, & u = v = 0, & p = p_{\infty}, & at\ r = \infty \end{array}\right\} \tag{8.88}$$

where R is the radius of the sphere and U_{∞} the uniform velocity far away from the sphere.

From the potential flow theory, the solution of Eq. (8.87) satisfying the required properties is

$$p - p_{\infty} = -\frac{A \cos \phi}{r^2} = -\frac{A_z}{r^3} \tag{8.89}$$

where A is a constant to be determined from the boundary conditions (Eq.(8.88)).

Substituting Eq.(8.89) into Eq. (8.85) one obtains

$$\left.\begin{array}{l} \nabla^2 w = \dfrac{A}{\mu}\left(\dfrac{3z^2}{r^5} - \dfrac{1}{r^3}\right) \\ \nabla^2 v = \dfrac{A}{\mu}\dfrac{3zy}{r^5} \\ \nabla^2 u = \dfrac{A}{\mu}\dfrac{3zx}{r^5} \end{array}\right\} \tag{8.90}$$

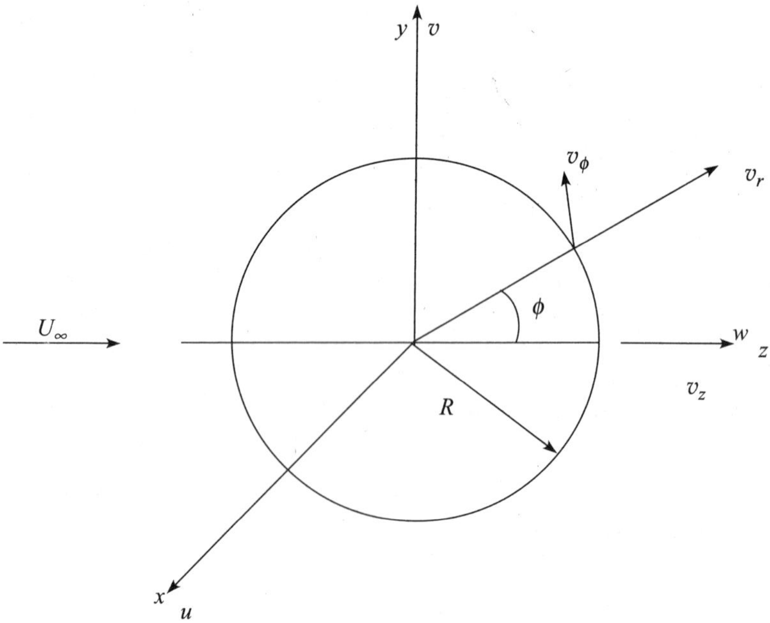

Fig. 8.23

The solutions of Eqs. (8.90) are beyond the scope of this text. The results of the solutions are

$$\left.\begin{aligned} w &= U_\infty\left[\frac{3}{4}\frac{Rz^2}{r^3}\left(\frac{R^2}{r^2}-1\right)+1-\frac{3}{4}\frac{R}{r}-\frac{1}{4}\frac{R^3}{r^3}\right] \\ v &= U_\infty\left[\frac{3}{4}\frac{Rzy}{r^3}\left(\frac{R^2}{r^2}-1\right)\right] \\ u &= U_\infty\left[\frac{3}{4}\frac{Rzx}{r^3}\left(\frac{R^2}{r^2}-1\right)\right] \end{aligned}\right\} \tag{8.91}$$

and the pressure is

$$p-p_\infty = -\frac{3}{2}\mu U_\infty R\left(\frac{z}{r^3}\right) \tag{8.92}$$

It can be readily verified that the velocity components in Eq. (8.91) satisfy Eq. (8.90) and the boundary conditions in Eq. (8.88). The pressure at the leading and trailing stagnation points of the sphere $(z = \mp R)$ are, respectively,

$$p_{\mp R} - p = \pm\frac{3}{2}\frac{\mu U_\infty}{R}$$

This indicates that the fluid exerts a force in the z direction. The total force (drag) acting on the sphere moving relative to an undisturbed stream along the z axis is equal to the surface integral of the normal pressure and the tangential shearing stresses acting on it. The normal pressure acting on the surface of the sphere is

$$(\tau_{rr})_{r=R} = \left[-(p-p_\infty)+2\mu\frac{\partial v_r}{\partial r}\right] \tag{8.93}$$

and the corresponding shearing stress is

$$\left(\tau_{r\phi}\right)_{r=R} = \mu\left[r\frac{\partial}{\partial r}\left(\frac{v_\phi}{r}\right)+\frac{1}{r}\frac{\partial v_r}{\partial \phi}\right]_{r=R} \tag{8.94}$$

The radial and tangential components of velocity v_r and v_ϕ respectively, can be calculated in terms of u, v, w, as follows:

$$\begin{aligned} v_r &= \left(u\frac{x}{\Gamma}+v\frac{y}{\Gamma}\right)\sin\phi + w\cos\phi \\ &= u\frac{x}{r}+v\frac{y}{r}+w\frac{z}{r} \end{aligned} \tag{8.95}$$

$$\begin{aligned} v_\phi &= -w\sin\phi+\left(u\frac{x}{\Gamma}+v\frac{y}{\Gamma}\right)\cos\phi \\ &= -w\frac{\Gamma}{r}+u\frac{xz}{r}+v\frac{yz}{r} \end{aligned} \tag{8.96}$$

where $\Gamma^2 = x^2 + y^2$. Substituting Eqs (8.91) into Eqs (8.95) and (8.96), we have

$$v_r = U_\infty \cos\phi\left(1-\frac{3}{2}\frac{R}{r}+\frac{1}{2}\frac{R^3}{r^3}\right) \tag{8.97}$$

$$v_\phi = -U_\infty \sin\phi\left(1-\frac{3R}{4r}-\frac{R^3}{4r^3}\right) \tag{8.98}$$

With the values of radial and tangential components of velocity in Eqs (8.85) and (8.86), the normal and shearing stresses acting on the surface of the sphere become [Eqs (8.93) to (8.94)]

$$(\tau_{r\phi})_{r=R} = -(p-p_\infty) = \frac{3}{2}\frac{\mu U_\infty}{R}\cos\phi \tag{8.99a}$$

and

$$(\tau_{r\phi})_{r=R} = -\frac{3}{2}\frac{\mu U_\infty}{R}\sin\phi \tag{8.99b}$$

The total drag on the sphere now can be calculated by integrating the z component of the normal and shearing stresses over the surface of the sphere as follows:

$$D = \iint(p-p_\infty)\cos\phi \mathrm{dA} + \iint \tau_{r\phi}\sin\phi\, dA$$

$$= 2\pi R^2 \int_{\pi}^{0} \left(-\frac{3}{2} \frac{\mu U_\infty}{R} \cos\phi \right) \cos\phi \sin\phi \, d\phi$$

$$+2\pi R^2 \int_{\pi}^{0} \left(-\frac{3}{2} \mu \frac{U_\infty}{R} \sin\phi \right) \sin^2 \phi \, d\phi \tag{8.100}$$

The expression given in Eq. (8.100) is known as the Stokes' formula for the drag of a sphere in which one-third of the values arises from the normal pressure forces and two-thirds from frictional forces. The coefficient of drag of the sphere at small Reynolds numbers is

$$C_D = \frac{D}{\left(\frac{\rho}{2}\right) U_\infty^2 \pi R^2} = \frac{24}{U_\infty (2R)/\upsilon} = \frac{24}{\text{Re}} \tag{8.101}$$

A comparison between Stokes' drag coefficient in Eq. (8.101) and experimental results is shown in Fig. 8.24. The approximate solution due to Stokes' is valid only for Re<1.

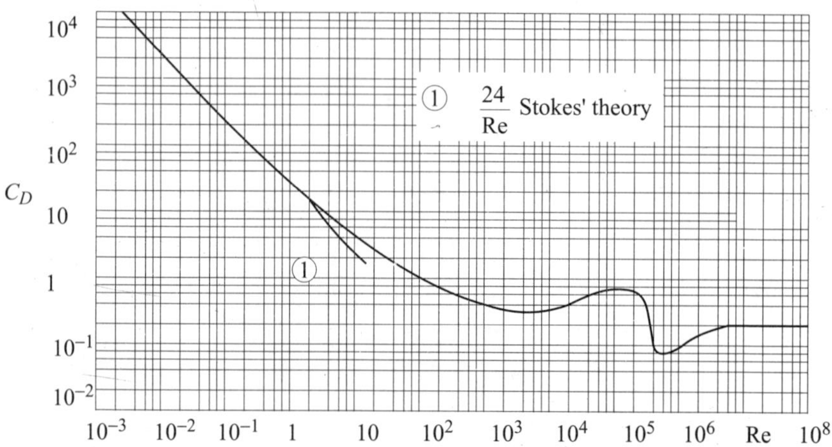

Fig. 8.24 Comparison between Stokes' drag coefficient and experimental drag coefficient

An important application of Stokes' law is the determination of viscosity of a viscous fluid by measuring the terminal velocity of a falling sphere. In this device, a sphere is dropped in a transparent cylinder containing the fluid under test. If the specific weight of the sphere is close to that of the liquid, the sphere will approach a small constant speed after being released in the fluid. Now we can apply Stokes' law for steady creeping flow around a sphere where the drag force on the sphere is given by Eq. (8.100).

With the sphere, falling at a constant speed, the acceleration is zero. This signifies that the falling body has attained terminal velocity and we can say that the sum of the buoyant force and drag force is equal to the weight of the body.

$$\frac{4}{3}\pi R^3 \rho_s g = \frac{4}{3}\pi R^3 \rho_l g + 6\pi\mu V_T R \tag{8.102}$$

where ρ_s is the density of the sphere, ρ_l is density of the liquid and V_T is the terminal velocity.

Solving for μ, we get

$$\mu = \frac{2}{9}\frac{gR^2}{V_T}(\rho_s - \rho_l) \tag{8.103}$$

The terminal velocity V_T, can be measured by observing the time for the sphere to cross a know distance betwwwn two points after its acceleration has ceased.

SUMMARY

- The Navier–Stokes equation, based on the conservation of momentum has been derived, which is valid for homogeneous and isotropic Newtonian and Stokesian fluids.
- The Navier–Stokes equation is often not amenable to an analytical solution due to the presence of non-linear inertia terms in it. However, there are some special situations where the non-linear inertia terms are reduced to zero. In such situations, exact solutions of the Navier–Stokes equation are obtainable. This includes the plane Poiseuille flow, the Couette flow, the flow through a straight pipe and the flow between two concentric rotating cylinders. All these flows are known as parallel flows where only one component of the velocity is non-trivial.
- Knowledge of the velocity field obtained through analytical methods permits calculation of shear stress, pressure drop and flow rate. Applications of the parallel flow theory to the measurement of viscosity and hydrodynamics of bearing lubrication are explained.

EXERCISES

8.1 Choose the correct answer:

(i) Bulk stress is equal to thermodynamic pressure

(a) if the second coefficient of viscosity is zero
(b) for incompressible flows
(c) for a compressible fluid with negligible second coefficient of viscosity
(d) if the bulk coefficient of viscosity is non-zero

(ii) Assumptions made in derivation of Navier–Stokes equations are

(a) continuum, incompressible flow, Newtonian fluid and μ = constant
(b) steady flow, incompressible flow, irrotational flow
(c) continuum, non-Newtonian fluid, incompressible flow
(d) continuum, Newtonian fluid, Stokes' hypothesis and isotropy

(iii) In a fully developed pipe flow
   (a) The pressure gradient is greater than the wall shear stress
   (b) The inertia force balances the wall shear stress
   (c) The pressure gradient balances the wall shear stress only and has a constant value
   (d) None of the above

(iv) In the case of fully developed flow through tubes
   (a) Darcy's friction factor is four times the skin friction coefficient
   (b) Darcy's friction factor and skin friction coefficients are same
   (c) Darcy's friction factor is double the skin friction coefficient
   (d) the skin friction coefficient is greater than the Darcy's friction factor.

(v) Based on the hydrodynamic theory of lubrication, state which of the following are correct.
   (a) The load bearing capacity remains unchanged so long either the slipper or the bearing moves in the same direction while the other is held fixed.
   (b) Reversing the direction of the movement of the slipper, bearing remaining fixed, does not cause any change in the load bearing capacity.
   (c) $u\frac{\partial u}{\partial x} >> \mu \frac{\partial^2 u}{\partial y^2}$
   (d) For a large film thickness, $h(x)$, the maximum pressure location shifts from the middle.

(vi) Observation on a spherical object falling in a liquid pool is the method of measuring viscosity by making use of Stokes' viscosity law. The falling body attains terminal velocity if
   (a) the weight of the falling body is more than the sum of the buoyancy force and the drag force
   (b) the drag force is equal to the buoyancy force
   (c) the buoyancy force is more than the drag force
   (d) the sum of the buoyancy force and the drag force is equal to the weight of the body.

8.2 (i) What is the basic difference between Euler's equations of motion and Navier–Stokes equations?

(ii) In case of flow through a straight tube of circular cross section with rotational symmetry, the axial component of velocity is the only non-trivial component and all the fluid particles move in the same direction only. Find out the average velocity and the maximum velocity within the tube. If Darcy–Weisbach equation for pressure drop over a finite length is given by $h_f = f(L/D)(V^2/2g)$, prove that $f = 64/\text{Re}$, where L is the length and D is the diameter of the tube.

8.3 What is the relationship between the average velocity and maximum velocity in case of parallel flow between two fixed parallel plates? What do you understand by inlet region and developed region?

Ans. ($U_{max} = 1.5\, U_{av}$)

8.4 Show that in case of a Couette flow, the shear stress at the horizontal mid-plane of the channel is independent of the pressure gradient imposed on the flow.

8.5 (i) Find out the total load and the frictional resistance on a block moving with a velocity U, over a horizontal plate separated by a thin layer of lubricating oil, the thickness of layer being h_1 and h_2 at the edges of the block which has a straight bottom.

(ii) Also show that the volume flow rate of lubricant is given by

$$Q = U\frac{h_1 h_2}{h_1 + h_2}$$

8.6 Oil flows between two parallel plates, one of which is at rest and the other moves with a velocity U. (i) If the pressure is decreasing in the direction of flow at the rate of 5 Pa/m, the dynamic viscosity is 0.05 kg/ms, the spacing of the horizontal plate is 0.04 m and the volumetric flow Q per unit width is 0.02 m^2/s, what is the velocity U? (ii) Calculate U if the pressure is increasing at a rate of 5 Pa/m in the direction of flow.

Ans. ((i) 0.97 m/s (ii) = 1.027 m/s)

8.7 Water flows between two very large, horizontal, parallel flat plates 20 mm apart. If the average velocity of water is 0.15 m/s, what is the shear stress (i) at the lower plate, and (ii) 5 mm and 10 mm above the lower plate? Assume $\mu = 1.1 \times 10^{-3}$ Ns/m^2.

Ans. ((i) 0.0495 N/m^2 (ii) 0.0248 N/m^2)

8.8 A Newtonian liquid flows slowly under gravity along an inclined flat surface that makes an angle θ with the horizontal plane. The film thickness is T and it is constant. The flow is two-dimensional. (i) Show that the fluid velocity u along the x (flow) direction is given by

$$u = \frac{g \sin \theta}{\nu} y\left(T - \frac{y}{2}\right)$$

(ii) Calculate the average velocity u_{av}, and the volumetric flow rate Q per unit width of the surface. The pressure within the fluid is a function of y alone, where y is the normal to the flow direction. The v component of velocity is trivial.

Ans. ($U_{av} = g \sin \theta\, T^2/3\nu$, $Q = g \sin \theta\, T^3/3\nu$)

8.9 A horizontal circular pipe of outer radius R_1, is placed concentrically inside another circular pipe of inner radius R_2. Considering fully developed laminar flow in the annular space between pipes show that the maximum velocity occurs at a radius R_0 given by

$$R_0 = \left[\frac{R_2^2 - R_1^2}{2 \ln (R_2/R_1)}\right]^{1/2}$$

8.10 The Reynolds number for flow of oil through a 5 cm diameter pipe is 1700. The kinematic viscosity, $\nu = 1.02 \times 10^{-6}$ m^2/s. What is the velocity at a point 0.625 cm away from the wall.

Ans. (0.03 m/s)

8.11 The velocity along the centre line of the Hagen–Poiseuille flow in a 0.1 m diameter pipe is 2 m/s. If the viscosity of the fluid is 0.07 kg/ms and its specific gravity is 0.92, calculate (i) the volumetric flow rate, (ii) shear stress of the fluid at the pipe wall, (iii) local skin friction coefficient, and (iv) the Darcy friction coefficient.

Ans. ((i) 7.854×10^{-3} m^3/s (ii) 5.6 N/m^2, (iii) 0.012 (iii) 0.048)

8.12 Kerosene at 10 °C flows steadily at 20 l/min through a 150 m long horizontal length of 5.5 cm diameter cast iron pipe. Compare the pressure drop of the kerosene flow with that of the same flow rate of benzene at 10 °C through the same pipe. For kerosene at 10 °C, $\rho = 820$ kg/m^3 and $\mu = 0.0025$ Ns/m^2 and for benzene $\rho = 899$ kg/m^3 and $\mu = 0.0008$ Ns/m^2. Why do you obtain greater pressure drop for benzene?

8.13 A viscous oil flows steadily between parallel plates. The fully developed velocity profile is given by

$$u = -\frac{h^2}{8\mu}\left(\frac{\partial p}{\partial x}\right)\left[1-\left(\frac{2y}{h}\right)^2\right]$$

where the total gap between the plates is $h = 3$ mm and y is the distance from the centre line. The viscosity of the oil is 0.5 Ns/m^2 and the pressure gradient is – 1200 N/m^2/m. Find the magnitude and direction of the shear stress on the upper plate, and the volumetric flow rate per metre width of the channel.

Ans. ((i) –1.80 N/m^2, (ii) 5.40×10^{-6} m^3/s m)

8.14 Consider steady, laminar, incompressible and fully developed flow of a viscous fluid parallel to the axis in the annular space between two coaxial cylinders having radii r_1 and r_2 respectively. A constant pressure gradient dp/dz drives the flow. Show that the volumetric flow rate is given by

$$Q = \frac{\pi r_1^4}{8\mu}\left(-\frac{dp}{dz}\right)\left[\left(\eta^4 - 1\right) - \frac{\left(\eta^2-1\right)^2}{\ln \eta}\right]$$

$$\left.\begin{aligned} u_{max} &= \frac{U(1+P)^2}{4P} \quad \text{for } P \geq 1 \\ u_{min} &= \frac{U(1+P)^2}{4P} \quad \text{for } P \leq 1 \end{aligned}\right\} \quad (8.42b)$$

where $\eta = r_2/r_1$

8.15 A fully developed laminar flow is taking place in the annulus between two concentric pipes. The inner pipe is stationary, and the outer pipe is moving in the axial direction with a velocity V_0. Assume the axial pressure gradient to be zero ($dp/dz = 0$). Find out a general expression for the shear stress as a function of radial coordinate. Also find out a general expression for the velocity profile $V_z(r)$.

Ans. $\left(\text{(i) } \tau = A/r, \text{ (ii) } V_z = V_0 \dfrac{\ln (r/r_i)}{\ln (r_o/r_i)}\right)$

8.16 Refer to Fig. 8.25. This problem illustrates the secret of the strength of cello-tape joints:

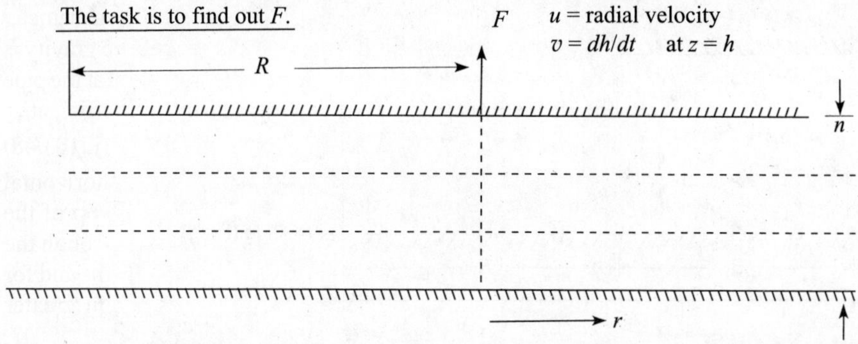

A disc of radius r is at a uniform distance h from a large flat plate, and the gap h is filled with a viscous liquid. The disc moves upwards at $v = dh/dt$, in the response to a central force F. Due to the symmetry of the problem (in the θ direction), the governing equations are

Continuity: $$\frac{1}{r}\frac{\partial}{\partial r}(ru) + \frac{\partial v}{\partial z} = 0 \qquad u = \textit{radial velocity}$$

Momentum
$$\begin{cases} u\dfrac{\partial u}{\partial r} + v\dfrac{\partial u}{\partial z} = -\dfrac{1}{\rho}\dfrac{\partial p}{\partial r} + \nu\left(\dfrac{\partial^2 u}{\partial r^2} + \dfrac{1}{r}\dfrac{\partial u}{\partial r} - \dfrac{u}{r^2} + \dfrac{\partial^2 u}{\partial z^2}\right) \\ u\dfrac{\partial v}{\partial r} + \nu\dfrac{\partial v}{\partial z} = -\dfrac{1}{\rho}\dfrac{\partial p}{\partial z} + \nu\left(\dfrac{\partial^2 v}{\partial r^2} + \dfrac{\partial^2 v}{\partial z^2}\right) \end{cases}$$

Hints:

(a) Show that if $h << R$ and $\frac{h}{\nu}\frac{dh}{dt} << 1$, inertia terms can be neglected in the momentum equations and we get the following:
Continuity equation remains unchanged
$$\frac{\partial p}{\partial r} = \mu\frac{\partial^2 u}{\partial z^2}; \frac{\partial p}{\partial z} = 0$$

(b) Solve these differential equations to get
$$F = \frac{3}{2}\pi\mu\frac{R^4}{h^3}\frac{dh}{dt}$$

(c) Estimate F if $R = 1$ cm, $\frac{dh}{dt} = 1$ mm/s, $h = 0.1$ mm, $\mu = 4.2 \times 10^{-6}$ N–cm^{-2}–s

8.17 The gap between the slider and guide of a step bearing as shown in Fig. 8.26 is uniform ($= h_1$) over length B_1 and also uniform ($= h_2 \neq h_1$)) over length B_2. Find the load that the bearing can carry.

Ans. ($W = 3\mu U B_1 B_2 (h_1 - h_2)(B_1 + B_2) / (B_1 h_2^3 + B_2 h_1^3)$)

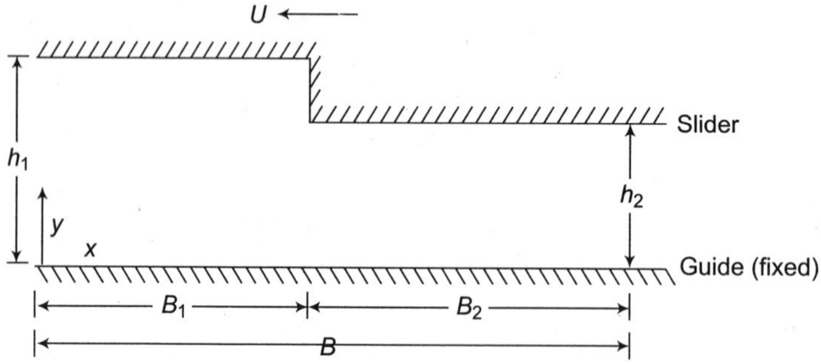

Fig. 8.26

8.18 Derive an expression for the terminal velocity of a sphere for Reynolds number (based on the diameter) much less than 1. Using the expression, find the settlement speed of blood cells in plasma, considering a radius of 3×10^{-3} mm for the cell, a difference of densities between cell and the plasma of 0.07×10^{-6} kg/mm^3 and a plasma viscosity of 1.27×10^{-3} Ns/m^2

Ans. (3.89 mm/hr)

9

LAMINAR BOUNDARY LAYERS

9.1 INTRODUCTION

The boundary layer of a flowing fluid is the thin layer close to the wall. In a flow field, viscous stresses are very prominent within this layer. Although the layer is thin, it is very important to know the details of flow within it. The main-flow velocity within this layer tends to zero while approaching the wall. Also the gradient of this velocity component in a direction normal to the surface is large as compared to the gradient of this component in the streamwise direction.

9.2 BOUNDARY LAYER EQUATIONS

In 1904, Ludwig Prandtl, the well-known German scientist, introduced the concept of boundary layer [1] and derived the equations for boundary layer flow by correct reduction of the Navier–Stokes equations. He hypothesised that for fluids having a relatively small viscosity, the effect of internal fictitious in the fluid is significant only in a narrow region surrounding the solid boundaries or bodies over which the fluid flows. Thus, close to the body is the boundary layer where shear stresses exert an increasingly larger effect on the fluid as one moves from free stream towards the solid boundary. However, outside the boundary layer where the effect of the shear stresses on the flow is small compared to values inside the boundary layer (since the velocity gradient $\partial u/\partial y$ is negligible), the fluid particles experience no vorticity, and therefore, the flow is similar to a potential flow. Hence, the *surface* at the boundary layer interface is a rather fictitious one dividing rotational and irrotational flow. Prandtl's model regarding the boundary layer flow is shown in Fig. 9.1. Hence with the exception of the immediate vicinity of the surface, the flow is frictionless (inviscid) and the velocity is U. In the region very near to the surface (in the thin layer), there is friction in the flow which signifies that the fluid is retarded until it adheres to the surface. The transition of the mainstream velocity from zero at the surface to full magnitude takes place across the boundary layer. Its thickness is δ which is a function of the coordinate direction x. The thickness is considered to be very small compared to the characteristic length L of the domain. In the normal direction, within the thin layer, the gradient $\partial u/\partial y$ is very large compared to the gradient in the flow direction $\partial u/\partial x$. The next step is to simplify the Navier–Stokes equations for steady two-dimensional laminar incompressible flows. Considering the Navier–Stokes equations together with the equation of continuity, the following dimensional form is obtained:

$$u\frac{\partial u}{\partial x}+v\frac{\partial u}{\partial y} = -\frac{1}{\rho}\frac{\partial p}{\partial x}+\frac{\mu}{\rho}\left[\frac{\partial^2 u}{\partial x^2}+\frac{\partial^2 u}{\partial y^2}\right] \tag{9.1}$$

$$u\frac{\partial v}{\partial x}+v\frac{\partial v}{\partial y} = -\frac{1}{\rho}\frac{\partial p}{\partial y}+\frac{\mu}{\rho}\left[\frac{\partial^2 v}{\partial x^2}+\frac{\partial^2 v}{\partial y^2}\right] \tag{9.2}$$

$$\frac{\partial u}{\partial x}+\frac{\partial v}{\partial y}=0 \tag{9.3}$$

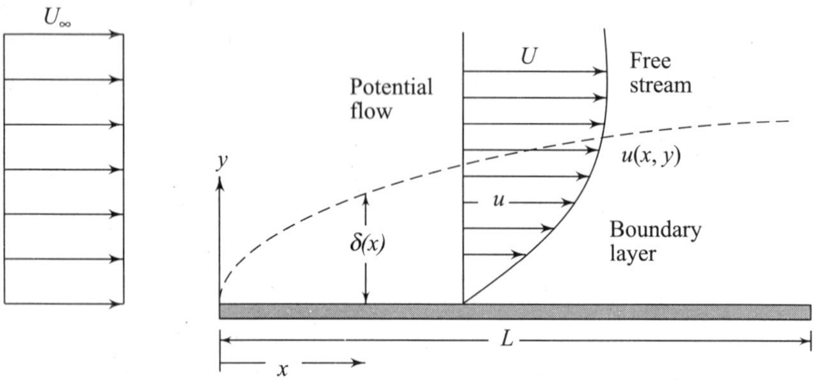

Fig. 9.1 Boundary layer on a flat plate

Here the velocity components u and v are acting along the streamwise x and normal y directions respectively. The static pressure is p, while ρ is the density and μ is the dynamic viscosity of the fluid.

The equations are now non-dimensionalised. The length and the velocity scales are chosen as L and U_∞ respectively. The non-dimensional variables are

$$u^* = \frac{u}{U_\infty},\ v^* = \frac{v}{U_\infty},\ p^* = \frac{p}{\rho U_\infty^2}$$

$$x^* = \frac{x}{L},\ y^* = \frac{y}{L}$$

where U_∞ is the dimensional free stream velocity and the pressure is non-dimensionalised by twice the dynamic pressure $p_d = (1/2)\ \rho\ U_\infty^2$. Using these non-dimensional variables, Eqs (9.1) to (9.3) become

$$u^*\frac{\partial u^*}{\partial x^*}+v^*\frac{\partial u^*}{\partial y^*} = -\frac{\partial p^*}{\partial x^*}+\frac{1}{\text{Re}}\left[\frac{\partial^2 u^*}{\partial x^{*2}}+\frac{\partial^2 u^*}{\partial y^{*2}}\right] \tag{9.4}$$

$$u^*\frac{\partial v^*}{\partial x^*}+v^*\frac{\partial v^*}{\partial y^*} = -\frac{\partial p^*}{\partial y^*}+\frac{1}{\text{Re}}\left[\frac{\partial^2 v^*}{\partial x^{*2}}+\frac{\partial^2 v^*}{\partial y^{*2}}\right] \tag{9.5}$$

$$\frac{\partial u^*}{\partial x^*} + \frac{\partial v^*}{\partial y^*} = 0 \tag{9.6}$$

where the Reynolds number,

$$\text{Re} = \frac{\rho U_\infty L}{\mu}$$

Let us now examine what happens to the u velocity as we go across the boundary layer. At the wall the u velocity is zero. The value of u on the inviscid side, i.e., on the free stream side beyond the boundary layer. is U. For the case of external flow over a flat plate, this U is equal to U_∞.

Based on the above, we can identify the following scales for the boundary layer variables:

Variable	*Dimensional scale*	*Non-dimensional scale*
u	U_∞	1
x	L	1
y	δ	$\varepsilon\,(=\delta/L)$

The symbol ε describes a value much smaller than 1. Now let us look at the order of magnitude of each individual terms involved in Eqs (9.4), (9.5) and (9.6). We start with the continuity Eq. (9.6). One general rule of incompressible fluid mechanics is that we are not allowed to drop any term from the continuity equation. From the scales of boundary layer variables, the derivative $\partial u^*/\partial x^*$ is of the order 1. The second term in the continuity equation $\partial v^*/\partial y^*$ should also be of the order 1. Now, what makes $\partial v^*/\partial y^*$ to have the order 1? Admittedly v^* has to be of the order ε because y^* becomes ε ($= \delta/L$) at its maximum. Next, consider Eq. (9.4). Inertia terms are of the order 1. Among the second order derivatives, $\partial^2 u^*/\partial x^{*2}$ is of the order 1 and $\partial^2 u^*/\partial y^{*2}$ contains a large estimate of $(1/\varepsilon^2)$. However, after multiplication with 1/Re, the sum of these two second order derivatives should produce at least one term which is of the same order of magnitude as the inertia terms. This is possible only if the Reynolds number (Re) is of the order of $1/\varepsilon^2$.

The order of Reynolds number can be determined in the following manner. While deriving the boundary layer equation, the basic assumption is that outside the boundary layer, the inertia force is >> viscous force. However, within the boundary layer, the inertia force and viscous forces are comparable. If inertia force and the viscous force components can be represented by $\rho u \dfrac{\partial u}{\partial x}$ and $\mu \dfrac{\partial^2 u}{\partial y^2}$ respectively, we can say that within the boundary layer

$$\rho u \frac{\partial u}{\partial x} \sim \mu \frac{\partial^2 u}{\partial y^2}$$

or

$$\rho U \frac{U}{L} \sim \mu \frac{U}{\delta^2}$$

or

$$\frac{\rho UL}{\mu} \sim \frac{L^2}{\delta^2}$$

Finally, it can be conclusively said that

$$\text{Re} \sim \frac{1}{\varepsilon^2}$$

It follows from the Eq. (9.4) that $-\partial p^*/\partial x^*$ will not exceed the order of 1 so as to be in balance with the remaining terms. Finally, Eqs (9.4), (9.5) and (9.6) can be rewritten as

$$u^*\frac{\partial u^*}{\partial x^*} + v^*\frac{\partial u^*}{\partial y} = -\frac{\partial p^*}{\partial x^*} + \frac{1}{\text{Re}}\left[\frac{\partial^2 u^*}{\partial x^{*2}} + \frac{\partial^2 u^*}{\partial y^{*2}}\right] \tag{9.4}$$

$$(1)\frac{(1)}{(1)} \quad (\varepsilon)\frac{(1)}{(\varepsilon)} \quad = \quad (1) \quad (\varepsilon^2)\left[\frac{(1)}{(1)} + \frac{1}{(\varepsilon^2)}\right]$$

$$u^*\frac{\partial v^*}{\partial x^*} + v^*\frac{\partial v^*}{\partial y^*} = -\frac{\partial p^*}{\partial y^*} + \frac{1}{\text{Re}}\left[\frac{\partial^2 v^*}{\partial x^{*2}} + \frac{\partial^2 v^*}{\partial y^{*2}}\right] \tag{9.5}$$

$$(1)\frac{(\varepsilon)}{(1)} \quad (\varepsilon)\frac{(\varepsilon)}{(\varepsilon)} \quad = \quad (?) \quad (\varepsilon^2)\left[\frac{(\varepsilon)}{(1)} + \frac{\varepsilon}{(\varepsilon^2)}\right]$$

$$\frac{\partial u^*}{\partial x^*} + \frac{\partial v^*}{\partial y^*} = 0 \tag{9.6}$$

$$\frac{(1)}{(1)} \quad \frac{(\varepsilon)}{(\varepsilon)}$$

As a consequence of the order of magnitude analysis $\frac{\partial^2 u^*}{\partial x^{*2}}$ can be dropped from the x direction momentum equation, because on multiplication with 1/Re it assumes the smallest order of magnitude. Now consider the y direction momentum Eq. (9.5). All the terms of this equation are of a smaller magnitude than those of Eq. (9.4). This equation can only be balanced if $\partial p^*/\partial y^*$ is of the same order of magnitude as other terms. Thus the y momentum equation reduces to

$$\frac{\partial p^*}{\partial y^*} = O(\varepsilon) \tag{9.7}$$

This means that the pressure across the boundary layer does not change. The pressure is impressed on the boundary layer, and its value is determined by hydrodynamic considerations. This also implies that the pressure p is only a function of x. The pressure forces on a body are solely determined by the inviscid flow outside the boundary layer. The application of Eq. (9.4) at the outer edge of boundary layer gives

$$u^* \frac{du^*}{dx^*} = -\frac{dp^*}{dx^*} \tag{9.8a}$$

In dimensional form, this can be written as

$$U \frac{dU}{dx} = -\frac{1}{\rho}\frac{dp}{dx} \tag{9.8b}$$

On integrating Eq. (9.8b), the well- known Bernoulli's equation is obtained,

$$p + \frac{1}{2}\rho U^2 = \text{a constant} \tag{9.9}$$

Finally, it can be said that by the order of magnitude analysis, the Navier–Stokes equations are simplified into equations given below.

$$u^* \frac{\partial u^*}{\partial x^*} + v^* \frac{\partial u^*}{\partial y^*} = -\frac{\partial p^*}{\partial x^*} + \frac{1}{\text{Re}}\frac{\partial^2 u^*}{\partial y^{*2}} \tag{9.10}$$

$$\frac{\partial p^*}{\partial y^*} = 0 \tag{9.11}$$

$$\frac{\partial u^*}{\partial x^*} + \frac{\partial v^*}{\partial y^*} = 0 \tag{9.12}$$

These are known as Prandtl's boundary layer equations. The available boundary conditions are

Solid surface

$$\left.\begin{array}{ll} & \text{at } y^* = 0, \quad u^* = 0 = v^* \\ \text{or} & \text{at } y = 0 \quad u = 0 = v \end{array}\right\} \tag{9.13}$$

Outer edge of boundary layer

$$\left.\begin{array}{ll} & \text{at } y^* = (\varepsilon) = \frac{\delta}{L}, \quad u^* = 1 \\ \text{or} & \text{at } y = \delta, u = U(x) \end{array}\right\} \tag{9.14}$$

The unknown pressure p in the x momentum equation can be determined from Bernoulli's Eq. (9.9), if the inviscid velocity distribution $U(x)$ is also known. The preceding derivations are related to a flat surface, but these can be easily extended to curved surfaces. While doing so, it is seen that Eqs (9.10) to (9.14) continue to be applicable only if the curvature does not change abruptly. However, the boundary layer equations are relatively easier to solve as compared to the Navier–Stokes equations and have been solved by various analytical and numerical techniques.

We solve the Prandtl boundary layer equations for $u^*(x, y)$ and $v^*(x, y)$ with U obtained from the outer inviscid flow analysis. The equations are solved by commencing at the leading edge of the body and moving downstream to the desired location. Note that the reduced momentum Eq. (9.10) is still non-linear. However, it does allow the no-slip boundary condition to be satisfied which constitutes a significant improvement over the potential flow analysis while solving real fluid flow problems. The *Prandtl boundary layer equations* are thus a simplification of the Navier–Stokes equations.

9.3 BLASIUS FLOW OVER A FLAT PLATE

The classical problem considered by H. Blasius was a two-dimensional, steady, incompressible flow over a flat plate at zero angle of incidence with respect to the uniform stream of velocity U_∞. The fluid extends to infinity in all directions from the plate. The physical problem is already illustrated in Fig. 9.1.

Blasius wanted to determine (a) the velocity field solely within the boundary layer, (b) the boundary layer thickness (δ), (c) the shear stress distribution on the plate, and (d) the drag force on the plate.

The Prandtl boundary layer equations in the case under consideration are

$$u\frac{\partial u}{\partial x} + v\frac{\partial u}{\partial y} = \nu\frac{\partial^2 u}{\partial y^2} \tag{9.15}$$

$$\frac{\partial u}{\partial x} + \frac{\partial v}{\partial y} = 0 \tag{9.3}$$

The boundary conditions are

$$\begin{aligned} &\text{at } y = 0, \quad u = v = 0 \\ &\text{at } y = \infty \quad u = U_\infty \end{aligned} \tag{9.16}$$

It may be mentioned that the substitution of the term $\left[-\frac{1}{\rho}\frac{\mathrm{d}p}{\mathrm{d}x}\right]$ in the original boundary layer momentum equation in terms of the free stream velocity produces $\left[U_\infty \frac{\mathrm{d}U_\infty}{\mathrm{d}x}\right]$ which is equal to zero. Hence the governing Eq. (9.15) does not contain any pressure-gradient term. However, the characteristic parameters of this problem are U_∞, ν, x, y, i.e.,

$$u = u\,(U_\infty, \nu, x, y)$$

Before we write down this relationship in terms of two non-dimensional parameters, we have to be acquainted with the *law of similarity* in boundary layer flows. It states that the u component of velocity with two velocity profiles of $u(x, y)$ at different x locations differ only by scale factors in u and y. Therefore, the velocity profiles $u(x, y)$ at all values of x can be made congruent if they are plotted in coordinates which have been made dimensionless with reference to the scale factors. The local free stream velocity $U(x)$ at section x is an obvious scale factor for u, because the dimensionless $u(x)$ varies between zero and unity with y at all sections. The scale factor for y, denoted by $g(x)$, is proportional to the local boundary layer thickness so that y itself varies between zero and unity. The principle of similarity demands that the velocity at two arbitrary x locations, namely, x_1 and x_2 should satisfy the equation

$$\frac{u[x_1, \{y/g(x_1)\}]}{U(x_1)} = \frac{u[x_2, \{y/g(x_2)\}]}{U(x_2)} \tag{9.17}$$

Now, for Blasius flow, it is possible to write

$$\frac{u}{U_\infty} = F\left(\frac{y}{\sqrt{\dfrac{\nu x}{U_\infty}}}\right) = F(\eta) \tag{9.18}$$

where $\eta \sim \dfrac{y}{\delta}$ and $\delta \sim \sqrt{\dfrac{\nu x}{U_\infty}}$

or more precisely,
$$\eta = \frac{y}{\sqrt{\dfrac{\nu x}{U_\infty}}} \tag{9.19}$$

The stream function can now be obtained in terms of the velocity components as

$$\psi = \int u \,\mathrm{d}y = \int U_\infty F(\eta) \sqrt{\frac{\nu x}{U_\infty}}\,\mathrm{d}\eta = \sqrt{U_\infty \nu x} \int F(\eta)\,\mathrm{d}\eta$$

or
$$\psi = \sqrt{U_\infty \nu x}\; f(\eta) + \text{constant} \tag{9.20}$$

where $\int F(\eta)\,\mathrm{d}\eta = f(\eta)$ and the constant of integration is zero if the stream function at the solid surface is set equal to zero.

Now, the velocity components and their derivatives are

$$u = \frac{\partial \psi}{\partial y} = \frac{\partial \psi}{\partial \eta}\cdot\frac{\partial \eta}{\partial y} = U_\infty f'(\eta) \tag{9.21a}$$

$$v = -\left(\frac{\partial \psi}{\partial x} + \frac{\partial \psi}{\partial \eta}\cdot\frac{\partial \eta}{\partial x}\right) = -\sqrt{U_\infty \nu}\left[\frac{1}{2}\cdot\frac{1}{\sqrt{x}} f(\eta) + \sqrt{x} f'(\eta)\left\{-\frac{1}{2}\frac{y}{\sqrt{\dfrac{\nu x}{U_\infty}}}\frac{1}{x}\right\}\right]$$

or
$$v = \frac{1}{2}\sqrt{\frac{\nu U_\infty}{x}}\,[\eta f'(\eta) - f(\eta)] \tag{9.21b}$$

$$\frac{\partial u}{\partial x} = U_\infty f''(\eta)\frac{\partial \eta}{\partial x} = U_\infty f''(\eta)\cdot\left[-\frac{1}{2}\cdot\frac{y}{\sqrt{\dfrac{\nu x}{U_\infty}}}\cdot\frac{1}{x}\right]$$

or
$$\frac{\partial u}{\partial x} = \frac{U_\infty}{2}\cdot\frac{\eta}{x}\cdot f''(\eta) \tag{9.21c}$$

$$\frac{\partial u}{\partial y} = U_\infty f''(\eta)\cdot\frac{\partial \eta}{\partial y} = U_\infty f''(\eta)\cdot\left[\frac{1}{\sqrt{\dfrac{\nu x}{U_\infty}}}\right]$$

or $$\frac{\partial u}{\partial y} = U_\infty \sqrt{\frac{U_\infty}{\nu x}}\, f''(\eta) \tag{9.21d}$$

$$\frac{\partial^2 u}{\partial y^2} = U_\infty \sqrt{\frac{U_\infty}{\nu x}}\, f'''(\eta) \left\{ \frac{1}{\sqrt{\frac{\nu x}{U_\infty}}} \right\}$$

or $$\frac{\partial^2 u}{\partial y^2} = \frac{U_\infty^2}{\nu x} f'''(\eta) \tag{9.21e}$$

Substituting (9.21) into (9.15), we have

$$-\frac{U_\infty^2}{2}\frac{\eta}{x}\cdot f'(\eta)\, f''(\eta) + \frac{U_\infty^2}{2x}[\eta f'(\eta) - f(\eta)]\, f''(\eta) = \frac{U_\infty^2}{x} f'''(\eta)$$

or $$-\frac{1}{2}\frac{U_\infty^2}{x} f(\eta)\, f''(\eta) = \frac{U_\infty^2}{x} f'''(\eta)$$

or $$2f'''(\eta) + f(\eta) f''(\eta) = 0 \tag{9.22}$$

This is known as Blasius equation. The boundary conditions as in Eq. (9.16), in combination with Eq. (9.21a) and (9.21b) become

$$\left.\begin{aligned} &\text{at } \eta = 0: \quad f(\eta) = 0, \quad f'(\eta) = 0 \\ &\text{at } \eta = \infty: \quad f'(\eta) = 1 \end{aligned}\right\} \tag{9.23}$$

Equation (9.22) is a third order non-linear differential equation. Blasius obtained the solution of this equation in the form of series expansion through analytical techniques which is beyond the scope of this text. However, we shall discuss a numerical technique to solve the aforesaid equation which can be understood rather easily.

It is to be observed that the equation for f does not contain x. Further boundary conditions at $x = 0$ and $y = \infty$ merge into the condition $\eta \rightarrow \infty, u/U_\infty = f' = 1$. This is the key feature of similarity solution.

We can rewrite Eq. (9.22) as three first order differential equations in the following way:

$$f' = G \tag{9.24a}$$

$$G' = H \tag{9.24b}$$

$$H' = -\frac{1}{2} f H \tag{9.24c}$$

Let us next consider the boundary conditions. The condition $f(0) = 0$ remains valid. Next, the condition $f'(0) = 0$ means that $G(0) = 0$. Finally, $f'(\infty) = 1$ gives us $G(\infty) = 1$. Note that the equations for f and G have initial values. However, the value for $H(0)$ is not known. Hence, we do not have a usual initial-value problem. Nevertheless, we handle this problem as an initial-value problem by choosing values of $H(0)$ and solving by numerical methods $f(\eta)$, $G(\eta)$, and $H(\eta)$. In general, the condition $G(\infty) = 1$ will not be satisfied for the function G arising from the numerical solution. We then choose other initial values of H so that eventually we find an $H(0)$ which results in $G(\infty) = 1$. This method is called the *shooting technique*.

In Eq. (9.24), the primes refer to differentiation w.r.t the similarity variable η. The integration steps following the Runge–Kutta method are given below:

$$f_{n+1} = f_n + \frac{1}{6}(k_1 + 2k_2 + 2k_3 + k_4) \tag{9.25a}$$

$$G_{n+1} = G_n + \frac{1}{6}(l_1 + 2l_2 + 2l_3 + l_4) \tag{9.25b}$$

$$H_{n+1} = H_n + \frac{1}{6}(m_1 + 2m_2 + 2m_3 + m_4) \tag{9.25c}$$

One moves from η_n to $\eta_{n+1} = \eta_n + h$. A fourth order accuracy is preserved if h is constant along the integration path, that is, $\eta_{n+1} - \eta_n = h$ for all values of n. The values of k, l and m are as follows:

For generality, let the system of governing equations be

$$f' = F_1(f, G, H, \eta),\ G' = F_2(f, G, H, \eta) \text{ and } H' = F_3(f, G, H, \eta).$$

Then,

$$k_1 = h\,F_1(f_n, G_n, H_n, \eta_n)$$
$$l_1 = h\,F_2(f_n, G_n, H_n, \eta_n)$$
$$m_1 = h\,F_3(f_n, G_n, H_n, \eta_n)$$

$$k_2 = h\,F_1\left\{\left(f_n + \frac{1}{2}k_1\right), \left(G_n + \frac{1}{2}l_1\right), \left(H_n + \frac{1}{2}m_1\right), \left(\eta_n + \frac{h}{2}\right)\right\}$$

$$l_2 = h\,F_2\left\{\left(f_n + \frac{1}{2}k_1\right), \left(G_n + \frac{1}{2}l_1\right), \left(H_n + \frac{1}{2}m_1\right), \left(\eta_n + \frac{h}{2}\right)\right\}$$

$$m_2 = h\,F_3\left\{\left(f_n + \frac{1}{2}k_1\right), \left(G_n + \frac{1}{2}l_1\right), \left(H_n + \frac{1}{2}m_1\right), \left(\eta_n + \frac{h}{2}\right)\right\}$$

In a similar way, k_3, l_3, m_3 and k_4, l_4, m_4 are calculated following standard formulae for the Runge–Kutta integration. For example, k_3 is given by

$$k_3 = hF_1\left\{\left(f_n + \frac{1}{2}k_2\right), \left(G_n + \frac{1}{2}l_2\right), \left(H_n + \frac{1}{2}m_2\right), \left(\eta_n + \frac{h}{2}\right)\right\}$$

The functions F_1, F_2 and F_3 are G, H, $-fH/2$, respectively. Then at a distance $\Delta\eta$ from the wall, we have

$$f(\Delta\eta) = f(0) + G(0)\,\Delta\eta \tag{9.26a}$$

$$G(\Delta\eta) = G(0) + H(0)\,\Delta\eta \tag{9.26b}$$

$$H(\Delta\eta) = H(0) + H'(0)\,\Delta\eta \tag{9.26c}$$

$$H'(\Delta\eta) = -\frac{1}{2}f(\Delta\eta)\,H(\Delta\eta) \tag{9.26d}$$

As mentioned earlier, $f''(0) = H(0) = \lambda$ is unknown. It must be determined such that the condition $f'(\infty) = G(\infty) = 1$ is satisfied. The condition at infinity is usually approximated at a finite value of η (around $\eta = 10$). The process of obtaining λ accurately involves iteration and may be calculated using the procedure described below.

For this purpose, consider Fig. 9.2(a) where the solutions of G versus η for two different values of $H(0)$ are plotted. The values of $G(\infty)$ are estimated from the G

curves and are plotted in Fig. 9.2(b). The value of $H(0)$ now can be calculated by finding the value $\tilde{H}(0)$ at which the line 1–2 crosses the line $G(\infty) = 1$. By using similar triangles, it can be said that

$$\frac{\tilde{H}(0) - H(0)_1}{1 - G(\infty)_1} = \frac{H(0)_2 - H(0)_1}{G(\infty)_2 - G(\infty)_1}$$

By solving this, we get $\tilde{H}(0)$. Next we repeat the same calculation as above by using $\tilde{H}(0)$ and the better of the two initial values of $H(0)$. Thus we get another improved value $\tilde{\tilde{H}}(0)$. This process may continue, that is, we use $\tilde{\tilde{H}}(0)$ and $\tilde{H}(0)$ as a pair of values to find more improved values for $H(0)$, and so forth. It should be always kept in mind that for each value of $H(0)$, the curve $G(\eta)$ versus η is to be examined to get the proper value of $G(\infty)$.

The functions $f(\eta), f'(\eta) = G$ and $f''(\eta) = H$ are plotted in Fig. 9.3. The velocity components, u and v inside the boundary layer can be computed from Eqs (9.21a) and (9.21b) respectively. Measurements to test the accuracy of theoretical results were carried out by many scientists. In his experiments, J. Nikuradse, found excellent agreement with the theoretical results with respect to velocity distribution (u/U_∞) within the boundary layer of a stream of air on a flat plate. However, some values of the velocity profile shape $f'(\eta) = u/U_\infty = G$ and $f''(\eta) = H$ are given in Table 9.1.

Table 9.1 Blasius Velocity Profile $G = u/U_\infty$ f and H after Schlichting [2]

η	f	G	H
0	0	0	0.33206
0.2	0.00664	0.06641	0.33199
0.4	0.02656	0.13277	0.33147
0.8	0.10611	0.26471	0.32739
1.2	0.23795	0.39378	0.31659
1.6	0.42032	0.51676	0.29667
2.0	0.65003	0.62977	0.26675
2.4	0.92230	0.72899	0.22809
2.8	1.23099	0.81152	0.18401
3.2	1.56911	0.87609	0.13913
3.6	1.92954	0.92333	0.09809
4.0	2.30576	0.95552	0.06424
4.4	2.69238	0.97587	0.03897
4.8	3.08534	0.98779	0.02187
5.0	3.28329	0.99155	0.01591
8.8	7.07923	1.00000	0.00000

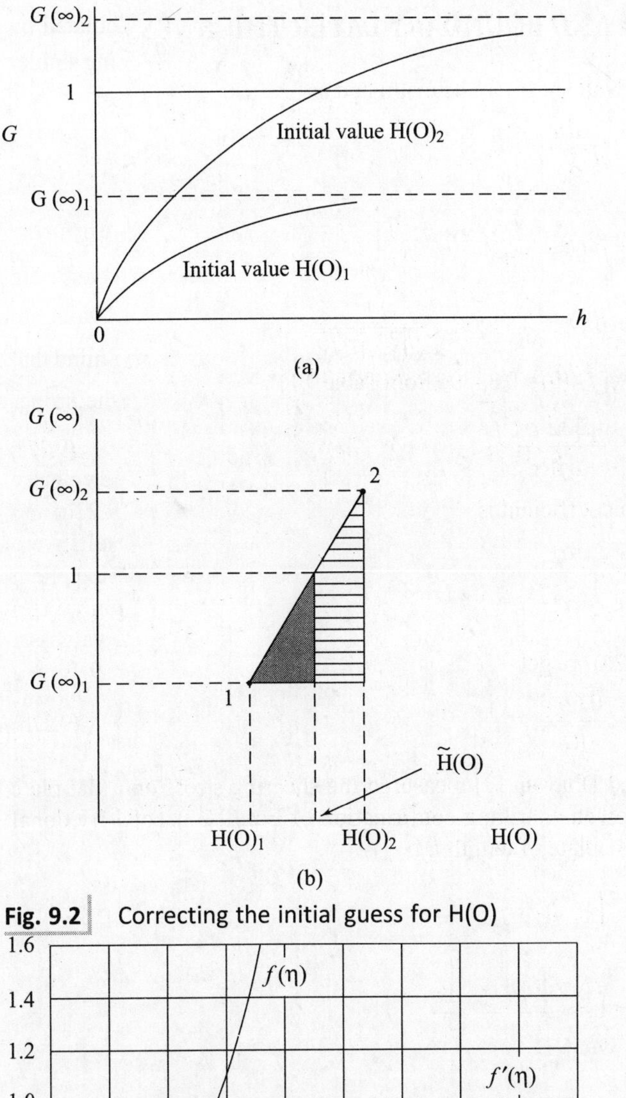

Fig. 9.2 Correcting the initial guess for H(O)

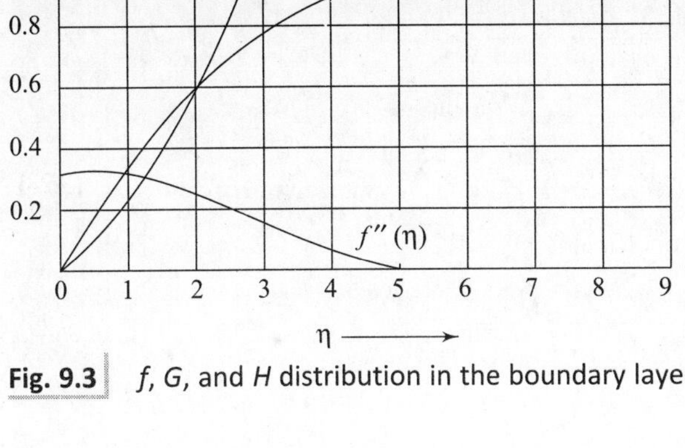

Fig. 9.3 *f*, *G*, and *H* distribution in the boundary layer

9.4 WALL SHEAR AND BOUNDARY LAYER THICKNESS

With the profile known, wall shear can be evaluated as

$$\tau_w = \mu \left.\frac{\partial u}{\partial y}\right|_{y=0}$$

or
$$\tau_w = \mu U_\infty \frac{\partial}{\partial \eta} f'(\eta) \cdot \left.\frac{\partial \eta}{\partial y}\right|_{\eta=0}$$

or
$$\tau_w = \mu U_\infty \times 0.33206 \times \frac{1}{\sqrt{(\nu x)/U_\infty}}$$

$$[f''(0) = 0.33206 \text{ from Table 9.1}]$$

or
$$\tau_w = \frac{0.332\, \rho U_\infty^2}{\sqrt{\text{Re}_x}} \tag{9.27a}$$

and the local skin friction coefficient is

$$C_{fx} = \frac{\tau_w}{\frac{1}{2}\rho U_\infty^2}$$

Substituting from (9.27a) we get

$$C_{fx} = \frac{0.664}{\sqrt{\text{Re}_x}} \tag{9.27b}$$

In 1951, Liepmann and Dhawan [3] measured the shearing stress on a flat plate directly. Their results showed a striking confirmation of Eq. (9.27).Total frictional force per unit width for the plate of length L is

$$F = \int_0^L \tau_w \, dx$$

or
$$F = \int_0^L \frac{0.332\, \rho U_\infty^2}{\sqrt{\frac{U_\infty}{\nu}}} \frac{dx}{\sqrt{x}}$$

or
$$F = \left[\frac{0.332\, \rho U_\infty^2}{\sqrt{U_\infty/\nu}} \times \frac{x^{1/2}}{\left(\frac{1}{2}\right)}\right]_0^L$$

or
$$F = 0.664 \times \rho U_\infty^2 \sqrt{\frac{\nu L}{U_\infty}} \tag{9.28}$$

and the average skin friction coefficient is

$$\overline{C}_f = \frac{F}{\frac{1}{2}\left(\rho U_\infty^2 L\right)} = \frac{1.328}{\sqrt{\text{Re}_L}} \tag{9.29}$$

where, $\mathrm{Re}_L = U_\infty L/\nu$.

Since u/U_∞ approaches 1.0 as $y \to \infty$, it is customary to select the boundary layer thickness δ as that point where u/U_∞ approaches 0.99. From Table 9.1, u/U_∞ reaches 0.99 at $\eta = 5.0$ and we can write

$$\delta \Big/ \sqrt{\left(\frac{\nu x}{U_\infty}\right)} \approx 5.0$$

or

$$\delta \approx 5.0 \sqrt{\left(\frac{\nu x}{U_\infty}\right)} = \frac{5.0\, x}{\sqrt{\mathrm{Re}_x}} \tag{9.30}$$

However, the aforesaid definition of boundary layer thickness is somewhat arbitrary, a physically more meaningful measure of boundary layer estimation is expressed through displacement thickness.

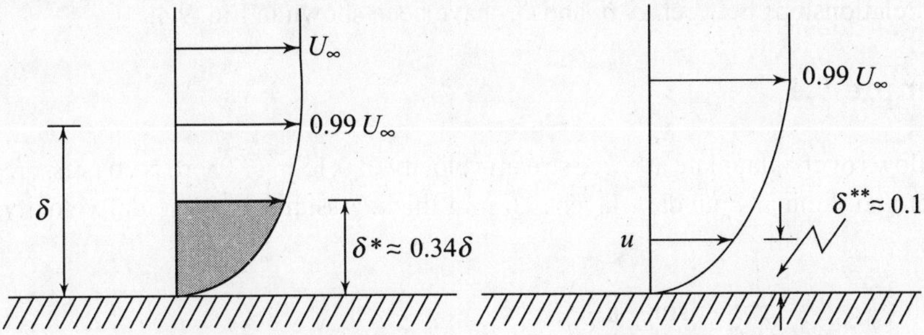

Fig. 9.4

Displacement thickness (δ^*) is defined as the distance by which the external potential flow is displaced outwards due to the decrease in velocity in the boundary layer.

$$U_\infty \delta^* = \int_0^\infty (U_\infty - u)\, \mathrm{d}y$$

Therefore,

$$\delta^* = \int_0^\infty \left(1 - \frac{u}{U_\infty}\right) \mathrm{d}y \tag{9.31}$$

Substituting the values of (u/U_∞) and η from Eqs (9.21a) and (9.19) into Eq. (9.31), we obtain

$$\delta^* = \sqrt{\frac{\nu x}{U_\infty}} \int_0^\infty (1 - f')\, \mathrm{d}\eta = \sqrt{\frac{\nu x}{U_\infty}} \lim_{\eta \to \infty} [\eta - f(\eta)]$$

or

$$\delta^* = 1.7208 \sqrt{\frac{\nu x}{U_\infty}} = \frac{1.7208\, x}{\sqrt{\mathrm{Re}_x}} \tag{9.32}$$

Following the analogy of the displacement thickness, a momentum thickness may be defined. *Momentum thickness* (δ^{**}) is defined as the loss of momentum in the boundary layer as compared with that of potential flow. Thus,

$$\rho U_\infty^2 \delta^{**} = \int_0^\infty \rho u (U_\infty - u)\, dy$$

or
$$\delta^{**} = \int_0^\infty \frac{u}{U_\infty}\left(1 - \frac{u}{U_\infty}\right) dy \tag{9.33}$$

With the substitution of (u/U_∞) and η from Eq. (9.21a) and (9.19), we can evaluate numerically the value of δ^{**} for a flat plate as

$$\delta^{**} = \sqrt{\frac{\nu x}{U_\infty}} \int_0^\infty f'(1 - f')\, d\eta$$

or
$$\delta^{**} = 0.664 \sqrt{\frac{\nu x}{U_\infty}} = \frac{0.664\, x}{\sqrt{\mathrm{Re}_x}} \tag{9.34}$$

The relationships between δ, δ^* and δ^{**} have been shown in Fig. 9.4.

Example 9.1

Water flows over a flat plate at a free stream velocity of 0.15 m/s. There is no pressure gradient and laminar boundary layer is 6 mm thick. Assume a sinusoidal velocity profile

$$\frac{u}{U_\infty} = \sin \frac{\pi}{2}\left(\frac{y}{\delta}\right)$$

For the flow conditions stated above, calculate the local wall shear stress and skin friction coefficient.

$$[\mu = 1.02 \times 10^{-3}\ \mathrm{kg/ms},\ \rho = 1000\ \mathrm{kg/m^3}]$$

Solution

$$\tau = \mu \frac{\partial u}{\partial y} = \frac{\mu U_\infty}{\delta} \cdot \frac{\partial (u/U_\infty)}{\partial (y/\delta)}$$

$$= \frac{\mu U_\infty}{\delta} \frac{\pi}{2} \cos\left(\frac{\pi}{2} \cdot \frac{y}{\delta}\right) = \frac{1.57\, \mu U_\infty}{\delta} \cos\left(\frac{\pi}{2} \cdot \frac{y}{\delta}\right)$$

$$\tau_w = \tau\big|_{y=0} = \frac{1.57\, \mu U_\infty}{\delta}$$

or
$$\tau_w = \frac{1.57 \times 1.02 \times 10^{-3} \times 0.15}{6 \times 10^{-3}} = 0.04\ \mathrm{N/m^2}$$

$$C_f = \frac{\tau_w}{\frac{1}{2} \rho U_\infty^2} = \frac{2 \times 0.04}{1000 \times (0.15)^2} = 3.5 \times 10^{-3}$$

9.5 MOMENTUM-INTEGRAL EQUATIONS FOR BOUNDARY LAYER

If we are to employ boundary layer concepts in real engineering designs, we need to devise approximate methods that would quickly lead to an answer even if the accuracy is somewhat less. Karman and Pohlhausen devised a simplified method by satisfying only the boundary conditions of the boundary layer flow rather than satisfying Prandtl's differential equations for each and every particle within the boundary layer. We shall discuss this method herein.

Consider the case of a steady, two-dimensional and incompressible flow, i.e., we shall refer to Eqs (9.10) to (9.14). Upon integrating the dimensional form of Eq. (9.10) with respect to $y = 0$ (wall) to $y = \delta$ (where δ signifies the interface of the free stream and the boundary layer), we obtain

$$\int_0^\delta \left(u\frac{\partial u}{\partial x} + v\frac{\partial u}{\partial y} \right) \mathrm{d}y = \int_0^\delta \left(-\frac{1}{\rho}\frac{\mathrm{d}p}{\mathrm{d}x} + \nu\frac{\partial^2 u}{\partial y^2} \right) \mathrm{d}y$$

or
$$\int_0^\delta u\frac{\partial u}{\partial x}\,\mathrm{d}y + \int_0^\delta v\frac{\partial u}{\partial y}\,\mathrm{d}y = \int_0^\delta -\frac{1}{\rho}\frac{\mathrm{d}p}{\mathrm{d}x}\,\mathrm{d}y + \int_0^\delta \nu\frac{\partial^2 u}{\partial y^2}\,\mathrm{d}y \tag{9.35}$$

The second term of the left-hand side can be expanded as

$$\int_0^\delta v\frac{\partial u}{\partial y}\,\mathrm{d}y = [vu]_0^\delta - \int_0^\delta u\frac{\partial v}{\partial y}\,\mathrm{d}y$$

or
$$\int_0^\delta v\frac{\partial u}{\partial y}\,\mathrm{d}y = U_\infty v_\delta + \int_0^\delta u\frac{\partial u}{\partial x}\,\mathrm{d}y \quad \left(\text{since } \frac{\partial u}{\partial x} = -\frac{\partial v}{\partial y}\right)$$

or
$$\int_0^\delta v\frac{\partial u}{\partial y}\,\mathrm{d}y = -U_\infty\int_0^\delta \frac{\partial u}{\partial x}\,\mathrm{d}y + \int_0^\delta u\frac{\partial u}{\partial x}\,\mathrm{d}y \tag{9.36}$$

Substituting Eq. (9.36) in Eq. (9.35) we obtain

$$\int_0^\delta 2u\frac{\partial u}{\partial x}\,\mathrm{d}y - U_\infty\int_0^\delta \frac{\partial u}{\partial x}\,\mathrm{d}y = -\int_0^\delta \frac{1}{\rho}\frac{\mathrm{d}p}{\mathrm{d}x}\,\mathrm{d}y - \nu\left.\frac{\partial u}{\partial y}\right|_{y=0} \tag{9.37}$$

Substituting the relation between $\frac{\mathrm{d}p}{\mathrm{d}x}$ and the free stream velocity U_∞ for the inviscid zone in Eq. (9.37), we get

$$\int_0^\delta 2u\frac{\partial u}{\partial x}\,\mathrm{d}y - U_\infty\int_0^\delta \frac{\partial u}{\partial x}\,\mathrm{d}y - \int_0^\delta U_\infty\frac{\mathrm{d}U_\infty}{\mathrm{d}x}\,\mathrm{d}y = -\left(\frac{\mu\left.\frac{\partial u}{\partial y}\right|_{y=0}}{\rho} \right)$$

or
$$\int_0^\delta \left(2u\frac{\partial u}{\partial x} - U_\infty\frac{\partial u}{\partial x} - U_\infty\frac{\mathrm{d}U_\infty}{\mathrm{d}x} \right)\mathrm{d}y = -\frac{\tau_w}{\rho}$$

which is reduced to

$$\int_0^{\delta} \frac{\partial}{\partial x} \{[u\,(U_\infty - u)]\,\mathrm{d}y\} + \frac{\mathrm{d}U_\infty}{\mathrm{d}x} \int_0^{\delta} (U_\infty - u)\,\mathrm{d}y = \frac{\tau_w}{\rho}$$

Since the integrals vanish outside the boundary layer, we are allowed to put $\delta = \infty$.

$$\int_0^{\infty} \frac{\partial}{\partial x} [u\,(U_\infty - u)]\mathrm{d}y + \frac{\mathrm{d}U_\infty}{\mathrm{d}x} \int_0^{\infty} (U_\infty - u)\,\mathrm{d}y = \frac{\tau_w}{\rho}$$

or
$$\frac{\mathrm{d}}{\mathrm{d}x} \int_0^{\infty} [u\,(U_\infty - u)]\mathrm{d}y + \frac{\mathrm{d}U_\infty}{\mathrm{d}x} \int_0^{\infty} (U_\infty - u)\,\mathrm{d}y = \frac{\tau_w}{\rho} \tag{9.38}$$

Substituting Eq. (9.31) and (9.33) in Eq. (9.38) we obtain

$$\frac{\mathrm{d}}{\mathrm{d}x}\left[U_\infty^2\,\delta^{**}\right] + \delta^{*}\,U_\infty\,\frac{\mathrm{d}U_\infty}{\mathrm{d}x} = \frac{\tau_w}{\rho} \tag{9.39}$$

Equation (9.39) is known as the momentum integral equation for two-dimensional incompressible laminar boundary layer. The same remains valid for turbulent boundary layers as well. Needless to say, the wall shear stress (τ_w) will be different for laminar and turbulent flows. The term $\left(U_\infty \dfrac{\mathrm{d}U_\infty}{\mathrm{d}x}\right)$ signifies spacewise acceleration of the free stream. Existence of this term means the presence of free stream pressure gradient in the flow direction. For example, we get finite value of $\left(U_\infty \dfrac{\mathrm{d}U_\infty}{\mathrm{d}x}\right)$ outsidethe boundary layer in the entrance region of a pipe or a channel. For external flows, the existence of $\left(U_\infty \dfrac{\mathrm{d}U_\infty}{\mathrm{d}x}\right)$ depends on the shape of the body. During the flow over a flat plate, $\left(U_\infty \dfrac{\mathrm{d}U_\infty}{\mathrm{d}x}\right) = 0$ and the momentum integral equation is reduced to

$$\frac{\mathrm{d}}{\mathrm{d}x}\left[U_\infty^2\,\delta^{**}\right] = \frac{\tau_w}{\rho} \tag{9.40}$$

9.6 SEPARATION OF BOUNDARY LAYER

It has been observed that the flow is reversed at the vicinity of the wall under certain conditions. The phenomenon is termed as separation of boundary layer. Separation takes place due to excessive momentum loss near the wall in a boundarylayer trying to move downstream against increasing pressure, i.e., $\mathrm{d}p/\mathrm{d}x > 0$, which is called *adverse pressure gradient*. Figure 9.5 shows the flow past a circular cylinder, in an infinite medium. Up to $\theta = 90°$, the flow area is like a constricted passage and the flow behaviour is like that of a nozzle. Beyond $\theta = 90°$ the flow area is diverged, therefore, the flow behaviour is very similar to a diffuser. This dictates the inviscid pressure distribution on the cylinder which is shown by a firm line in Fig. 9.5. Here p_∞ and U_∞ are the pressure and velocity in the free stream and p is the local pressure on the cylinder.

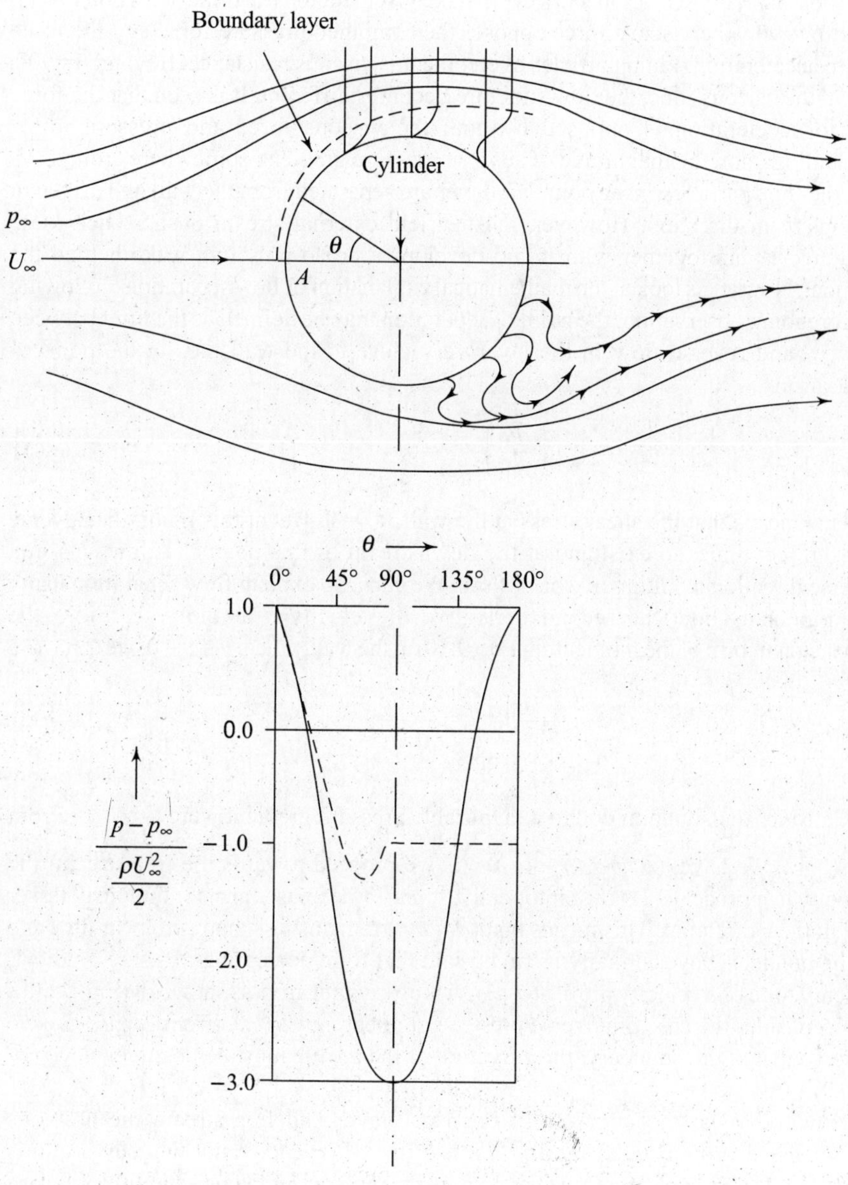

Fig. 9.5 Flow separation and formation of wake behind a circular cylinder

Consider the forces in the flow field. It is evident that in the inviscid region, the pressure force and the force due to streamwise acceleration are acting in the same direction (pressure gradient being negative/favourable) until $\theta = 90°$. Beyond $\theta = 90°$, the pressure gradient is positive or adverse. Due to the adverse pressure gradient the pressure force and the force due to acceleration will be opposing each other in the inviscid zone of this part. So long as no viscous effect is considered, the situation does not cause any sensation. However, in the viscid region (near the solid boundary), up to $\theta = 90°$, the viscous force opposes the combined pressure force and the force due to acceleration. Fluid particles overcome this viscous resistance. Beyond $\theta = 90°$, within the viscous zone, the flow structure becomes different. It is seen that the force due to acceleration is opposed by both the viscous force and pressure force. Depending upon the magnitude of adverse pressure gradient, somewhere around $\theta =$ 90°, the fluid particles, in the boundary layer are separated from the wall and driven in the upstream direction. However, the far field external stream pushes back these separated layers together with it and develops a broad pulsating wake behind the cylinder. Now let us look at the mathematical explanation of flow separation. Following the foregoing observation, the point of separation may be defined as the limit between forward and reverse flow in the layer very close to the wall, i.e., at the point of separation

$$\left(\frac{\partial u}{\partial y}\right)_{y=0} = 0 \tag{9.41}$$

This means that the shear stress at the wall, $\tau_w = 0$. But at this point, the adverse pressure continues to exist and at the downstream of this point the flow acts in a reverse direction resulting in a back flow. We can also explain flow separation using the argument about the second derivative of velocity u at the wall. From the dimensional form of the momentum Eq. (9.10) at the wall, where $u = v = 0$, we can write

$$\left(\frac{\partial^2 u}{\partial y^2}\right)_{y=0} = \frac{1}{\mu}\frac{dp}{dx} \tag{9.42}$$

Consider the situation due to a favourable pressure gradient were $\frac{dp}{dx} < 0$. From Eq. (9.42) we have, $(\partial^2 u/\partial y^2)_{\text{wall}} < 0$. As we proceed towards the free stream, the velocity u approaches U_∞ aysmptotically, so $\partial u/\partial y$ decreases at a continuously lesser rate in the y direction. This means that $(\partial^2 u/\partial y^2)$ remains less than zero near the edge of the boundary layer. Finally it can be said that for a decreasing pressure gradient, the curvature of a velocity profile $(\partial^2 u/\partial y^2)$ is always negative as shown in (Fig. 9.6(a). Next consider the case of adverse pressure gradient, $\partial p/\partial x > 0$. From Eq. (9.42), we observe that at the boundary, the curvature of the profile must be positive (since $\partial p/\partial x > 0$).

However, near the interface of boundary layer and free stream the previous argument regarding $\partial u/\partial y$ and $\partial^2 u/\partial y^2$ still holds good and the curvature is negative. Thus we observe that for an adverse pressure gradient, there must exist a point for which $\partial^2 u/\partial y^2 = 0$. This point is known as *the point of inflection* of the velocity profile in the boundary layer as shown in Fig. 9.6(b). However, point of separation means $\partial u/\partial y = 0$ at the wall. In addition, Eq. (9.42) depicts $\partial^2 u/\partial y^2 > 0$ at the wall since can only occur due to adverse pressure gradient. But we have already seen that at the edge of the boundary layer, $\partial^2 u/\partial y^2 < 0$. It is therefore, clear that if there is a point of separation, there must exist a point of inflection in the velocity profile.

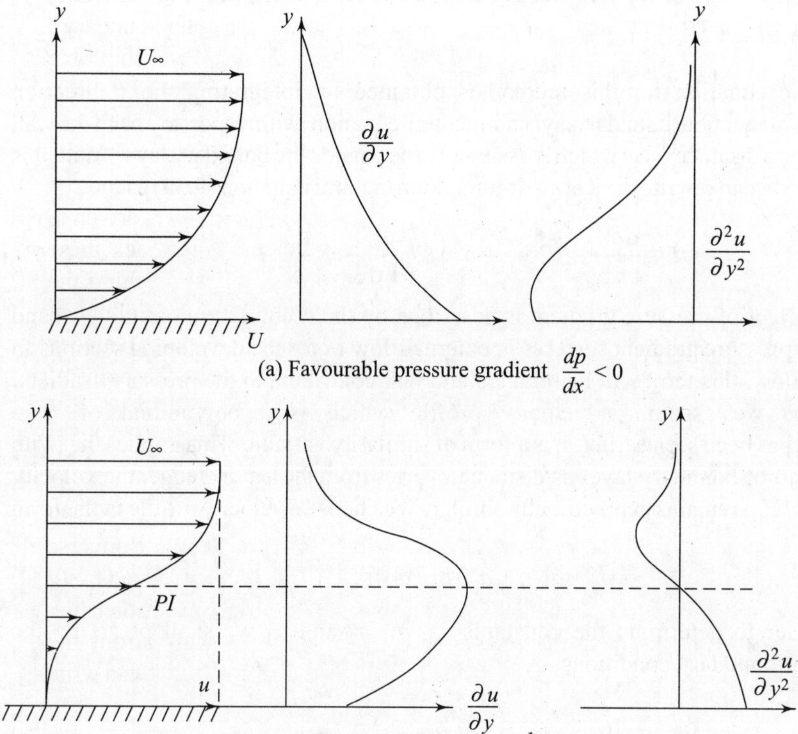

(a) Favourable pressure gradient $\frac{dp}{dx} < 0$

(b) Adverse pressure gradient, $\frac{dp}{dx} > 0$

Fig. 9.6 Velocity distribution within a boundary layer

Let us reconsider the flow past a circular cylinder and continue our discussion on the wake behind a cylinder. The pressure distribution which was shown by the firm line in Fig. 9.5 is obtained from the potential flow theory. However, somewhere near $\theta = 90°$ (in experiments it has been observed to be at $\theta = 81°$), the boundary layer detaches itself from the wall. Meanwhile, pressure in the wake remains close to separation-point-pressure since the eddies (formed as a consequence of the retarded layers being carried together with the upper layer through the action of shear) cannot convert rotational kinetic energy into pressure head. The actual pressure distribution is shown by the dotted line in Fig. 9.5. Since the wake zone pressure is less than that of the forward stagnation point (pressure at point *A* in Fig. 9.5), the cylinder experiences a drag force which is basically attributed to the pressure difference. The drag force, brought about by the pressure difference is known as *form drag* whereas the shear stress at the wall gives rise to *skin friction drag*. Generally, these two drag forces together are responsible for resultant drag on a body.

9.7 KARMAN–POHLHAUSEN APPROXIMATE METHOD FOR SOLUTION OF MOMENTUM INTEGRAL EQUATION OVER A FLAT PLATE

The basic equation for this method is obtained by integrating the x direction momentum equation (boundary layer momentum equation) with respect to y from the wall (at $y = 0$) to a distance $\delta(x)$ which is assumed to be outside the boundary layer. With this notation, we can rewrite the Karman momentum integral equation (9.39) as

$$U_\infty^2 \frac{d\delta^{**}}{dx} + \left(2\delta^{**} + \delta^*\right) U_\infty \frac{dU_\infty}{dx} = \frac{\tau_w}{\rho} \tag{9.43}$$

The effect of pressure gradient is described by the second term on the left-hand side. For pressure gradient surfaces in external flow or for the developing sections in internal flow, this term will be retained and will contribute to the pressure gradient. However, we assume a velocity profile which is a polynomial of $\eta = y/\delta$. As it has been seen earlier, η is a form of similarity variable. This implies that with the growth of boundary layer as distance x varies from the leading edge, the velocity profile (u/U_∞) remains geometrically similar. We choose a velocity profile in the form

$$\frac{u}{U_\infty} = a_0 + a_1\,\eta + a_2\,\eta^2 + a_3\,\eta^3 \tag{9.44}$$

In order to determine the constants a_0, a_1, a_2 and a_3 we shall prescribe the following boundary conditions:

at $\quad y = 0, \quad u = 0 \quad$ or $\quad$ at $\eta = 0, \quad \dfrac{u}{U_\infty} = 0 \quad$ (9.45a)

at $\quad y = 0, \quad \dfrac{\partial^2 u}{\partial y^2} = 0 \quad$ or $\quad$ at $\eta = 0, \quad \dfrac{\partial^2}{\partial \eta^2}(u/U_\infty) = 0 \quad$ (9.45b)

at $\quad y = \delta, \quad u = U_\infty \quad$ or $\quad$ at $\eta = 1, \quad \dfrac{u}{U_\infty} = 1 \quad$ (9.45c)

at $\quad y = \delta, \quad \dfrac{\partial u}{\partial y} = 0 \quad$ or $\quad$ at $\eta = 1, \quad \dfrac{\partial (u/U_\infty)}{\partial \eta} = 0 \quad$ (9.45d)

These requirements will yield

$$a_0 = 0, \quad a_2 = 0, \quad a_1 + 3a_3 = 0 \quad \text{and} \quad a_1 + a_3 = 1$$

Finally, we obtain the following values for the coefficients in Eq. (9.44),

$$a_0 = 0, \quad a_1 = \frac{3}{2} \quad a_2 = 0 \quad \text{and} \quad a_3 = -\frac{1}{2}$$

and the velocity profile becomes

$$\frac{u}{U_\infty} = \frac{3}{2}\eta - \frac{1}{2}\eta^3 \tag{9.46}$$

For flow over a flat plate, $\frac{\mathrm{d}p}{\mathrm{d}x} = 0$, hence $U_\infty \frac{\mathrm{d}U_\infty}{\mathrm{d}x} = 0$ and the governing Eq. (9.43) reduces to

$$\frac{\mathrm{d}\delta^{**}}{\mathrm{d}x} = \frac{\tau_w}{\rho U_\infty^2} \tag{9.47}$$

Again from Eq. (9.33), the momentum thickness is

$$\delta^{**} = \int_0^\infty \frac{u}{U_\infty}\left(1 - \frac{u}{U_\infty}\right)\mathrm{d}y$$

or

$$\delta^{**} = \int_0^\delta \frac{u}{U_\infty}\left(1 - \frac{u}{U_\infty}\right)\mathrm{d}y$$

or

$$\delta^{**} = \delta\int_0^1 \left(1 - \frac{3}{2}\eta + \frac{1}{2}\eta^3\right)\left(\frac{3}{2}\eta - \frac{1}{2}\eta^3\right)\mathrm{d}\eta$$

or

$$\delta^{**} = \frac{39}{280}\delta$$

The wall shear stress is given by

$$\tau_w = \mu \left.\frac{\partial u}{\partial y}\right|_{y=0}$$

or

$$\tau_w = \mu\left[\frac{\partial}{\delta\,\partial\eta}\left\{U_\infty\left(\frac{3}{2}\eta - \frac{1}{2}\eta^3\right)\right\}\right]_{\eta=0}$$

or

$$\tau_w = \frac{3\mu U_\infty}{2\delta}$$

Substituting the values of δ^{**} and τ_w in Eq. (9.47), we get

$$\frac{39}{280}\frac{\mathrm{d}\delta}{\mathrm{d}x} = \frac{3\mu U_\infty}{2\delta\,\rho U_\infty^2}$$

or

$$\int \delta\,\mathrm{d}\delta = \int \frac{140}{13}\frac{\mu}{\rho U_\infty}\mathrm{d}x + C_1$$

or

$$\frac{\delta^2}{2} = \frac{140}{13}\frac{\nu x}{U_\infty} + C_1 \tag{9.48}$$

where C_1 is any arbitrary unknown constant.

The condition at the leading edge (at $x = 0$, $\delta = 0$) yields

$$C_1 = 0$$

Finally ,we obtain

$$\delta^2 = \frac{280}{13}\frac{\nu x}{U_\infty} \tag{9.49}$$

or

$$\delta = 4.64\sqrt{\frac{\nu x}{U_\infty}}$$

or

$$\delta = \frac{4.64x}{\sqrt{\mathrm{Re}_x}} \tag{9.50}$$

This is the value of boundary layer thickness on a flat plate. Although the method is an approximate one, the result is found to be reasonably accurate. The value is slightly lower than the exact solution of laminar flow over a flat plate given by Eq. (9.30). As such, the accuracy depends on the order of the velocity profile. It may be mentioned that instead of Eq. (9.44), we can as well use a fourth order polynomial as

$$\frac{u}{U_\infty} = a_0 + a_1\eta + a_2\eta^2 + a_3\eta^3 + a_4\eta^4 \tag{9.51}$$

In addition to the boundary conditions in Eq. (9.45), we shall require another boundary condition

at

$$y = \delta, \ \frac{\partial^2 u}{\partial y^2} = 0 \quad \text{or} \quad \text{at } \eta = 1, \ \frac{\partial^2 (u/U_\infty)}{\partial \eta^2} = 0$$

To determine the constants as $a_0 = 0$, $a_1 = 2$, $a_2 = 0$, $a_3 = -2$ and $a_4 = 1$. Finally the velocity profile will be

$$\frac{u}{U_\infty} = 2\eta - 2\eta^3 + \eta^4$$

Subsequently, for a fourth order profile the growth of boundary layer is given by

$$\delta = \frac{5.83\,x}{\sqrt{\mathrm{Re}_x}} \tag{9.52}$$

This is also very close to the value of the exact solution.

Example 9.2

Air at standard conditions flows over a flat plate. The freestream speed is 3 m/sec. Find δ and τ_w at $x = 1$ m from the leading edge (assume a cubic velocity profile). For air, $\nu = 1.5 \times 10^{-5}$ m^2/s and $\rho = 1.23$ kg/m^3.

Solution

Applying the results developed in Section 9.7 for cubic velocity profile and the growth of boundary layer, we can write

$$\frac{U}{U_\infty} = \frac{3}{2}\eta - \frac{1}{2}\eta^3, \text{ where } \eta = \frac{y}{\delta} \text{ at any } x$$

and
$$\frac{\delta}{x} = \frac{4.64}{\sqrt{\text{Re}_x}}$$

For air with $\nu = 1.5 \times 10^{-5}$ m^2/s, the local Reynolds number at x is

$$\text{Re}_x = \frac{U_\infty x}{\nu} = \frac{3 \times 1}{1.5 \times 10^{-5}} = 2 \times 10^5$$

$$\delta = \frac{4.64\,x}{\sqrt{\text{Re}_x}} = \frac{4.64 \times 1}{\sqrt{2 \times 10^5}} \text{ m} = 0.0103 \text{ m} = 10.3 \text{ mm}$$

$$\tau_w = \mu \frac{\partial u}{\partial y}\bigg|_{y=0} = \frac{\mu U_\infty}{\delta} \cdot \frac{d}{d\eta}\left[\frac{3}{2}\eta - \frac{1}{2}\eta^3\right]_{\eta=0}$$

$$\tau_w = \frac{3\mu U_\infty}{2\delta} = \frac{3\rho\nu U_\infty}{2\delta}$$

or
$$\tau_w = \frac{3 \times 1.23 \times 1.5 \times 10^{-5} \times 3}{2 \times 0.0103}$$

$$\tau_w = 8.06 \times 10^{-3} \text{ N/m}^2$$

Example 9.3

Air moves over a flat plate with a uniform free stream velocity of 10 m/s. At a position 15 cm away from the front edge of the plate, what is the boundary layer thickness? Use a parabolic profile in the boundary layer. For air, $\nu = 1.5 \times 10^{-5}$m^2/s and $\rho = 1.23$ kg/m^3.

Solution

For a parabolic profile let us take

$$\frac{u}{U_\infty} = a + by + cy^2$$

The boundary conditions are

at $y = 0, \quad u = 0$

at $y = \delta, \quad u = U_\infty$

at $y = \delta, \quad \dfrac{\partial u}{\partial y} = 0$

Evaluating the constants, we get

$$\frac{u}{U_\infty} = 2\left(\frac{y}{\delta}\right) - \left(\frac{y}{\delta}\right)^2 = 2\eta - \eta^2$$

Now,
$$\tau_w = \mu \frac{\partial u}{\partial y}\bigg|_{y=0} = \frac{\mu U_\infty}{\delta} \cdot \frac{\partial (u/U_\infty)}{\partial (y/\delta)}\bigg|_{\eta=0}$$

$$= \frac{\mu U_\infty}{\delta} \cdot \frac{\mathrm{d}(2\eta - \eta^2)}{\mathrm{d}\eta}\bigg|_{\eta=0} = \frac{2\mu U_\infty}{\delta}$$

Applying momentum integral equation,

$$\tau_w = \rho U_\infty^2 \frac{\mathrm{d}\delta}{\mathrm{d}x} \int_0^1 \frac{u}{U_\infty}\left(1 - \frac{u}{U_\infty}\right) \mathrm{d}\eta$$

$$\frac{2\mu U_\infty}{\delta} = \rho U_\infty^2 \frac{\mathrm{d}\delta}{\mathrm{d}x} \int_0^1 (2\eta - \eta^2)(1 - 2\eta + \eta^2)\, \mathrm{d}\eta$$

$$\frac{2\mu U_\infty}{\delta \rho U_\infty^2} = \frac{\mathrm{d}\delta}{\mathrm{d}x} \int_0^1 (2\eta - 5\eta^2 + 4\eta^3 - \eta^4)\, \mathrm{d}\eta$$

or
$$\frac{2\mu}{\delta \rho U_\infty} = \frac{2}{15} \frac{\mathrm{d}\delta}{\mathrm{d}x}$$

or
$$\delta \,\mathrm{d}\delta = \frac{15\mu}{\rho U_\infty} \mathrm{d}x$$

$$\frac{\delta^2}{2} = \frac{15\mu}{\rho U_\infty} x + C$$

It is assumed that at $x = 0$, $\delta = 0$, which yields $C = 0$.
Thus,

$$\delta = \sqrt{\frac{30\mu}{\rho U_\infty} x}$$

or
$$\frac{\delta}{x} = \sqrt{\frac{30\mu}{\rho U_\infty x}} = \frac{5.48}{\sqrt{\mathrm{Re}_x}}$$

In this problem,
$$\mathrm{Re}_x = \frac{10 \times 15 \times 10^{-2}}{1.5 \times 10^{-5}} = 1 \times 10^5$$

$$\delta = \frac{5.48}{\sqrt{\mathrm{Re}_x}} \times 15 \text{ cm} = 0.259 \text{ cm}$$

or
$$\delta = 2.59 \text{ mm}$$

Example 9.4

Air moves over a 10 m long flat plate. The transition from laminar to turbulent flow takes place between Reynolds numbers of 2.5×10^6 and 3.6×10^6. What are the minimum and maximum distance from the front edge of the plate along which one expect laminar flow in the boundary layer? The free stream velocity is 30 m/s and $\nu = 1.5 \times 10^{-5}$ m²/s.

Solution

We can see that the range of Reynolds numbers for laminar flow is 2.5×10^6 to 3.6×10^6

For the lower limit,

$$2.15 \times 10^6 = \frac{30x}{1.5 \times 10^{-5}}$$

or $$x_{min} = 1.075 \text{ m}$$

For the upper limit,

$$3.6 \times 10^{-6} = \frac{30x}{1.5 \times 10^{-5}}$$

or $$x_{max} = 1.8 \text{ m}$$

Example 9.5

Water at 15 °C flows over a flat plate at a speed of 1.2 m/s. The plate is 0.3 m long and 2 m wide. The boundary layer on each surface of the plane is laminar. Assume the velocity profile is approximated by a linear expression for which $\frac{\delta}{x} = \frac{3.46}{\sqrt{Re_x}}$.

Determine the drag force on the plate. For water $\nu = 1.1 \times 10^{-6}$ m²/s, $\rho = 1000$ kg/m³.

Solution

On a flat plate, the drag is due to skin friction acting on each side of the plate

$$F_D = 2 \int_0^L \tau_{wx} b \, dx$$

For linear profile, $$\frac{u}{U_\infty} = \frac{y}{\delta} \quad \text{and} \quad \tau_w = \mu \left. \frac{\partial u}{\partial y} \right|_{y=0}$$

or $$\tau_w = \frac{\mu U_\infty}{\delta} \cdot \left. \frac{\partial (U/U_\infty)}{\partial (y/\delta)} \right|_{y=0} = \frac{\mu U_\infty}{\delta}$$

$$F_D = 2\int_0^L \frac{\mu U_\infty}{\delta} b\,dx = 2\int_2^L \frac{\mu U_\infty}{3.46}\sqrt{\frac{U_\infty}{\nu x}}\cdot b\,dx$$

$$= \frac{2\mu U_\infty}{3.46}\sqrt{\frac{U_\infty}{\nu}}\, b\int_2^L \frac{1}{x^{(1/2)}}dx$$

$$= \frac{2\mu U_\infty b}{3.46}\sqrt{\frac{U_\infty}{\nu}}\left[2x^{\frac{1}{2}}\right]_o^L$$

$$= \frac{4\mu U_\infty b}{3.46}\sqrt{\frac{U_\infty L}{\nu}} = \frac{\mu U_\infty b}{3.46}\sqrt{\mathrm{Re}_L}$$

$$\mathrm{Re}_L = \frac{U_\infty L}{\nu} = \frac{1.2\times 0.3}{1.1\times 10^{-6}} = 3.27\times 10^5$$

Therefore, $(\mathrm{Re}_L)^{1/2} = 572$

Thus, $$F_D = \frac{4\times 1.1\times 10^{-3}\times 1.2\times 2\times 572}{3.46} = 1.745\ \mathrm{N}$$

Example 9.6

Air is flowing over a thin flat plate which is 1 m long and 0.3 m wide. At the leading edge, the flow is assumed to be uniform and $U_\infty = 30$ m/s. The flow condition is independent of z (see Fig. 9.7). Using the control volume *abcd*, calculate the mass flow rate across the surface *ab*. Determine the magnitude and direction of the x component of force required to hold the plate stationary. The velocity profile at *bc* is given by

$$\frac{U}{U_\infty} = 2\left(\frac{y}{2}\right) - \left(\frac{y}{\delta}\right)^2$$

and $\delta = 4$ mm. Density of air = 1.23 kg/m^3 and $\nu = 1.5\times 10^{-5}$ m^2/s.

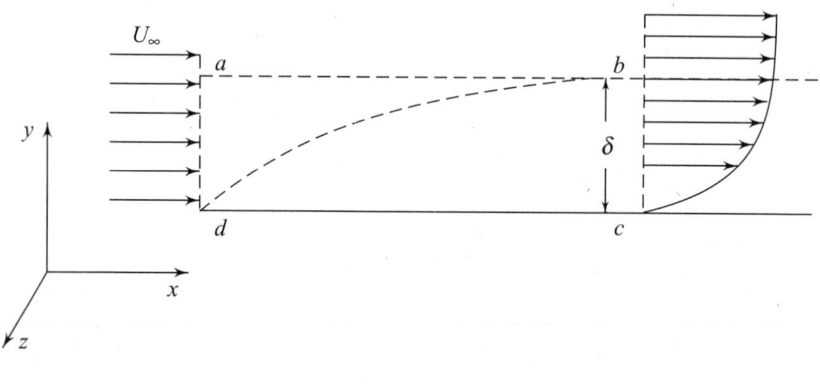

Fig. 9.7 Control volume of interest on flat plate

Solution

At *bc*, $\frac{U}{U_\infty} = 2\eta - \eta^2;\ \eta = y/\delta$

Steady state continuity equation is given by

$$\int_S \rho \vec{V} \cdot d\vec{A} = 0$$

or
$$-\rho U_\infty b\delta + \int_0^\delta \rho u\, b\, dy + \dot{m}_{ab} = 0$$

but $\int_0^\delta \rho u\, b\, dy = \rho U_\infty b\delta \int_0^1 (2\eta - \eta^2) d\eta = \rho U_\infty b\delta \left[\eta^2 - \frac{\eta^3}{3}\right]_0^1 = \frac{2}{3}\rho U_\infty b\delta$

$$m_{ab} = \rho U_\infty b\delta - \frac{2}{3}\rho U_\infty b\delta = \frac{1}{3}\rho U_\infty b\delta$$

or
$$\dot{m}_{ab} = \frac{1}{3} \times 1.23 \times 30 \times 0.3 \times 0.004 = 0.0147 \text{ kg/s}$$

Steady state momentum equation is given by

$$\int_{CS} u \rho \mathbf{\vec{V}} \cdot d\mathbf{\vec{A}} = F_{sx}$$

or
$$R_x = u_{da}\{-\rho U_\infty b\delta\} + u_{ab} m_{ab} + \int_0^\delta u \rho u b\, dy$$

But $u_{da} = u_{ab} = U_\infty$, and

$$\int_0^\delta u \rho u\, b\, dy = \rho U_\infty^2 b\delta \int_0^1 (2\eta - \eta^2)^2 d\eta$$

$$= \rho U_\infty^2 b\delta \left[\frac{4}{3}\eta^2 - \eta^4 + \frac{\eta^5}{3}\right]_0^1 = \frac{8}{15}\rho U_\infty^2 b\delta$$

Thus,
$$R_x = -\rho U_\infty^2 b\delta + \frac{1}{3}\rho U_\infty^2 b\delta + \frac{8}{15}\rho U_\infty^2 b\delta = -\frac{2}{15}\rho U_\infty^2 b\delta$$

$$R_x = -\frac{2}{15}\{1.23 \times (3)^2 \times 0.3 \times 0.004\}$$

$= -0.177$ N (the force must be applied to the CV by the plate)

Hence,
$$F_x = R_x = 0.177 \text{ N (to the left)}$$

9.8 INTEGRAL METHOD FOR NON-ZERO PRESSURE GRADIENT FLOWS

A wide variety of 'integral methods' in this category have been discussed by Rosenhead [4]. The Thwaites method [5] is found to be a very elegant method, which is an extension of the method due to Holstein and Bohlen [6]. We shall discuss the Holstein–Bohlen method in this section.

This is an approximate method for solving boundary layer equations for two-dimensional generalised flows. The integrated Eq. (9.39) for laminar flow with pressure gradient can be written as

$$\frac{\mathrm{d}}{\mathrm{d}x}[U^2\delta^{**}] + \delta^* U\frac{\mathrm{d}U}{\mathrm{d}x} = \frac{\tau_w}{\rho}$$

or

$$U^2\frac{\mathrm{d}\delta^{**}}{\mathrm{d}x} + (2\delta^{**} + \delta^*)U\frac{\mathrm{d}U}{\mathrm{d}x} = \frac{\tau_w}{\rho} \tag{9.53}$$

The velocity profile of the boundary layer is considered to be a fourth-order polynomial in terms of the dimensionless distance $\eta = y/\delta$, and is expressed as

$$u/U = a\eta + b\eta^2 + c\eta^3 + d\eta^4$$

The boundary conditions are

$$\eta = 0: u = 0, v = 0 \qquad \frac{\nu}{\delta^2}\frac{\partial^2 u}{\partial \eta^2} = \frac{1}{\rho}\frac{\mathrm{d}p}{\mathrm{d}x} = -U\frac{\mathrm{d}U}{\mathrm{d}x}$$

$$\eta = 1: u = U, \qquad \frac{\partial u}{\partial \eta} = 0, \ \frac{\partial^2 u}{\partial \eta^2} = 0$$

A dimensionless quantity, may be introduced as

$$\lambda = \frac{\delta^2}{\nu}\frac{\mathrm{d}U}{\mathrm{d}x} \tag{9.54}$$

The following relations are obtained

$$a = 2 + \frac{\lambda}{6}, \quad b = -\frac{\lambda}{2}, \quad c = -2 + \frac{\lambda}{2}, \quad d = 1 - \frac{\lambda}{6}$$

Now, the velocity profile can be expressed as

$$u/U = F(\eta) + \lambda G(\eta), \tag{9.55}$$

where

$$F(\eta) = 2\eta - 2\eta^3 + \eta^4, \quad G(\eta) = \frac{1}{6}\,\eta(1-\eta)^3$$

The shear stress $\tau_w = \mu(\partial u/\partial y)_{y=0}$ is given by

$$\frac{\tau_w \delta}{\mu U} = 2 + \frac{\lambda}{6} \tag{9.56}$$

We use the following dimensionless parameters,

$$L=\frac{\tau_w\delta^{**}}{\mu U} \tag{9.57}$$

$$K=\frac{(\delta^{**})^2}{\nu}\frac{dU}{dx} \tag{9.58}$$

$$H=\delta^{*}/\delta^{**} \tag{9.59}$$

The integrated momentum Eq. (9.53) reduces to

$$U\frac{d\delta^{**}}{dx}+\delta^{**}(2+H)\frac{dU}{dx}=\frac{\nu L}{\delta^{**}}$$

or

$$U\frac{d}{dx}\left[\frac{(\delta^{**})^2}{\nu}\right]=2[L-K(H+2)] \tag{9.60}$$

The parameter L is related to the skin friction and K is linked to the pressure gradient. If we take K as the independent variable, L and H can be shown to be the functions of K since

$$\frac{\delta^{*}}{\delta}=\int_0^1[1-F(\eta)-\lambda G(\eta)]d\eta=\frac{3}{10}-\frac{\lambda}{120} \tag{9.61}$$

$$\frac{\delta^{**}}{\delta}=\int_0^1(F(\eta)+\lambda G(\eta)(1-F(\eta)-\lambda G(\eta))d\eta$$

$$=\frac{37}{315}-\frac{\lambda}{945}-\frac{\lambda^2}{9072} \tag{9.62}$$

$$K=\frac{[\delta^{**}]^2}{\delta^2}\lambda=\lambda\left(\frac{37}{315}-\frac{\lambda}{945}-\frac{\lambda^2}{9072}\right)^2 \tag{9.63}$$

Therefore,

$$L=\left(2+\frac{\lambda}{6}\right)\frac{\delta^{**}}{\delta}=\left(2+\frac{\lambda}{6}\right)\left(\frac{37}{315}-\frac{\lambda}{945}-\frac{\lambda^2}{9072}\right)=f_1(K)$$

$$H=\frac{\delta^{*}}{\delta^{**}}=\frac{(3/10)-(\lambda/120)}{(37/315)-(\lambda/945)-(\lambda^2/9072)}=f_2(K)$$

The right-hand side of Eq. (9.60) is thus a function of K alone. Walz[7] pointed out that this function can be approximated with a good degree of accuracy by a linear function of K so that

$$2[L-K(H+2)]=a-bK$$

Equation (9.60) can now be written as

$$\frac{\mathrm{d}}{\mathrm{d}x}\left(\frac{U[\delta^{**}]^2}{\nu}\right) = a-(b-1)\,\frac{U[\delta^{**}]^2}{\nu}\,\frac{1}{U}\frac{\mathrm{d}U}{\mathrm{d}x}$$

Solution of this differential equation for the dependent variable $(U[\delta^{**}]^2/\nu)$ subject to the boundary condition $U = 0$ when $x = 0$, gives

$$\frac{U[\delta^{**}]^2}{\nu} = \frac{a}{U^{b-1}}\int_0^x U^{b-1}\,\mathrm{d}x$$

With $a = 0.47$ and $b = 6$, the approximation is particularly close between the stagnation point and the point of maximum velocity. Finally, the value of the dependent variable is

$$[\delta^{**}]^2 = \frac{0.47\,\nu}{U^6}\int_0^x U^5\,\mathrm{d}x \tag{9.64}$$

By taking the limit of Eq. (9.64), according to L'Hospital's rule, it can be shown that

$$[\delta^{**}]^2|_{x=0} = 0.47\,\nu/6U'(0)$$

This corresponds to $K = 0.0783$. It may be mentioned that $[\delta^{**}]$ is not equal to zero at the stagnation point. If $([\delta^{**}]^2/\nu)$ is determined from Eq. (9.64), $K(x)$ can be obtained from Eq. (9.58). Table 9.2 gives the necessary parameters for obtaining results, such as velocity profile and shear stress τ_w. The approximate method can be applied successfully to a wide range of problems.

As mentioned earlier, K and λ are related to the pressure gradient and the shape factor. Introduction of K and λ in the integral analysis enables extension of Karman–Pohlhausen method for solving flows over curved geometry. However, the analysis is not valid for the geometries, where $\lambda < -12$ and $\lambda > +12$.

Table 9.2 Auxiliary functions after Holstein and Bohlen [6]

λ	K	$f_1(K)$	$f_2(K)$
12	0.0948	2.250	0.356
10	0.0919	2.260	0.351
8	0.0831	2.289	0.340
7.6	0.0807	2.297	0.337
7.2	0.0781	2.305	0.333
7.0	0.0767	2.309	0.331

λ	K	$f_1(K)$	$f_2(K)$
6.6	0.0737	2.318	0.328
6.2	0.0706	2.328	0.324
5.0	0.0599	2.361	0.310
3.0	0.0385	2.427	0.283
1.0	0.0135	2.508	0.252
0	0	2.554	0.235
−1	−0.0140	2.604	0.217
−3	−0.0429	2.716	0.179
−5	−0.0720	2.847	0.140
−7	−0.0999	2.999	0.100
−9	−0.1254	3.176	0.059
−11	−0.1474	3.383	0.019
−12	−0.1567	3.500	0

9.9 ENTRY FLOW IN A DUCT

Growth of boundary layer has a remarkable influence on flow through a constant area duct or pipe. Let us consider a flow entering a pipe with uniform velocity. The boundary layer starts growing on the wall at the entrance of the pipe. Gradually it becomes thicker in the downstream and the flow becomes fully developed when the boundary layers from the wall meet at the axis of the pipe (Fig. 9.8). The velocity profile is nearly rectangular at the entrance and it gradually changes to a parabolic profile at the fully developed region. Before the boundary layers from the periphery meet at the axis, there prevails a core region which is uninfluenced by viscosity. Since the volume-flow must be same for every section and the boundary layer thickness increases in the flow direction, the inviscid core accelerates, and there is a corresponding fall in pressure. It can be shown that for laminar incompressible flows, the velocity profile approaches the parabolic profile through a distance *Le* from the entry of the pipe, which is given by

$$\frac{Le}{D} \approx 0.05\,\text{Re}, \qquad \text{where } \text{Re} = \frac{U_{\text{av}} D}{\nu}$$

For a Reynolds number of 2000, this distance, which is often referred to as the entrance length is about 100 pipe-diameters. For turbulent flows, the entrance region is shorter, since the turbulent boundary layer grows faster.

At the entrance region, the velocity gradient is steeper at the wall, causing a higher value of shear stress as compared to a developed flow. In addition, momentum flux across any section in the entrance region is higher than that typically at the inlet due to the change in shape of the velocity profile. Arising out of these, an additional pressure drop is brought about at the entrance region as compared to the pressure drop in the fully developed region.

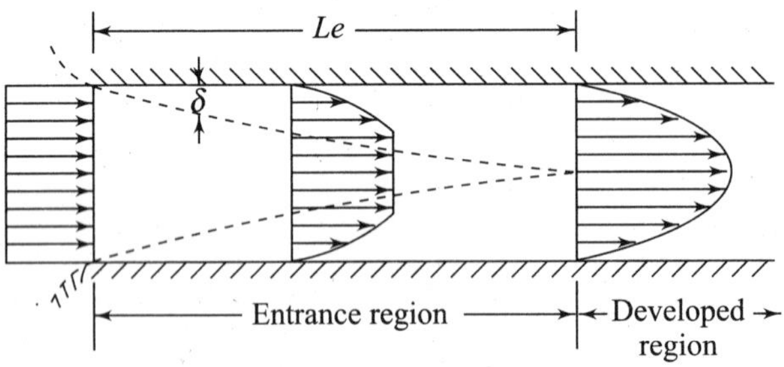

Fig. 9.8 Development of boundary layer in the entrance region of a duct

9.10 CONTROL OF BOUNDARY LAYER SEPARATION

It has already been seen that the total drag on a body is attributed to form drag and skin friction drag. In some flow configurations, the contribution of form drag becomes significant. In order to reduce the form drag, the boundary layer separation should be prevented or delayed so that somewhat better pressure recovery takes place and the form drag is reduced considerably. There are some popular methods for this purpose which are stated as follows:

(i) By giving the profile of the body a streamlined shape as shown in Fig. 9.9. This has an elongated shape in the rear part to reduce the magnitude of the pressure gradient. The optimum contour for a streamlined body is the one for which the wake zone is very narrow and the form drag is minimum.

(ii) The injection of fluid through porous wall can also control the boundary layer separation. This is generally accomplished by blowing high energy fluid particles tangentially from the location where separation would have taken place otherwise. This is shown in Fig. 9.10. The injection of fluid promotes turbulence and thereby increases skin friction. But the form drag is reduced considerably due to suppression of flow separation and this reduction can be of significant magnitude so as to ignore the enhanced skin friction drag.

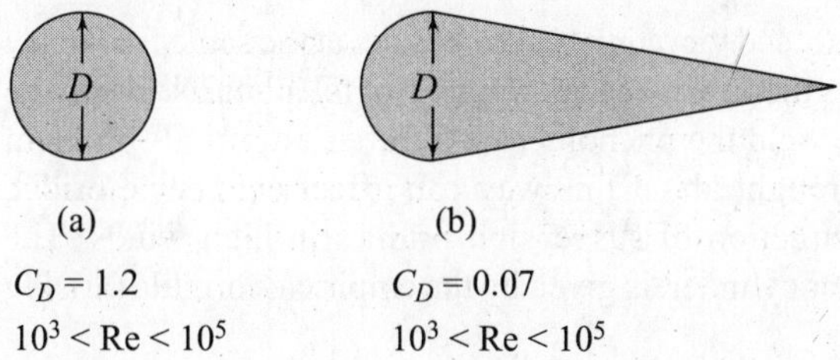

Fig. 9.9 Reduction of drag coefficient (C_D) by giving the profile a streamlined shape

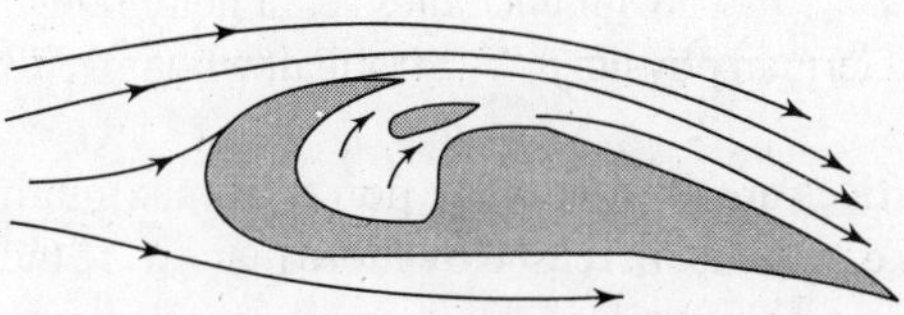

Fig. 9.10 Boundary layer control by blowing

9.11 MECHANICS OF BOUNDARY LAYER TRANSITION

One of the interesting problems in fluid mechanics is the physical mechanism of transition from laminar to turbulent flow. The problem evolves about the generation of both steady and unsteady vorticity near a body, its subsequent molecular diffusion, its kinematic and dynamic convection and redistribution downstream, and the resulting feedback on the velocity and pressure fields near the body. We can perhaps realise the complexity of the transition problem by examining the behaviour of a real flow past a cylinder.

Figure 9.11 (a) shows the flow past a cylinder for a very low Reynolds number (~1). The flow smoothly divides and reunites around the cylinder. At a Reynolds number of about 4, the flow separates in the downstream and the wake is formed by two symmetric eddies. The eddies remain steady and symmetrical but grow in size up to a Reynolds number of about 40 as shown in Fig. 9.11(b).

When the Reynolds number crosses 40, oscillation in the wake induces asymmetry and finally the wake starts shedding vortices into the stream. This situation is termed as onset of periodicity as shown in Fig. 9.11(c) and the wake keeps on undulating up to a Reynolds number of 90. As the Reynolds number further increases, the eddies are shed alternately from a top and bottom of the cylinder and the regular pattern of alternately shed clockwise and counter-clockwise vortices form *Von Karman vortex street* as in Fig. 9.11(d). However, periodicity is eventually induced in the flow field with the vortex-shedding phenomenon. The periodicity is characterised by the frequency of vortex shedding, *f*.

Each time a vortex is shed from the cylinder, a circulation is produced and consequently an unbalanced lateral force acts on the cylinder. The shedding of vortices alternately from the top and bottom of the cylinder produces alternating lateral forces which may cause a forced vibration of the cylinder at the same frequency of the shedding frequency. This is the cause of the 'singing' of telephone wires in the breeze. These alternating forces yield the phenomena of 'flutter', a serious problem with the aircrafts. Vibrations brought about this way can affect chimneys, bridge piers and they have caused destruction of suspension bridges in high winds. The shedding frequency, f, of a circular cylinder, is given by the empirical formula (arising out of experimental data)

$$\frac{fD}{U_\infty} = 0.198\left(1-\frac{19.7}{\text{Re}}\right) \tag{9.65}$$

The formula works well for $250 < \text{Re} < 2 \times 10^5$. The dimensionless parameter fD/U_∞ is known as the *Strouhal number* after the German physicist, V. Strouhal, who in 1878 first investigated the 'singing' of wires.

If the cylinder itself oscillates, the vortices are shed near the points of maximum displacements of the cylinder and the wake width is increased by more than twice the magnitude. The lateral force is increased and f decreases. Under various operating conditions the oscillation frequency interacts with the vortex formation frequency differently. Depending upon the flow situation, the wake comprises either a '2S' mode, representing two single vortices formed per cycle or the '2P' mode, whereby two vortex pairs are formed per cycle. For some flows (P + S) mode is also visible.

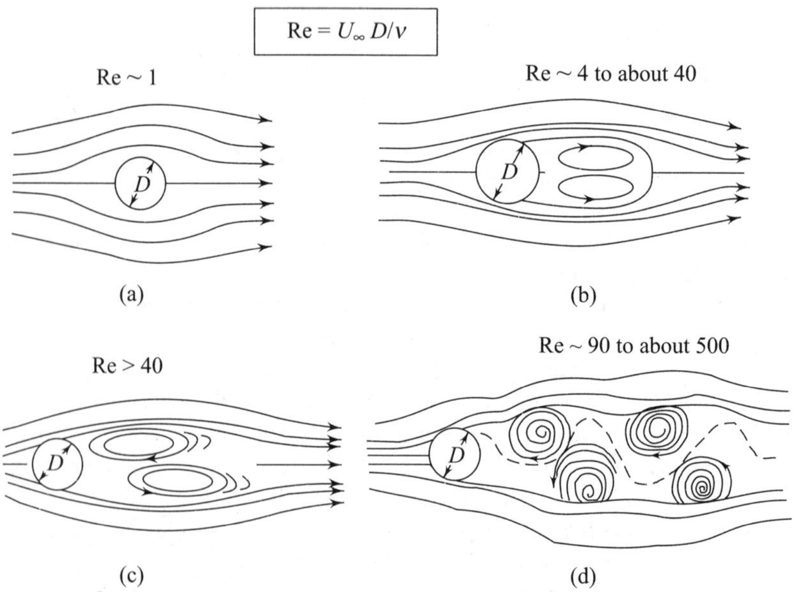

Fig. 9.11 Influence of Reynolds number on wake-zone aerodynamics

At high Reynolds numbers, the large angular velocities and the rates of shear associated with the individual vortices induce random turbulence close to the cylinder. The undesirable vibration caused by a cylinder can be eliminated by attaching a longitudinal fin to the downstream side. These are called wake splitters or splitter plates. If the length of the wake-splitter is greater than one diameter of the cylinder, interaction between the vortices is stopped and shedding ceases. Sometimes tall towers or chimneys have helical projections, similar to screw threads in order to induce asymmetrical three-dimensional flows that destroy alternate shedding.

An understanding of the transitional flow processes helps in practical problems either by improving procedures for predicting positions or for determining methods of advancing or retarding the transition position.

The critical value at which the transition occurs in pipe flow is $\text{Re}_{cr} = 2300$. The actual value depends upon the disturbance in flow. Some experiments have shown the critical Reynolds number to reach as high as 40,000. The precise upper bound is no known, but the lower bound appears to be $\text{Re}_{cr} = 2300$. Below this value, the flow remains laminar even when subjected to strong disturbances.

For $2300 \leq \text{Re}_{cr} \leq 2600$, the flow alternates randomly between laminar and partially turbulent. Near the centerline, the flow is more laminar than turbulent, whereas near the wall, the flow is more turbulent than laminar. For flow over a flat plate, turbulent regime is observed between Reynolds numbers ($U_\infty x/\nu$) of 3.5×10^5 and 10^6.

9.12 SEVERAL EVENTS OF TRANSITION

Transitional flow consists of several events as shown in Fig. 9.12. Let us consider the sequence of events:

1. Region of instability of small wavy disturbances Consider a laminar flow over a flat plate aligned with the flow direction (Fig. 9.12). It has been seen that in the presence of an adverse pressure gradient, at a high Reynolds number (water velocity approximately 9-cm/s), two-dimensional waves appear. These waves can be made visible by a method known as tellurium method. In 1929, Tollmien and Schlichting predicted that the waves would form and grow in the boundary layer. These waves are called *Tollmien-Schlichting* waves.

2. Three-dimensional waves and vortex formation Disturbances in the free stream or oscillations in the upstream boundary layer can generate wave growth, which has a variation in the spanwise direction. This leads an initially two-dimensional wave to a three-dimensional form. In many such transitional flows, periodicity is observed in the spanwise direction. This is accompanied by the appearance of vortices whose axes lie in the direction of flow.

3. Peak-valley development with streamwise vortices As the three-dimensional wave propagates downstream, the boundary layer flow develops into a complex streamwise vortex system. However, within this vortex system, at some spanwise location, the velocities fluctuate violently. These locations are called peaks and the neighbouring locations of the peaks are valleys (Fig. 9.13).

4. Vorticity concentration and shear layer development At the spanwise locations corresponding to the peak, the instantaneous streamwise velocity profiles demonstrate the following. Often, an inflexion is observed on the velocity profile. The inflectional profile appears and disappears once after each cycle of the basic wave.

5. Breakdown The instantaneous velocity profiles produce high shear in the outer region of the boundary layer. The velocity fluctuations develop from the shear layer at a higher frequency than that of the basic wave. However, these velocity fluctuations have a strong ability to amplify any slight three-dimensionality, which is already present in the flow field. As a result, a staggered vortex pattern evolves with the streamwise wavelength twice the wavelength of *Tollmien–Schlichting wavelength*. The spanwise wavelength of these structures is about one-half of the streamwise value. This is known as breakdown. Klebanoff et al. [8] refer to the high frequency fluctuations as hairpin eddies.

6. Turbulent-spot development The hairpin-eddies travel at a speed grater than that of the basic (primary) waves. As they travel downstream, eddies spread in the spanwise direction and towards the wall. The vortices begin a cascading breakdown into smaller vortices. In such a fluctuating state, intense local changes occur at random locations in the shear layer near the wall in the form of turbulent spots. Each spot grows almost linearly with the downstream distance. The creation of spots is considered as the main event of transition.

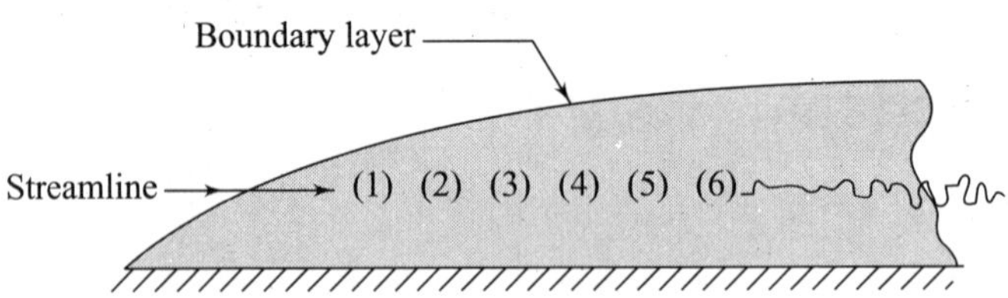

Fig. 9.12 Sequence of event involved in transition

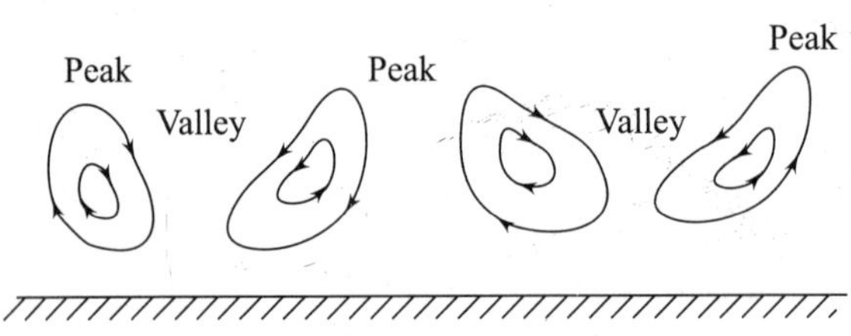

Fig. 9.13 Cross-stream view of the streamwise vortex system

SUMMARY

- The boundary layer is the thin layer of fluid adjacent to the solid surface. Phenomenologically, the effect of viscosity is very prominent within this layer.
- The main stream velocity undergoes a change from zero at the solid-surface to the full magnitude through the boundary layer.
- Effectively, the boundary layer theory is a complement to the inviscid flow theory.
- The governing equation for the boundary layer can be obtained through correct

reduction of the *Navier Stokes equations* within the thin layer referred above. There is no variation in pressure in the y direction within the boundary layer.

- The pressure is impressed on the boundary layer by the outer inviscid flow, which can be calculated using *Bernoulli's equation*.
- The boundary layer equation is a second order non-linear partial differential equation. The exact solution of this equation is known as the *similarity solution*. For the flow over a flat plate, the similarity solution is often referred to as *Blasius solution*. Complete analytical treatment of this solution is beyond the scope of this text. However, the momentum integral equation can be derived from the boundary layer equation which is amenable to analytical treatment.
- The solutions of the momentum integral equation are called approximate solutions of the boundary layer equation.
- The boundary layer equations are valid up to the point of separation. At the point of separation, the flow gets detached from the solid surface due to excessive adverse pressure gradient.
- Beyond the point of separation, the flow reversal produces eddies. During flow past bluff-bodies, the desired pressure recovery does not take place in a separated flow and the situation gives rise to *pressure drag* or *form drag*.

REFERENCES

1. Prandtl, L. Über Flussigkeitsbewegung bei sehr kleiner Reibung, *Proc. Third Intern. Math. Congress, Heidelberg*, pp.484–491, 1904.
2. Schlichting, H., *Boundary Layer Theory*, Seventh Edition, McGraw-Hill, New York, 1987.
3. Liepmann, H.W. and Dhawan, S., "Direct Measurements of Local Skin Friction in Low-speed and High-speed Flow", *Proc. First US National Congress Appl. Mech.*, 869, 1951.
4. Resenhead, L. (Ed), *Laminar Boundary Layers*, Oxford University Press, London, 1963.
5. Thwaites, B., "Approximate Calculation of the Laminar Boundary Layer", *Aeronaut Quart.*, Vol. 1, pp .245–280, 1949.
6. Holstein, H. and Bohlen, T., Ein einfaches Verfahren zur Berechnung Laminarer Reibungs-schichten die dem Naherungsverfahren von K. Pohlhausen genugen, *Lilienthal-Bericht*, S. 10, 5–16, 1940.
7. Walz, A., Ein neuer Anstaz Fur das Geswindigkeitsprofil der laminaren Reibungsschicht, *Lilienthal-Bericht*, 141,pp. 8–12, 1941.
8. Klebanoff, P.S., Tidstrom, K.D., and Sargent, L.M., "The Three-Dimensional Nature of Boundary Layer Instability", *J. Fluid Mech.*, Vol. 12, pp. 1–34, 1962.

EXERCISES

9.1 Choose the correct answer:

(i) The laminar boundary layer thickness on a flat plate varies as

(a) $x^{(-1/2)}$ (b) $x^{(4/5)}$ (c) $x^{(1/2)}$ (d) x^2

(ii) The turbulent boundary layer thickness on a flat plate varies as

(a) $x^{(+1/2)}$ (b) $x^{(4/5)}$ (c) $x^{(1/7)}$ (d) $x^{(6/7)}$

(iii) The injection of air through a porous wall can control the boundary layer separation on the upper curved surface of an aerofoil wing. The injection of fluid also promotes turbulence. The final result is
- (a) increase in the skin friction and decrease in the form drag
- (b) increase in the form drag and decrease in the skin friction
- (c) decrease in both the skin friction and form drag
- (d) increase in both the skin friction and form drag

(iv) In the entrance region of a pipe, the boundary layer grows and the inviscid core accelerates. This is accompanied by a
- (a) rise in pressure
- (b) constant pressure gradient
- (c) fall in pressure in the flow direction
- (d) pressure pulse

(v) Flow separation is caused by
- (a) reduction of pressure to vapour pressure
- (b) a negative pressure gradient
- (c) a positive pressure gradient
- (d) the boundary layer thickness reducing to zero.

(vi) At the point of separation
- (a) shear stress is zero
- (b) velocity is negative
- (c) pressure gradient is negative
- (d) shear stress is maximum

(vii) Choose the expression for the momentum thickness of an incompressible boundary layer

(a) $\dfrac{5.0\,x}{\sqrt{Re_x}}$ (b) $\int_0^\infty \left(1-\dfrac{u}{U_\infty}\right)dy$

(c) $\int_0^\infty \dfrac{u}{U_\infty}\left(1-\dfrac{u}{U_\infty}\right)dy$ (d) $\int_0^\infty (u/U_\infty)\,dy$

(viii) For cross flow over a circular cylinder at a Reynolds number $\left[\text{Re} = \dfrac{U_\infty D}{\nu}\right]$ greater than 200, the wake is
- (a) at a pressure equal to the forward stagnation point
- (b) at a pressure lower than the forward stagnation point
- (c) the principal cause of skin friction
- (d) at a pressure higher than the forward stagnation point

9.2 Two students are asked to solve the Blasius flow over a flat plate to determine the variation of boundary layer thickness as a function of the Reynolds number. One student solves the problem by similarity method and arrives at the result $\dfrac{\delta}{x} = \dfrac{5.0}{\sqrt{Re_x}}$ The other student chooses to solve the problem by using the momentum-integral equation and Karman-Pohlhausen method and finds that $\frac{\delta}{x} = \frac{4.64}{\sqrt{Re_x}}$ Which of the two results is expected to be closer to the experimental results and why?

9.3 A scientist claims that a highly viscous flow around a body can generate the same flow patterns as the flow of an inviscid and incompressible fluid around that body. According to our understanding, the Reynolds number for the first flow is very small, while the Reynolds number for the second flow can be taken to be infinity. Do you think it is possible to get the same flow patterns for the two extreme values of Reynolds number? Use mathematical analysis to prove or disprove the Scientist's claim.

9.4 Nikuradse, a student of Prandtl, obtained experimental data for laminar flow over a flat plate placed at zero angle of attack. His measurements suggest:

$$\frac{u}{U} = a\left(\frac{y}{\delta}\right) + b\left(\frac{y}{\delta}\right)^3$$

where y is the perpendicular distance from the plate. The local velocity is u. Evaluate a and b from the physical boundary conditions. Obtain the expressions for the boundary layer thickness δ, displacement thickness δ^*, momentum thickness δ^{**}, and the shear stress τ_w on the surface of the plate.

Ans. (($a = 3/2$, $b = -1/2$, $\delta/x = 4.64/(\mathrm{Re}_x)^{0.5}$, ($\delta^*/x = 1.73\,(\mathrm{Re}_x)^{0.5}$, $\tau_w = 0.323\,\mu\,(\mathrm{Re}_x)^{0.5}\,U/x$)

9.5 Given the choice between cos Ay and sin Ay as velocity profiles, which one would you prefer? To determine the choice of the profile, find the displacement thickness, momentum thickness, wall shear stress and boundary layer thickness, from the momentum integral equation for flow over a flat plate.

Ans. (sin Ay, $\delta^* = 0.363\,\delta$, $\delta^{**} = 0.137\,\delta$, $\tau_w = \pi\mu\,U_\infty/2\delta$, $\delta/x = 4.8/(\mathrm{Re}_x)^{0.5}$

9.6 In boundary layer theory, a boundary layer can be characterized by any of the following quantities (i) boundary layer thickness (ii) displacement thickness (iii) momentum thickness. How do these quantities differ in their physical as well as mathematical definitions? For the flow over a flat plate, which of these is expected to have the highest value at a given location on the wall, and which the lowest?

9.7 Reduce the Prandtl's boundary layer equations to a simpler form than that given by equations (9.10) – (9.12) for the following cases:

(a) Flow over a flat plate.

(b) The case $\tau_{yx} = C_1$ (a constant)

(c) The case where velocity (v) is directly proportional to kinematic viscosity (ν)

(d) Solve the Prandtl's boundary layer equations for $v = \nu$ assuming pressure gradient $\frac{\partial p}{\partial x} = 0$.

Ans. ((a) $u\frac{\partial u}{\partial x} + v\frac{\partial u}{\partial y} = \nu\frac{\partial^2 u}{\partial y^2}$, (b) $p + \frac{1}{2}\rho u^2 = -\frac{C_1}{\nu}\int v\,dx$,

(c) $\frac{d^2 u}{dy^2} - a\frac{du}{dy} = \frac{1}{\mu}\frac{dp}{dx}$, (d) $u = \frac{C_1}{a}\exp(ay) + C_2$)

9.8 For the laminar flow over a flat plate, the experiments confirm the velocity profile $\frac{u}{U}=\frac{3}{2}\left(\frac{y}{\delta}\right)-\frac{1}{2}\left(\frac{y}{\delta}\right)^3$. For the turbulent flow over a flat plate, the experimental observations over a range of Reynolds number suggest $\frac{u}{U}=\left(\frac{y}{\delta}\right)^{1/7}$. Find the ratio of (δ^*/δ) for laminar and turbulent cases.

Ans. $((\delta^*/\delta)_{\text{lam}}=3/8, (\delta^*/\delta)_{\text{tur}}=1/8)$

9.9 Assuming the velocity profile $\frac{u}{U_\infty}=\tanh\frac{y}{a(x)}$ for the boundary layer over a flat plate at zero incidence, find the relations for δ, δ^* δ^{**} and τ_w. Check whether the profile satisfies all the boundary conditions.
Note: $a(x)\neq\delta(x)$ where δ is such that at $y=\delta$, $u=0.99\,U_\infty$

Ans. $((\delta/x=6.76/(\text{Re}_x)^{0.5}, \delta^*/x=1.77/(\text{Re}_x)^{0.5}\ \delta^{**}/x=0.783/(\text{Re}_x)^{0.5},$
$\tau_w=\mu U_\infty(\text{Re}_x)^{0.5}/2.553\,x)$

9.10 An approximate expression for the velocity profile in a steady, two - dimensional incompressible boundary layer is

$$\frac{u}{U_\infty}=1-e^{-\eta}+k\left(1-e^{-\eta}-\sin\frac{\pi\eta}{6}\right),\quad \text{for } 0\le\eta\le 3$$

$$=1-e^{-\eta}-ke^{-\eta},\quad \text{for}\quad \eta\ge 3$$

where $\eta=y/\delta(x)$. Show that the profile satisfies the following boundary conditions:

$$\text{at } y=0,\ u=0$$

$$\text{at } y=\infty,\ u=U_\infty,\ \frac{\partial u}{\partial y}=\frac{\partial^2 u}{\partial y^2}=0$$

Also find out k from an appropriate boundary condition.

Ans. $(k=(\delta^2/\nu)(\mathrm{d}U_\infty/\mathrm{d}x)-1)$

9.11 Water of kinematic viscosity $\nu=1.02\times10^{-6}\ \text{m}^2/\text{s}$ is flowing steadily over a smooth flatplate at zero angle of attack with a velocity 1.6 m/s. The length of the plate is 0.3 m. Calculate (i) the thickness of the boundary layer at 15 cm from the leading edge (ii) the rate of growth of boundary layer at 15 cm from the leading edge, and (iii) the total drag coefficient on one side of the plate. Assume a parabolic velocity profile.

Ans. $(\delta=1.7\ \text{mm}, \mathrm{d}\delta/\mathrm{d}x=5.625\times10^{-3}, \overline{C}_f=1.935\times10^{-3})$

9.12 Air flows steadily over a smooth flat plate at a velocity of 4.39 m/s. The length of the plate is 12.2 m in the direction of the flow. The kinematic viscosity and density of air are $1.34\times10^{-5}\ \text{m}^2/\text{s}$ and $1.031\ \text{kg/m}^3$ respectively. Find (a) the boundary layer thickness at 15.24 cm from the leading edge and (b) the drag coefficient. Assume a parabolic velocity profile.

Ans. ((a) 3.41×10^{-5} m, (b) 5.94×10^{-3})

9.13 Use the Prandtl's boundary layer equations and show that the velocity profile for a laminar flow past a flat plate has an infinite radius of curvature on the surface of the plate.

9.14 What do you mean by the "point of separation" of a boundary layer? How will the velocity gradient $\frac{\partial u}{\partial y}$ and the second gradient $\frac{\partial^2 u}{\partial y^2}$ vary within the boundary layer at the point of separation? Show the variation graphically. Here u is the velocity along the wall and y is the co-ordinate perpendicular to the wall.

9.15 Water flows between two parallel walls as shown in Fig. 9.14. The velocity is uniform at the entrance and core region. Beyond a distance Le downstream from the entrance, the flow becomes fully developed so that the velocity varies over the entire width $2h$ of the channel. In the boundary layer region, velocity varies as $u = U(x)\left(\frac{y}{\delta}\right)^2$, where $\delta = \alpha\sqrt{x}$; α being a constant. Determine the acceleration on the axis of symmetry for $0 \le x \le Le$.

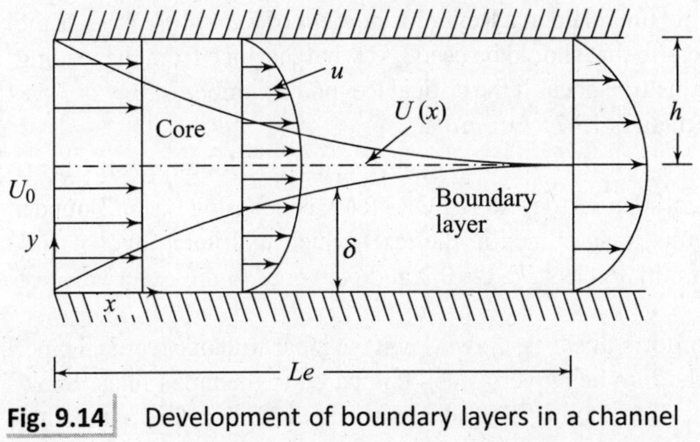

Fig. 9.14 Development of boundary layers in a channel

Ans. ($a = (U_0^2/\,3Le)\,(x/Le)^{-1/2}/(1-(2/3)(x/Le)^{0.5})^3$)

9.16 Consider the laminar boundary layer on a flat plate with uniform suction velocity V_0 as shown in Fig. 9.15.

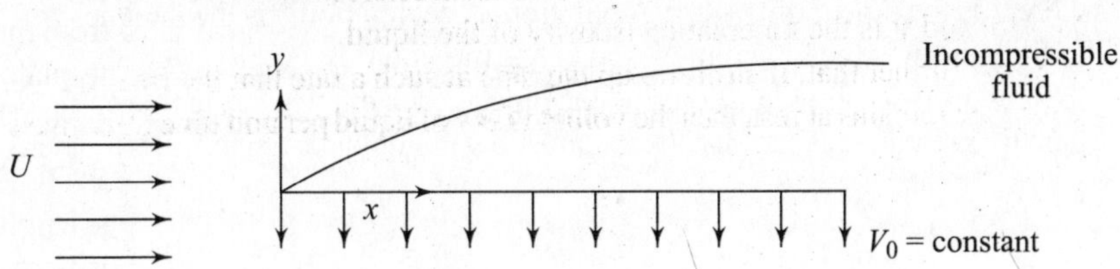

Fig. 9.15 Flow on a flat plate with uniform suction velocity

Far down the plate (large x), a fully developed situation may be shown to exist in which the velocity distribution does not vary with x. Find the velocity distribution in this region, as well as the wall shear. The governing equations are

$$\frac{\partial u}{\partial x} + \frac{\partial v}{\partial y} = 0 \quad \text{and} \quad u\frac{\partial u}{\partial x} + v\frac{\partial v}{\partial y} = -\frac{1}{\rho}\frac{dp}{dx} + \nu\frac{\partial^2 u}{\partial y^2}$$

The boundary conditions are at $y = 0$, $u = 0$, $v = V_0$ and $u(\infty) = U$

Ans. ($u = U\{1 - \exp(-V_0 y/\nu)\}$, $\tau_w = \rho U V_0$)

9.17 Consider a laminar boundary layer on flat plate with a velocity profile given by

$$\frac{u}{U} = \frac{3}{2}\eta - \frac{1}{2}\eta^3, \qquad \text{where, } \eta = y/\delta$$

For this profile $\dfrac{\delta}{x} = \dfrac{4.64}{\text{Re}_x}$.

Determine an expression for the local skin friction coefficient in terms of distance and flow properties.

Ans. ($C_f = 0.647\,(\text{Re}_x)^{0.5}$)

9.18 A low-speed wind tunnel is provided with air supply up to a speed of 50 m/s at 20 °C. One needs to study the behaviour of the boundary layer over a flat plate kept inside the wind tunnel up to a Reynolds number of $\text{Re}_x = 10^8$. What is the minimum plate length that should be used? At what distance from the leading edge would the transition occur if the critical Reynolds number is $\text{Re}_{x,c} = 3.5 \times 10^5$? At 25 °C, ν of air is 15.7×10^{-6} m²/s.

Ans. ($x_{\text{min}} = 31.4$ m, $x_{\text{cr}} = 0.109$ m)

9.19 Perform a numerical solution (develop a FORTRAN code) using Eq. (9.24) and a Runge–Kutta method (as outlined in the text) which will iterate the Blasius equation from an initial guess $H(0) = 0.2$ and converge to the exact value of $H(0) = 0.4696$.

9.20 A liquid film of uniform thickness flows down the inner wall of a vertical pipe, the thickness of the film being very small compared to the pipe radius. Show that, in the absence of any appreciable shear force on the free surface of the film, the volume flow of liquid per unit time, Q_1, is given by

$$Q_1 = \frac{2\pi r g t^3}{3\nu}$$

where, r is the pipe radius, g the gravitational acceleration, t is the thickness of the film and ν is the kinematic viscosity of the liquid.

Show further that, if air flows up the pipe at such a rate that the free surface of the film remains at rest, then the volume flow of liquid per unit time Q, is given by

$$Q \cong \frac{Q_1}{4}\left(1 - \frac{1}{\rho g}\frac{dp}{dx}\right).$$

where $\frac{\mathrm{d}p}{\mathrm{d}x}$ is the pressure drop along the pipe, ρ is the density of the liquid and other symbols are as defined above.

9.21 The velocity distribution in the laminar boundary layer is of the form

$$\frac{u}{U_e} = F(\eta) + \lambda G(\eta)$$

where, $$F(\eta) = \frac{3}{2}\eta - \frac{\eta^3}{2}\ ;\ G(\eta) = \frac{\eta}{4} - \frac{\eta^2}{2} + \frac{\eta^3}{4};$$

$$\eta = \frac{y}{\delta} \text{ and } \lambda = \frac{\delta^2}{\nu}\frac{\mathrm{d}Ue}{\mathrm{d}x}$$

Find the value of λ when the flow is on the point of separating and show that then the displacement thickness will be half the boundary layer thickness.

9.22 Show that the shape factor (H) has the value $\approx$ 2.6 for the boundary layer flow over a flat plate. Also calculate the position where the flow is critical for flow velocity of 3.048 m/s and kinematic viscosity 9.29×10^{-7} m^2/s. Given that at the critical location Reynolds Number (based on distance from the leading edge surface) is related to shape factor (H) by $-\log(R_{\text{critical}}) = \text{H}$.

9.23 Determine the distance downstream from the bow of a ship moving at 3.9 m/s relative to still water at which the boundary layer will become turbulent. Also find the boundary layer thickness and total friction drag coefficient for this portion of the surface of the ship. Given the kinematic viscosity = 1.124×10^{-6} m^2/s.

Ans. (0.144 m, 1.02×10^{-3} m, 1.88×10^{-3})

10

TURBULENT FLOW

10.1 INTRODUCTION

A turbulent motion is an irregular motion. The irregularity associated with turbulence is such that it can be described by the laws of probability and turbulent fluid motion can be considered as an irregular condition of flow in which various quantities (such as velocity components and pressure) show a random variation with time and space in such a way that the statistical average of those quantities can be quantitatively expressed.

An irregular motion is associated with random fluctuations. It is postulated that the fluctuations inherently come from disturbances (such as roughness of a solid surface) and they may be either dampened out due to viscous damping or may grow by drawing energy from the free stream. At a Reynolds number less than the critical, the kinetic energy of flow is not enough to sustain random fluctuation against the viscous damping and in such cases laminar flow continues to exist. At a somewhat higher Reynolds number than the critical Reynolds number, the kinetic energy of flow supports the growth of fluctuations and transition to turbulence takes place.

10.2 CHARACTERISTICS OF TURBULENT FLOW

In view of the preceding discussion, the most important characteristic of turbulent motion is the fact that velocity and pressure at a point fluctuate with time in a random manner (Fig. 10.1).

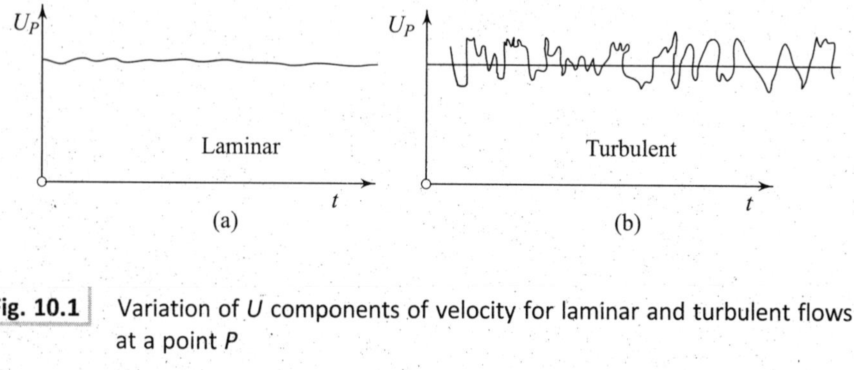

Fig. 10.1 Variation of U components of velocity for laminar and turbulent flows at a point P

The mixing in turbulent flow is more due to these fluctuations. As a result we can see more uniform velocity distributions in turbulent pipe flows as compared to laminar flows (see Fig. 10.2).

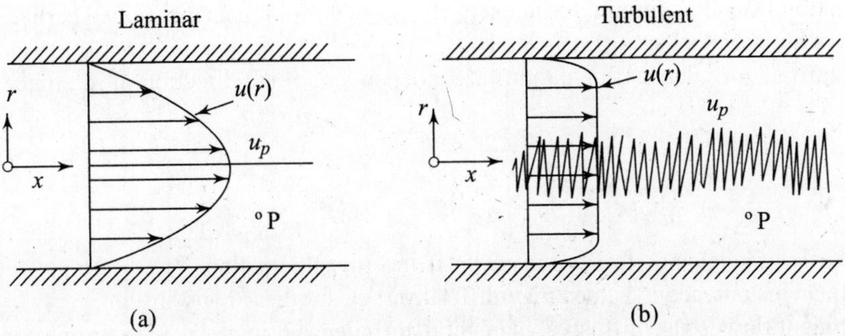

Fig. 10.2 Comparison of velocity profiles in a pipe for (a) laminar and (b) turbulent flows

Turbulence can be generated by frictional forces at the confining solid walls or by the flow of layers of fluids with different velocities over one another. The turbulence generated in these two ways are considered to be different. Turbulence generated and continuously affected by fixed walls is designated as *wall turbulence*, and turbulence generated by two adjacent layers of fluid in the absence of walls is termed as *free turbulence*.

One of the effects of viscosity on turbulence is to make the flow more homogeneous and less dependent on direction. If the turbulence has the same structure quantitatively in all parts of the flow field, the turbulence is said to be *homogeneous*. The turbulence is called *isotropic* if its statistical features have no directional preference and perfect disorder persists. Its velocity fluctuations are independent of the axis of reference, i.e., invariant to axis rotation and reflection. Isotropic turbulence is by its definition always homogeneous. In such a situation, the gradient of the mean velocity does not exist. The mean velocity is either zero or constant throughout. However, when the mean velocity has a gradient, the turbulence is called *anisotropic*.

A little more discussion on homogeneous and isotropic turbulence is needed at this stage. The term 'homogeneous turbulence' implies that the velocity fluctuations in the system are random. The average turbulent characteristics are independent of the position in the fluid, i.e., invariant to axis translation.

Consider the root mean square velocity fluctuations:

$$u' = \sqrt{\overline{u^2}}\ , v' = \sqrt{\overline{v^2}}\ , w' = \sqrt{\overline{w^2}}$$

In homogeneous turbulence, the rms values of u', v' and w' can all be different, but each value must be constant over the entire turbulent field. It is also to be understood that even if the rms fluctuation of any component, say u's are constant over the entire field, the instantaneous values of u may differ from point to point at any instant.

In addition to its homogeneous nature, if the velocity fluctuations are independent of the axis of reference, i.e., invariant to axis rotation and reflection, the situation leads to isotropic turbulence, which by definition as mentioned earlier, is always homogeneous.

In isotropic turbulence, fluctuations are independent of the direction of reference and

$$\sqrt{\overline{u^2}} = \sqrt{\overline{v^2}} = \sqrt{\overline{w^2}}$$

or

$$u' = v' = w'$$

Again, it is of relevance to say that even if the rms fluctuations at any point are same, their instantaneous values may differ from each other at any instant.

Turbulent flow is also diffusive. In general, turbulence brings about better mixing of a fluid and produces an additional diffusive effect. The term 'eddy-diffusion' is often used to distinguish this effect from molecular diffusion. The effects caused by mixing are as if the viscosity is increased by a factor of 100 or more. At a large Reynolds number there exists a continuous transport of energy from the free stream to the large eddies. Then, from the large eddies smaller eddies are continuously formed. Near the wall, the smallest eddies dissipate energy and destroy themselves.

10.3 LAMINAR–TURBULENT TRANSITION

For turbulent flow over a flat plate, the boundary layer starts out as laminar flow at the leading edge and subsequently, the flow turns into transition flow and very shortly thereafter turns into turbulent flow. The turbulent boundary layer continues to grow in thickness, with a small region below it, called a *viscous sublayer*. In this *sublayer*, the flow is well behaved, just as the laminar boundary layer (Fig. 10.3).

A careful observation further suggests that at a certain axial location, the laminar boundary layer tends to become unstable. Physically this means that the disturbances in the flow grow in amplitude at this location.

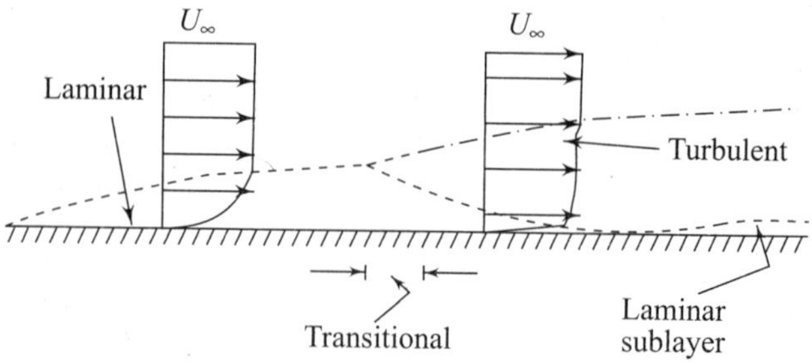

Fig. 10.3 Laminar-turbulent transition

Free stream turbulence, wall roughness and acoustic signals may be among the sources of such disturbances. Transition to turbulent flow is thus initiated with the instability in laminar flow. The possibility of instability in boundary layer was felt by Prandtl as early as 1912. The theoretical analysis of Tollmien and Schlichting showed that unstable waves could exist if the Reynolds number was 575. The Reynolds number was defined as

$$\mathrm{Re} = U_\infty \delta^* / \nu$$

where U_∞ is the free stream velocity and δ^* is the displacement thickness. Taylor developed an alternate theory, which assumed that the transition is caused by a momentary separation at the boundary layer associated with the free stream turbulence. In a pipe flow the initiation of turbulence is usually observed at Reynolds numbers ($U_\infty D/\nu$) in the range of 2000 to 2700. The development starts with a laminar profile, undergoes a transition, changes over to turbulent profile and then stays turbulent thereafter (Fig. 10.4). The length of development is of the order of 25 to 40 diameters of the pipe. The mechanisms related to the growth and the decay of turbulence in a flow field are indeed an advanced topic and beyond the scope of this text. However, interested readers may like to refer to Tennekes and Lumley and Hinze for advanced knowledge.

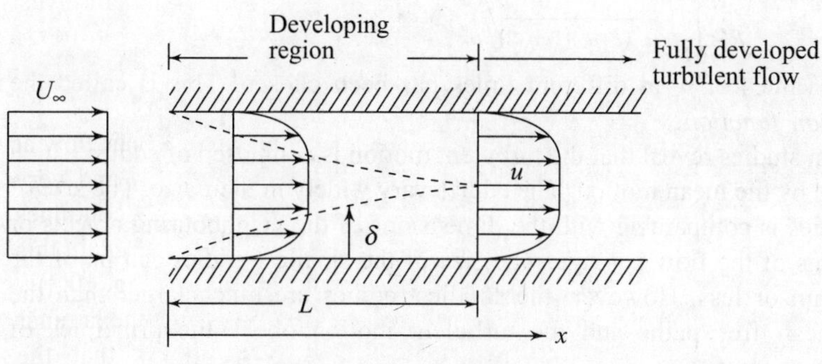

Fig. 10.4 Development of turbulent flow in a circular duct

10.4 CORRELATION FUNCTIONS

A statistical correlation can be applied to fluctuating velocity terms in turbulence. Turbulent motion is by definition eddying motion. Notwithstanding the circulation strength of the individual eddies, a high degree of correlation exists between the velocities at two points in space, if the distance between the points is smaller than the diameter of the eddy.

Consider a statistical property of a random variable (velocity) at two points separated by a distance r. An Eulerian correlation tensor (nine terms) at the two points can be defined by

$$\mathbf{Q} = \overline{\mathbf{u}(x)\mathbf{u}\,(x+r)}$$

In other words, the dependence between the two velocities at two points is measured by the correlations, i.e., the time averages of the products of the quantities measured at two points. The correlation of the u' components of the turbulent velocity of these two points is defined as

$$\overline{u'(x)u'(x+r)}$$

It is conventional to work with the non-dimensional form of the correlation, such as

$$R(r) = \frac{\overline{u'(x)u'(x+r)}}{\left(\overline{u'^2}(x)\overline{u'^2}(x+r)\right)^{1/2}}$$

A value of $R(r)$ of unity signifies a perfect correlation of the two quantities involved and their motion is in phase. Negative value of the correlation function implies that the time averages of the velocities in the two correlated points have different signs. Figure 10.5 shows typical variations of the correlation R with increasing separation r.

To describe the evolution of a fluctuating function $u'(t)$, we need to know the manner in which the value of u' at different times are related. For this purpose a correlation function

$$R(\tau) = \overline{u'(t)u'(t+\tau)}/\overline{u'^2}$$

between the values of u' at different times has been chosen. This is called the *autocorrelation function*.

Correlation studies reveal that the turbulent motion is composed of eddies which are convected by the mean motion. The eddies vary widely in their size. The size of the large eddies is comparable with the dimensions of the neighbouring objects or the dimensions of the flow passage. The size of the smallest eddies can be of the order of 1 mm or less. However, the smallest eddies are much larger than the molecular mean free paths and the turbulent motion obeys the principles of continuum mechanics.

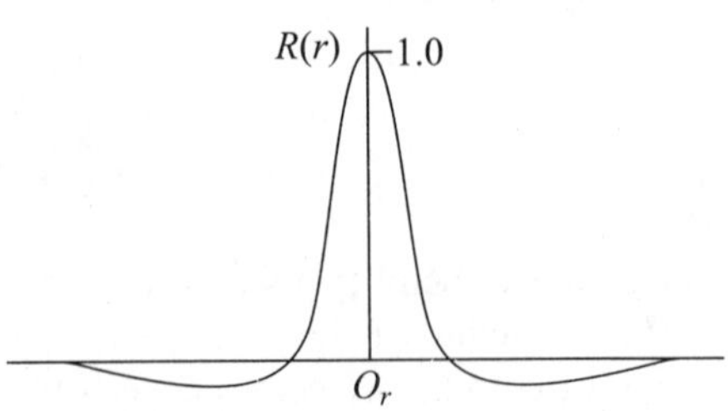

Fig. 10.5 Variation of R with the distance of separation r

10.5 MEAN MOTION AND FLUCTUATIONS

In 1883, O. Reynolds conducted experiments with pipe flow by feeding into the stream a thin thread of liquid dye. For low Reynolds numbers, the dye traced a straight line and did not disperse. With increasing velocity, the dye thread got mixed in all directions and the flowing fluid appeared to be uniformly coloured in the downstream. It was conjectured that on the main motion in the direction of the axis, there existed a superimposed motion all along the main motion at right angles to it. The superimposed motion causes exchange of momentum in transverse direction and the velocity distribution over the cross section is more uniform than in laminar flow. This description of turbulent flow which consists of superimposed streaming and fluctuating (eddying) motion is referred to as *Reynolds decomposition of turbulent flow*.

Here we shall discuss different descriptions of mean motion. Generally, for Eulerian velocity u, the following two methods of averaging could be obtained:

(i) Time average for a stationary turbulence

$$\overline{u}^t(x_0) = \lim_{t_1 \to \infty} \frac{1}{2t_1} \int_{-t_1}^{t_1} u(x_0, t)\, \mathrm{d}t$$

(ii) Space average for a homogeneous turbulence

$$\overline{u}^s(t_0) = \lim_{x \to \infty} \frac{1}{2x} \int_{-x}^{x} u(x, t_0)\, \mathrm{d}x$$

For a stationary and homogeneous turbulence, it is assumed that the two averages lead to the same result: $\overline{u}^t = \overline{u}^s$ and the assumption is known as the *ergodic hypothesis*.

In our analysis, average of any quantity will be evaluated as a *time average*. We take t_1 as a finite interval. This interval must be larger than the timescale of turbulence. Needless to say that it must be small compared with the period t_2 of any slow variation (such as periodicity of the mean flow) in the flow field that we do not consider to be *chaotic* or *turbulent*.

Thus, for a parallel flow, it can be written that the axial velocity component is

$$u(y, t) = \overline{u}(y) + u'(\Gamma, t) \tag{10.1}$$

As such, the time mean component $\overline{u}(y)$ determines whether the turbulent motion is steady or not. Let us look at two different turbulent motions described in Fig. 10.6 (a) and (b). The symbol Γ signifies any of the space variables.

While the motion described by Fig. 10.6(a) is for a turbulent flow with steady mean velocity, Fig. 10.6(b) shows an example of turbulent flow with unsteady mean velocity. The time period of the high-frequency fluctuating component is t_1, whereas the time period for the unsteady mean motion is t_2 and for obvious reason $t_2 >> t_1$. Even if the bulk motion is parallel, the fluctuation u' being random varies in all directions. Now let us look at the continuity equation

$$\frac{\partial u}{\partial x}+\frac{\partial v}{\partial y}+\frac{\partial w}{\partial z}=0$$

Invoking Eq. (10.1) in the above expression, we get

$$\frac{\partial \bar{u}}{\partial x}+\frac{\partial u'}{\partial x}+\frac{\partial v}{\partial y}+\frac{\partial w}{\partial z}=0 \quad (10.2)$$

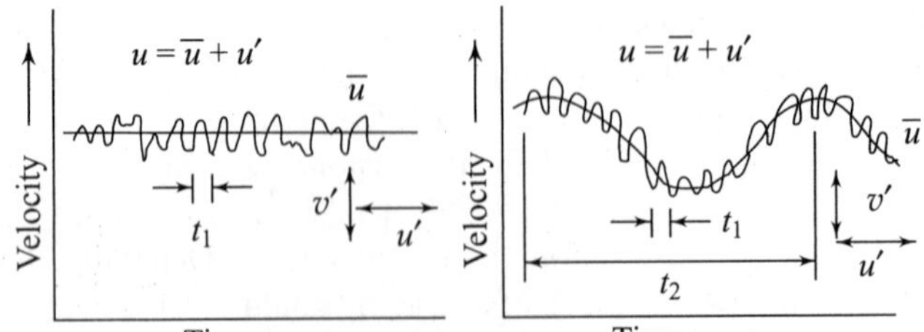

(a) Steady (b) Unsteady

Fig. 10.6 Steady and unsteady mean motions in a turbulent flow

Since $\dfrac{\partial u'}{\partial x} \neq 0$, Eq. (10.2) depicts that *y* and *z* components of velocity exist even for the parallel flow if the flow is turbulent. We can write

$$\left.\begin{aligned} u(y,t) &= \bar{u}(y) + u'(\Gamma, t) \\ v &= 0 + v'(\Gamma, t) \\ w &= 0 + w'(\Gamma, t) \end{aligned}\right\} \quad (10.3)$$

However, the fluctuating components do not bring about the bulk displacement of a fluid element. The instantaneous displacement is $u'\,dt$, and that is not responsible for the bulk motion. We can conclude from the above

$$\int_{-T}^{T} u'\,dt = 0, \qquad (t_1 < T \leq t_2)$$

Due to the interaction of fluctuating components, macroscopic momentum transport takes place. Therefore, interaction effect between two fluctuating components over a long period is non-zero and this can be expressed as

$$\int_{-T}^{T} u'\,v'\,dt \neq 0$$

We take time average of these two integrals and write

$$\overline{u'} = \frac{1}{2T}\int_{-T}^{T} u'\,dt = 0 \quad (10.4a)$$

and $$\overline{u'v'} = \frac{1}{2T}\int_{-T}^{T} u'v'\,dt \neq 0 \tag{10.4b}$$

Now, we can make a general statement with any two fluctuating parameters, say, f' and g', as

$$\overline{f'} = \overline{g'} = 0 \tag{10.5a}$$

$$\overline{f'g'} \neq 0 \tag{10.5b}$$

The time averages of the spatial gradients of the fluctuating components also follow the same laws, and they can be written as

$$\left.\begin{aligned} &\overline{\frac{\partial f'}{\partial s}} = \overline{\frac{\partial^2 f'}{\partial s^2}} = 0 \\ \text{and} \quad &\overline{\frac{\partial (f'g')}{\partial s}} \neq 0 \end{aligned}\right\} \tag{10.6}$$

The *intensity of turbulence* or *degree of turbulence* in a flow is described by the relative magnitude of the root mean square value of the fluctuating components with respect to the time-averaged main velocity. The mathematical expression is given by

$$I = \sqrt{\frac{1}{3}\left(\overline{u'^2} + \overline{v'^2} + \overline{w'^2}\right)} \Big/ U_\infty \tag{10.7a}$$

The degree of turbulence in a wind tunnel can be brought down by introducing screens of fine mesh at the bell mouth entry. In general, at a certain distance from the screens, the turbulence in a wind tunnel becomes isotropic, i.e., the mean oscillation in the three components are equal,

$$\overline{u'^2} = \overline{v'^2} = \overline{w'^2}$$

In this case, it is sufficient to consider the oscillation u' in the direction of flow and to put

$$I = \sqrt{\overline{u'^2}} \Big/ U_\infty \tag{10.7b}$$

This simpler definition of turbulence intensity is often used in practice even in cases when turbulence is not isotropic.

Following Reynolds decomposition, it is suggested to separate the motion into a mean motion and a fluctuating or eddying motion. Denoting the time average of the u component of velocity by $\overline{u}$ and fluctuating component as u', we can write the following:

$$u = \overline{u} + u', \quad v = \overline{v} + v', \quad w = \overline{w} + w', \quad p = \overline{p} + p' \tag{10.8}$$

By definition, the time averages of all quantities describing fluctuations are equal to zero.

$$\overline{u'} = 0, \quad \overline{v'} = 0, \quad \overline{w'} = 0, \quad \overline{p'} = 0 \tag{10.9}$$

The fluctuations u', v', and w' influence the mean motion $\bar{u}$, $\bar{v}$ and $\bar{w}$ in such a way that the mean motion exhibits an apparent increase in the resistance to deformation. In other words, the effect of fluctuations is an apparent increase in viscosity or macroscopic momentum diffusivity.

We shall state some rules of mean time-averages here. If f and g are two dependent variables and if s denotes any one of the independent variables x, y, z, t, then

$$\overline{\bar{f}} = \bar{f}; \quad \overline{f+g} = \bar{f} + \bar{g}; \quad \overline{\bar{f}\cdot g} = \bar{f}\cdot\bar{g};$$

$$\overline{\frac{\partial f}{\partial s}} = \frac{\partial \bar{f}}{\partial s}; \quad \overline{\int f \, \mathrm{d}s} = \int \bar{f} \, \mathrm{d}s \tag{10.10}$$

10.6 DERIVATION OF GOVERNING EQUATIONS FOR TURBULENT FLOW

For incompressible flows, the Navier–Stokes equations can be rearranged in the form

$$\rho\left[\frac{\partial u}{\partial t} + \frac{\partial (u^2)}{\partial x} + \frac{\partial (uv)}{\partial y} + \frac{\partial (uw)}{\partial z}\right] = -\frac{\partial p}{\partial x} + \mu \nabla^2 u \tag{10.11a}$$

$$\rho\left[\frac{\partial v}{\partial t} + \frac{\partial (uv)}{\partial x} + \frac{\partial (v^2)}{\partial y} + \frac{\partial (vw)}{\partial z}\right] = -\frac{\partial p}{\partial y} + \mu \nabla^2 v \tag{10.11b}$$

$$\rho\left[\frac{\partial w}{\partial t} + \frac{\partial (uw)}{\partial x} + \frac{\partial (vw)}{\partial y} + \frac{\partial (w^2)}{\partial z}\right] = -\frac{\partial p}{\partial z} + \mu \nabla^2 w \tag{10.11c}$$

and

$$\frac{\partial u}{\partial x} + \frac{\partial v}{\partial y} + \frac{\partial w}{\partial z} = 0 \tag{10.12}$$

Let us express the velocity components and pressure in terms of time-mean values and corresponding fluctuations. In continuity equation, this substitution and subsequent time averaging will lead to

$$\overline{\frac{\partial(\bar{u} + u')}{\partial x}} + \overline{\frac{\partial(\bar{v} + v')}{\partial y}} + \overline{\frac{\partial(\bar{w} + w')}{\partial z}} = 0$$

or

$$\left(\frac{\partial \bar{u}}{\partial x} + \frac{\partial \bar{v}}{\partial y} + \frac{\partial \bar{w}}{\partial z}\right) + \left(\frac{\partial \overline{u'}}{\partial x} + \frac{\partial \overline{v'}}{\partial y} + \frac{\partial \overline{w'}}{\partial z}\right) = 0$$

Since

$$\frac{\partial \overline{u'}}{\partial x} = \frac{\partial \overline{v'}}{\partial y} = \frac{\partial \overline{w'}}{\partial z} = 0 \quad \text{[From Eq. (10.6)]}$$

We can write

$$\frac{\partial \bar{u}}{\partial x} + \frac{\partial \bar{v}}{\partial y} + \frac{\partial \bar{w}}{\partial z} = 0 \tag{10.13a}$$

From Eqs (10.13a) and (10.12), we obtain

$$\frac{\partial u'}{\partial x}+\frac{\partial v'}{\partial y}+\frac{\partial w'}{\partial z}=0 \tag{10.13b}$$

It is evident that the time-averaged velocity components and the fluctuating velocity components, each satisfy the continuity equation for incompressible flow. Let us imagine a two-dimensional flow in which the turbulent components are independent of the z direction. Eventually, Eq. (10.13b) tends to become

$$\frac{\partial u'}{\partial x}=-\frac{\partial v'}{\partial y} \tag{10.14}$$

On the basis of condition (10.14), it is postulated that if at an instant there is an increase in u' in the x direction, it will be followed by an increase in v' in the negative y direction. In other words, $\overline{u'v'}$ is non-zero and negative. This is another important consideration within the framework of mean-motion description of turbulent flows.

Invoking the concepts of (10.8) into the equations of motion (10.11a, b, c), we obtain expressions in terms of mean and fluctuating components. Now, forming time averages and considering the rules of (10.10), we discern the following. The terms which are linear, such as $\partial u'/\partial t$ and $\partial^2 u'/\partial x^2$ vanish when they are averaged [from (10.6)]. The same is true for the mixed terms like $\bar{u}\cdot u'$, or $\bar{u}\cdot v'$, but the quadratic terms in the fluctuating components remain in the equations. After averaging, they form $\overline{u'^2}$, $\overline{u'v'}$, etc.

For example, if we perform the aforesaid exercise on the x momentum equation, we shall obtain

$$\rho\left[\frac{\overline{\partial(\bar{u}+u')}}{\mathrm{d}t}+\frac{\overline{\partial(\bar{u}+u')^2}}{\partial x}+\frac{\overline{\partial(\bar{u}+u')(\bar{v}+v')}}{\partial y}+\frac{\overline{\partial(\bar{u}+u')(\bar{w}+w')}}{\partial z}\right]$$

$$=-\frac{\overline{\partial(\bar{p}+p')}}{\partial x}+\mu\left[\frac{\overline{\partial^2(\bar{u}+u')}}{\partial x^2}+\frac{\overline{\partial^2(\bar{u}+u')}}{\partial y^2}+\frac{\overline{\partial^2(\bar{u}+u')}}{\partial z^2}\right]$$

or $$\rho\left\{\frac{\partial\bar{u}}{\partial t}^{0}+\frac{\partial\overline{u'}}{\partial t}^{0}+\frac{\partial(\bar{u}^2+\overline{u'^2})}{\partial x}+\frac{\partial(\bar{u}\cdot\bar{v}+\overline{u'v'})}{\partial y}+\frac{\partial(\bar{u}\cdot\bar{w}+\overline{u'w'})}{\partial z}\right\}$$

$$=-\frac{\partial\bar{p}}{\partial x}-\frac{\partial\overline{p'}}{\partial x}^{0}+\mu\left[\frac{\partial^2\bar{u}}{\partial x^2}+\frac{\partial^2\bar{u}}{\partial y^2}+\frac{\partial^2\bar{u}}{\partial z^2}+\left(\frac{\partial^2\overline{u'}}{\partial x^2}^{0}+\frac{\partial^2\overline{u'}}{\partial y^2}^{0}+\frac{\partial^2\overline{u'}}{\partial z^2}^{0}\right)\right]$$

or $$\rho\left\{\frac{\partial(\bar{u}^2)}{\partial x}+\frac{\partial(\bar{u}\cdot\bar{v})}{\partial y}+\frac{\partial(\bar{u}\cdot\bar{w})}{\partial z}\right\}=-\frac{\partial\bar{p}}{\partial x}+\mu\nabla^2\bar{u}$$

$$-\rho\left[\frac{\partial}{\partial x}\overline{u'^2}+\frac{\partial}{\partial y}\overline{u'v'}+\frac{\partial}{\partial z}\overline{u'w'}\right]$$

Introducing simplifications arising out of continuity Eq. (10.13a), we shall obtain

$$\rho\left[\bar{u}\frac{\partial\bar{u}}{\partial x}+\bar{v}\frac{\partial\bar{u}}{\partial y}+\bar{w}\frac{\partial\bar{u}}{\partial z}\right]=-\frac{\partial\bar{p}}{\partial x}+\mu\nabla^2\bar{u}$$

$$-\rho\left[\frac{\partial}{\partial x}\overline{u'^2}+\frac{\partial}{\partial y}\overline{u'v'}+\frac{\partial}{\partial z}\overline{u'w'}\right]$$

Performing a similar treatment on y and z momentum equations, finally we obtain the momentum equations in the form

$$\rho\left[\bar{u}\frac{\partial\bar{u}}{\partial x}+\bar{v}\frac{\partial\bar{u}}{\partial y}+\bar{w}\frac{\partial\bar{u}}{\partial z}\right]=-\frac{\partial\bar{p}}{\partial x}+\mu\nabla^2\bar{u}$$

$$-\rho\left[\frac{\partial}{\partial x}\overline{u'^2}+\frac{\partial}{\partial y}\overline{u'v'}+\frac{\partial}{\partial z}\overline{u'w'}\right] \tag{10.15a}$$

$$\rho\left[\bar{u}\frac{\partial\bar{v}}{\partial x}+\bar{v}\frac{\partial\bar{v}}{\partial y}+\bar{w}\frac{\partial\bar{v}}{\partial z}\right]=-\frac{\partial\bar{p}}{\partial y}+\mu\nabla^2\bar{v}$$

$$-\rho\left[\frac{\partial}{\partial x}\overline{u'v'}+\frac{\partial}{\partial y}\overline{v'^2}+\frac{\partial}{\partial z}\overline{v'w'}\right] \tag{10.15b}$$

$$\rho\left[\bar{u}\frac{\partial\bar{w}}{\partial x}+\bar{v}\frac{\partial\bar{w}}{\partial y}+\bar{w}\frac{\partial\bar{w}}{\partial z}\right]=-\frac{\partial\bar{p}}{\partial z}+\mu\nabla^2\bar{w}$$

$$-\rho\left[\frac{\partial}{\partial x}\overline{u'w'}+\frac{\partial}{\partial y}\overline{v'w'}+\frac{\partial}{\partial z}\overline{w'^2}\right] \tag{10.15c}$$

The left-hand side of Eqs (10.15a)–(10.15c) are essentially similar to the steady-state Navier–Stokes equations if the velocity components u, v and w are replaced by $\bar{u}$, $\bar{v}$ and $\bar{w}$ respectively. The same argument holds good for the first two terms on the right-hand side of Eqs (10.15a)–(10.15c). However, the equations contain some additional terms which depend on turbulent fluctuations of the stream. These additional terms can be interpreted as components of a stress tensor. Now, the resultant surface force per unit area due to these terms may be considered as

$$\rho\left[\bar{u}\frac{\partial\bar{u}}{\partial x}+\bar{v}\frac{\partial\bar{u}}{\partial y}+\bar{w}\frac{\partial\bar{u}}{\partial z}\right]=-\frac{\partial\bar{p}}{\partial x}+\mu\nabla^2\bar{u}$$

$$+\left[\frac{\partial}{\partial x}\sigma'_{xx}+\frac{\partial}{\partial y}\tau'_{yx}+\frac{\partial}{\partial z}\tau'_{zx}\right] \tag{10.16a}$$

$$\rho\left[\bar{u}\frac{\partial\bar{v}}{\partial x}+\bar{v}\frac{\partial\bar{v}}{\partial y}+\bar{w}\frac{\partial\bar{v}}{\partial z}\right]=-\frac{\partial\bar{p}}{\partial y}+\mu\nabla^2\bar{v}$$

$$+\left[\frac{\partial}{\partial x}\tau'_{xy}+\frac{\partial}{\partial y}\sigma'_{yy}+\frac{\partial}{\partial z}\tau'_{zy}\right] \tag{10.16b}$$

$$\rho\left[\bar{u}\frac{\partial \bar{w}}{\partial x}+\bar{v}\frac{\partial \bar{w}}{\partial y}+\bar{w}\frac{\partial \bar{w}}{\partial z}\right]=-\frac{\partial \bar{p}}{\partial z}+\mu\nabla^2\bar{w}$$

$$+\left[\frac{\partial}{\partial x}\tau'_{xz}+\frac{\partial}{\partial y}\tau'_{yz}+\frac{\partial}{\partial z}\sigma'_{zz}\right] \tag{10.16c}$$

Comparing Eqs (10.15) and (10.16), we can write

$$\begin{bmatrix}\sigma'_{xx} & \tau'_{xy} & \tau'_{xz}\\ \tau'_{xy} & \sigma'_{yy} & \tau'_{yz}\\ \tau'_{xz} & \tau'_{yz} & \sigma'_{zz}\end{bmatrix}=-\rho\begin{bmatrix}\overline{u'^2} & \overline{u'v'} & \overline{u'w'}\\ \overline{u'v'} & \overline{v'^2} & \overline{v'w'}\\ \overline{u'w'} & \overline{v'w'} & \overline{w'^2}\end{bmatrix} \tag{10.17}$$

It can be said that the mean velocity components of turbulent flow satisfy the same Navier–Stokes equations of laminar flow. However, for the turbulent flow, the laminar stresses must be increased by additional stresses which are given by the stress tensor (10.17). These additional stresses are known as apparent stresses of turbulent flow or *Reynolds stresses*. Since turbulence is considered as eddying motion and the aforesaid additional stresses are added to the viscous stresses due to mean motion in order to explain the complete stress field, it is often said that the apparent stresses are caused by eddy viscosity. The total stresses are now

$$\left.\begin{aligned}\sigma_{xx}&=-\bar{p}+2\mu\frac{\partial \bar{u}}{\partial x}-\rho\overline{u'^2}\\ \tau_{xy}&=\mu\left(\frac{\partial \bar{u}}{\partial y}+\frac{\partial \bar{v}}{\partial x}\right)-\overline{\rho u'v'}\end{aligned}\right] \tag{10.18}$$

and so on. The apparent stresses are much larger than the viscous components, and the viscous stresses can even be dropped in many actual calculations.

It may be mentioned here that the terms, such as, $\frac{\partial \bar{u}}{\partial t}$, $\frac{\partial \bar{v}}{\partial t}$ and $\frac{\partial \bar{w}}{\partial t}$ were dropped considering the mean motion is not varying with time, confirming the paradigm of stationary random. However, there are flows where the mean motions are time varying and under such a situation, one has to retain these terms in the time-averaged Navier–Stokes equations (10.16). The governing momentum equations, for such flows are called Unsteady Reynolds Averaged Navier–Stokes (URANS) equations.

Example 10.1

Prove that

$$\overline{\int_{t-T/2}^{t+T/2}\phi\,\mathrm{d}\xi}=\int_{t-T/2}^{t+T/2}\bar{\phi}\,\mathrm{d}\xi$$

where ϕ is a constinuous function of ξ.

Solution

$$\overline{\int_{t-T/2}^{t+T/2} \phi \, \mathrm{d}\xi} = \frac{1}{T} \int_{t-T/2}^{t+T/2} \left[\int_{t-T/2}^{t+T/2} \phi \, \mathrm{d}\xi \right] \mathrm{d}t$$

Changing the order of integration, we can write

or $$\overline{\int_{t-T/2}^{t+T/2} \phi \, \mathrm{d}\xi} = \int_{t-T/2}^{t+T/2} \frac{1}{T} \left[\int_{t-T/2}^{t+T/2} \phi \, \mathrm{d}t \right] \mathrm{d}\xi$$

or $$\overline{\int_{t-T/2}^{t+T/2} \phi \, \mathrm{d}\xi} = \int_{t-T/2}^{t+T/2} \bar{\phi} \, \mathrm{d}\xi$$

10.7 TURBULENT BOUNDARY LAYER EQUATIONS

For a two-dimensional flow ($w = 0$) over a flat plate, the thickness of turbulent boundary layer is assumed to be much smaller than the axial length and the order of magnitude analysis (refer to Chapter 9) may be applied. As a consequence, the following inferences are drawn:

(a) $\dfrac{\partial \bar{p}}{\partial y} = 0$, (b) $\dfrac{\partial \bar{p}}{\partial x} = \dfrac{\mathrm{d}\bar{p}}{\mathrm{d}x}$

(c) $\dfrac{\partial^2 \bar{u}}{\partial x^2} << \dfrac{\partial^2 \bar{u}}{\partial y^2}$, and

(d) $\dfrac{\partial}{\partial x}\left(-\rho \overline{u'^2}\right) << \dfrac{\partial}{\partial y}\left(-\rho \overline{u'v'}\right)$

The turbulent boundary layer equation together with the equation of continuity becomes

$$\frac{\partial \bar{u}}{\partial x} + \frac{\partial \bar{v}}{\partial y} = 0 \quad (10.19)$$

$$\bar{u}\frac{\partial \bar{u}}{\partial x} + \bar{v}\frac{\partial \bar{v}}{\partial y} = -\frac{1}{\rho}\frac{\mathrm{d}\bar{p}}{\mathrm{d}x} + \frac{\partial}{\partial y}\left[\nu \frac{\partial \bar{u}}{\partial y} - \overline{u'v'}\right] \quad (10.20)$$

A comparison of Eq. (10.20) with laminar boundary layer Eq. (9.10) depicts that u, v and p are replaced by the time-averaged values $\bar{u}$, $\bar{v}$ and $\bar{p}$, and laminar viscous force per unit volume $\dfrac{\partial(\tau_l)}{\partial y}$ is replaced by $\dfrac{\partial}{\partial y}(\tau_l + \tau_t)$, where $\tau_l = \mu \dfrac{\partial \bar{u}}{\partial y}$ is the laminar shear stress and $\tau_t = -\rho \overline{u'v'}$ is the turbulent stress.

10.8 BOUNDARY CONDITIONS

All the components of apparent stresses vanish at the solid walls and only stresses which act near the wall are the viscous stresses of laminar flow. The boundary conditions, to be satisfied by the mean velocity components, are similar to laminar flow. A very thin layer next to the wall behaves like a near wall region of the laminar flow. This layer is known as the *laminar sublayer* and its velocities are such that the viscous forces dominate over the inertia forces. No turbulence exists in it (see Fig. 10.7). For a developed turbulent flow over a flat plate, in the near wall region, inertial effects are insignificant, and we can write from Eq. (10.20),

$$\nu \frac{\partial^2 \bar{u}}{\partial y^2} - \frac{\partial \overline{(u' v')}}{\partial y} = 0$$

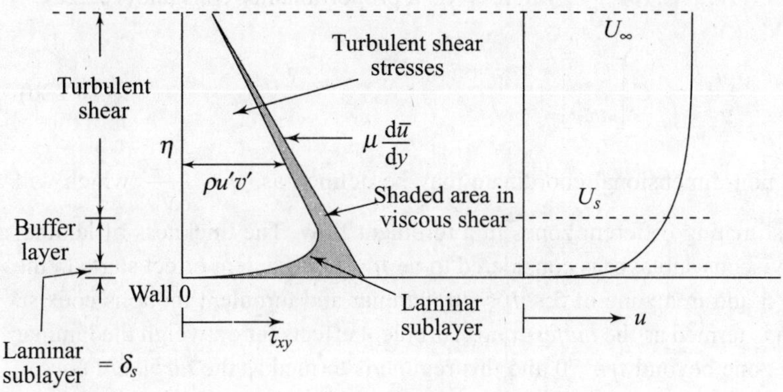

Fig. 10.7 Different zones of a turbulent flow past a wall

which can be integrated as, $\frac{\nu \, \partial \bar{u}}{\partial y} - \overline{u' v'}$ = constant

Again, as we know that the fluctuating components, do not exist near the wall, the shear stress on the wall is purely viscous and it follows that

$$\nu \frac{\partial \bar{u}}{\partial y}\bigg|_{y=0} = \frac{\tau_w}{\rho}$$

However, the wall shear stress in the vicinity of the laminar sublayer is estimated as

$$\tau_w = \mu \left[\frac{U_s - 0}{\delta_s - 0} \right] = \mu \frac{U_s}{\delta_s} \tag{10.21a}$$

where U_s is the fluid velocity at the edge of the sublayer. The flow in the sublayer is specified by a velocity scale (characteristic of this region). We define the friction velocity,

$$u_\tau = \left[\frac{\tau_w}{\rho}\right]^{1/2} \tag{10.21b}$$

as our velocity scale. Once u_τ is specified, the structure of the sublayer is specified. It has been confirmed experimentally that the turbulent intensity distributions are scaled with u_τ. For example, maximum value of the $\overline{u'^2}$ is always about $8u_\tau^2$. The relationship between u_τ and U_s can be determined from Eqs (10.21a) and (10.21b) as

$$u_\tau^2 = \nu \frac{U_s}{\delta_s}$$

Let us assume $\quad U_s = \bar{C} U_\infty$.

Now we can write

$$u_\tau^2 = \bar{C} \nu \frac{U_\infty}{\delta_s} \quad \text{where } \bar{C} \text{ is a proportionality constant} \tag{10.22a}$$

or

$$\frac{\delta_s u_\tau}{\nu} = \bar{C}\left[\frac{U_\infty}{u_\tau}\right] \tag{10.22b}$$

Hence, a non-dimensional coordinate may be defined as, $\eta = \dfrac{y u_\tau}{\nu}$ which will help us in estimating different zones in a turbulent flow. The thickness of laminar sublayer or viscous sublayer is considered to be $\eta \approx 5$. Turbulent effect starts in the zone of $\eta > 5$ and in a zone of $5 < \eta < 70$, laminar and turbulent motions coexist. This domain is termed as the *buffer zone*. Turbulent effects far outweigh the laminar effect in the zone beyond $\eta = 70$ and this regime is termed as the *turbulent core*.

For flow over a flat plate, the turbulent sheat stress $(-\rho\overline{u'v'})$ is constant throughout in the y direction and this becomes equal to τ_w at the wall. In the event of flow through a channel, the turbulent shear stress $(-\rho\overline{u'v'})$ varies with y and it is possible to write

$$\frac{\tau_t}{\tau_w} = \frac{\zeta}{h} \tag{10.22c}$$

where the channel is assumed to have a height $2h$ and ζ is the distance measured from the centreline of the channel $(= h - y)$. Figure 10.7 explains such variation of turbulent stress.

10.9 SHEAR STRESS MODELS

In analogy with the coefficient of viscosity for laminar flow, J. Boussinesq introduced a mixing coefficient μ_τ, for the Reynolds stress term by invoking

$$\tau_t = -\overline{\rho u'v'} = \mu_t \frac{\partial \bar{u}}{\partial y}$$

Now the expressions for shearing stresses are written as

$$\tau_l = \rho \nu \frac{\partial u}{\partial y}, \tau_t = \mu_t \frac{\partial \bar{u}}{\partial y} = \rho \nu_t \frac{\partial \bar{u}}{\partial y}$$

such that the equation

$$\bar{u}\frac{\partial \bar{u}}{\partial x} + \bar{v}\frac{\partial \bar{u}}{\partial y} = -\frac{1}{\rho}\cdot\frac{\partial p}{\partial x} + \frac{\partial}{\partial y}\left\{\nu \frac{\partial \bar{u}}{\partial y} - \overline{u'v'}\right\}$$

may be written as

$$\bar{u}\frac{\partial \bar{u}}{\partial x} + \bar{v}\frac{\partial \bar{u}}{\partial y} = -\frac{1}{\rho}\cdot\frac{\partial p}{\partial x} + \frac{\partial}{\partial y}\left\{(\nu + \nu_t)\frac{\partial \bar{u}}{\partial y}\right\} \tag{10.23}$$

The term ν_t is known as *eddy viscosity* and the model is known as *eddy viscosity* model. The difficulty in using Eq. (10.23) can be discussed herein. The value of ν_t is not known. The term ν is a property of the fluid whereas ν_t is attributed to random fluctuations and is not a property of the fluid. However, it is necessary to find out empirical relations between ν_t and the mean velocity. We shall discuss one such well-known relation between the aforesaid apparent or eddy viscosity and the mean velocity components in the following subsection.

10.9.1 Prandtl's Mixing Length Hypothesis

Let us consider a fully developed turbulent boundary layer (Fig. 10.3). The streamwise mean velocity varies only from streamline to streamline. The main flow direction is assumed parallel to the x axis (Fig. 10.8).

The time-average components of velocity are given by $\bar{u} = \bar{u}\ (y)$, $\bar{v} = 0$, $\bar{w} = 0$. The fluctuating component of transverse velocity v' transports mass and momentum across a plane at y_1 from the wall. The shear stress due to the fluctuation is

given by $$\tau_t = -\rho\, \overline{u'v'} = \mu_t \frac{\partial \bar{u}}{\partial y} \tag{10.24}$$

A lump of fluid, which comes to the layer y_1 from a layer $(y_1 - l)$ has a positive value of v'. If the lump of fluid retains its original momentum then its velocity at its current location y_1 is smaller than the velocity prevailing there. The difference in velocities is then

$$\Delta u_1 = \bar{u}\ (y_1) - \bar{u}\ (y_1 - l) \approx l\left(\frac{\partial \bar{u}}{\partial y}\right)_{y_1} \tag{10.25}$$

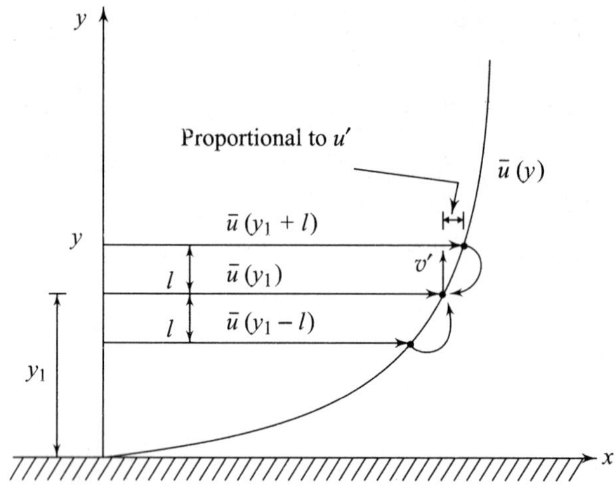

Fig. 10.8 One-dimensional parallel flow and Prandtl's mixing length hypothesis

The above expression is obtained by expanding the function $\bar{u}\,(y_1 - l)$ in a Taylor series and neglecting all higher order terms and higher order derivatives. As it is said, l is a small length scale known as Prandtl's mixing length. Prandtl proposed that the transverse displacement of any fluid particle is, on an average, l. Let us consider another lump of fluid with a negative value of v'. This is arriving at y_1 from $(y_1 + l)$. If this lump retains its original momentum, its mean velocity at the current lamina y_1 will be somewhat more than the original mean velocity of y_1. This difference is given by

$$\Delta u_2 = \bar{u}\,(y_1 + l) - \bar{u}\,(y_1) \approx l\left(\frac{\partial \bar{u}}{\partial y}\right)_{y_1} \qquad (10.26)$$

The velocity differences caused by the transverse motion can be regarded as the turbulent velocity components at y_1. We calculate the time average of the absolute value of this fluctuation as

$$\overline{|u'|} = \frac{1}{2}\left(|\Delta u_1| + |\Delta u_2|\right) = l\left|\left(\frac{\partial \bar{u}}{\partial y}\right)_{y_1}\right| \qquad (10.27)$$

Suppose these two lumps of fluid meet at a layer y_1. The lumps will collide with a velocity $2u'$ and diverge. This proposes the possible existence of transverse velocity component in both directions with respect to the layer at y_1. Now, suppose that the two lumps move away in a reverse order from layer y_1 with a velocity $2u'$. The empty space will be filled from the surrounding fluid creating transverse velocity components which will again collide at y_1. Keeping in mind this argument and the physical explanation accompanying Eqs. (10.14), we may state that

$$\overline{|v'|} \sim \overline{|u'|}$$

or $$|v'| = \text{const}\ |u'| = (\text{const})\ l \left|\frac{\partial \bar{u}}{\partial y}\right|$$

along with the condition that the moment at which u' is positive, v' is more likely to be negative and conversely when u' is negative. Possibly, we can write at this stage

$$\overline{u'v'} = -C_1\ \overline{|u'|}\ \overline{|v'|}$$

or $$\overline{u'v'} = -C_2\, l^2 \left(\frac{\partial \bar{u}}{\partial y}\right)^2 \tag{10.28}$$

where C_1 and C_2 are different proportionality constants. However, the constant C_2 can now be included in still unknown mixing length and Eq. (10.28) may be rewritten as

$$\overline{u'v'} = -l^2 \left(\frac{\partial \bar{u}}{\partial y}\right)^2$$

For the expression of turbulent shearing stress τ_t, we may write

$$\tau_t = -\rho\ \overline{u'v'} = \rho\, l^2 \left(\frac{\partial \bar{u}}{\partial y}\right)^2 \tag{10.29}$$

After comparing this expression with the eddy viscosity concept and Eq. (10.24), we may arrive at a more precise definition,

$$\tau_t = \rho\, l^2 \left|\frac{\partial \bar{u}}{\partial y}\right| \left(\frac{\partial \bar{u}}{\partial y}\right) = \mu_t \frac{\partial \bar{u}}{\partial y} \tag{10.30a}$$

where the apparent viscosity may be expressed as

$$\mu_t = \rho\, l^2 \left|\frac{\partial \bar{u}}{\partial y}\right| \tag{10.30b}$$

and the apparent kinematic viscosity is given by

$$\nu_t = l^2 \left|\frac{\partial \bar{u}}{\partial y}\right| \tag{10.30c}$$

The decision of expressing one of the velocity gradients of Eq. (10.29) in terms of its modulus as $\left|\frac{\partial \bar{u}}{\partial y}\right|$ was made in order to assign a sign to τ_t according to the sign of $\frac{\partial \bar{u}}{\partial y}$. It may be mentioned that the apparent viscosity and consequently, the mixing length are not properties of fluid. They are dependent on turbulent fluctuation. However, our problem is still not resolved. How to determine the value of l, the mixing length? Several correlations, using experimental results for τ_t have been proposed to determine l.

However, so far the most widely used value of mixing length in the regime of isotropic turbulence is given by

$$l = \chi y \tag{10.31}$$

where y is the distance from the wall and χ is known as von Karman constant (≈ 0.4).

10.10 UNIVERSAL VELOCITY DISTRIBUTION LAW AND FRICTION FACTOR IN DUCT FLOWS FOR VERY LARGE REYNOLDS NUMBERS

For flows in a rectangular channel at very large Reynolds numbers the laminar sublayer can practically be ignored. The channel may be assumed to have a width $2h$ and the x axis will be placed along the bottom wall of the channel. We shall consider a turbulent stream along a smooth flat wall in such a duct and denote the distance from the bottom wall by y, while $u(y)$ will signify the velocity. In the neighbourhood of the wall, we shall apply

$$l = \chi y$$

According to Prandtl's assumption, the turbulent shearing stress will be

$$\tau_t = \rho \chi^2 y^2 \left(\frac{\partial \bar{u}}{\partial y} \right)^2 \tag{10.32}$$

At this point, Prandtl introduced an additional assumption which like a plane Couette flow takes a constant shearing stress throughout, i.e,

$$\tau_t = \tau_w \tag{10.33}$$

where τ_w denotes the shearing stress at the wall. Invoking once more the friction velocity $u_\tau = \left[\frac{\tau_w}{\rho} \right]^{1/2}$, we obtain

$$u_\tau^2 = \chi^2 y^2 \left(\frac{\partial \bar{u}}{\partial y} \right)^2 \tag{10.34}$$

or

$$\frac{\partial \bar{u}}{\partial y} = \frac{u_\tau}{\chi y} \tag{10.35}$$

On integrating we find

$$\bar{u} = \frac{u_\tau}{\chi} \ln y + C \tag{10.36}$$

Despite the fact that Eq. (10.36) is derived on the basis of the friction velocity in the neighbourhood of the wall because of the assumption that $\tau_w = \tau_t =$ constant, we shall use it for the entire region. At $y = h$ (at the horizontal mid-plane of the channel), we have $\bar{u} = U_{max}$. The constant of integration is eliminated by considering

$$U_{max} = \frac{u_\tau}{\chi} \ln h + C$$

or
$$C = U_{max} - \frac{u_\tau}{\chi} \ln h$$

Substituting C in Eq. (10.36), we get

and
$$\frac{U_{max} - \bar{u}}{u_\tau} = \frac{1}{\chi} \ln\left(\frac{h}{y}\right) \tag{10.37}$$

Equation (10.37) is known as universal velocity defect law of Prandtl and its distribution has been shown in Fig. 10.9.

Here, we have seen that the friction velocity u_τ is a reference parameter for velocity. We shall now discuss the problem with $(\bar{u}/u_\tau)$ and $\eta\ (= y\, u_\tau/\nu)$ as parameters. Equation (10.36) can be rewritten once again for this purpose as

$$\frac{\bar{u}}{u_\tau} = \frac{1}{\chi} \ln y + C$$

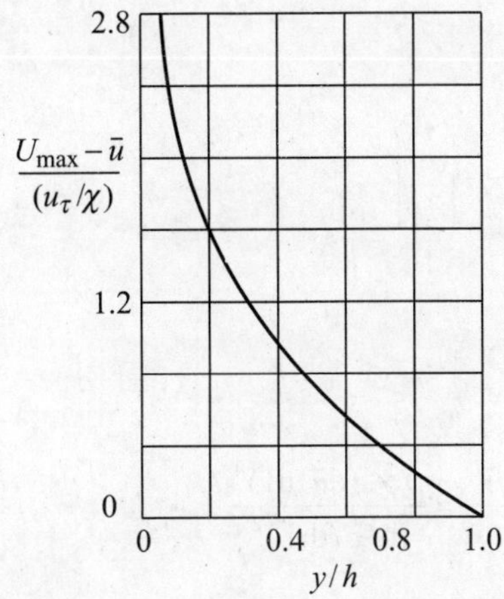

Fig. 10.9 Distribution of universal velocity defect law of Prandtl in a turbulent channel flow

The no-slip condition at the wall cannot be satisfied with a finite constant of integration. This is expected that the appropriate condition for the present problem should be that $\bar{u} = 0$ at a very small distance $y = y_0$ from the wall. Hence, Eq. (10.36) becomes

$$\frac{\bar{u}}{u_\tau} = \frac{1}{\chi} (\ln y - \ln y_0) \tag{10.38}$$

The distance y_0 is of the order of magnitude of the thickness of the viscous layer. Now we can write Eq. (10.38) as

$$\frac{\bar{u}}{u_\tau} = \frac{1}{\chi} \left[\ln y \frac{u_\tau}{\nu} - \ln \beta \right]$$

or
$$\frac{\bar{u}}{u_\tau} = A_1 \ln \eta + D_1 \tag{10.39}$$

where $A_1 = (1/\chi)$, the unknown β is included in D_1.

Equation (10.39) is generally known as the *universal velocity* profile because of the fact that it is applicable from moderate to a very large Reynolds number. However, the constants A_1 and D_1 have to be found out from experiments.

Example 10.2

The well-known scientist, Theodore von Karman, suggested the mixing length to be $l = \chi \left| \dfrac{\mathrm{d}\bar{u}/\mathrm{d}y}{\mathrm{d}^2\bar{u}/\mathrm{d}y^2} \right|$. Using this relation drive the velocity profile near the wall of a flat-plate boundary layer flow.

Solution

We know

$$\tau_t = \mu_t \frac{\partial \bar{u}}{\partial y}, \text{ where } \mu_t = \rho l^2 \left| \frac{\partial \bar{u}}{\partial y} \right|$$

So,
$$\tau_t = \rho l^2 \left| \frac{\partial \bar{u}}{\partial y} \right|^2$$

Substituting von Karman's suggestion, we get

$$\tau_t = \frac{\rho \chi^2 (\mathrm{d}\bar{u}/\mathrm{d}y)^2 (\mathrm{d}\bar{u}/\mathrm{d}y)^2}{(\mathrm{d}^2\bar{u}/\mathrm{d}y^2)^2} = \frac{\rho \chi^2 (\mathrm{d}\bar{u}/\mathrm{d}y)^4}{(\mathrm{d}^2\bar{u}/\mathrm{d}y^2)^2}$$

or
$$\left(\frac{\mathrm{d}\bar{u}}{\mathrm{d}y}\right)^4 = \frac{\tau_t}{\rho} \frac{1}{\chi^2} \left(\frac{\mathrm{d}^2\bar{u}}{\mathrm{d}y^2}\right)^2$$

Assuming $\tau_t = \tau_w$ and considering $u_\tau = \sqrt{\tau_w/\rho}$

$$\left(\frac{\mathrm{d}\bar{u}}{\mathrm{d}y}\right)^4 = \frac{u_\tau^2}{\chi^2} \left(\frac{\mathrm{d}^2\bar{u}}{\mathrm{d}y^2}\right)^2$$

Taking the square root, and applying physical argument that $(\mathrm{d}\bar{u}/\mathrm{d}y)$ cannot be imaginary, we obtain

$$\left(\frac{\mathrm{d}\bar{u}}{\mathrm{d}y}\right)^2 = \pm \frac{u_\tau}{\chi} \left(\frac{\mathrm{d}^2\bar{u}}{\mathrm{d}y^2}\right)$$

Let
$$m = \mathrm{d}\bar{u}/\mathrm{d}y$$

then $$dm/dy = \pm \frac{\chi}{u_\tau} m^2$$

Integration yields,

$$-\frac{1}{m} = \pm \frac{\chi}{u_\tau} y + C_1$$

Using $m \Rightarrow \infty$ as $y \Rightarrow 0$, we get $C_1 = 0$

Then, $$\frac{d\bar{u}}{dy} = \frac{u_\tau}{\chi y}, \qquad \text{since } \frac{d\bar{u}}{dy} \geq 0$$

Integrating, we obtain,

$$\bar{u} = \frac{u_\tau}{\chi} \ln y + C_2$$

At some value $y = y_0$, $\bar{u} = 0$

Invoking this, $$C_2 = -\frac{u_\tau}{\chi} \ln y_0$$

Thus, $$\frac{\bar{u}}{u_\tau} = \frac{1}{\chi} \cdot \ln (y - y_0)$$

Let us substitute $y_0 = \beta \frac{\nu}{u_\tau}$ order of which is same as viscous sublayer and β is an arbitrary constant.

Thus, we shall get

$$\frac{\bar{u}}{u_\tau} = \frac{1}{\chi}\left(\ln \frac{u_\tau y}{\nu} - \ln \beta\right)$$

or $$\frac{\bar{u}}{u_\tau} = A_1 \ln \eta + D_1$$

This is the universal velocity profile.

It may be mentioned that A_1 and D_1 are near-universal constants for turbulent flow past smooth impermeable walls. The parameter D_1 varies with the pressure gradient. The original pipe-flow measurements by Prandt' s student, J. Nikuradse, suggests $\chi = 0.4$ (thereby $A_1 = 2.5$) and $D_1 = 5.5$. The determination of correct χ for various geometrical configurations is a subject of research even today.

The universal velocity profile is not only valid for channel flows; it retains the same functional relationship for circuilar pipes as well. It may be mentioned that even without the assumption of having a constant shear throughout, the universal velocity profile can be derived. The following example makes it clear.

Example 10.3

Using Karman's relation $l = \chi \left| \dfrac{d\bar{u}/dy}{d^2\bar{u}/dy^2} \right|$, show that the universal velocity distribution in a fully developed channel flow is given by

$$\frac{U_{max} - \bar{u}}{u_\tau} = -\frac{1}{\chi}\left[\ln\left(1 - \sqrt{\frac{y}{h}}\right) + \sqrt{\frac{y}{h}}\right]$$

where, $2h$ is the height of the channel, y is the distance measured from the centre line of the channel and χ is an empirical constant. The pressure gradient in flow direction is (dp/dx).

Solution

From Reynolds equation, we get

$$\frac{\partial(\tau_t)}{\partial y} = \frac{\partial \bar{p}}{\partial x}$$

or
$$\tau_t = \left(\frac{\partial \bar{p}}{\partial x}\right) y + C_1$$

At $y = 0$, $\tau_t = 0$, that makes $C_1 = 0$

Thus,
$$\tau_t = \left(\frac{\partial \bar{p}}{\partial x}\right) y$$

and
$$\tau_w = \left(\frac{\partial \bar{p}}{\partial x}\right) h$$

we get
$$\frac{\tau_t}{\tau_w} = \frac{y}{h}$$

From Karman's relation, we can write

$$\tau_t = \frac{\rho \chi^2 (\partial \bar{u}/\partial y)^4}{\left(\partial^2 \bar{u}/\partial y^2\right)^2}$$

Then
$$\tau_w = \frac{h \rho \chi^2 (\partial \bar{u}/\partial y)^4}{y\left(\partial^2 \bar{u}/\partial y^2\right)^2}$$

$$u_\tau^2 = \frac{h \chi^2 (\partial \bar{u}/\partial y)^4}{y\left(\partial^2 \bar{u}/\partial y^2\right)^2}$$

Thus,
$$\frac{\partial^2 \bar{u}}{\partial y^2} = \pm \frac{\chi}{u_\tau}\sqrt{\frac{h}{y}}\left(\left|\frac{\partial \bar{u}}{\partial y}\right|\right)^2$$

Substituting for $m = \frac{\partial \bar{u}}{\partial y}$ and integrating,

$$-\frac{1}{m} = \pm\, 2\frac{\chi}{u_\tau}\sqrt{h\,y} + C_2$$

at $y = h, m \to \infty$ and $C_2 = \pm\, 2\,\frac{\chi}{u_\tau}\, h$

Now, we know that $\frac{\partial \bar{u}}{\partial y} \leq 0$ for $y > 0$ and we write

$$\mathrm{d}\bar{u} = -\frac{u_\tau}{2\,\chi\, h}\int \frac{\mathrm{d}y}{1 - \sqrt{y/h}}$$

Substituting for $\xi = 1 - \sqrt{\frac{y}{h}}$ and integrating,

$$\bar{u} = \frac{u_\tau}{\chi}[\ln \xi - \xi] + C_3$$

or

$$\bar{u} = \frac{u_\tau}{\chi}\left[\ln\left(1 - \sqrt{\frac{y}{h}}\right) - \left(1 - \sqrt{\frac{y}{h}}\right)\right] + C_3$$

at $y = 0, \bar{u} = U_{\max}$

$$C_3 = U_{\max} + \frac{u_\tau}{\chi}$$

Finally,

$$\frac{U_{\max} - \bar{u}}{u_\tau} = \frac{1}{\chi}\left[\ln\left(1 - \sqrt{\frac{y}{h}}\right) - \sqrt{\frac{y}{h}}\right]$$

Some Important Considertions for High Reynolds Number Pipe Flows

As discussed earlier, the experiments performed by Nikuradse showed that Eq. (10.39) is in good agreement with experimental results. Based on Nikurads' s and Reichardt' s experimental data, the empirical constants of Eq. (10.39) can be determined for a smooth pipe as

$$\frac{\bar{u}}{u_\tau} = 2.5 \ln \eta + 5.5 \tag{10.40}$$

There is a serious lacuna in this expression; one cannot extend this up to the wall. Actually the profile drops linearly to zero very close to the wall within thicknes too small to be seen. For the key to the profile shape, we are indebted to the physical insight of Prandtl and von-Karman. They concluded that the profile consist of an inner and outer layer, plus an intermediate overlap layer between the two. The outer layer (turbulent zone) is represented well by Eq. (10.40).

The velocity distribution for the outer layer (turbulent zone) has been shown through curve (b) in Fig. 10.10.

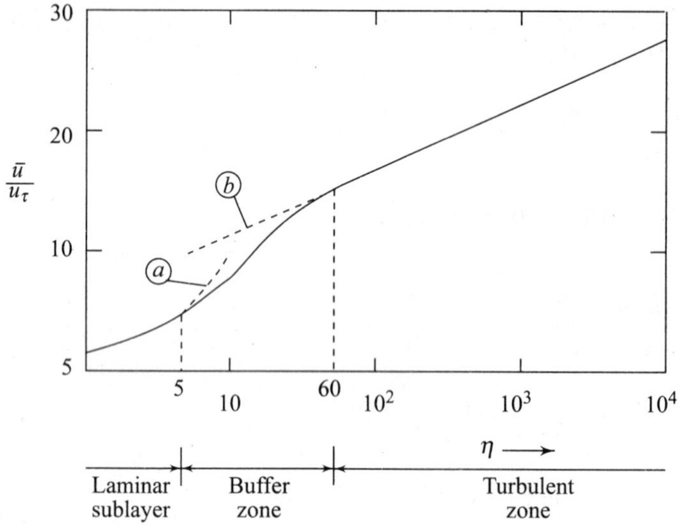

Fig. 10.10 The universal velocity distribution law for smooth pipes

However, the corresponding friction factor concerning Eq. (10.40) is

$$\frac{1}{\sqrt{f}} = 2.0 \log_{10} (\text{Re}\sqrt{f}) - 0.8 \tag{10.41}$$

As mentioned earlier, the universal velocity profile does not match very close to the wall where the viscous shear predominates the flow. Prandtl and von Karman suggested a modification for the *laminar sublayer* and the *buffer zone* which are

$$\frac{\bar{u}}{u_\tau} = \eta = \frac{u_\tau y}{\nu} \quad \text{for } \eta < 5.0 \tag{10.42}$$

and

$$\frac{\bar{u}}{u_\tau} = 11.5 \log_{10} \frac{u_\tau y}{\nu} - 3.0 \quad \text{for } 5 < \eta < 60 \tag{10.43}$$

Equation (10.42) has been shown through curve (a) in Fig. 10.10.

It may be worthwhile to mention here that a surface is said to be hydraulically smooth as long as

$$0 \le \frac{\varepsilon_p u_\tau}{\nu} \le 5 \tag{10.44}$$

where ε_p is the average height of the protrusions inside the pipe.

Physically, the above expression means that for smooth pipes protrusions will not be extended outside the laminar sublayer. If protrusions exceed the thickness of the laminar sublayer, it is conjectured (also justified though experimental verification) that some additional frictional resistance will contribute to pipe friction due to the form drag experienced by the protrusions in the boundary layer. In rough pipes, experiments indicate that the velocity profile may be expressed as:

$$\frac{\bar{u}}{u_\tau} = 2.5 \ln \frac{y}{\varepsilon_p} + 8.5 \tag{10.45}$$

At the centre-line, the maximum velocity is expressed as

$$\frac{U_{max}}{u_\tau} = 2.5 \ln \frac{R}{\varepsilon_p} + 8.5 \tag{10.46}$$

Note that ν no longer appears with R and ε_p. This means that for completely rough zone of turbulent flow, *the profile is independent of Reynolds number and a strong function of pipe roughness*. However, for pipe roughness of varying degrees, the recommendation due to Colebrook and White works well. Their formula is

$$\frac{1}{\sqrt{f}} = 1.74 - 2.0 \log_{10} \left[\frac{\varepsilon_p}{R} + \frac{18.7}{\mathrm{Re}\sqrt{f}} \right] \tag{10.47}$$

where R is the pipe radius.

For $\varepsilon_p \to 0$, this equation produces the result of the smooth pipes (Eq. (10.41)). For $\mathrm{Re} \to \infty$, it gives the expression for friction factor for a completely rough pipe at a very high Reynolds number which is given by

$$f = \frac{1}{\left(2 \log \dfrac{R}{\varepsilon_p} + 1.74 \right)^2} \tag{10.48}$$

Turbulent flow through pipes has been investigated by many researchers because of its enormous practical importance.

In the next section, we shall discuss, in detail the velocity distribution and other important aspects of turbulent pipe flows.

10.11 FULLY-DEVELOPED TURBULENT FLOW IN A PIPE FOR MODERATE REYNOLDS NUMBERS

The entry length of a turbulent flow is much shorter than that of a laminar flow, J. Nikuradse determined that a fully developed profile for turbulent flow can be observed after an entry length of 25 to 40 diameters. We shall focus herein our attention to fully-developed turbulent flow. Considering a fully developed turbulent pipe flow (Fig. 10.11) we can write

$$2 \pi R \tau_w = -\left(\frac{dp}{dx} \right) \pi R^2 \tag{10.49}$$

or
$$\left(-\frac{dp}{dx}\right) = \frac{2\tau_w}{R} \tag{10.50}$$

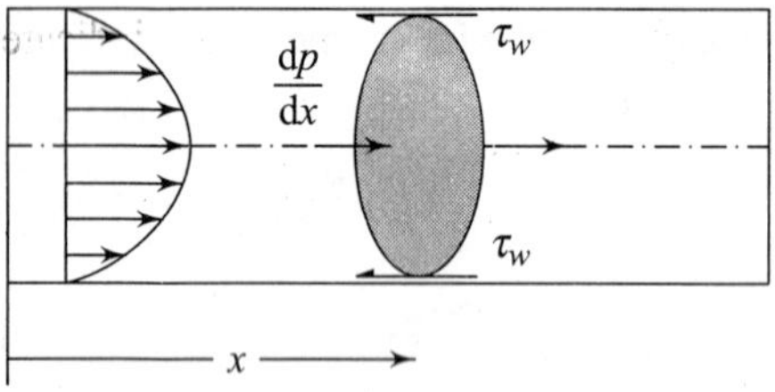

Fig. 10.11 Fully developed turbulent pipe flow

It can be said that in a fully developed flow, the pressure gradient balances the wall shear stress only and has a constant value at any x. However, the friction factor (Darcy friction factor) is defined in a fully developed flow as

$$-\left(\frac{dp}{dx}\right) = \frac{\rho f U_{av}^2}{2D} \tag{10.51}$$

Comparing Eq. (10.50) with Eq. (10.51), we can write

$$\tau_w = \frac{f}{8}\rho U_{av}^2 \tag{10.52}$$

H. Blasius conducted a critical survey of available experimental results and established the empirical correlation for the above equation as

$$f = 0.3164\ \mathrm{Re}^{-0.25}, \text{ where } \mathrm{Re} = \rho U_{av} D/\mu \tag{10.53}$$

It is found that the Blasius's formula is valid in the range of Reynolds number of Re $\le 10^5$. At the time when Blasius compiled the experimental data, results for higher Reynolds numbers were not available. However, later, J. Nikuradse carried out experiments with the laws of friction in a very wide range of Reynolds numbers, $4 \times 10^3 \le \mathrm{Re} \le 3.2 \times 10^6$. The velocity profile in this range follows:

$$\frac{u}{\bar{u}} = \left[\frac{y}{R}\right]^{1/n} \tag{10.54}$$

where $\bar{u}$ is the time mean velocity at the pipe centre and y is the *distance from the wall*. The exponent n varies slightly with Reynolds number. In the range of Re ~ 10^5, n is 7.

The ratio of $\bar{u}$ and U_{av} for the aforesaid profile is found out by considering the volume flow rate Q as

$$Q = \pi R^2 U_{av} = \int_0^R 2\pi r\, u\, dr$$

or
$$\pi R^2 U_{av} = 2\pi \bar{u} \int_R^0 (R - y)\,(y/R)^{1/n}\,(-dy)$$

or $$\pi R^2 U_{av} = 2\pi \bar{u} \left[\frac{n}{n+1} \left(R^{\frac{n-1}{n}} \; y^{\frac{n+1}{n}} \right) - \frac{n}{2n+1} \left(y^{\frac{2n+1}{n}} \; R^{-\frac{1}{n}} \right) \right]_0^R$$

or $$\pi R^2 U_{av} = 2\pi \bar{u} \left[R^2 \frac{n}{n+1} - \frac{n}{2n+1} R^2 \right]$$

or $$\pi R^2 U_{av} = 2\pi R^2 \bar{u} \left[\frac{n^2}{(n+1)(2n+1)} \right]$$

or $$\frac{U_{av}}{\bar{u}} = \frac{2n^2}{(n+1)(2n+1)} \tag{10.55a}$$

Now, for different values of n (for different Reynolds numbers) we shall obtain different values of $U_{av}/\bar{u}$ from Eq. (10.55a). On substitution of Blasius resistance formula (10.53) in Eq. (10.52), the following expression for the shear stress at the wall can be obtained:

$$\tau_w = \frac{0.3164}{8} \text{Re}^{-0.25} \; \rho U_{av}^2$$

or $$\tau_w = 0.03955 \, \rho U_{av}^2 \left(\frac{\nu}{2R \, U_{av}} \right)^{1/4}$$

or $$\tau_w = 0.03325 \, \rho U_{av}^{7/4} \left(\frac{\nu}{R} \right)^{1/4}$$

or $$\tau_w = 0.03325 \, \rho \left(\frac{U_{av}}{\bar{u}} \right)^{7/4} (\bar{u})^{7/4} \left(\frac{\nu}{R} \right)^{1/4}$$

For $n = 7$, $U_{av}/\bar{u}$ becomes equal to 0.8. Substituting $U_{av}/\bar{u} = 0.8$ in the above equation, we get

$$\tau_w = 0.03325 \, \rho \, (0.8)^{1/4} (\bar{u})^{7/4} (\nu/R)^{1/4}$$

Finally it produces $$\tau_w = 0.0225 \, \rho (\bar{u})^{7/4} (\nu/R)^{1/4} \tag{10.55b}$$

or $$u_\tau^2 = 0.0225 \, (\bar{u})^{7/4} \left(\frac{\nu}{R} \right)^{1/4}$$

where u_τ is friction velocity. However, u_τ^2 may be split into $u_\tau^{7/4}$ and $u_\tau^{1/4}$ and we obtain

$$\left(\frac{\bar{u}}{u_\tau} \right)^{7/4} = 44.44 \left(\frac{u_\tau R}{\nu} \right)^{1/4}$$

or $$\frac{\bar{u}}{u_\tau} = 8.74 \left(\frac{u_\tau R}{\nu} \right)^{1/7} \tag{10.56a}$$

Now we can assume that the above equation is not only valid at the pipe axis ($y = R$) but also at any distance from the wall y and a general form is proposed as

$$\frac{\bar{u}}{u_\tau} = 8.74\left(\frac{yu_\tau}{\nu}\right)^{1/7} \tag{10.56b}$$

In conclusion, it can be said that $(1/7)^{th}$ power velocity distribution law (10.56b) can be derived from Blasius's resistance formula (10.53). Equation (10.55b) gives the shear stress relationship in the pipe the flow at a moderate Reynolds number, i.e, $Re \le 10^5$. Unlike very high Reynolds number flow, here laminar effect cannot be neglected and the laminar sublayer brings about remarkable influence on the outer zones.

It is worth mentioning that the friction factor for pipe flows f, defined by Eq. (10.53) is valid for a specific range of Reynolds number and for a particular surface condition. The experimental results for a wide range of Reynolds numbers and variety of pipe roughness can be summarised through the *Moody diagram*, shown in Chapter 11.

10.12 SKIN FRICTION COEFFICIENT FOR BOUNDARY LAYERS ON A FLAT PLATE

Calculations of skin friction drag on lifting surface and on aerodynamic bodies are somewhat similar to the analyses of skin friction on a flat plate. Because of zero pressure gradient, the flat plate at zero incidence is easy to consider. In some of the applications cited above, the pressure gradient will differ from zero but the skin friction will not be dramatically different as long as there is no separation.

We begin with the momentum integral equation for flat plate boundary layer which is valid for both laminar and turbulent flow.

$$\frac{d}{dx}\left(U_\infty^2\,\delta^{**}\right) = \frac{\tau_w}{\rho} \tag{10.57a}$$

Invoking the definition of C_{fx} $\left(C_{fx} = \dfrac{\tau_w}{\frac{1}{2}\rho U_\infty^2}\right)$, Eq. (10.57a) can be rewritten as

$$C_{fx} = 2\frac{d\delta^{**}}{dx} \tag{10.57b}$$

Due to the similarity in the laws of wall, correlations of the previous section may be applied to the flat plate by substituting δ for R and U_∞ for the time mean velocity at the pipe centre. The rationale for using the turbulent pipe flow results in the situation of a turbulent flow over a flat plate is to consider that the time mean velocity, at the centre of the pipe is analogous to the free stream velocity, both the velocities being defined at the edge of boundary layer thickness.

Finally, the velocity profile will be [following Eq. (10.54)]

$$\frac{u}{U_\infty} = \left[\frac{y}{\delta}\right]^{1/7} \qquad \text{for } Re \le 10^5 \tag{10.58}$$

If we evaluate momentum thickness with this profile, we shall obtain

$$\delta^{**} = \int_0^\delta \left(\frac{y}{\delta}\right)^{1/7}\left[1-\left(\frac{y}{\delta}\right)^{1/7}\right] dy = \frac{7}{72}\delta \tag{10.59}$$

It is worth mentioning that Prandtl suggested the usage of the turbulent pipe flow profile for the bounday layers over flat plates, on the grounds stated above. Although Eq. (10.58) satisfactorily describes the velocity distribution in most of the boundary layer, it fails to satisfy the boundary condition at the wall surface because $\partial u/\partial y = (1/7)\,(U_\infty\, \delta^{-1/7}\, y^{-6/7}) = \infty$ at $y = 0$. Actually the viscous sublayer exists immediately adjacent to the wall. And it is so thin that the velocity variation within the sublayer may be taken as linear and tangential to the one-seventh profile at the point where the viscous sublayer merges with the turbulent part of the boundary layer (Fig 10.12).

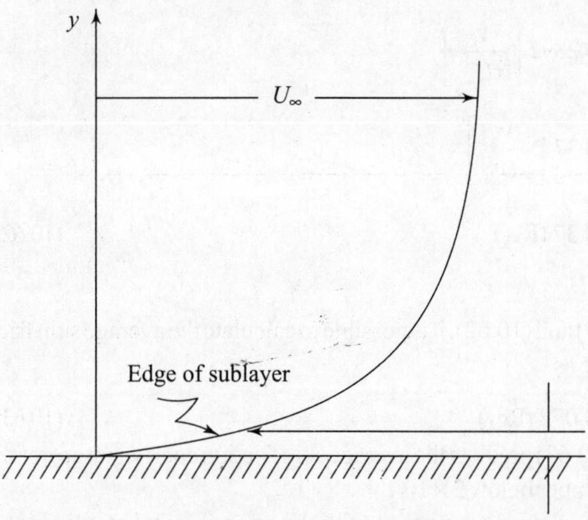

Fig. 10.12 Turbulent boundary layer on a flat plate

Consequently, the law of shear stress (in range of Re $\leq 10^5$) for the flat plate is found out by making use of the pipe flow expression of Eq. (10.55b) as

$$\tau_w = 0.0225\, \rho(\bar{u})^{7/4}\left(\frac{\nu}{R}\right)^{1/4}$$

or
$$\frac{\tau_w}{\rho(\bar{u})^2} = 0.0225\left[\frac{\nu}{R\bar{u}}\right]^{1/4}$$

Substituting U_∞ for $\bar{u}$ and δ for R in the above expression, we get

or
$$\frac{\tau_w}{\rho U_\infty^2} = 0.0225\left[\frac{\nu}{\delta U_\infty}\right]^{1/4} \tag{10.60}$$

Once again substituting Eqs (10.59) and (10.60) in Eq. (10.57), we obtain

$$\frac{7}{72}\cdot\frac{d\delta}{dx} = 0.0225\left[\frac{\nu}{\delta U_\infty}\right]^{1/4}$$

or $$\delta^{1/4}\frac{d\delta}{dx} = 0.2314\left[\frac{\nu}{U_\infty}\right]^{1/4}$$

or $$\delta^{5/4} = 0.2892\, x\left(\frac{\nu}{U_\infty}\right)^{1/4} + C \qquad (10.61)$$

For simplicity, if we assume that the turbulent boundary layer grows from the leading edge of the plate we shall be able to apply the boundary conditions $x = 0$, $\delta = 0$, which will yield $C = 0$, and Eq. (10.61) will become

$$(\delta/x)^{5/4} = 0.2892\left[\frac{\nu}{xU_\infty}\right]^{1/4}$$

or $$\frac{\delta}{x} = 0.37\left[\frac{\nu}{xU_\infty}\right]^{1/5}$$

or $$\frac{\delta}{x} = 0.37\,(\mathrm{Re}_x)^{-1/5} \qquad (10.62)$$

where $\mathrm{Re}_x = (U_\infty x)/\nu$

From Eqs (10.57b), (10.59) and (10.62), it is possible to calculate the average skin friction coefficient on a flat plate as

$$\bar{C}_f = 0.072\,(\mathrm{Re}_L)^{-1/5} \qquad (10.63)$$

It can be shown that Eq. (10.63) predicts the average skin friction coefficient correctly in the regime of Reynolds number below 2×10^6.

This result is found to be in good agreement with the experimental results in the range of Reynolds number between 5×10^5 and 10^7 which is given by

$$\bar{C}_f = 0.074\,(\mathrm{Re}_L)^{-1/5} \qquad (10.64)$$

Equation (10.64) is a widely accepted correlation for the average value of turbulent skin friction coefficient on a flat plate.

With the help of Nikuradse's experiments, Schlichting obtained the semi-empirical equation for the average skin friction coefficient as

$$\bar{C}_f = \frac{0.455}{(\log \mathrm{Re})^{2.58}} \qquad (10.65)$$

Equation (10.65) was derived assuming the flat plate to be completely turbulent over its entire length. In reality, a portion of it is laminar from the leading edge to some downstream position. For this purpose, it was suggested to use

$$\bar{C}_f = \frac{0.455}{(\log \mathrm{Re})^{2.58}} - \frac{A}{\mathrm{Re}} \qquad (10.66a)$$

where A has various values depending on the value of Reynolds number at which the transition takes place. If the transition is assumed to take place around a Reynolds number of 5×10^5, the average skin friction correlation of Schlichling can be written as

$$\overline{C}_f = \frac{0.455}{(\log \text{Re})^{2.58}} - \frac{1700}{\text{Re}} \tag{10.66b}$$

All that we have presented so far are valid for a smooth plate. Schlichting used a logarithmic expression for turbulent flow over a rough surface and derived

$$\overline{C}_f = \left(1.89 + 1.62 \log \frac{L}{\varepsilon_p}\right)^{-2.5} \tag{10.67}$$

Example 10.4

During flow over a flat plate the laminar boundary layer undergoes a transition to turbulent boundary layer as the flow proceeds in the downstream. It is observed that a parabolic laminar profile is finally changed into a 1/7th power law velocity profile in the turbulent regime. Find out the ratio of turbulent and laminar boundary layers if the momentum flux within the boundary layer remains constant.

Solution

Assume width of the boundary layers be a. Then momentum flux is

$$A = \int u \, \rho \, u \, a \, dy = \rho \, U_\infty^2 \, a \, \delta \int \left(\frac{u}{U_\infty}\right)^2 d\eta$$

where $\eta = y/\delta$

For laminar flow, $\dfrac{u}{U_\infty} = 2\,\eta - \eta^2$

$$A_{\text{lam}} = \rho U_\infty^2 \, a \, \delta_{\text{lam}} \int_0^1 (4\eta^2 - 4\eta^3 + \eta^4) \, d\eta$$

$$= \rho U_\infty^2 \, a \, \delta_{\text{lam}} \left[\frac{4}{3}\eta^3 - \eta^4 + \frac{\eta^5}{5}\right]_0^1$$

$$= \frac{8}{15} \rho U_\infty^2 \, a \, \delta_{\text{lam}}$$

For 1/7th power law turbulent profile,

$$\frac{\overline{u}}{U_\infty} = \eta^{1/7}$$

$$A_{\text{turb}} = \rho U_\infty^2 \, a \, \delta_{\text{turb}} \int_0^1 (\eta^{1/7})^2 \, d\eta$$

$$= \rho U_\infty^2 \, a \, \delta_{\text{turb}} \int_0^1 \eta^{2/7} \, d\eta$$

$$= \rho U_\infty^2 \, a \, \delta_{\text{turb}} \left[\frac{7}{9} \eta^{9/7} \right]_0^1 = \frac{7}{9} \rho U_\infty^2 \, a \, \delta_{\text{turb}}$$

Comparing the momentum fluxes,

$$\frac{\delta_{\text{turb}}}{\delta_{\text{lam}}} = \frac{72}{105}$$

It is to be noted that generally the turbulent boundary layer grows faster than the laminar boundary layer when a completely turbulent flow is considered from the leading edge. However the present result is valid at transition for a constant momentum flow.

Example 10.5

Air ($\rho = 1.23$ kg/m^3 and $\nu = 1.5 \times 10^{-5}$ m^2/s) is flowing over a flat plate. The free stream speed is 15 m/s. At a distance of 1 m from the leading edge, calculate δ and τ_w for (i) completely laminar flow and (ii) completely turbulent flow for a 1/7th power law velocity profile.

Solution

Applying the results developed in Chapters 9 and 10, we can write for parabolic velocity profile (laminar flow)

$$\frac{\delta}{x} = \frac{5.48}{\sqrt{\text{Re}_x}} \quad \text{and} \quad \tau_w = \mu \left. \frac{\partial u}{\partial y} \right|_{y=0}$$

$$\text{Re}_x = \frac{U x}{\nu} = \frac{15 \times 1}{1.5 \times 10^{-5}} = 1.0 \times 10^6$$

$$\delta = \frac{5.48}{\sqrt{1.0 \times 10^6}} \times 1 \text{ m} = 5.48 \text{ mm}$$

$$\tau_w = \mu \left. \frac{\partial u}{\partial y} \right|_{y=0} = \frac{\mu U_\infty}{\delta} \cdot \frac{d}{d\eta} [2\eta - \eta^2]_{\eta=0}$$

$$\tau_w = \frac{2 \times 1.23 \times 1.5 \times 10^{-5} \times 15}{0.00548} = 0.101 \text{ N/m}^2$$

For turbulent flow,

$$\frac{\delta}{x} = \frac{0.37}{(\text{Re}_x)^{1/5}} \quad \text{(from Eq. 10.62)}$$

or $$\delta = \frac{0.370}{(1.0 \times 10^6)^{1/5}} \times 1\ \text{m} = 23.34\ \text{mm}$$

or $$\delta/x = 0.0233$$

$$\tau_w = 0.0225\, \rho U_\infty^2 \left(\frac{\nu}{U_\infty \delta}\right)^{1/4} \quad \text{(from Eq. 10.60)}$$

$$\tau_w = 0.0225 \times 1.23 \times (15)^2 \left(\frac{\nu}{U_\infty x} \cdot \frac{x}{\delta}\right)^{1/4}$$

or $$\tau_w = 0.0225 \times 1.23 \times (15)^2 \left[\frac{1}{1.0 \times 10^6} \times \frac{1}{0.0233}\right]^{1/4}$$

$$= 0.502\ \text{N/m}^2$$

The turbulent boundary layer has a larger shear stress than the laminar boundary layer.

SUMMARY

- Turbulent motion is an irregular motion of fluid particles in a flow field. However, for homogeneous and isotropic turbulence, the flow field can be described by time-mean motions and fluctuating components. This is called Reynolds decomposition of turbulent flow.
- In a three-dimensional flow field, the velocity components and the pressure can be expressed in terms of the time-averages and the corresponding fluctuations. Substitution of these dependent variables in the Navier–Stokes equations for incompressible flow and subsequent time-averaging yield the governing equations for the turbulent flow. The mean velocity components of turbulent flow satisfy the same Navier–Stokes equations for laminar flow. However, for the turbulent flow, the laminar stresses are increased by additional stresses arising out of the fluctuating velocity components. These additional stresses are known as apparent stresses of turbulent flow or Reynolds stresses.
- In analogy with the laminar shear stresses, the turbulent shear stresses can be expressed in terms of mean velocity gradients and a mixing coefficient known as eddy viscosity. The eddy viscosity (ν_t) can be expressed as $\nu_t = l^2 \left|\frac{d\bar{u}}{dy}\right|$, where l is known as Prandtl's mixing length.
- For a homogeneous and isotropic turbulence, the most widely used value of mixing length is given by $l = \chi y$. In this expression, y is the distance from the wall and χ is known as the von Karman constant (≈ 0.4). For high Reynolds number, fully developed turbulent duct flows, the velocity profile is given by

$$\frac{\bar{u}}{u_\tau} = A_1 \ln \eta + D_1$$

where $\bar{u}$ is the time-mean velocity at any η $(= yu_\tau/\nu)$ and u_τ is the friction velocity given by $\sqrt{\tau_w/\rho}$. The constants A_1 and D_1 are determined from experiments which are 2.5 and 5.5, respectively, for smooth pipes. The corresponding friction factor (f) is given by

the expression $\frac{1}{\sqrt{f}} = 2.0 \log_{10} (\text{Re}\sqrt{f}\,) - 0.8$.

- However, for pipe roughness of varying degree, the following recommendation of Colebrook and White works well:

$$\frac{1}{\sqrt{f}} = 1.74 - 2.0 \log_{10}\left[\frac{\varepsilon_p}{R} + \frac{18.7}{\text{Re}\sqrt{f}}\right]$$

where ε_p/R is pipe roughness.

- In the range of Re $\leq 10^5$, the velocity distribution in a smooth pipe is given by

$$\frac{\bar{u}}{u_\tau} = 8.74\left(\frac{yu_\tau}{\nu}\right)^{1/7}$$

- The friction factor in this regime is given by Blasius as

$$f = 0.3164\ (\text{Re})^{-0.25}$$

- The growth of boundary layer for turbulent flow over a flat plate is given by

$$\frac{\delta}{x} = 0.37\ (\text{Re}_x)^{-1/5}$$

- The expression for the average skin friction coefficient on the entire plate of length L has been determined as

$$\bar{C}_f = 0.072\ (\text{Re}_L)^{-1/5}$$

- This result is found to be in good agreement with the experimental results in the range of $5 \times 10^5 < \text{Re} < 10^7$ which is given by

$$\bar{C}_f = 0.074\ (\text{Re}_L)^{-1/5}$$

- For turbulent flow over a rough plate, the average skin friction coefficient is given by

$$\bar{C}_f = \left(1.89 + 1.62 \log \frac{L}{\varepsilon_p}\right)^{-2.5}$$

REFERENCES

1. Tennekes, H., and Lumley, J.L, *A First Course in Turbulence*, the MIT Press, Cambridge, Massachusetts, 1972.
2. Hinze, J.O., *Turbulence*, McGraw-Hill Book Company, New York, 1987.

EXERCISES

10.1 Only write down the option (true/false) or the choice (a, b, c or d) or the appropriate conditions.

(i) For flow through pipes, due to the same pressure gradient, the turbulent velocity profile will be more uniform than the laminar velocity profile. (True/False)

(ii) If the mean velocity has a gradient, the turbulence is called isotropic. (True/False)

(iii) $\dfrac{\partial u'}{\partial x} + \dfrac{\partial v'}{\partial y} = 0$ for a turbulent flow signifies

(a) conservation bulk momentum transport

(b) increase in u' in the positive x direction will be followed by increase in v' in negative y direction

(c) turbulence is anisotropic

(d) turbulence is isotropic

(vi) In a turbulent pipe flow the initiation of turbulence is usually observed at a Reynolds number (based on pipe diameter) of

(a) 3.5×10^5 (b) 2×10^6

(c) between 2000 and 2700 (d) 5000

(v) A turbulent boundary is thought to be comprised of laminar sublayer, a buffer layer and a turbulent zone. The velocity profile outside the laminar sublayer is described by a

(a) parabolic profile (b) cubic profile

(c) linear profile (d) logarithmic profile

(vi) A laminar boundary layer is less likely to separate than a turbulent boundary layer. (True/False)

10.2 Show that, with the help of both the mixing length hypothesis due to Prandtl and mixing length law due to Karman (given in Example 10.2), the universal velocity profile near the wall in case of a fully developed turbulent flow through a circular pipe can be expressed as

$$\frac{U_{\max} - \bar{u}}{u_\tau} = \frac{1}{\kappa} \ln\left(\frac{R}{R-r}\right)$$

where r is the radius of the pipe and κ is a constant.

10.3 Calculate power required to move a flat plate, 8 m long and 3 m wide in water ($\rho = 1000$ kg/m^3, $\mu = 1.02 \times 10^{-3}$ kg/ms) at 8 m/s for the following cases:

(i) the boundary layer is turbulent over the entire surface of the plate

(ii) the transition takes place at Re = 5×10^5.

Ans. ((i) 12.536×10^3 W (ii) 12.518×10^3 W)

10.4 The transition Reynolds number in a pipe flow based on U_{av} is approximately 2300. How can this value be extrapolated for the flow over a flat plate if U_∞ in the flat plate case is analogous to U_{max} in the pipe and δ is analogous to pipe radius R?

Ans. ($Re_x = 2.116 \times 10^5$)

10.5 A plate 50 cm long and 2.5 m wide moves in water at a speed of 15 m/s. Estimate its drag if the transition takes place at Re = 5×10^5 for (i) a smooth wall, and (ii) a rough wall, $\varepsilon_p = 0.1$ mm. For water, $\rho = 1000$ kg/m^3 and $\mu = 1.02 \times 10^{-3}$ kg/ms.

Ans. ((i) 411.405 W (ii) 806.168 W)

10.6 Estimate the power required to move a flat plate, 15 m long and 4 m wide, in oil ($\rho = 800$ kg/m^3, $\nu = 10^{-5}$m^2/s) at 4m/s, under the following cases:

(a) The boundary layer is assumed laminar over the entire surface of the plate.
(b) Transition to turbulence occurs at Re = 3 * 10^5 and plate is smooth.
(c) The boundary layer is turbulent over the entire plate which is smooth.
(d) The boundary layer is turbulent over the entire rough plate with

$$\frac{u_\infty \varepsilon_p}{\nu} = 10^3.$$

Ans. ((a) 1665.5 Watt (b) 9486 Watt (c) 10023.94 Watt, (d) 17200 Watt)

10.7 In turbulent flat-plate flow, the wall shear stress is given by the formula

$$\tau_w = 0.0225\ \rho U_\infty^2 \left[\frac{\nu}{\delta U_\infty}\right]^{1/4}$$

Two important equations concerning the 1/7th power law velocity profiles are

$$C_{fx} = 2\frac{d\delta^{**}}{dx} \quad \text{and} \quad \delta^{**} = \frac{7}{72}\delta$$

From the above three equations, find the final expression for skin friction coefficient (C_{fx}).

Ans. ($C_{fx} = 0.0576\,(Re_x)^{-1/5}$)

10.8 Water flows at a rate of 0.05 m^3/s in a 20 cm diameter cast iron pipe ($\varepsilon_p/D = 0.0007$). What is the head (pressure) loss per kilometer of the pipe? For water, $\rho = 1000$ kg/m^3, $\nu = 1.0 \times 10^{-6}$ m^2/s. Use Moody's chart.

Ans. (12.2 m)

10.9 Air ($\rho = 1.2$ kg/m^3 and $\nu = 1.5 \times 10^{-5}$ m^2/s) flows at a rate of 2.5 m^3/s in a 30 cm × 60 cm steel rectangular duct ($\varepsilon_p = 4.6 \times 10^{-5}$ m). What is the pressure drop per 50 m of the duct? Use Moody's chart.

Ans. (217 pa)

10.10 Water is being transported through a rough pipe line ($u_\tau \varepsilon_p/\nu = 100$), 1 km long with maximum velocity of 4 m/s. If the Reynolds number is 1.5×10^6, find out the diameter of the pipe and power required to maintain the flow. For water, $\rho =$ 1000 kg/m^3, $\nu = 1.0 \times 10^{-6}$ m^2/s.

Ans. ($D = 0.454$ m, $P = 134.113$ kW)

10.11 Water ($\rho = 1000$ kg/m^3, $\nu = 2 \times 10^{-6}$ m^2/s) is transported through a horizontal pipeline, 800 m long, with a maximum velocity of 3 m/s. If the Reynolds number is 10°, find the diameter of the pipe (with and without the use of Moody Diagram).

Also calculate the thickness of laminar sub-layer and the buffer layer, and the power required to maintain the flow. Calculate your results for a fully rough pipe with $\frac{u_\infty \varepsilon_p}{\nu} = 100$.

Ans. (Diameter of the pipe 0.8 m, laminar sub-layer thickness 0.1 mm, buffer layer thickness 1.3 mm, power required 50250 W)

10.12 Modify the friction drag coefficient given by Eq. (10.64) as $\bar{C}_f = 0.074\ (\text{Re}_L)^{-1/5} - A/\text{Re}_L$. Let the flow be laminar up to a distance X_{cr} from the leading edge and turbulent for $X_{cr} \le x \le L$. Consider the transition to occur at $\text{Re}_x = 5\times10^5$.

Ans. (A=1700)

10.13 Air flows over a smooth flat plate at a velocity of 4.4 m/s. The density of air is 1.029 kg/m^3 and the kinematic viscosity is 1.35×10^{-5} m^2/s. The length of the plate is 12 m in the direction of flow. Calculate (i) the boundary layer thickness at 16 cm and 12 m respectively, from the leading edge and (ii) the drag coefficient for the entire plate surface (one side) considering turbulent flow.

Ans. ((i) 3.5×10^{-3} m, and 0.0207 m (ii) $\bar{C}_f = 3.554\times10^{-3}$)

10.14 The velocity distribution for a laminar boundary layer flow is given by $\frac{u}{u_e} = \sin\left(\frac{\pi}{2}\cdot\frac{y}{\delta}\right)$. The velocity at $y = k$ is given by u_k. It is assumed that the small roughness of height k will not generate eddies to disturb the boundary layer if $\frac{u_k k}{\nu}$ is less than about 5.0. Show that at a distance x from the leading edge, the maximum permissible roughness height for the boundary layer to remain undistributed is given by $\frac{k}{c} = \frac{A}{(\text{Re})^{3/4}}\left(\frac{x}{c}\right)^{1/4}$ where, $\text{Re} = \frac{u_e c}{\nu}$, c is the total length of the plate and A is a constant.

10.15 Find the frictional drag on the top and sides of a box-shaped moving van 2.4 m wide, 3.0 m high, and 10.5 m long travelling at 100 km/hr through air ($\nu = 1.4 \times 10^{-6}$ m^2/s). Assume that the vehicle has a rounded nose so that the flow does not separate from the top and side. Also assume that a turbulent boundary layer starts immediately at the leading edge.

Also, find the thickness of the boundary layer and the shear stress at the trailing edge.

Ans. (Drag = 105.9 N, B.L. = 0.136 m, Shear stress = 0.904 Pa)

11

VISCOUS FLOWS THROUGH PIPES

11.1 INTRODUCTION

Fully developed laminar and turbulent flows through pipes of uniform cross section have already been discussed in Sections 8.3.4 and 10.11 respectively. While a complete analytical solution for the equation of motion in case of a laminar flow is available, even the advanced theories in the analyses of turbulent flow depend at some point on experimentally derived information. One of the most important items of information that a hydraulic engineer needs is the power required to force fluids at a certain steady rate through a pipe or pipe network system. This information is furnished in practice through some routine solution of pipe flow problems with the help of available empirical and theoretical information. This chapter deals with the typical approaches to the solution of pipe flow problems in practice.

11.2 CONCEPT OF FRICTION FACTOR IN A PIPE FLOW

The friction factor in the case of a pipe flow has already been mentioned in Section 8.3.4. A little elaborate discussion on the friction factor or friction coefficient is still needed for the sake of its use in different practical problems. Skin friction coefficient for a fully developed flow through a closed duct is defined as

$$C_f = \frac{\tau_w}{(1/2)\rho V^2} \tag{11.1}$$

where, V is the average velocity of flow given by $V = Q/A$, Q and A are the volume flow rate through the duct and the cross-sectional area of the duct respectively. From a force balance of a typical fluid element (Fig. 11.1) in course of its flow through a duct of constant cross-sectional area, we can write

$$\tau_w = \frac{\Delta p^* A}{S L} \tag{11.2}$$

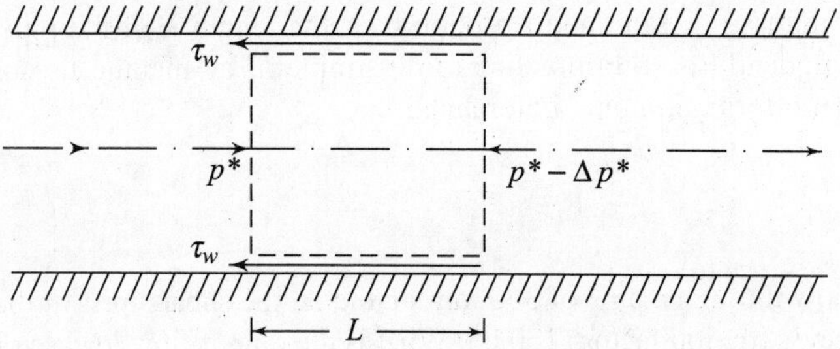

Fig. 11.1 Force balance of a fluid element in the course of flow through a duct

where, τ_w is the shear stress at the wall and Δp^* is the piezometric pressure drop over a length of L. A and S are respectively the cross-sectional area and wetted perimeter of the duct. Substituting the expression (11.2) in Eq. (11.1), we have,

$$C_f = \frac{\Delta p^* A}{S L (1/2)\rho V^2}$$

$$= \frac{1}{4}\frac{D_h}{L}\frac{\Delta p^*}{(1/2)\rho V^2} \tag{11.3}$$

where, $D_h = 4A/S$ and is known as the *hydraulic diameter*. In case of a circular pipe, $D_h = D$, the diameter of the pipe. The coefficient C_f defined by Eqs (11.1) or (11.3) is known as *Fanning's friction factor*. To do away with the factor 1/4 in the Eq. (11.3), Darcy defined a friction factor f as

$$f = \frac{D_h}{L}\frac{\Delta p^*}{(1/2)\rho V^2} \tag{11.4}$$

Comparison of Eqs (11.3) and (11.4) gives $f = 4C_f$. Equation (11.4) can be written for a pipe flow as

$$f = \frac{D}{L}\frac{\Delta p^*}{(1/2)\rho V^2} \tag{11.5}$$

Equation (11.5) is written in a different fashion for its use in the solution of pipe flow problems in practice as

$$\Delta p^* = f\frac{L}{D}\frac{1}{2}\rho V^2 \tag{11.6a}$$

or in terms of head loss (energy loss per unit weight)

$$h_f = \frac{\Delta p^*}{\rho g} = f\frac{L}{D}(V^2/2g) \tag{11.6b}$$

where, h_f represents the loss of head due to friction over the length L of the pipe. Equation (11.6b) is frequently used in practice to determine h_f by making use of theoretical or empirical information on f beforehand.

Example 11.1

In a fully developed flow through a pipe of 300 mm diameter, the shear stress at the wall is 50 Pa. The Darcy's friction factor f is 0.05. What is the rate of flow in case of (i) water flowing through the pipe and (ii) oil of specific gravity 0.70 flowing through the pipe?

Solution

Darcy's friction factor f is defined (see Eq. 11.1 to 11.4) as

$$f = 4 \times \frac{\tau_w}{\frac{1}{2}\rho V^2}$$

where, τ_w is the wall shear stress and V is the average flow velocity.

Therefore, $$V = \sqrt{\frac{8\,\tau_w}{\rho f}}$$

and, flow rate $$Q = V \times \pi R^2 \text{ (where } R \text{ is the pipe radius)}$$

(i) For water flowing through the pipe

$$Q = \pi \times (0.3)^2 \sqrt{\frac{8 \times 50}{10^3 \times 0.05}}$$

$$= 0.8 \text{ m}^3/\text{s}$$

(ii) For oil flowing through the pipe

$$Q = \pi \times (0.3)^2 \sqrt{\frac{8 \times 50}{0.70 \times 10^3 \times 0.05}}$$

$$= 0.96 \text{ m}^3/\text{s}$$

11.3 VARIATION OF FRICTION FACTOR

In case of a laminar fully developed flow through pipes, the friction factor f, is found from the exact solution of the Navier–Stokes equation as discussed in Section 8.4.3. It is given by

$$f = \frac{64}{\text{Re}} \tag{11.7}$$

It has also been discussed in Sections 10.10 and 10.11 that in case of a turbulent flow, friction factor depends on both the Reynolds number and the roughness of the pipe surface.

Sir Thomas E. Stanton (1865–1931) first started conducting experiments on a number of pipes of various diameters and materials and with various fluids. Afterwards, a German engineer Nikuradse carried out experiments on flows through pipes in a very wide range of Reynolds number. A comprehensive documentation of the experimental and theoretical investigations on the laws of friction in pipe flows has been made in the form of a diagram, as shown in Fig. 11.2, by L.F. Moody to show the variation of friction factor f, with the pertinent governing parameters, namely, the Reynolds number of flow and the relative roughness ε/D of the pipe. This diagram is known as Moody's diagram which is employed till today as the best means for predicting the values of f.

Roughness in commercial pipes is due to the protrusions at the surface which are random both in size and spacing. However, the commercial pipes are specified by the average roughness which is the measure of some average height of the protrusions. This equivalent average roughness is determined from the experimental comparisons of flow rate and pressure drop in a commercial pipe with that of a pipe with artificial roughness created by gluing grains of sand of uniform size to the wall. Friction factor f, in laminar flow, as given by Eq. (11.7), is independent of the roughness of the pipe wall, unless the roughness is so great that the irregularities make an appreciable change in diameter of the pipe. Beyond a Reynolds number of 2000, i.e., in turbulent region, the flow depends on the roughness of the pipe. Figure 11.2 depicts that the friction factor f, at a given Reynolds number, in the turbulent region, depends on the relative roughness, defined as the ratio of average roughness to the diameter of the pipe, rather than the absolute roughness. For moderate degree of roughness, a pipe acts as a smooth pipe up to a value of Re where the curve of f versus Re for the pipe coincides with that of a smooth pipe. This zone is known as the *smooth zone of flow*. The region where f versus Re curves (Fig. 11.2) become horizontal showing that f is independent of Re, is known as the *rough zone* and the intermediate region between the smooth and rough zone is known as the *transition zone*. The position and extent of all these zones depend on the relative roughness of the pipe. In the smooth zone of flow, the laminar sublayer becomes thick, and hence, it covers appreciably the irregular surface protrusions. Therefore all the curves for smooth flow coincide. With increasing Reynolds number, the thickness of sublayer decreases and hence the surface bumps protrude through it. The higher is the roughness of the pipe, the lower is the value of Re at which the curve of f versus Re branches off from smooth pipe curve (Fig. 11.2). In the rough zone of flow, the flow resistance is mainly due to the form drag of those protrusions. The pressure drop in this region is approximately proportional to the square of the average velocity of flow. Thus f becomes independent of Re in this region.

In practice, there are three distinct classes of problems relating to flow through a single pipe line as follows:

(i) The flow rate and pipe diameter are given. One has to determine the loss of head over a given length of pipe and the corresponding power required to maintain the flow over that length.

(ii) The loss of head over a given length of a pipe of known diameter is given. One has to find out the flow rate and the transmission of power accordingly.

(iii) The flow rate through a pipe and the corresponding loss of head over a part of its length are given. One has to find out the diameter of the pipe.

In the first category of problems, the friction factor f, is found out explicitly from the given values of flow rate and pipe diameter. Therefore, the loss of head h_f and the power required P can be calculated by the sraightforward application of Eq. (11.6b). A typical example of this category of problems is given below:

Example 11.2

Determine the loss of head in friction when water at 15 °C flows through a 300 m long galvanised steel pipe of 150 mm diameter at 0.05 m^3/s. (Kinematic viscosity of water at 15 °C = 1.14×10^{-6} m^2/s. Average surface roughness for galvanised steel = 0.15 mm). Also calculate the pumping power required to maintain the above flow.

Solution

Average velocity of flow $V = \dfrac{0.05}{(\pi/4)(0.15)^2} = 2.83$ m/s

Therefore, Reynolds number $\mathrm{Re} = \dfrac{VD}{\nu} = \dfrac{2.83 \times 0.15}{1.14 \times 10^{-6}} = 3.72 \times 10^5$

Relative roughness $\varepsilon/D = 0.15/150 = 0.001$

From Fig. 11.2, $f = 0.02$

Hence, using Eq. (11.6b)

$$h_f = 0.02 \frac{300}{0.15} \frac{(2.83)^2}{2 \times 9.81} = 16.33 \text{ m}$$

Power required to maintain a flow at the rate of Q under a loss of head of h_f is given by

$$\begin{aligned} P &= \rho g h_f Q \\ &= 10^3 \times 9.81 \times 16.33 \times 0.05 \text{ W} \\ &= 8 \text{ kW} \end{aligned}$$

In the second and third category of problems, both the flow rate and the pipe diameter are not known beforehand to determine the friction factor. Therefore the problems in these categories cannot be solved by the straightforward application of Eq. (11.6b), as shown in Example 11.2 above. A method of iteration is suggested in this case where a guess is first made regarding the value of f. With the guess value of f the flow rate or the pipe diameter, whichever is unknown in the problem, is found out as a first approximation using the Eq. (11.6b). Then the guess value of f is updated with the new value of Reynolds number found from the approximate value of flow rate or pipe diameter as calculated. The problem is repeated till a legitimate convergence in f is achieved. Examples of this typical method dealing with the problems belonging to categories (ii) and (iii), as mentioned above, are given below.

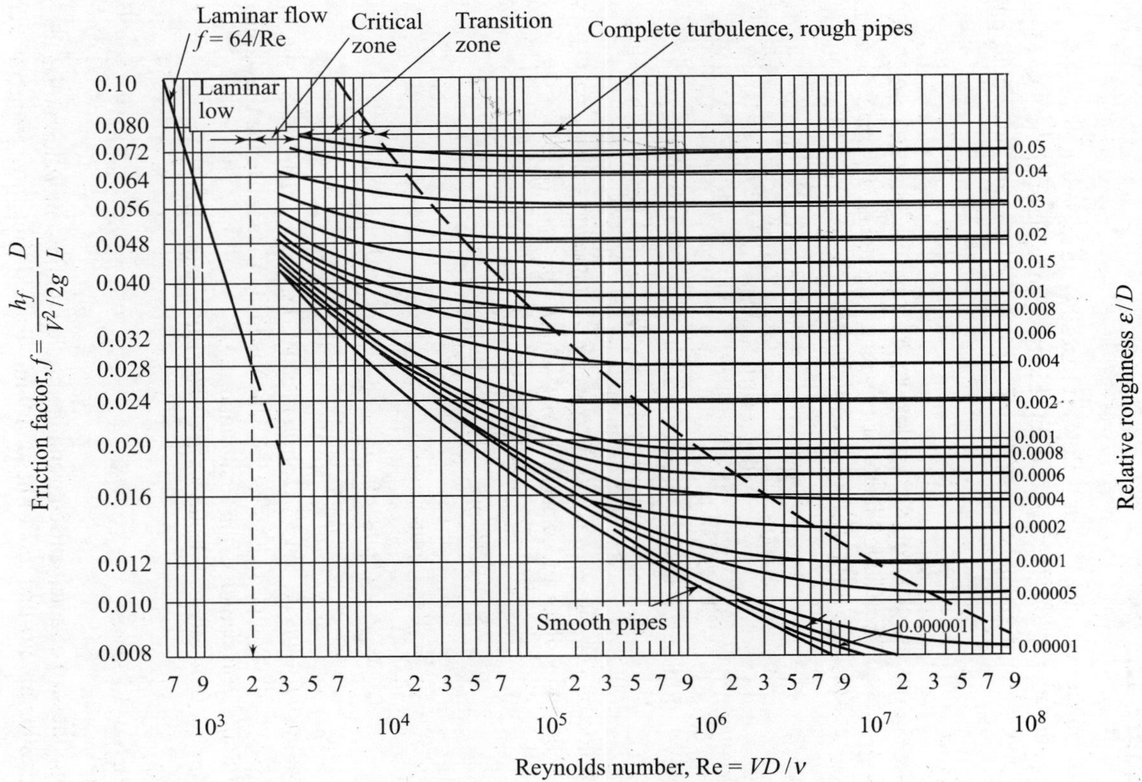

Values of ε for new pipes

Riveted steel	1–10 mm
Concrete	0.3–3 mm
Cast iron	0.25 mm
Galvanised iron	0.15 mm
Mild steel	0.045 mm
Drawn tubing, brass, lead, glass	0.0015 mm

Fig.11.2 Friction factors for pipes (adopted from Trans. ASME, 66, 672, 1944)

Example 11.3

Oil of kinematic viscosity 10^{-5} m^2/s flows at a steady rate through a cast iron pipe of 100 mm diameter and of 0.25 mm average surface roughness. If the loss of head over a pipe length of 120 m is 5 m of the oil, what is the flow rate through the pipe?

Solution

Since the velocity is unknown, Re is unknown. Relative rougness $\varepsilon/D = 0.25/100 = 0.0025$. A guess of the friction factor at this relative roughness is made from Fig. 11.2 as $f = 0.026$ Then Eq. (11.6b) gives a first trial

$$5 = 0.026\,\frac{120}{0.10}\,\frac{V^2}{2\times 9.81}$$

when, $$V = 1.773 \text{ m/s}$$

Then, $$\text{Re} = \frac{1.773\times 0.10}{10^{-5}} = 1.773\times 10^4$$

The value of Re, with ε/D as 0.0025, gives $f = 0.0316$ (Fig. 11.2). The second step of iteration involves a recalcualtion of V with $f = 0.0316$, as

$$5 = 0.0316\times\frac{120}{0.1}\,\frac{V^2}{2\times 9.81}$$

which gives $$V = 1.608 \text{ m/s}$$

and $$\text{Re} = \frac{1.608\times 0.10}{10^{-5}} = 1.608\times 10^4$$

The value of f at this Re (Fig. 11.2) becomes 0.0318. The relative change between the two successive values of f is 0.63% which is insignificant. Hence the value of $V = 1.608$ m/s is accepted as the final value.
Therefore, the flow rate $Q = 1.608\times(\pi/4)\times(0.10)^2 = 0.013$ m^3/s

Example 11.4

Determine the size of a galvanised iron pipe needed to transmit water a distance of 180 m at 0.085 m^3/s with a loss of head of 9 m. (Take kinematic viscosity of water $\nu = 1.14\times 10^{-6}$ m^2/s, and the average surface roughness for galvanised iron = 0.15 mm).

Solution

From Eq. (11.6b),

$$9 = f\,\frac{180}{D}\left(\frac{0.085}{\pi D^2/4}\right)^2\frac{1}{2\times 9.81}$$

which gives $$D^5 = 0.012\,f \qquad (11.8)$$

and $$\text{Re} = \frac{0.085\,D}{(\pi D^2/4)\times 1.14\times 10^{-6}}$$

$$= 9.49\times 10^4\,\frac{1}{D} \qquad (11.9)$$

First, a guess in f is made as 0.024.

Then from Eq. (11.8) $D = 0.196\,\text{m}$

and from Eq. (11.9) $\text{Re} = 4.84\times 10^5$

The relative roughness $\varepsilon/D = \dfrac{0.15}{0.196}\times 10^{-3} = 0.00076$

With the values of Re and ε/D, the updated value of f is found from Fig. 11.2 as 0.019. With this value of f as 0.019, a recalculation of D and Re from Eqs (11.8) and (11.9) gives $D = 0.187$ m, $\text{Re} = 5.07\times 10^5$. ε/D becomes $(0.15/0.187)\times 10^{-3} = 0.0008$. The new values of Re and $\in/D$ predict $f = 0.0192$ from Fig. 11.2. This value of f differs negligibly (by 1%) from the previous value of 0.019. Therefore the calculated diameter $D = 0.187$ m is accepted as the final value.

11.4 ENERGY CONSIDERATIONS IN PIPE FLOW

The discussions made in this chapter so far implicate that there is invariably an energy loss to overcome viscous resistances in pipe flow. In order to assess the underlying consequences from a control volume energy balance perspective, we consider a control volume (CV) as depicted in Fig. 11.3. For applying Reynolds transport theorem (Eq. (5.3)) for energy conservation as applicable to the identified CV, we set $N = E$ (where E is the total energy of the system = Kinetic energy + Potential energy + Internal energy $= m\dfrac{V^2}{2} + mgz + mi$; i being the internal energy per unit mass) and $\eta = e$ (where $e = \dfrac{V^2}{2} + gz + i$, which is energy per unit mass). Accordingly, Eq. (5.3) becomes

$$\left.\frac{dE}{dt}\right|_{\text{system}} = \frac{\partial}{\partial t}\int_{CV}\rho e\, d\forall + \int_{CS}\rho e\left(\vec{V}_r\cdot\hat{n}\right)dA \qquad (11.10)$$

We further consider steady flow through a pipe so that $\dfrac{\partial}{\partial t}\displaystyle\int_{CV}\rho e\, d\forall = 0$ and furthermore, we assume a stationary CV, so that $\vec{V}_r = \vec{V}$. Considering these two assumptions, Eq. (11.10) simplifies to

$$\left.\frac{dE}{dt}\right|_{\text{system}} = \int_{CS}\rho e\left(\vec{V}\cdot\hat{n}\right)dA \qquad (11.10a)$$

For describing Eq. (11.10a) in terms of heat-work interactions, we refer to the first law of thermodynamics for a system which states that the heat δQ added to a system, minus the work done δW by the system, equals to the change in its energy E that depends only upon the initial and final states of the system. In mathematical form

$$\delta Q - \delta W = dE \tag{11.11a}$$

Equation (11.11a) can be expressed on the time rate basis as

$$\dot{Q} - \dot{W} = \left.\frac{dE}{dt}\right|_{\text{system}} \tag{11.11b}$$

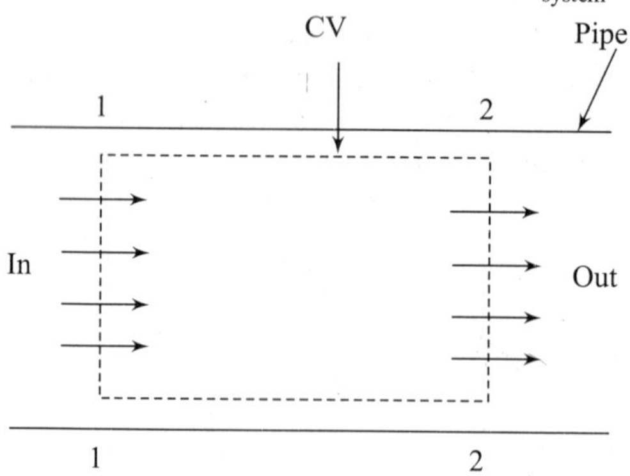

Fig. 11.3 A control volume used to derive energy equation in pipe flow

Substituting Eq. (11.11b) into Eq. (11.10a), we get

$$\dot{Q} - \dot{W} = \int_{CS} \rho e\left(\vec{V}\cdot\hat{n}\right)dA \tag{11.12}$$

The rate of work done by the fluid can be expressed as (refer Fig. 11.3)

$$\dot{W} = -\int_{CS} p v \rho\left(\vec{V}.\hat{n}\right)dA \tag{11.13}$$

Equation is the specific volume Eq. (11.13) follows from the consideration that there is a work associated with the occurrence of flow in presence of pressure, which is called as flow energy or flow work. At the inlet, work is input to the control volume, which by the present sign convention, implicates negative work done by the control volume. Reverse is the case at the flow outlet. It is important to mention here that there is no work done to overcome the shear stress of the wall as there is no displacement of fluid at the wall.

In many traditional texts, the rate of work done represented by Eq. (11.13) is clubbed up with the rate of energy flow term depicted by the right-hand side of Eq. (11.12). In that case, the net rate of work done, $\dot{W}$, as appearing in the left hand side of Eq. (11.12) should be taken as zero, since the same effect should not be taken twice.

Combining Eqs (11.12) and (11.13), one can write

$$\dot{Q} = \int_{out} \rho\left(\frac{p}{\rho} + \frac{V^2}{2} + gz + i\right) V dA - \int_{in} \rho\left(\frac{p}{\rho} + \frac{V^2}{2} + gz + i\right) V dA \tag{11.14}$$

The average velocity at across section in a flowing stream is defined on the basis of the volumetric flow rate as

$$\bar{V} = \frac{\int_A V dA}{A}$$

The kinetic energy per unit mass of the fluid is usually expressed as $\alpha\left(\bar{V}^2/2\right)$, where α is known as the kinetic energy correction factor. This correction factor is introduced so that the kinetic energy becomes expressible in terms of the average velocity over a section on adjustment with this factor. Therefore, we can write (noting that rate of flow kinetic energy $\dot{m}\alpha\frac{\bar{V}^2}{2} = (\rho A\bar{V})\alpha\frac{\bar{V}^2}{2}$)

$$\alpha\frac{1}{2}\rho\bar{V}^3 A = \int_A \frac{1}{2}\rho V^3 dA$$

Thus,
$$\alpha = \frac{\int \rho V^3 dA}{\rho\bar{V}^3 A} \tag{11.15a}$$

For a constant density flow,

$$\alpha = \frac{\int V^3 dA}{\bar{V}^3 A} \tag{11.15b}$$

In case of a laminar fully developed constant density flow through a circular pipe, the value of α becomes 2. This can be derived straightway from Eq. (11.15b) by noting that $\frac{V}{\bar{V}} = 2\left(1 - \frac{r^2}{R^2}\right)$ (Eq. (8.55) and $dA = 2\pi r dr$. For a turbulent flow through a pipe, the value of α usually varies from 1.01 to 1.15. In the absence of a prior knowledge about the velocity distribution, the value of α, in most of the practical analyses, is taken as unity for turbulent flows. The rationale behind this argument is that the profile of average velocity for turbulent flow through a pipe is virtually uniform due to efficient mixing, resulting in only marginal deviation of the local average velocities from the mean averaged velocity.

Assuming that pressure is not varying at a given cross section, and denoting the Subscripts 1 and 2 for inlet and outlet, respectively, (refer to Fig. 11.3), Eq. (11.14), with the aid of Eq. (11.15a), can be written as

$$\dot{Q}_{CV} = \frac{p_2}{\rho}\dot{m}_{out} + \alpha_1\dot{m}_{out}\frac{\bar{V}_1^2}{2} + gz_2\dot{m}_{out} + i_2\dot{m}_{out} - \frac{p_1}{\rho}\dot{m}_{in} - \alpha_1\dot{m}_{in}\frac{\bar{V}_1^2}{2} + gz_1\dot{m}_{in} + i_1\dot{m}_{in}$$

or
$$\frac{p_1}{\rho} + \alpha_1\frac{\bar{V}_1^2}{2} + gz_1 = \frac{p_2}{\rho} + \alpha_2\frac{\bar{V}_2^2}{2} + gz_2 + \left[(i_2 - i_1) - \frac{\dot{Q}_{CV}}{\dot{m}}\right] \tag{11.16}$$

Equation (11.16) is a statement of energy equation in pipe flow.

part of the mechanical energy converted into the intermolecular energy is termed as the *loss of energy*, since our attention is focussed only on the mechanical energy of the fluid.

Apart from the losses due to friction between the solid surface and the fluid layer past it, the loss of mechanical energy is also incurred when the path of the fluid is suddenly changed in course of its flow through a closed duct due to any abrupt change in the geometry of the duct. In long ducts, these losses are very small compared to the frictional loss, and hence they are often termed as minor losses. But minor losses may, however, outweigh the frictional loss in short pipes or ducts. The source of these losses is usually confined to a very short length of the duct, but the turbulence produced may persist for a considerable distance downstream. A few such minor losses are discussed below.

11.5.1 Losses Due to Sudden Enlargement

If the cross section of a pipe with fluid flowing through it, is abruptly enlarged (Fig. 11.4(a)) at a certain place, the fluid emerging from the smaller pipe is unable to follow the abrupt deviation of the boundary. The streamline takes a typical diverging pattern as shown in Fig. 11.4(a). This creates pockets of turbulent eddies in the corners resulting in the dissipation of mechanical energy into intermolecular energy.

The basic mechanism of this type of loss is similar to that of losses due to separation, in case of flow of fluid against an adverse pressure gradient. Here the fluid flows against an adverse pressure gradient. The upstream pressure p_1 at section *a–b* is lower than the downstream pressure p_2 at section *e–f* since the upstream velocity V_1 is higher than the downstream velocity V_2 as a consequence of continuity. The fluid particles near the wall due to their low kinetic energy cannot overcome the adverse pressure hill in the direction of flow and hence follow up the reverse path under the favourable pressure gradient (from p_2 to p_1). This creates a zone of recirculating flow with turbulent eddies near the wall of the larger tube at the abrupt change of cross section, as shown in Fig. 11.4(a), resulting in a loss of total mechanical energy. For high values of Reynolds number, usually found in practice, the velocity in the smaller pipe may be assumed sensibly uniform over the cross section. Due to the vigorous mixing caused by the turbulence, the velocity becomes again uniform at a far downstream section *e–f* from the enlargement (approximately 8 times the larger diameter). A control volume *abcdefgha* is considered (Fig. 11.4(a)) for which the momentum theorem can be written as

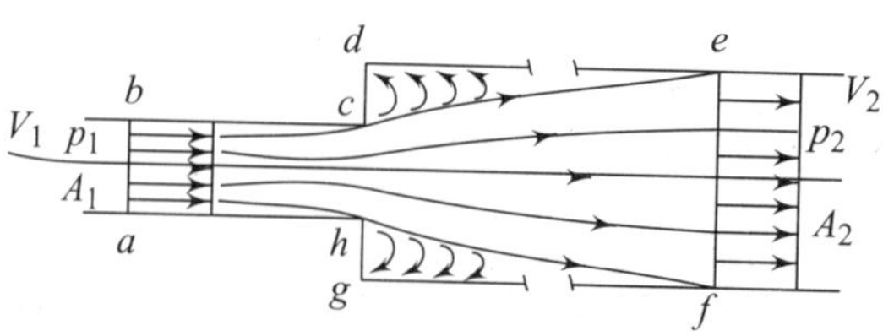

Fig. 11.4 (a) Flow through abrupt but finite enlargement

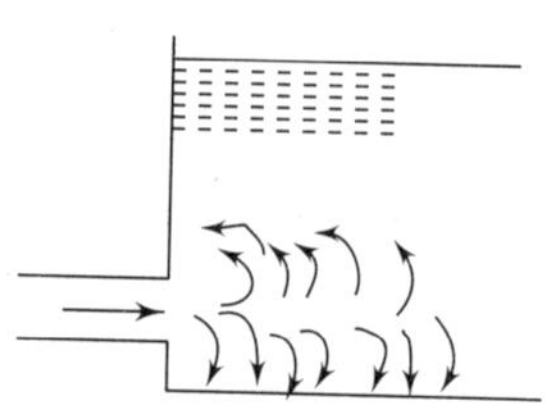

Fig. 11.4 (b) Flow at infinite enlargement (Exit loss)

$$p_1A_1 + p'(A_2 - A_1) - p_2A_2 = \rho Q(V_2 - V_1) \tag{11.18}$$

where A_1, A_2 are the cross-sectional areas of the smaller and larger parts of the pipe respectively, Q is the volumetric flow rate and p' is the mean pressure of the eddying fluid over the annular face, *gd*. It is known from experimental evidence that $p' = p_1$ Hence the Eq. (11.18) becomes

$$(p_2 - p_1)A_2 = \rho Q(V_1 - V_2) \tag{11.19}$$

From the equation of continuity,

$$Q = V_2 A_2 \tag{11.20}$$

With the help of Eq. (11.20), Eq. (4.19) becomes

$$p_2 - p_1 = \rho V_2 (V_1 - V_2) \tag{11.21}$$

Applying Bernoulli's equation between sections *ab* and *ef* in consideration of the flow to be incompressible and the axis of the pipe to be horizontal, we can write

$$\frac{p_1}{\rho} + \frac{V_1^2}{2} = \frac{p_2}{\rho} + \frac{V_2^2}{2} + gh_L$$

or

$$\frac{p_2 - p_1}{\rho} = \frac{V_1^2 - V_2^2}{2} - gh_L \tag{11.22}$$

where h_L is the loss of head. Substituting $(p_2 - p_1)$ from Eq. (11.21) into Eq. (11.22), we obtain

$$h_L = \frac{(V_1 - V_2)^2}{2g} = \frac{V_1^2}{2g}[1 - (A_1/A_2)]^2 \tag{11.23}$$

In view of the assumptions made, Eq. (11.23) is subjected to some inaccuracies, but experiments show that for coaxial pipes they are within only a few per cent of the actual values.

11.5.2 Exit Loss

If, in Eq. (11.23), $A_2 \to \infty$, then the head loss at an abrupt enlargement tends to $V_1^2/2g$. The physical resembleness of this situation is the submerged outlet of a pipe discharging into a large reservoir as shown in Fig. 11.4(b). Since the fluid velocities are arrested in a large reservoir, the entire kinetic energy at the outlet of the pipe is dissipated into the intermolecular energy of the reservoir through the creation of turbulent eddies. In such circumstances, the loss is usually termed as exit loss for the pipe and equals to the velocity head at the discharge end of the pipe.

11.5.3 Losses Due to Sudden Contraction

An abrupt contraction is geometrically the reverse of an abrupt enlargement (Fig. 11.5). Here also the streamlines cannot follow the abrupt change of geometry and hence gradually converge from an upstream section of the larger tube. However, immediately downstream of the junction of area contraction, the cross-sectional area of the stream tube becomes the minimum and less than that of the smaller pipe.

This section of the stream tube is known as *the vena contracta*, after which the stream widens again to fill the pipe. The velocity of flow in the converging part of the stream tube from Section 1–1 to Section *c–c* (vena contracta) increases due to continuity and the pressure decreases in the direction of flow accordingly in compliance with the Bernoulli's theorem. In an accelerating flow, under a favourable pressure gradient, losses due to separation cannot take place. But in the decelerating part of the flow from Section *c–c* to Section 2–2, where the stream tube expands to fill the pipe, losses take place in the similar fashion as occur in case of a sudden geometrical enlargement. Hence eddies are formed between the vena contracta *c–c* and the downstream Section 2–2. The flow pattern after the vena contracta is similar to that after an abrupt enlargement, and the loss of head is thus confined between Section *c–c* to Section 2–2. Therefore, we can say that the losses due to contraction is not for the contraction itself, but due to the expansion followed by the contraction. Following Eq. (11.23), the loss of head in this case can be written as

$$h_L = \frac{V_2^2}{2g}\,[(A_2/A_c) - 1]^2 = \frac{V_2^2}{2g}\,[(1/C_c) - 1]^2 \tag{11.24}$$

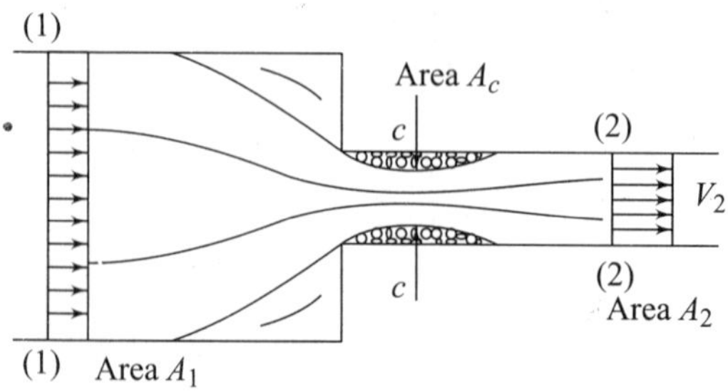

Fig. 11.5 Flow through a sudden contraction

where A_c represents the cross-sectional area of the vena contracta, and C_c is the coefficient of contraction defined by

$$C_c = A_c/A_2 \tag{11.25}$$

Equation (11.24) is usually expressed as

$$h_L = K(V_2^2/2g) \tag{11.26}$$

where

$$K = [(1/C_c) - 1]^2 \tag{11.27}$$

Although the area A_1 is not explicitly involved in the Eq. (11.24), the value of C_c depends on the ratio A_2/A_1. For coaxial circular pipes and at fairly high Reynolds numbers, Table 11.1 gives representative values of the coefficient *K*.

Table 11.1

A_2/A_1	0	0.04	0.16	0.36	0.64	1.0
K	0.5	0.45	0.38	0.28	0.14	0

11.5.4 Entry Loss

As $A_1 \to \infty$, the value of K in the Eq. (11.26) tends to 0.5 as shown in Table 11.1. This limiting situation corresponds to the flow from a large reservoir into a sharp-edged pipe, provided the end of the pipe does not protrude into the reservoir (Fig. 11.6(a)). The loss of head at the entrance to the pipe is therefore given by 0.5 $(V_2^2/2g)$ and is known as *entry loss*. A protruding pipe (Fig. 11.6(b)) causes a greater loss of head, while on the other hand, if the inlet of the pipe is well rounded (Fig. 11.6(c)), the fluid can follow the boundary without separating from it, and the entry loss is much reduced and even may be zero depending upon the rounded geometry of the pipe at its inlet.

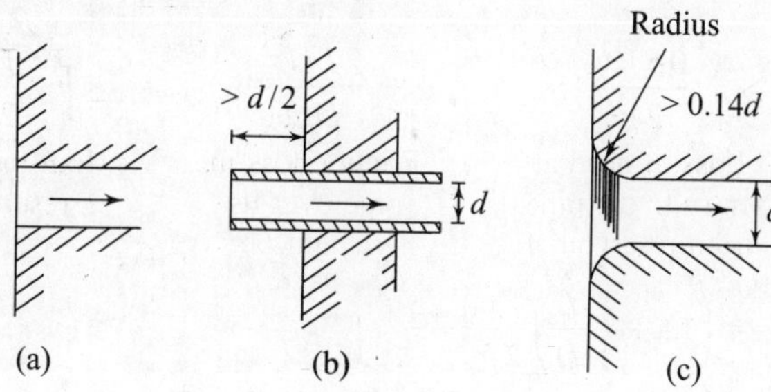

Usually accepted values of K: (a) 0.5, (b) 1.0, (c) 0

Fig. 11.6 Flow from a reservoir to a sharp edged pipe

11.6 CONCEPT OF FLOW POTENTIAL AND FLOW RESISTANCE

Consider the flow of water from one reservoir to another as shown in Fig. 11.7. The two reservoirs A and B are maintained with constant levels of water. The difference between these two levels is ΔH as shown in the figure. Therefore water flows from reservoir A to reservoir B. Application of Bernoulli's equation between two points A and B at the free surfaces in the two reservoirs gives

$$\frac{p_{\text{atm}}}{\rho g} + H_A + Z_A = \frac{p_{\text{atm}}}{\rho g} + H_B + Z_B + h_f$$

or

$$\Delta H = (Z_A + H_A) - (Z_B + H_B) = h_f \qquad (11.28)$$

where h_f is the loss of head in the course of flow from A to B. Therefore, Eq. (11.28) states that under steady state, the head causing flow ΔH becomes equal to the total loss of head due to the flow. Considering the possible hydrodynamic losses, the total loss of head h_f, can be written in terms of its different components as

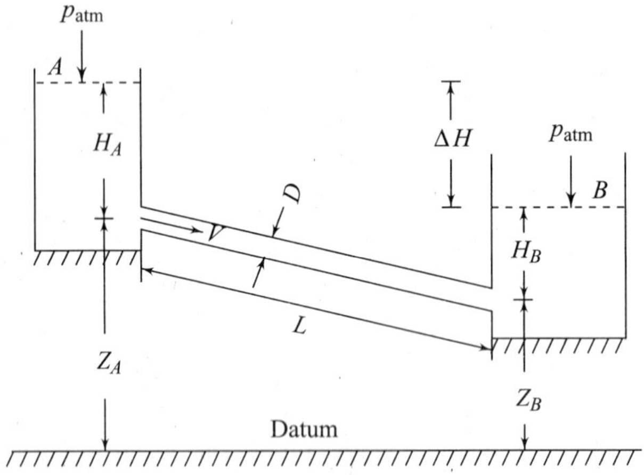

Fig. 11.7 Flow of liquid from one reservoir to another

$$h_f = \underbrace{\frac{0.5\,V^2}{2g}}_{\substack{\text{Loss of head at}\\ \text{entry to the pipe}\\ \text{from reservoir } A}} + \underbrace{f\,\frac{L}{D}\frac{V^2}{2g}}_{\substack{\text{Friction loss in}\\ \text{pipe over its}\\ \text{length } L}} + \underbrace{\frac{V^2}{2g}}_{\substack{\text{Exit loss to the}\\ \text{reservoir } B}}$$

$$= \left(1.5 + f\,\frac{L}{D}\right)\frac{V^2}{2g} \tag{11.29}$$

where V is the average velocity of flow in the pipe. The velocity V, in the above equation is usually substituted in terms of flow rate Q, since, under steady state, the flow rate remains constant throughout the pipe even if its diameter changes.

Therefore, we replace V in Eq. (11.29) as $V = 4Q/\pi D^2$ and finally get

$$h_f = \left[8\left(1.5 + f\,\frac{L}{D}\right)\frac{1}{\pi^2 D^4 g}\right]Q^2$$

or

$$h_f = R\,Q^2 \tag{11.30}$$

where

$$R = \left[\frac{8}{\pi^2 D^4 g}\left(1.5 + f\,\frac{L}{D}\right)\right] \tag{11.31}$$

The term R is defined as the *flow resistance*. In a situation where f becomes independent of Re, the flow resistance expressed by Eq. (11.31) becomes simply a function of the pipe geometry. With the help of Eq. (11.28), Eq. (11.30) can be written as

$$\Delta H = RQ^2 \tag{11.32}$$

ΔH in Eq. (11.32) is the head causing the flow and is defined as the difference in flow potentials between A and B.

This equation is comparable to the voltage-current relationship in a purely resistive electrical circuit. In a purely resistive electrical circuit, $\Delta V = r\,i$, where ΔV is the voltage or electrical potential difference across a resistor whose resistance is r and the electrical current flowing through it is i. The difference however is that while the voltage drop in an electrical circuit is linearly proportional to current, the difference in the flow potential in a fluid circuit is proportional to the square of the flow rate. Therefore, the fluid flow system as shown in Fig. 11.7 and described by Eq. (11.32) can be expressed by an equivalent electrical network system as shown in Fig. 11.18.

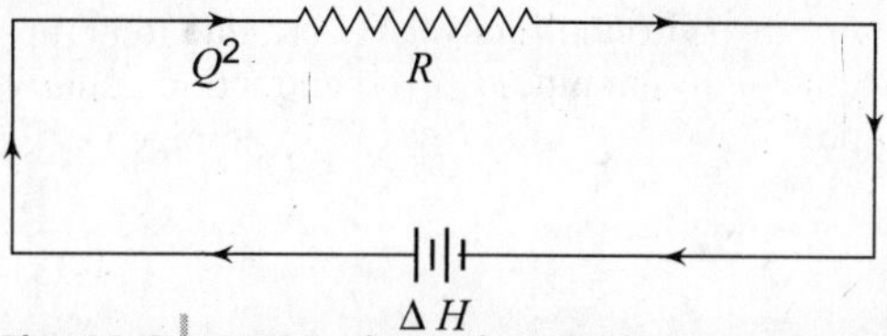

Fig. 11.8 Equivalent electrical network system for a simple pipe flow problem shown in Fig. 11.7

11.7 FLOW THROUGH BRANCHED PIPES

In several practical situations, flow takes place under a given head through different pipes joined together either in series or in parallel or in a combination of both of them.

11.7.1 Pipes in Series

If a pipeline is joined to one or more pipelines in continuation, these are said to constitute pipes in series. A typical example of pipes in series is shown in Fig. 11.9. Here three pipes *A*, *B* and *C* are joined in series.

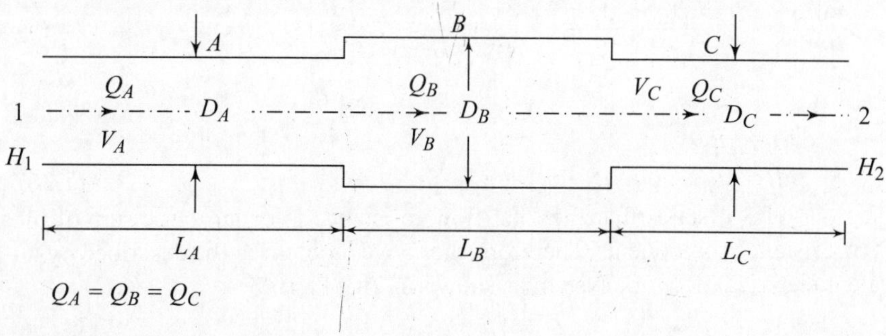

Fig. 11.9 Pipes in series

In this case, rate of flow Q remains same in each pipe. Hence,

$$Q_A = Q_B = Q_C = Q$$

If the total head available at Section 1 (at the inlet to pipe *A*) is H_1 which is greater than H_2, the total head at Section 2 (at the exit of pipe *C*), then the flow takes place from 1 to 2 through the system of pipelines in series. Application of energy equation between Sections 1 and 2 gives

$$H_1 - H_2 = h_f$$

where, h_f is the loss of head due to the flow from 1 to 2. Recognising the minor and major losses associated with the flow, h_f can be written as

$$h_f = \underbrace{f_A \frac{L_A}{D_A}\frac{V_A^2}{2g}}_{\text{Friction loss in pipe } A} + \underbrace{\frac{(V_A - V_B)^2}{2g}}_{\text{Loss due to enlargement at entry to pipe } B} + \underbrace{f_B \frac{L_B}{D_B}\frac{V_B^2}{2g}}_{\text{Friction loss in pipe } B} + \underbrace{\left(\frac{1}{C_c} - 1\right)^2 \frac{V_C^2}{2g}}_{\text{Loss due to abrupt contraction at entry to pipe } C}$$

$$+ f_C \frac{L_C}{D_C}\frac{V_c^2}{2g} \tag{11.33}$$

Friction loss in pipe C

The subscripts *A*, *B* and *C* refer to the quantities in pipe *A*, *B* and *C* respectively. C_c is the coefficient of contraction.

The flow rate Q satisfies the equation,

$$Q = \frac{\pi D_A^2}{4} V_A = \frac{\pi D_B^2}{4} V_B = \frac{\pi D_C^2}{4} V_C \tag{11.34}$$

Velocities V_A, V_B and V_C in Eq. (11.33) are substituted from Eq. (11.34), and we get

$$h_f = \Bigg[\underbrace{\frac{8}{g\pi^2} f_A \frac{L_A}{D_A^5}}_{R_1} + \underbrace{\frac{8}{g\pi^2}\left(1 - \frac{D_A^2}{D_B^2}\right)^2 \frac{1}{D_A^4}}_{R_2} + \underbrace{\frac{8}{g\pi^2} f_B \frac{L_B}{D_B^5}}_{R_3}$$

$$+ \underbrace{\frac{8}{g\pi^2}\left(\frac{1}{C_C} - 1\right)^2 \frac{1}{D_C^4}}_{R_4} + \underbrace{\frac{8}{g\pi^2} f_c \frac{L_C}{D_C^5}}_{R_5}\Bigg] Q^2 \tag{11.35}$$

or $\qquad h_f = RQ^2$

where, $\qquad R = R_1 + R_2 + R_3 + R_4 + R_5 \qquad$ (11.36)

Equation (11.36) states that the total flow resistance is equal to the sum of the different resistance components. Therefore, the above problem can be described by an equivalent electrical network system as shown in Fig. 11.10.

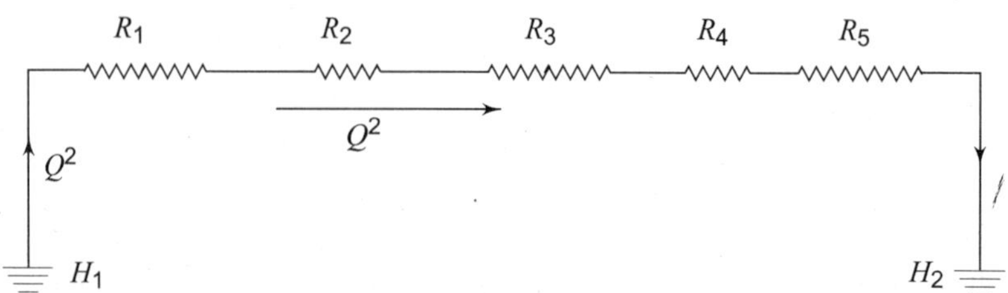

Fig. 11.10 Equivalent electrical network system for flow through pipes in series

11.7.2 Pipes in Parallel

When two or more pipes are connected, as shown in Fig. 11.11, so that the flow divides and subsequently comes together again, the pipes are said to be in parallel. In this case (Fig. 11.11), equation of continuity gives

$$Q = Q_A + Q_B \tag{11.37}$$

where, Q is the total flow rate and Q_A and Q_B are the flow rates through pipes A and B respectively. Loss of head between the locations 1 and 2 can be expressed by applying energy equation either through the path 1–A–2 or 1–B–2. Therefore, we can write

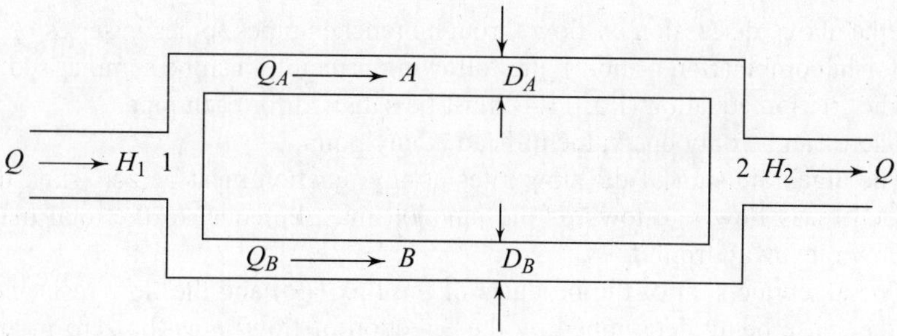

Fig. 11.11 Pipes in parallel

$$H_1 - H_2 = f_A \frac{L_A}{D_A}\frac{V_A^2}{2g} = \frac{8L_A}{\pi^2 D_A^5\, g} f_A\, Q_A^2$$

and
$$H_1 - H_2 = f_B \frac{L_B}{D_B}\frac{V_B^2}{2g} = \frac{8L_B}{\pi^2 D_B^5\, g} f_B\, Q_B^2$$

Equating the above two expressions, we get

$$Q_A^2 = \frac{R_B}{R_A} Q_B^2 \tag{11.38}$$

where,
$$R_A = \frac{8\, L_A}{\pi^2 D_A^5\, g} f_A$$

$$R_B = \frac{8\, L_B}{\pi^2 D_B^5\, g} f_B$$

Equations (11.37) and (11.38) give

$$Q_A = \frac{K}{1+K} Q, Q_B = \frac{1}{1+K} Q \tag{11.39}$$

where,
$$K = \sqrt{R_B / R_A} \tag{11.40}$$

The flow system can be described by an equivalent electrical circuit as shown in Fig. 11.12.

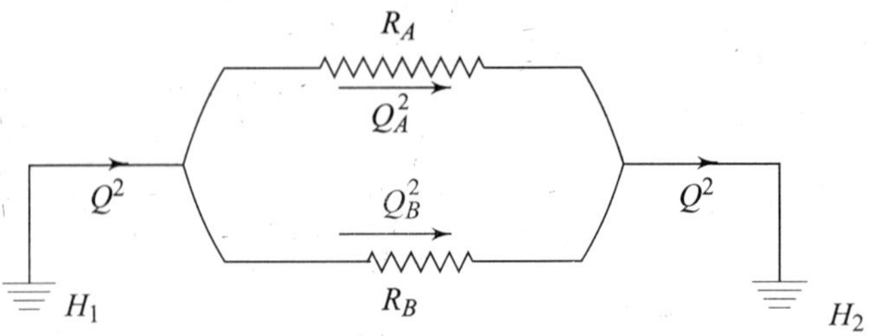

Fig. 11.12 Equivalent electrical network system for flow through pipes in parallel

From the above discussion on flow through branched pipes (pipes in series or in parallel, or in combination of both), the following principles can be summarised:

(i) The friction equation (Eq. 11.4) must be satisfied for each pipe.
(ii) There can be only one value of head at any point.
(iii) The algebraic sum of the flow rates at any junction must be zero, i.e., the total mass flow rate towards the junction must be equal to the total mass flow rate away from it.
(iv) The algebraic sum of the products of the flux (Q^2) and the flow resistance (the sense being determined by the direction of flow) must be zero in any closed hydraulic circuit.

The principles (iii) and (iv) can be written analytically as

$$\Sigma Q = 0 \text{ at a node (Junction)} \tag{11.41}$$

$$\Sigma R|Q|Q = 0 \text{ in a loop} \tag{11.42}$$

While Eq. (11.41) implies the principle of continuity in a hydraulic circuit, Eq. (11.42) is referred to as pressure equation of the circuit.

Example 11.6

Three pipes of 400 mm, 200 mm and 300 mm diameters and having lengths of 400 m, 200 m and 300 m, respectively are connected in series to make a compound pipe. The ends of this compound pipe are connected with two tanks whose difference in water levels is 16 m, as shown in (Fig.11.13(a)). If the friction factor f, for all the pipes is same and equal to 0.02, determine the discharge through the compound pipe neglecting first the minor losses and then including them. Draw the equivalent electrical network system. (Take coefficient of contraction = 0.6.)

Solution

Application of energy equation between sections A and B (Fig. 11.13 (a)) gives

$$\frac{p_{\text{atm}}}{\rho g} + 0 + 16 = \frac{p_{\text{atm}}}{\rho g} + 0 + 0 + h_f$$

or

$$h_f = 16 \text{ m} \tag{11.43}$$

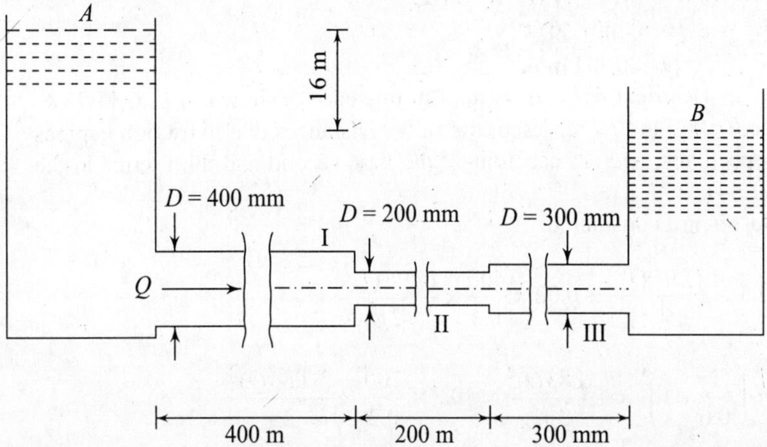

(a) Flow of water through pipes in series

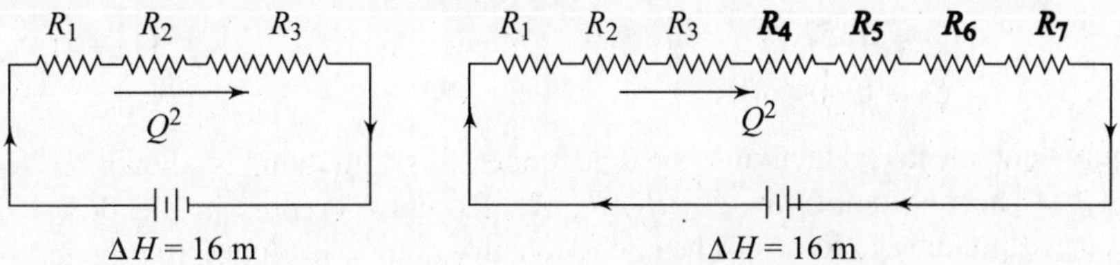

(i) Without minor loss (ii) With minor loss

(b) Equivalent electrical network system for flow through pipes in series

Fig. 11.13

Let Q be the volumetric rate of discharge through the pipelines. Then,

the velocity of flow in pipe I (Fig. 11.13a) $= \dfrac{4Q}{\pi(0.4)^2} = 7.96\ Q$

the velocity of flow in pipe II (Fig. 11.13a) $= \dfrac{4Q}{\pi(0.2)^2} = 31.83\ Q$

the velocity of flow in pipe III (Fig. 11.13a) $= \dfrac{4Q}{\pi(0.3)^2} = 14.15\ Q$

When minor losses are not considered, the loss of head h_f, in the course of flow from A to B constitutes of the friction losses in three pipes only, and can be written as

$$h_f = \left[0.02 \times \frac{400}{0.4} \times \frac{(7.96)^2}{2g} + 0.02 \times \frac{200}{0.2} \times \frac{(31.83)^2}{2g} + 0.02 \times \frac{300}{0.3} \times \frac{(14.15)^2}{2g}\right] Q^2$$

$$= 1301.46\ Q^2 \tag{11.44}$$

Equating Eq. (11.43) with Eq. (11.44) we have,

$$16 = 1301.46\,Q^2$$

which gives $$Q = 0.111 \text{ m}^3/\text{s}$$

The equivalent electrical network system in this case is shown in Fig. 11.13 b. The resistances R_1, R_2, and R_3 represent the flow resistances due to friction in pipes I, II and III respectively, and are accordingly the first, second and third terms in the RHS of Eq. (11.44).

When minor losses are considered,

$$h_f = 16 = 0.5 \times \frac{(7.96Q)^2}{2g} + 0.02 \times \frac{400}{0.4} \times \frac{(7.96Q)^2}{2g}$$

$$+ \left(\frac{1}{0.6} - 1\right)^2 \times \frac{(31.83Q)^2}{2g} + 0.2 \times \frac{200}{0.2} \times \frac{(31.83Q)^2}{2g}$$

$$+ \frac{(31.83Q - 14.15Q)^2}{2g} + 0.2 \times \frac{300}{0.3} \times \frac{(14.15Q)^2}{2g} + \frac{(14.15Q)^2}{2g}$$

$$= 1352\,Q^2 \tag{11.45}$$

which gives $$Q = 0.109 \text{ m}^3/\text{s}$$

The equivalent electrical network system, under this situation, is shown in Fig. 11.13 (b). The resistances R_1, R_2, R_3, R_4, R_5, R_6 and R_7 represent the flow resistances corresponding to losses of head due to entry at pipe I, friction in pipe I, contraction at entrance to pipe II, friction in pipe II, expansion at entrance to pipe III, friction in pipe III, exit from pipe III respectively.

Example 11.7

Two reservoirs 5.2 km apart are connected by a pipeline which consists of a 225 mm diameter pipe for the first 1.6 km, sloping at 5.7 m per km. For the remaining distance, the pipe diameter is 150 mm laid at a slope of 1.9 m per km. The levels of water above the pipe openings are 6 m in the upper reservoir and 3.7 m in the lower reservoir. Taking $f = 0.024$ for both the pipes and $C_c = 0.6$, calculate the rate of discharge through the pipeline.

Solution

The connections of pipelines are shown in Fig. 11.14. From the given conditions of pipe slopings,

$$h_1 = 5.7 \times 1.6 = 9.12 \text{ m}$$

$$h_2 = 1.9 \times 3.6 = 6.84 \text{ m}$$

Therefore, the length of the first pipe $$L_1 = \sqrt{(1.6 \times 10^3)^2 + (9.12)^2 \text{ m}}$$

$$= 1.6 \text{ km}$$

and, the length of the second pipe $\quad L_2 = \sqrt{(3.6 \times 10^3)^2 + (6.84)^2}\ \text{m}$

$$= 3.6\,\text{km}$$

Applying energy equation between the sections *A* and *B* taking the horizontal plane through the pipe connection in the lower reservoir as datum (Fig. 11.14), we can write

$$\frac{p_{\text{atm}}}{\rho g} + 6 + 9.12 + 6.84 = \frac{p_{\text{atm}}}{\rho g} + 3.7 + h_f \quad \text{or} \quad h_f = 18.26\ \text{m}$$

where, h_f is the total loss of head in the flow. Considering all the losses in the path of flow, we can write

$$h_f = \underset{\text{Entrance loss to pipe 1}}{0.5\frac{V_1^2}{2g}} + \underset{\text{Friction loss in pipe 1}}{0.024 \times \frac{1.6 \times 10^3}{0.225} \cdot \frac{V_1^2}{2g}} + \underset{\text{Loss due to contraction at the entrance to pipe 2}}{\left(\frac{1}{0.6} - 1\right)^2 \frac{V_2^2}{2g}}$$

$$+ \underset{\text{Friction loss in pipe 2}}{0.024 \times \frac{3.6 \times 10^3}{0.150} \frac{V_2^2}{2g}} + \underset{\text{Exit loss from pipe 2 to lower reservoir}}{\frac{V_2^2}{2g}}$$

or
$$18.26 = 171.17\frac{V_1^2}{2g} + 577.44\frac{V_2^2}{2g} \tag{11.46}$$

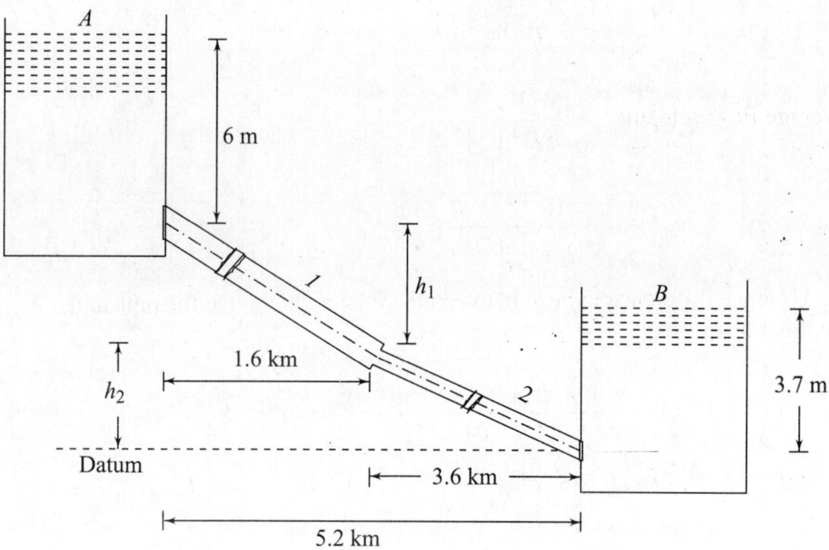

Fig. 11.14 Flow of water from a upper reservoir to a lower one through pipes in series

where, V_1 and V_2 are the average flow velocities in pipe I and II, respectively. If Q is the rate of discharge, then

$$V_1 = \frac{4Q}{\pi(0.225)^2} = 25.15\,Q$$

$$V_2 = \frac{4Q}{\pi(0.15)^2} = 56.59\,Q$$

Inserting the expressions for V_1 and V_2 in Eq. (11.46), we have

$$18.26 = \left[\frac{171.17 \times (25.15)^2}{2 \times 9.81} + \frac{577.44 \times (56.59)^2}{2 \times 9.81}\right] Q^2$$

which gives $\quad Q = 0.0135\ \text{m}^3/\text{s}$

Example 11.8

A pipeline of 0.6 m in diameter is 1.5 km long. In order to augment the discharge, another parallel line of the same diameter is introduced in the second half of the length. Neglecting minor losses, find the increase in discharge if $f = 0.04$. The head at inlet is 30 m over that at the outlet.

Solution

Initially, for the single pipe, the discharge is calculated from the relationship

$$\Delta H = h_f = f\,\frac{L}{D}\,\frac{V^2}{2g}$$

The average flow velocity $V = \dfrac{4Q}{\pi D^2}$

Hence, $$\Delta H = \frac{16}{\pi^2 \times 2 \times g}\,\frac{fL}{D^5}\,Q^2$$

(where ΔH is the difference in head between the inlet and outlet at the pipe and h_f is the frictional head loss).

or $$Q^2 = \frac{30 \times \pi^2 \times 2 \times 9.81 \times (0.6)^5}{16 \times 0.04 \times 1500}$$

or $$Q = 0.686\ \text{m}^3/\text{s}$$

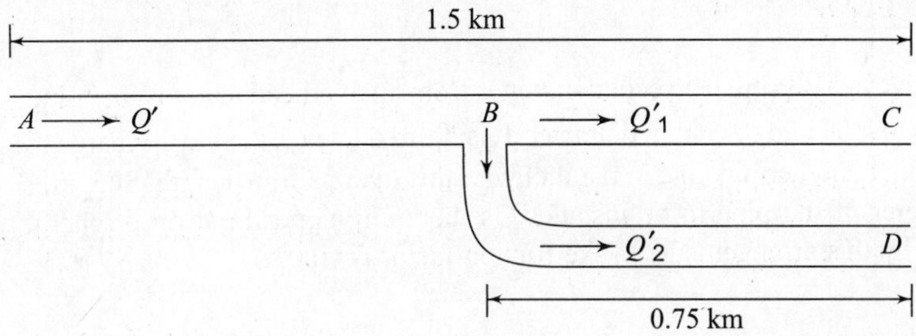

Fig. 11.15 Flow through a compound pipe

Let Q' be the discharge through the first half of the pipe when another parallel line of the same diameter is introduced to the second half of the length, as shown in Fig. 11.15. If Q'_1 and Q'_2 are the flow rates through the two branched pipes in parallel, then from continuity,

$$Q' = Q'_1 + Q'_2$$

We can write for the two parallel paths BC and BD

$$H_B - H_C = \frac{0.04 \times 0.75}{0.6 \times 2 \times 9.81}\left[\frac{4}{\pi (0.6)^2}\right]^2 Q_1'^2$$

$$H_B - H_D = \frac{0.04 \times 0.75}{0.6 \times 2 \times 9.81}\left[\frac{4}{\pi (0.6)^2}\right]^2 Q_2'^2$$

At outlet, $H_C = H_D$

Therefore, we get from the above two equations along with the equation of continuity

$$Q'_1 = Q'_2 = Q'/2$$

Applying energy equation between A and C through the hydraulic path ABC, we have

$$30 = \frac{0.04 \times 0.75 \times 10^3}{0.6 \times 2 \times 9.81}\left[\frac{4}{\pi (0.6)^2}\right]^2 Q'^2$$

$$+ \frac{0.04 \times 0.75 \times 10^3}{0.6 \times 2 \times 9.81}\left[\frac{4}{\pi (0.6)^2}\right]^2 \left(\frac{Q'}{2}\right)^2$$

$$= 39.85\, Q'^2$$

which gives $Q' = 0.868 \text{ m}^3/\text{s}$

Therefore, the increase in the rate of discharge by the new arrangement becomes

$$Q' - Q = 0.868 - 0.686 = 0.182 \text{ m}^3/\text{s}$$

which is $0.182 \times 100/0.686 = 26.4\%$ of the initial rate of discharge.

Example 11.9

A pipeline conveys 8.33 litre per second of water from an overhead tank to a building. The pipe is 2 km long and 0.15 m in diameter. It is desired to increase the discharge by 30% by installing another pipeline in parallel with this over half the length. Suggest a suitable diameter of the pipe to be installed. Is there any upper limit on discharge augmentation by this arrangement? (Take friction factor $f = 0.03$.)

Solution

The height H, of the overhead tank above the building can be determined from the conditions with a single pipe.

$$H = h_f = 0.03\frac{2000}{0.15}\left[\frac{(4\times 0.00833)}{\pi(0.15)^2}\right]^2\frac{1}{2\times 9.81} = 4.53 \text{ m}$$

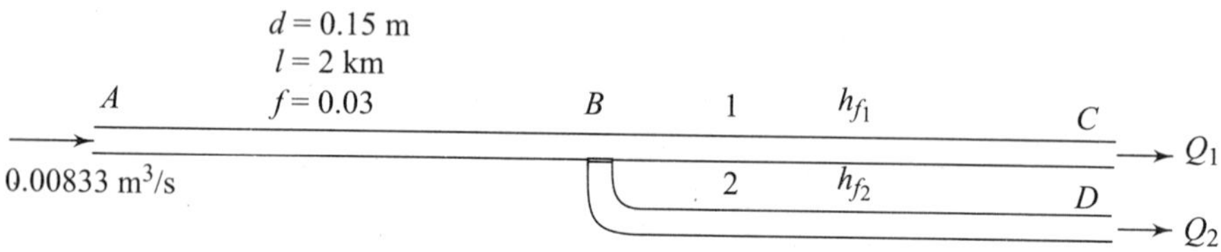

Fig. 11.16 Flow through a compound pipe

In the new plan as shown in Fig. 11.16

$$h_f = 4.53 = h_{f_{AB}} + h_{f_{BC}} \tag{11.47}$$

again,

$$h_{f_{BC}} = h_{f_{BD}} = \frac{f L_1}{2g\,d_1}\left[\frac{4Q_1}{\pi(d_1)^2}\right]^2 = \frac{f L_2}{2g\,d_2}\left[\frac{4Q_2}{\pi(d_2)^2}\right]^2$$

Here, $L_1 = L_2 = 1000$ m

Therefore, $$(Q_1/Q_2)^2 = (d_1/d_2)^5 \tag{11.48}$$

$$h_{f_{AB}} = \frac{0.03\times 1000}{2g\,(0.15)}\left[\frac{4Q}{\pi(0.15)^2}\right]^2$$

Therefore, Eq. (11.47) can be written as

$$\frac{0.03\times 1000\times 8Q^2}{9.81\times\pi^2\times(0.15)^5} + \frac{0.03\times 1000\times 8\times Q_1^2}{9.81\times\pi^2\times(0.15)^5} = 4.53 \tag{11.49}$$

In this case, $Q = 1.3 \times 0.00833 = 0.0108 \text{ m}^3/\text{s}$

Then, from Eq. (11.49), we get

$$Q_1^2 = 0.00014 - (0.0108)^2$$

which gives $Q_1 = 0.0048 \text{ m}^3/\text{s}$

From continuity, $Q_2 = 0.0108 - 0.0048 = 0.006 \text{ m}^3/\text{s}$

From Eq. (11.48), we have $d_2 = \left(\dfrac{0.006}{0.0048}\right)^{2/5} \times 0.15$

$$= 0.164 \text{ m}$$

It can be observed from Eq. (11.41) that

$$Q_1^2 = 0.00014 - Q^2$$

or

$$Q^2 = 0.00014 - Q_1^2$$

Now Q will be maximum when $\dot{Q}_1$ will be minimum. For a physically possible situation, the minimum value of $\dot{Q}_1$ will be zero. Therefore, the maximum value of Q will be

$$Q_{max} = \sqrt{0.00014} = 0.0118 \text{ m}^3/\text{s}$$

which is 41.6% more than the initial value. The case ($Q_1 = 0$, $Q = 0.0118 \text{ m}^3/\text{s}$) corresponds to a situation of an infinitely large branched pipe, i.e., $d_2 \to \infty$.

Example 11.10

Two points A and B at the ground level are supplied equal quantity of water through branched pipes each 200 mm in diameter and 10 m long. Water supply is made from an overhead tank whose water level above the ground is 12 m, and the length and diameter of the pipe up to the junction point O are 14 m and 500 mm, respectively. The point O is also on the ground level as shown in Fig. 11.17. The connection of a new pipe of 200 mm diameter and 20 m length is to be made from O to C. The friction factor f, for all the pipes is 0.016. Pipelines A and B are provided valves for controlling the flow rates.

Calculate (i) the flow rates at A and B when the valves are fully open, before C was connected, (ii) the flow rates at A, B and C with the valves fully open, (iii) the valve resistance coefficients on pipelines A and B so as to obtain equal flow rates at A, B and C, and the value of such flow rates (Neglect entry and bend losses). Assume kinetic energy correction factor as unity.

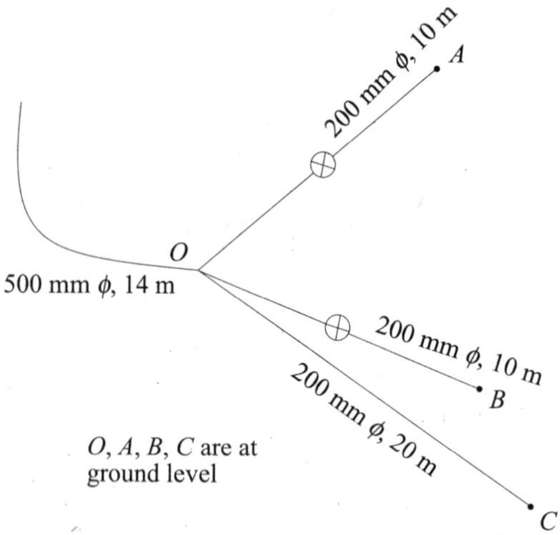

Fig. 11.17 Supply of water from an overhead tank through branched pipes

Solution

(i) Let the flow rate through the main pipe from the overhead tank to the junction O be Q and those through the pipes OA and OB are Q_1 and Q_2. From continuity,

$$Q_1 + Q_2 = Q$$

Since length, diameter and friction factor for the pipes OA and OB are equal,

$$Q_1 = Q_2 = Q/2$$

Velocity in the main pipe from the tank to the point $O = \dfrac{4Q}{\pi(0.5)^2}$

$$= 5.09\,Q$$

Velocity in the pipe $OA = \dfrac{2Q}{\pi(0.2)^2} = 15.92\,Q$

Applying energy equation between a section at the water level in the overhead tank and the section A through the path connecting the main pipe and the pipe OA, we can write

$$12 = 0.016\,\frac{14}{0.5}\,\frac{1}{2g}(5.09Q)^2 + 0.016\,\frac{10}{0.2}\,\frac{1}{2g}(15.92Q)^2$$

or $\qquad 12 = 10.92\,Q^2$

which gives $\qquad Q = 1.05\ \text{m}^3/\text{s}.$

Hence flow rates at A and B are

$$Q_1 = Q_2 = \frac{1.05}{2} = 0.525 \text{ m}^3/\text{s}$$

(ii) Let Q be the flow rate in the main pipe and Q_1, Q_2, Q_3 be the flow rates through the pipes OA, OB and OC, respectively.

From continuity, $$Q = Q_1 + Q_2 + Q_3 \quad (11.50)$$

If the discharge pressures at A, B and C are equal, then the sum of the frictional loss and the velocity head (or the exit loss) through each pipe OA, OB and OC must be equal. Hence we can write

$$\left(1 + 0.016\frac{10}{0.2}\right)\frac{16}{\pi^2 (0.2)^4} Q_1^2 = \left(1 + 0.016\frac{10}{0.2}\right)\frac{16}{\pi^2 (0.2)^4} Q_2^2$$

$$= \left(1 + 0.016\frac{20}{0.2}\right)\frac{16}{\pi^2 (0.2)^4} Q_3^2$$

which gives $$Q_2 = Q_1 \quad (11.51)$$

and $$Q_3 = 0.832\, Q_1 \quad (11.52)$$

Therefore, from Eqs (11.50), (11.51) and (11.52), we get

$$Q = 2.832\, Q_1$$

Applying energy equation between a section at the water level in the overhead tank and the section A through the hydraulic path connecting the main pipe and the pipe OA, we can write

$$12 = 0.016\, \frac{14}{0.5}\frac{1}{2g}\left(\frac{4 \times 2.832 Q_1}{\pi (0.5)^2}\right)^2 + 0.016\frac{10}{0.2}\frac{1}{2g}\left[\frac{4Q_1}{\pi (0.2)^2}\right]^2$$

$$= 46.06\, Q_1^2$$

which gives $Q_1 = 0.51 \text{ m}^3/\text{s}$

and from (11.51) $Q_2 = 0.51 \text{ m}^3/\text{s}$

from (11.52) $Q_3 = 0.42 \text{ m}^3/\text{s}$

from (11.50) $Q = 1.44 \text{ m}^3/\text{s}$

(iii) Let Q_1 be the flow rate through OA, OB and OC. Then the flow rate through the main pipe $Q = 3\, Q_1$.

Since the diameter of the pipes OA, OB and OC are same, the average velocity of flow through these pipes will also be the same. Let this velocity be V_1.

Then, $$V_1 = \frac{4Q_1}{\pi (0.2)^2} = 31.83\, Q_1$$

Velocity through the main pipe $V = \dfrac{4 \times 3Q_1}{\pi (0.5)^2} = 15.28\, Q_1$

Let K be the valve resistance coefficient in pipe OA or OB.

Equating the total losses through two parallel hydraulic paths OC and any one of OA and OB, we have

$$\left(0.016 \times \frac{10}{0.2} + K + 1\right)\frac{(31.83)^2}{28} Q_1^2 = \left(0.016 \times \frac{20}{0.2} + 1\right)\frac{(31.83)^2}{2g} Q_1^2$$

or $\qquad 1.8 + K = 2.6$

Hence $\qquad K = 0.8$

Applying energy equation between a section at the water level in the overhead tank and the section A through the path connecting the main pipe and OA, we have

$$12 = 0.016\frac{14}{0.5}\frac{(15.28)^2}{2g} Q_1^2 + \left[0.016\frac{10}{0.2} + 0.8 + 1\right]\frac{(31.83)^2}{2g} Q_1^2$$

$$= 139.76\, Q_1^2$$

which gives $\qquad Q_1 = 0.293 \text{ m}^3/\text{s}$

and hence, $\qquad Q = 3 \times 0.293 = 0.879 \text{ m}^3/\text{s}$

Example 11.11

Two reservoirs are connected through a 300 mm diameter pipe line, 1000 m long as shown in Fig. 11.18. At a point B, 300 m from the reservoir A, a valve is inserted on a short branch line which discharges to the atmosphere. The valve may be regarded as a rounded orifice 75 mm diameter, $C_d = 0.65$. If friction factor f for all the pipes is 0.013, calculate the rate of discharge to the reservoir C when the valve at B is fully opened. Estimate the leakage through the short pipe line at B. Assume kinetic energy correction factor as unity.

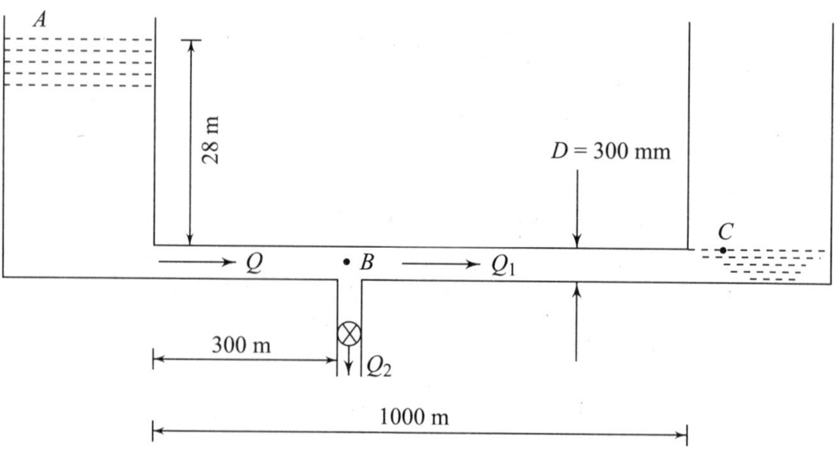

Fig. 11.18 Flow of water between two reservoir through a pipe with a bypass discharge to atmosphere

Solution

Let the flow rate through the first 300 m of the pipe be Q and the flow rates through the next 700 m of the pipe and the short branch line containing the valve be Q_1 and Q_2 respectively.

From continuity, $\quad Q = Q_1 + Q_2$

Velocity in the pipe $\quad BC = \dfrac{4Q_1}{\pi(0.3)^2} = 14.15\, Q_1$

Applying energy equation between sections B and C,

$$\frac{p_B}{\rho g} + \frac{(14.15)^2}{2g} Q_1^2 = \frac{p_{\text{atm}}}{\rho g} + 0.013 \frac{700}{0.3} \frac{(14.15)^2}{2g} Q_1^2 + \frac{(14.15)^2}{2g} Q_1^2$$

or

$$\frac{p_B - p_{\text{atm}}}{\rho g} = 309.55\, Q_1^2 \tag{11.53}$$

The discharge through the valve acting as an orifice can be written as

$$Q_2 = 0.65 \times \pi \left(\frac{0.075}{4}\right)^2 \sqrt{\frac{2(p_B - p_{\text{atm}})}{\rho}} \tag{11.54}$$

Using Eqs (11.53) and (11.54), we have

$$Q_2 = 0.65 \times \pi \left(\frac{0.075}{4}\right)^2 \sqrt{2 \times 309.55 \times 9.81}\; Q_1$$

$$= 0.224\, Q_1$$

Hence, $\quad Q = 1.224\, Q_1$

Applying energy equation between A and C through the path ABC, we have,

$$28 = \left(0.5 + 0.013 \frac{300}{0.3}\right) \frac{(14.15 \times 1.224)^2}{2g} Q_1^2$$

$$+ \left(1 + 0.013 \frac{700}{0.3}\right) \frac{(14.15)^2}{2g} Q_1^2$$

$$= 526.16\, Q_1^2$$

which gives $\quad Q_1 = 0.231 \text{ m}^3/\text{s}$

$$Q_2 = 0.224 \times 0.231 = 0.052 \text{ m}^3/\text{s}$$

Example 11.12

Two reservoirs open to the atmosphere are connected by a pipe 800 m long. The pipe goes over a hill whose height is 6 m above the level of water in the upper reservoir. The pipe diameter is 300 mm and friction factor $f = 0.032$. The difference in water levels in the two reservoirs is 12.5 m. If the absolute pressure of water anywhere in the pipe is not allowed to fall below 1.2 m of water in order to prevent vapour formation, calculate the length of pipe in the portion between the upper reservoir and the hill sumit, and also the discharge through the pipe. Neglect bend losses. Draw the equivalent electrical network system. Assume kinetic energy correction factor as unity.

Solution

Let the length of pipe upstream of C be L_1 and that of the downstream be L_2 (Fig. 11.19(a)).

It is given $\quad L_1 + L_2 = 800$ m

Considering the entry, friction and exit losses,

the total loss from A to C $= h_{f1} = \left(0.5 + \dfrac{0.032\,L_1}{0.3}\right)\dfrac{V^2}{2g}$ (11.55)

the total loss from C to B $= h_{f2} = \left(1 + \dfrac{0.032\,L_2}{0.3}\right)\dfrac{V^2}{2g}$

Therefore, the total loss from A to $B = h_f = \left(0.5 + \dfrac{0.032 \times 800}{0.3} + 1\right)\dfrac{V^2}{2g}$

$$= 86.83\,\frac{V^2}{2g}$$

Applying energy equation between A and B, we have

$$\Delta H = h_f$$

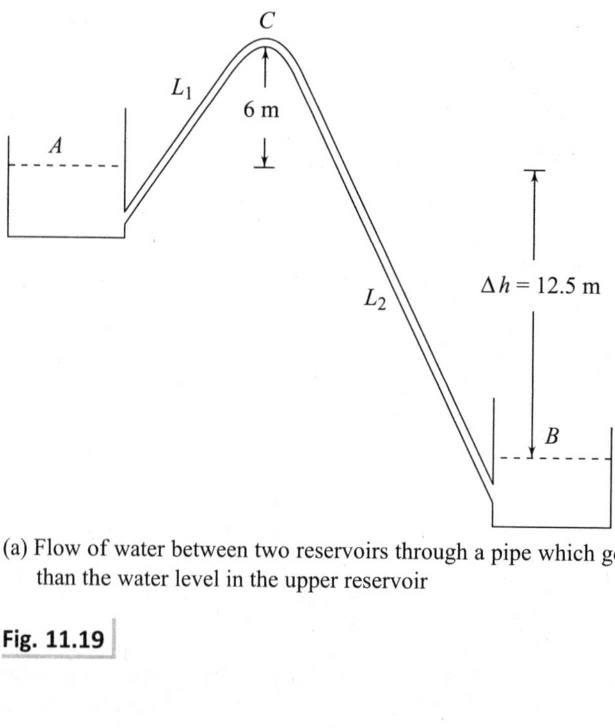

(a) Flow of water between two reservoirs through a pipe which goes over a height more than the water level in the upper reservoir

Fig. 11.19

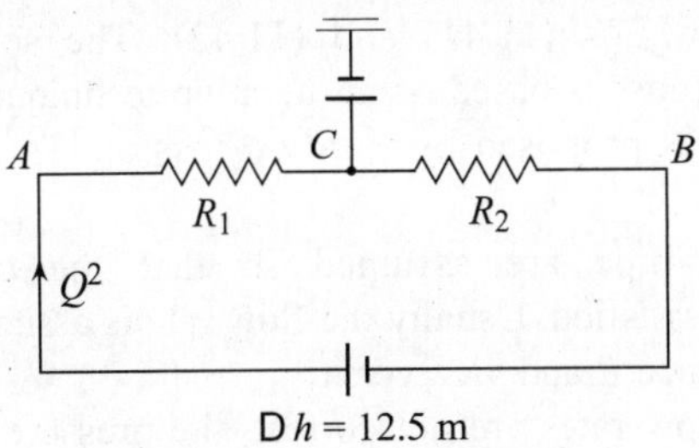

(b) Equivalent electrical network of pipe flow problem of Example 11.12

Fig. 11.19

or
$$12.5 = 86.83\,\frac{V^2}{2g}$$

which gives
$$V = \sqrt{\frac{12.5 \times 2 \times 9.81}{86.83}} = 1.68 \text{ m/s}$$

Applying energy equation between A and C, we have

$$\frac{p_{atm}}{\rho g} = \frac{p_c}{\rho g} + \frac{V^2}{2g} + 6 + h_{f1} \qquad (11.56)$$

With the atmospheric pressure

$$p_{atm} = 760 \text{ mm of Hg}$$

$$= \frac{760 \times 13.6}{1000} = 10.34 \text{ m of water,}$$

Equation (11.56) becomes

$$10.34 = 1.2 + 6 + \frac{(1.68)^2}{2 \times 9.81} + h_{f1}$$

which gives
$$h_{f1} = 2.99 \text{ m}$$

Using the value of h_{f1} = 2.99 m, and V = 1.68 m/s in Eq. (11.55) we get

$$(0.5 + 0.107\,L_1)\,\frac{(1.68)^2}{2 \times 9.81} = 2.99$$

or
$$0.5 + 0.107\,L_1 = 20.78$$

which gives
$$L_1 = 189.53 \text{ m}$$

Rate of discharge through the pipe

$$Q = \frac{\pi}{4}\,(0.3)^2 \times 1.68 = 0.119 \text{ m}^3\text{/s}$$

The equivalent electrical network of the system is shown in Fig. 11.19 (b).

11.7.3 Pipe Network: Solution by Hardy–Cross Method

The distribution of water supply in practice is often made through a pipe network comprising a combination of pipes in series and parallel. The flow distribution in a pipe network is determined from Eqs (11.41) and (11.42). The solution of Eqs (11.41) and (11.42) for the purpose is based on an iterative technique with an initial guess in Q. The method was proposed by Hardy–Cross and is described below:

(a) The flow rates in each pipe are assumed so that the continuity (Eq. 11.41) at each node is satisfied. Usually the flow rate is assumed more for smaller values of resistance R and vice versa.

(b) If the assumed values of flow rates are not correct, the pressure equation (Eq. (11.42) will not be satisfied. The flow rate is then altered based on the error in satisfying the Eq. (11.42).

Let Q_0 be the correct flow in a path whereas the assumed flow be Q. The error dQ in flow is then defined as

$$Q = Q_0 + \mathrm{d}Q \tag{11.57}$$

Let

$$h = R|Q|Q \tag{11.58a}$$

and

$$h' = R|Q_0|Q_0 \tag{11.58b}$$

Then according to Eq. (11.46)

$$\Sigma h' = 0 \quad \text{in a loop} \tag{11.59a}$$

and

$$\Sigma h = e \quad \text{in a loop} \tag{11.59b}$$

Where e is defined to be the error in pressure equation for a loop with the assumed values of flow rate in each path.

From Eqs (11.59a) and (11.59b) we have

$$\Sigma (h - h') = e$$

or

$$\Sigma\, \mathrm{d}h = e \tag{11.60}$$

Where dh (= $h - h'$) is the error in pressure equation for a path. Again from Eq. (11.58a), we can write

$$\frac{\mathrm{d}h}{\mathrm{d}Q} = 2R|Q|$$

or

$$\mathrm{d}h = 2R|Q|\mathrm{d}Q \tag{11.61}$$

Substituting the value of dh from Eq. (11.61) in Eq. (11.60) we have

$$\Sigma\, 2R|Q|\mathrm{d}Q = e$$

Considering the error dQ to be the same for all hydraulic paths in a loop, we can write

$$\mathrm{d}Q = \frac{e}{\Sigma 2R|Q|} \tag{11.62}$$

The Eq. (11.62) can be written with the help of Eqs (11.58a) and (11.59b) as

$$\mathrm{d}Q = \frac{\Sigma R|Q|Q}{\Sigma 2R|Q|} \tag{11.63}$$

The error in flow rate dQ is determined from Eq. (11.63) and the flow rate in each path of a loop is then altered according to Eq. (11.57). The procedure is repeated unless a reasonable convergence is achieved to get the correct flow rates.

The Hardy–Cross method can also be applied to a hydraulic circuit containing a pump or a turbine. The pressure equation (Eq. (11.42)) is only modified in consideration of a head source (pump) or a head sink (turbine) as

$$-\Delta H + \Sigma R|Q|Q = 0 \tag{11.64}$$

where ΔH is the head delivered by a source in the circuit. Therefore, the value of ΔH to be substituted in Eq. (11.64) will be positive for a pump and negative for a turbine.

The application of Hardy–Cross method in a pipe network is illustrated in the following example.

Example 11.13

A pipe network with two loops is shown in Fig. 11.20. Determine the flow in each pipe for an inflow of 5 units at the junction *A* and outflows of 2.0 units and 3.0 units at junctions *D* and *C*, respectively. The resistance *R*, for different pipes are shown in the figure.

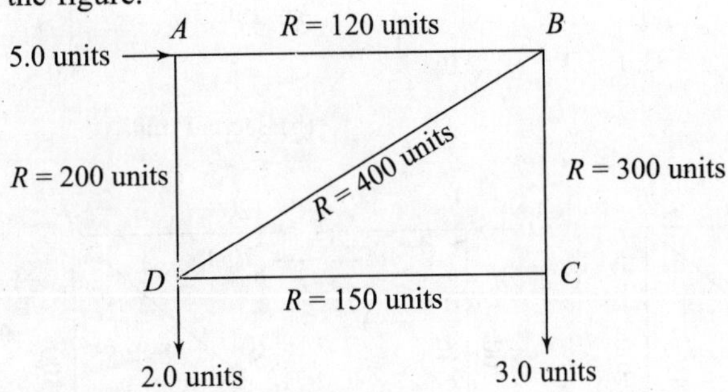

Fig. 11.20 A pipe network

Solution

Flow direction is assumed positive clockwise for both the loops *ABD* and *BCD*. The iterative solutions based on the Hardy–Cross method has been made. The five trials have been made and the results of each trial is shown in Fig. 11.21; for each trial, dQ is calculated from Eq. (11.31). After the fifth trial, the error dQ is so small that it changes the flow only in the third place of decimal. Hence the calculation has not been continued beyond the fifth trial.

First trial:

Loop ABD		*Loop BCD*	
$R\|Q\|Q$	$2R\|Q\|$	$R\|Q\|Q$	$2R\|Q\|$
$120 \times 2^2 = 480$	$2 \times 120 \times 2 = 480$	$300 \times (1.2)^2 = 432$	$2 \times 300 \times 1.2 = 720$
$400 \times (0.8)^2 = 256$	$2 \times 400 \times 0.8 = 640$	$-150 \times (1.8)^2 = -486$	$2 \times 150 \times 1.8 = 540$
$-200 \times 3^2 = -1800$	$2 \times 200 \times 3 = 1200$	$-400 \times (0.8)^2 = -256$	$2 \times 400 \times 0.8 = 640$
$\Sigma R\|Q\|Q = -1064 2\Sigma R\|Q\| = 2320$		$\Sigma R\|Q\|Q = -310 2\Sigma R\|Q\| = 1900$	
$dQ = \dfrac{\Sigma R\|Q\|Q}{\Sigma 2R\|Q\|}$		$dQ = \dfrac{\Sigma R\|Q\|Q}{\Sigma 2R\|Q\|}$	
$= \dfrac{-1064}{2320} = -0.46$		$= \dfrac{-300}{1900} = -0.16$	

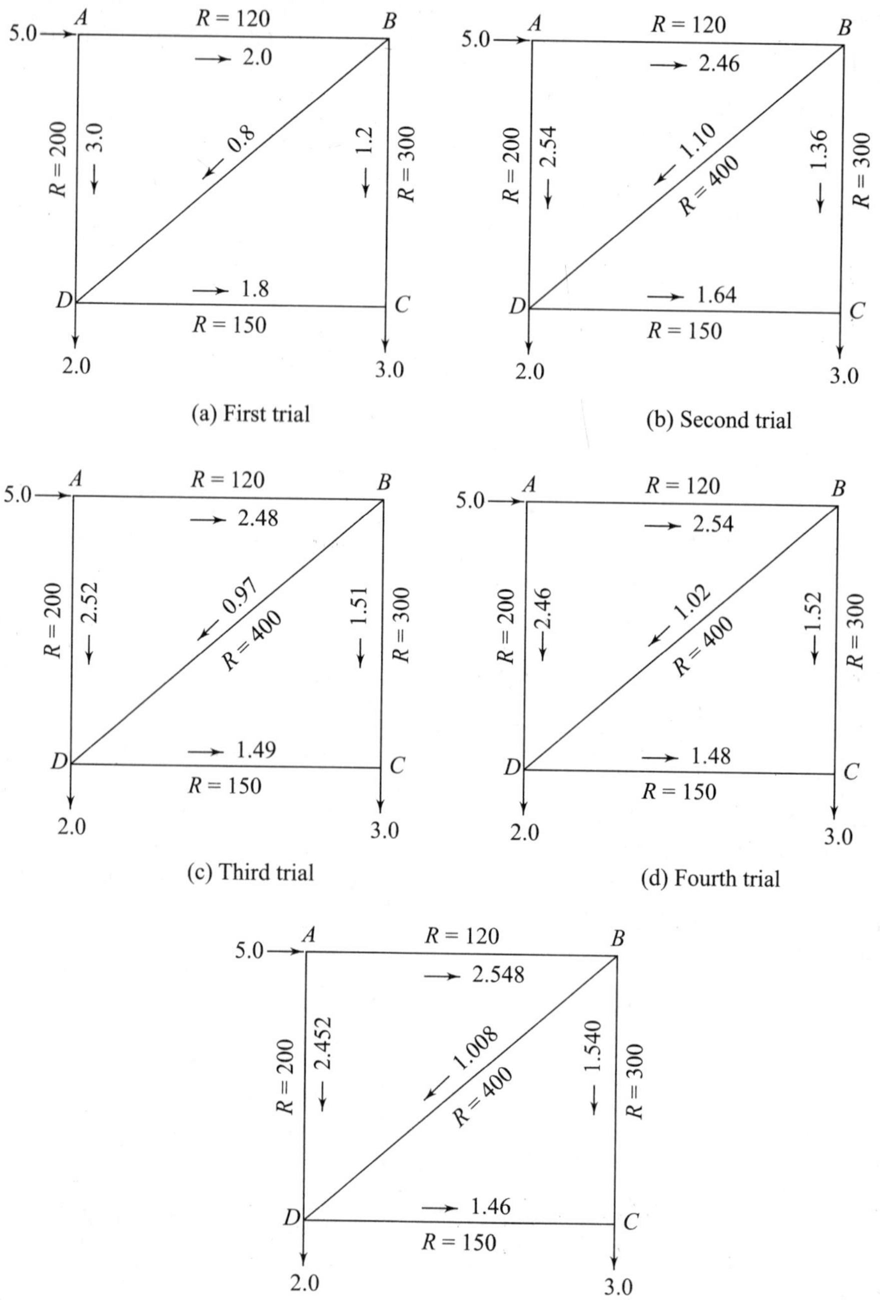

Fig. 11.21 Flow distribution in a pipe network after different trials for Example 11.13

Second trial:

Loop ABD		*Loop BCD*	
$R\|Q\|Q$	$2R\|Q\|$	$R\|Q\|Q$	$2R\|Q\|$
$120 \times (2.46)^2$ $= 726.19$	$2 \times 120 \times 2.46$ $= 590.40$	$300 \times (1.36)^2$ $= 554.88$	$2 \times 300 \times 1.36$ $= 816$
$400 \times (1.10)^2$ $= 484.00$	$2 \times 400 \times 1.10$ $= 880.00$	$-150 \times (1.64)^2$ $= -403.44$	$2 \times 150 \times 1.64$ $= 492$
$-1200 \times (2.54)^2$ $= -1290.32$	$2 \times 200 \times 2.54$ $= 1016.00$	$-400 \times (1.10)^2$ $= -484.00$	$2 \times 400 \times 1.10$ $= 880$
$\Sigma R\|Q\|Q = -50.13$	$2\Sigma R\|Q\| = 2486.40$	$\Sigma R\|Q\|Q = -332.56$	$2\Sigma R\|Q\| = 2188$
$dQ = \dfrac{\Sigma R\|Q\|Q}{2\Sigma R\|Q\|} = \dfrac{-50.13}{2486.40} = -0.02$		$dQ = \dfrac{\Sigma R\|Q\|Q}{2\Sigma R\|Q\|} = \dfrac{-332.56}{2188} = -0.15$	

Third trial:

Loop ABD		*Loop BCD*	
$R\|Q\|Q$	$2R\|Q\|$	$R\|Q\|Q$	$2R\|Q\|$
$120 \times (2.48)^2$ $= 738.05$	$2 \times 120 \times 2.48$ $= 595.20$	$300 \times (1.51)^2$ $= 684.03$	$2 \times 300 \times 1.51$ $= 906.00$
$400 \times (0.97)^2$ $= 376.36$	$2 \times 400 \times 0.97$ $= 776.00$	$-150 \times (1.49)^2$ $= -333.01$	$2 \times 150 \times 1.49$ $= 447.00$
$-200 \times (2.52)^2$	$2 \times 200 \times 2.52$	$-400 \times (0.97)^2$	$2 \times 400 \times 0.97$
$= -1270.08$	$= 1008.00$	$= -376.36$	$=776.00$
$\Sigma R\|Q\|Q = -155.67$	$2\Sigma R\|Q\| = 2379.20$	$\Sigma R\|Q\|Q = -25.34$	$2\Sigma R\|Q\| = 2129$
$dQ = \dfrac{\Sigma R\|Q\|Q}{\Sigma 2R\|Q\|} = \dfrac{-155.67}{2379.67} = -0.06$		$dQ = \dfrac{\Sigma R\|Q\|Q}{\Sigma 2R\|Q\|} = \dfrac{-25.34}{2129} = -0.01$	

Fourth trial:

Loop ABD		*Loop BCD*	
$R\|Q\|Q$	$2R\|Q\|$	$R\|Q\|Q$	$2R\|Q\|$
$120 \times (2.54)^2$ $= 774.20$	$2 \times 120 \times 2.54$ $= 609.60$	$300 \times (1.52)^2$ $= 693.12$	$2 \times 300 \times 1.52$ $= 912.00$
$400 \times (1.02)^2$ $= 416.16$	$2 \times 400 \times 1.02$ $= 816.00$	$-150 \times (1.48)^2$ $= -328.56$	$2 \times 150 \times 1.48$ $= 444.00$

$-200\times(2.46)^2$	$2\times200\times2.46$	$-400\times(1.02)^2$	$2\times400\times1.02$
$=-1210.32$	$=984.00$	$=-416.16$	$=816.00$
$\Sigma R\|Q\|Q=-19.96$	$2\Sigma R\|Q\|=2409.60$	$\Sigma R\|Q\|Q=-51.6$	$2\Sigma R\|Q\|=2172$
$dQ=\dfrac{\Sigma R\|Q\|Q}{\Sigma 2R\|Q\|}$		$dQ=\dfrac{\Sigma R\|Q\|Q}{\Sigma 2R\|Q\|}$	
$=\dfrac{-19.96}{2409.60}$		$=\dfrac{-51.6}{2172}$	
$=-0.008$		$=-0.02$	

Fifth trial:

Loop ABD		*Loop BCD*	
$R\|Q\|Q$	$2R\|Q\|$	$R\|Q\|Q$	$2R\|Q\|$
$120\times(2.58)^2$	$2\times120\times2.58$	$300\times(1.54)^2$	$2\times300\times1.54$
$=779.08$	$=619.20$	$=711.48$	$=924.00$
$400\times(1.008)^2$	$2\times400\times1.008$	$-150\times(1.46)^2$	$2\times150\times1.46$
$=406.42$	$=806.40$	$=-319.74$	$=438.00$
$-200\times(2.452)^2$	$2\times200\times2.452$	$-400\times(1.08)^2$	$2\times400\times1.008$
$=-1202.46$	$=980.80$	$=-406.42$	$=806.40$
$\Sigma R\|Q\|Q=-16.96$	$2\Sigma R\|Q\|=2406.40$	$\Sigma R\|Q\|Q=-14.68$	$2\Sigma R\|Q\|=2168.40$
$dQ=\dfrac{\Sigma R\|Q\|Q}{\Sigma 2R\|Q\|}$		$dQ=\dfrac{\Sigma R\|Q\|Q}{\Sigma 2R\|Q\|}$	
$=\dfrac{-16.96}{2406.40}$		$=\dfrac{-14.68}{2168.40}$	
$=-0.007$		$=-0.007$	

11.8 FLOW THROUGH PIPES WITH SIDE TAPPINGS

In course of flow through a pipe, a fluid may be withdrawn from the side tappings along the length of the pipe as shown in Fig. 11.22. If the side tappings are very closely spaced, the loss of head over a given length of pipe can be obtained as shown below.

The rate of flow through the pipe, under this situation, decreases in the direction of flow due to side tappings. Therefore, the average flow velocity at any section of the pipe is not constant. The frictional head loss dh_f, over a small length dx of the pipe at any section can be written as

$$dh_f = f\frac{dx}{D}\frac{V_x^2}{2g} \tag{11.65}$$

where, V_x is the average flow velocity at that section. If the side tappings are very close together, Eq. (11.65) can be integrated to determine the loss of head due to friction over a given length L of the pipe, provided, V_x can be replaced in terms of the length of the pipe. Let us consider, for this purpose, a Section 1–1 at the upstream just after which the side tappings are provided. If the tappings are uniformly and closely spaced, so that the fluid is removed at a uniform rate q per unit length of the pipe, then the volume flow rate Q_x at a distance x from the inlet Section 1–1 can be written as

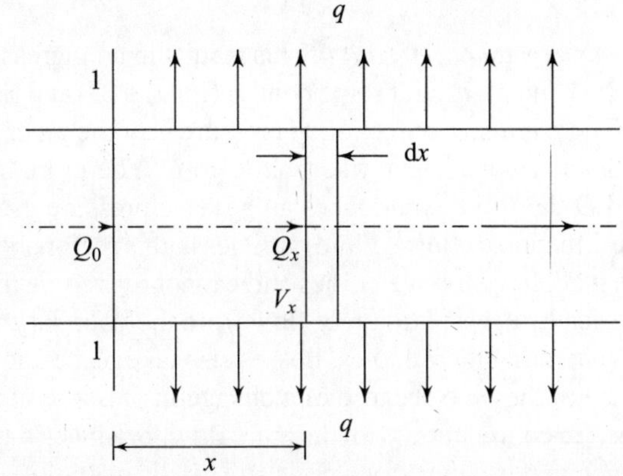

Fig. 11.22 Flow through pipes with side tappings

$$Q_x = Q_0 - qx$$

where, Q_0 is the volume flow rate at Sec. 1–1. Hence,

$$V_x = \frac{4Q_x}{\pi D^2} = \frac{4Q_0}{\pi D^2}\left(1 - \frac{q}{Q_0}x\right) \tag{11.66}$$

Substituting V_x from Eq. (11.66) into Eq. (11.65), we have,

$$\mathrm{d}h_f = \frac{16Q_0^2 f}{2\pi^2 D^5 g}\left(1 - \frac{q}{Q_0}x\right)^2 \mathrm{d}x \tag{11.67}$$

Therefore, the loss of head due to friction over a length L, is given by

$$h_f = \int_0^L \mathrm{d}h_f = \frac{8Q_0^2 f L}{\pi^2 D^5 g}\left(1 - \frac{q}{Q_0}L + \frac{1}{3}\frac{q^2}{Q_0^2}L^2\right) \tag{11.68a}$$

Here, the friction factor f, has been assumed to be constant over the length L, of the pipe. If the entire flow at Sec. 1-1 is drained off over the length L, then,

$$Q_0 - qL = 0 \quad \text{or} \quad \frac{q}{Q_0} = \frac{1}{L}$$

Equation (11.68(a)), under this situation, becomes

$$h_f = \frac{8}{3}\frac{Q_0^2 f L}{\pi^2 D^5 g} = \frac{1}{3}f\frac{L}{D}\left(\frac{4Q_0}{\pi D^2}\right)^2 \frac{1}{2g} = \frac{1}{3}f\frac{L}{D}V_0^2\frac{1}{2g} \tag{11.68b}$$

where, V_0 is the average velocity of flow at the inlet Section 1–1.

Equation (11.68b) indicates that the loss of head due to friction over a length L of a pipe, where the entire flow is drained off uniformly from the side tappings, becomes one third of that in a pipe of same length and diameter, but without side tappings.

11.9 LOSSES IN PIPE BENDS

Bends are provided in pipes to change the direction of flow through it. An additional loss of head, apart from that due to fluid friction, takes place in the course of flow through pipe bend. The fluid takes a curved path while flowing through a pipe bend as shown in Fig. 11.23. Whenever a fluid flows in a curved path, there must be a force acting radially inwards on the fluid to provide the inward acceleration, known as *centripetal acceleration*. This results in an increase in pressure near the outer wall of the bend, starting at some point A (Fig. 11.23) and rising to a maximum at some point B. There is also a reduction of pressure near the inner wall giving a minimum pressure at C and a subsequent rise from C to D. Therefore between A and B and between C and D the fluid experiences an adverse pressure gradient (the pressure increases in the direction of flow). Fluid particles in this region, because of their close proximity to the wall, have low velocities and cannot overcome the adverse pressure gradient and this leads to a separation of flow from the boundary and consequent losses of energy in generating local eddies. Losses also take place due to a secondary flow in the radial plane of the pipe because of a change in pressure in the radial depth of the pipe. This flow, in conjunction with the main flow, produces a typical spiral motion of the fluid which persists even for a downstream distance of fifty times the pipe diameter from the central plane of the bend. This spiral motion of the fluid increases the local flow velocity and the velocity gradient at the pipe wall, and therefore results in a greater frictional loss of head than that which occurs for the same rate of flow in a straight pipe of the same length and diameter.

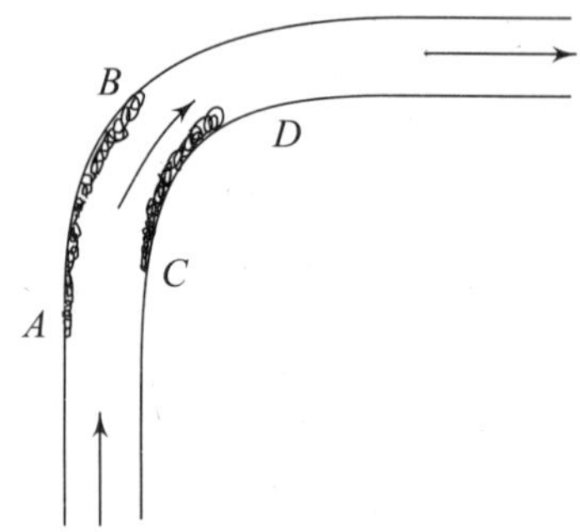

Fig. 11.23 Flow through pipe bend

The additional loss of head (apart from that due to usual friction) in flow through pipe bends is known as bend loss and is usually expressed as a fraction of the velocity head as $KV^2/2g$, where V is the average velocity of flow through the pipe. The value of K depends on the total length of the bend and the ratio of radius of curvature of the bend and pipe diameter R/D. The radius of curvature R is usually taken as the radius of curvature of the centre line of the bend. The factor K varies slightly with Reynolds number Re in the typical range of Re encountered in practice, but increases with surface roughness.

11.10 LOSSES IN PIPE FITTINGS

An additional loss of head takes place in the course of flow through pipe fittings like valves, couplings and so on. In general, more restricted the passage is, greater is the loss of head. For turbulent flow, the losses are proportional to the square of the average flow velocity and are usually expressed by $KV^2/2g$, where V is the average velocity of flow. The value of K depends on the exact shape of the flow passages. Typical values of K are given in Table 11.2. Since the eddies generated by fittings persist for some distance downstream, the total loss of head caused by two fittings close together is not necessarily the same as the sum of the losses which each alone would cause.

These losses are sometimes expressed in terms of an equivalent length of an unobstructed straight pipe in which an equal loss would occur for the same average flow velocity. That is

$$K\frac{V^2}{2g} = f\frac{L_e}{D}\frac{V^2}{2g} \quad \text{or} \quad \frac{L_e}{D} = \frac{K}{f} \tag{11.69}$$

where L_e represents the equivalent length which is usually expressed in terms of the pipe diameter as given by Eq. (11.69). Thus L_e/D depends upon the friction factor f, and therefore on the Reynolds number and roughness of the pipe.

Table 11.2 Approximate Loss Coefficients K for Commercial Pipe Fittings

Type and position of fittings		*Values of K*
Globe valve,	wide open	10
Gate valve,	wide open	0.2
	three-quarters open	1.15
	half open	5.6
	quarter open	24
Pump foot valve		1.5
90° elbow (threaded)		0.9
45° elbow (threaded)		0.4
Side outlet of *T* junction		1.8

Example 11.14

A pump requires 50 kW to supply water at a rate of 0.2 m^3/s to an overhead tank. The pipe connecting the delivery end of the pump to the overhead tank is 120 m long and 300 mm in diameter and has a friction factor $f = 0.02$. A valve is inserted in the delivery pipe to control the flow rate. The loss coefficient of the valve under wide open condition is 5.0. Water is supplied from a reservoir 2 m below the horizontal level of the pump though a suction pipe 6 m long and 400 mm in diameter having $f = 0.03$. Determine the maximum height from the plane of the pump at which the overhead tank can be placed under this situation. (Take the efficiency of the pump $\eta = 80\%$). Assume kinetic energy correction factor as unity.

Solution

Let H be the height of the overhead tank from the pump

p_d be the pressure at the delivery side of the pump

p_s be the pressure at the suction side of the pump.

The average velocity of flow in the delivery pipe

$$V_d = \frac{4 \times 0.2}{\pi \times (0.3)^2} = 2.83 \text{ m/s}$$

The average velocity of flow in the suction pipe

$$V_s = \frac{4 \times 0.2}{\pi \times (0.4)^2} = 1.59 \text{ m/s}$$

Applying energy equation between a section at the inlet to the delivery pipe and a point at the water surface in the overhead tank where the pressure is atmospheric, we have,

$$\frac{p_d}{\rho g}+\frac{(2.83)^2}{2g} = \frac{p_{atm}}{\rho g}+H+\left(0.02\times\frac{120}{0.3}+5.0+1.0\right)\times\frac{(2.83)^2}{2g}$$

or
$$\frac{p_d - p_{atm}}{\rho g} = H + 5.31 \tag{11.70}$$

Applying energy equation between a section on the water surface in the supply reservoir and a section at the end of the suction pipe connecting the pump, we can write,

$$\frac{p_{atm}}{\rho g} = \frac{p_s}{\rho g}+2+\left(1+0.5+0.03\times\frac{6.0}{0.4}\right)\times\frac{(1.59)^2}{2g}$$

or
$$\frac{p_{atm} - p_s}{\rho g} = 2.25 \tag{11.71}$$

From Eqs (11.70) and (11.71), we get

$$\frac{p_d - p_s}{\rho g} = H+7.56$$

Power delivered by the pump to water = 50 × 0.8 = 40 kW

Therefore, we can write,

$$0.2 \times (p_d - p_s) = 40 \times 10^3$$

or
$$0.2 \times 10^3 \times 9.81\,(H + 7.56) = 40 \times 10^3$$

which gives, $H = 12.83$ m

Example 11.15

Water flows through a pipe line of 300 mm in diameter and 20 km long in a horizontal plane. At a point *B*, the pipe is branched off into two parallel pipes each of 150 mm diameter and 3.5 km long as shown in Fig. 11.24. In one of the these pipes, water is completely drained off from side tappings at a constant rate of 0.01 litre/s per metre length of the pipe. Determine the flow rate and loss of head in the main pipe. (Take friction factor for all the pipes as 0.012.)

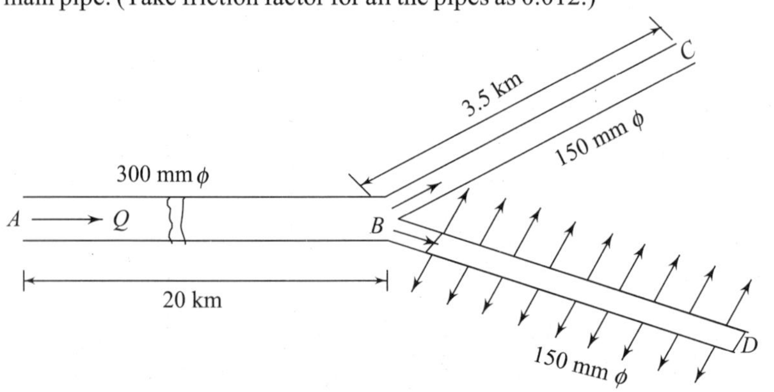

Fig. 11.24 Flow through a branched pipe with side tappings

Solution

Let Q be the flow rate through the pipe *AB* and be divided at *B* into Q_1 and Q_2 for the pipes *BC* and *BD* respectively. Then from continuity,

$$Q = Q_1 + Q_2$$

Since the entire flow at inlet to the pipe *BD* is drained off through side tappings at a constant rate of 0.01 litre per metre length,

$$Q_2 = 0.01 \times 3500 = 35 \text{ litre/s} = 0.035 \text{ m}^3\text{/s}$$

Hence, average velocity at inlet to pipe *BD*

$$= \frac{4 \times 0.035}{\pi (0.15)^2} = 1.98 \text{ m/s}$$

The loss of head in *BD* can be written with the help of Eq. (11.68b) as

$$h_{fBD} = \frac{1}{3} \times 0.012 \times \frac{3500}{0.15} \times \frac{(1.98)^2}{2 \times 9.81} = 18.65 \text{ m}$$

Since *B* is a common point and *C* and *D* are at the same horizontal level and have the same pressure which is equal to that of the atmosphere, the loss of head in the parallel pipes *BC* and *BD* are equal.

Therefore, $h_{fBC} = h_{fBD} = 18.65$ m (11.72)

Average flow velocity in pipe $BC = \dfrac{4Q_1}{\pi (0.15)^2} = 56.59\, Q_1$

Equating h_{fBC} given by Eq. (11.72) with the different losses taking place in pipe *BC*, we can write

$$h_{fBC} = 18.65 = 0.012 \times \frac{3500}{0.15} \times \frac{(56.59)^2}{2g} Q_1^2 + \frac{(56.59)^2}{2g} Q_1^2$$

$$= 45865.6\, Q_1^2$$

which gives $Q_1 = 0.02 \text{ m}^3\text{/s}$

Hence, $Q = 0.035 + 0.02 + 0.055 \text{ m}^3\text{/s}$

Velocity in the main pipe $AB = \dfrac{4 \times 0.055}{\pi (0.3)^2} = 0.78$ m/s

The loss of head in the main pipe $AB = 0.012 \times \dfrac{20000}{0.30} \dfrac{(0.78)^2}{2g} = 24.81$ m

11.11 POWER TRANSMISSION BY A PIPELINE

In certain occasions, hydraulic power is transmitted by conveying the fluid through a pipeline. For example, water from a reservoir at a high altitude is often conveyed by a pipeline to an impulse hydraulic turbine in a hydroelectric power station. The hydrostatic head of water is thus transmitted by a pipeline. Let us analyse the efficiency of power transmission under the situation.

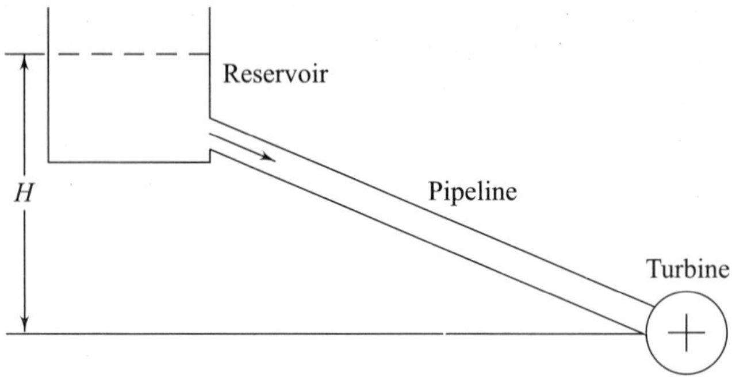

Fig. 11.25 Transmission of hydraulic power by a pipeline to a turbine

The potential head of water in the reservoir = H (the difference in the water level in the reservoir and the turbine centre) (Fig. 11.25)

The head available at the pipe exit (or at the turbine entry) = $H_E = H - h_f$

where h_f is the loss of head in the pipeline due to friction.

Assuming that the friction coefficient and other loss coefficients are constant, we can write

$$h_f = RQ^2$$

where Q is the volume flow rate and R is the hydraulic resistance of the pipeline. Therefore, the power available P at the exit of the pipeline becomes

$$P = \rho g Q H_E = \rho g Q(H - RQ^2)$$

For P to be maximum, for a given head H, dP/dQ should be zero. This gives

$$H - 3RQ^2 = 0$$

or
$$RQ^2 = h_f = \frac{H}{3} \tag{11.73}$$

[d^2P/dQ^2 is always negative which shows that P has only a maximum value (not a minimum) with Q].

From Eq. (11.73), we can say that maximum power is obtained when one third of the head available at the source (reservoir) is lost due to friction in the flow.

The efficiency of power transmission η_p is defined as

$$\eta_p = \frac{\rho g Q(H - RQ^2)}{\rho g Q H}$$

$$= 1 - \frac{RQ^2}{H} \tag{11.74}$$

The efficiency η_p equals to unity for the trivial case of $Q = 0$. For flow to commence $RQ^2 \leq H$ and hence η_p is a monotonically decreasing function of Q from a maximum value of unity to zero. The zero value of η_p corresponds to the situation given by $RQ^2 = H$ (or, $Q = \sqrt{H/R}$) when the head H available at the reservoir is totally lost to overcome friction in the flow through the pipe. The efficiency of transmission at the condition of maximum power delivered is obtained by substituting RQ^2 from Eq. (11.73) in Eq. (11.74) as

$$\eta_{\text{at } P = P_{\max}} = 1 - \frac{H/3}{H}$$

$$= \frac{2}{3}$$

Therefore, the maximum power transmission efficiency through a pipeline is 67%.

SUMMARY

- Fanning's friction coefficient C_f, for a flow through a closed duct is defined in terms of shear stress at the wall as $C_f = \tau_w/(1/2)\rho V^2$, and in terms of piezometric pressure drop Δp^* over a length L, as $C_f = \frac{1}{4}(D_h/L)\Delta p^*/(1/2)\rho V^2$. Darcy's friction factor f, is defined as $f = 4C_f$.
- Loss of head in a pipe flow is expressed in terms of Darcy's friction factor f as $h_f = f(L/D)(V^2/2g)$.
- Friction factor in case of a laminar fully developed flow through pipes is found from the exact solution of Navier–Stokes equation and is given by $f = 64/\text{Re}$. Friction factor in a turbulent flow depends on both the Reynolds number of flow and the roughness at pipe surface.
- Loss of head due to fluid friction may also be estimated by applying energy conservation equation between two sections of flow. This loss can be attributed to irreversible conversion of viscous dissipation into intermolecular form of energy and any possible heat transfer from the system to the surroundings through the pipe walls.
- The head causing the flow is known as flow potential. Flows, in practice, takes place through several pipes joined together either in series or in parallel or in a combination of both of them. The flow through a pipe network system has to overcome the pipe friction and other resistances due to minor losses. The relationship between the head causing the flow ΔH and the flow rate Q can be expressed as $\Delta H = RQ^2$, where R is the flow resistance in the hydraulic path. This equation is analogous to the voltage-current relationship in a purely resistive electrical circuit. Therefore, the pipe flow system can be described by an equivalent electrical network system.
- The loss of head due to friction over a length L of a pipe, where the entire flow is drained off uniformly from the side tappings, becomes one-third of that in a pipe of the same length and diameter, but without side tappings.
- An additional head loss over that due to pipe friction takes place in a flow through pipe bends and pipe fittings like valves, couplings and so on.

  The hydraulic power can be transmitted by a pipeline. For a maximum power transmission, the head lost due to friction in the flow equals to one-third of the head at source to be transmitted. The maximum power trasmission efficiency is 2/3 (67%).

EXERCISES

11.1 Under what circumstances is the friction factor for the flow through a pipe of constant diameter
 (a) inversely proportional to the Reynolds number
 (b) dependent on the relative roughness only
 (c) independent of the relative roughness

11.2 Choose the correct answer:
 (i) Friction loss through a pipe flow implies
  (a) loss of energy due to the coefficient of friction between the material of the pipe and the fluid
  (b) loss due to dynamic coefficient of friction
  (c) loss of flow rate in a pipe due to surface roughness
  (d) loss of energy due to surface roughness
  (e) loss of momentum due to surface roughness
 (ii) For pipes arranged in series
  (a) the flow may be different in different pipes
  (b) the head loss per unit length must be more in a smaller pipe
  (c) the velocity must be the same in all pipes
  (d) the head loss must be the same in all pipes
  (e) the flow rate must be the same in all pipes
 (iii) In parallel pipe systems
  (a) the pipes must be placed geometrically parallel to each other
  (b) the flow must be the same in all pipes
  (c) the head loss per unit length must be the same for all pipes
  (d) the head loss across each of the parallel pipes must be the same
  (e) None of the above

11.3 A 200 mm diameter pipe 200 m long discharges oil from a tank into the atmosphere. At the midpoint of the pipe length, the pressure is one and a half times the atmospheric pressure. The specific gravity of the oil is 0.9. The friction factor $f = 0.03$. Calculate (i) the discharge rate of oil and (ii) the pressure in the tank at the inlet of the pipe.

Ans. (0.086 m^3/s, 206.64 kN/m^2)

11.4 Calculate the power required to pump sulphuric acid (viscosity 0.04 Ns/m^2 and specific gravity 1.83) at 45 litre/s from a supply tank through a glass-lined 150 mm diameter pipe, 18 m long, into a storage tank. The liquid level in the storage tank is 6 m above from that in the supply tank. For laminar flow $f = 64/Re$ and for turbulent flow $f = 0.0056\,(1 + 100Re^{-1/3})$. Take all losses into account

Ans. (6.12 kW)

11.5 The total head at inlet to a pipe network system is 20 m of water more than that at its outlet. Compare the rate of discharge of water, if the network system consists of (i) three pipes each 700 m long but of diameters 450 mm, 300 mm and 600 mm respectively in the order from inlet to outlet, (ii) the same three pipes in parallel. Assume friction factor for all the pipes to be 0.01, and the coefficient of contraction $C_c = 0.6$.

Ans. ((a) 0.263 m^3/s, (b) 2.73 m^3/s)

11.6 Water flows from a tank A to a tank B. The difference in water level between the two tanks is 7 m. The tanks are connected by a 30.5 m of 300 mm diameter pipe ($f = 0.02$) followed by the 30.5 m of 150 mm diameter pipe ($f = 0.015$). There are two 90° bends in each pipe ($k = 0.50$ each), the coefficient of contraction $C_c = 0.75$. If the junction of the two pipes is 5 m below the top water level, find the pressure heads (in gauge) in 300 mm and 150 mm pipe at the junction.

Ans. (4.76 m, 3.56 m)

11.7 There is a sudden increase in the diameter of a pipe from d_1 to d_2. What would be the ratio d_2/d_1 if the minor loss is independent of the direction of flow? Assume coefficient of contraction $C_c = 0.6$.

Ans. $\left(\sqrt{3}\right)$

11.8 Show that the loss of head Δh due to friction for a laminar flow in a diffuser of round cross section and with a small taper angle α is given by

$$\Delta h = 64\, \mu Q\, (d_2^3 - d_1^3)/[3\, \pi\, \rho\, g\, d_1^3\, d_2^3 \tan(\alpha/2)]$$

where, Q is the rate of volumetric discharge, μ and ρ are the viscosity and density of the fluid respectively, d_1 and d_2 are the diameters of the diffuser at its inlet and outlet respectively. Assume that Poiseuille's law is valid for each element of the diffuser length.

11.9 A single pipe 400 mm in diameter and 400 m long conveys water at the rate of 0.5 m^3/s. Find the increase in discharge if another pipe of 200 m long and 200 mm in diameter is joined parallel with the existing pipe over half of its length. Friction factor for all the pipes is same.

Ans. (0.04 m^3/s)

11.10 Three piping systems (I), (II) and (III) are studied (Fig. 11.26). Take $f = 0.012$ for all the pipes. Indicate which one has the greatest and which one has the lowest capacity under a given head.

(Greatest - II, Lowest - III)

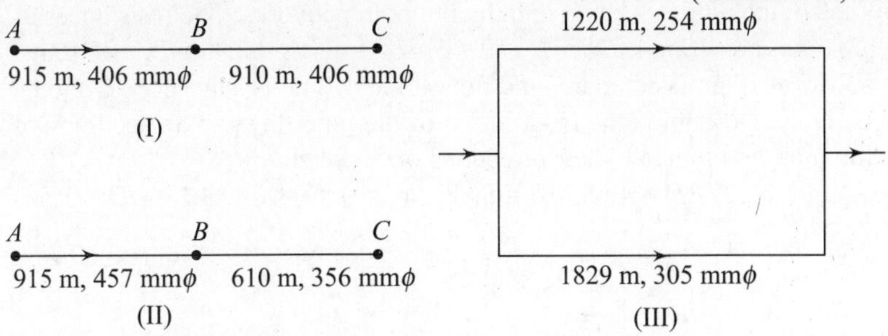

Fig. 11.26 Different piping systems

11.11 A pump delivers water through two pipes laid in parallel. One pipe is 100 mm in diameter and 45 m long and discharges to atmosphere at a level of 6 m above the pump outlet. The other pipe, 150 mm in diameter and 60 m long, discharges to atmosphere at a level of 8 m above the pump outlet. The two pipes are connected to a junction immediately after the pump. The inlet to the pump is 600 mm below the level of its outlet. Taking the datum level as that of the pump inlet, determine the total head at the pump outlet if the flow rate through it is 0.037 m^3/s. Take friction factor for the pipes $f = 0.032$, and neglect losses at the pipe junction.

Ans. (9.64 m)

11.12 Water flows out of a reservoir through a horizontal pipe 500 mm in diameter and 400 m long. The level of water in the reservoir is 10 m. Due to partial closure of the pipe at the discharge end by an obstruction, the flow velocity through the pipe is 3 m/s, and the pressure loss per unit length is 135 N/m^3. Calculate the pipe friction factor and the loss coefficient of the obstruction. Estimate the flow velocity when the resistance is withdrawn completely. Neglect entry loss, but account for the exit velocity head.

Ans. ($f = 0.015$, $k = 8.8$, 3.88 m/s)

11.13 A pipe system consists of three pipes connected in series (i) 300 m long, 150 mm in diameter (ii) 150 m long, 100 mm in diameter and (iii) 250 m long, 200 mm in diameter. Determine the equivalent length of a 125 mm diameter pipe. (Take friction factor $f = 0.02$, coefficient of contraction $C_c = 0.6$).

Ans. (620.4 m)

11.14 Two reservoirs are connected by three cast iron pipes in series. The length and diameter of the pipes are $L_1 = 600$ m, $D_1 = 0.3$ m, $L_2 = 900$ m, $D_2 = 0.4$ m, $L_3 = 1500$ m and $D_3 = 0.45$ m, respectively. Find out the Reynolds number in each of the pipes. The density and viscosity of water are 1000 kg/m^3 and 1.1×10^{-3} Ns/m^2, respectively. The friction factor in each pipe may be approximated as 0.02. The loss due to expansion at the junctions between Pipe 1 and Pipe 2 as well as between Pipe 2 and Pipe 3 may be neglected. The discharge is 0.11 m^3/s. Determine the difference in elevation between the top surfaces of the reservoirs. Include the entry loss to Pipe 1 and exit loss between Pipe 3 and the adjacent reservoir.

Ans. (8.426 m)

11.15 A ring main consists of a quadrilateral network *ABCD* and a triangular network *ADE*, the pipe *AD* being common to both networks. The resistances of the pipelines are $AB = 4$, $BC = 2$, $CD = 5$, $DA = 4$, $AE = 2$, and $DE = 3$ units. Let a flow of 10 units enter at *E* and flows of 3, 4, 3 units leave at *B*, *C* and *D*, respectively. Determine the magnitudes of the pipe flows to an accuracy of 0.1flow unit and indicate their directions on a sketch.

Ans. ($EA = 5.32$, $ED = 4.68$, $AB = 3.82$, $BC = 0.82$, $DC = 3.18$, $AD = 1.50$)

12

FLOWS WITH A FREE SURFACE

12.1 INTRODUCTION

The flow of a fluid is not always required to be bounded on all sides by solid surfaces as discussed in the previous chapters. The flow of liquids, under certain circumstances, may take place when the uppermost boundary is the free surface of the liquid itself. Then the cross section of flow is not determined entirely by the solid boundaries. The controlling parameters of flow in this case are different from those in the case of flow through closed ducts.

Flow with a free surface takes place in open channels. The free surface is subjected only to atmospheric pressure which is constant. Therefore the flow is caused by the weight of the fluid. It has been discussed earlier on several occasions, that a uniform flow through a closed duct takes place due to a drop in the *Piezometric pressure* p^* $(= p + \rho gz)$. But for an open channel, a uniform flow is caused by the second term ρgz since the static pressure remains constant in the direction of flow. Natural streams, rivers, artificial canals, irrigation ditches and flumes are examples of open channels in practice. Pipe lines or tunnels which are not completely full of liquid also have the essential features of open channels.

12.2 FLOW IN OPEN CHANNELS

12.2.1 Geometrical Terminologies

Depth of Flow h The depth of flow h at any section (Fig. 12.1) is the vertical distance of the bed of the channel from the free surface at that section.

Top Breadth B It is the breadth of the channel section at the free surface (Fig. 12.1).

The Water Area A The water area is the flow cross-sectional area perpendicular to the direction of flow.

The Wetted Perimeter P The wetted perimeter P is the perimeter of the solid boundary in contact with the liquid.

Hydraulic Radius Rh The hydraulic radius R_h is defined as $R_h = A/P$.

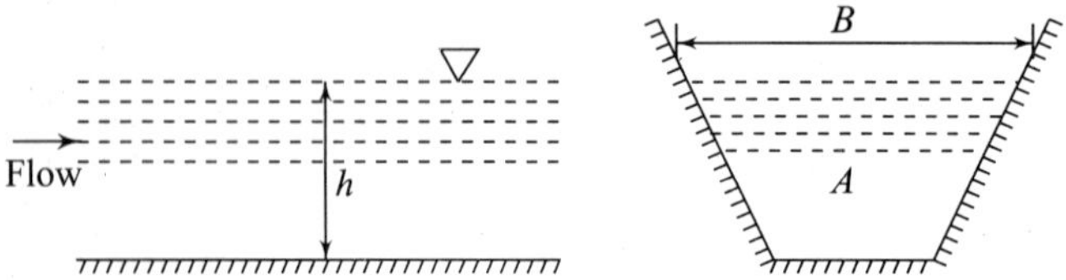

Fig. 12.1 Geometry of a straight channel

12.2.2 Types of Flow in Open Channels

The flow in an open channel may be uniform or non-uniform, steady or unsteady, laminar or turbulent.

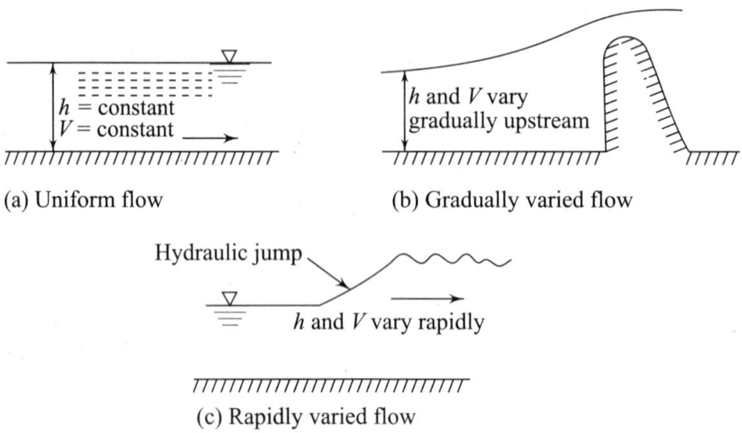

Fig. 12.2 Uniform and non-uniform flow in an open channel

Uniform Flow Uniform flow occurs in a channel when the cross section and depth of flow do not change along the length of the channel. This is characterised by the liquid surface being parallel to the base of the channel (Fig. 12.2(a)).Under this circumstance, the velocity of liquid does not change either in magnitude or direction from one section to another in the part of the channel under consideration.

Non-uniform Flow Flow in which the liquid surface is not parallel to the base of the channel (Fig. 12.2(b)) is said to be non-uniform or varied, since the depth of flow continuously varies from one section to another. This flow occurs in a channel which is shaped irregularly and also in a prismatic channel when depth and velocity vary. The change in depth may be gradual or rapid according to which a non-uniform flow is termed as a gradually varied flow (Fig. 12.2(b)) or a rapidly varied flow (Fig. 12.2(c)). In a gradually varied flow, the degree of non-uniformity is small and gradual. This may extend upstream to a considerable distance due to some control structure, e.g., spillway of a dam, as shown in (Fig. 12.2(b)). In a rapidly varied flow, the change in depth and velocity is rather abrupt or takes place within a short distance. Boundary frictional losses are small and the head loss arises mainly from eddy formation. Such a flow is observed in a hydraulic jump (Fig. 12.2(c)), which will be discussed later in Section 12.4 of this chapter.

Steady or Unsteady Flow Flow is termed steady or unsteady according to whether the velocity at a point in the channel is invariant with time or not. Unsteady non-uniform flow is more common in practice. It occurs when a sluice gate is operated in a dam or during a tidal bore. Non-uniform flow always occurs in short channels because a certain length of channel is required for the establishment of uniform flow. Analysis of unsteady non-uniform flows is more complicated and difficult as compared to that of a steady uniform flow.

The flow is an open channel may be either laminar or turbulent depending upon the relative magnitudes of the viscous and inertia forces. Reynolds number Re, as the criterion of transition from laminar to turbulent flow, is defined in this case as Re = $V_{av}\, l/\nu$, where V_{av} is the average flow velocity at any cross section l, the characteristic length is usually the hydraulic radius R_h (= A/P) and ν is the kinematic viscosity of the liquid. The lower critical value of Reynolds number below which the flow is always laminar is 600. Flows in open channels are usually turbulent in practice. Laminar flow may be observed in small grooves in domestic draining boards set at a small slope.

Another important classification of an open channel flow is made on the basis of whether a small disturbance in the flow can travel upstream or not. This depends on the flow velocity and is characterised by the magnitude of Froude number Fr. When Froude number is less than 1.0, any small disturbance can travel against the flow and affects the upstream condition, and the flow is described as tranquil. When Froude number is greater than 1, a small disturbance cannot propagate upstream but is washed downstream, and the flow is said to be rapid. When Froude number is exactly equal to 1.0, the flow is said to be critical. Further discussion on tranquil and rapid flow has been made in Section 12.2.7. Therefore, a complete description of flow consists of four characteristics. The flow may be

(a) Either uniform or non-uniform
(b) Either steady or unsteady
(c) Either laminar or turbulent
(d) Either tranquil or rapid

12.2.3 Application of Bernoulli's Equation in Open Channels

Bernoulli's equation can be well applied to flow through an open channel, since no restriction to flow between boundaries of a particular kind was made in the derivation of this equation. Bernoulli's equation can be written for a steady incompressible and inviscid flow as

$$\frac{p}{\rho g} + \frac{V^2}{2g} + z = \text{constant along a streamline}$$

In case of a flow through a channel where the streamlines are sensible, straight and parallel or a little curved in nature (for a gradually varied flow), there is only a hydrostatic variation of pressure over the cross-section This implies that the pressure at any point in the stream is governed only by its depth below the free surface. Consider the flow through a straight channel, as shown in Fig. 12.3. Pressure head ($p/\rho g$) at any point in the channel is therefore the vertical height of the free surface from the point. Hence, the sum of pressure head ($p/\rho g$) and potential head (z) at any point becomes equal to the height of the free surface, at the cross section containing the point, above a horizontal datum of reference. (Fig. 12.3). Bernuolli's equation is thus simplified to the following form:

The height of liquid surface above datum + $(V^2/2g)$ = constant (12.1)

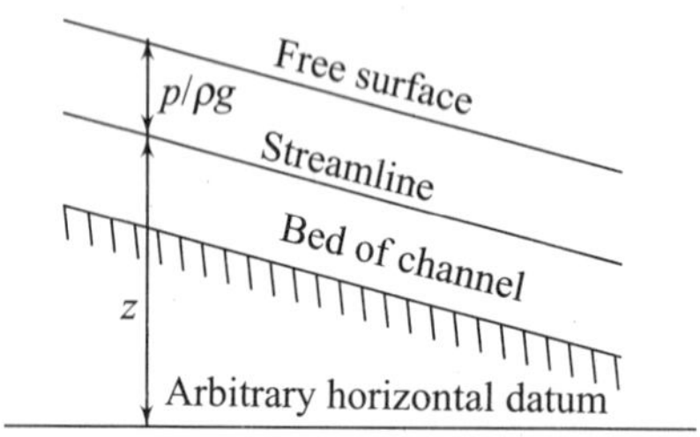

Fig. 12.3 Representation of pressure head and potential head in flow through a straight channel

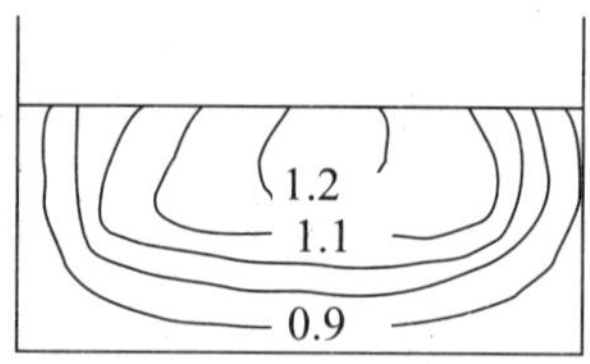

Fig. 12.4 Contours of constant velocity in a rectangular channel

provided that friction is negligible. If it is assumed that, at the section considered, the velocity is same at all streamlines, then Eq. (12.1) is valid for the entire stream. In practice, however, a uniform distribution of velocity over a section is never achieved. The actual velocity distribution in an open channel is influenced both by the solid boundaries and by the free surface. The irregularities in the boundaries of an open channel are usually very large and greatly influence the velocity distribution. A typical velocity distribution for a channel of rectangular section is shown in Fig. 12.4. The maximum velocity usually occurs at a point slightly below the free surface. The numeric in Fig. 12.4 represents the ratio of actual velocity to the velocity at free surface. In practice, when liquid flows from one section to another, friction converts a part of the mechanical energy into intermolecular energy and this part of energy is regarded to be lost. If this loss of mechanical energy per unit weight between Sections 1 and 2 (Fig. 12.5) is h_f, then for steady flow, mechanical energy balance equation (Eq. 12.1) between the two sections can be written as

$$\left(\text{Height of liquid surface above a horizontal datum of reference}\right) + \frac{V_1^2}{2g} = \left(\text{Height of liquid surface above a horizontal datum of reference}\right) + \frac{V_2^2}{2g} + h_f$$

or
$$h_1 + z_1 + \frac{V_1^2}{2g} = h_2 + z_2 + \frac{V_2^2}{2g} + h_f \qquad (12.2a)$$

where V_1 and V_2 are the average flow velocities over the cross sections at 1 and 2 respectively.

To take account of non-uniformity of velocity over the cross section Eq. (12.2a) may be written as

$$h_1 + z_1 + \alpha_1 \frac{V_1^2}{2g} = h_2 + z_2 + \alpha_2 \frac{V_2^2}{2g} + h_f \qquad (12.2b)$$

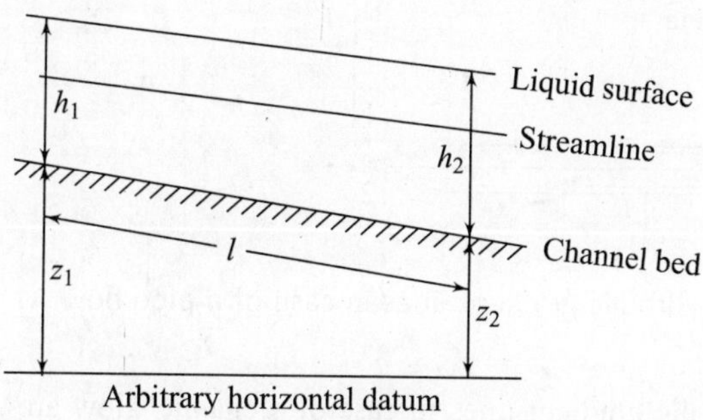

Fig. 12.5 Representation of total head at two sections in a channel flow

12.2.4 Energy Gradient and Hydraulic Gradient Lines

The concept of energy and hydraulic gradient lines is not restricted to channel flows, rather it is referred, in general, to all kinds of flows through closed or open ducts. The energy gradient line is the contour of the total mechanical energy per unit weight or the total head ($z + p/\rho g + V^2/2g$) at a cross section, as ordinate against the distance along the flow. The hydraulic gradient line is obtained by plotting the sum of potential and pressure heads ($z + p/\rho g$) as ordinate against the same abscissa (the distance along the flow). Thus, the hydraulic gradient line is the contour of the free surface in an open channel. The *hydraulic gradient line in any kind of flow* can be constructed by subtracting the velocity head $V^2/2g$ from the energy gradient line at every Section The energy and hydraulic gradient lines are illustrated in case of a pipe flow and flow through a straight channel in Figs 12.6 and 12.7 respectively.

Figure 12.6 shows a flow of fluid through a pipe one end of which is attached to a reservoir maintained with a constant height of water, and the other end to a converging nozzle that increases the velocity at the expense of pressure. The energy gradient line, as shown, fall gradually and continuously due to the frictional head loss in the pipeline and the nozzle. The hydraulic gradient line always runs below the energy gradient line, difference being the velocity head at the corresponding section. Since the velocity increases in the nozzle, the hydraulic gradient line falls sharply in that region. Both the energy gradient and hydraulic gradient lines meet on the reservoir surface where the velocity is negligible. If a pump or a turbine is fitted in the system, the energy gradient line would show an abrupt rise across the pump by an amount equal to the head developed, or an abrupt fall across the turbine by an amount equal to the head extracted. The hydraulic gradient line on the other hand, may show an abrupt rise or fall in a pipeline, if there occurs a sudden enlargement or contraction of the pipe at any section.

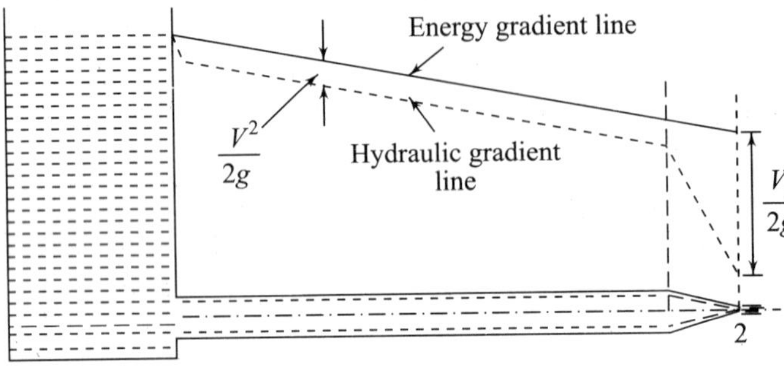

Fig. 12.6 Energy gradient and hydraulic gradient lines in case of a pipe flow

The energy gradient and hydraulic gradient lines in case of a channel flow are shown in Fig. 12.7. The hydraulic gradient line in this situation is the liquid surface itself. In case of uniform flow, the depth of the bed h, and accordingly the average velocity and kinetic energy correction factor, remain the same along the direction of flow in the channel ($h_1 = h_2$, $V_1 = V_2$, $\alpha_1 = \alpha_2$). As a consequence, the energy gradient line, the hydraulic gradient line or the liquid surface and the channel bed run parallel to each other. The loss of mechanical energy due to friction per unit length of the bed becomes h_f/l, where h_f is the total loss of head over the length of the channel l. The quantity h_f/l is termed as the *energy gradient* since it corresponds to the slope of the energy gradient line. For a uniform flow through a channel, the energy gradient becomes equal to the geometrical gradient of the channel bed and of the liquid surface.

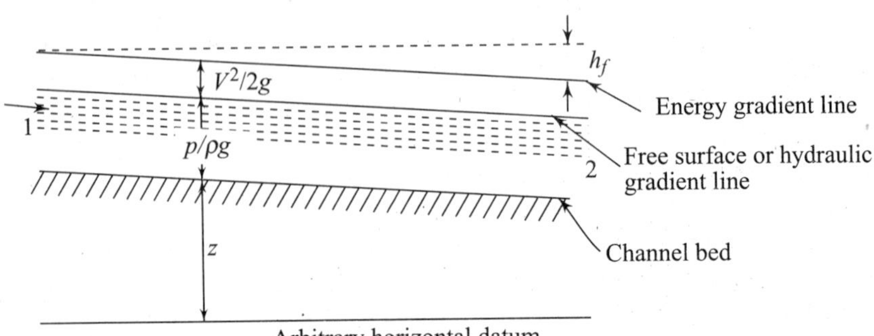

Fig. 12.7 Energy gradient and hydraulic gradient lines in case of a channel flow

12.2.5 Steady Uniform Flow—the Chezy Equation

A relationship between the average flow velocity and pressure drop in a steady uniform flow through a straight channel will now be developed. Let us consider a control volume *abcd* in a straight channel as shown in Fig. 12.8. The hydrostatic pressure forces at the surfaces *ab* and *cd* balance each other. The other forces acting on the control volume are the component of weight along the flow direction and the shear force at the solid boundary. Since the flow is steady and uniform, the rate of momentum influx to the control volume at *ab* is equal to the rate of momentum efflux from it at *cd*. Therefore, the net rate of momentum efflux from the control volume is zero. Hence, applying the momentum theorem to the control volume for a steady and uniform flow, we can write, as follows:

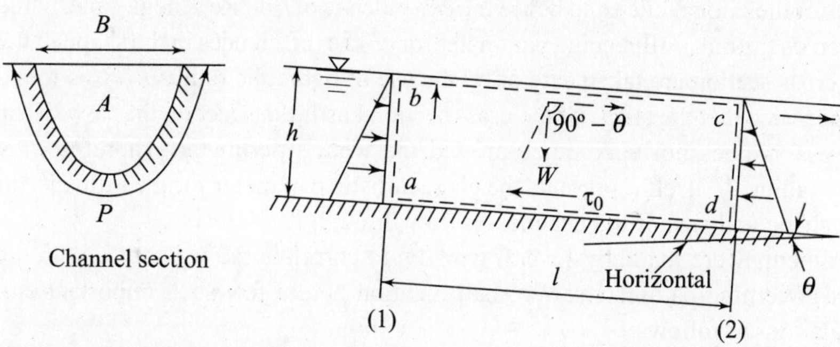

Fig. 12.8 Application of momentum theorem on a control volume in a uniform flow through a straight channel

The net force acting on the control volume in the direction of flow = Net rate of momentum efflux from the control volume = 0

or $\quad W \sin\theta - \tau_0 Pl = 0$

or $\quad \rho g Al \sin\theta = \tau_0 Pl$

or
$$\tau_0 = \rho g (A/P) \sin\theta = \rho g R_h S \tag{12.3}$$

where, τ_0 is the average shear stress at the solid boundary and S (= sin θ) is the slope of the bed of the channel. The hydraulic radius $R_h = A/P$ as defined in Section 12.2.1.

An expression to substitute τ_0 in terms of the average flow velocity V is needed. This is done by expressing τ_0 in terms of Fanning's friction factor f, as

$$\tau_0 = \frac{1}{2}\rho V^2 f \tag{12.4}$$

Moreover, in almost all cases of practical interest, the Reynolds number of flow in an open channel is sufficiently high where the shear stress at the boundary is proportional to the square of the average velocity and hence f remains constant. Combining Eq. (12.3) with (12.4), we have

$$V = (2g/f)^{1/2} (R_h S)^{1/2}$$

or
$$V = c\,(R_h S)^{1/2} \tag{12.5}$$

This is the well-known *Chezy equation*. The parameter $c = (2g/f)^{1/2}$ is called the *Chezy's coefficient* and has the dimension $L^{1/2}\,T^{-1}$. Although the validity of Eq. (12.5) has been experimentally verified for uniform flow only, it can also be used with reasonable accuracy for gradually varied flows. An expression similar to Eq. (12.3) can be derived for a non-uniform flow also.

Variation of Chezy Coefficient To determine the velocity V, from Chezy equation (Eq. 12.5), one has to know the value of c, the Chezy coefficient. In case of flow through pipes, as described in Chapter 11, the friction factor f, depends on both the Reynolds number Re, and on the relative roughness ε/d of the solid surface. Thus, Chezy's coefficient may be expected to depend on both Re and ε/R_h (R_h is the hydraulic radius), and also on the shape and size of the channel. The flow in open channels are fully turbulent in practice and hence the dependence of c on Re is negligible, while ε/R_h becomes the only influencing parameter for c. The differences in the shape of the channel cross section are taken care of by the use of hydraulic radius R_h. It is found from experience that the shape of the cross section has little effect on the flow, if the shear stress τ_0 does not vary much around the wetted perimeter. Therefore, the hydraulic radius R_h, itself represents the characteristic parameter for the influence of both the shape and size of the channel on flow through it.

Experiments were made by several workers to correlate the value of c with the pertinent governing parameters. We shall mention here a few such important empirical relations as follows:

$$c = \frac{23 + \dfrac{1}{n} + \dfrac{0.00155}{S}}{1 + \left(23 + \dfrac{0.00155}{S}\right)\dfrac{n}{R_h}} \tag{12.6}$$

[*Ganguillet–Kulter (G.K.) formula*]

$$c = (1/n)\,R_h^y \quad \text{[\textit{Pavlovskii formula}]} \tag{12.7}$$

where $y = 2.5\,n - 0.13 - 0.75\,R_h(n - 0.1)$

$$c = (1/n)R_h^{1/6} \quad \text{[\textit{Manning's formula}]} \tag{12.8}$$

The parameter n in all these formulae is the *roughness coefficient*. The hydraulic radius R_h has to be substituted in metre to get the value of c in $m^{1/2}$ s from the above relations. The simplest expression amongst all is the Eq. (12.8) due to Manning. The values of roughness coefficient n for a straight channel are shown in Table 12.1. It is interesting to note that all the formulae (Eq. 12.6, 12.7 and 12.8) give the same value for the Chezy's coefficient $c = 1/n$ at the unit hydraulic radius $R_h = 1$. Inserting the value of c from Eq. (12.8) into Eq. (12.5), the expression for velocity can be written as

$$V = (1/n)\,R_h^{2/3}\,S^{1/2} \tag{12.9}$$

This equation is widely used in calculating the flow velocity in an open channel because of its simplest form and yet good agreement with experiments.

Table 12.1 Values of Manning's Roughness Coefficient *n* for Straight Uniform Channels

Type of surface	*n*
Smooth cement, planed timber	0.010
Rough timber, canvas	0.012
Cast iron, good ashler masonry, brick work	0.013
Vertified clay, asphalt, good concrete	0.015
Rubble masonry	0.018
Firm gravel	0.020
Canals and rivers in good condition	0.025
Canals and rivers in bad condition	0.035

Example 12.1

The depth of a uniform steady flow of water in a 1.22 m wide rectangular cement lined channel laid on a slope of 4 m in 10000 m, is 610 mm. Find the rate of discharge using Manning's value for *c* (Chezy coefficient).

Solution

We have to use Eq. (12.9) for the present purpose.

Here, $$R_h = \frac{1.22 \times (0.61)}{1.22 + 2 \times 0.61} = 0.305\,\text{m} = 305\,\text{mm}$$

$$S(\text{slope}) = \frac{4}{10000} = 0.0004$$

n (from the Table 12.1) = 0.01
Therefore, from Eq. (12.9)

$$V = \frac{1}{0.01}\,(0.305)2/3\,(0.0004)^{1/2}$$

or $$Q = V \cdot A = \frac{1.22 \times 0.61}{0.01}\,(0.305)^{2/3}\,(0.0004)^{1/2}$$

$$= 0.674\,\text{m}^3$$

Example 12.2

A trapezoidal channel, having a bottom width of 6.096 m and side slopes 1 to 1, flows 1.219 m deep on a slope of 0.0009. Find the rate of uniform discharge. Take n (roughness coefficient) = 0.025.

Solution

Here A, the cross-sectional area of flow (Fig. 12.9)

$$= \frac{1}{2}[6.096 + 6.096 + 2 \times 1.219] \times 1.219$$

$$= (6.096 + 1.219) \times 1.219 = 8.917\ \text{m}^2$$

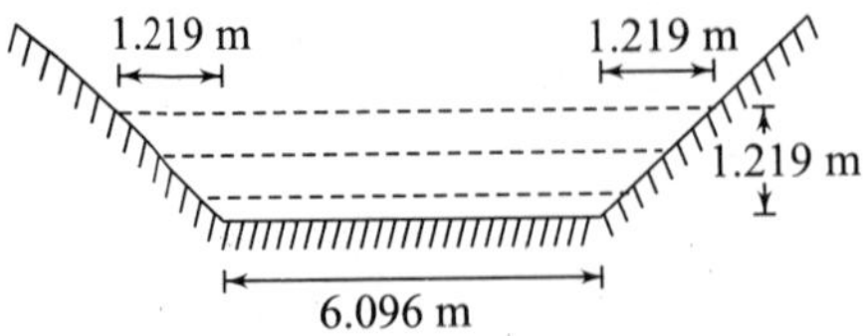

Fig. 12.9 A trapezoidal channel

Wetted perimeter $\quad P = 6.096 + 2(1.219/\cos 45°)$

$$= 6.096 + 2 \times 1.219 \times (2)^{1/2}$$

$$= 9.544\ \text{m}$$

Therefore, $\quad R_h = 8.917/9.544 = 0.934\ \text{m}$

Now we apply Eq. (12.9) to get the discharge as

$$Q = \frac{8.917 \times (0.934)^{2/3}(0.0009)^{1/2}}{0.025} = 10.22\ \text{m}^3/\text{s}$$

Example 12.3

How deep will water flow at the rate of 6.79 m^3/s in a rectangular channel 6.1 m wide, laid on a slope of 0.0001? Use $n = 0.0149$.

Solution

Let the depth be h

Then A (cross-sectional areas of flow) = $6.1 \times h$

P (wetted perimeter) = $6.1 + 2h$

Therefore, R_h (Hydraulic radius) = $\dfrac{6.1 \times h}{6.1 + 2h}$

By making use of Eq. (12.9),

$$6.79 = \frac{6.1h}{0.0149}\left(\frac{6.1 \times h}{6.1 + 2h}\right)^{2/3}(0.0001)^{1/2}$$

or
$$1.66 = h\left(\frac{6.1h}{6.1+2h}\right)^{2/3} \tag{12.10}$$

The value of h is found out from this equation by the method of successive trails. Equation (12.10) is therefore written, for this purpose, as

$$h = 1.66\left(\frac{6.1+2h}{6.1h}\right)^{2/3} \tag{12.11}$$

For a first trial, let us put $h = 1.50$ in the RHS. of Eq. (12.11) and get

$$h^1 = 1.65$$

Superscript on h indicates the number of trials.

Now we put h^1 in the RHS of Eq. (12.11) for the second trial to get a new value of h as

$$h^2 = 1.59$$

Putting this value of h in Eq. (12.11) we obtain

$$h^3 = 1.61$$

Putting the new value of h again in Eq. (12.11) we obtain

$$h^4 = 1.60$$

The difference between the two successive values of h now becomes 0.62%. Therefore, we can write the final value of the depth h as 1.60 m.

Example 12.4

Show that the vertical distribution of velocity is parabolic for a uniform laminar flow in a wide open channel with constant slope and depth of flow.

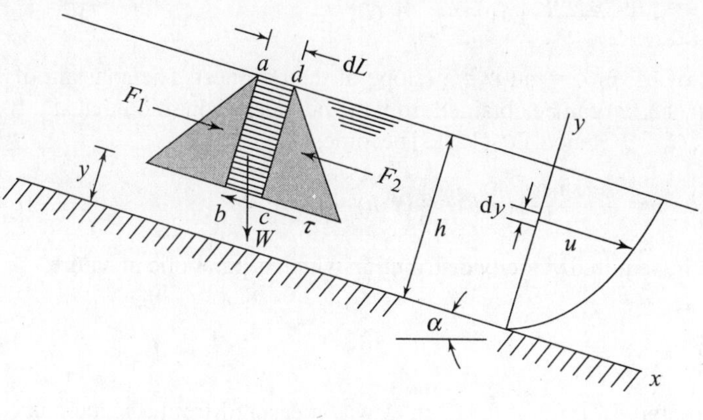

Fig. 12.10 A uniform laminar flow in an open channel with constant slope and depth of flow

Solution

Let the depth of flow be h (Fig. 12.10). A control volume *abcda* of length dL and of width B (the width of the channel) is taken as shown in the figure. Now we have to apply the momentum theorem to this control volume.

The forces acting on the surfaces *ab* and *cd* are the hydrostatic pressure forces as shown in the figure.

Let F_1 and F_2 be the hydrostatic pressure forces on these two surfaces *ab* and *cd* respectively.

Therefore the net force acting on the control volume in the direction of flow can be written as

$$F_x = F_1 - F_2 + \rho g(h - y)\, \mathrm{d}L\, B \sin\alpha - \tau\, \mathrm{d}L\, B$$

Since

$$F_1 = F_2$$

$$F_x = \rho\, g(h - y)\, \mathrm{d}L\, B \sin\alpha - \tau\, \mathrm{d}L\, B$$

For a steady uniform flow, the momentum coming into the control volume across the face *ab* is equal to that leaving from the control volume across the face *cd*. Therefore the net rate of momentum efflux from the control volume is zero.

Hence, we can write, from the momentum theorem applied to the control volume *abcda*,

$$F_x = \rho\, g(h - y)\, \mathrm{d}L\, B \sin\alpha - \tau\, \mathrm{d}L\, B = 0$$

which gives,

$$\tau = \rho\, g(h - y) \sin\alpha \qquad (12.12)$$

For a laminar flow,

$$\tau = \mu \frac{\mathrm{d}u}{\mathrm{d}y}$$

Substituting the expression of τ in the Eq. (12.12), we get

$$\mathrm{d}u = \frac{\rho g}{\mu}(h - y) \sin\alpha\, \mathrm{d}y$$

or

$$u = \frac{\rho g \sin\alpha}{\mu}\left(hy - \frac{y^2}{2}\right) + C_1 \qquad (12.13)$$

For small values of α, $\sin\alpha = \tan\alpha = S$ (slope of the channel). The constant of integration C_1 in Eq. (12.13) can be obtained from the boundary condition that at $y = 0$, $u = 0$, which gives $C_1 = 0$. Hence, Eq. (12.13) becomes

$$u = \frac{\rho g \sin\alpha}{\mu}\left[(y/h) - \frac{1}{2}(y/h)^2\right] \qquad (12.14)$$

Equation (12.14) is the required velocity distribution which is parabolic in nature.

Example 12.5

In a hydraulics laboratory, a flow of 0.412 m^3/s was measured from a rectangular channel flowing 1.22 m wide and 0.61 m deep. If the slope of the channel was 0.0004, find its roughness factor using Manning's formula.

Solution

Here,

$$A = 1.22 \times (0.61)$$
$$= 0.7442\,\text{m}^2$$
$$P = 1.22 + 2 \times 0.61 = 2.44$$

Therefore, $R_h = A/P = 0.305\,\text{m}$

Using Eq. (12.9)

$$Q = 0.412 = \frac{1.22 \times 0.610}{n} (0.305)^{2/3} (0.0004)^{1/2}$$

which gives $n = 0.0163$

12.2.6 Optimum Hydraulic Cross Section

With the help of Manning's equation [Eq. (12.9)] for the velocity of flow, we can write the expression for volumetric discharge rate through an open channel as

$$Q = \frac{A}{n} R_h^{2/3} S^{1/2} = \frac{A^{5/3} S^{1/2}}{n P^{2/3}} \tag{12.15}$$

A typical application of the above equation in the design of artificial canal for uniform flow is the economical proportioning of its cross section. It may be observed from Eq. (12.15) that the discharge rate Q, is maximum when the wetted perimeter is minimum for a given flow area. The most efficient cross section, from the hydraulic point of view, is semicircular as it has the least wetted perimeter among all sections with the same flow area. It is desirable to use such a section not only for the sake of obtaining the maximum discharge for a given cross-sectional area, but for the sake of economy due to the fact that a minimum wetted perimeter requires a minimum of lining material. The cross-sectional area of a channel under this condition is known as *the optimum hydraulic cross section.* The condition is characterised by the maximum value of the hydraulic radius $R_h = A/P$. Although a semi circular channel has the maximum hydraulic mean radius and it is built from prefabricated sections, the semicircular shape is impractical for other forms of construction. Trapezoidal sections on the other hand, are very popular. We should therefore find out the condition for maximum hydraulic mean radius for a trapezoidal section as follows:

Let us consider a trapezoidal section, as shown in Fig. 12.11.

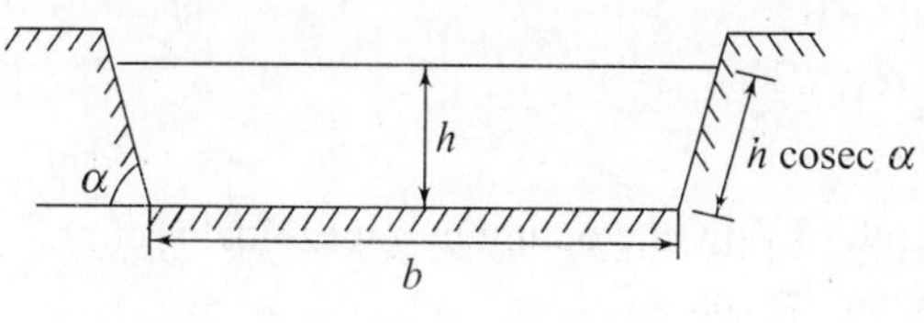

Fig. 12.11 Section of a trapezoidal channel

Cross-sectional area of flow $A = bh + h^2 \cos \alpha$

Wetted perimeter $P = b + 2h \operatorname{cosec} \alpha$

Since $b = (A/h) - h \cot \alpha$,

$$R_h = \frac{A}{P} = \frac{A}{(A/h) - h\cot\alpha + 2h\operatorname{cosec}\alpha} \tag{12.16}$$

For a given value of A, the expression is a maximum when its denominator is a minimum. This is found from the consideration

$$\frac{\mathrm{d}}{\mathrm{d}h}[(A/h) - h\cot\alpha + 2h\operatorname{cosec}\alpha] = 0$$

or
$$-(A/h^2) - \cot\alpha + 2\operatorname{cosec}\alpha = 0$$

or
$$A = h^2(2\operatorname{cosec}\alpha - \cot\alpha) \tag{12.17}$$

The second derivative, $2A/h^3$, is clearly positive and so the condition is indeed for a minimum of the denominator of Eq. (12.16), and hence for a maximum of R_h. Substituting the value of A from Eq. (12.17) in the expression for R_h, i.e., into Eq. (12.16), we have

$$R_h = \frac{h^2(2\operatorname{cosec}\alpha - \cot\alpha)}{h(2\operatorname{cosec}\alpha - \cot\alpha) - h\cot\alpha + 2h\operatorname{cosec}\alpha}$$

$$= \frac{h^2(2\operatorname{cosec}\alpha - \cot\alpha)}{2h(2\operatorname{cosec}\alpha - \cot\alpha)} = \frac{h}{2} \tag{12.18}$$

In other words, for maximum efficiency, a trapezoidal channel should be so proportioned that its hydraulic mean radius is half the central depth of flow. Since a rectangle is a special case of a trapezium (with $\alpha = 90°$), the optimum proportions for a rectangular section is given by $R_h = h/2$, and from Eq. (12.17), $A = 2h^2$ which finally gives that the width of the rectangle $B = 2h^2/h = 2h$.

If, instead of depth of flow, the side slope is varied to give the optimum cross section, i.e. maximum R_h, then we can find the required condition as

$$\frac{\mathrm{d}}{\mathrm{d}\alpha}[(A/h) - h\cot\alpha + 2h\operatorname{cosec}\alpha] = 0$$

or
$$h\operatorname{cosec}\alpha(\operatorname{cosec}\alpha - 2\cot\alpha) = 0$$

Since
$$h \neq 0,$$

$$\operatorname{cosec}\alpha - 2\cot\alpha = 0$$

or
$$\cos\alpha = 1/2$$

which gives
$$\alpha = 60° \tag{12.19}$$

This concludes that, for a given depth of flow the optimum trapezoidal section, given by maximum R_h, is half of a regular hexagon.

Example 12.6

(i) Determine the most efficient section of trapezoidal channel, $n = 0.025$, to carry 12.74 m^3/s. To prevent scouring, the maximum velocity is to be 0.92 m/s and the side slopes of the trapezoidal channel are 1 vertical to 2 horizontal. (ii) What slope S of the channel is required?

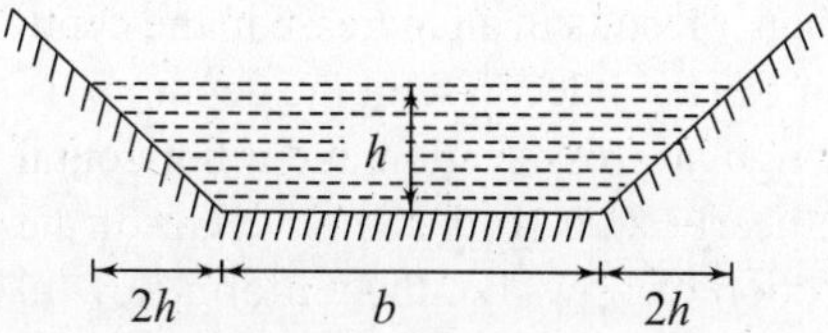

Fig. 12.12 A trapezoidal channel

Solution

(i) It is known from Eq. (12.18) that for the most efficient section (the minimum wetted perimeter for a given discharge) of a trapezoidal channel

$$R_h = h/2$$

where R_h is the hydraulic radius and h is the depth of flow (Fig. 12.12). Hence we can write,

$$R_h = h/2 = A/P = \frac{bh + 2(h/2)(2h)}{b + 2h(5)^{1/2}}$$

or
$$b = 2h\,(5)^{1/2} - 4h \quad (12.20)$$
$$= 0.472\,h$$

where b is the width at the base (Fig. 12.12). Again, from continuity, the cross-sectional area to accommodate the maximum velocity is given by

$$A = 12.74/0.92 = bh + 2h^2$$

or
$$b = (13.85 - 2h^2)/h \quad (12.21)$$

Equating (12.20) and (12.21), we get

$$h = 2.37\text{ m} \quad \text{and} \quad b = 1.12\text{ m}$$

(ii) Using Manning's equation, i.e. Eq., (12.9), for this trapezoidal channel with $b = 1.12$ m, $h = 2.37$ m and $n = 0.025$, we can write

$$0.92 = \frac{(2.37/2)^{2/3}\ S^{1/2}}{0.025}$$

or
$$S = 0.00042$$

12.2.7 Propagation of Waves by Small Disturbances in Open Channels

Any temporary disturbance of a free surface produces waves; for example, a stone dropped into a pond, removal or insertion of an obstruction like sudden opening or closing of a sluice gate in a river causes waves which are propagated upstream and downstream of the source of disturbances. The depth of water in a channel is considered to be shallow when it is small as compared to the length of a wave on its surface. Again, a wave is termed as a positive wave when it results in an increase in the depth of stream, and is termed as a negative one if it causes a decrease in the depth.

We consider an open channel with a rectangular cross section and a horizontal base. The slope of the bed is assumed to be nearly zero so that the weight of the liquid has a negligible component in the direction of flow. Let the uniform flow in the channel, represented by velocity V_1 and depth h_1, (Fig. 12.13(a)) be disturbed by a small disturbance, for example, the closing of a gate downstream, so that a positive wave travels upstream with a constant velocity C (relative to the bed of the channel). Due to the disturbance, the changes in downstream conditions of the flow are considered in a sense that a short distance downstream of the wave the flow has again become uniform with altered values of velocity and depth as V_2 and h_2, respectively (Fig. 12.13(a)).

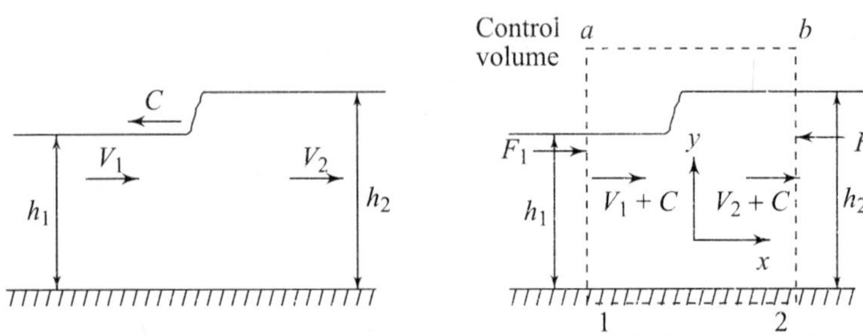

(a) Propagation of a wave in an open channel

(b) A typical control volume in the analysis of propagation velocity of a surface wave in a channel flow

Fig. 12.13

The change in velocity from V_1 to V_2 caused by the passage of the wave is the result of a net force acting on the fluid. The magnitude of this force can be found out by applying the momentum theorem to a control volume enclosing the wave. Such a control volume $1ab21$ shown in Fig. 12.13(b) is taken for our analysis. In order of make the flow steady, the frame of reference is chosen where the surge wave is stationary while the fluid upstream and downstream the wave moves with velocity $V_1 + C$ and $V_2 + C$, respectively. In other words, we can say, that the coordinate axes for the analysis, are considered to be fixed with the moving wave. The net force acting on the control volume in the x direction is due to the forces acting on the surfaces 1a and 2b. The forces acting on these surfaces (Fig. 12.13(b)) are the hydrostatic pressure forces and can be written as

$$F_1 = \frac{\rho g h_1^2}{2} \text{ and } F_2 = \frac{\rho g h_2^2}{2}$$

Here the width of the channel has been considered to be unity.
Therefore, the net force in the x direction on the control volume,

$$= \frac{\rho g h_1^2 - \rho g h_2^2}{2} \tag{12.22}$$

The net rate of x momentum efflux from the control volume

$$= \rho Q\,(V_2 - V_1) \tag{12.23}$$

where Q is the rat of volumetric discharge
From continuity,

$$Q = (V_1 + C)h_1 = (V_2 + C)h_2 \tag{12.24a}$$

which gives $$V_2 = (V_1 + C)\,\frac{h_1}{h_2} - C \tag{12.24b}$$

For a steady flow, the momentum theorem as applied to the control volume 1ab2 gives

$$\frac{\rho g}{2}\,(h_1^2 - h_2^2) = \rho Q\,(V_2 - V_1) \tag{12.25}$$

Substituting for Q and V_2 from Eqs (12.24a) and (12.24b) respectively into Eq. (12.25) we get

$$\frac{\rho g}{2}\,(h_1^2 - h_2^2) = \rho(V_1 + C)h_1\left[(V_1 + C)\frac{h_1}{h_2} - C - V_1\right]$$

$$= \rho(V_1 + C)^2\,\frac{h_1}{h_2}\,(h_1 - h_2)$$

which gives, $$V_1 + C = (gh_2)^{1/2}\left[\frac{1 + h_2/h_1}{2}\right]^{1/2} \tag{12.26}$$

If the height of the wave is considered to be small, which is usually true for waves created by small disturbances, then, $h_2 = h_1 = h$, and Eq. (12.26) can be written as

$$V_1 + C = (gh)^{1/2} \tag{12.27}$$

This equation implies that the velocity of the wave relative to the undisturbed liquid is $(gh)^{1/2}$. Though this derivation applies only to waves propagated in rectangular channels, it can be shown that, for channels with different types of cross section, the velocity of propagation of a small wave is $(g\bar{h})^{1/2}$ relative to the undisturbed liquid, where $\bar{h}$ is the mean depth given by

$$\bar{h} = \frac{\text{Area of cross section}}{\text{Width of the liquid surface}} = \frac{A}{B} \tag{12.28}$$

Equation (12.27) gives the velocity of a wave whose height is small. A larger wave will be propagated with a higher velocity than that given by the Eq. (12.27). Moreover, the height of the wave does not remain constant over an appreciable distance due to frictional effects. In the derivation of Eq. (12.27), the effect of friction has been assumed to be negligible for the control volume 1ab2 (Fig. 12.13(b)), since the distance between the sections 1*a* and 2*b* are considered to be very small. The present analysis is, however, valid for a shallow depth.

12.2.8 Specific Energy and Alternative Depth of Flow

Definition of Specific Energy In the definition of total energy of a flowing fluid, a reference horizontal datum is chosen arbitrarily so that the potential energy of a fluid element is prescribed from the datum. In a channel flow, the sum of the pressure head, $p/\rho g$ and the potential head, z, (measured from any horizontal datum) is equivalent to the height of free surface above the datum, and the total energy is therefore equivalent to this height plus the height corresponding to velocity head $V^2/2g$ as already explained in Fig. 12.7 in Section 12.2.4. Specific energy of a fluid element at any point in a channel flow is defined as its total energy per unit weight where the component potential energy is measured from the base or bed of the channel as the datum. Therefore, specific energy E_s, at any section of the channel if given by

$$E_s = h + \frac{V_{av}^2}{2g} \tag{12.29}$$

where V_{av} represents the average flow velocity. If A and Q are the cross sectional area and rate of volumetric flow respectively at the section considered, then

$$V_{av} = Q/A \tag{12.30a}$$

Again, if the width of the channel at that section is b, then

$$A = b\,h \tag{12.30b}$$

With the help of Eqs (12.30a) and (12.30b), Eq. (12.29) can be written as

$$E_s = h + \left(\frac{q^2}{2g}\right)\frac{1}{h^2} \tag{12.31}$$

where $\qquad q = Q/b$

Equation (12.31) relates the *specific energy* with the depth of flow and the discharge per unit width. Out of the three variables E_s, q and h, any two can vary independently and the third one becomes dependent by Eq. (12.31). Our particular interest centres around the instances in which q is constant while h and E_s vary, i.e., how the specific energy varies with depth of flow for a given rate of discharge. If E_s is plotted against h for a constant value of q, we get a curve, as shown in Fig. 12.14, which is known as the *specific energy diagram*. At small values of h, the second term on the right-hand side of Eq. (12.31) becomes predominant over the first one and then E_s becomes an inverse function of h with $E_s \rightarrow \propto$ as $h \rightarrow 0$. Therefore, this part of the specific energy curve becomes asymptotic to the E_s axis. Conversely, as h increases, the velocity becomes smaller and the second term $(q^2/2g)\,\dfrac{1}{h^2}$ becomes insignificant compared to the first term h and therefore E_s varies directly with h in this region and finally becomes asymptotic to the line $E_s = h$. Between these two extremes, there is clearly a minimum value of E_s. The depth of flow at which the minimum value of E_s occurs is known as *critical depth* h_c. The value of $E_{s_{\min}}$ and h_c can be found out as follows:

For E_s to be minimum, we can write from Eq. (12.31)

$$\partial E_s/\partial h = 1 + \frac{q^2}{2g}(-2/h^3) = 0$$

which gives $\qquad h = (q^2/g)^{1/3}$

This value of h is the critical depth h_c and hence we can write

$$h_c = (q^2/g)^{1/3} \tag{12.32}$$

The corresponding minimum value of E_s is obtained by substituting q in terms of h_c from Eq. (12.32) into Eq. (12.31) as

$$E_{s_{\min}} = h_c + (h_c^3/2h_c^2) = \frac{3}{2}\,h_c \tag{12.33}$$

We now examine another interesting case where h and q vary while specific energy E_s is kept constant. Again, with the help of Eq. (12.31), the curve of h against q for constant E_s is drawn as shown in Fig. 12.15. Here we observe that q reaches a maximum at a given value of h which indicates a maximum discharge for a given specific energy. To obtain this condition, we first write the Eq. (12.31) in a form

$$q^2 = 2gh^2\,(E_s - h)$$

For maximum discharge,

$$2q\,\frac{\partial q}{\partial h} = 2g(2E_s\,h - 3h^2) = 0$$

which gives

$$h = \frac{2}{3}\,E_s \tag{12.34}$$

From Eqs (12.33) and (12.34) we conclude that, at the critical depth, either the discharge is maximum for a given specific energy or the specific energy is minimum for a given discharge.

Critical Velocity The velocity of flow at the critical depth is known as *critical velocity* in case of a channel flow. Since the velocity of flow $V = Q/bh = q/h$, the critical velocity V_c may be determined from Eq. (12.32) as

$$V_c = q/h_c = \frac{(gh_c^3)^{1/2}}{h_c} = (gh_c)^{1/2} \tag{12.35}$$

Though the expressions for critical depth and critical velocity have been derived here for a rectangular channel, the same results can be obtained for a channel with any shape of section provided the mean depth $\bar{h}$ as defined by Eq. (12.28) is used in place of depth of flow h for a rectangular cross Section

Physical Implication of Critical Velocity and Definition of Tranquil and Rapid Flow The most important outcome of critical velocity is that it separates two distinct types of flow—one in which the velocity is less than the critical value, and the other in which the velocity exceeds the critical value. We find that for each value of E_s other than the minimum (Fig. 12.14), and for each value of q other than the maximum (Fig. 12.15), there are two possible values of h, one greater and one less than h_c (although 12.31) is a cubic in h, the third root is always negative and is therefore physically meaningless). These two values of h are known as *alternative depths*. When $h < h_c$, the flow velocity V is greater than V_c and when $h > h_c$, V is less than V_c. Before examining the physical significance of these two regimes of flow given by $V > V_c$ and $V < V_c$, we first find the physical implication of the critical velocity. We have shown in Section 12.2.7 that the velocity of propagation (relative to the undisturbed liquid) of a small surface wave in a shallow liquid equals to $(gh)^{1/2}$, where h is the mean depth of flow in case of varying cross section, or simply the depth of flow, in case of a rectangular cross section throughout the channel. A surface wave can be caused by any small disturbance to the flow. Hence the surface wave can be considered as a messenger, propagated against the flow, for the liquid upstream to be informed about the disturbances downstream so that it can change its behaviour accordingly. The absolute velocity of surface wave propagating upstream will be $(gh^{1/2} - V)$, which is positive when $V < gh^{1/2}$ and negative when $V > gh^{1/2}$. This implies physically that when the flow velocity is less than the critical velocity, the surface wave will have the opportunity to reach the upstream and to influence the upstream liquid by the disturbances downstream. On the other hand, when the flow velocity is greater than the critical velocity, the surface wave cannot propagate upstream and hence the information about events downstream is never conveyed upstream. When the flow velocity is equal to the critical velocity, a small wave which tries to travel upstream cannot progress, since $(gh^{1/2} - V)$ becomes zero. The wave under this situation is known as a standing wave.

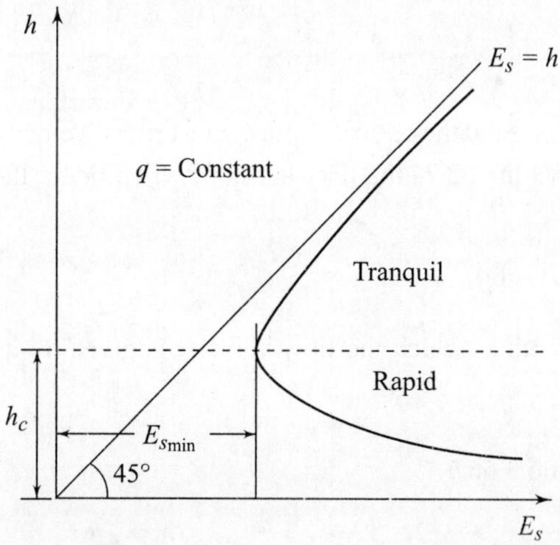

Fig. 12.14 Variation of specific energy with the depth of flow for a given discharge

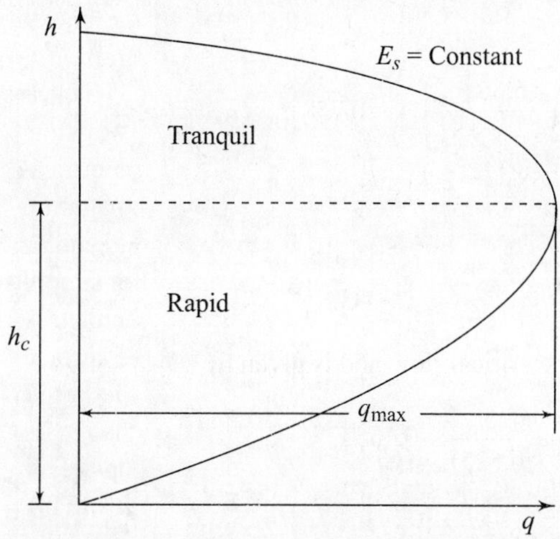

Fig. 12.15 Variation of discharge with depth of flow for a given specific energy

These three regimes of flow can be characterised by a dimensionless parameter defined as the ratio of flow velocity to the critical velocity $V/(gh)^{1/2}$. We have already seen in Section 6.2 that this dimensionless term is known as Froude number Fr, where $\text{Fr} = V/(gh)^{1/2}$.

Flow in which the velocity V, is less than the critical velocity $(gh)^{1/2}$, i.e., when Froude number $\text{Fr} < 1$, is referred to as tranquil flow. Flow in which the velocity V is greater than the critical velocity, i.e., when $\text{Fr} > 1$, is termed as rapid or shooting flow. The flow in which the velocity is equal to the critical velocity, i.e., when $\text{Fr} = 1$ is known as critical flow.

Example 12.7

A rectangular channel carries 5.66 m^3/s. Find the critical depth h_c and critical velocity V_c for (i) a width of 3.66 m and (ii) a width of 2.74 m, (iii) what slope will produce the critical velocity in (i) if $n = 0.020$?

Solution

(i) Critical depth is defined as the depth at which the flow velocity is given by its critical value as

$$V_c = (gh_c)^{1/2}$$

again, $$V_c = Q/bh_c = 5.66/3.66\, h_c$$

Therefore, $$5.66/3.66\, h_c = (gh_c)^{1/2}$$

or $$h_c = \left[\frac{5.66 \times 5.66}{3.66 \times 3.66 \times 9.81}\right]^{1/3} = 0.625 \text{ m}$$

Now, $$V_c = (9.81 \times 0.625)^{1/2} = 2.48 \text{ m/s}$$

(ii) When the width is 2.74 m

$$h_c = \left[\frac{5.66 \times 5.66}{2.74 \times 2.74 \times 9.81}\right]^{1/3} = 0.758 \text{ m}$$

and, $$V_c = (9.81 \times 0.758)^{1/2} = 2.73 \text{ m/s}$$

(iii) Applying Eq. (12.9), we can write

$$V_c = \frac{R_c^{2/3}\, S^{1/2}}{n}$$

where R_c is the hydraulic radius at the critical flow and is given by

$$R_c = \frac{3.66 \times 0.625}{(3.66 + 2 \times 0.625)} = 0.466$$

Hence, $$2.48 = \frac{(0.466)^{2/3}}{0.02}\, S^{1/2}$$

which gives $$S = 0.0068$$

Example 12.8

A rectangular channel, 9.14 m wide, carries 7.64 m^3/s when flowing 914 mm deep. (i) What is the specific energy? (ii) Is the flow tranquil or rapid?

Solution

(i) We know from Eq. (12.31) that

$$E_s = h + \left(\frac{q^2}{2g}\right)\frac{1}{h^2}$$

or $$E_s = 0.914 + \frac{1}{2 \times 9.81 \times (0.914)^2}\left(\frac{7.64}{9.14}\right)^2$$

$$= 0.957\,\text{m}$$

(ii) From Eq. (12.32),

$$h_c = \left[\left(\frac{7.64}{9.14}\right)^2 \frac{1}{9.81}\right]^{1/3} = 0.415\,\text{m} = 415\,\text{mm}$$

Therefore, the flow is tranquil since the depth of flow is grater than the critical depth.

Example 12.9

A trapezoidal channel has a bottom width of 6.1 m and side slope of 2 horizontal to 1 vertical. When the depth of water is 1.07 m, the flow is 10.47 m^3/s. (i) What is the specific energy of flow? (ii) Is the flow tranquil or rapid?

Solution

The cross sectional area of flow,

$$A = 6.1(1.07) + 2\left(\frac{1}{2}\right)(1.07)(2.14) = 8.82\,\text{m}^2$$

From Eq. (12.31), we can write

$$E_s = h + \frac{1}{2g}(Q/A)^2$$

where Q is the volumetric flow rate.

Hence, $$E_S = 1.07 + \frac{1}{2 \times 9.81}\left(\frac{10.47}{8.82}\right)^2 = 1.14\,\text{m}$$

To determine the critical depth, we have to first find out a similar relation as given in Eq. (12.32) for a channel whose width varies with the depth. For this purpose, we start with Eq. (12.31) as

$$E_S = h + \frac{1}{2g}(Q/A)^2$$

where Q is the flow rate and A is the cross-sectional area. At critical condition, i.e., for minimum specific energy.

$$\frac{\mathrm{d}E_S}{\mathrm{d}h} = 1 + \frac{Q^2}{2g}\left(-\frac{2}{A^3}\frac{\mathrm{d}A}{\mathrm{d}h}\right) = 0$$

substituting $\mathrm{d}A = B'\,\mathrm{d}h$ (B' is the width at the water surface), we get

$$(Q^2 B')/(g A_c^3) = 1$$

or $$Q^2/g = A_c^3/B'$$

From the geometry of the channel, $A_c = 6.1\,h_c + 2\,h_c^2$

and $$B' = 6.1 + 4\,h_c$$

(where h_c is the critical depth and A_c is the corresponding cross-sectional area of flow)
Therefore,

$$(10.47)^2/9.81 = (6.1\,h_c + 2\,h_c^2)^3/(6.1 + 4\,h_c)$$

Solving by trial, $h_c = 0.625\,\text{m}$

Since the actual depth exceeds the critical one, the flow is tranquil.

12.3 FLOW IN CLOSED CIRCULAR CONDUITS ONLY PARTLY FULL

Flows in closed conduits partly full are usually encountered in practice, namely, in drains and sewers. Since the liquid has a free surface inside the conduits, the flow is governed by the principles of channel flow. There are however some special characteristic features of the flow which result from the convergence of the boundary to the top.

Let us consider a circular conduit of diameter d, partly full of liquid flowing through it. Let the angle subtended by the free surface at the centre of the conduit be 2θ as shown in Fig. 12.16(a).

The area of cross section of the liquid

$$A = \frac{d^2\theta}{4} - 2\left(\frac{1}{2}\frac{d}{2}\sin\theta\frac{d}{2}\cos\theta\right)$$

$$= \frac{d^2}{4}\left(\theta - \frac{1}{2}\sin 2\theta\right) \qquad (12.36)$$

The wetted perimeter $P = d\theta$

Therefore, the hydraulic radius $R_h = A/P = \dfrac{d}{4}\left(1 - \dfrac{1}{2}\dfrac{\sin 2\theta}{\theta}\right)$

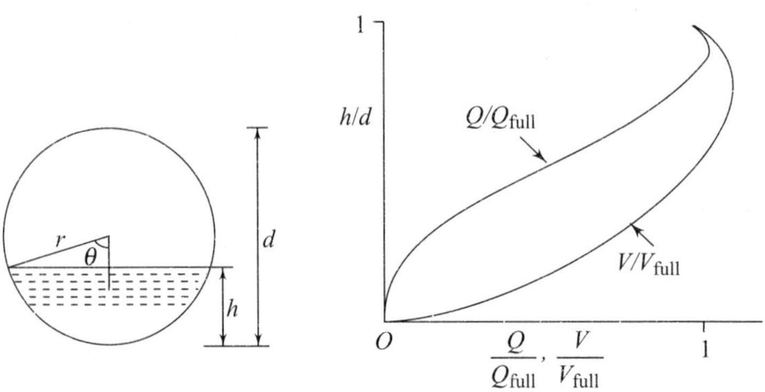

(a) Flow through a closed conduit partly full

(b) Variation of discharge and velocity with depth of flow for a closed conduit

Fig. 12.16

The rate of discharge may be calculated from *Manning's equation* (Eq. 12.9) as

$$Q = \frac{d^2}{4}\left(\theta - \frac{1}{2}\sin 2\theta\right)\left(\frac{1}{n}\right)S^{1/2}\left\{\frac{d}{4}\left(1 - \frac{1}{2}\frac{\sin 2\theta}{\theta}\right)\right\}^{2/3}$$

$$= K\left(\theta - \frac{\sin 2\theta}{2}\right)\left(1 - \frac{\sin 2\theta}{2\theta}\right)^{2/3} \tag{12.37}$$

where the constant $K = \frac{d^{8/3}}{4^{5/3}}\frac{S^{1/2}}{n}$ (12.38)

The rate of discharge for the conduit flowing full can be obtained by putting $\theta = \pi$ in Eq. (12.37) as

$$Q_{full} = K\pi$$

The rate of discharge Q is usually expressed in a dimensionless form as

$$\frac{Q}{Q_{full}} = \frac{1}{\pi}\left(\theta - \frac{\sin 2\theta}{2}\right)\left(1 - \frac{\sin 2\theta}{2\theta}\right)^{2/3} \tag{12.39}$$

In a similar fashion we can also write,

$$\frac{V}{V_{full}} = \left(1 - \frac{\sin 2\theta}{2\theta}\right)^{2/3} \tag{12.40}$$

The depth of flow h (Fig. (12.16(a)) can be expressed in a dimensionless from h/d as

$$\frac{h}{d} = \frac{1}{2} - \frac{1}{2}\cos\theta \tag{12.41}$$

The variations of Q/Q_{full}, and V/V_{full} with h/d are shown in Fig. 12.16(b). The maximum value of Q/Q_{full} is found to be (from Eq. (12.39)) 1.08, at $h/d = 0.94$. This indicates that the rate of discharge through a conduit is more in case of conduit partly full with $h/d = 0.94$ than that in the case of the conduit flowing full. Similarly, it is found from Eq. (12.40) that the maximum value of $V/V_{full} = 1.14$ at $h/d = 0.81$. The physical explanation for this can be attributed to the typical variation of Chezy's coefficient with the hydraulic radius R_h, in Manning's formula. However, the values are based on the assumption that Manning's roughness coefficient n is independent of the depth of flow. In practice, n tends to decrease with increasing flow depth. For this reason, the experimental results differ slightly from the theoretical values with constant n, and show the maximum discharge and velocity at $h/d = 0.97$ and 0.83, respectively. Under fluctuating condition of discharge, in practice, low velocity may cause deposition of solids in the conduit whereas high velocity at large depths of flow may cause excessive scour. This is rectified to some extent by changing the shape of the conduit from circular to oval - or egg-shaped sections as shown in Fig. 12.17.

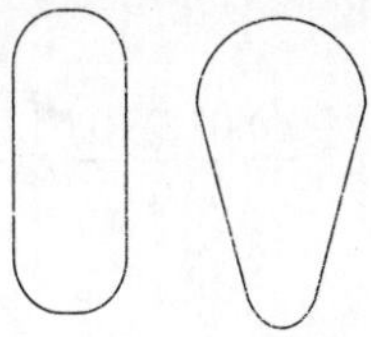

Fig. 12.17 Commonly used sections for fluctuating flows through conduits partly full

Example 12.10

A circular culvert has a capacity of 0.5 m^3/s when flowing full. Velocity should not be less than 0.7 m/s if the depth is one-fourth of the diameter. Assuming uniform flow, find the diameter and the slope, taking Manning's roughness coefficient $n = 0.012$.

Solution

Putting $h/d = 1/4$ in Eq. (12.41), we get

$$\frac{1}{4} = \frac{1}{2} - \frac{1}{2}\cos\theta$$

or $$\cos\theta = \frac{1}{2}$$

which gives $$\theta = \pi/3 \text{ radians.}$$

From Eq. (12.40), we get

$$\frac{V}{V_{\text{full}}} = \left[1 - \frac{\sin 2\pi/3}{2\pi/3}\right]^{2/3}$$

$$= 0.70$$

Hence, $$V_{\text{full}} = \frac{V}{0.70} = \frac{0.70}{0.70} = 1 \text{ m/s}$$

From continuity $$Q_{\text{full}} = \frac{\pi}{4} d^2 V_{\text{full}}$$

or $$0.5 = \frac{\pi}{4} d^2 \times 1$$

which gives d, the diameter for the culvert = 0.798 m. When flowing full, the hydraulic radius

$$R_h = A/P = d/4 = 0.798/4 = 0.1995 \text{ m}$$

From Eq. (12.9)

$$1 = \frac{(0.1995)^{2/3}}{0.012} S^{1/2}$$

or $$S = \frac{(0.012)(0.012)}{(0.1995)^{4/3}} = 0.0012$$

12.4 HYDRAULIC JUMP

A sudden transition from a rapid flow to a tranquil flow is known as *hydraulic jump*. A rapid flow, in practice may occur due to the release of liquid in a channel at high velocity under a sluice gate or at the foot of a steep spillway. If this flow has to be decelerated to a uniform tranquil flow due to some obstruction downstream or by the roughness of the boundary of a long channel with a mild slope, then the only possible way is a sudden change from rapid to tranquil flow at some location rather than a gradual transition via the critical condition. This can be explained in the following way:

A deceleration of flow is accompanied by an increase in the depth of flow which, in the regime of rapid flow decreases the specific energy (Fig. 12.14). If this increase in depth continues beyond the critical value then the specific energy has to increase (Fig. 12.14) which is not possible under the circumstances without any addition of energy from outside. Hence the specific energy may only decrease. Therefore a demand from a rapid flow to a uniform tranquil flow due to some resistance downstream in a channel is met only through a sudden transition before the critical condition is reached. This is known as hydraulic jump. It represents a typical discontinuity in the flow (Fig. 12.18(a)) during which the usual specific energy-depth of flow relation is invalid. The process of hydraulic jump is highly irreversible and is shown by the path 1-2 in Fig. (12.18(b)). The hydraulic jump results in the formation of eddies and turbulences which are responsible for the loss of mechanical energy h_j (Fig. 12.15(b)). The most important task in this context is to determine the relationship between the depths of flow before and after the hydraulic jump.

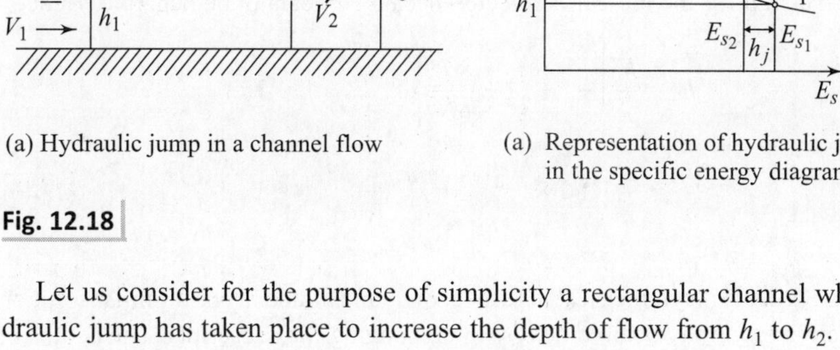

(a) Hydraulic jump in a channel flow

(a) Representation of hydraulic jump in the specific energy diagram

Fig. 12.18

Let us consider for the purpose of simplicity a rectangular channel where a hydraulic jump has taken place to increase the depth of flow from h_1 to h_2. The jump may be considered as a standing wave through which the change has occurred. The from Eq. (12.26), we can write, putting $C = 0$,

$$V_1 = (gh_2)^{1/2} \left(\frac{1 + h_2/h_1}{2} \right)^{1/2}$$

$$= (gh_1)^{1/2} \, (h_2/h_1)^{1/2} \left(\frac{1 + h_2/h_1}{2} \right)^{1/2}$$

Therefore,

$$\frac{V_1}{(gh_1)^{1/2}} = \mathrm{Fr}_1 = \left[\frac{h_2}{h_1}\frac{(1 + h_2/h_1)}{2}\right]^{1/2} \tag{12.42}$$

Since $h_2 > h_1$, the right-hand side of this equation is greater than unity. This concludes that a hydraulic jump, *Froude number* before the jump is greater than unity and hence the flow is rapid. We can also calculate the Froude number after the jump as

$$\mathrm{Fr}_2 = \frac{V_2}{(gh_2)^{1/2}} = \left[\frac{1 + h_2/h_1}{2}\right]^{1/2}\left(\frac{h_1}{h_2}\right)$$

$$= \left[\frac{h_1}{h_2}\frac{(1 + h_1/h_2)}{2}\right]^{1/2} \tag{12.43}$$

For $h_1/h_2 < 1$, the right-hand side of this equation is less than unity which concludes that the Froude number after the jump is less than unity and the flow becomes tranquil.

A rearrangement of Eq. (12.42) gives

$$h_2^2 + h_1 h_2 - \frac{2V_1^2 h_1}{g} = 0$$

If we put $V_1 = q/h_1$, where q is the discharge per unit width, we get

$$h_1 h_2^2 + h_1^2 h_2 - 2q^2/g = 0 \tag{12.44}$$

which gives

$$h_2 = -\frac{h_1}{2}(\pm)\left[\left(\frac{h_1^2}{4} + \frac{2q^2}{gh_1}\right)\right]^{1/2} \tag{12.45}$$

The negative sign for the radical is rejected because h_2 cannot be negative. Hence,

$$h_2 = \frac{h_1}{2}\left[-1 + \left\{1 + \frac{8q^2}{gh_1^3}\right\}^{1/2}\right]$$

or

$$\frac{h_2}{h_1} = \frac{1}{2}\,[(1 + 8\,\mathrm{Fr}_1^2)^{1/2} - 1] \tag{12.46}$$

(Since, $8\,q^2/gh_1^3 = 8\,V_1^2/gh_1 = 8\,\mathrm{Fr}_1^2$)

Equation (12.44) is symmetrical in respect of h_1 and h_2 and hence a similar solution for h_1 in terms of h_2 may be obtained by interchanging the subscripts. The depths of flow on both sides of a hydraulic jump are termed as the *conjugate depths* for the jump.

Loss of Mechanical Energy in Hydraulic Jump The loss of mechanical energy that takes place in a hydraulic jump is calculated by the application of energy equation (Bernoulli's equation). If the loss of total head in the jump is h_j as shown in Fig.

12.18(b), then we can write by the application of Bernoulli's equation between Sections 1 and 2 (Fig. 12.18(a)) neglecting the slope of the channe!,

$$h_1 + (V_1^2/2g) = h_2 + (V_2^2/2g) + h_j$$

or
$$h_j = h_1 - h_2 + \frac{V_1^2 - V_2^2}{2g}$$

$$= h_1 - h_2 + \frac{q^2}{2g}\left(\frac{1}{h_1^2} - \frac{1}{h_2^2}\right) \tag{12.47}$$

(Since from continuity, $q = V_1 h_1 = V_2 h_2$)
From Eq. (12.44), we can write

$$\frac{q^2}{2g} = \frac{h_1 h_2^2 + h_1^2 h_2}{4}$$

Invoking this relation into Eq. (12.47), we get

$$h_j = h_1 - h_2 + \left(\frac{h_1 h_2^2 + h_2^2 h_2}{4}\right)\left(\frac{1}{h_1^2} - \frac{1}{h_2^2}\right)$$

which finally gives

$$h_j = \frac{(h_2 - h_1)^3}{4 h_1 h_2} \tag{12.48}$$

The loss of head h_j amounts to be the part of mechanical energy that is being dissipated into intermolecular energy as a result of the creation of eddies and turbulences in the wave. Friction at the boundaries make a negligible contribution to it. This dissipation of energy results in a little rise in the liquid temperature. The hydraulic jump is a very effective means of reducing unwanted energy in a stream which is usually generated by the rapid discharge from a steep spillway to the channel.

Example 12.11

A rectangular channel, 6.1 m wide, carries 11.32 m^3/s and discharges onto a 6.1 m wide apron having no slope with a mean velocity of 6.1 m/s. (i) What is the height of the hydraulic jump? (ii) What energy is absorbed (lost) in the jump?

Solution

(i) $V_1 = 6.1$ m/s
q_1 (the rate of discharge per unit width) = 11.32/6.1

= 1.86 m^3/sm width

Therefore, $h_1 = q_1/V_1 = 1.86/6.1 = 0.305$ m

and $\text{Fr}_1 = V_1/(gh_1)^{1/2} = 6.1/(9.81 \times 0.305)^{1/2} = 3.53$

using Eq. (12.46),

$$h_2/h_1 = \frac{1}{2}\,[\{1 + 8(3.53)^2\}^{1/2} - 1]$$

from which $h_2 = 1.38\,\text{m}$

Hence, the height of the hydraulic jump = 1.38 – 0.305 = 1.075 m
using Eq. (12.48)

Loss of head in the hydraulic jump $$h_j = \frac{(1.07)^3}{4 \times 0.305 \times 1.38}$$

$$= 0.73\,\text{m}$$

Therefore the loss of total energy per second $= \rho g Q h_j$

$$= \frac{9.81 \times 10^3 \times (11.32) \times (0.73)}{10^3} = 81.06\,\text{kW}$$

Example 12.12

A control sluice spanning, the entry to a 3.5 m wide rectangular channel, admits 5.5 m^3/s of water with a uniform velocity of 4.14 m/s. Explain under what conditions a hydraulic jump will be formed and, assuming that these conditions exist, calculate (i) the height of the jump, and (ii) power dissipated in the jump.

Solution

The upstream depth of flow is

$$h_1 = \frac{5.5}{3.5 \times 4.14} = 0.379\,\text{m}$$

The upstream Froude number, $\text{Fr}_1 = \dfrac{4.14}{(9.81 \times 0.379)^{1/2}} = 2.15$

For the hydraulic jump to occur, the downstream flow must be tranquil and the depth of flow at downstream must satisfy the Eq. (12.46).
Therefore,

$$h_2 = \frac{0.379}{2}\,[\{1 + 8(2.15)^2\}^{1/2} - 1] = 0.978\,\text{m}$$

(i) Therefore, the height of the jump

$$\Delta h = (h_2 - h_1) = (0.978 - 0.379) = 0.6\,\text{m}$$

(ii) The loss of head in the jump is found out from Eq. (12.48) as

$$h_j = \frac{(0.6)^3}{4 \times 0.978 \times 0.379} = 0.146\,\text{m}$$

The rate of dissipation of energy $= \rho g Q h_j$

$$= 10^3 \times 9.81 \times 5.5 \times 0.146\,\text{W} = 7.66\,\text{kW}$$

12.5 OCCURRENCE OF CRITICAL CONDITIONS

We have discussed so far the nature of tranquil, rapid and critical flows. We have also seen that the transition from rapid to tranquil flow occurs through a hydraulic jump. Now it is important to know that under what conditions the critical flow occurs. The location where the critical flow occurs is called the *control section.* The following situations show the occurrence of critical conditions.

Change of Slope of Channel Bed Critical flow occurs when a tranquil flow changes to a rapid one. One such situation is illustrated in Fig. 12.19 which shows a long prismatic channel of mild slope connected to another long channel of steep slope with identical cross section. At a large distance from the junction, there will be uniform tranquil flow in the mild channel and uniform rapid flow in the steep channel. The depths in the channels will be the normal depths corresponding to the respective slope and rate of flow. The transition from tranquil to rapid flow will be non-uniform and must pass through the critical condition that occurs at the junction. If the change of the slope is abrupt, an appreciable curvature of the streamlines takes place near the junction. This will not justify the assumption of a hydrostatic variation of pressure at the Section This may result in the occurrence of critical condition given by the flow velocity $(gh)^{1/2}$ not exactly at the junction of the two slopes, but slightly upstream of it.

The discharge of liquid from a long channel of steep slope to a long channel of mild slope requires the flow to change from rapid to tranquil. This transition takes place abruptly through a hydraulic jump near the junction point.

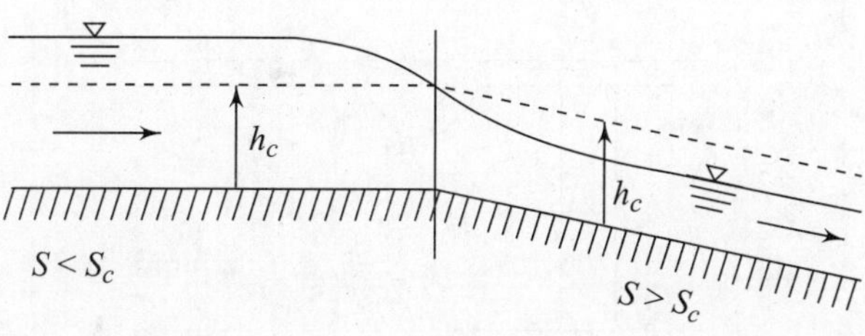

Fig. 12.19 Transition from tranquil to rapid flow

Flow over a Spillway and in a Channel with a Rise in its Bed The critical flow may occur even in a channel with a constant slope. A rise in the channel floor or bed may bring about a critical flow. Flow over a spillway (Fig. 12.20(a)) and flow in a channel with a rise in its bed caused by some obstruction (Fig. 12.20(b)) will pass through a critical condition.

In the case of a flow through a rectangular channel of constant width, the total mechanical energy per unit weight at any cross section is usually written as

$$H = h + \frac{q^2}{2gh^2} + z$$

where h is the depth of flow and z is the elevation of the bed from any reference horizontal datum.

Neglecting the effect of friction, we can write

$$\frac{\mathrm{d}H}{\mathrm{d}x} = \frac{\mathrm{d}h}{\mathrm{d}x} - \frac{q^2}{gh^3}\frac{\mathrm{d}h}{\mathrm{d}x} + \frac{\mathrm{d}z}{\mathrm{d}x} = 0 \qquad (12.49)$$

where, x is the distance measured along the direction of flow.

Since $\qquad q = Q/b = Vbh/b = Vh,$

$$\frac{\mathrm{d}h}{\mathrm{d}x} - \frac{V^2}{gh}\frac{\mathrm{d}h}{\mathrm{d}x} + \frac{\mathrm{d}z}{\mathrm{d}x} = 0$$

Recalling that $\mathrm{Fr} = V/(gh)^{1/2}$, we have

$$\frac{\mathrm{d}h}{\mathrm{d}x}(1 - \mathrm{Fr}^2) + \frac{\mathrm{d}z}{\mathrm{d}x} = 0 \qquad (12.50)$$

In case of flow over a spillway, (Fig. 12.20(a)), $\mathrm{d}z/\mathrm{d}x = 0$ at the crest, and since $\mathrm{d}h/\mathrm{d}x \neq 0$, $(1 - \mathrm{Fr}^2) = 0$. This gives $\mathrm{Fr} = 1$, i.e., the *critical condition* at the crest.

As another example for the occurrence of critical flow, we consider a rise in channel bed caused by some obstruction or gradual transition as shown in (Fig. 12.17(b)).

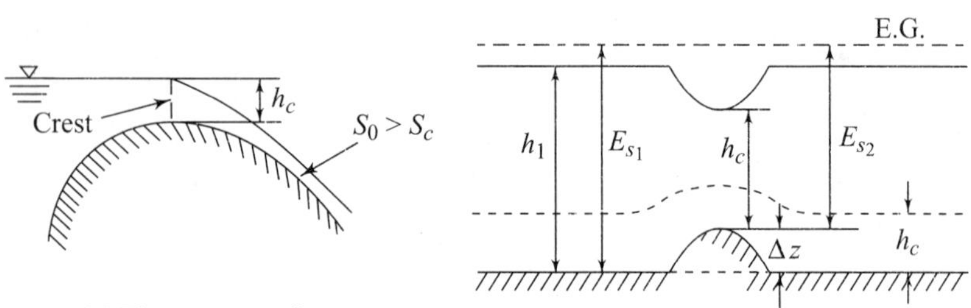

(a) Flow over a spilway

(b) Flow over an obstruction

Fig. 12.20

We consider a tranquil flow upstream of the hump or the obstruction. Therefore, the approach velocity to the obstruction is below the critical one and let the uniform depth upstream be h_1 and the corresponding specific energy be E_{s1}. If E_{s2} is the specific energy at the crest of the hump, then for a steady flow and a constant width of channel (i.e., q is constant), E_{s1} and E_{s2} satisfy the relation

$$E_{s_2} = E_{s_1} - \Delta z$$

At any value of $E_{s_2} > E_{s_c}$ there are two possible depths corresponding to E_{s_2}. To have the state at the crest with the lower depth out of the two, the specific energy of the flow upstream the hump should pass through a minimum and then increase. This is possible only if the bed could rise above the level of the hump and then drop. Therefore, under the present situation, the possible state is with the higher depth corresponding to E_{s_2} and the flow remains tranquil over the hump and the surface of water falls under this situation, since, from Eq. 12.50, $dh/dx < 0$ when $dz/dx > 0$ for $Fr < 1$. At the crest of the hump, $dz/dx = 0$ and $dh/dx = 0$, and hence according to Eq. (12.50), Fr may or may not be equal to unity. Therefore we can say that the critical condition may or may not exist at the crest in general. As the height of the step is raised, i.e., Δz is increased $E_{s_1} - E_{s_2}$ increases until E_{s_2} corresponds to the critical specific energy E_{s_c}. Any further rise in Δz will maintain critical flow over the step.

SUMMARY

- Flow with a free surface is caused by the weight of the flowing fluid. Flow in open channels is an example of such a flow. A uniform flow through an open channel is characterised by the liquid surface being parallel to the base of the channel whose cross section is same along the length of the channel. In a non-uniform flow, the liquid surface is not parallel to the base of the channel.
- Energy gradient line is the contour of total head (total mechanical energy per unit weight) at a cross section as ordinate against the distance along the flow as abscissa. The hydraulic gradient line is the contour of the sum of potential and pressure heads as ordinate against the distance along the flow as abscissa.
- The relationship between the average flow velocity and pressure drop in a steady uniform flow through a straight channel is given by the well known Chezy equation as $V = c\,(R_h\, S_b)^{1/2}$. The Chezy coefficient c, includes the friction factor f, and depends on the surface roughness and the hydraulic radius of the channel. The simplest and widely used empirical relation, in this regard, is given by $c = (1/n)\, R_h^{1/6}$ and is known as *Manning's formula*, where n is the roughness coefficient.
- The optimum hydraulic cross section of a channel is characterised by the maximum value of the hydraulic radius. For a trapezoidal section, this condition is satisfied when the hydraulic radius becomes equal to half the central depth of flow.
- Specific energy of a fluid element at any point in a channel flow is defined as its total energy per unit weight where the component potential energy is measured from the base of the channel. At critical depth given by $h_c = (q^2/g)^{1/3}$, the specific energy of flow is a minimum for a given discharge, or the discharge is a maximum for a given specific energy.

- The velocity of flow at the critical depth is known as critical velocity and is given by $V_c = (gh_c)^{1/2}$. Flow in which the velocity is less than the critical velocity is known as tranquil flow, while the flow with a velocity greater than the critical velocity is referred to as rapid flow. A small disturbance in the open channels propagates upstream as a surface wave with a velocity (relative to the undisturbed fluid) equal to $(gh)^{1/2}$. Therefore, a surface wave caused by any disturbance downstream can propagate upstream in a tranquil flow, while it cannot do so in a rapid flow.
- A sudden transition from rapid flow to a tranquil one is known as hydraulic jump and takes place through an abrupt discontinuity in the flow. The loss of head in a hydraulic jump is given by $(h_2 - h_1)^3/4h_1 h_2$, here h_1 and h_2 are the depths of flow before and after the jump respectively.

EXERCISES

12.1 Choose the correct answer:

(i) In an open-channel flow, the free surface, the hydraulic gradient, and energy gradient lines are such that
  (a) the three of them coincide
  (b) the first two coincide
  (c) the last two must remain parallel
  (d) the first and the last must remain parallel
  (e) the three of them are different but parallel

(ii) A small disturbance in the rapid flow in an open channel
  (a) can propagate both upstream and downstream
  (b) can propagate neither upstream nor downstream
  (c) cannot propagate upstream

(iii) For a critical flow in an open channel
  (a) specific energy is maximum for a given flow
  (b) shear stress is maximum at the bed surface
  (c) the flow is minimum for a given specific energy
  (d) the specific energy is minimum for a given flow

(iv) A hydraulic jump must occur when
  (a) the flow is rapid
  (b) the depth is less than the critical depth
  (c) the slope is mild or level
  (d) the flow is increased in a given channel
  (e) the bed slope changes from steep to mild

12.2 The breadth of a rectangular channel is twice its depth. Assuming the Chezy coefficient c to be 55 $m^{1/2}/s$, find the cross-sectional dimensions of the channel and the slope to satisfy the conditions that the discharge when flowing full should be 0.8 m^3/s, and the velocity when flowing half-full should be 0.6 m/s.

Ans. ($h = 0.74$ m, $B = 1.48$ m, $s = 0.00048$)

12.3 A channel of symmetrical trapezoidal section, 900 mm deep and with top and bottom widths 1.8 m and 600 mm, respectively, carries water at a depth of 600 mm. If the channel slopes uniformly at 1 in 2600 and Chezy's coefficient is 60 $m^{1/2}/s$, calculate the steady rate of flow in the channel.

Ans. (0.38 m^3/s)

12.4 An open channel of trapezoidal section with 5 m width at the base and with side slope of 2 horizontal: 1 vertical has a bed slope of 1 in 3000. It is found that when the flow is 8.5 m^3/s, the depth of water in the channel is 1.5 m. Calculate the flow rate when the depth is 1 m assuming the validity of Manning's formula.

Ans. (4.0 m^3/s)

12.5 A long channel of trapezoidal section is constructed from rubble masonry at a bed slope of 1 in 7000. The sides slope at $\tan^{-1}$ 1.5 to the horizontal and the required flow rate is 2.8 m^3/s. Determine the base width of the channel if the maximum depth is 1 m (use Table 12.1 for roughness coefficient of the channel).

Ans.(4.46 m)

12.6 A trapezoidal channel with a bottom width of 1.5 m and side slopes of 2 horizontal: 1 vertical has a bed slope of 1 in 3000. If the depth of water flowing through the channel is 2.5 m, what is the average shear stress at the boundary?

Ans. (4.19 N/m^2)

12.7 A trapezoidal canal with side slopes of 1 horizontal: 1 vertical and bed slopes of 0.00035 discharges water at the rate of 24 m^3/s. Determine the base width and depth of flow if the shear stress at the boundary is not to exceed 6 N/m^2. Take Manning's roughness factor $n = 0.028$.

Ans. (6.64 m, 2.66 m)

12.8 A sewer pipe is to be laid at a slope of 1 in 8100 to carry a maximum discharge of 600 litres/s when the depth of water is 75% of the vertical diameter. Find the diameter of this pipe if the value of Manning's roughness factor $n = 0.025$.

Ans. (1.79 m)

12.9 A circular conduit is to satisfy the following conditions: Capacity when flowing full, 0.13 m^3/s, velocity when the depth is one quarter the diameter, not less than 600 mm/s. Assuming uniform flow, determine the diameter and the slope if Chezy's coefficient $c = 58$ $m^{1/2}/s$.

Ans. (442 mm, 0.0016)

12.10 Determine the dimensions of the most economical trapezoidal concrete channel with a bed slope of 1 in 4000 and a side slope of 1 vertical to 2 horizontal to carry water at the rate of 0.15 m^3/s. Take Manning's $n = 0.015$.

Ans. ($b = 0.19$ m, $h = 0.41$ m)

12.11 In a long rectangular channel 3 m wide, the specific energy is 1.8 m and the rate of flow is 12 m^3/s. Calculate two possible depths of flow and the corresponding Froude numbers. If Manning's roughness factor $n = 0.014$, what is the critical slope for this discharge?

Ans. (1.03 m, 1.36 m, 1.22, 0.80, 0.0039)

12.12 For a constant specific energy of 2 m., what maximum flow may occur in a rectangular channel 3 m wide?

Ans. (14.48 m^3/s)

12.13 A horizontal rectangular channel of constant width has a sluice gate installed in it. At a position of 1 m opening, the velocity of water is 10 m/s. Determine whether a jump can occur, and if so,

(a) the height downstream
(b) the loss of the head in the jump
(c) the ratio of Froude numbers across it.

Ans. ((a) 4.04 m, (b) 1.74 m, (c) 8.12)

12.14 In a rectangular channel of 0.6 m wide, a jump occurs where the Froude number is 3. The depth after the jump is 0.6 m. Estimate the loss of head and the power dissipated due to the jump.

Ans. (0.22 m, 0.78 kW)

13

APPLICATIONS OF UNSTEADY FLOWS

13.1 INTRODUCTION

Although most of the engineering problems are steady or quasi-steady in nature, there are certain classes of problems in practice where the phenomenon of unsteady flow becomes significant. In an unsteady flow, velocity, pressure, density, etc., at a particular point change with time. Such variations pose considerable difficulties in solving unsteady flow problems. Problems of unsteady flow may be put into the following three broad categories according to the rate at which the changes in hydrodynamic parameters occur:

(i) *Slow changes* of flow where the velocity changes slowly so that the temporai acceleration can be neglected. An example of this category of problems is the continuous filling or emptying of a reservoir as discussed in Section 5.8 of Chapter 5.

(ii) *Rapid changes* of flow causing the temporal acceleration to be important. Examples of this category of problems are oscillations of liquids in U tubes and between reservoirs, flows in positive displacement pumps and in hydraulic and pneumatic servo-mechanisms.

(iii) *Very fast* changes of flow, arising from sudden opening or closing of a valve, so that density changes considerably and elastic force becomes significant.

The present chapter discusses a few unsteady flow problems of engineering importance.

13.2 INERTIA PRESSURE AND ACCELERATIVE HEAD

Whenever any fluid element undergoes acceleration, either positive or negative, it must be acted upon by a net external force. This force corresponds to a difference in Piezometric pressure across the fluid element. This pressure difference is known as the *inertia pressure*.

Let us consider, as a simple case, a stream tube of length L, and uniform cross-sectional area A. The velocity of fluid flowing through it is considered to be uniform both across a section and along the flow. Let the velocity of flow at any instant be V. Therefore, the mass of the fluid concerned is ρAL and the force causing the acceleration, according to Newton's second law, is the product of mass and acceleration. The acceleration here is the temporal acceleration. Hence, the force causing acceleration equals to $\rho AL\,(\partial V/\partial t)$. If this force arises because of a difference in the piezometric pressure Δp_i between the upstream and downstream ends of the tube, then

$$\Delta p_i A = \rho AL \frac{\partial V}{\partial t}$$

or

$$\Delta p_i = \rho L \frac{\partial V}{\partial t} \tag{13.1}$$

Δp_i, as defined by Eq. (13.1), is known as the inertia pressure (difference in piezometric pressure responsible for fluid acceleration). The corresponding head can be written as

$$h_i = \frac{\Delta p_i}{\rho g} = \frac{L}{g}\frac{\partial V}{\partial t} \tag{13.2}$$

where h_i is known as *inertia head* or *accelerative head.*

Energy Equation with Accelerative Head While deriving Bernoulli's equation in Section 4.3.1 of Chapter 4, we considered the flow to be steady. If the unsteady term of the Euler's equation, i.e., the temporal derivative of the velocity is taken care of in the derivation of Bernoulli's equation, then we can arrive at a modified form of the Bernoulli's equation for an unsteady but incompressible flow as

$$\frac{1}{g}\int \frac{\partial V}{\partial t}\,\mathrm{d}S + \frac{V^2}{2g} + \frac{p}{\rho g} + z = C \tag{13.3}$$

where C is a constant along a streamline. The first term in Eq. (13.3) represents the accelerative head. Therefore, Bernoulli's equation between two points 1 and 2 along a streamline can be written, for an unsteady flow, along with the consideration of friction loss as

$$\frac{p_1}{\rho g} + \frac{V_1^2}{2g} + z_1 = \frac{p_2}{\rho g} + \frac{V_2^2}{2g} + z_2 + h_f + h_i$$

where h_f is the head loss due to friction, and

$$h_i = \frac{1}{g}\int_1^2 \frac{\partial V}{\partial t}\,\mathrm{d}S$$

We shall now describe a few applications of unsteady flow problems in practice.

13.3 ESTABLISHMENT OF FLOW

The initiation of flow in a pipeline is governed by inertia pressure. Let us consider a pipe of uniform cross section and of length L, to convey liquid from a reservoir as shown in Fig. 13.1. The reservoir maintains a constant height of liquid above the pipe connection to the reservoir. The pipe has a valve at its downstream end which is initially closed, and the pressure downstream the valve is constant. When the valve is opened, the difference in Piezometric pressure between the ends of the pipeline is applied to the static liquid column in it. Since at this moment, viscous and other resistive forces are zero because of no movement of the liquid, this inertia pressure force, being the net external force, tries to accelerate the liquid column to a maximum. As soon as the flow initiates, the viscous and other types of resistive forces, if any, arise and gradually become prominent with the increase in velocity and eventually balance the pressure force to establish a steady state. Therefore we see that the flow within the pipe increases from zero to a steady value determined by the frictional and other losses in the pipe. Even if the valve could be opened instantaneously, the fluid would not reach its steady state velocity instantaneously. The attainment of a steady flow in the pipeline after the instantaneous opening of a valve at its downstream is known as *the establishment of flow*. An analytical expression for the response characteristic of the liquid column to the steady state can be derived as follows.

Let the loss of head in the pipeline be represented by $KV^2/2g$, where V is the instantaneous average velocity at any section which remains same in the direction of flow. The term $KV^2/2g$ includes both the frictional head loss and the minor losses (entry loss, valve loss, etc.). We can write the Bernoulli's equation in consideration of accelerative head between points 1 and 2 (Fig. 13.1) as

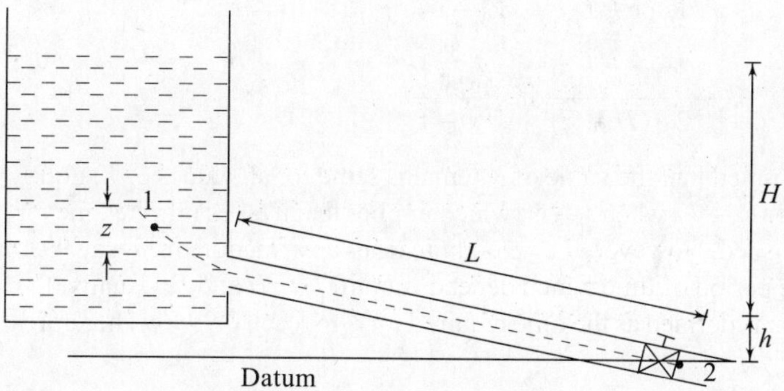

Fig. 13.1 Establishment of flow in a pipeline

$$\frac{V_1^2}{2g} + \frac{p_1}{\rho g} + (z+h) = \frac{V^2}{2g} + \frac{p_2}{\rho g} + h_f + \frac{1}{g}\int_1^2 \frac{\partial V}{\partial t}\,\mathrm{d}S \tag{13.4}$$

$V_1 << V$ for much larger cross-sectional area of the reservoir as compared to that of the pipeline, and

$$p_1 = p_{\text{atm}} + \rho g(H - z)$$
$$p_2 = p_{\text{atm}} \text{ (atmospheric pressure)}$$
$$h_f = KV^2/2g$$

Therefore, we have from Eq. (13.4)

$$(H+h) - \frac{V^2}{2g} = \frac{L}{g}\frac{\partial V}{\partial t} + \frac{KV^2}{2g} \tag{13.5}$$

Since the velocity is a function of time only, the partial derivative of V in Eq. (13.5) is changed to a total derivative and we get an ordinary differential equation as

$$\frac{dV}{dt} = \frac{1}{L}\left[g(H+h) - (1+K)\frac{V^2}{2}\right] \tag{13.6}$$

Let V_0 be the steady state velocity. Then, applying Bernoulli's equation between 1 and 2, at steady state, we have,

$$(H+h) = (1+K)\frac{V_0^2}{2g}$$

substituting the value of $(H+h)$ into Eq. (13.6), we get

$$\frac{dV}{dt} = \frac{1+K}{2L}(V_0^2 - V^2)$$

On integrating the equation we have

$$t = \frac{2L}{1+K}\int_0^V \frac{dV}{(V_0^2 - V^2)}$$

$$= \frac{L}{V_0(1+K)} \ln \frac{V_0+V}{V_0-V}$$

or

$$t = \frac{LV_0}{2g(H+h)} \ln \frac{V_0+V}{V_0-V} \tag{13.7}$$

Here we have assumed that the value of K remains same for all values of V. Equation (13.7) shows that $V \to V_0$ when $t \to \infty$, which implies that it takes infinite time for the flow to be established. However, the velocity reaches any fraction of V_0, say 99% of V_0, within a finite period of time which depends upon V_0, L, H and h. Usually, the time of establishment is defined as the time required for V to reach 0.99 V_0. Therefore, we get from Eq. (13.7),

$$t_{\text{establishment}} = \frac{LV_0}{2g(H+h)} \ln\left(\frac{1.99}{0.01}\right)$$

$$= 0.27\frac{LV_0}{(H+h)} \tag{13.8}$$

Example 13.1

A straight pipe 600 m in length, and 1m in diameter, with a constant friction factor $f = 0.025$, and a sharp inlet, leads from a reservoir where a constant level is maintained at 25 m above the pipe outlet which is initially closed by a globe valve ($K = 10$). If the valve is suddenly opened, find the time required to attain 90% of steady-state discharge.

Solution

This problem is an example of the straightforward application of Eq. (13.7) which gives the time for establishment of steady flow in a pipe. By making use of this equation for the present problem, we have

$$t = \frac{600 \times V_0}{(2 \times 9.81 \times 25)} \ln \frac{1.9}{0.1}$$

Steady state velocity V_0 is found out by the application of Bernoulli's equation, at steady state, between a point on the free surface of water in the reservoir and a point on the discharge plane after the valve, as

$$25 = \frac{V_0^2}{2g}\left(0.5 + \frac{0.025 \times 600}{1} + 10 + 1\right)$$

or
$$V_0 = \left[\frac{2 \times 9.81 \times 25}{26.5}\right]^{1/2}$$

Putting this value of V_0 in the above equation, we have

$$t = \frac{600}{(2 \times 9.81 \times 25 \times 26.5]^{1/2}} \ln \frac{1.9}{0.1}$$
$$= 15.5\text{s}$$

Example 13.2

A valve at the outlet end of a pipe 1m in diameter and 600 m long is rapidly opened. The pipe discharges to atmosphere and the Piezometric head at the inlet end of the pipe is 23 m (relative to the outlet level). The head loss through the open valve is 10 times the velocity head in the pipe, other minor losses amount to twice the velocity head, and f, the friction factor is assumed constant at 0.020. What is the velocity after 12 sec?

Solution

We first determine the steady-state velocity V_0 by the application of Bernoulli's equation, at steady state, between a point at the inlet end of the pipe and a point at its outlet end as

$$23 = \frac{V_0^2}{2g}\left[\frac{0.020 \times 600}{1} + 10 + 2\right]$$

Therefore,
$$V_0 = \left[\frac{2 \times 9.81 \times 23}{24}\right]^{1/2} = 4.34 \text{ m/s}$$

Let the velocity after 12 sec be V. Then, from Eq. (13.7) we can write

$$12 = \frac{600 \times 4.34}{2 \times 9.81 \times 23} \ln\left[\frac{(1+x)}{(1-x)}\right]$$

(where $x = V/V_0$)

Hence,
$$\ln\left[\frac{(1+x)}{(1-x)}\right] = 2.08$$

or $$\frac{(1+x)}{(1-x)} = 8$$

which gives $$x = 7/9$$

Therefore, $$V = \frac{7 \times 4.34}{9} = 3.37 \text{ m/s}$$

13.4 OSCILLATION IN A U-TUBE

(A) *Frictionless Liquid Column* Let us consider the oscillation of an inviscid liquid in a U-tube of internal diameter d, as shown in Fig. 13.2(a). Let l be the length of the liquid column.

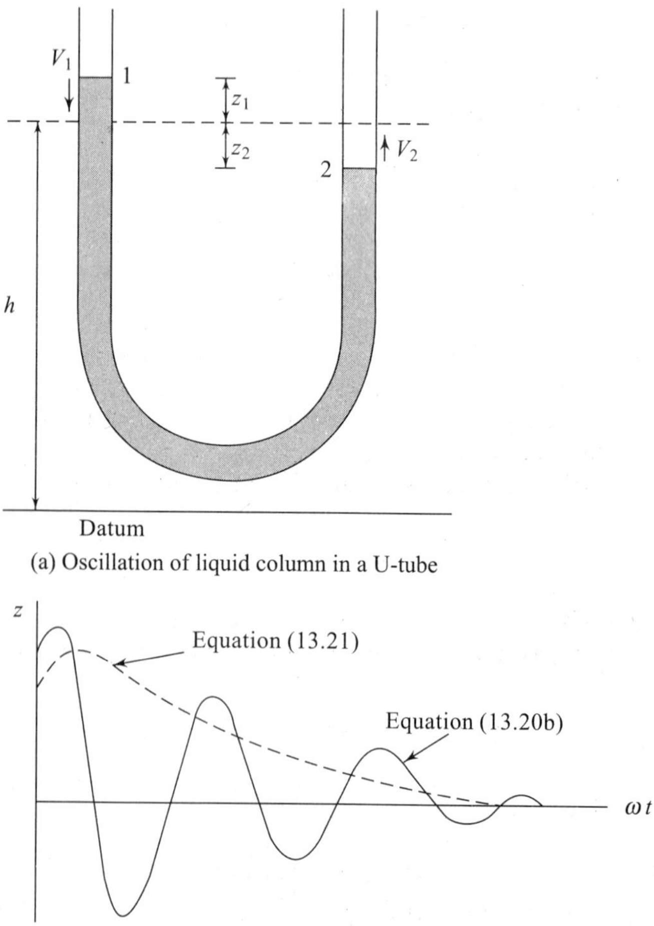

(a) Oscillation of liquid column in a U-tube

(b) Response characteristics with laminar resistance

Fig. 13.2

When the liquid is in equilibrium, the height of liquid column in both the limbs from a datum line is denoted by h. Let us consider, after the equilibrium of the liquid column being somehow disturbed, an instant when the meniscus in the left limb is coming down with a velocity V_1, while that in right limb is going up with a velocity V_2 as shown in Fig. 13.2(a). Since, the tube is uniform in cross section,

$$V_1 = V_2 = V \tag{13.9a}$$

and

$$z_1 = z_2 = z \tag{13.9b}$$

where V and z represent the velocity of liquid column in the U-tube and the displacement of liquid level from its equilibrium position in either limb respectively.

The Bernouli's equation for unsteady flow between the points 1 and 2 (Fig. 13.2(a)) can be written in the present case as

$$\frac{p_{\text{atm}}}{\rho g} + \frac{V^2}{2g} + (h + z) = \frac{p_{\text{atm}}}{\rho g} + \frac{V^2}{2g} + (h - z) + \frac{1}{g}\int_1^2 \frac{\mathrm{d}V}{\mathrm{d}t}\,\mathrm{d}S \tag{13.10}$$

or

$$\frac{\mathrm{d}V}{\mathrm{d}t} - \frac{2g}{l}z = 0 \tag{13.11}$$

Since z is diminishing with time at the instant considered, we can write

$$V = -\frac{\mathrm{d}z}{\mathrm{d}t}$$

Hence,

$$\frac{\mathrm{d}V}{\mathrm{d}t} = -\frac{\mathrm{d}^2 z}{\mathrm{d}t^2}$$

Therefore, we have from Eq. (13.11)

$$\frac{\mathrm{d}^2 z}{\mathrm{d}t^2} + \frac{2g}{l}z = 0 \tag{13.12}$$

The solution of Eq. (13.12) is

$$z = A\cos(2g/l)^{1/2}t + B\sin(2g/l)^{1/2}t \tag{13.13}$$

To determine the constants A and B, initial conditions are taken as

at $t = 0$; $z = z_0$ (the maximum displacement from the equilibrium position)

and, $\mathrm{d}z/\mathrm{d}t = 0$

which gives $A = z_0$ and $B = 0$

Therefore, Eq. (13.13) becomes

$$z = z_0 \cos\left(\frac{2g}{l}\right)^{1/2} t \tag{13.14}$$

This equation implies that the liquid column executes an undamped periodic oscillation with an amplitude z_0 and a time period of $2\pi(l/2g)^{1/2}$.

(B) *Viscous Fluid* If we consider the viscous effects in the oscillation of liquid columns, the Bernoulli's equation between 1 and 2 can be written as

$$\frac{p_{atm}}{\rho g} + \frac{V^2}{2g} + (h+z) = \frac{p_{atm}}{\rho g} + \frac{V^2}{2g} + (h-z) + h_f + \frac{1}{g}\int_1^2 \frac{dV}{dt}\, dS \tag{13.15}$$

where h_f is the frictional head loss in the tube due to the motion of the liquid column, and can be expressed in terms of velocity head as

$$h_f = \frac{f l V^2}{2gd}$$

If we consider the flow to be laminar, friction factor f, can be written as

$$f = \frac{64}{\text{Re}} = \frac{64\nu}{Vd}$$

Hence,
$$h_f = \frac{32\nu l}{gd^2} V$$

Invoking this value into Eq. (13.15), we get

$$\frac{dV}{dt} + \frac{32\nu}{d^2} V - \frac{2g}{l} z = 0$$

Substituting
$$V = -\frac{dz}{dt}$$

and
$$\frac{dV}{dt} = -\frac{d^2 z}{dt^2}$$

we have,

$$\frac{d^2 z}{dt^2} + \frac{32\nu}{d^2}\frac{dz}{dt} + \frac{2g}{l} z = 0 \tag{13.16}$$

The differential equation corresponds to a damped oscillatory system. The general solution of the equation can be written as

$$z = Ae^{C_1 t} + Be^{C_2 t} \tag{13.17}$$

The values of C_1 and C_2 are the roots of the equation

$$m^2 + \frac{32\nu}{d^2} m + \frac{2g}{l} = 0$$

where m is a general variable.
Hence,

$$C_1 = -\frac{16\nu}{d^2} + \left[\left(\frac{16\nu}{d^2}\right)^2 - \left(\frac{2g}{l}\right)\right]^{1/2} \tag{13.18a}$$

and
$$C_2 = -\frac{16\nu}{d^2} - \left[\left(\frac{16\nu}{d^2}\right)^2 - \left(\frac{2g}{l}\right)\right]^{1/2} \tag{13.18b}$$

putting
$$a = \frac{16\nu}{d^2} \tag{13.19a}$$

$$\omega^2 = \frac{2g}{l} \tag{13.19b}$$

and
$$\zeta = a/\omega = \frac{16\nu}{d^2}(l/2g)^{1/2} \tag{13.19c}$$

we can write

$$C_1 = [-\zeta + (\zeta^2 - 1)^{1/2}]\omega$$

and
$$C_2 = [-\zeta - (\zeta^2 - 1)^{1/2}]\omega$$

The nature of the solution of Eq. (13.16) depends on three conditions: whether the damping factor (a) $\zeta < 1$, (b) $\zeta > 1$ and (c) $\zeta = 1$.

(a) When $\zeta < 1$ (light damping), the general solution of Eq. (13.16) is written as a special form of Eq. (13.17) as

$$z = Ae^{-\zeta\omega t}\sin[(1-\zeta^2)^{1/2}\,\omega t + \phi] \tag{13.20a}$$

The amplitude A and the phase difference ϕ are found from the initial conditions. If we assume the initial conditions as

at
$$t = 0, \quad z = z_0 \quad \text{and} \quad dz/dt = 0$$

we get from Eq. (13.20a),

$$A = \frac{z_0}{(1-\zeta^2)^{1/2}}$$

and
$$\phi = \tan^{-1}\left[\frac{(1-\zeta^2)^{1/2}}{\zeta}\right]$$

Equation (13.20a) can then be written as

$$z = \frac{z_0}{(1-\zeta^2)^{1/2}}\,e^{-\zeta\omega t}\sin\left[(1-\zeta^2)^{1/2}\,\omega t + \tan^{-1}\frac{(1-\zeta^2)^{1/2}}{\zeta}\right] \tag{13.20b}$$

The time period of oscillation is

$$T = \frac{2\pi}{\omega(1-\zeta^2)^{1/2}} \tag{13.20c}$$

The flow under this situation oscillates with diminishing amplitudes (Fig. 13.2b), because of the exponential damping term, and eventually comes to rest.

(b) When $\zeta > 1$ (large damping) the Eq. (13.17) can be written as

$$z = A\exp[\{-\zeta + (\zeta^2 - 1)^{1/2}\}\omega t] + B\exp[\{-\zeta - (\zeta^2 - 1)^{1/2}\omega t] \tag{13.21}$$

with the initial conditions as $z = 0$, $\dfrac{dz}{dt} = 0$ at $t = 0$ we have

$$A = \frac{z_0}{2}\left[1 + \frac{\zeta}{(\zeta^2 - 1)^{1/2}}\right]$$

and
$$B = \frac{z_0}{2}\left[1 - \frac{\zeta}{(\zeta^2 - 1)^{1/2}}\right]$$

The flow under this situation does not oscillate, rather asymptotically reaches the equilibrium position as shown in Fig. 13.2(b).

(c) When $\zeta = 1$ (critical damping), the solution of Eq. (13.16) becomes

$$z = (A + Bt)\, e^{-\zeta \omega t} \tag{13.22a}$$

with the same initial conditions as described above in (a) and in (b), we get

$$A = z_0,\; B = \zeta\, \omega z_0$$

Hence Eq. (13.22a) becomes

$$z = z_0 (1 + \zeta \omega\, t)\, e^{-\zeta \omega t} \tag{13.22b}$$

The motion, under this situation is in transition, i.e., it changes from oscillatory to non-oscillatory types.

Example 13.3

A 20 mm diameter U-tube contains liquid column of length 4 m. The kinematic viscosity of the liquid is 8×10^{-6} m^2/s. If the liquid column oscillates, find the time period of oscillation assuming the flow to be laminar. Find also the ratio of two successive amplitudes.

Solution

The differential equation for oscillation of a liquid column in a U-tube (in consideration of flow to be laminar) is given by Eq. (13.16), and the nature of its solution depends upon the value of damping factor given by

$$\zeta = \left(\frac{16\nu}{d^2} \sqrt{\frac{l}{2g}} \right)$$

Here,

$$\zeta = \frac{16 \times 8 \times 10^{-6}}{(20)^2 \times 10^{-6}} \sqrt{\frac{4}{2 \times 9.81}} = 0.14$$

Since $\zeta < 1$, the oscillatory flow in the present case is represented by the Eq. (13.20a), and hence the time period is given by the Eq. (13.20c) as

$$T = \frac{2\,\pi}{\{1 - (0.14)^2\}^{1/2}} \sqrt{\frac{4}{2 \times 9.81}} = 2.86 \text{ s}$$

The ratio of two successive amplitudes can be written with the help of Eq. (13.20a) as

$$\frac{z_{(t)}}{z_{(t+T)}} = e^{\zeta \omega T} = \exp\left(0.14\sqrt{2 \times 9.81/4} \times 2.86\right) = 2.43$$

13.5 DAMPED OSCILLATION BETWEEN TWO RESERVOIRS

We now consider the oscillation of a viscous liquid column between two prismatic reservoirs connected by a long pipeline as shown in Fig. 13.3. The flow in the pipeline is assumed to be turbulent so that the head loss becomes proportional to the square of the velocity. Let us assume that the reservoirs are of uniform cross-sectional area A_1 and A_2. The pipeline is of uniform circular cross section of diameter d_p and area a_p.

The total length of the pipeline is l as shown in Fig. 13.3. Let the height of the liquid levels, under equilibrium position, from a reference datum be h.

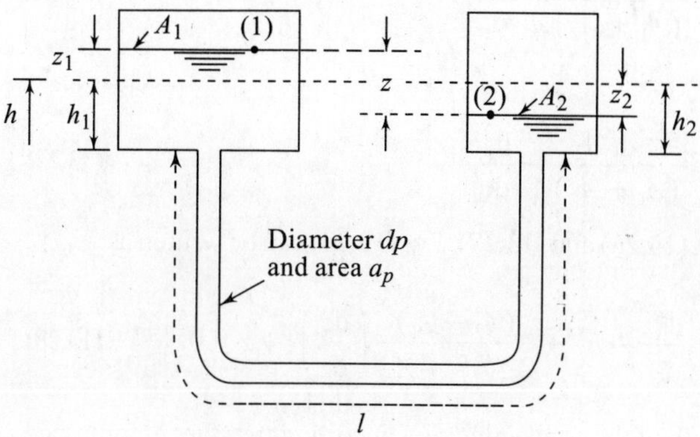

Fig. 13.3 Oscillation of liquid column between two reservoirs connected with a pipline

Applying Bernoulli's equation between the liquid levels (1) and (2); [Fig. 13.3] when the liquid column is in motion, we have,

$$\frac{p_{\text{atm}}}{\rho g} + \frac{V_1^2}{2g} + h + z_1 = \frac{p_{\text{atm}}}{\rho g} + \frac{V_2^2}{2g} + h - z_2 + h_f + \frac{1}{g}\int_1^2 \frac{dV}{dt}\, dS \tag{13.23}$$

Let V be the velocity at a distance S from the surface 1 along a streamline, where the cross-sectional area is A which may be a_p, A_1 or A_2 depending upon the distance S, and the liquid level in reservoir A_1 be moving down with a velocity V_1, while that in reservoir A_2 be moving up with a velocity V_2.

From continuity,

$$V_1 A_1 = V_2 A_2 = VA \tag{13.24}$$

Again, from Kinematic condition,

$$V_1 = -\frac{dz_1}{dt} \tag{13.25a}$$

$$V_2 = -\frac{dz_2}{dt} \tag{13.25b}$$

and from geometrical condition $z = z_1 + z_2$ (13.25c)

Equations (13.24), (13.25a), (13.25b) and (13.25c) give

$$\frac{dz_1}{dt} = \frac{A_2}{A_1 + A_2}\frac{dz}{dt} \tag{13.26a}$$

and
$$\frac{dz_2}{dt} = \frac{A_1}{A_1 + A_2}\frac{dz}{dt} \tag{13.26b}$$

Again,
$$V = V_1\frac{A_1}{A} = -\frac{A_1}{A}\frac{dz_1}{dt}$$
$$= -\frac{A_1 A_2}{(A_1 + A_2)A}\frac{dz}{dt}$$

Therefore,
$$\frac{dV}{dt} = -\frac{A_1 A_2}{(A_1 + A_2)A}\frac{d^2 z}{dt^2} \tag{13.27}$$

With the help of Eqs (13.25), (13.26) and (13.27), Eq. (13.23) can be written as

$$\frac{A_1 A_2}{g(A_1 + A_2)}\left(\int_1^2 \frac{dS}{A}\right)\frac{d^2 z}{dt^2} - h_f + \frac{(A_2 - A_1)}{2g(A_2 + A_1)}\left(\frac{dz}{dt}\right)^2 + z = 0 \tag{13.28}$$

If l_e is the equivalent length of the connecting pipe incorporating the minor losses, then the total head loss h_f, can be written as

$$h_f = \frac{f l_e}{2 g d_p}V^2 = \frac{f l_e}{2 g d_p}\frac{A_1^2 A_2^2}{(A_1 + A_2)^2 a_p^2}\left(\frac{dz}{dt}\right)^2 \tag{13.29a}$$

Again,
$$\int_1^2 \frac{dS}{A} = \frac{h_1 + z_1}{A_1} + \frac{l}{a_p} + \frac{h_2 - z_2}{A_2} \cong \frac{l}{a_p} \tag{13.29b}$$

(Since A_1 and A_2 are much larger than a_p)

With the help of Eqs (13.29a) and (13.29b), Eq. (13.28) can be written as

$$\frac{d^2 z}{dt^2} - M\left(\frac{dz}{dt}\right)^2 + Nz = 0 \tag{13.30}$$

where,
$$M = \frac{f l_e}{2 d_p\, l\, a_p}\frac{A_1 A_2}{(A_1 + A_2)} - \frac{a_p}{2l}\frac{(A_2 - A_1)}{A_1 / A_2}$$

and
$$N = \frac{g\, a_p}{l}\frac{(A_2 - A_1)}{A_1 A_2}$$

The Eq. (13.30) is a non-linear ordinary differential equation in z. The non-linearity arises due to the term $\left(\frac{dz}{dt}\right)^2$. This equation can be solved numerically for z with suitable initial conditions. Fourth order Runge Kutta method is best adopted for this purpose. However, an analytical solution for the first derivative of z, i.e. $\frac{dz}{dt}$ can be obtained. By substituting $y = \left(\frac{dz}{dt}\right)^2$, in Eq. (13.30) we get

$$\frac{dy}{dz} - 2\,My + 2\,Nz = 0$$

the solution of which is

$$y = \left(\frac{dz}{dt}\right)^2 = \frac{N}{2\,M^2}(2Mz + 1) + C\,e^{2Mz} \tag{13.31}$$

If we put the initial condition

$$z = z_0,\ \frac{dz}{dt} = 0 \quad \text{at} \quad t = 0$$

we get,

$$C = \frac{-N}{2\,M^2}(2Mz_0 + 1)\exp(-2\,Mz_0)$$

Therefore, Eq. (13.31) becomes,

$$\frac{dz}{dt} = \pm\left[\frac{N}{2M^2}[(2Mz + 1) - (2\,Mz_0 + 1)\exp\{2\,M(z - z_0)\}]\right]^{1/2} \tag{13.32}$$

To find the time-displacement (z vs t) relationship, Eq. (13.32) has to be solved numerically with the initial condition as $z = z_0$ at $t = t_0$.

13.6 WATER HAMMER

In the preceding sections, we considered unsteady problems where though the changes in velocity were high to make the acceleration head as significant as the velocity head, but at the same time were too low to cause the compressibility effect on the liquid. We now consider the category of unsteady flow phenomena where the change in velocity is so rapid that the compressibility effect of the liquid becomes prominent and hence the elastic forces are important. As a result, a change in pressure does not take place instantaneously throughout the fluid. This means that if a change in pressure is caused by a change in velocity at any location, this change is not sensed immediately by the entire fluid—rather this is sensed by the propagation of a pressure wave with a finite velocity. The problem assumes importance in fields like hydroelectric plants where the flow of water in a pipeline is required to be decreased suddenly by manipulating a valve downstream. This causes a phenomenon like knocking of the pipe system due to repeated up and down motion of a pressure wave within the pipe. It is also our common experience that when a domestic water tap is turned off very quickly, a heavy knocking sound is heard and the entire pipe vibrates. This typical phenomenon is known as water hammer. The name is perhaps a little unfortunate because, not only water, but any liquid in a pipe under such situation will cause the phenomenon of water hammer.

Instantaneous Closure of a Valve For a detailed physical explanation of the above phenomenon of water hammer, let us consider a simple situation where a long pipeline discharging water from a reservoir is fitted with a valve at its end as shown in Fig. 13.4(a). The uniform flow velocity in the pipe is considered to be V_0.

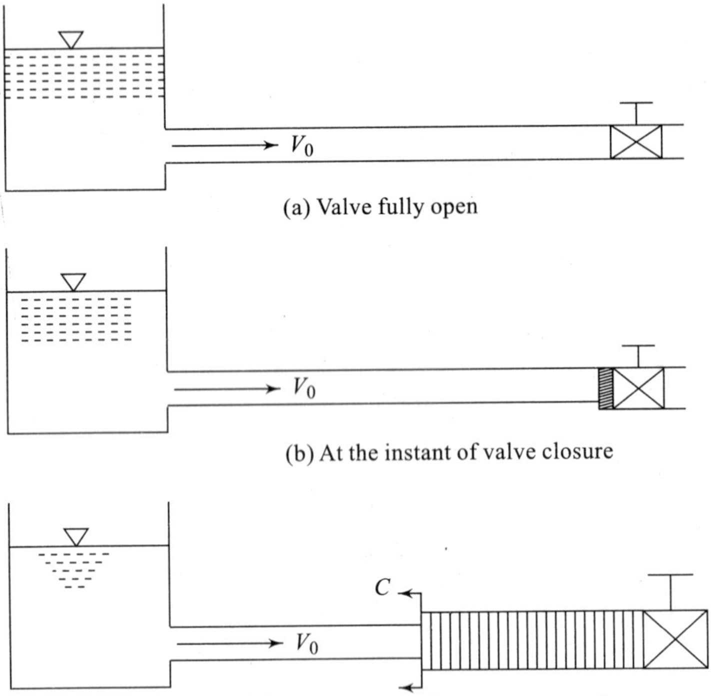

(a) Valve fully open

(b) At the instant of valve closure

(c) Propagation of pressure wave after the valve closure

Fig. 13.4 Effect of instantaneous valve closure

We assume that the valve is closed instantaneously to stop the discharge from the pipeline. An instantaneous closure of a valve is not possible in practice; an extremely rapid closure may be made at the best. However, the concept of instantaneous valve closure makes the explanation simple for a basic physical understanding of the problem. If the liquid is fully incompressible, then the instantaneous closure of the valve will cause the entire liquid in the pipe to come to rest instantaneously. But any liquid, in fact, is compressible to some extent and so its constituent particles do not decelerate instantaneously. Therefore even an instantaneous closure of the valve cannot make the entire column of fluid stationary at once.

Only the fluid particles adjacent to the valve will be stopped instantaneously, and the other would come to rest later (Fig. 13.4(b)). While the flow near the valve is stopped completely, the fluid far away from the valve still moves with a velocity V_0 and compresses the fluid adjacent to the valve increasing its pressure and density. This way, fluid column comes to rest layer by layer from valve end to the reservoir (Fig. 13.4(c)). The kinetic energy of the liquid coming to rest is transformed partly into elastic energy of liquid by compression and partly into elastic energy of pipe due to its expansion. The process of deceleration and subsequent pressure rise of the liquid column due to the valve closure is conceived by the propagation of a pressure wave upstream as a message that is generated at the valve end. As the pressure wave moves upstream, the fluid downstream, though which it has moved, comes to rest and the portion of the pipe downstream expands, depending upon its rigidity, due to rise in pressure of the fluid. The fluid upstream, where the pressure wave is yet to reach, is still in motion with the velocity V_0. The velocity with which the pressure wave moves upstream is very high compared to the velocity of the liquid. The increase in pressure head of the liquid, and the velocity of propagation of pressure wave are the two important parameters to be determined in analysing any water hammer problem.

Velocity of Pressure Wave Figure 13.5(a) shows a pipe in which liquid flowing from left to right with a velocity V_0, is brought to rest by a pressure wave *XX* moving from right to left.

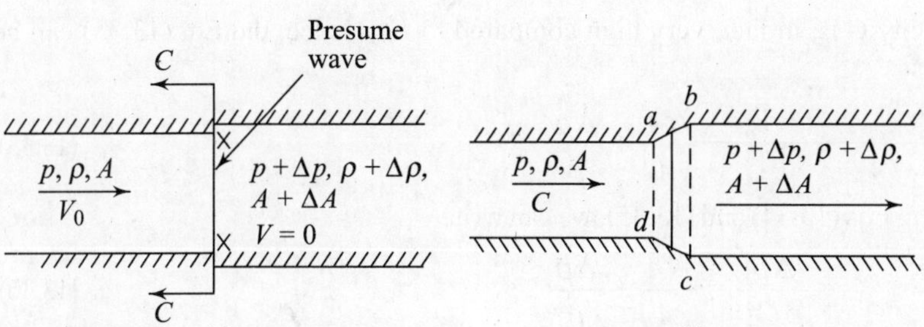

(a) Propagation of a pressure wave in a pipe flow

(b) Model of a control volume analysis in determining the wave velocity

Fig. 13.5

Let the pressure and density of the undisturbed liquid left of the wave be p and ρ respectively, and the cross-sectional area of the pipe be A. After the wave has passed, these quantities become $p + \Delta p$, $\rho + \Delta\rho$ and $A + \Delta A$ respectively as shown in Fig. 13.5(a). Let the velocity of propagation of the pressure wave be C relative to the flowing liquid, and hence, $C–V_0$ with respect to the stationary pipe. The conditions will appear steady if we refer to coordinate axes moving with the wave, which, in other words, means to consider a system where a velocity $C–V_0$ in an opposite direction to that of the wave is superimposed on the flow to bring the wave front stationary as illustrated in Fig. 13.5(b). Here the wave will appear to be stationary while the fluid from left approaches with a velocity C and moves away with a velocity $C–V_0$ after crossing the wave. Now we apply the continuity and momentum equations for a steady flow to an elemental control volume *abcd* across the wave front as shown in Fig. 13.5(b).

Continuity Equation

$$A\rho\, C = (A + \Delta A)(\rho + \Delta\rho)(C - V_o)$$

or

$$A\rho\, C = (A\rho + \rho\,\Delta A + A\,\Delta\rho)(C - V_o)$$

(neglecting the higher order term $\Delta A\,\Delta\rho$)

or

$$A\rho V_o = (C - V_o)(\rho\,\Delta A + A\,\Delta\rho)$$

Dividing both the sides by $A\rho\,(C - V_o)$, we get

$$\frac{V_o}{C - V_o} = \frac{\Delta A}{A} + \frac{\Delta \rho}{\rho} \tag{13.33}$$

Momentum Equation Neglecting the wall shear force, we can apply the momentum theorem to the control volume *abcd* as

$$A\rho C\,[(C - V_o) - C] = p(A + \Delta A) - (p + \Delta p)(A + \Delta A)$$

or $$A\rho\, CV_0 = \Delta p\, A \text{ (the higher order term } \Delta p \Delta A \text{ is neglected)}$$

or $$V_o/C = \frac{\Delta p}{\rho C^2} \tag{13.34}$$

The velocity C is, in fact, very high compared to V_o. Hence, the Eq. (13.33) can be written as

$$\frac{V_o}{C} = \frac{\Delta A}{A} + \frac{\Delta \rho}{\rho} \tag{13.35}$$

Comparing Eqs (13.34) and (13.35), we can write

$$\frac{\Delta p}{\rho C^2} = \frac{\Delta A}{A} + \frac{\Delta \rho}{\rho} \tag{13.36}$$

The change in density of a fluid is related to its change in pressure through the bulk modulus of elasticity E (Eq. (1.22) in Chapter 1) as

$$\Delta p = E\,\frac{\Delta \rho}{\rho} \tag{13.37}$$

Substituting the value of $\Delta\rho/\rho$ from Eq. (13.37) into Eq. (13.36) we have

$$\frac{\Delta p}{\rho C^2} = \frac{\Delta A}{A} + \frac{\Delta p}{E}$$

or $$C^2 = \frac{\Delta p}{\rho\left(\dfrac{\Delta A}{A} + \dfrac{\Delta p}{E}\right)} = \frac{E/\rho}{1 + \dfrac{E}{\Delta p}\left(\dfrac{\Delta A}{A}\right)}$$

Hence, $$C = \left[\frac{E/\rho}{1 + \dfrac{E}{\Delta p}\left(\dfrac{\Delta A}{A}\right)}\right]^{1/2} \tag{13.38}$$

The quantity $\Delta A/A$ in Eq. (13.38) is found out in consideration of the elasticity of the pipe. It is assumed that the pipe is subjected to circumferential hoop stress σ_t but negligible longitudinal stress. Then we can write

$$\frac{\Delta A}{A} = \frac{2\Delta d}{d} = \frac{2\sigma_t}{E_p} \tag{13.39}$$

where σ_t is the hoop stress and E_p is the elasticity of the pipe material. For a circular pipe in which the thickness t of the wall is small compared to the diameter d, the hoop stress is given by

$$\sigma_t = \frac{\Delta pd}{2t}$$

Therefore, from Eq. (13.39)

$$\frac{\Delta A}{A} = \frac{\Delta pd}{tE_p} \qquad (13.40)$$

Inserting the expression of $\Delta A/A$ from Eq. (13.40) into Eq. (13.38), we have

$$C = \left[\frac{E/\rho}{1 + Ed/E_p t}\right]^{1/2} \qquad (13.41)$$

For a rigid pipe, the quantity Ed/E_pt is small compared to unity, and hence the Eq. (13.41) can be written as

$$C = [E/\rho]^{1/2} \qquad (13.42)$$

The quantity $(E/\rho)^{1/2}$ corresponds to the speed of sound through an elastic medium. Therefore, Eq. (13.42) implies that the speed of pressure wave relative to the flowing liquid is equal to the local acoustic speed through the liquid. Taking the value of E for water at 20 °C as 2.2×10^6 kN/m^2 and $\rho = 10^3$ kg/m^3, the value of C from Eq. (13.42) is found to be 1482 m/s. Other liquids give figures of the same order. Let us calculate the value of C from Eq. (13.41) in consideration of pipe elasticity. For a steel pipe, $E_p = 2 \times 10^8$ kN/m^2. Considering the diameter and thickness of the pipe to be 75 mm and 6 mm respectively, we have

$$C = \left[\frac{2.2 \times 10^9/10^3}{1 + \dfrac{2.2 \times 10^9 \times 0.075}{2 \times 10^{11} \times 0.006}}\right]^{1/2}$$

$$= 1390.7 \text{ m/s}$$

Hence we see that the variation in the value of C calculated from Eqs (13.41) and (13.42) is marginal as compared to their absolute values. In fact, the values of C are much in excess of any liquid flow velocity encountered in practice. Therefore, the Eq. (13.42) is used to determine the value of C for all practical purposes.

Reflection of Waves and Pressure Fluctuation We have so long discussed how a pressure wave is generated at the valve end due its instantaneous closure and is transmitted upstream by decelerating and pressurising the liquid column in the pipe. If the pipe is not of infinite length, the reflection of pressure wave at the reservoir and valve ends causes a periodic fluctuation of pressure at any location in the pipe. This is illustrated in Fig. 13.6.

Let us assume, for the sake of simplicity, that the flow is inviscid. When the valve is closed instantaneously, a pressure wave moves upstream with a velocity C relative to the liquid as discussed earlier. The wave, as it progresses, brings the liquid to rest increasing its pressure (Fig. 13.6(b)). Let us consider the initial pressure to be p_0 and the corresponding pressure head to be h_o ($= p_o/\rho g$). The increase in presure head of the liquid due to the propagation of pressure wave upstream can be found from Eq. (13.34) as $\Delta h = \Delta p/\rho g = CV_0/g$. Therefore, after a time $t = l/C$ (Fig. 13.6(c)), where l is the length of the pipe, the whole pipe is filled with high pressure liquid (the pressure head being more than the original one by an amount CV_0/g) at rest.

The situation illustrated in Fig. 13.6(c) is unstable since there occurs a discontinuity of pressure at the reservoir end, because the liquid is at original pressure in the reservoir unlike in the pipe where it is at increased pressure. What happens, in this situation, is that the liquid begins to flow from the pipe back into the reservoir so as to equalise the liquid pressure in the pipe to the original value existing in the reservoir. This is conceived by the propagation of a reflected pressure wave from reservoir end towards the valve end. The action of this reflected wave from the reservoir end is to superimpose a negative pressure head, $-\Delta h$ of same magnitude of CV_0/g on the existing positive pressure head Δh and to set a velocity of the liquid towards the reservoir. When the pressure wave reaches the valve end at $t = 2\ l/C$, the entire liquid in the pipe is at original pressure and is moving with a velocity V_o towards the reservoir. The pipe diameter is also back to its original value. This condition, as depicted in Fig. 13.6e, is similar to that at $t = 0$ (Fig. 13.6(a)) except that the liquid velocity V_0 is in the opposite direction.

As liquid tries to maintain its inertia of motion, i.e., its velocity V_0 towards the reservoir end (Fig. 13.6(e)), the decompression of the liquid column in the pipe takes place. Therefore, the pressure of the liquid in the pipe falls below its original value. This decrease in pressure in the liquid column again starts from the valve end and progresses gradually towards the reservoir end. The fall in pressure in the entire liquid column is thus conceived by the propagation of a negative pressure wave as a reflected wave from the valve end. The magnitude of the reflected wave is same as that of the incident wave, and the sign remains unchanged.

At time $t = 3l/C$, when the negative pressure wave reaches the reservoir end, the entire fluid in the pipe is at rest and at a pressure head lower than the original one by an amount of Δh (Fig. 13.6(g)). This is again an unstable situation due to pressure discontinuity at the reservoir end and causes a flow of liquid from the reservoir end to the valve end to equalise the pressure in the liquid, i.e., to destroy the negative pressure head of the liquid in the pipe. This process is again depicted by the propagation of a positive pressure wave Δh, from the reservoir end towards the valve, and at time $t = 4l/C$ this pressure wave will reach the valve end when the pressure of the entire liquid column in the pipe is again at its original value and the velocity is V_0 towards the valve.

Therefore, we observe that after a time period of $t = 4\ l/C$, the initial condition of the liquid in the pipe, i.e., the condition at the instant when the valve was closed (at $t = 0$), is reached (Fig. 13.6(i)). This complete cycle of events is repeated and, in the absence of friction, would be repeated indefinitely, with the same period of time $4\ l/C$ and with undiminishing intensity of pressure waves.

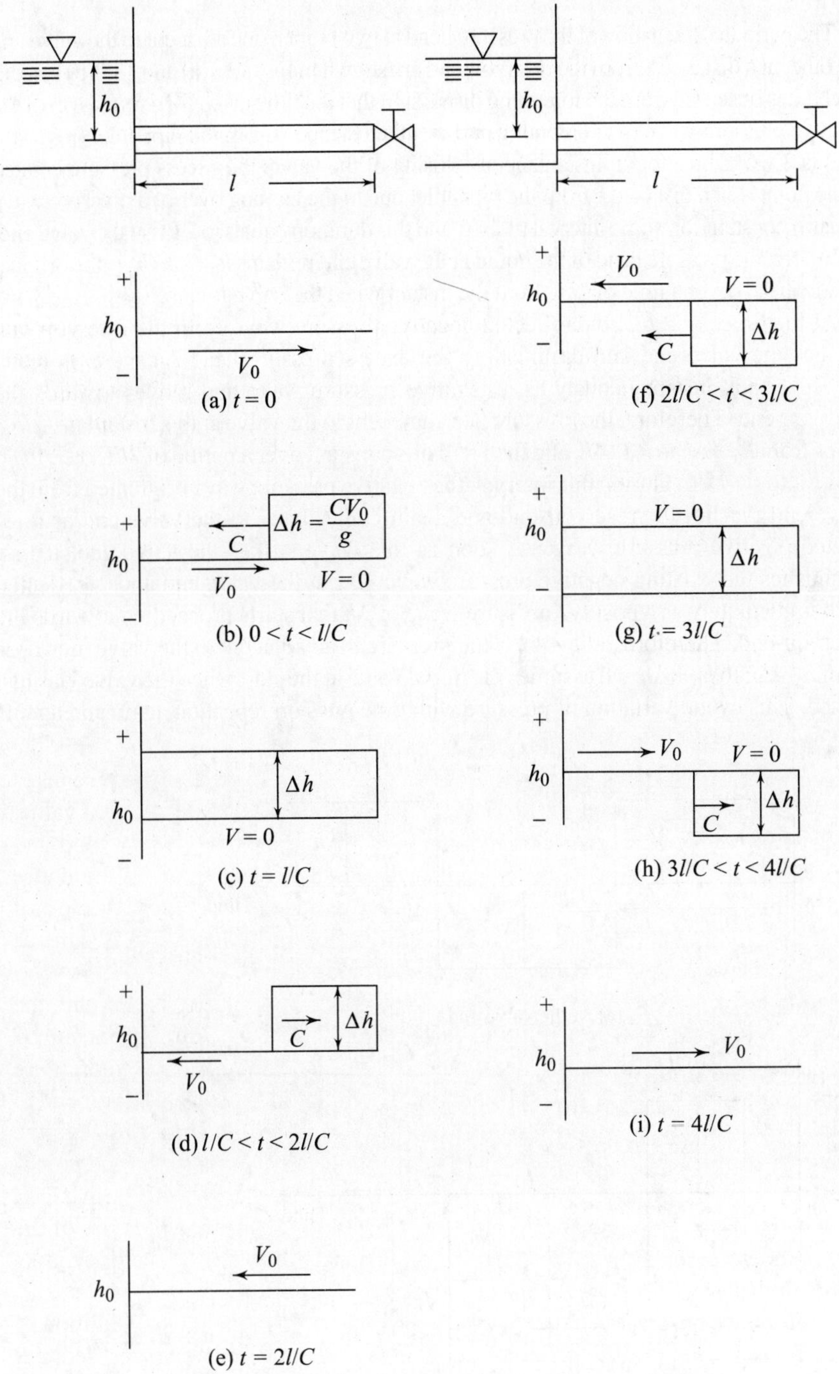

Fig. 13.6 Temporal histories of pressure head along a pipe length after an instantaneous closure of valve (inviscid fluid)

The periodic fluctuation of the pressure head at two points, one adjacent to the valve and the other at a distance x from the reservoir end are shown in Fig. 13.7(a) and 13.7(b) respectively. It is observed from the foregoing discussion that the time taken for a round trip of the positive pressure wave over any point, say A, at a distance x from the pipe inlet (reservoir end) is $2\,x/C$. Thus, for an instantaneous closing of the valve, the excess pressure created at the point A at a distance x from the pipe inlet due to the passing over of a pressure wave remains constant for a time interval of $2x/C$ and this duration equals to $2\,l/C$ at the valve end. Therefore, the pressure head of the liquid at the valve end remains $h_0 + \Delta h$ (h_0 is the original presure head) over a time of $2l/C$ from the instant when the valve is closed.

At the time $t = 2\,l/C$, the reflected negative pressure wave from the reservoir end reaches the valve end and diminishing the excess pressure head Δh there, is again reflected back instantaneously as a negative pressure wave and moves towards the reservoir end. Therefore, the pressure head adjacent to the valve at this instant, $t = 2l/C$, drops from $h_0 + \Delta h$ to $h_0 - \Delta h$, and then remains constant over a period of $2l/C$, i.e., from $t = 2\,l/C$ to $t = 4\,l/C$. During this interval, the negative pressure wave originated from the valve end reaches the reservoir end and again comes back to the valve end as a reflected positive pressure wave. As soon as this wave strikes the valve end, it first diminishes the existing negative pressure wave $-\Delta h$ at the valve end and is reflected back immediately as a positive pressure wave of Δh that starts proceeding towards the reservoir end. Therefore, at $t = 4\,l/C$ the pressure head adjacent to the valve increases from $h_0 - \Delta h$ to $h_0 + \Delta h$ and assumes the initial value at the start when the valve was just closed. This cyclic variation of pressure with time goes on repeating again and again.

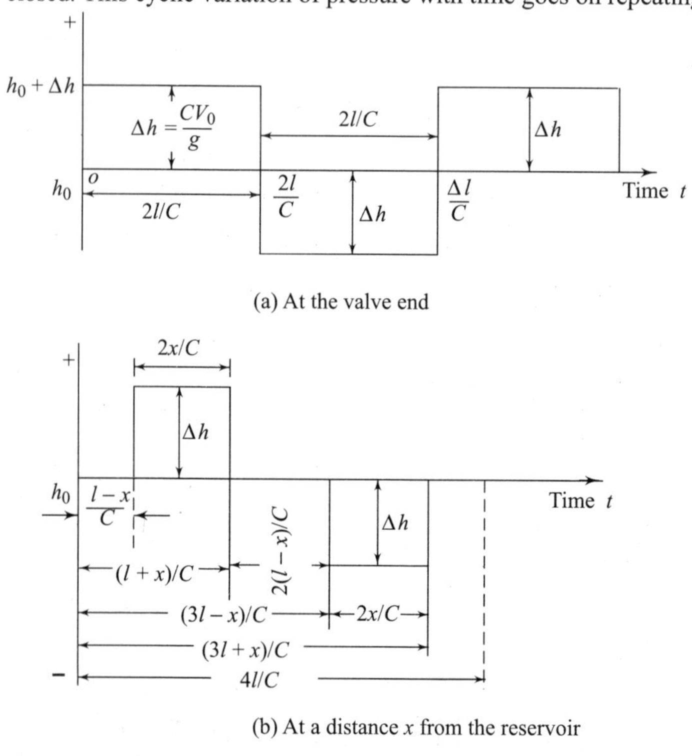

Fig. 13.7 Pressure-time diagram for instantaneous valve closure

Figure 13.7(b) shows the pressure-time diagram for a point at a distance x from the reservoir. In this case, the pressure at the point remains at its original value from the instant the value is closed ($t = 0$) until the positive pressure wave, originating from the valve end, reaches there after a time $t = (l - x)/C$. Therefore, at $t = (l - x)/C$, the pressure head changes to $h_o + \Delta h$ and remains the same for a period of $2\,x/C$ which is the time required for the round trip of the pressure wave to the reservoir and back to that point. At time $t = (l + x)/C$ (Fig. 13.7(b)), the pressure head changes from $h_o + \Delta h$ to h_o, the original pressure, and remains at this value for a period of $2(l - x)/C$ during which the negative pressure wave reaching the valve end is again reflected back to the point. At this instant, given by $t = (3\,l - x)/C$, the pressure head at the point falls from h_o to $h_o - \Delta h$ and remains at this value for a period of $2\,x/C$ until the negative pressure wave, after reaching the reservoir end, is reflected back as a positive pressure wave to that point. Therefore, at $t = (3l + x)/C$, the pressure head increases instantaneously from $h_o - \Delta h$ to h_o. After a time of $l - x/C$ from then the positive wave reaches the valve end, when the situation in the entire pipe is identical to that of the initial one when the valve was just closed.

The effect of friction on the pressure-time diagram for a point at the valve end is shown in Fig. 13.8. Due to the viscous dissipation of energy, the amplitude of pressure wave is reduced in each reflection and hence the oscillations of the pressure wave is damped. The interesting feature is that while the excess pressure over the time period $2l/C$ remains constant in the case without friction, it changes when frictional effect is considered. When the velocity of fluid is reduced, so is the head lost to friction. Therefore, the head available at the downstream end of the pipe consequently rises somewhat as layer after layer of the fluid is slowed down. This effect is transmitted back from each layer in turn with a velocity C, and so the full effect is not felt at the valve until a time $2l/C$ after its closure. In Fig. 13.8, this effect is indicated by the upward slope of the line *ab*. During the second time interval of $2l/C$, velocity and pressure amplitudes have reversed their signs, and thus the line slopes downwards. The frictional effect is usually neglected since the friction head is small compared to the head produced by the water hammer. However, it is always safer to design a pipeline assuming the initial head at the valve to be the same as in the reservoir, and thus neglecting subsequent frictional effects.

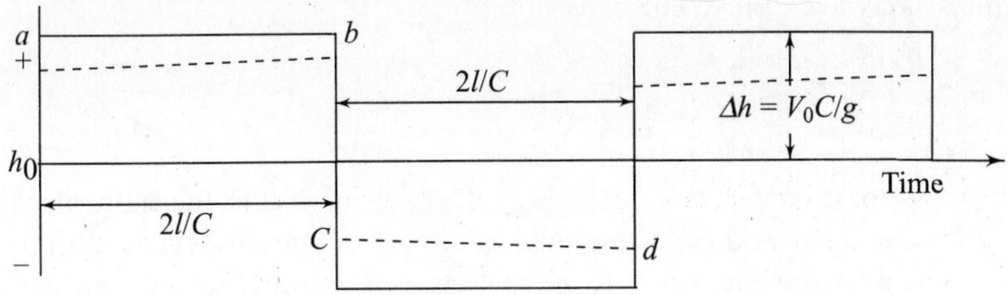

Fig. 13.8 Effect of friction on pressure-time diagram at valve end for instantaneous valve closure

Rapid and Slow Closure of the Valve Our discussion has so far been based on the instantaneous closure of the valve which means that the time taken for the valve to be fully closed is zero. But this is practically impossible, and therefore, some time must elapse for the complete closure of the valve. If this time interval of valve closure is equal to or less than $2l/C$, then results are not essentially different from that discussed from an instantaneous valve closure. Therefore, when the time for the valve to be fully closed is less than or equal to $2l/C$, the closing of valve is known as *rapid closure*. In rapid closure, though the pressure head at the valve is gradually built up as the valve is closed, the maximum pressure head reached for an inviscid fluid, however is the same and equals to CV_o/g as with the instantaneous closure. This is because the conversion of entire kinetic energy of fluid to its strain energy (or pressure energy) is completed before any reflected wave reaches the valve end. If on the other hand, the time for complete closure of the valve is greater than $2\,l/C$, then before the entire kinetic energy being converted into strain energy to raise the pressure head to its maximum value of CV_o/g, a reflected wave of negative pressure arrives to reduce the pressure head at the valve end. This situation is termed as *slow closure* of valve. Therefore, we see that the maximum pressure rise depends on whether the time during which the valve is closed is greater or less than $2\,l/C$. When the time of valve closure is much longer than $2\,l/C$, the effect of compressibility may be neglected. Thus we can summarise the above discussion as follows:

t_c(time taken for valve closure) = 0 (instantaneous closure)

$\leq 2l/C$ (rapid closure)

$> 2l/C$ (slow closure)

$\gg 2l/C$ (slow closure where compressibility effect and subsequent phenomenon of water hammer can be neglected)

When a valve is rapidly closed ($t_c < 2l/C$), the whole length of the pipe is not subjected to peak pressure. Let, the length x_o of the pipe from the reservoir end be subjected to reduced pressure while the remaining portion, $(l - x_o)$ up to the valve end be subjected to peak pressure head CV_o/g. The value of x_o depends upon the value of t_c, the time of valve closure, and can be obtained by equating the time for the peak pressure to be generated up to the length $(l - x_o)$ with the time for the first reflected negative pressure wave to reach there as

$$t_c + \frac{l - x_0}{C} = \frac{l}{C} + \frac{x_0}{C}$$

or

$$x_o = Ct_c/2$$

When $t_c = 0$, i.e., for instantaneous closure, $x_o = 0$, which means that the entire pipe is subjected to maximum pressure. The essential feature in the analysis of water hammer problems due to a rapid or slow closure of the valve is to assume that the movement of the valve does not take place continuously, rather in series of discrete steps of instantaneous partial closure occurring at equal intervals of $2l/C$ or a sub-multiple of $2l/C$. Between these discrete steps, the valve is assumed stationary. Each of these steps generates its own particular wave which is similar in form to those depicted in Fig. 13.7(a) and 13.7(b). We can calculate the increase in pressure head due to the first step of instantaneous closure by assuming that the velocity is reduced from V_0 to V_1 in this step. The momentum equation for a control volume circumscribing the pressure wave in a steady state, under this situation, can be written as

$$\rho C\,[(C - V_0 + V_1) - C] = -\Delta p$$

or

$$\rho C(V_0 - V_1) = \Delta p$$

Hence,

$$\Delta h = \frac{\Delta p}{\rho g} = \frac{C(V_0 - V_1)}{g} \tag{13.43}$$

For a rapid closure, the total pressure head at the valve end at the instant of its complete closure is given by

$$\Sigma\,\Delta h = \frac{CV_0}{g}$$

This is because no reflected wave returns back to the valve before it is completely closed. Determination of V_1 and the pressure head developed for the first step is made as follows.

Let the initial pressure head and the pressure head after the first step of partial valve closure be h_0 and h_1 respectively. Then Eq. (13.43) can be written as

$$h_1 - h_0 = \frac{C(V_0 - V_1)}{g} \tag{13.44}$$

Another relation between h_1 and V_1 is required if either is to be calculated. Let us consider that the valve discharges into atmosphere, and it is regarded as similar to an orifice with a constant coefficient of discharge C_d. Therefore, we can write from continuity

$$AV_1 = Q = C_d A_v (2gh_1)^{1/2} \tag{13.45a}$$

where A_v is the area of valve opening after the first step of closure and A is the cross-sectional area of pipe where the fluid velocity is V_1. Equation (13.45a) can be written as

$$V_1 = B\,(h_1)^{1/2} \tag{13.45b}$$

where,

$$B = C_d (A_v/A)(2g)^{1/2}$$

The factor B is usually known as the *valve opening factor* or *area coefficient*. It should be noted that C_d is not necessarily constant, and therefore the variation of B with the valve setting has to be determined by experiment for each design of valve. Simultaneous solution of Eq. (13.44) and (13.45b) gives the values of V_1 and h_1. Calculations are usually carried out step by step for each discrete step of partial closure of the valve.

Surge Tanks In many practical situations, problems associated with water hammer may be overcome by the use of a surge tank. One such situation occurs in hydroelectric power stations. In hydroelectric installations, the turbine is supplied with water via a long pipeline or a tunnel cut through rock known as *penstock*. If the electric power taken from the generator which is mechanically coupled to the turbine is suddenly altered the turbine tends to change its speed. However this speed is kept constant to maintain the constancy in cycle frequency in the power line, by altering the water flow rate to the turbine through the operation of a valve at its inlet. This is known as *governing of turbines* and the mechanism, through which it is automatically done, is known as *governor*. Therefore, it is the consequent acceleration or deceleration of water in the pipeline which may give rise to water hammer.

The minimisation of water hammer is of utmost importance because the large pressure fluctuations not only produce a harmful effect on the pipeline but also impede the governing. By using a surge tank in the pipeline at a convenient place near the turbine, the adverse effect of water hammer can be restricted to a shorter length of the penstock. Such an arrangement is shown in Fig. 13.9.

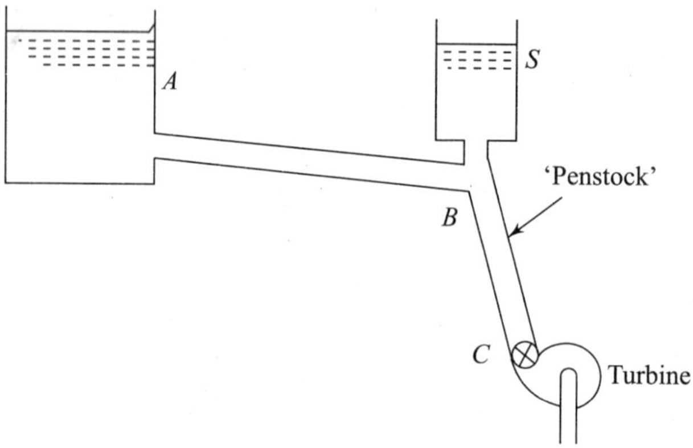

Fig. 13.9 A simple surge tank

The simplest type of surge tank is an open vertical cylinder *S* (Fig. 13.9) of large diameter. It may be constructed of steel, or tunnelled in rock, and should be as close to the turbine as possible. The upstream pipeline *AB* is of small slope, and the top of the surge tank *S* is higher than the water level in the reservoir *A*. When there is a sudden reduction in load on the turbines, the rate of flow of water to the turbines is decreased through the governing mechanism. But the rate of flow in the line *AB* cannot fall at once to the required new value. What happens, under this situation, is that the temporary surplus of water goes into the surge tank *S* and the rise in water level in the surge tank then creates a hydrostatic head which decelerates the water in pipe *AB*. In case the required deceleration is very high, water is allowed to overflow from the top of the surge tank so that the head in the surge tank does not increase indefinitely. Thus a gradual deceleration of water in pipe *AB* takes place. Therefore, a much shorter length of pipe *BC* is now subjected to water hammer effects due to partial closure of the valve *C*. Therefore the pipe *BC* must be constructed strong enough to withstand the increased pressure.

Another important feature of a surge tank is that it provides a reverse supply of water to make up a temporary deficiency of flow down the pipe *AB* when the demand at the turbines is increased. If the load on the turbines is suddenly increased, a sudden acceleration of the water column in the supply pipe is required. The excessive drop in pressure at the turbines, under this situation, is controlled by supplying water from the surge tank and thus meeting up the demand. As the water level in the surge tank is drawn down, the difference in head along *AB* is increased, and so the water there is gradually accelerated until the rate of flow in *AB* equals to that required by the turbine.

We present here an analysis of the reduction of flow rate in a hydroelectric installation with a simple cylindrical surge tank as shown in Fig. 13.10. The part *AB* of the pipe is free from water hammer effects since it has two open reservoirs at its ends. Therefore, the flow in this part is treated as a simple inertia problem similar to that discussed in Section 13.2. The flow in pipe *BC* is subjected to water hammer.

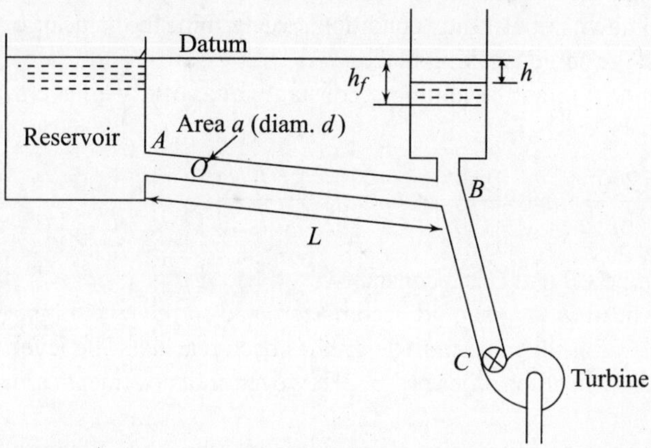

Fig. 13.10 Working principle of a cylindrical surge tank

At any instant, we can write from continuity,

$$aV = A\left[-\frac{dh}{dt}\right] + Q \tag{13.46}$$

where,

A = Cross-sectional area of surge tank
a = Cross-sectional area of upstream pipeline *AB*
V = Average velocity (over a cross-section) in pipe *AB*
h = Depth of water level in surge tank below that of the reservoir, which is taken as datum
Q = Rate of volume flow through the pipe *BC* to the turbine

Under steady condition, the level in the surge tank h would be constant and would equal to the frictional head loss h_f due to flow from the reservoir to the surge tank through pipe AB. But at any instant while the surge is taking place, the level in the surge tank goes up fro. . its steady state level, and thus an additional head of $(h_f - h)$ is available to decelerate the liquid in pipe AB. If the area of the surge tank is considered to be large compared to a, the area of pipe AB, then the frictional head and the head required to decelerate the liquid in the surge tank can be neglected compared to that required for decelerating the liquid in pipe AB. Therefore, we can write, according to Eq. (13.2),

$$h_f - h = \frac{L}{g}\left(-\frac{\mathrm{d}V}{\mathrm{d}t}\right) \tag{13.47}$$

Instantaneous values of h and V can be found out from simultaneous solution of Eqs (13.46) and (13.47). It is difficult to have a closed form solution if friction factor f in determining h_f is not constant. However, numerical integrations of Eqs (13.46) and (13.47) are possible, with a known initial steady condition, to determine the value of h and V at every instant while the surge is taking place.

For a special case when $Q = 0$, and assuming a constant value of f, we have a solution for V as

$$V^2 = \frac{2gd}{4fL}\left(h + \frac{ad}{4fA}\right) + C\exp\left(\frac{4fAh}{ad}\right)$$

where d is the diameter of pipe AB and C is a constant.

Under steady condition, head $(h_f - h)$ should become zero and so the level in the tank should fall immediately after the maximum height has been reached. The level then oscillates about the steady position where $h = h_f$. However, the movements are damped out by friction.

We can conclude from the above discussion that a surge tank has the following two distinct functions:

(i) *Minimisation of water hammer effect* in the pipelines leading from penstock to the turbines.

(ii) *Taking up the surplus water* when the load is reduced and meeting up with the extra water when the load is increased.

A simple cylindrical surge tank has the disadvantage in a sense that these two effects are in no way separated, and hence it becomes a little sluggish in operation. Tanks of different designs with varying cross section along the height and with overflow devices or damping arrangements such as a restriction in the entrance are incorporated in practice.

Example 13.4

Determine the maximum time for rapid valve closure on a pipeline 600 mm in diameter, 450 m long, made of steel ($E = 207 \times 10^6$ kN/m^2) with a wall thickness of 12.5 mm. The pipe contains benzene of specific gravity 0.88, $E = 1.035 \times 10^6$ kN/m^2 flowing at 0.85 m^3/s. The pipe is not restricted longitudinally.

Solution

The maximum time for a rapid valve closure is given by

$$t_{max} = \frac{2l}{C}$$

where l is the length of the pipe and C is the velocity (relative to flow of liquid) of pressure wave created by the valve closure.
C is given according to Eq. (13.41) as

$$C = \left[\frac{E/\rho}{1 + (Ed/E_p t)}\right]^{1/2}$$

$$= \left[\frac{1.035 \times 10^9 / 0.88 \times 10^3}{1 + \left(\frac{1.035 \times 600}{207 \times 12.5}\right)}\right]^{1/2}$$

or $$C = 974 \text{ m/s}$$

Hence, $$t_{max} = \frac{2 \times 450}{974} = 0.924 \text{ s}$$

Example 13.5

Water has to flow uniformly at the rate of 0.20 m^3/s though a pipe of 200 mm diameter. Calculate the minimum thickness of the pipe that has to be provided if, for a sudden stoppage of flow, the pipe should not be stressed more than 5×10^4 kN/m^2. (Take E for water = 2×10^6 kN/m^2 and E for the pipe material = 120×10^6 kN/m^2)

Solution

The velocity of flow V through the pipe is given by

$$V = \frac{4 \times (0.20)}{\pi (0.2)^2}$$

$$= 6.37 \text{ m/s}$$

The velocity of pressure wave created due to valve closure is determined using Eq. (13.41) as

$$C = \sqrt{\frac{(2 \times 10^9 / 10^3)}{1 + (2 \times 0.2/120\, t)}}$$

(where t is the thickness of the pipe)

$$= \sqrt{\frac{2 \times 10^6}{1 + 0.0033/t}}$$

$$= \sqrt{\frac{2t}{t+0.0033}} \times 10^3 \text{ m/s}$$

Now, $$\Delta p = \rho CV = 6.37\sqrt{\frac{2t}{t+0.0033}} \times 10^6 \text{ N/m}^2$$

Again, from the consideration of stress in the pipe wall

$$t = \frac{\Delta p d}{2\sigma} = \frac{6.37 \times 0.2}{2 \times 5 \times 10^7}\sqrt{\frac{2t}{t+0.0033}} \times 10^6$$

$$= 0.01274\sqrt{\frac{2t}{t+0.0033}}$$

or $6161t^2 + 20.33t - 2 = 0$

This equation of t gives one positive root of $t = 0.016$ as the feasible solution. Therefore, $t = 0.016$ m $= 16$ mm.

Example 13.6

A uniform U-tube has two vertical limbs open to atmosphere and connected by a horizontal middle part. The left and right limbs are filled with liquids of length l_1, l_2 and density ρ_1, ρ_2 respectively. The liquid columns meet in the horizontal part of the tube. Calculate the frequency of oscillation under gravity, neglecting viscous effect.

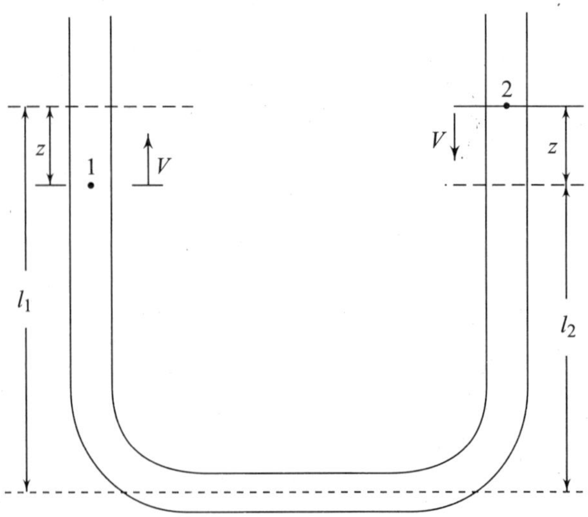

Fig. 13.11 Oscillations of two liquid columns in a U-tube

Solution

At equilibrium position, the heights of liquid columns in left and right limbs from the horizontal base of the manometer are l_1 and l_2 respectively (Fig. 13.11). Let us consider an instant of oscillation when the liquid column in the left limb moves upward while that in the right limb moves downward as shown in Fig. 13.11. By the application of Bernoulli's equation for unsteady flow between points 1 and 2 (Fig. 13.11) we get,

$$\rho_1 \frac{V^2}{2} + p_{atm} + \rho_1 g(l_1 - z) = \rho_2 \frac{V^2}{2} + p_{atm} + \rho_2 g(l_2 + z) + \int_1^2 \rho \frac{\partial V}{\partial t} \mathrm{d}s$$

[Displacement of the liquid columns and hence their velocities are equal since the cross-sectional area of two limbs are considered to be the same]

or $$(\rho_2 - \rho_1)\frac{V^2}{2} + (\rho_2 l_2 - \rho_1 l_1)g + (\rho_2 + \rho_1)gz + (\rho_1 l_1 + \rho_2 l_2)\frac{\mathrm{d}V}{\mathrm{d}t} = 0 \quad (13.48)$$

Equating the hydrostatic pressures at the base of the manometer in the equilibrium position of the liquids, we have

$$\rho_2 l_2 g = \rho_1 l_1 g \quad (13.49)$$

With the help of Eq. (13.49) and writing

$$V = \frac{\mathrm{d}z}{\mathrm{d}t} \text{ and } \frac{\mathrm{d}V}{\mathrm{d}t} = \frac{\mathrm{d}^2 z}{\mathrm{d}t^2}$$

We have from Eq. (13.48)

$$\frac{\mathrm{d}^2 z}{\mathrm{d}t^2} + \frac{(\rho_2 - \rho_1)}{2(\rho_2 l_2 + \rho_1 l_1)}\left(\frac{\mathrm{d}z}{\mathrm{d}t}\right)^2 + \frac{g(\rho_1 + \rho_2)}{2(\rho_2 l_2 + \rho_1 l_1)} z = 0$$

For small values of $(\mathrm{d}z/\mathrm{d}t)^2$ (when amplitudes of oscillation are small compared to the lengths of the liquid columns), this equation becomes

$$\frac{\mathrm{d}^2 z}{\mathrm{d}t^2} + \frac{g(\rho_1 + \rho_2)}{(\rho_2 l_2 + \rho_1 l_1)} z = 0$$

Hence, the frequency of oscillation becomes

$$\omega = \left[\frac{g(\rho_1 + \rho_2)}{(\rho_2 l_2 + \rho_1 l_1)}\right]^{1/2}$$

Example 13.7

A cast iron pipe of 300 mm diameter and 8 mm thick is 1500 m long. The pipe is to convey 200 litre/s of water.

(i) Estimate the maximum time of closure of a valve at the downstream end that would be recognised as a rapid closure.

(ii) What is the peak water hammer pressure produced by rapid closure?

(iii) What is the length of the pipe subjected to peak water hammer pressure if the time of closure is 2.0 s?

[For water $E = 2200$ MPa; for cast iron $E = 80 \times 10^9$ Pa]

Solution

(i) The velocity of the pressure wave due to valve closure is determined according to Eq. (13.41) as

$$C = \sqrt{\frac{2.2 \times 10^9}{10^3\left(1 + \dfrac{2.2 \times 0.3}{80 \times 8 \times 10^{-3}}\right)}}$$

$$= 1041 \text{ m/s}$$

The maximum time of valve closure to be recognised as a rapid one is given by

$$t_{\max} = \frac{2l}{C} = \frac{2 \times 1500}{1041} = 2.88 \text{ s}$$

(ii) The velocity of flow through the pipe

$$V_0 = \frac{0.2 \times 4}{\pi \times (0.3)^2} = 2.83 \text{ m/s}$$

Therefore, the peak pressure due to rapid closure

$$p_{\max} = \rho C V_0 = 10^3 \times 1041 \times 2.83 \text{ Pa}$$

$$= 2.95 \text{ MPa}$$

(iii) Let the length of the pipe from the valve end which will be subjected to peak pressure be x. Then equating the time for the peak pressure to be generated upto the length x from the valve end with the time for the first reflected negative pressure wave to reach there, we have

$$\frac{x}{C} + 2 = \frac{l}{C} + \frac{l - x}{C}$$

or

$$\frac{x}{1041} + 2 = \frac{1500}{1041} + \frac{1500 - x}{1041}$$

which gives

$$x = 459 \text{ m}$$

Example 13.8

A 400 mm steel pipe is 2000 m long and conveys 100 litre/s of water with a static head of 200 m at the downstream end of the pipe. If a valve at the downstream end is closed in 6 s, estimate the stress in the pipe wall at the valve. The pipe thickness is 5 mm. [For water $E = 2.2 \times 10^9$ Pa; for steel $E = 2.2 \times 10^{11}$ Pa]

Use an approximate expression to calculate the maximum rise in pressure head for a slow closure as $\Delta p_s = \dfrac{2l}{TC} \cdot \Delta p_r$, where Δp_s and Δp_r are the peak rises in pressure due to slow and rapid closure respectively. l, C and T are the length of the pipe, the wave velocity and the time of valve closure respectively]

Solution

The wave velocity C is given by

$$C=\sqrt{\frac{2.2\times10^9}{10^3\left(1+\dfrac{2.2\times0.4}{2.2\times10^2\times5\times10^{-3}}\right)}}$$

$$= 1105 \text{ m/s}$$

Velocity of flow, $\quad V_o=\dfrac{0.1\times4}{\pi\times(0.4)^2}=0.796 \text{ m/s}$

The peak rise in water hammer pressure due to rapid closure,

$$\Delta p_r=\rho CV_o=10^3\times1105\times0.796=879 \text{ kPa}$$

The maximum time of rapid closure, $T_{\max}=\dfrac{2\times2000}{1105}=3.62 \text{ s}$

Since the time of closure is 6 s which is greater than 3.62 s, the present situation corresponds to a slow closure.

The rise in pressure head at the valve end due to the slow closure is given by

$$\Delta p_s=879\times\frac{3.62}{6}=530 \text{ kPa}$$

Therefore, the rise in total pressure $\Delta p=\Delta p_s+\Delta p_{\text{static}}$

$$= 530\times10^3+200\times10^3\times9.81 \text{ Pa}$$

$$= 2.5 \text{ MPa}$$

Therefore the stress σ is determined as

$$\sigma=\frac{\Delta pd}{2t}=\frac{2.5\times10^6\times(0.4)}{2\times5\times10^{-3}}=1\times10^8 \text{ N/m}^2=100 \text{ MN/m}^2$$

SUMMARY

- The temporal acceleration in an unsteady flow becomes important when the change in velocity is rapid. In a very fast change of flow, arising from sudden opening or closing of valve, the density of fluid changes considerably and the elastic force becomes significant.
- The difference in the Piezometric pressure, causing a uniform temporal acceleration of a liquid column, is known as *the inertia pressure* and the corresponding head is known as the inertia head, which is given by (L/g) $(\partial V/\partial t)$, where L is the length of the liquid column being accelerated.
- Oscillation of an inviscid liquid column in a U-tube shows an undamped periodic motion with a time period of $2\pi(l/2g)^{1/2}$, where l is the length of the liquid column. The nature of oscillation of a viscous liquid column in a U-tube depends upon the kind of flow and damping factor. For a laminar flow, the oscillation is of damped periodic in nature with diminishing am-

plitude when the damping factor is less than unity. The flow is a non-oscillatory type reaching the equilibrium position asymptotically or a transitory one changing from oscillatory to non-oscillatory types depending upon whether the damping factor is greater than unity or equals to unity respectively.

- When the flow in a pipe line is suddenly reduced by closing a valve downstream, a phenomenon like knocking of the pipe system takes place due to repeated up and down motion of a pressure wave within the pipe. This phenomenon is known as water hammer. The disturbance created at the valve end, due to its closure, propagates upstream as a messenger in the form of a pressure wave with a velocity C (relative to the liquid medium) which equals to $[(E/\rho) / (1 + Ed/E_p t)]^{1/2}$. The rise in pressure head due to deceleration of the liquid to rest by the instantaneous closure of a valve is given by CV_o/g. The valve closure is said to be rapid when the time of closing the valve is less than or equal to $2l/C$ (l being the length of the pipe), so that the maximum rise in pressure head at the valve end becomes equal to $C\,V_0/g$. The valve closure is considered to be slow when the time of closing the valve is greater than $2l/C$ and under this situation the maximum rise in pressure head at the valve end becomes less than $C\,V_o/g$ due to the arrival of a reflected wave of negative pressure head from the reservoir end.
- The problem of water hammer in the penstock in a hydroelectric power station is circumvented by the use of a surge tank.

EXERCISES

13.1 Choose the correct answer:

(i) A long pipe connected to a water tank, providing a constant head, has a valve at its downstream end which is suddenly opend. If t_1 is the time to reach 90% of the steady state flow determined by neglecting friction and other losses, and t_2 is the corresponding time obtained by including friction and other losses, then

(a) $t_2 > t_1$ (b) $t_2 = t_1$ (c) $t_2 < t_1$ (d) $t_2 \geq t_1$
(e) $t_2 \leq t_1$

(ii) The propagation velocity of a pressure wave in a rigid pipe carrying a fluid of density ρ and viscosity μ varies as

(a) ρ (b) $\sqrt{1/\rho}$ (c) ρ/μ (d) $\sqrt{\rho}$

(iii) The downstream valve of a pipe conveying a liquid at steady rate is closed during a time interval of l/C, where l is the length of the pipe and C is the wave velocity relative to liquid in the pipe. Under this situation, the peak water hammer pressure would be experienced

(a) only at the valve end
(b) by one fourth length of the pipe from the valve end
(c) by half of the pipe length
(d) by the full pipe length

(iv) In a pipe of 4000 m long carrying oil, the velocity of propagation of a pressure wave is 500 m/s. A valve at the downstream end is closed suddenly. At the mid point of the pipeline, the peak water hammer pressure will exist for a duration of

(a) 1.0s (b) 2.0s (c) 4.0s (d) 8.0s

(v) A surge tank is provided in a hydroelectric power station to

(a) reduce frictional losses in the system

(b) reduce water hammer problem in the penstock

(c) increase the net head across the turbine

13.2 A 200 mm diameter and 2000 m long pipe leads from a large reservoir to an outlet which is 30 m below the water level in the reservoir. If a valve at the pipe outlet is suddenly opened, find the time required to reach (i) 50% and (ii) 90% of steady state discharge. Assume the friction factor, $f = 0.02$ and minor losses (excluding the exit loss) as $10\,(V^2/2g)$.

Ans. (6.23s, 16.71s)

13.3 Two reservoirs with a constant difference of 15 m in their free water surface are connected by a 200 mm diameter pipe of length 500 m and $f = 0.020$. The minor losses in the pipe (including the exit loss) can be taken as 10 times the velocity head in the pipe. If a valve controlling the flow is suddenly opened, (i) find the time for 95% of the steady flow to be established, and (ii) find the flow at the end of 10s from the opening of the valve.

Ans. (13.78s, 0.06 m^3/s)

13.4 Determine the error in calculating the excess pressure of water hammer in a steel pipe carrying water with an inner diameter $d = 15$ mm and a wall thickness $t = 2$ mm if the elasticity of the material of the pipe wall is disregarded. Take $E = 2.07 \times 10^5$ MN/m^2 for steel and $E = 2.2 \times 10^3$ MN/m^2 for water.

Ans. (3.92%)

13.5 A steel pipe 300 mm in diameter and 1500 m long conveys crude oil having a specific gravity of 0.8 and a bulk modulus of elasticity 1520 MPa. The rate of discharge of oil is 0.08 m^3/s. A valve at the downstream end of the pipe is completely closed in 2s. If the thickness of the pipe is 20 mm, calculate the additional stress in the pipe due to the valve operation. (For steel pipe, modulus of elasticity $= 2.07 \times 10^5$ MPa).

Ans. (8.88 MPa)

13.6 A steel pipeline of 1200 m long, 500 mm in diameter has a wall thickness of 5 mm. The pipe discharges water at the rate of 0.1 m^3/s. The static head at the outlet is 200 m of water. If the working stress of steel is 0.1 kN/mm^2, calculate the minimum time of closure of a downstream valve. For water: $E = 2.2 \times 10^3$ MPa and for steel: $E = 2.07 \times 10^5$ MPa.

Ans. (32.17s)

13.7 A valve at the end of a pipe 600 m long is closed in five equal steps each of $2\,l/C$, where $C = 1200$ m/s (the wave velocity relative to liquid in the pipe). The initial head at the valve which discharges to atmosphere, is 100 m and the initial velocity in the pipe is 1 m/s. Neglecting the frictional effects, determine the head at the valve after 1, 2 and 3 s.

Ans. (116.64 m, 111.5 m, 87.14 m)

13.8 Show that, if the friction loss in a pipeline is proportional to the square of the velocity, the oscillatory motion of the level in a simple, open, cylindrical surge tank following complete shut-down of the turbines in a hydroelectric plant is given by an equation of the form

$$=0\frac{d^2H'}{dt'^2}+\alpha\left(\frac{dH'}{dt'}\right)+\beta H'=0$$

where $H' = H/l$, $t' = t\sqrt{\frac{g}{l}}$, with H being the instantaneous depth of water level in surge tank below that of the reservoir, l the length of the pipeline from reservoir to surge tank and t the time, α and β are the dimensionless constants. Find the values of α and β for surge tank whose diameter is 10 times more than that of the pipeline and the length to diameter ratio of the pipeline is 200. (Take friction factor $f = 0.02$)

Ans. ($\alpha = -200$, $\beta = 0.01$)

14

COMPRESSIBLE FLOW

14.1 INTRODUCTION

Compressible flow is often called variable density flow. For the flow of all liquids and for the flow of gases under certain conditions, the density changes are so small that assumption of constant density remains valid (see Chapter 1).

Consider a small element of fluid of volume $\varv$. The pressure exerted on the element by the neighbouring fluid is p. If the pressure is now increased by an amount dp, the volume of the element will correspondingly be reduced by the amount d$\varv$. The compressibility of the fluid, K, is thus defined as

$$K = -\frac{1}{\varv}\frac{d\varv}{dp} \tag{14.1}$$

However, when a gas is compressed, its temperature increases. Therefore, the above mentioned definition of compressibility is not complete unless the temperature condition is specified. If the temperature is maintained at a constant level, the isothermal compressibility is defined as

$$K_T = -\frac{1}{\varv}\left(\frac{d\varv}{dp}\right)_T \tag{14.2}$$

Compressibility is a property of fluids. Liquids have very low value of compressibility (for example, compressibility of water is 5×10^{-10} m^2/N at 1 atm under isothermal condition), while gases have very high compressibility (for example, compressibility of air is 10^{-5} m^2/N at 1 atm under isothermal condition). If the fluid element is considered to have unit mass, $\varv$ is the specific volume (volume per unit mass) and the density is $\rho = 1/\varv$. In terms of density, Eq. (14.1) becomes

$$K = \frac{1}{\rho}\frac{d\rho}{dp} \tag{14.3}$$

We can also say that for a change in pressure, dp, the change in density is

$$d\rho = \rho K dp \tag{14.4}$$

So far we have thought about a fluid and its property–compressibility. If we also consider the fluid motion, we shall appreciate that the flows are initiated and maintained by changes in pressure on the fluid. It is also known that high pressure gradient is responsible for high speed flow. However, for a given pressure gradient (dp), the change in density of a liquid will be smaller than the change in density of a gas (as seen in Eq. (14.4)). So, for flow of gases, moderate to high pressure gradients lead to substantial changes in the density. Due to such pressure gradients, gases flow with high velocity. Such flows, where ρ is a variable, are known as compressible flows.

If we recapitulate Chapter 1, we can say that the proper criterion for a nearly incompressible flow is a small Mach number,

$$\text{Ma} = \frac{V}{a} << 1 \tag{14.5}$$

where V is the flow velocity and a is the speed of sound in the fluid. For small Mach numbers, changes in fluid density are small everywhere in the flow field. In this chapter we shall treat compressible flows which have Mach numbers greater than 0.3 and exhibit appreciable density changes.

The Mach number is the most important parameter in compressible flow analysis. Aerodynamicists make a distinction between different regions of Mach numbers in the following ways:

- Ma < 0.3: incompressible flow; change in density is negligible.
- 0.3 < Ma < 0.8: subsonic flow; density changes are significant but shock waves do not appear.
- 0.8 < Ma < 1.2: transonic flow; shock waves appear and divide the subsonic and supersonic regions of the flow. Transonic flow is characterized by mixed regions of locally subsonic and supersonic flow.
- 1.2 < Ma < 3.0: supersonic flow; flow field everywhere is above acoustic speed. Shock waves appear and across the shock wave, the streamline changes direction discontinuously.
- 3.0 < Ma: hypersonic flow; where the temperature, pressure and density of the flow increase almost explosively across the shock wave.

The above five categories of flow are appropriate to external aerodynamics. For internal flow, it is to be studied whether the flow is subsonic (Ma < 1) or supersonic (Ma > 1). The effect of change in area on velocity changes in subsonic and supersonic regime is of considerable interest. By and large, in this chapter we shall mostly focus our attention to internal flows. The material in this chapter is inspired by the two-volume classical book on compressible flows by A.H. Shapiro[see Ref.1].

14.2 THERMODYNAMIC RELATIONS OF PERFECT GASES

14.2.1 Perfect Gas

Compressible flow calculations can be made by assuming the fluid to be a perfect gas. A perfect gas is one in which intermolecular forces are neglected. The equation of state for a perfect gas can be derived from the kinetic theory. It was synthesised from laboratory experiments by Robert Boyle, Jacques Charles, Joseph Gay-Lussac and John Dalton. However, for a perfect gas, it can be written

$$p\forall = MRT \tag{14.6}$$

where p is pressure (N/m^2), $\mathcal{V}$ is the volume of the system (m^3), M is the mass of the system (kg), R is the specific gas constant (J/kg K) and T is the temperature (K). This equation of state can be written as

$$p\upsilon = RT \tag{14.7}$$

where υ is the specific volume (m^3/kg). We can also write

$$p = \rho RT \tag{14.8}$$

where ρ is the density (kg/m^3).

In another approach, which is particularly useful in chemically reacting systems, the equation of state is written as

$$p\mathcal{V} = N\mathcal{R}T \tag{14.9}$$

where N is the number of moles in the system, and $\mathcal{R}$ is the universal gas constant which is same for all gases. It may be recalled that a mole of a substance is that amount which contains a mass equal to the molecular weight of the gas and which is identified with the particular system of units being used. For example, in case of oxygen (O_2), 1 kilogram-mole (or kg · mol) has a mass of 32 kg. Because the masses of different molecules are in the same ratio as their molecular weights, 1 mol of different gases always contains the same number of molecules, i.e., 1 kg-mol always contains 6.02×10^{26} molecules, independent of the species of the gas. Dividing Eq. (14.9) by the number of moles of the system yields

$$p\mathcal{V}^1 = \mathcal{R}T \tag{14.10}$$

If, Eq. (14.9) is divided by the mass of the system, we can write

$$p\upsilon = \eta\mathcal{R}T \tag{14.11}$$

where υ is the specific volume as before and η is the mole-mass ratio (kg-mol/kg). Also, Eq. (14.9) can be divided by system volume, which results in

$$p = C\mathcal{R}T \tag{14.12}$$

where C is the concentration (kg-mol/m^3).

The equation of state can also be expressed in terms of particles. If N_A is the number of particles in a mole (Avogadro constant, which for a kilogram-mole is 6.02×10^{26} particles), from Eq. (14.12) we obtain

$$p = (N_A C)\left(\frac{\mathcal{R}}{N_A}\right)T \tag{14.13}$$

In the above equation, $(N_A C)$ is the number density, i.e., number of particles per unit volume and $(\mathcal{R}/N_A)$ is the gas constant per particle, which is nothing but the Boltzmann constant.

Finally, Eq. (14.13) can be written as

$$p = nkT \tag{14.14}$$

where n is the number density and k is the Boltzmann constant.

So far, we have come across different forms of equations of state for perfect gas. They are necessarily same. A closer look depicts that there are a variety of gas constants. They are categorised as

1. *Universal gas constant*: When the equation deals with moles, it is in use. It is same for all the gases.

$$\mathcal{R} = 8314 \text{ J/ (kg-mol-K)}$$

2. *Characteristic gas constant*: When the equation deals with mass, the characteristic gas constant (R) is used. It is a gas constant per unit mass and it is different for different gases. As such $R = \mathcal{R}/M$, where M is the molecular weight. For air at standard conditions,

$$R = 287 \text{ J/(kg-K)}$$

3. *Boltzmann constant*: When the equation deals with particles, Boltzmann constant is used. It is a gas constant per particle.

$$k = 1.38 \times 10^{-23} \text{ J/K}$$

However, the question is how accurately one can apply the perfect gas theory? It has been experimentally determined that at low pressures (1 atm or less) and at high temperature (273 K and above), the value of (pv/RT) for most pure gases differs with unity by a quanity less than one per cent. It is also understood that at very cold temperatures and high pressures the molecules are densely packed. Under such circumstances, the gas is defined as real gas and the perfect gas equation of state is replaced by van der Waals equation which is

$$\left(p + \frac{a}{v^2}\right)(v - b) = RT \tag{14.15}$$

where a and b are constants and depend on the type of the gas. In conclusion, it can be said that for a wide range of applications related to compressible flows, the temperatures and pressures are such that the equation of state for the perfect gas can be applied with a high degree of confidence.

14.2.2 Internal Energy and Enthalpy

Microscopic view of a gas is a collection of particles in random motion. Energy of a particle can consist of translational energy, rotational energy, vibrational energy and electronic energy. All these energies summed over all the particles of the gas, form the internal energy, e, of the gas.

Let, us imagine a gas is in equilibrium. Equilibrium signifies gradients in velocity, pressure, temperature and chemical concentrations do not exist. Let e be the internal energy per unit mass. Then the enthalpy, h, is defined per unit mass, as $h = e + pv$, and we know that

$$\left.\begin{aligned} e &= e(T, v) \\ h &= h(T, p) \end{aligned}\right\} \tag{14.16}$$

If the gas is not chemically reacting and the intermolecular forces are neglected, the system can be called as a thermally perfect gas, where internal energy and enthalpy are functions of temperature only. One can write

$$\left.\begin{aligned} e &= e(T) \\ h &= h(T) \\ \mathrm{d}e &= c_v \, \mathrm{d}T \\ \mathrm{d}h &= c_p \, \mathrm{d}T \end{aligned}\right\} \tag{14.17}$$

If the specific heats are constant it can be called as a calorically perfect gas where

$$\left.\begin{aligned} e &= c_v T \\ h &= c_p T \end{aligned}\right\} \tag{14.18}$$

In most of the compressible flow applications, the pressure and temperatures are such that the gas can be considered as calorically perfect. However, for calorically perfect gases, we can accept constant specific heats and write

$$c_p - c_v = R \tag{14.19}$$

and the specific heats at constant pressure and constant volume are defined as

$$c_p = \left(\frac{\partial h}{\partial T}\right)_p$$

$$c_v = \left(\frac{\partial e}{\partial T}\right)_v \tag{14.20}$$

From Eq. (14.19), one can write

$$1 - \frac{c_v}{c_p} = \frac{R}{c_p} \tag{14.21}$$

We also know that $c_p/c_v = \gamma$. We can rewrite Eq. (14.21) as

$$1 - \frac{1}{\gamma} = \frac{R}{c_p}$$

or

$$c_p = \frac{\gamma R}{\gamma - 1} \tag{14.22}$$

In a similar way, from Eq. (14.19) we can write

$$c_v = \frac{R}{\gamma - 1} \tag{14.23}$$

14.2.3 First Law of Thermodynamics

Let us imagine a system with a fixed mass of gas. If δq amount of heat is added to the system across the system-boundary and if δw is the work done on the system by the surroundings, then there will be an eventual change in internal energy of the system which is denoted by d*e* and we can write

$$\mathrm{d}e = \delta q + \delta w \tag{14.24}$$

This is the First law of thermodynamics. Here, d*e* is an exact differential and its value depends only on *initial and final states of the system*. However, δq and δw are dependent on the process. A process signifies the way by which heat can be added and the work is done on the system. Here, we shall be interested in the isentropic process which is a combination of adiabatic (no heat is added to or taken away from the system) and reversible process (occurs through successive stages, each stage consists of an infinitesimal small gradient). In an isentropic process, entropy of a system remains same.

14.2.4 Entropy and the Second Law of Thermodynamics

Equation (14.24) does not tell us about the direction (i.e., a hot body with respect to its surrounding will gain temperature or cool down) of the process. To determine the proper direction of a process, we define a new state variable, the entropy, which is

$$ds = \frac{\delta q_{rev}}{T} \tag{14.25}$$

where s is the entropy of the system, δq_{rev} is the heat added reversibly to the system and T is the temperature of the system. Entropy is a state variable and it can be connected with any type of process, reversible or irreversible. An effective value of δq_{rev} can always be assigned to relate initial and end points of an irreversible process, where the actual amount of heat added is δq. One can write

$$ds = \frac{\delta q}{T} + ds_{irrev} \tag{14.26}$$

It states that the change in entropy during a process is equal to the actual heat added divided by the temperature plus a contribution from the irreversible dissipative phenomena. The dissipative phenomena always increase the entropy,

$$ds_{irrev} \geq 0 \tag{14.27}$$

Significance of the 'greater than' sign is understood. The 'equal to' sign represents a reversible process. A combination of Eqs (14.26) and (14.27) yields,

$$ds \geq \frac{\delta q}{T} \tag{14.28}$$

If the process is adiabatic, $\delta q = 0$, Eq. (14.28) yields,

$$ds \geq 0 \tag{14.29}$$

Equations (14.28) and (14.29) are the expressions for the second law of thermodynamics. The second law tells us in what direction the process will take place. The direction of a process is such that the change in entropy of the system plus surrounding is always positive or zero (for a reversible adiabatic process). In conclusion, it can be said that the second law governs the direction of a natural process.

For a reversible process, it can be said (see Nag [2]) that $\delta w = -p d\mathcal{v}$, where $d\mathcal{v}$ is change in volume and from the first law of thermodynamics it can be written as

$$\delta q - p d\mathcal{v} = de \tag{14.30}$$

If the process is reversible, we use the definition of entropy in the form $\delta q_{rev} = T\,ds$, then Eq. (14.30) becomes

$$T\,ds - p d\mathcal{v} = de$$

or

$$T ds = de + p d\mathcal{v} \tag{14.31}$$

Another form can be obtained in terms of enthalpy. For example, by definition

$$h = e + p\mathcal{v}$$

Differentiating, we obtain

$$dh = de + p d\mathcal{v} + \mathcal{v} dp \tag{14.32}$$

Combining Eqs (14.31) and (14.32), we have

$$Tds = dh - v dp \tag{14.33}$$

Equations (14.31) and (14.33) are termed as first *Tds* equation and second *Tds* equation, respectively.

For a thermally perfect gas, we have $dh = c_p dT$ (from Eq. 14.20) and we can substitute this in Eq. (14.33) to obtain

$$ds = c_p \frac{dT}{T} - \frac{v dp}{T} \tag{14.34}$$

Further substitution of $pv = RT$ into Eq. (14.34) yields

$$ds = c_p \frac{dT}{T} - R\frac{dp}{p} \tag{14.35}$$

Integrating Eq. (14.35) between states 1 and 2,

$$s_2 - s_1 = \int_{T_1}^{T_2} c_p \frac{dT}{T} - R \ln \frac{p_2}{p_1} \tag{14.36}$$

If c_p is a variable, we shall require gas tables; but for constant c_p, we obtain the analytic expression

$$s_2 - s_1 = c_p \ln \frac{T_2}{T_1} - R \ln \frac{p_2}{p_1} \tag{14.37}$$

In a similar way, starting with Eq. (14.31) and making use of the relation $de = c_v dT$, the change in entropy can also be obtained as

$$s_2 - s_1 = c_v \ln \frac{T_2}{T_1} + R \ln \frac{v_2}{v_1} \tag{14.38}$$

14.2.5 Isentropic Relation

An isentropic process has already been described as reversible-adiabatic. For an adiabatic process $\delta q = 0$, and for a reversible process, $ds_{irrev} = 0$. From Eq. (14.26), we can see that for an isentropic process, $ds = 0$. However, in Eq. (14.37), substitution of isentropic condition yields,

$$c_p \ln \frac{T_2}{T_1} = R \ln \frac{p_2}{p_1}$$

or

$$\ln \frac{p_2}{p_1} = \frac{c_p}{R} \ln \frac{T_2}{T_1}$$

or

$$\frac{p_2}{p_1} = \left(\frac{T_2}{T_1}\right)^{c_p/R} \tag{14.39}$$

Substituting Eq. (14.22) in Eq. (14.39), we get

$$\frac{p_2}{p_1} = \left(\frac{T_2}{T_1}\right)^{\frac{\gamma}{\gamma-1}} \tag{14.40}$$

In a similar way, from Eq. (14.38)

$$0 = c_v \ln \frac{T_2}{T_1} + R \ln \frac{v_2}{v_1}$$

$$\ln \frac{v_2}{v_1} = -\frac{c_v}{R} \ln \frac{T_2}{T_1}$$

or
$$\frac{v_2}{v_1} = \left(\frac{T_2}{T_1}\right)^{-c_v/R} \tag{14.41}$$

Substituting Eq. (14.23) in Eq. (14.41), we get

$$\frac{v_2}{v_1} = \left(\frac{T_2}{T_1}\right)^{\frac{-1}{\gamma-1}} \tag{14.42}$$

From our known relationship of $\rho_2/\rho_1 = v_1/v_2$, we can write

$$\frac{\rho_2}{\rho_1} = \left(\frac{T_2}{T_1}\right)^{\frac{1}{\gamma-1}} \tag{14.43}$$

Combining Eq. (14.40) with Eq. (14.43), we find

$$\frac{p_2}{p_1} = \left(\frac{\rho_2}{\rho_1}\right)^{\gamma} = \left(\frac{T_2}{T_1}\right)^{\frac{\gamma}{\gamma-1}} \tag{14.44}$$

14.3 SPEED OF SOUND

A pressure pulse in an incompressible flow behaves like that in a rigid body. A displaced particle displaces all the particles in the medium. In a compressible fluid, on the other hand, displaced mass compresses and increases the density of the neighbouring mass which in turn increases the density of the adjoining mass and so on. Thus, a disturbance in the form of an elastic wave or a pressure wave travels through the medium. If the amplitude of the elastic wave is infinitesimal, it is termed as acoustic wave or sound wave.

Figure 14.1(a) shows an infinitesimal pressure pulse propagating at a speed "a" towards still fluid ($V = 0$) at the left. The fluid properties ahead of the wave are p, T and ρ, while the properties behind the wave are $p + dp$, $T + dT$ and $\rho + d\rho$. The fluid velocity dV is directed toward the left following wave but much slower.

In order to make the analysis steady, we superimpose a velocity "**a**" directed towards right, on the entire system (Fig. 14.1(b)). The wave is now stationary and the fluid appears to have velocity "**a**" on the left and (**a** – dV) on the right. The flow in Fig. 14.1 (b) is now steady and one dimensional across the wave. Consider an area A on the wave front. A mass balance gives

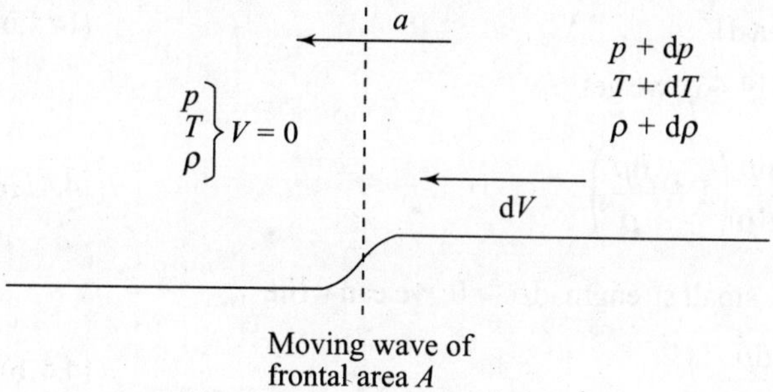

(a) Wave propagating into still fluid

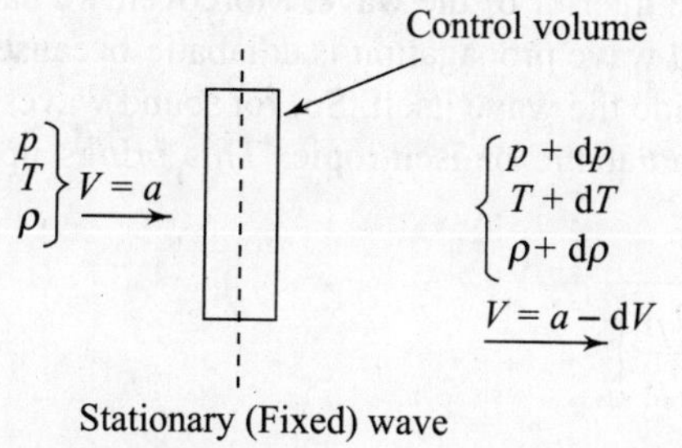

(b) Stationary wave

Fig. 14.1 Propagation of a sound wave

$$\rho A a = (\rho + \mathrm{d}\rho) A (\mathbf{a} - \mathrm{d}V)$$

or
$$\mathrm{d}V = \mathbf{a}\left[\frac{\mathrm{d}\rho}{\rho + \mathrm{d}\rho}\right] \tag{14.45}$$

This shows that $\mathrm{d}V > 0$ if $d\rho$ is positive. A compression wave leaves behind a fluid moving in the direction of the wave (Fig. 14.1(a)). Equation (14.45) also signifies that the fluid velocity on the right is much smaller than the wave speed "**a**". Within the framework of infinitesimal strength of the wave (sound wave), this "**a**" itself is very small.

Now, let us apply the momentum balance on the same control volume in Fig. 14.1 (b). It says that the net force in the x direction on the control volume equals the rate of outflow of x momentum minus the rate of inflow of x momentum. In symbolic form, this yields

$$pA - (p + \mathrm{d}p) A = A\rho\, a (a - \mathrm{d}V) - (A\, \rho\, a) a$$

In the above expression, $A\rho a$ is the mass flow rate. The first term on the right-hand side represents the rate of outflow of x momentum and the second term represents the rate of inflow of x momentum. Simplifying the momentum equation, we get

$$dp = \rho a \, dV \tag{14.46}$$

Combining Eqs (14.45) and (14.46), we get

$$a^2 = \frac{dp}{d\rho}\left(1 + \frac{d\rho}{\rho}\right) \tag{14.47a}$$

In the limit of infinitesimally small strength, $d\rho \to 0$, we can write

$$a^2 = \frac{dp}{d\rho} \tag{14.47b}$$

Notice that in the limit of infinitesimally strength of sound wave, there are no velocity gradients on either side of the wave. Therefore, the frictional effects (irreversible) are confined to the interior of the wave. Moreover, we can appreciate that the entire process of sound wave propagation is adiabatic because there is no temperature gradient except inside the wave itself. So, for sound waves, we can see that the process is reversible adiabatic or isentropic. This brings up the correct expression for the sound speed

$$a = \sqrt{\left(\frac{\partial p}{\partial \rho}\right)_s} \tag{14.48}$$

For a perfect gas, by using of p/ρ^γ = constant, and $p = \rho RT$, we deduce the speed of sound as

$$a = \sqrt{\frac{\gamma p}{\rho}} = \sqrt{\gamma RT} \tag{14.49}$$

For air at sea level and at a temperature of 15 °C, it gives $a = 340$ m/s.

14.4 PRESSURE FIELD DUE TO A MOVING SOURCE

Consider a point source emanating infinitesimal pressure disturbances in a still fluid, in which the speed of sound is "a". If the point disturbance, is stationary then the wave fronts are concentric spheres. This is shown in Fig. 14.2(a), where the wave fronts at intervals of Δt are shown.

Now suppose that source moves to the left at speed $U < a$. Figure 14.2(b) shows four locations of the source, 1 to 4, at equal intervals of time Δt, with point 4 being the current location of the source. At point 1, the source emanated a wave which has spherically expanded to a radius $3a\Delta t$ in an interval of time $3\Delta t$. During this time the source has moved to the location 4 at a distance of $3U\Delta t$ from point 1. The figure also shows the locations of the wave fronts emitted while the source was at points 2 and 3, respectively.

When the source speed is supersonic ($U > a$) as shown in Fig. 14.2(c), the point source is ahead of the disturbance and an observer in the downstream location is unaware of the approaching source. The disturbance emitted at different points of time are enveloped by an imaginary conical surface known as the Mach cone. The half angle of the cone, α, is known as the Mach angle and is given by

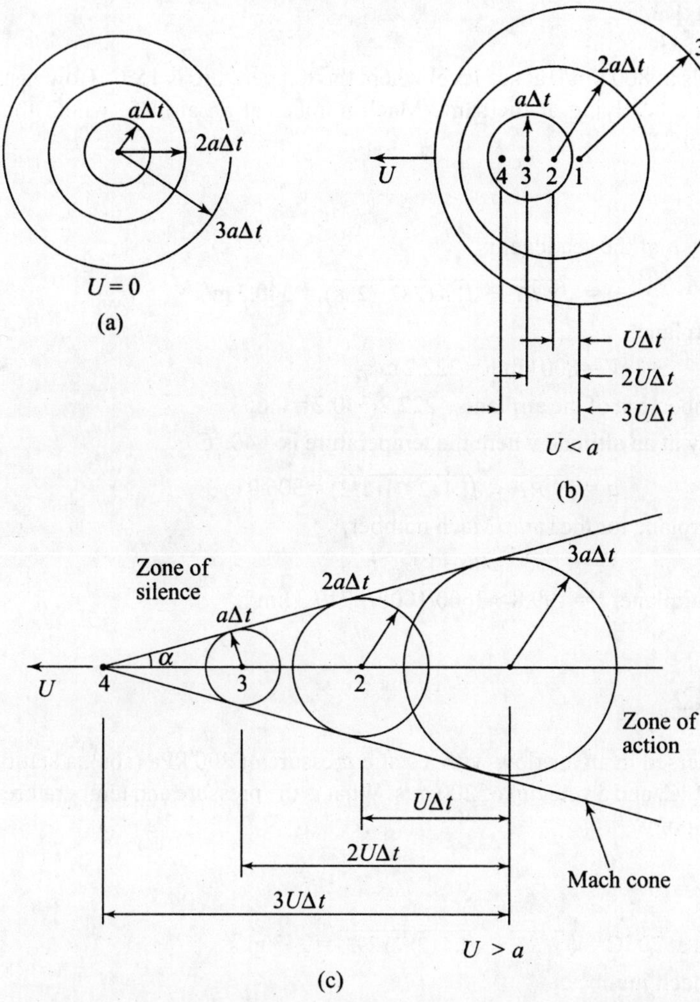

Fig. 14.2 Wave fronts emitted from a point source in a still fluid when the source speed U is (a) $U = 0$, (b) $U < a$, and (c) $U > a$

$$\sin \alpha = \frac{a\Delta t}{U\Delta t} = \frac{1}{\text{Ma}}$$

or

$$\alpha = \sin^{-1} (1/\text{Ma})$$

Since the disturbances are confined to the cone, the area within the cone is known as *the zone of action* and the area outside the cone is *the zone of silence*. An observer does not feel the effects of the moving source till the Mach cone covers his position.

Example 14.1

An airplane travels at 800 km/h at sea level where the temperature is 15°C. How fast would the airplane be flying at the same Mach number at an altitude where the temperature is – 40 °C?

Solution

The sonic velocity a at the sea level is

$$a = \sqrt{\gamma RT} = \sqrt{1.4\,(287)\,(288)} = 340.2 \text{ m/s}$$

Velocity of the airplane,

$$V = 800 \text{ km/h} = 222.2 \text{ m/s}$$

So, the Mach number, Ma of the airplane = 222.2/340.2 = 0.653

The sonic velocity at an altitude where the temperature is – 40 °C

$$a = \sqrt{\gamma RT} = \sqrt{1.4\,(287)\,(233)} = 306.0 \text{ m/s}$$

Velocity of the airplane for the same Mach number,

$$V = 0.653 \times 306 = 199.8 \text{ m/s}$$

or velocity of the airplane, $V = 199.8 \times 3600/1000 = 719.3$ km/h

Example 14.2

An object is immersed in an air flow with a static pressure of 200 kPa (abs), a static temperature of 20 °C, and a velocity of 200 m/s. What is the pressure and temperature at the stagnation point?

Solution

Velocity of sound at 20 °C = $\sqrt{\gamma RT} = \sqrt{1.4\,(287)\,293} = 343$ m/s

Corresponding Mach number,

$$\text{Ma} = 200/343 = 0.583$$

Stagnation temperature, $T_0 = (293)\,[1 + 0.2 \times (0.583)^2]$

$$= 293 \times 1.068 = 312.9\,K = 39.9\,°\text{C}$$

Stagnation pressure, $p_0 = (200)\,(1.068)^{3.5} = 251.8$ kPa

14.5 BASIC EQUATIONS FOR ONE-DIMENSIONAL FLOW

Having had an exposure to the speed of sound, we begin our study of a class of compressible flows that can be treated as one-dimensional flow. Such a simplification is meaningful for flow through ducts where the centreline of the ducts does not have a large curvature and the cross section of the ducts does not vary abruptly. For one-dimensional assumption, the flow can be studied by ignoring the variation of velocity and other properties across the cross-normal direction of the flow. However, these distributions are taken care of by assigning an average value over the cross section

(Fig. 14.3). The area of the duct is taken as $A(x)$ and the flow properties are taken as $p(x)$, $\rho(x)$, $V(x)$, etc. The forms of the basic equations in a one-dimensional compressible flow are discussed next.

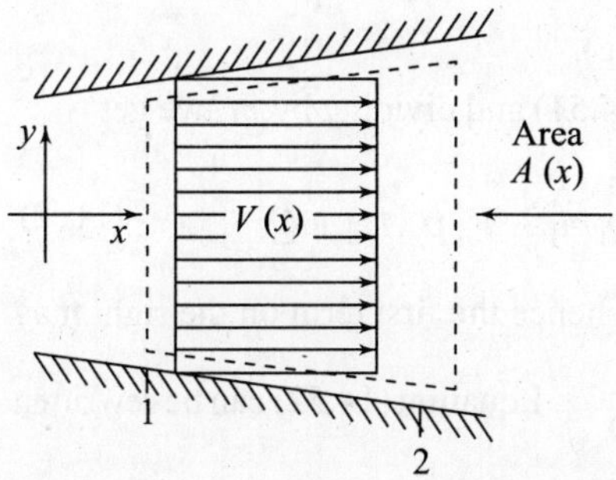

Fig. 14.3 One-dimensional approximation

Continuity Equation For steady one-dimensional flow, the equation of continuity is

$$\rho(x)\, V(x)\, A(x) = \dot{m} = \text{constant}$$

Differentiating, we get

$$\frac{d\rho}{\rho} + \frac{dV}{V} + \frac{dA}{A} = 0 \tag{14.50}$$

Energy Equation Let us consider a control volume within the duct shown by dotted lines in Fig. 14.3. The first law of thermodynamics for a control volume fixed in space is

$$\frac{d}{dt}\iiint \rho\left(e+\frac{V^2}{2}\right) d\mathcal{V} + \iint\left(e+\frac{V^2}{2}\right)\rho\, V\cdot dA$$
$$= \iint V\cdot(\tau\cdot dA) - \iint q\cdot dA \tag{14.51}$$

where $\frac{V^2}{2}$ is the kinetic energy per unit mass. The first term on the left-hand side signifies the rate of change of energy (internal + kinetic) within the control volume, and the second term depicts the flux of energy out of control surface. The first term on the right-hand side represents the work done on the control surface, and the second term on the right means the heat transferred through the control surface. Here q is the heat flux per unit area per unit time. It may be mentioned that dA is directed along the outward normal.

We shall assume steady state so that the first term on the left-hand side of Eq. (14.51) is zero. Writing $m = \rho_1 V_1 A_1 = \rho_2 V_2 A_2$ (where the subscripts are for Sections 1 and 2), the second term on the left of Eq. (14.51) yields

$$\iint\left(e+\frac{V^2}{2}\right)\rho V\cdot dA = \dot{m}\left[\left(e_2+\frac{V_2^2}{2}\right)-\left(e_1+\frac{V_1^2}{2}\right)\right]$$

The work done on the control surfaces is

$$\iint V \cdot (\tau \cdot \mathrm{d}A) = V_1 p_1 A_1 - V_2 p_2 A_2$$

The rate of heat transfer to the control volume is

$$-\iint q \cdot \mathrm{d}A = Q\,\dot{m}$$

where Q is the heat added per unit mass (in J/kg).

Invoking all the aforesaid relations in Eq. (14.51) and dividing by $\dot{m}$, we get

$$e_2 + \frac{V_2^2}{2} - e_1 - \frac{V_1^2}{2} = \frac{1}{\dot{m}}[V_1 p_1 A_1 - V_2 p_2 A_2] + Q \tag{14.52}$$

We know that the density ρ is given by $\dot{m}/VA$, hence the first term on the right may be expressed in terms of v $\left(\text{specific volume; } \frac{1}{\rho}\right)$. Equation (14.52) can be rewritten as

$$e_2 + \frac{V_2^2}{2} - e_1 - \frac{V_1^2}{2} = p_1 v_1 - p_2 v_2 + Q \tag{14.53}$$

It is understood that $p_1 v_1$ is the work done (per unit mass) by the surrounding in pushing fluid into the control volume. Following a similar argument, $p_2 v_2$ is the work done by the fluid inside the control volume on the surroundings in pushing fluid out of the control volume. Equation (14.53) may be reduced to a simpler fcrm. Noting that $h = e + pv$, we obtain

$$h_2 + \frac{V_2^2}{2} = h_1 + \frac{V_1^2}{2} + Q \tag{14.54}$$

This is energy equation, which is valid even in the presence of friction or non-equilibrium conditions between Sections 1 and 2. It is evident that the sum of enthalpy and kinetic energy remains constant in an adiabatic flow. Enthalpy performs a similar role that internal energy performs in a non-flowing system. The difference between the two types of systems is the flow work pv required to push the fluid through a section.

Bernoulli and Euler Equations For inviscid flows, the steady form of the momentum equation is the Euler equation,

$$\frac{\mathrm{d}p}{\rho} + V\,\mathrm{d}V = 0 \tag{14.55}$$

Integrating along a streamline, we get the Bernoulli's equation for a compressible flow as

$$\int \frac{\mathrm{d}p}{\rho} + \frac{V^2}{2} = \text{constant} \tag{14.56}$$

For adiabatic frictionless flows the Bernoulli's equation is identical to the energy equation. To appreciate this, we have to remember that this is an isentropic flow, so that the *T*d*s* equation is given by

$$T\mathrm{d}s = \mathrm{d}h - v\,\mathrm{d}p$$

which yields $$dh = \frac{dp}{\rho}$$

Then the Euler equation (14.55) can also be written as

$$VdV + dh = 0$$

Needless to say that this is identical to the adiabatic form of the energy Eq. (14.54). The merger of the momentum and energy equation is attributed to the elimination of one of the flow variables due to constant entropy.

Momentum Principle for a Control Volume For a finite control volume between Sections 1 and 2 (Fig. 14.3), the momentum principle is

$$p_1A_1 - p_2A_2 + F = \dot{m}\,V_2 - \dot{m}\,V_1$$

or $$p_1A_1 - p_2A_2 + F = \rho_2V_2^2A_2 - \rho_1V_1^2\,A_1 \qquad (14.57)$$

where F is the x component of resultant force exerted on the fluid by the walls. The momentum principle, Eq. (14.57), is applicable even when there are frictional dissipative processes within the control volume.

14.6 STAGNATION AND SONIC PROPERTIES

The stagnation values are useful reference conditions in a compressible flow. Suppose the properties of a flow (such as T, p, ρ, etc.) are known at a point. The stagnation properties at a point are defined as those which are to be obtained if the local flow were imagined to cease to zero velocity isentropically. The stagnation values are denoted by a subscript zero. Thus, the stagnation enthalpy is defined as

$$h_0 = h + \frac{1}{2}V^2$$

For a perfect gas, this yields,

$$c_pT_0 = c_pT + \frac{1}{2}V^2 \qquad (14.58)$$

which defines the stagnation temperature. It is meaningful to express the ratio of (T_0/T) in the form

$$\frac{T_0}{T} = 1 + \frac{V^2}{2c_pT} = 1 + \frac{\gamma - 1}{2}\cdot\frac{V^2}{\gamma RT}$$

or $$\frac{T_0}{T} = 1 + \frac{\gamma - 1}{2}\mathrm{Ma}^2 \qquad (14.59)$$

If we know the local temperature (T) and Mach number (Ma), we can find out the stagnation temperature T_0. Consequently, isentropic relations can be used to obtain stagnation pressure and stagnation density as

$$\frac{p_0}{p} = \left(\frac{T_0}{T}\right)^{\frac{\gamma}{\gamma-1}} = \left[1 + \frac{\gamma - 1}{2}\mathrm{Ma}^2\right]^{\frac{\gamma}{\gamma-1}} \qquad (14.60)$$

$$\frac{\rho_0}{\rho} = \left(\frac{T_0}{T}\right)^{\frac{1}{\gamma-1}} = \left[1 + \frac{\gamma-1}{2}\mathrm{Ma}^2\right]^{\frac{1}{\gamma-1}} \tag{14.61}$$

In general, the stagnation properties can vary throughout the flow field. However, if the flow is adiabatic, then $h + \frac{V^2}{2}$ is constant throughout the flow (Eq. 14.54). It follows that the h_0, T_0, and a_0 are constant throughout an adiabatic flow, even in the presence of friction. It is understood that all stagnation properties are constant along an isentropic flow. If such a flow starts from a large reservoir where the fluid is practically at rest, then the properties in the reservoir are equal to the stagnation properties everywhere in the flow (Fig. 14.4).

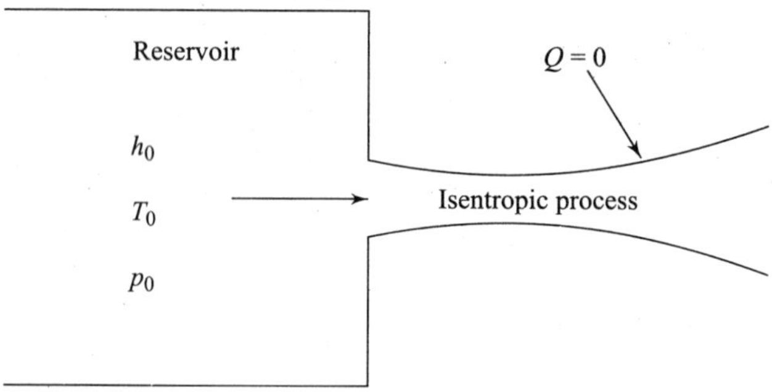

Fig. 14.4 An isentropic process starting from a reservoir

There is another set of conditions of comparable usefulness where the flow is sonic, Ma = 1.0. These sonic, or critical properties are denoted by asterisks: p^*, ρ^*, a^*, and T^*. These properties are attained if the local fluid is imagined to expand or compress isentropically until it reachers Ma = 1.

We have already discussed that the total enthalpy, hence T_0, is conserved so long the process is adiabatic, irrespective of frictional effects. In contrast, the stagnation pressure p_0 and density ρ_0 decrease if there is friction.

From Eq. (14.58), we note that

$$V^2 = 2c_p(T_0 - T)$$

or

$$V = \left[\frac{2\gamma R}{\gamma-1}(T_0 - T)\right]^{\frac{1}{2}} \tag{14.62a}$$

is the relationship between the fluid velocity and local temperature (T), in an adiabatic flow. The flow can attain a maximum velocity of

$$V_{max} = \left[\frac{2\gamma R T_0}{\gamma-1}\right]^{\frac{1}{2}} \tag{14.62b}$$

As it has already been stated, the unity Mach number, Ma = 1, condition is of special significance in compressible flow, and we can now write from Eqs. (14.59), (14.60) and (14.61),

$$\frac{T_0}{T^*} = \frac{1+\gamma}{2} \tag{14.63a}$$

$$\frac{p_0}{p^*} = \left(\frac{1+\gamma}{2}\right)^{\frac{\gamma}{\gamma-1}} \tag{14.63b}$$

$$\frac{\rho_0}{\rho^*} = \left(\frac{1+\gamma}{2}\right)^{\frac{1}{\gamma-1}} \tag{14.63c}$$

For diatomic gases, like air $\gamma = 1.4$, the numerical values are

$$\frac{T^*}{T_0} = 0.8333, \quad \frac{p^*}{p_0} = 0.5282, \quad \text{and} \quad \frac{\rho^*}{\rho_0} = 0.6339$$

The fluid velocity and acoustic speed are equal at sonic condition and is

$$V^* = a^* = [\gamma R T^*]^{1/2} \tag{14.64a}$$

or

$$V^* = \left[\frac{2\gamma}{\gamma+1} R T_0\right]^{\frac{1}{2}} \tag{14.64b}$$

We shall employ both stagnation conditions and critical conditions as reference conditions in a variety of one-dimensional compressible flows.

14.6.1 Effect of Area Variation on Flow Properties in Isentropic Flow

In considering the effect of area variation on flow properties in isentropic flow, we shall concern ourselves primarily with the velocity and pressure. We shall determine the effect of change in area A, on the velocity V, and the pressure p.

From Eq. (14.55), we can write

$$\frac{\mathrm{d}p}{\rho} + d\left(\frac{V^2}{2}\right) = 0$$

or

$$\mathrm{d}p = -\rho V \,\mathrm{d}V$$

Dividing by ρV^2, we obtain

$$\frac{\mathrm{d}p}{\rho V^2} = -\frac{\mathrm{d}V}{V} \tag{14.65}$$

A convenient differential form of the continuity equation can be obtained from Eq. (14.50) as

$$\frac{\mathrm{d}A}{A} = -\frac{\mathrm{d}V}{V} - \frac{\mathrm{d}\rho}{\rho}$$

Substituting from Eq. (14.65),

$$\frac{\mathrm{d}A}{A} = \frac{\mathrm{d}p}{\rho V^2} - \frac{\mathrm{d}\rho}{\rho}$$

or

$$\frac{\mathrm{d}A}{A} = \frac{\mathrm{d}p}{\rho V^2}\left[1 - \frac{V^2}{\mathrm{d}p/\mathrm{d}\rho}\right] \tag{14.66}$$

Invoking the relation (14.47b) for isentropic process in Eq. (14.66), we get

$$\frac{dA}{A} = \frac{dp}{\rho V^2}\left[1 - \frac{V^2}{a^2}\right] = \frac{dp}{\rho V^2}[1 - \text{Ma}^2] \tag{14.67}$$

From Eq. (14.67), we see that for Ma < 1 an area change causes a pressure change of the same sign, i.e., positive dA means positive dp for Ma < 1. For Ma > 1, an area change causes a pressure change of opposite sign.

Again, substituting from Eq. (14.65) into Eq. (14.67), we obtain

$$\frac{dA}{A} = -\frac{dV}{V}[1 - \text{Ma}^2] \tag{14.68}$$

From Eq. (14.68), we see that Ma < 1 an area change causes a velocity change of opposite sign, i.e., positive dA means negative dV for Ma < 1. For Ma > 1, an area change causes a velocity change of same sign.

These results are summarised in Fig. 14.5, and the relations (14.67) and (14.68) lead to the following important conclusions about compressible flows:

(i) At subsonic speeds (Ma < 1) a decrease in area increases the speed of flow. A subsonic nozzle should have a convergent profile and a subsonic diffuser should possess a divergent profile. The flow behaviour in the regime of Ma < 1 is therefore qualitatively the same as in incompressible flows.

(ii) In supersonic flows (Ma > 1), the effect of area changes are different. According to Eq. (14.68), a supersonic nozzle must be built with an increasing area in the flow direction. A supersonic diffuser must be a converging channel. Divergent nozzles are used to produce supersonic flow in missiles and launch vehicles.

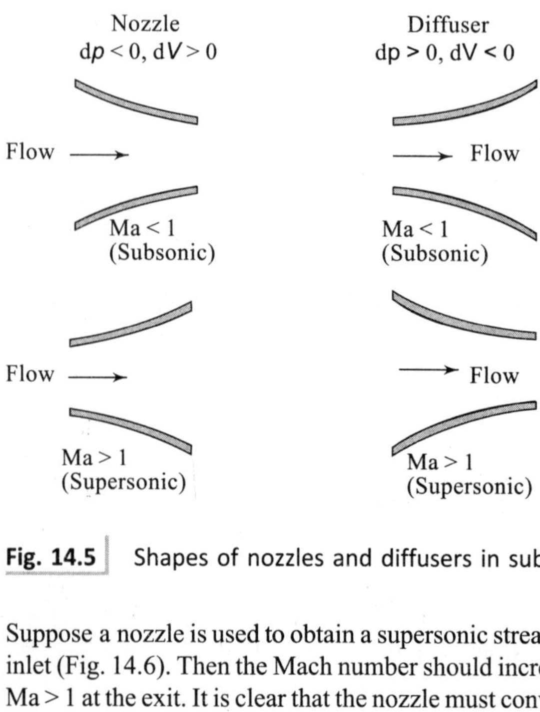

Fig. 14.5 Shapes of nozzles and diffusers in subsonic and supersonic regimes

Suppose a nozzle is used to obtain a supersonic stream staring from low speeds at the inlet (Fig. 14.6). Then the Mach number should increase from Ma = 0 near the inlet to Ma > 1 at the exit. It is clear that the nozzle must converge in the subsonic portion and diverge in the supersonic portion. Such a nozzle is called a *convergent-divergent nozzle*. A convergent-divergent nozzle is also called a *de Laval nozzle*, after Carl G.P. de Laval who first used such a configuration in his steam turbines in the late nineteenth century. From Fig. 14.6 it is clear that the Mach number must be unity at the throat, where the area is neither increasing nor decreasing. *This is consistent with Eq. (14.68) which shows that dV can be non-zero at the throat only if* Ma = 1. It also follows that the sonic velocity can be achieved only at the throat of a nozzle or a diffuser.

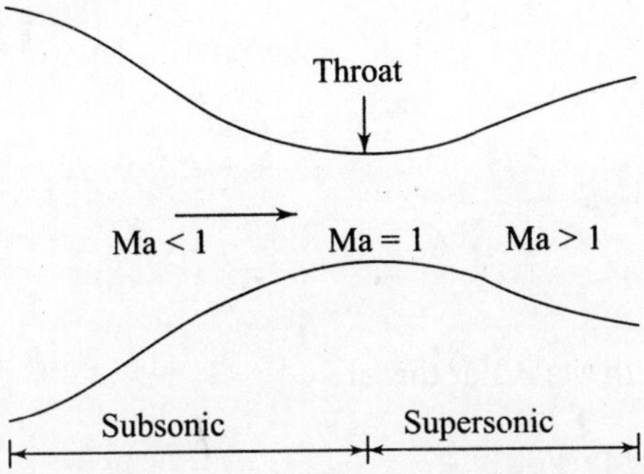

Fig. 14.6 A convergent-divergent nozzle

The condition, however, does not restrict that Ma must necessarily be unity at the throat. According to Eq. (14.68), a situation is possible where Ma ≠ 1 at the throat if dV = 0 there. For example, the flow in a convergent-divergent duct may be subsonic everywhere with Ma increasing in the convergent portion and decreasing in the divergent portion with Ma ≠ 1 at the throat (see Fig. 14.7). The first part of the duct is acting as a nozzle, whereas the second part is acting as a diffuser. Alternatively, we may have a convergent-divergent duct in which the flow is supersonic everywhere with Ma decreasing in the convergent part and increasing in the divergent part and again Ma ≠ 1 at the throat (see Fig. 14.8).

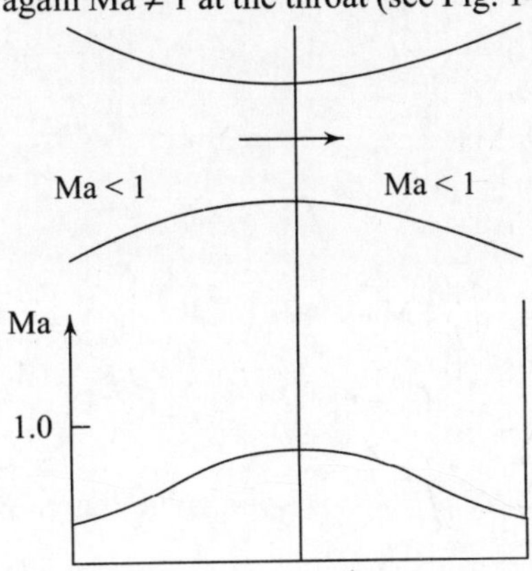

Fig. 14.7 Convergent-divergent duct with Ma ≠ 1 at throat

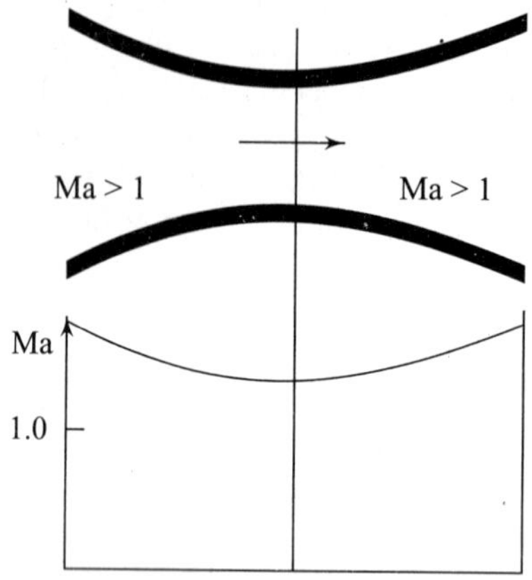

Fig. 14.8 Convergent-divergent duct with $\text{Ma} \neq 1$ at throat

14.6.2 Isentropic Flow in a Converging Nozzle

Let us consider the mass flow rate of an ideal gas through a converging nozzle. If the flow is isentropic, we can write

$$\dot{m} = \rho A V$$

or
$$\frac{\dot{m}}{A} = \frac{p}{RT} . a \,\text{Ma} \quad \text{[invoking Eqs (14.5) and (14.8)]}$$

or
$$\frac{\dot{m}}{A} = \frac{p}{RT} \cdot \sqrt{\gamma RT} \cdot \text{Ma}$$

or
$$\frac{\dot{m}}{A} = \frac{p}{\sqrt{T}} \cdot \sqrt{\frac{\gamma}{R}} \cdot \text{Ma}$$

or
$$\frac{\dot{m}}{A} = \frac{p}{p_0} \cdot p_0 \cdot \sqrt{\frac{T_0}{T}} \sqrt{\frac{1}{T_0}} \sqrt{\frac{\gamma}{R}} \cdot \text{Ma}$$

or
$$\frac{\dot{m}}{A} = \left(\frac{T_0}{T}\right)^{\frac{-\gamma}{\gamma-1}} \cdot \left(\frac{T_0}{T}\right)^{\frac{1}{2}} \frac{p_0}{\sqrt{T_0}} \cdot \sqrt{\frac{\gamma}{R}} \cdot \text{Ma}$$

[invoking Eq. (14.44)]

or
$$\frac{\dot{m}}{A} = \sqrt{\frac{\gamma}{R}} \cdot \frac{p_0 \text{Ma}}{\sqrt{T_0}} \left(\frac{T_0}{T}\right)^{-\frac{(\gamma+1)}{2(\gamma-1)}}$$

or
$$\frac{\dot{m}}{A} = \sqrt{\frac{\gamma}{R}} \cdot \frac{p_0 \mathrm{Ma}}{\sqrt{T_0}} \cdot \frac{1}{\left[1+\frac{\gamma-1}{2}\mathrm{Ma}^2\right]^{\frac{(\gamma+1)}{2(\gamma-1)}}} \tag{14.69}$$

In the expression (14.69), p_0, T_0, γ and R are constant. The discharge per unit area $\frac{\dot{m}}{A}$ is a function of Ma only. There exists a particular value of Ma for which $(\dot{m}/A)$ is maximum. Differentiating with respect to Ma and equating it to zero, we get

$$\frac{\mathrm{d}(\dot{m}/A)}{\mathrm{dMa}} = \sqrt{\frac{\gamma}{R}} \cdot \frac{p_0}{\sqrt{T_0}} \cdot \frac{1}{\left[1+\frac{\gamma-1}{2}\mathrm{Ma}^2\right]^{\frac{(\gamma+1)}{2(\gamma-1)}}} + \sqrt{\frac{\gamma}{R}} \cdot \frac{p_0 \mathrm{Ma}}{\sqrt{T_0}}$$

$$\left[-\frac{(\gamma+1)}{2(\gamma-1)}\left\{1+\frac{\gamma-1}{2}\mathrm{Ma}^2\right\}^{\frac{-(\gamma+1)}{2(\gamma-1)}-1}\left\{\frac{\gamma-1}{2}2\mathrm{Ma}\right\}\right]=0$$

or
$$1-\frac{\mathrm{Ma}^2(\gamma+1)}{2\left\{1+\frac{\gamma-1}{2}\mathrm{Ma}^2\right\}}=0$$

or $\mathrm{Ma}^2(\gamma+1) = 2+(\gamma-1)\mathrm{Ma}^2$

or $\mathrm{Ma} = 1$

So, discharge is maximum when $\mathrm{Ma} = 1$.

We know that $V = a\mathrm{Ma} = \sqrt{\gamma RT}\,\mathrm{Ma}$. By logarithmic differentiation, we get

$$\frac{\mathrm{d}V}{V} = \frac{\mathrm{dMa}}{\mathrm{Ma}} + \frac{1}{2}\frac{\mathrm{d}T}{T} \tag{14.70}$$

We also know that

$$\frac{T}{T_0} = \left[1+\frac{\gamma-1}{2}\mathrm{Ma}^2\right]^{-1} \tag{14.59 repeated}$$

By logarithmic differentiation, we get

$$\frac{\mathrm{d}T}{T} = -\frac{(\gamma-1)\mathrm{Ma}^2}{1+\frac{(\gamma-1)}{2}\mathrm{Ma}^2} \cdot \frac{\mathrm{dMa}}{\mathrm{Ma}} \tag{14.71}$$

From Eqs (14.70) and (14.71), we get

$$\frac{dV}{V} = \frac{d\text{Ma}}{\text{Ma}}\left[1 - \frac{\{(\gamma-1)/2\}\text{Ma}^2}{1+\frac{(\gamma-1)}{2}\text{Ma}^2}\right]$$

$$\frac{dV}{V} = \frac{1}{1+\frac{(\gamma-1)}{2}\text{Ma}^2}\cdot\frac{d\text{Ma}}{\text{Ma}} \tag{14.72}$$

From Eqs (14.68) and (14.72) we get

$$\frac{dA}{A}\frac{1}{(\text{Ma}^2-1)} = \frac{1}{1+\frac{\gamma-1}{2}\text{Ma}^2}\cdot\frac{d\text{Ma}}{\text{Ma}}$$

$$\frac{dA}{A} = \frac{(\text{Ma}^2-1)}{1+\frac{(\gamma-1)}{2}}\cdot\frac{d\text{Ma}}{\text{Ma}} \tag{14.73}$$

By substituting $Ma = 1$ in Eq. (14.73), we get $dA = 0$ or A = constant. Some $Ma = 1$ can occur only at the throat and nowhere else, and this happens only when the discharge is maximum. When Ma = 1, the discharge is maximum and the nozzle is said to be choked. The properties at the throat are termed as critical properties which are already expressed through Eq. (14.63a), (14.63b) and (14.63c). By substituting Ma = 1 in Eq. (14.69), we get

$$\frac{\dot{m}}{A^*} = \sqrt{\frac{\gamma}{R}}\cdot\frac{p_0}{\sqrt{T_0}}\cdot\frac{1}{\left[\frac{(\gamma+1)}{2}\right]^{\frac{\gamma+1}{2(\gamma-1)}}} \tag{14.74}$$

(as we have earlier designated critical or sonic conditions by a superscript asterisk). Dividing Eq. (14.74) by Eq. (14.69) we obtain

$$\frac{A}{A^*} = \frac{1}{\text{Ma}}\left[\left\{\frac{2}{\gamma+1}\right\}\left\{1+\frac{(\gamma-1)}{2}\text{Ma}^2\right\}\right]^{\frac{\gamma+1}{2(\gamma-1)}} \tag{14.75}$$

From Eq. (14.75) we see that a choice of Ma gives a unique value of A/A^*. The variation of A/A^* with Ma is shown in Fig. 14.9. Note that the curve is double valued; that is, for a given value of A/A^* (other than unity), there are two possible values of Mach number. This signifies the fact that the supersonic nozzle is diverging.

The values of T_0/T, p_0/p, ρ_0/ρ and A/A^* at a point can be determined from Eqs. T_2/T_1 (14.59) – (14.61) and Eq. (14.75), if the local Mach number is known. For $\gamma = 1.4$, these values can be tabulated and the table is known as the Isentropic table. The sample values of the isentropic table are shown in Table 14.1 (see Babu [3]).

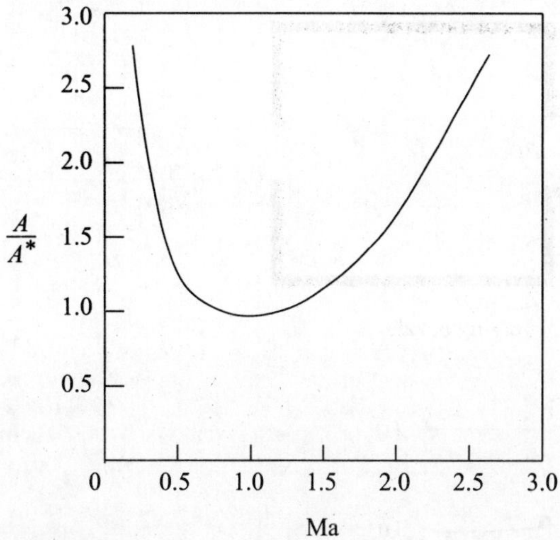

Fig. 14.9 Variation of A/A* with Ma in isentropic flow for $\gamma = 1.4$

14.6.3 Pressure Distribution and Choking in a Converging Nozzle

Let us first consider a convergent nozzle as shown in Fig. 14.10(a). Figure 14.10(b) shows the pressure ratio p/p_0 along the length of the nozzle. The inlet conditions of the gas are at the stagnation state (p_0, T_0) which are constants. The pressure at the exit plane of the nozzle is denoted by p_E and the back pressure is p_B which can be varied by the adjustment of the valve. At the condition $p_0 = p_E = p_B$, there shall be no flow through the nozzle. The pressure is p_0 throughout, as shown by condition (i) in Fig. 14.10(b). As p_B is gradually reduced, the flow rate shall increase. The pressure will decrease in the direction of flow as shown by condition (ii) in Fig. 14.10(b). The exit plane pressure p_E shall remain equal to p_B so long as the maximum discharge condition is not reached. Condition (iii) in Fig. 14.10(b) illustrates the pressure distribution in the maximum discharge situation. When ($\dot{m}$/A) attains its maximum value, given by substituting Ma = 1 in Eq. (14.69), p_E is equal to p^*. Since the nozzle does not have a diverging section, further reduction in back pressure p_B will not accelerate the flow to supersonic condition. As a result, the exit pressure p_E shall continue to remain at p^* even though p_B is lowered further. The convergent-nozzle discharge against the variation of back pressure is shown in Fig. 14.11. As it has been pointed out earlier, the maximum value of ($\dot{m}/A$) at Ma = 1 is stated as the choked flow. With a given nozzle, the flow rate cannot be increased further. Thus neither the nozzle exit pressure, nor the mass flow rate are affected by lowering p_B below p^*.

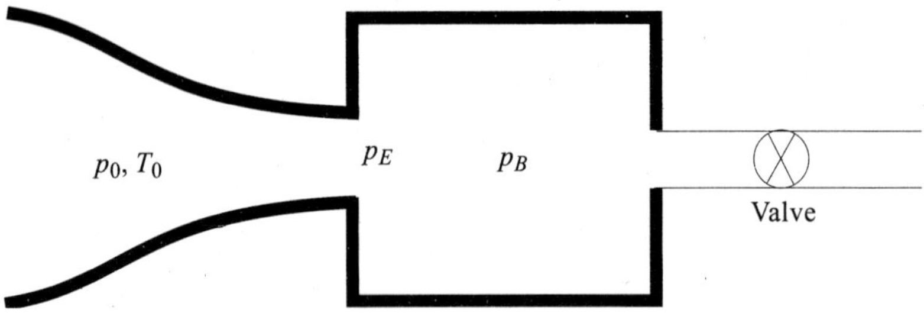

(a) Compressible flow through a converging nozzle

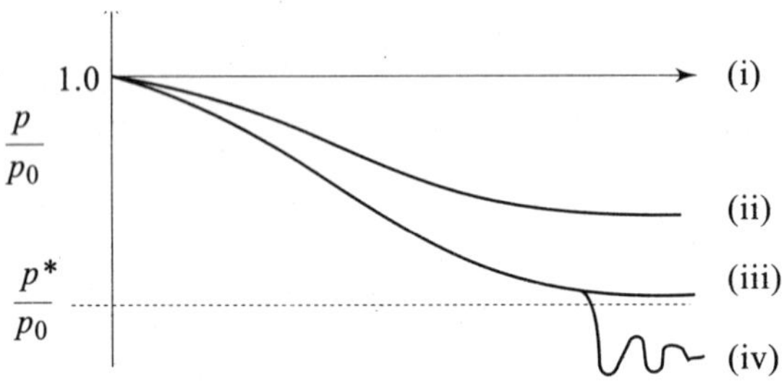

(b) Pressure distribution along a converging nozzle for different values of back pressure

Fig. 14.10

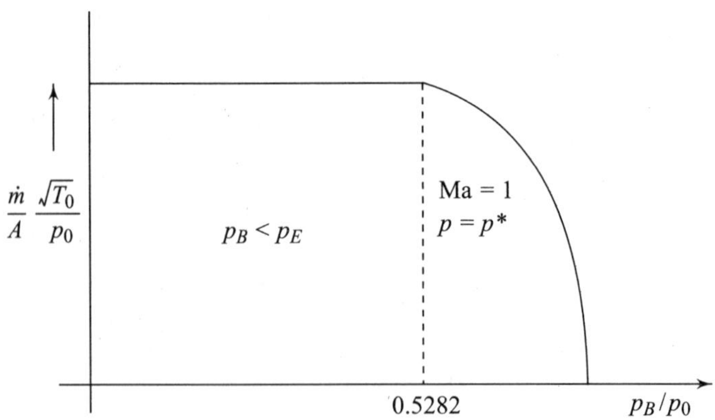

Fig. 14.11 Mass flow rate and the variation of back pressure in a converging nozzle

However for p_B less than p^*, the flow leaving the nozzle has to expand to match the lower back pressure as shown by condition (iv) in Fig. 14.10(b). This expansion process is three-dimensional and the pressure distribution cannot be predicted by a one-dimensional theory. Experiments reveal that a series of shocks form in the exit stream, resulting in an increase in entropy.

14.6.4 Isentropic Flow in a Converging-Diverging Nozzle

Now consider the flow in a convergent-divergent nozzle (Fig. 14.12). The upstream stagnation conditions are assumed constant; the pressure in the exit plane of the nozzle is denoted by p_E; the nozzle discharges to the back pressure, p_B. With the value initially closed, there is no flow through the nozzle; the pressure is constant at p_0. Opening the valve slightly produces the pressure distribution shown by curve (i). Completely subsonic flow is discerned. Then p_B is lowered in such a way that sonic condition is reached at the throat (ii). The flow rate becomes maximum for a given nozzle and the stagnation conditions. On further reduction of the back pressure, the flow upstream of the throat does not respond. However, if the back pressure is reduced further (cases (iii) and (iv)), the flow initially becomes supersonic in the diverging section, but then adjusts to the back pressure by means of a normal shock standing inside the nozzle. In such cases, the position of the shock moves downstream as p_B is decreased, and for curve (iv) the normal shock stands right at the exit plane. The flow in the entire divergent portion up to the exit plane is now supersonic. When the back pressure is reduced even further (v), there is no normal shock anywhere within the nozzle, and the jet pressure adjusts to p_B by means of oblique shock waves outside the exit plane. A converging diverging nozzle is generally intended to produce supersonic flow near the exit plane. If the back pressure is set at (vi), the flow will be isentropic throughout the nozzle, and supersonic at nozzle exit. Nozzles operating at $p_B = p_{VI}$ (corresponding to curve (vi) in Fig. 14.12) are said to be at design conditions. Rocket-propelled vehicles use converging-diverging nozzles to accelerate the exhaust gases to the maximum possible velocity to produce high thrust.

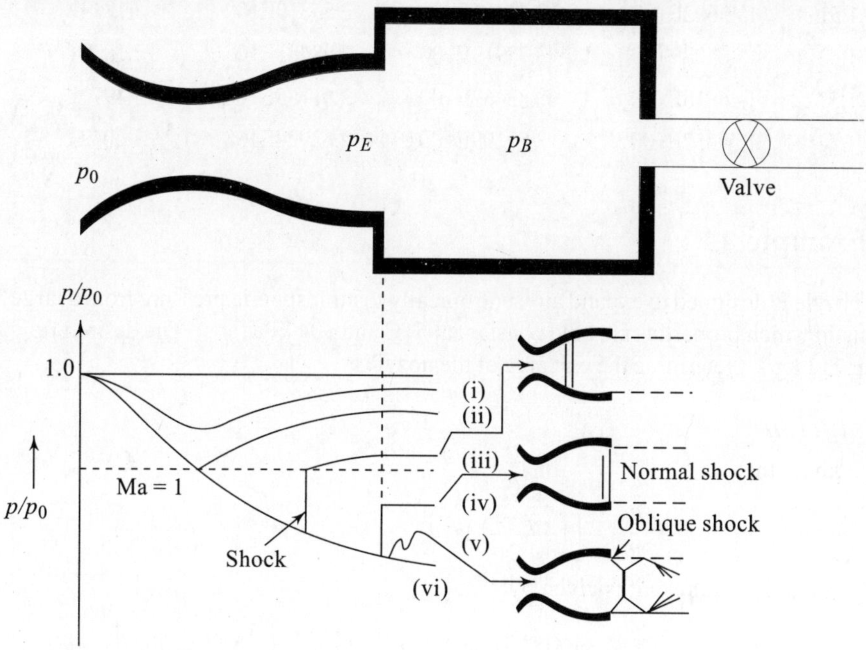

Fig. 14.12 Pressure distribution along a converging-diverging nozzle for different values of back pressure p_B

Table 14.1 Isentropic Table

Ma	$\frac{T_0}{T}$	$\frac{P_0}{P}$	$\frac{\rho_0}{\rho}$	$\frac{A}{A^*}$
0.00	1.00000E+00	1.00000E+00	1.00000E+00	∞
0.01	1.00002E+00	1.00007E+00	1.00005E+00	57.87384
0.02	1.00008E+00	1.00028E+00	1.00020E+00	28.94213
0.03	1.00018E+00	1.00063E+00	1.00045E+00	19.30054
0.11	1.00242E+00	1.00850E+00	1.00606E+00	5.29923
0.12	1.00288E+00	1.01012E+00	1.00722E+00	4.86432
0.13	1.00338E+00	1.01188E+00	1.00847E+00	4.49686
0.14	1.00392E+00	1.01379E+00	1.00983E+00	4.18240
0.15	1.00450E+00	1.01584E+00	1.01129E+00	3.91034
0.57	1.06498E+00	1.24651E+00	1.17045E+00	1.22633
0.58	1.06728E+00	1.25596E+00	1.17678E+00	1.21301
0.59	1.06962E+00	1.26562E+00	1.18324E+00	1.20031
0.60	1.07200E+00	1.27550E+00	1.18984E+00	1.18820
1.51	1.45602E+00	3.72465E+00	2.55810E+00	1.18299
1.52	1.46208E+00	3.77919E+00	2.58481E+00	1.18994
1.53	1.46818E+00	3.83467E+00	2.61185E+00	1.19702
1.54	1.47432E+00	3.89109E+00	2.63924E+00	1.20423

Example 14.3

A nozzle is designed to expand air isentropically to atmospheric pressure from a large tank in which properties are held constant at 5 °C and 304 kPa (abs). The desired flow rate is 1 kg/s. Determine the exit area of the nozzle.

Solution

We know that

$$\frac{p_0}{p_e} = \left(1 + \frac{\gamma - 1}{2}\text{Ma}_e^2\right)^{\gamma/\gamma-1}$$

Mach number at the exit is given by

$$\text{Ma}_e = \left[\frac{2}{\gamma - 1}\left\{\left(\frac{p_0}{p_e}\right)^{(\gamma-1)/\gamma} - 1\right\}\right]^{0.5}$$

$$\text{Ma}_e = 1.36$$

Since $\text{Ma}_e > 1.0$, the nozzle is converging-diverging. Again, we know

$$\frac{T_0}{T_e} = 1 + \frac{(\gamma - 1)}{2}\text{Ma}_e^2$$

$$T_e = \frac{T_0}{1 + \frac{(\gamma-1)}{2}\text{Ma}_e^2} = \frac{278}{1 + 0.2(1.36)^2} = 203\text{ K}$$

$$\rho_e = \frac{p_e}{RT_e} = \frac{101 \times 10^3}{287 \times 203} = 1.73\text{ kg/m}^3$$

$$V_e = \text{Ma}_e\, a_e = \text{Ma}_e\,(\gamma\, RT_e)^{0.5} = 1.36\,(1.4 \times 287 \times 203)^{0.5}$$
$$= 388\text{ m/s}$$

We also know that $\dot{m} = \rho_e V_e A_e$; so the exit area A_e is

$$A_e = \dot{m}/\rho_e V_e = 1.0/1.73 \times 388 = 1.49 \times 10^{-3}\text{ m}^2.$$

Example 14.4

Air at an absolute pressure 60.0 kPa and 27 °C enters a passage at 486 m/s. The cross sectional area at the entrance is 0.02 m^2. At Section 2, further downstream, the pressure is 78.8 kPa (abs). Assuming isentropic flow, calculate the Mach number at Section 2. Also, identify the type of the nozzle.

Solution

For isentropic flow, $p_{01} = p_{02} = p_0 = $ constant

At Section 1, $\text{Ma}_1 = V_1/a_1$;

the sonic velocity, $a_1 = (\gamma RT)^{0.5} = (1.4 \times 287 \times 300)^{0.5}$

$$= 347\text{ m/s}$$

So, $\text{Ma}_1 = 486/347 = 1.40$

Now, $$P_{01} = p_1\left(1 + \frac{\gamma - 1}{2}\text{Ma}_1^2\right)^{\gamma/(\gamma-1)}$$

$$= 60(1 + 0.2\,(1.40)^2)^{3.5} = 191\text{ kPa}$$

Again, we can write $$\frac{p_{02}}{p_2} = \left(1 + \frac{\gamma - 1}{2}\text{Ma}_2^2\right)^{\gamma/(\gamma-1)}$$

and $p_{02} = p_{01}$

So, $$\mathrm{Ma}_2 = \left\{\frac{2}{\gamma-1}\left[\left(\frac{p_{01}}{p_2}\right)^{(\gamma/(\gamma-1))} - 1\right]\right\}^{0.5}$$

$$\mathrm{Ma}_2 = 1.2$$

Since $\mathrm{Ma}_2 < \mathrm{Ma}_1$ and $\mathrm{Ma}_2 > 1.0$, the flow passage from 1 to 2 is a supersonic diffuser.

Example 14.5

A supersonic diffuser decelerates air isentropically from a Mach number of 3 to a Mach number of 1.4. If the static pressure at the diffuser inlet is 30.0 kPa (abs), calculate the static pressure rise in the diffuser and the ratio of inlet to outlet area of the diffuser.

Solution

For isentropic flow, $p_{01} = p_{02} = p_{03}$

Now, $$\frac{p_0}{p_1} = \left(1 + \frac{\gamma-1}{2}\mathrm{Ma}_1^2\right)^{\gamma/(\gamma-1)}$$

so, $$\frac{p_2}{p_1} = \frac{p_2}{p_0} \times \frac{p_0}{p_1} = \frac{\left(1 + \frac{\gamma-1}{2}\mathrm{Ma}_1^2\right)^{\gamma/(\gamma-1)}}{\left(1 + \frac{\gamma-1}{2}\mathrm{Ma}_2^2\right)^{\gamma/(\gamma-1)}}$$

$$= \frac{[1 + 0.2(3.0)^2]^{3.5}}{[1 + 0.2(1.4)^2]^{3.5}} = 11.5$$

Now, $p_2 - p_1 = 11.5, p_1 - p_1 = 10.5 \times 30.0\ \mathrm{kPa} = 315\ \mathrm{kPa}$ is the static pressure rise in the diffuser.

Again, from continuity, $\rho_1 V_1 A_1 = \rho_2 V_2 A_2$

or $$\frac{A_1}{A_2} = \rho_2 V_2 / \rho_1 V_1$$

We also know that $p/\rho^\gamma =$ constant and $\rho_2/\rho_1 = (p_2/p_1)^{1/\gamma}$

From the definition of Mach number, we can write $\mathrm{Ma} = V/a$ and $a = \sqrt{\gamma RT}$

Now, $$V_2/V_1 = \mathrm{Ma}_2\, a_2/\mathrm{Ma}_1\, a_1 = \frac{\mathrm{Ma}_2}{\mathrm{Ma}_1}\left(\frac{T_2}{T_1}\right)^{0.5}$$

Since T_0 is constant, $$\frac{T_2}{T_1} = \frac{T_2}{T_0} \times \frac{T_0}{T_1} = \frac{1 + \frac{\gamma-1}{2}\mathrm{Ma}_1^2}{1 + \frac{\gamma-1}{2}\mathrm{Ma}_2^2}$$

Finally, we get

$$\frac{A_1}{A_2}=\frac{\rho_2}{\rho_1}\frac{V_2}{V_1}=\left(\frac{p_2}{p_1}\right)^{1/\gamma}\times\left(\frac{\text{Ma}_2}{\text{Ma}_1}\right)\times\left[\frac{1+\frac{\gamma-1}{2}\text{Ma}_1^2}{1+\frac{\gamma-1}{2}\text{Ma}_2^2}\right]^{1/2}$$

$$=(11.5)^{0.714}\times\frac{1.4}{3.0}\times\left[\frac{1+0.2\,(3.0)^2}{1+0.2\,(1.4)^2}\right]^{1/2}$$

or $$A_1/A_2=3.79$$

So, the area ratio is 3.79.

Example 14.6

Air flows isentropically through a converging nozzle attached to a large tank where the absolute pressure is 171 kPa and the temperature is 27 °C. At the inlet section, the Mach number is 0.2. The nozzle discharges into the atmosphere through an area 0.015 m^2. Determine the magnitude and direction of the force that must be applied to hold the nozzle in place.

Solution

Refer to Fig. 14.13.

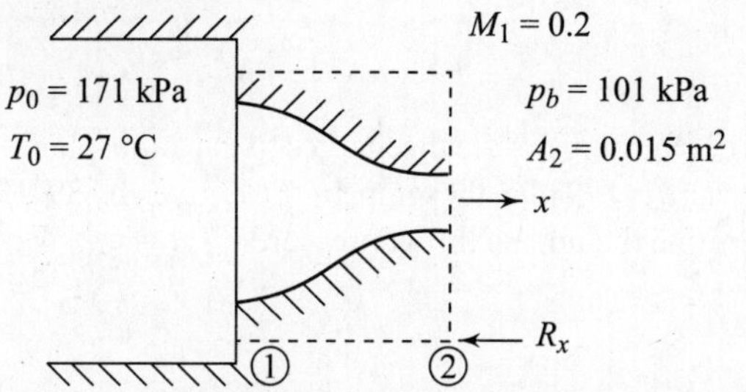

Fig. 14.13 Magnitude and direction of force required to keep the nozzle in place

$$\text{Ma}_2=\left[\frac{2}{\gamma-1}\left(\frac{p_0}{p_{\text{th}}}\right)^{(\gamma-1)/\gamma}-1\right]^{0.5}=\left[\frac{2}{0.4}\left(\frac{171}{101}\right)^{0.286}-1\right]^{0.5}$$

$$=0.901$$

So, the flow is not choked

$$T_2=T_0\Big/\left[1+\frac{\gamma-1}{2}\text{Ma}_2^2\right]=300/[1+0.2(0.901)^2]=258\text{ K}$$

$$V_2=\text{Ma}_2 a_2=\text{Ma}_2\,(\gamma R T_2)^{0.5}=0.901\,(1.4\times 287\times 258)^{0.5}$$

$$=290\text{ m/s}$$

$$\rho_2 = \frac{p_2}{RT_2} = \frac{101\times10^3}{287\times258} = 1.36\ \text{kg/m}^3$$

$$\dot{m} = \rho_2 V_2 A_2 = 1.36\times290\times0.016 = 5.92\ \text{kg/s}$$

$$T_1 = T_0\bigg/\left[1+\frac{\gamma-1}{2}\text{Ma}_1^2\right] = 300/\{1+0.2\,(0.2)^2\} = 298\ \text{K}$$

$$V_1 = \text{Ma}_1\, a_1 = \text{Ma}_1\,(\gamma RT_1)^{0.5} = 0.2\,(1.4\times287\times298)^{0.5}$$

$$= 69.2\ \text{m/s}$$

$$p_1 = p_0\bigg/\left[1+\frac{\gamma-1}{2}\text{Ma}_1^2\right]^{\gamma/(\gamma-1)} = 171/[1+0.2\,(0.2)^2]^{3.5}$$

$$= 166\ \text{kPa}$$

$$\rho_1 = p_1/(RT_1) = 166\times10^3/(287\times298) = 1.94\ \text{kg/m}^3$$

$$A_1 = \dot{m}/\rho_1 V_1 = 5.92/(1.94\times69.2) = 0.044\ \text{m}^2$$

$$R_x = p_1 A_1 - p_2 A_2 - p_{\text{atm}}(A_1 - A_2) - \dot{m}(V_2 - V_1)$$

$$= p_{1g} A_1 - p_{2g} A_2 - \dot{m}\,(V_2 - V_1)$$

$$= (166-101)\times10^3\times0.044 - 5.92\,(290-69.2)$$

$$R_x = 1560\ \text{N (to the left)}$$

Example 14.7

Air flowing isentropically through a converging nozzle discharges to the atmosphere. At any section where the absolute pressure is 179 kPa, the temperature is given by 39 °C and the air velocity is 177 m/s. Determine the nozzle throat pressure.

Solution

Refer to Fig. 14.14.

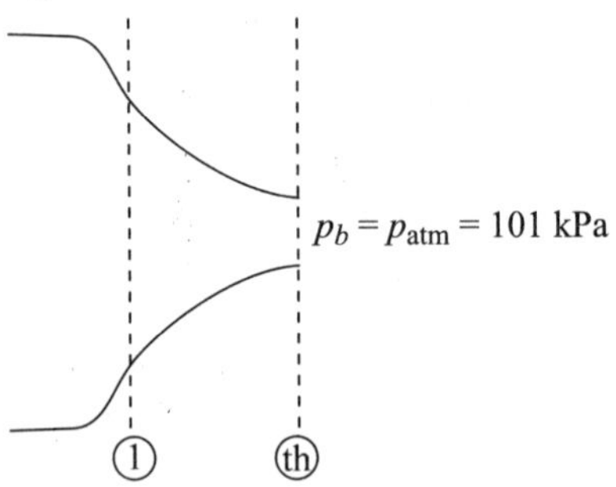

$T_1 = 39$ °C
$p_1 = 179$ kPa
$V_1 = 177$ m/sec

Fig. 14.14 Pressure, temperature and velocity are specified at any section of a converging nozzle

The nozzle will be choked ($Ma_{th} = 1.0$) if $p_b/p_0 = 0.528$

$$Ma_1 = V_1/a_1;\ a_1 = \sqrt{\gamma R T_1} = (1.4 \times 287 \times 312)^{0.5}$$

$$= 354 \text{ m/s}$$

$$Ma_1 = V_1/a_1 = 177/354 = 0.5$$

$$\frac{p_0}{p_1} = \left(1 + \frac{\gamma - 1}{2} Ma_1^2\right)^{\gamma/(\gamma-1)}$$

$$p_0 = 179\,(1 + 0.2\,(0.5)^2)^{3.5}$$

or $$p_0 = 212 \text{ kPa}$$

So, $$p_b/p_0 = 101/212 = 0.476 \text{ which is less than } 0.528$$

For $$Ma_{th} = 1.0,\ p_{th}/p_0 = 0.528$$

$$P_{th} = 0.528 \times p_0 = 0.528 \times 212 = 112 \text{ kPa}$$

Example 14.8

Air flows steadily and isentropically in a converging-diverging nozzle. At the throat, the air is at 140 kPa (abs), and at 60 °C. The throat cross-sectional area is 0.05 m^2. At a certain section in the diverging part of the nozzle, the pressure is 70.0 kPa (abs). Calculate the velocity and area of the this section.

Solution

Refer to Fig. 14.15.

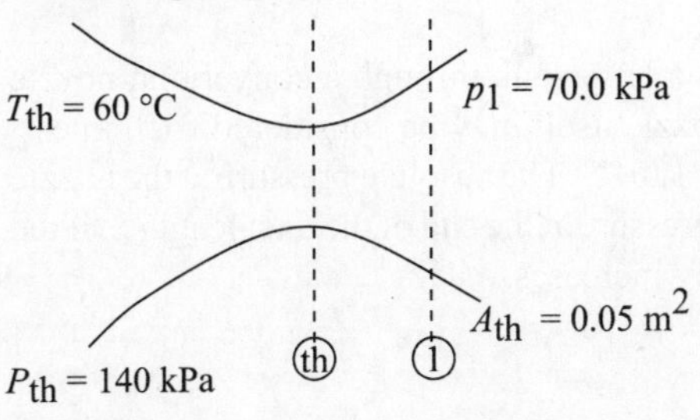

Fig. 14.15 Flow in a converging-diverging nozzle and conditions at a diverging section

Since $p_1 < p_{th}$, flow downstream of throat is supersonic and $Ma_{th} = 1.0$

$$p_0 = p_{th}\left[1 + \frac{\gamma - 1}{2} Ma_{th}^2\right]^{\gamma/(\gamma-1)}$$

$$= 140\,[1 + 0.2(1.0)^2]^{3.5}$$

$$= 265 \text{ kPa}$$

$$T_0 = T_{\text{th}}\left[1+\frac{\gamma-1}{2}\text{Ma}_{\text{th}}^2\right] = 333\,[1+0.2\,(1.0)] = 400\ \text{K}$$

$$V_{\text{th}} = \text{Ma}_{\text{th}}\, a_{\text{th}} = \text{Ma}_{\text{th}}\,(\rho\, RT)^{0.5}$$
$$= 1.0\,(1.4\times 287\times 333)^{0.5} = 366\ \text{m/s}$$

$$\text{Ma}_1 = \left[\frac{2}{\gamma-1}\left(\frac{p_0}{p_1}\right)^{(\gamma-1)/\gamma} - 1\right]^{0.5}$$

$$= \left[\frac{2}{0.4}\left(\frac{265}{70}\right)^{0.286} - 1\right]^{0.5} = 1.52$$

$$T_1 = \frac{T_0}{1+\frac{\gamma-1}{2}\text{Ma}_1^2} = \frac{400}{1+0.2(1.52)^2} = 274\ \text{K}$$

$$V_1 = \text{Ma}_1\, a_1 = 1.52\,(91.4\times 287\times 274)^{0.5} = 504\ \text{m/s}$$

$$\dot{m} = \rho_{\text{th}}\, V_{\text{th}}\, A_{\text{th}} = \rho_1\, V_1\, A_1$$

$$A_1 = \frac{\rho_{\text{th}}}{\rho_1}\cdot\frac{V_{\text{th}}}{V_1}\cdot A_{\text{th}} = \left(\frac{p_{\text{th}}}{p_1}\right)^{\frac{1}{\gamma}}\cdot\frac{V_{\text{th}}}{V_1}\cdot A_{\text{th}}$$

$$A_1 = \left(\frac{140}{70}\right)^{0.714}\times\frac{366}{504}\times 0.05 = 0.0596\ \text{m}^2$$

Example 14.9

Air flows steadily and adiabatically from a large tank through a converging nozzle connected to a constant area duct. The nozzle itself may be considered frictionless. Air in the tank is at $p = 1.00$ MPa (abs), $T = 125$ °C. The absolute pressure at the nozzle exit (duct inlet) is 784 kPa. Determine the pressure at the end of the duct length L, if the temperature there is 65 °C, and the entropy increases.

Solution

Refer to Fig. 14.16.

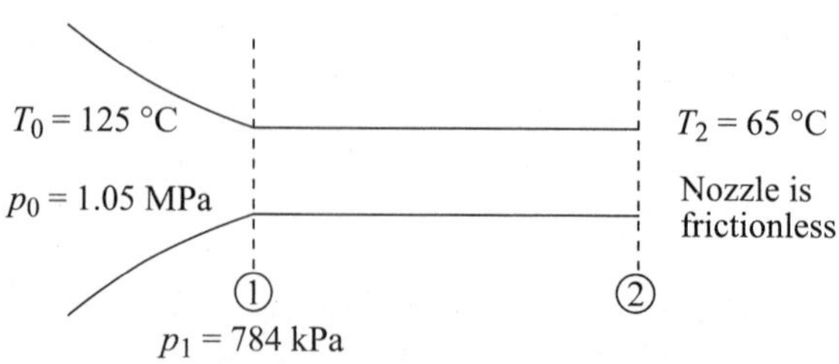

Fig. 14.16 Flow from a tank through a nozzle connected to a duct

$$\text{Ma}_1 = \left[\frac{2}{\gamma-1}\left\{\left(\frac{p_0}{p_1}\right)^{(\gamma-1)/\gamma} - 1\right\}\right]^{0.5} = \left[\frac{2}{0.4}\left(\frac{398}{T_1} - 1\right)\right]^{0.5}$$

$$= 0.60$$

$$T_1 = \frac{T_0}{1 + \frac{\gamma-1}{2}\text{Ma}_1^2} = \frac{398}{1 + 0.2\,(0.60)^2} = 317\text{ K},\ T_2 = 338\text{ K}$$

Again,
$$T_0 = \text{constant and } \text{Ma}_2 = \left[\frac{2}{\gamma-1}\left(\frac{T_0}{T_2}\right) - 1\right]^{0.5} = 0.942$$

$$V_2 = \text{Ma}_2\, a_2 = 0.942\,(1.4 \times 287 \times 338)^{0.5} = 347\text{ m/s}$$

$$V_1 = \text{Ma}_1\, a_1 = 0.60\,(1.4 \times 287 \times 371)^{0.5} = 232\text{ m/s}$$

$$\rho_1 = p_1/(RT_1) = 784 \times 10^3/(287 \times 371) = 7.36\text{ kg/m}^3$$

$$\rho_2 = \frac{V_1}{V_2}\rho_1 = 4.92\text{ kg/m}^3$$

$$p_2 = \rho_2 R T_2 = 4.92 \times 287 \times 338 = 477\text{ kPa}$$

$$T\text{d}s = \text{d}h - v\,\text{d}p = c_p\,\text{d}T - \frac{1}{\rho}\text{d}p$$

$$s_2 - s_1 = \int_{s_1}^{s_2} \text{d}s = c_p \ln\frac{T_2}{T_1} - R\ln\frac{p_2}{p_1}$$

$$= 10\ln(338/371) - 287\ln(477/784) = 49.5\text{ J/kg K}$$

14.7 NORMAL SHOCKS

Shock waves are highly localised irreversibilities in the flow. Within the distance of a mean free path, the flow passes from a supersonic to a subsonic state, the velocity decreases suddenly and the pressure rises sharply. To be more specific, a shock is said to have occurred if there is an abrupt reduction of velocity in the downstream in course of a supersonic flow in a passage or around a body. Normal shocks are substantially perpendicular to the flow and oblique shocks are inclined at other angles. Shock formation is possible for confined flows as well as for external flows. Normal shock and oblique shock may mutually interact to make another shock pattern. Different type of shocks are presented in Fig. 14.17.

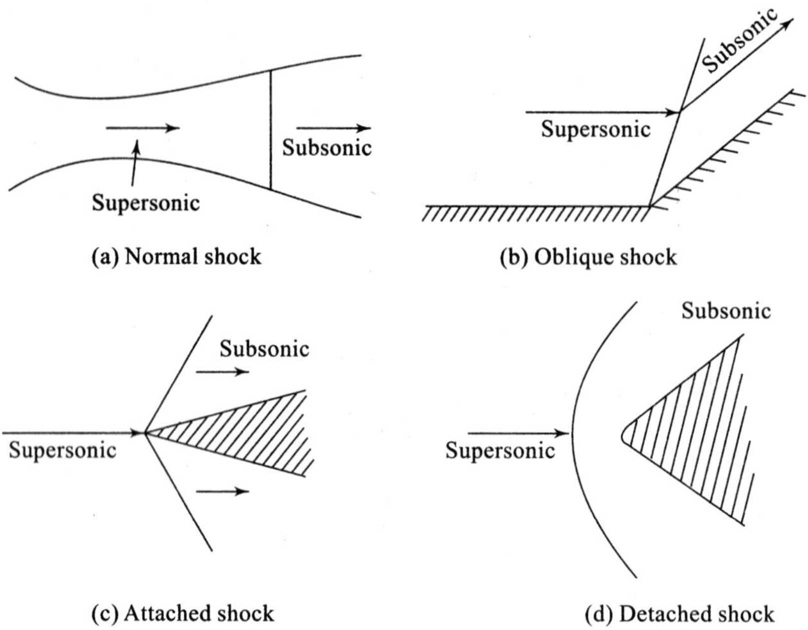

Fig. 14.17 Different type of shocks

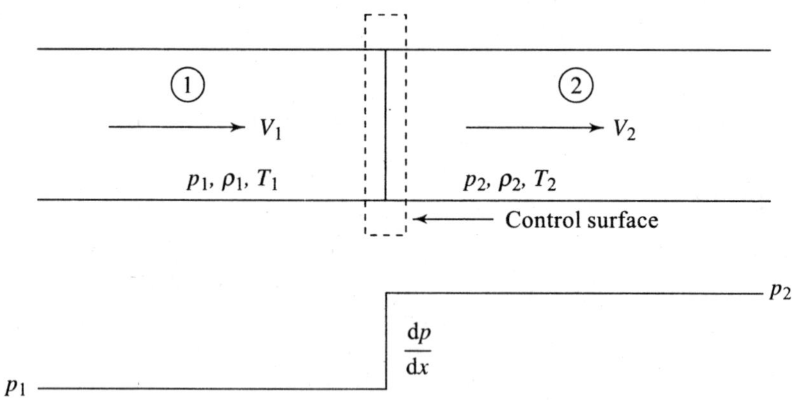

Fig. 14.18 One-dimensional normal shock

Figure 14.18 shows a control surface that includes a normal shock. The fluid is assumed to be in thermodynamic equilibrium upstream and downstream of the shock, the properties of which are designated by the subscripts 1 and 2, respectively.

Continuity equation can be written as

$$\frac{\dot{m}}{A} = \rho_1 V_1 = \rho_2 V_2 = G \tag{14.76}$$

where G is the mass velocity kg/m^{2}s.
From momentum equation, one can write

$$p_1 - p_2 = \frac{\dot{m}}{A}(V_2 - V_1) = \rho_2 V_2^2 - \rho_1 V_2^2$$

or
$$p_1 + \rho_1 V_1^2 = p_2 + \rho_2 V_2^2 \tag{14.77a}$$

or
$$F_1 = F_2 \tag{14.77b}$$

where $F = p + \rho V^2$ can be termed as *impulse function*.

The energy equation may be written as

$$h_1 + \frac{V_1^2}{2} = h_2 + \frac{V_2^2}{2} = h_{01} = h_{02} = h_0 \tag{14.78}$$

where h_0 is stagnation enthalpy.
From the second law of thermodynamics, it may be written as

$$s_2 - s_1 \geq 0 \tag{14.79}$$

But Eq. (14.79) is of little help in calculating actual entropy change across the shock. To calculate the entropy change, we have

$$T\mathrm{d}s = \mathrm{d}h - v\mathrm{d}p \tag{14.33 repeated}$$

For an ideal gas we can write

$$\mathrm{d}s = c_p \frac{\mathrm{d}T}{T} - R\frac{\mathrm{d}p}{p}$$

For constant specific heat, this equation can be integrated to give

$$s_2 - s_1 = c_p \ln \frac{T_2}{T_1} - R \ln \frac{p_2}{p_1} \tag{14.80}$$

For an ideal gas the equation of state can be written as

$$p = \rho RT \tag{14.8 repeated}$$

Equations (14.76), (14.77a), (14.78), (14.80) and (14.8) are the governing equations for the flow of an ideal gas through normal shock. If all the properties at state '1' (upstream of the shock) are known, then we have six unknowns ($T_2, p_2, \rho_2, V_2, h_2, s_2$) in these five equations. However, we have known relationship between h and T [Eq. (14.17)] for an ideal gas which is given by $\mathrm{d}h = c_p\,\mathrm{d}T$. For an ideal gas with constant specific heats,

$$\Delta h = h_2 - h_1 = c_p(T_2 - T_1) \tag{14.81}$$

Thus, we have the situation of six equations and six unknowns.

If all the conditions at state '1' (immediately upstream of the shock) are known, how many possible states '2' (immediate downstream of the shock) are there? The mathematical answer indicates that there is a unique state '2' for a given state '1'. Before describing the physical picture and precise location of these two states let us introduce Fanno line and Rayleigh line flows.

14.7.1 Fanno Line Flows

If we consider a problem of frictional adiabatic flow through a duct, the governing Eqs (14.76), (14.78), (14.80) (14.8) and (14.81) are valid between any two points '1' and '2'. Equation (14.77a) requires to be modified in order to take into account the frictional force, R_x, of the duct wall on the flow and we obtain

$$R_x + p_1 A - p_2 A = \dot{m} V_2 - \dot{m} V_1 \tag{14.82}$$

So, for a frictional flow, we thus have a situation of six equations and seven unknowns. If all the conditions of '1' are known, how many possible states '2' are there? Mathematically, we get number of possible states '2'. With an infinite number of possible states '2' for a given state '1', what do we observe if all possible states '2' are plotted on a *T-s* diagram? The locus of all possible states '2' reachable from state '1' is a continuous curve passing through state '1'. However, the question is how to determine this curve? Perhaps the simplest way is to assume different values of T_2. For an assumed value of T_2, the corresponding values of all other properties at '2' and R_x can be determined.

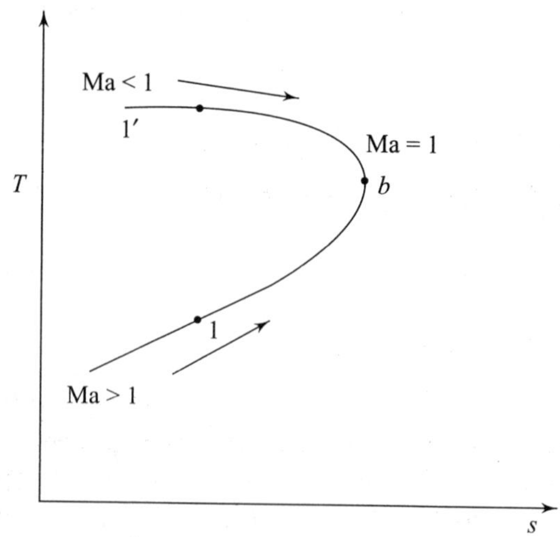

Fig. 14.19 Fanno line representation of constant area adiabatic flow

The locus of all possible downstream states is called the Fanno line and is shown in Fig. 14.19. Point '*b*' corresponds to maximum entropy where the flow is sonic. This point splits the Fanno line into subsonic (upper)and supersonic (lower) portions. If the inlet flow is supersonic and corresponds to point 1 in Fig. 14.15, then friction causes the downstream flow to move closer to point '*b*' with a consequent decrease of Mach number towards unity. Each point on the curve between point 1 and '*b*' corresponds to a certain duct length *L*. As *L* is made larger, the conditions at the exit move closer to point '*b*'. Finally, for a certain value of *L*, the flow becomes sonic. Any further increase in *L* is not possible without a drastic revision of the inlet conditions. Consider the alternative case where the inlet flow is subsonic, say, given the point 1′ in Fig. 14.19. As *L* increases, the exit conditions move closer to point '*b*'. If *L* is increased to a sufficiently large value, then point '*b*' is reached and the flow at the exit becomes sonic. The flow is again choked and any further increase in *L* is not possible without an adjustment of the inlet conditions.

14.7.2 Rayleigh Line Flows

If we consider the effects of heat transfer on a frictionless compressible flow through a duct, the governing Eq. (14.76), (14.77a), (14.80), (14.8) and (14.81) are valid between any two points '1' and '2'. Equation (14.78) requires to be modified in order to account for the heat transferred to the flowing fluid per unit mass, δQ, and we obtain

$$\delta Q = h_{02} - h_{01} \tag{14.83}$$

So, for frictionless flow of an ideal gas in a constant area duct with heat transfer, we have again a situation of six equations and seven unknowns. If all conditions at state '1' are known, how many possible states '2' are there? Mathematically, there exists infinite number of possible states '2'. With an infinite number of possible states '2' for a given state '1', what do we observe if all possible states '2' are plotted on a *T*-*s* diagram? The locus of all possible states '2' reachable from state '1' is a continuous curve passing through state '1'. Again, the question arises as to how to determine this curve? The simplest way to go about this problem is to assume different values of T_2. For an assumed value of T_2, the corresponding values of all other properties at '2' and δQ can be determined. The results of these calculations are shown on the *T*-*s* plane in Fig. 14.20. The curve in Fig. 14.20 is called the Rayleigh line.

At the point of maximum temperature (point '*c*' in Fig. 14.20), the value of Mach number for an ideal gas is $1/\sqrt{\gamma}$. At the point of maximum entropy, the Mach number is unity. On the upper branch of the curve, the flow is always subsonic and it increases monotonically as we proceed to the right along the curve. At every point on the lower branch of the curve, the flow is supersonic, and it decreases monotonically as we move to the right along the curve. Irrespective of the initial Mach number, with heat addition, the flow state proceeds to the right and with heat rejection, the flow state proceeds to the left along the Rayleigh line. For example, let us consider a flow which is at an initial state given by 1 on the Rayleigh line in fig. 14.20. If heat is added to the flow, the conditions in the downstream region 2 will move close to point "*b*". The velocity reduces due to increase in pressure and density, and Ma approaches unity. If δQ is increased to a sufficiently high value, then point '*b*' will be reached and flow in region 2 will be sonic. The flow is again choked, and any further increase in δQ is not possible without an adjustment of the initial condition. The flow cannot become subsonic by any further increase in δQ.

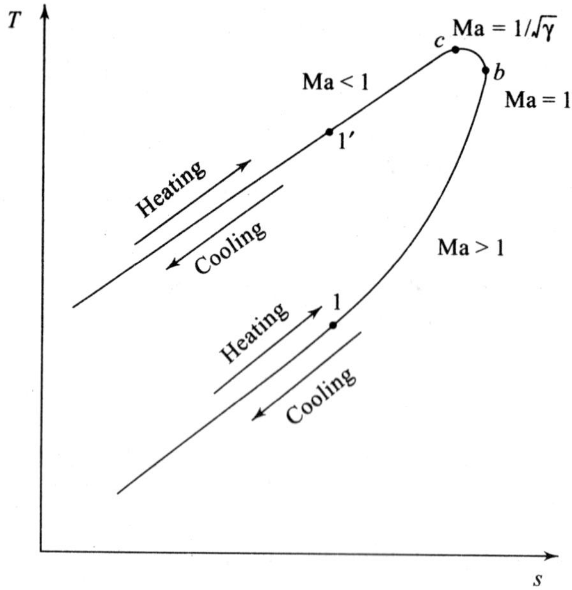

Fig. 14.20 Rayleigh line representation of frictionless flow in a constant area duct with heat transfer

14.7.3 Physical Picture of Flow through a Normal Shock

It is possible to obtain physical picture of flow through a normal shock by employing some of the ideas of Fanno line and Rayleigh line flows. Flow through a normal shock must satisfy Eqs (14.76), (14.77a), (14.78), (14.80), (14.8) and 14.81). Since all the condition of state '1' are known, there is no difficulty in locating state '1' on *T-s* diagram. In order to draw a Fanno line curve through state '1', we require a locus of mathematical states that satisfy Eqs (14.76), (14.78), (14.80), (14.8) and (14.81). The Fanno line curve does not satisfy Eq. (14.77a). A Rayleigh line curve through state '1' gives a locus of mathematical states that satisfy Eqs (14.76), (14.77a), (14.80), (14.8) and (14.81). The Rayleigh line does not satisfy Eq. (14.78). Both the curves on a same *T-s* diagram are shown in Fig. 14.21. As we have already pointed out, the normal shock should satisfy all the six equations stated above. At the same time, for a given state '1', the end state '2' of the normal shock must lie on both the Fanno line and Rayleigh line passing through state '1'. Hence, the intersection of the two lines at state '2' represents the conditions downstream from the shock. In Fig. 14.21, the flow through the shock is indicated as transition from state '1' to state '2'. This is also consistent with directional principle indicated by the second law of thermodynamics, i.e. $s_2 > s_1$. From Fig. 14.21, it is also evident that the flow through a normal shock signifies a change of speed from supersonic to subsonic. Normal shock is possible only in a flow which is initially supersonic.

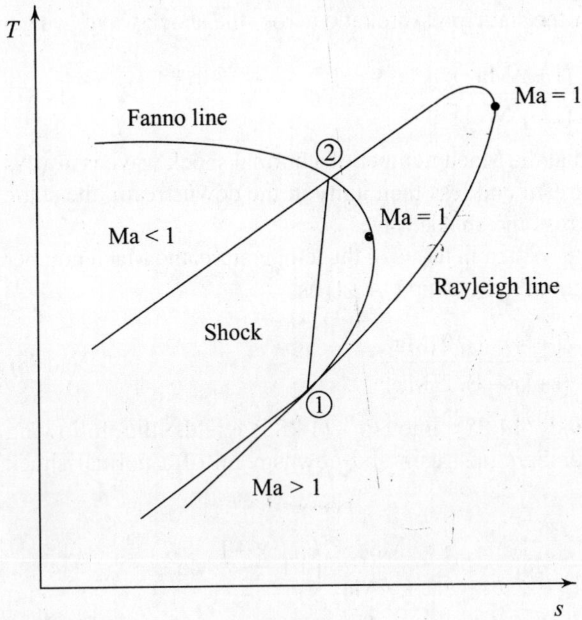

Fig. 14.21 Intersection of Fanno line and Rayleigh line and the solution for normal shock condition

14.7.4 Calculation of Flow Properties Across a Normal Shock

The easiest way to analyse a normal shock is to consider a control surface around the wave as shown in Fig. 14.18. The continuity equation (14.76), the momentum equation (14.77) and the energy equation (14.78) have already been discussed earlier. The energy equation can be simplified for an ideal gas as

$$T_{01} = T_{02} \tag{14.84}$$

By making use of the equation for the speed of sound (14.49) and the equation of state for ideal gas (14.8), the continuity equation can be rewritten to include the influence of Mach number as

$$\frac{p_1}{RT_1}\,\mathrm{Ma}_1\sqrt{\gamma RT_1} = \frac{p_2}{RT_2}\,\mathrm{Ma}_2\sqrt{\gamma RT_2} \tag{14.85}$$

The Mach number can be introduced in momentum equation in the following way:

$$\rho_2 V_2^2 - \rho_1 V_1^2 = p_1 - p_2$$

$$p_1 + \frac{p_1}{RT_1} V_1^2 = p_2 + \frac{p_2}{RT_2} V_2^2$$

$$p_1(1+\gamma \mathrm{Ma}_1^2) = p_2(1+\gamma \mathrm{Ma}_2^2) \tag{14.86}$$

Rearranging this equation for the static pressure ratio across the shock wave, we get

$$\frac{p_2}{p_1} = \frac{(1+\gamma\,\mathrm{Ma}_1^2)}{(1+\gamma\,\mathrm{Ma}_2^2)} \tag{14.87}$$

As we have already seen that the Mach number of a normal shock wave is always greater than unity in the upstream and less than unity in the downstream, the static pressure always increases across the shock wave.

The energy equation can be written in terms of the temperature and Mach number using the stagnation temperature relationship (14.84) as

$$\frac{T_2}{T_1} = \frac{\{1+[\gamma-1)/2]\mathrm{Ma}_1^2}{\{1+[\gamma-1)/2]\mathrm{Ma}_2^2} \tag{14.88}$$

Substituting Eqs (14.87) and (14.88) into Eq. (14.85) yields the following relationship for the Mach numbers upstream and downstream of a normal shock wave:

$$\frac{\mathrm{Ma}_1}{1+\gamma\,\mathrm{Ma}_1^2}\left(1+\frac{\gamma-1}{2}\mathrm{Ma}_1^2\right)^{1/2} = \frac{\mathrm{Ma}_2}{1+\gamma\,\mathrm{Ma}_2^2}\left(1+\frac{\gamma-1}{2}\mathrm{Ma}_2^2\right)^{1/2} \tag{14.89}$$

Then, solving this equation for Ma_2 as a function of Ma_1, we obtain two solutions. One solution is trivial, $\mathrm{Ma}_1 = \mathrm{Ma}_2$, which signifies no shock across the control volume. The other solution is

$$\mathrm{Ma}_2^2 = \frac{(\gamma-1)\,\mathrm{Ma}_1^2+2}{2\gamma\,\mathrm{Ma}_1^2-(\gamma-1)} \tag{14.90}$$

$\mathrm{Ma}_1 = 1$ in Eq. (14.90) results in $\mathrm{Ma}_2 = 1$. Equations (14.87) and (14.88) also show that there would be no pressure or temperature increase across the shock. In fact, the shock wave corresponding to $\mathrm{Ma}_1 = 1$ is the sound wave across which, by definition, pressure and temperature changes are infinitesimal. Therefore, it can be said that the sound wave represents a degenerated normal shock wave.

14.7.5 Rankine–Hugoniot Relation

Equation (14.87) describes the static pressure ratio across the normal shock. On substituting for Ma_2 from Eq. (14.90), we get

$$\frac{p_2}{p_1} = \frac{2\gamma\,\mathrm{Ma}_1^2}{\gamma+1} - \frac{\gamma-1}{\gamma+1} \tag{14.91}$$

The pressure ratio tends to infinity with the rising value of upstream Mach number. Eq. (14.88) describes the temperature ratio across the normal shock. Substitution of Ma_2 from from Eq. (14.90), we get

$$\frac{T_2}{T_1} = \frac{\left[2+(\gamma-1)\,\mathrm{Ma}_1^2\right]\left[2\gamma\,\mathrm{Ma}_1^2-(\gamma-1)\right]}{(\gamma+1)^2\,\mathrm{Ma}_1^2} \tag{14.92}$$

The limiting value of T_2/T_1, as $\mathrm{Ma}_1 \to \infty$ is

$$\frac{T_2}{T_1} = \lim_{\text{Ma}_1 \to \infty} \frac{\left[\left(2/\text{Ma}_1^2\right) + (\gamma - 1)\right]\left[2\gamma \text{Ma}_1^2 - (\gamma - 1)\right]}{(\gamma + 1)^2}$$

or $\dfrac{T_2}{T_1} \to \infty$

The ratio of densities after and before the shock is obtained by combining Eqs. (14.91) and (14.92)

$$\frac{\rho_2}{\rho_1} = \left(\frac{p_2}{p_1}\right)\left(\frac{T_1}{T_2}\right)$$

or
$$\frac{\rho_2}{\rho_1} = \frac{(\gamma+1)\,\text{Ma}_1^2}{2+(\gamma-1)\,\text{Ma}_1^2} \tag{14.93}$$

As, $\text{Ma}_1 \to \infty$, the limiting value of the density ratio is

$$\lim_{\text{Ma}_1 \to \infty}\left(\frac{\rho_2}{\rho_1}\right) = \lim_{\text{Ma}_1 \to \infty} \frac{(\gamma+1)}{\left[\dfrac{2}{\text{Ma}_1^2} + (\gamma-1)\right]}$$

$$= \frac{(\gamma+1)}{(\gamma-1)} = 6 \text{ for } \gamma = 1.4$$

Therefore even though the pressure and temperature ratios tend to infinity, the density ratio remains finite for increasing value of upstream Mach number. The ratios of pressures and densities on the downstream and upstream of a normal shock are plotted in Fig. 14.22 and the plot is known as **Rankine–Hugoniot** curve.

14.7.6 Strength of a Shock

The strength of a shock may be defined as the rise in pressure across the shock as compared to the upstream pressure. Thus,

$$S = \frac{p_2 - p_1}{p_1} = \frac{p_2}{p_1} - 1$$

Invoking Eq. (14.91), we get,

$$S = \frac{2\gamma \text{Ma}_1^2 - (\gamma-1) - (\gamma+1)}{\gamma+1}$$

or
$$S = \frac{2\gamma\left(\text{Ma}_1^2 - 1\right)}{\gamma+1} \tag{14.94}$$

The interpretation of Eq. (14.94) is as the following. A shock wave has vanishing strength as the upstream Mach number is close to unity The phenomenon is reduced to isentropic expansion as illustrated in Fig. 14.22. The pressure density ratios of a shock wave (at $Ma_1 \sim 1$) coincide with those of an isentropic flow. In other words, the weak shock waves are acoustic waves.

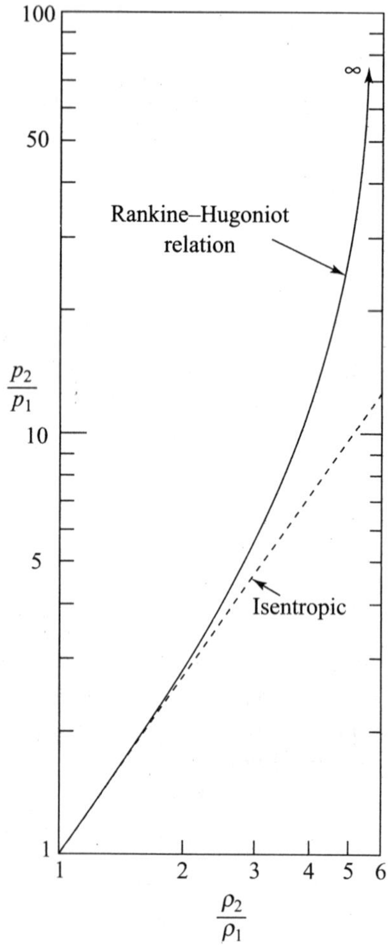

Fig. 14.22 Rankine–Hugoniot curve and isentropic expansion ($\gamma = 1.4$)

14.8 OBLIQUE SHOCK

The discontinuities in supersonic flows do not always exist as normal to the flow direction. There are oblique shocks which are inclined with respect to the flow direction. Let us refer to the shock structure on an obstacle, as depicted qualitatively in Fig.14.23. The segment of the shock immediately in front of the body behaves like a normal shock. Oblique shock is formed as a consequence of the bending of the shock in the free-stream direction. Sometimes in a supersonic flow through a duct, viscous effects cause the shock to be oblique near the walls, the shock being normal only in the core region. The shock is also oblique when a supersonic flow is made to change direction near a sharp corner.

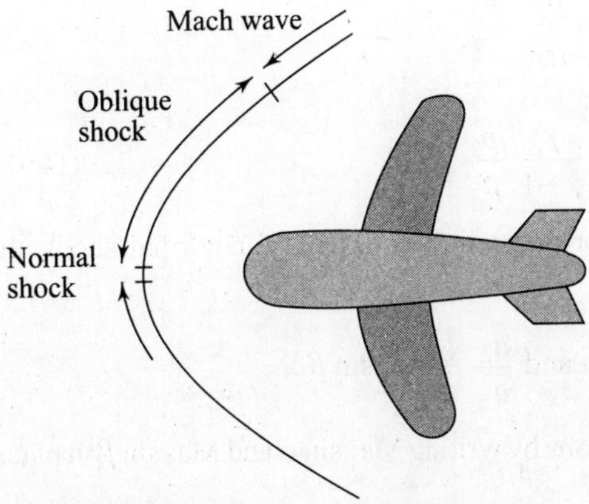

Fig. 14.23 Normal and oblique shock in front of an obstacle

The same relationships derived earlier for the normal shock are valid for the velocity components normal to the oblique shock. The oblique shock continues to bend in the downstream direction until the Mach number of the velocity component normal to the wave is unity. At that instant, the oblique shock degenerates into a so-called Mach wave across which changes in flow properties are infinitesimal.

Let us consider a two-dimensional oblique shock as shown in Fig. 14.24.

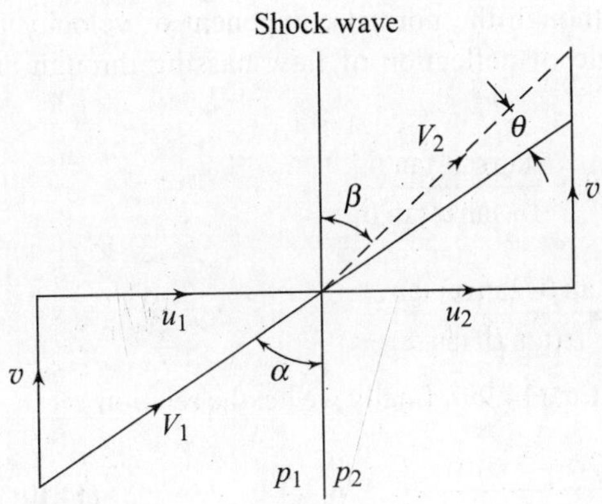

Fig. 14.24 Two-dimensional oblique shock

In analysing flow through such a shock, it may be considered as a normal shock on which a velocity v (parallel to the shock) is superimposed. The change across shock front is determined in the same way as for the normal shock. The equations for mass, momentum and energy conservation are, respectively,

$$\rho_1 u_1 = \rho_2 u_2 \tag{14.95}$$

$$\rho_1 u_1 (u_1 - u_2) = p_2 - p_1 \tag{14.96}$$

$$\frac{\gamma}{\gamma - 1} \cdot \frac{p_1}{\rho_1} + \frac{V_1^2}{2} = \frac{\gamma}{\gamma - 1} \cdot \frac{p_2}{\rho_2} + \frac{V_2^2}{2}$$

or

$$\frac{u_1^2}{2} + \frac{\gamma}{\gamma - 1} \cdot \frac{p_1}{\rho_1} = \frac{u_2^2}{2} + \frac{\gamma}{\gamma - 1} \cdot \frac{p_2}{\rho_2} \tag{14.97}$$

These equations are analogous to corresponding equations for normal shock. In addition to these, we have

$$\frac{u_1}{a_1} = \text{Ma}_1 \sin \alpha \text{ and } \frac{u_2}{a_2} = \text{Ma}_2 \sin \beta$$

Then modifying normal shock relations by writing $\text{Ma}_1 \sin \alpha$ and $\text{Ma}_2 \sin \beta$ in place of Ma_1 and Ma_2, we obtain

$$\frac{p_2}{p_1} = \frac{2\gamma \,\text{Ma}_1^2 \sin^2 \alpha - \gamma + 1}{\gamma + 1} \tag{14.98}$$

$$\frac{u_2}{u_1} = \frac{\rho_1}{\rho_2} = \frac{\tan \beta}{\tan \alpha} = \frac{\gamma - 1}{\gamma + 1} + \frac{2}{(\gamma + 1)\,\text{Ma}_1^2 \sin^2 \alpha} \tag{14.99}$$

$$\text{Ma}_2^2 \sin^2 \beta = \frac{2 + (\gamma - 1)\,\text{Ma}_1^2 \sin^2 \alpha}{1 + \tan^2 \alpha (\tan \beta / \tan \alpha)} \tag{14.100}$$

Note that although $\text{Ma}_2 \sin \beta < 1$, Ma_2 may be greater than 1. So the flow behind an oblique shock may be supersonic although the normal component of velocity is subsonic. In order to obtain the angle of deflection of flow passing through an oblique shock, we use the relation

$$\tan \theta = \tan (\alpha - \beta) = \frac{\tan \alpha - \tan \beta}{1 + \tan \alpha \tan \beta}$$

$$= \frac{\tan \alpha - (\tan \beta / \tan \alpha) \tan \alpha}{1 + \tan^2 \alpha (\tan \beta / \tan \alpha)}$$

Having substituted ($\tan \beta / \tan \alpha$) from Eq. (14.99), finally we get the relation

$$\tan \theta = \frac{\text{Ma}_1^2 \sin 2\alpha - 2 \cot \alpha}{\text{Ma}_1^2 (\gamma + \cos 2\alpha) + 2} \tag{14.101}$$

Sometimes, a design is done in such a way that an oblique shock is allowed instead of a normal shock. The losses for the case of oblique shock are much less than those of normal shock. This is the reason for making the nose angle of the fuselage of a supersonic aircraft small.

Example 14.10

A normal shock wave takes place during the flow of air at a Mach number of 1.8. The static pressure and temperature of the air upstream of the shock wave are 100 kPa (abs) and 15 °C. Determine the Mach number, pressure and temperature downstream of the shock.

Solution

Making use of Eq. (14.90), the Mach number downstream of the shock can be calculated as

$$\mathrm{Ma}_2^2 = \frac{(0.4)\,(1.8)^2 + 2}{(2.8)\,(1.8)^2 - 0.4} = 0.38;\ \text{or}\ \mathrm{Ma}_2 = 0.616$$

Equations (14.87) and (14.88) provide the downstream pressure and temperature.

$$p_2 = p_1\left(\frac{1+\gamma\,\mathrm{Ma}_1^2}{1+\gamma\,\mathrm{Ma}_2^2}\right) = 100\left(\frac{1+(1.4)\,(1.8)^2}{1+(1.4)\,(0.616)^2}\right) = 361\ \text{kPa}$$

$$T_2 = T_1\left(\frac{1+[(\gamma-1)/2]\,\mathrm{Ma}_1^2}{1+[\gamma-1)/2]\,\mathrm{Ma}_2^2}\right) = 288\left(\frac{1+(0.2)(3.24)}{1+(0.2)(0.38)}\right) = 288$$

$$= 441\ \text{K}$$

SUMMARY

- Fluid density varies significantly due to a large Mach number (Ma = V/a) flow. This leads to a situation where continuity and momentum equations must be coupled to the energy equation and the equation of state to solve for the four unknowns, namely, p, ρ, T and V.
- The stagnation enthalpy and hence, T_0 are conserved in isentropic flows. The effect of area variation on flow properties in an isentropic flow is of great significance. This reveals the phenomenon of choking (maximum mass flow) at the sonic velocity in the throat of a nozzle. At choked condition, the ratio of the throat pressure to the stagnation pressure is constant and it is equal to 0.528 for $\gamma = 1.4$. A nozzle is basically a converging or converging-diverging duct where the kinetic energy keeps increasing at the expense of static pressure. A diffuser has a reversed geometry where pressure recovery takes place at the expense of kinetic energy. At supersonic velocities, the normal-shock wave appears across which the gas discontinuously reverts to the subsonic conditions.
- In order to understand the effect of non-isentropic flow conditions, an understanding of constant area duct flow with friction and heat transfer is necessary. These are known as Fanno line flows and Rayleigh line flows, both of which entail choking of the exit flow. The conditions before and after a normal shock are defined by the points of intersection of Fanno and Rayleigh lines on a T-s diagram.

- If a supersonic flow is made to change its direction, the oblique shock is evolved. The oblique shock continues to bend in the downstream direction until the Mach number of the velocity component normal to the wave is unity.

REFERENCES

1. A.H. Shapiro, *The Dynamics and Thermodynamics of Compressible Fluid Flow*, 2 Vols, The Ronald Press, New York, 1953.
2. P.K. Nag, *Engineering Thermodynamics*, Second Edition, Tata McGraw-Hill, New Delhi, 1995.
3. V. Babu, *Fundametals of Gas Dynamics*, Ane Books Pvt Ltd, 2008.

EXERCISES

14.1 Choose the correct answer:

(i) Select the expression that does not give the speed of a sound wave ($\gamma = c_p/c_v$)

(a) $\sqrt{\gamma\, p/\rho}$ (b) $\sqrt{\gamma\, \rho/p}$ (c) $\sqrt{\partial p/\partial \rho}$ (d) $\sqrt{\gamma\, RT}$

(ii) Shock waves are highly localised irreversibilities in the flow. Within the distance of a mean free path, the flow passes from a

(a) supersonic to a subsonic state
(b) subsonic to a supersonic state
(c) subsonic state to a sonic state
(d) supersonic to a hypersonic state

(iii) The compressible flow upstream of a shock is always

(a) supersonic (b) subsonic
(c) sonic (d) none of these

(iv) Fluid is flowing through a duct with a Mach number equal to 1.2. An increase in cross-sectional area in the downstream will cause an

(a) decrease in velocity (b) increase in velocity
(c) increase in static pressure (d) choked flow situation

(v) In a steady, adiabatic flow (it is not known whether reversible or not) of a compressible fluid

(a) the stagnation temperature may vary throughout the flow field
(b) the stagnation pressure and stagnation density may change
(c) the stagnation temperature and stagnation density remain constant.

14.2 An airplane is capable of flying with a Mach number of 0.8. What can be the maximum speed of the airplane (i) at the sea level where temperature is 15 °C, and (ii) at the high altitude where the temperature is – 55 °C?

Ans. ((i) 272.13 m/s (ii) 236.76 m/s)

14.3 Air is at rest ($p = 101$ kPa, $T = 288$ K) in a chamber. It is expanded isentropically. What is the Mach number when the velocity becomes 200 m/s? What is the velocity when the speed becomes sonic? Also find out the maximum attainable speed. *Ans.* (0.587, 340 m/s, 760 m/s)

14.4 Oxygen flow from a reservoir in which the temperature is at 200 °C and the pressure is at 300 kPa (abs). Assuming isentropic flow, calculate the velocity, pressure and temperature where the Mach number is 0.8. For oxygen, $\gamma = 1.4$, $R = 260$ J/kg K. *Ans.* (312.5 m/s, 196.8 kPa, 419.3 K)

14.5 One problem in creating high Mach number flows is condensation of the oxygen component in air when the temperature reaches 50 K. If the temperature of a reservoir is 300 K and the flow is isentropic, at what Mach number will condensation of oxygen take place? *Ans.* (Ma = 5.0)

14.6 A venturimeter with throat diameter 20 mm is installed in a pipe line of 60 mm to measure air flow rate. The inlet side pressure and temperature are 400 kPa (abs) and 298 K. The throat pressure is 300 kPa (abs). The flow in the venturimeter is considered frictionless and without heat transfer. Estimate the mass flow rate of air.

14.7 Air flows steadily and isentropically into an aircraft inlet at a rate of 100 kg/s. At a section where the area is 0.464 m^2, the Mach number, temperature and absolute pressure are found to be 3, – 60 °C and 15.0 kPa. Determine the velocity and cross-sectional area downstream where $T = 138$ °C. Sketch the flow passage.

Ans. ($V_2 = 610$ m/s, $A_2 = 0.129$ m^2)

14.8 Air flows steadily and isentropically through a passage. At Section 1 where the cross-sectional area is 0.02 m^2, the air is at 40.0 kPa (abs), 60 °C, and the Mach number is 2.0. At a section 2 downstream, the velocity is 519 m/s. Calculate the Mach number at Section 2. Sketch the shape of the passage between Sections 1 and 2.

Ans. ($Ma_2 = 1.2$)

14.9 Air flows from a large tank ($p = 650$ kPa (abs), $T = 550$°C) through a converging nozzle, with a throat area of 600 mm^2, and discharges to the atmosphere. Determine the rate of mass flow under isentropic condition in the nozzle.

Ans. (0.548 kg/s)

14.10 Air enters a converging-diverging nozzle with negligible velocity at an absolute pressure of 1.0 MPa and a temperature of 60 °C. If the flow is isentropic and the exit temperature is – 11 °C, what is the Mach number at the exit?

Ans. (1.16)

14.11 Air is to be expanded through a converging-diverging nozzle by a frictionless adiabatic process from a pressure of 1.10 MPa (abs) and a temperature of 115 °C to a pressure of 141 kPa (abs). Determine the throat and exit areas for a well-designed shockless nozzle if the mass flow rate is 2 kg/s.

Ans. (8.86×10^{-4} m^2, 1.5×10^{-3} m^2)

14.12 Air, at a stagnation pressure of 7.2 MPa (abs) and a stagnation temperature of 1100 K, flows isentropically through a converging-diverging nozzle having a throat area of 0.01 m^2. Determine the velocity at the downstream section where the Mach number is 4.0. Also find out the mass flow rate.

Ans. (1300 m/s, 87.4 kg/s)

14.13 A normal shock wave exists in a 500 m/s stream of nitrogen with a static temperature of – 40 °C and static pressure of 70 kPa. Calculate the Mach number, pressure, and temperature downstream of the wave and entropy increase across the wave. For nitrogen, $\gamma = 1.4$, $R = 297$ J/kg K.

Ans. ($Ma_2 = 0.665$, $p_2 = 200$ kPa, $T_2 = 325$ K, $\Delta s = 34.1$ J/kg K)

15

PRINCIPLES OF FLUID MACHINES

15.1 INTRODUCTION

A fluid machine is a device which converts the energy stored by a fluid into mechanical energy or vice versa. The energy stored by a fluid mass appears in the form of potential, kinetic and intermolecular energy. Mechanical energy, on the other hand, is usually transmitted by a rotating shaft. Machines using liquid (mainly water, for almost all practical purposes) are termed as hydraulic machines. In this chapter we shall discuss, in general, the basic fluid mechanical principles governing energy transfer in a fluid machine and also a brief description of the different kinds of hydraulic machines along with their performances. Discussion on machines using air or other gases is beyond the scope of this chapter.

15.2 CLASSIFICATION OF FLUID MACHINES

Fluid machines may be classified under the following different categories.

15.2.1 Classification based on Direction of Energy Conversion

A device in which the kinetic, potential or intermolecular energy held by the fluid is converted in the form of mechanical energy by a rotating member is known as a *turbine*. The machines, on the other hand, where the mechanical energy from moving parts is transferred to a fluid to increase its stored energy by increasing either its pressure or velocity are known as *pumps*, *compressors*, *fans* or *blowers*.

15.2.2 Classification based on Principle of Operation

The machines whose functioning depends essentially on the change of volume of a certain amount of fluid within the machine are known as *positive displacement machines*. The word positive displacement comes from the fact that there is a physical displacement of the boundary of a certain fluid mass as a closed system. This principle is utilised in practice by the reciprocating motion of a piston within a cylinder while entrapping a certain amount of fluid in it. Therefore, the word 'reciprocating' is commonly used with the name of machines of this kind. A machine producing mechanical energy is known as a reciprocating engine while a machine developing energy of the fluid from mechanical energy is known as a reciprocating pump or a reciprocating compressor.

The machines, whose functioning basically depends on the principle of fluid dynamics, are known as *rotodynamic machines*. They are distinguished from positive displacement machines in requiring relative motion between the fluid and the moving part of the machine. The rotating element of the machine usually consisting of a number of vanes or blades, is known as the rotor or the impeller while the fixed part is known as the stator.

For turbines, the work is done by the fluid on the rotor, while, in case of pumps, compressors, fans or blowers, the work is done by the rotor on the fluid element. Depending upon the main direction of the fluid path in the rotor, the machine is termed as a *radial flow* or an *axial flow machine*. In a radial flow machine, the main direction of flow in the rotor is radial while in an axial flow machine, it is axial. For radial flow turbines, the flow is towards the centre of the rotor, while, for pumps and compressors, the flow is away from the centre. Therefore, radial flow turbines are sometimes referred to as radially *inward flow machines* and radial flow pumps as radially *outward flow machines*. Examples of such machines are the Francis turbines and the centrifugal pump or compressors. Examples of axial flow machines are Kaplan turbines and axial flow compressors. If the flow is partly radial and partly axial, the term *'mixed flow machine'* is used.

15.2.3 Classification based on Fluid Used

Fluid machines use either liquid or gas as the working fluid depending upon the purpose. A machine transferring mechanical energy of the rotor to the energy of fluid is termed as a pump when it uses liquid, and is termed as a compressor or a fan or a blower, when it uses gas. A compressor is a machine where the main objective is to increase the static pressure of a gas. Therefore, the mechanical energy held by the fluid is mainly in the form of pressure energy. Fans or blowers, on the other hand, mainly cause a high flow of gas, and hence utilise the mechanical energy of the rotor to increase mostly the kinetic energy of the fluid. In these machines, the change in static pressure is quite small.

For all practical purposes, the liquid used by the turbines producing power is water, and therefore, they are termed as *water turbines* or *hydraulic turbines*. Turbines handling gases in practical fields are usually referred to as *steam turbines*, *gas turbines*, or *air turbines* depending upon whether they use steam, gas (a mixture of air and products of burnt fuel in the air) or air.

15.3 ROTODYNAMIC MACHINES

In this section, we shall discuss the basic principle of rotodynamic machines and the performance of different kinds of those machines. The important element of a rotodynamic machine, in general, is a rotor consisting of a number of vanes or blades. There always exists a relative motion between the rotor vanes and the fluid. The fluid has a component of velocity and hence of momentum in a direction tangential to the rotor. While flowing through the rotor, tangential velocity and hence the momentum changes.

The rate at which this tangential momentum changes corresponds to a tangential force on the rotor. In a turbine, the tangential momentum of the fluid is reduced and therefore work is done by the fluid to the moving rotor. But in the case of pumps and compressors, there is an increase in the tangential momentum of the fluid and therefore work is absorbed by the fluid from the moving rotor.

15.3.1 Basic Equation of Energy Transfer in Rotodynamic Machines

The basic equation of fluid dynamics relating to energy transfer is same for all rotodynamic machines and is a simple form of Newton's laws of motion applied to a fluid element traversing a rotor. Here we shall make use of the momentum theorem as applicable to a fluid element while flowing through fixed and moving vanes. Figure 15.1 represents diagrammatically a rotor of a generalised fluid machine,

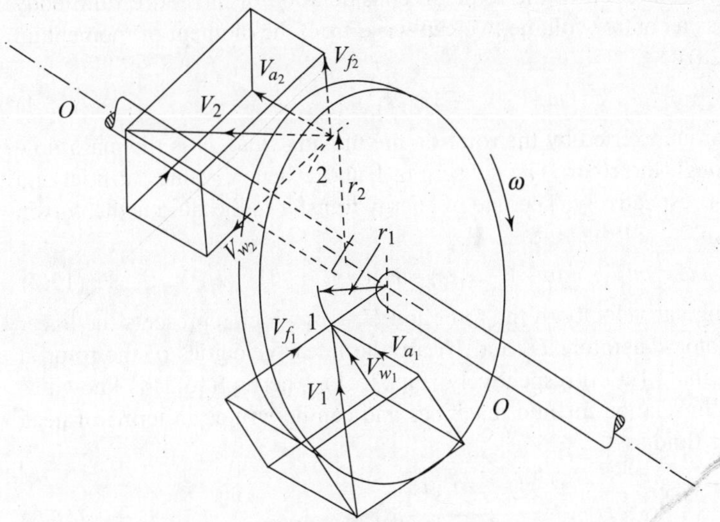

Fig. 15.1 Components of flow velocity in a generalised fluid machine

where 0–0 is the axis of rotation and ω the angular velocity. Fluid enters the rotor at 1, passes through the rotor by any path and is discharged at 2. The points 1 and 2 are at radii r_1 and r_2 from the centre of the rotor, and the directions of fluid velocities at 1 and 2 may be at any arbitrary angle. For the analysis of energy transfer due to fluid flow in this situation, we assume the following:

(a) The flow is steady, i.e., the mass flow rate is constant across any section (no storage or depletion of fluid mass in the rotor).

(b) The heat and work interactions between the rotor and its surroundings take place at a constant rate.

(c) The velocity is uniform over any area normal to the flow. This means that the velocity vector at any point is representative of the total flow over a finite area. This condition also implies that there is no leakage loss, and the entire fluid is undergoing the same process.

The velocity at any point may be resolved into three mutually perpendicular components as shown in Fig. 15.1. The axial component of velocity V_a is directed parallel to the axis of rotation, the radial component V_f is directed radially through the axis of rotation, while the tangential component V_w is directed at right angles to the radial direction and along the tangent to the rotor at that part.

The change in magnitude of the axial velocity components through the rotor causes a change in the axial momentum. This change gives rise to an axial force, which must be taken by a thrust bearing to the stationary rotor casing. The change in magnitude of radial velocity causes a change in momentum in the radial direction. However, for an axisymmetric flow, this does not result in any net radial force on the rotor. In case of a non-uniform flow distribution over the periphery of the rotor in practice, a change in momentum in the radial direction may result in a net radial force which is carried as a journal load. The tangential component V_w only has an effect on the angular motion of the rotor. In consideration of the entire fluid body within the rotor as a control volume, we can write from the moment of momentum theorem (Eq. 5.28))

$$T = m\,(V_{w_2} r_2 - Vw_1 r_1) \tag{15.1}$$

where T is the torque exerted by the rotor on the moving fluid, m is the mass flow rate of fluid through the rotor. The subscripts 1 and 2 denote values at inlet and outlet of the rotor respectively. The rate of energy transfer to the fluid is then given by

$$E = T\omega = m(V_{w_2} r_2\omega - V_{w_1} r_1\omega) = m(V_{w_2} U_2 - V_{w_1} U_1) \tag{15.2}$$

where ω is the angular velocity of the rotor and $U = \omega r$ which represents the linear velocity of the rotor. Therefore U_2 and U_1 are the linear velocities of the rotor at points 2 (outlet) and 1 (inlet) respectively (Fig. 15.1). The Eq. (15.2) is known as Euler's equation in relation to fluid machines and can be written in terms of head gained 'H' by the fluid as

$$H = \frac{V_{w_2} U_2 - V_{w_1} U_1}{g} \tag{15.3}$$

In usual convention relating to fluid machines, the head delivered by the fluid to the rotor is considered to be positive and vice versa. Therefore, Eq. (15.3) is written with a change in the sign on the right-hand side in accordance with the sign convention as

$$H = \frac{V_{w_1} U_1 - V_{w_2} U_2}{g} \tag{15.4}$$

Components of Energy Transfer It is worth mentioning in this context that either of the Eqs (15.2) and (15.4) is applicable regardless of changes in density or components of velocity in other directions. Moreover, the shape of the path taken by the fluid in moving from inlet to outlet is of no consequence. The expression involves only the inlet and outlet conditions. A rotor, the moving part of a fluid machine, usually consists of a number of vanes or blades mounted on a circular disc. Figure 15.2(a) shows the velocity triangles at the inlet and outlet of a rotor. The inlet and outlet portions of a rotor vane are only shown as a representative of the whole rotor.

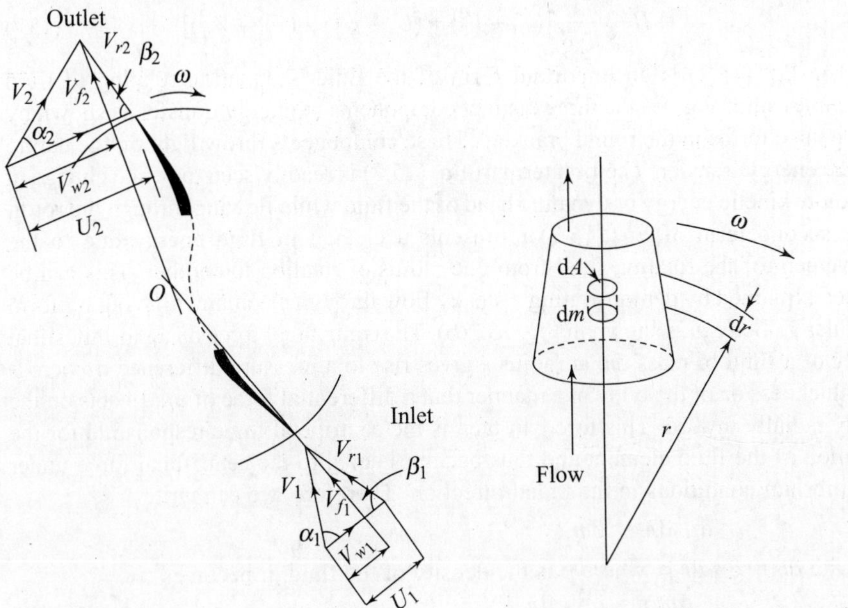

(a) Velocity triangles for a generalised rotor vane

(b) Centrifugal effect in a flow of fluid with rotation

Fig. 15.2

Vector diagrams of velocities at inlet and outlet correspond to two velocity triangles, where V_r is the velocity of fluid relative to the rotor and α_1, α_2 are the angles made by the directions of the absolute velocities at the inlet and outlet respectively with the tangential direction, while β_1 and β_2 are the angles made by the relative velocities with the tangential direction. The angles β_1 and β_2 should match with vane or blade angles at inlet and outlet respectively for a smooth, shockless entry and exit of the fluid to avoid undesirable losses. Now we shall apply a simple geometrical relation as follows:

From the inlet velocity triangle,

$$V_{r_1}^2 = V_1^2 + U_1^2 - 2\, U_1 V_1 \cos \alpha_1 = V_1^2 + U_1^2 - 2\, U_1 V_{w_1}$$

or $$U_1 V_{w_1} = \frac{1}{2}(V_1^2 + U_1^2 - V_{r_1}^2) \tag{15.5}$$

Similarly, from the outlet velocity triangle,

$$V_{r_2}^2 = V_2^2 + U_2^2 - 2\, U_2 V_2 \cos \alpha_2 = V_2^2 + U_2^2 - 2\, U_2 V_{w_2}$$

or $$U_2 V_{w_2} = \frac{1}{2}(V_2^2 + U_2^2 - V_{r_2}^2) \tag{15.6}$$

Invoking the expressions of $U_1 V_{w_1}$ and $U_2 V_{w_2}$ in Eq. (15.4), we get H (work head, i.e., energy per unit weight of fluid, transferred between the fluid and the rotor) as

$$H = \frac{1}{2g}[(V_1^2 - V_2^2) + (U_1^2 - U_2^2) + (V_{r_2}^2 - V_{r_1}^2)] \qquad (15.7)$$

The Eq. (15.7) is an important form of the Euler's equation relating to fluid machines since it gives the three distinct components of energy transfer as shown by the pair of terms in the round brackets. These components throw light on the nature of the energy transfer. The first term of Eq. (15.7) is readily seen to be the change in absolute kinetic energy or dynamic head of the fluid while flowing through the rotor. The second term of Eq. (15.7) represents a change in fluid energy due to the movement of the rotating fluid from one radius of rotation to another. This can be better explained by demonstrating a steady flow through a container having uniform angular velocity ω as shown in Fig. 15.2(b). The centrifugal force on an infinitesimal body of a fluid of mass dm at radius r gives rise to a pressure difference dp across the thickness dr of the body in a manner that a differential force of dp dA acts on the body radially inward. This force, in fact, is the centripetal force responsible for the rotation of the fluid element and thus becomes equal to the centrifugal force under equilibrium conditions in the radial direction. Therefore, we can write

$$dp \cdot dA = dm\, \omega^2 r$$

with dm = dA dr ρ, where ρ is the density of the fluid, it becomes

$$dp/\rho = \omega^2 r\, dr$$

For a reversible flow (flow without friction) between two points, say, 1 and 2, the work done per unit mass of the fluid (i.e., the flow work) can be written as

$$\int_1^2 \frac{dp}{\rho} = \int_1^2 \omega^2 r\, dr = \frac{\omega^2 r_2^2 - \omega^2 r_1^2}{2} = \frac{U_2^2 - U_1^2}{2}$$

This work is, therefore, done on or by the fluid element due to its displacement from radius r_1 to radius r_2 and hence becomes equal to the energy held or lost by it. Since the centrifugal force field is responsible for this energy transfer, the corresponding head (energy per unit weight) $U^2/2g$ is termed as the centrifugal head. The transfer of energy due to a change in centrifugal head $[(U_2^2 - U_1^2)/2g]$ causes a change in the static head of the fluid.

The third term represents a change in the static head due to a change in fluid velocity relative to the rotor. This is similar to what happens in case of a flow through a fixed duct of variable cross-sectional area. Regarding the effect of flow area on fluid velocity V_r relative to the rotor, a converging passage in the direction of flow through the rotor increases the relative velocity ($V_{r2} > V_{r1}$) and hence decreases the static pressure. This usually happens in case of turbines. Similarly, a diverging passage in the direction of flow through the rotor decreases the relative velocity ($V_{r2} < V_{r1}$) and increases the static pressure as occurs in case of pumps and compressors.

The fact that the second and third terms of Eq. (15.7) correspond to a change in static head can be demonstrated analytically by deriving Bernoulli's equation in the frame of the rotor.

In a rotating frame, the momentum equation for the flow of a fluid, assumed 'inviscid' can be written as

$$\rho\left[\frac{\partial \vec{v}}{\partial t} + \vec{v}\cdot\nabla\vec{v} + 2\vec{\omega}\times\vec{v} + \vec{\omega}\times(\vec{\omega}\times\vec{r})\right] = -\nabla p$$

where $\vec{v}$ is the fluid velocity relative to the coordinate frame rotating with an angular velocity $\vec{\omega}$.

We assume that the flow is steady in the rotating frame so that $\dfrac{\partial \vec{v}}{\partial t} = 0$. We choose a cylindrical coordinate system (r, θ, z) with the z axis along the axis of rotation. Then the momentum equation reduces to

$$\vec{v} \cdot \nabla \vec{v} + 2\omega \vec{i}_z \times \vec{v} - \omega^2 r \vec{i}_r = -\frac{1}{\rho} \nabla p$$

where, $\vec{i}_z$ and $\vec{i}_r$ are the unit vectors along z and r directions respectively. Let $\vec{i}_s$ be a unit vector in the direction of $\vec{v}$ and s be a coordinate along the stream line. Then we can write

$$v \frac{\partial v}{\partial s} \vec{i}_s + v^2 \frac{\partial \vec{i}_s}{\partial s} + 2\omega v \vec{i}_z \times \vec{i}_s - \omega^2 r \vec{i}_r = -\frac{1}{\rho} \nabla p$$

Taking scalar product with $\vec{i}_s$ it becomes

$$v \frac{\partial v}{\partial s} - \omega^2 r \frac{\partial r}{\partial s} = -\frac{1}{\rho} \frac{\partial p}{\partial s}$$

We have used $\vec{i}_s \cdot \dfrac{\partial \vec{i}_s}{\partial s} = 0$. With a little rearrangement, we have

$$\frac{\partial}{\partial s} \left(\frac{1}{2} v^2 - \frac{1}{2} \omega^2 r^2 + \frac{p}{\rho} \right) = 0$$

Since v is the velocity relative to the rotating frame we can replace it by V_r. Further $\omega r = U$ is the linear velocity of the rotor. Integrating the momentum equation from inlet to outlet along a streamline we have

$$\frac{1}{2} (V_{r_2}^2 - V_{r_1}^2) - \frac{1}{2} (U_2^2 - U_1^2) + \frac{p_2 - p_1}{\rho} = 0$$

or

$$\frac{1}{2} (U_1^2 - U_2^2) + \frac{1}{2} (V_{r_2}^2 - V_{r_1}^2) = \frac{p_1 - p_2}{\rho} \tag{15.8}$$

Therefore, we can say, with the help of Eq. (15.8), that the last two terms of Eq. (15.7) represent a change in the static head of fluid.

Energy Transfer in Axial Flow Machines For an axial flow machine, the main direction of flow is parallel to the axis of the rotor, and hence the inlet and outlet points of the flow do not vary in their radial locations from the axis of rotation. Therefore, $U_1 = U_2$ and the equation of energy transfer [Eq. (15.7)] can be written, under this situation, as

$$H = \frac{1}{2g} [(V_1^2 - V_2^2) + (V_{r_2}^2 - V_{r_1}^2)] \tag{15.9}$$

Hence, change in the static head in the rotor of an axial flow machine is only due to the flow of fluid through the variable area passage in the rotor.

Radially Outward and Inward Flow Machines For radially outward flow machines, $U_2 > U_1$; hence the fluid gains in static head, while, for a radially inward flow machine, $U_2 < U_1$, the fluid loses its static head. Therefore, in radial flow pumps or compressors the flow is always directed radially outward, and in a radial flow turbine it is directed radially inward.

Impulse and Reaction Machines The relative proportion of energy transfer obtained by the change in static head and by the change in dynamic head is one of the important factors for classifying fluid machines. The machine for which the change in static head in the rotor is zero is known as an *impulse machine*. In these machines, the energy transfer in the rotor takes place only by the change in the dynamic head of the fluid. The parameter characterising the proportion of changes in the dynamic and static head in the rotor of a fluid machine is known as the degree of reaction and is defined as the ratio of energy transfer by the change in the static head to the total energy transfer in the rotor.

Therefore, the degree of reaction,

$$R = \frac{\frac{1}{2g}\left[\left(U_1^2 - U_2^2\right) + \left(Vr_2^2 - Vr_1^2\right)\right]}{H} \tag{15.10}$$

For an impulse machine, $R = 0$, because there is no change in static pressure in the rotor. It is difficult to obtain a radial flow impulse machine, since the change in the centrifugal head is obvious there. Nevertheless, an impulse machine of radial flow type can be conceived by having a change in the static head in one direction contributed by the centrifugal effect, and an equal change in the other direction contributed by the change in the relative velocity. However, this has not been established in practice. Thus for an axial flow impulse machine $U_1 = U_2$, $V_{r_1} = V_{r_2}$. For an impulse machine, the rotor can be made open, that is, the velocity V_1 can represent an open jet of fluid flowing through the rotor, which needs no casing. A very simple example of an impulse machine is a paddle wheel rotated by the impingement of water from a stationary nozzle as shown in Fig. 15.3(a).

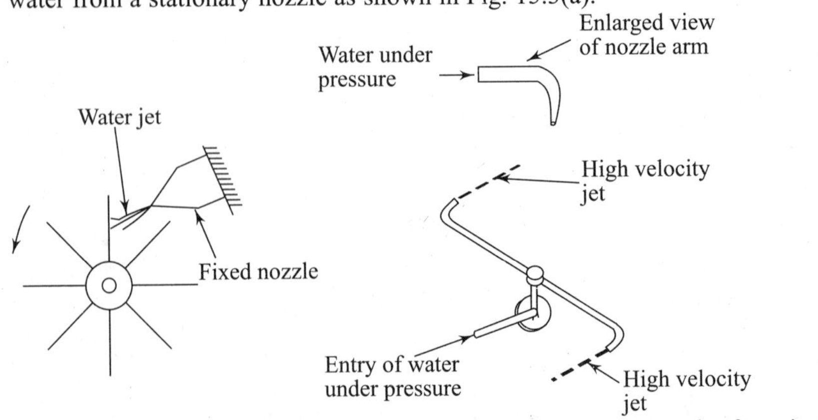

(a) Paddle wheel as an example of impulse turbine

(b) Lawn sprinkler as an example of reaction turbine

Fig. 15.3

A machine with any degree of reaction must have an enclosed rotor so that the fluid cannot expand freely in all directions. A simple example of a reaction machine can be shown by the familiar lawn sprinkler, in which water comes out (Fig. 15.3(b)) at a high velocity from the rotor in a tangential direction. The essential feature of the rotor is that water enters at high pressure and this pressure energy is transformed into kinetic energy by a nozzle which is a part of the rotor itself.

In the earlier example of impulse machine (Fig. 15.3(a)), the nozzle is stationary and its function is only to transform pressure energy to kinetic energy and finally this kinetic energy is transferred to the rotor by a pure impulse action. The change in momentum of the fluid in the nozzle gives rise to a reaction force but as the nozzle is held stationary, no energy is transferred by it. In the case of a lawn sprinkler (Fig. 15.3(b)), the nozzle, being a part of the rotor, is free to move and, in fact, rotates due to the reaction force caused by the change in momentum of the fluid and hence the word *reaction machine* follows.

Efficiencies The concept of efficiency of any machine comes from the consideration of energy transfer and is defined, in general, as the ratio of useful energy delivered to the energy supplied. Two efficiencies are usually considered for fluid machines—the hydraulic efficiency concerning the energy transfer between the fluid and the rotor, and the overall efficiency concerning the energy transfer between the fluid and the shaft. The difference between the two represents the energy absorbed by bearings, glands, couplings, etc. or, in general, by pure mechanical effects which occur between the rotor itself and the point of actual power input or output.

Therefore, for a pump or compressor,

$$\eta_{\text{hydraulic}} = \eta_h = \frac{\text{Useful energy gained by the fluid at final discharge}}{\text{Mechanical energy supplied to rotor}} \tag{15.11a}$$

$$\eta_{\text{overall}} = \frac{\text{Useful energy gained by the fluid at final discharge}}{\text{Mechanical energy supplied to shaft at coupling}} \tag{15.11b}$$

For a turbine,

$$\eta_h = \frac{\text{Mechanical energy delivered by the rotor}}{\text{Energy available from the fluid}} \tag{15.12a}$$

$$\eta_{\text{overall}} = \frac{\text{Mechanical energy in output shaft at coupling}}{\text{Energy available from the fluid}} \tag{15.12b}$$

The ratio of rotor and shaft energy is represented by the mechanical efficiency η_m.

Hence,

$$\eta_m = \frac{\eta_{\text{overall}}}{\eta_h} \tag{15.13}$$

15.3.2 Principle of Similarity and Dimensional Analysis in Rotodynamic Machines

The principle of similarity is a consequence of nature for any physical phenomenon. The concept of similarity and dimensional analysis related to the problems of fluid flow, in general, has been discussed in Chapter 6. By making use of this principle, it becomes possible to predict the performance of one machine from the results of tests on a geometrically similar machine, and also to predict the performance of the same machine under conditions different from the test conditions. For fluid machines, geometrical similarity must apply to all significant parts of the system, viz., the rotor, the entrance and discharge passages and so on. Machines which are geometrically similar form a homologous series. Therefore, the members of such a series, having a common shape are simply enlargements or reductions of each other. If two machines are kinematically similar, the velocity vector diagrams at inlet and outlet of the rotor of one machine must be similar to those of the other. Geometrical similarity of the inlet and outlet velocity diagrams is, therefore, a necessary condition for dynamic similarity.

Let us now apply dimensional analysis to determine the dimensionless parameters, i.e., the π terms as the criteria of similarity. For a machine of a given shape, and handling compressible fluid, the relevant variables are given in Table 15.1.

Table 15.1 Variable physical parameters of fluid machine

Variable physical parameters	*Dimensional formula*
D = Any physical dimension of the machine as a measure of the machine's size, usually the rotor diameter	L
Q = Volume flow rate through the machine	L^3T^{-1}
N = Rotational speed (rev./min.)	T^{-1}
H = Difference in head (energy per unit weight) across the machine. This may be either gained or given by the fluid depending upon whether the machine is a pump or a turbine respectively	L
ρ = Density of the fluid	ML^{-3}
μ = Viscosity of the fluid	$ML^{-1}T^{-1}$
E = Coefficient of elasticity of the fluid	$ML^{-1}T^{-2}$
g = Acceleration due to gravity	LT^{-2}
P = Power transferred between the fluid and the rotor (the difference between P and H is taken care of by the hydraulic efficiency η_h)	ML^2T^{-3}

In almost all fluid machines, flow with a free surface does not occur, and the effect of gravitational force is negligible. Therefore, it is more logical to consider the energy per unit mass gH as the variable rather than H alone so that acceleration due to gravity g does not appear as a separate variable. Therefore, the number of separate variables becomes eight: D, Q, N, gH, ρ, μ, E and P. Since the number of fundamental dimensions required to express these variables are three, the number of independent π terms (dimensionless terms), becomes five. Using Buckingham's π theorem with D, N and ρ as the repeating variables, the expressions for the π terms are obtained as

$$\pi_1 = \frac{Q}{ND^3}, \quad \pi_2 = \frac{gH}{N^2D^2}, \pi_3 = \frac{\rho ND^2}{\mu}, \pi_4 = \frac{P}{\rho N^3D^5}, \pi_5 = \frac{E/\rho}{N^2D^2}$$

We shall now discuss the physical significance and usual terminologies of the different π terms.

All lengths of the machine are proportional to D, and all areas to D^2. Therefore, the average flow velocity at any section in the machine is proportional to Q/D^2. Again, the peripheral velocity of the rotor is proportional to the product ND. The first π term can be expressed as

$$\pi_1 = \frac{Q}{ND^3} = \frac{Q/D^2}{ND} \propto \frac{\text{Fluid velocity } V}{\text{Rotor velocity } U}$$

Thus, π_1 represents the condition for kinematic similarity, and is known as *the capacity coefficient* or *discharge coefficient*. The second π term π_2 is known as the *head coefficient* since it expresses the head H in dimensionless form. Considering the fact that $ND \propto$ rotor velocity, the term π_2 becomes gH/U^2, and can be interpreted as the ratio of fluid head to kinetic energy of the rotor. Dividing π_2 by the square of π_1 we get

$$\frac{\pi_2}{\pi_1^2} = \frac{gH}{(Q/D^2)^2} \propto \frac{\text{Total fluid energy per unit mass}}{\text{Kinetic energy of the fluid per unit mass}}$$

The term π_3 can be expressed as $\rho\,(ND)D/\mu$ and thus represents the Reynolds number with rotor velocity as the characteristic velocity. Again, if we make the product of π_1 and π_3, it becomes $\rho\,(Q/D^2)D/\mu$ which represents the Reynold's number based on fluid velocity. Therefore, if π_1 is kept same to obtain kinematic similarity, π_3 becomes proportional to the Reynolds number based on fluid velocity.

The term π_4 expresses the power P in dimensionless form and is therefore known as *power coefficient*. Combination of π_4, π_1 and π_2 in the form of $\pi_4/\pi_1\pi_2$ gives $P/\rho QgH$. The term ρQgH represents the rate of total energy given up by the fluid, in case of turbine, and gained by the fluid in case of pump or compressor. Since P is the power transferred to or from the rotor. Therefore $\pi_4/\pi_1\,\pi_2$ becomes the hydraulic efficiency η_h for a turbine and $1/\eta_h$ for a pump or a compressor.

From the fifth π term, we get

$$\frac{1}{\sqrt{\pi_5}} = \frac{ND}{\sqrt{E/\rho}}$$

Multiplying π_1 on both sides, we get

$$\frac{\pi_1}{\sqrt{\pi_5}} = \frac{Q/D^2}{\sqrt{E/\rho}} \propto \frac{\text{Fluid velocity}}{\text{Local acoustic velocity}}$$

Therefore, we find that $\pi_1/\sqrt{\pi_5}$ represents the well known *Mach number*.

For a fluid machine handling incompressible fluid, the term π_5 can be dropped. Moreover, if the effect of liquid viscosity on the performance of fluid machines is neglected or regarded as secondary, (which is often sufficiently true for certain cases or over a limited range) the term π_3 can also be dropped. Then the relationship between the different dimensionless variables (π terms) can be expressed as

$$f\left[\frac{Q}{ND^3}, \frac{gH}{N^2D^2}, \frac{P}{\rho N^3 D^5}\right] = 0 \tag{15.14}$$

or, with another arrangement of the π terms,

$$\phi\left[\eta_h, \frac{gH}{N^2D^2}, \frac{P}{\rho N^3 D^5}\right] = 0 \tag{15.15}$$

If data obtained from tests on a model machine are plotted so as to show the variation of dimensionless parameters $\frac{Q}{ND^3}, \frac{gH}{N^2D^2}, \frac{P}{\rho N^3 D^5}$ with one another, then the graphs are applicable to any machine in the same homologous series. The curves for other homologous series would naturally be different. Therefore one set of relationship or curves of the π terms would be sufficient to describe the performance of all the members of one series.

The performance or operating conditions for a turbine handling a particular fluid are usually expressed by the values of N, P and H, and for a pump by N, Q and H. It is important to know the range of these operating parameters covered by a machine of a particular shape (homologous series). Such information enables us to select the type of machine best suited to a particular application, and thus serves as a starting point in its design. Therefore a parameter independent of the size of the machine D is required which will be the characteristic of all the machines of a homologous series. A parameter involving N, P and H but not D is obtained by dividing $(\pi_4)^{1/2}$ by $(\pi_2)^{5/4}$. Let this parameter be designated by K_{s_T} as

$$K_{s_T} = \frac{(P/\rho N^3 D^5)^{1/2}}{(gH/N^2D^2)^{5/4}} = \frac{NP^{1/2}}{\rho^{1/2}(gH)^{5/4}} \tag{15.16}$$

Similarly, a parameter involving N, Q and H but not D is obtained by dividing $(\pi_1)^{1/2}$ by $(\pi_2)^{3/4}$ and is represented by K_{s_P} as

$$K_{s_P} = \frac{(Q/ND^3)^{1/2}}{(gH/N^2D^2)^{3/4}} = \frac{NQ^{1/2}}{(gH)^{3/4}} \tag{15.17}$$

Since the dimensionless parameters K_{s_T} and K_{s_P} are found as a combination of basic π terms, they must remain same for complete similarity of flow in machines of a homologous series. Therefore, a particular value of K_{s_T} or K_{s_P} relates all the combinations of N, P and H or N, Q and H for which the flow conditions are similar in the machines of that homologous series. Interest naturally centres on the conditions for which the efficiency is maximum. For turbines, the values of N, P and H, and for pumps and compressors, the values of N, Q and H are usually quoted for which the machines run at maximum efficiency.

The machines of a particular homologous series, that is, of a particular shape, correspond to a particular value of K_s for their maximum efficient operation. Machines of different shapes have, in general, different values of K_s. Thus the parameter K_s (K_{s_T} or K_{s_P}) is referred to as the *shape factor* of the machines. Considering the fluids used by the machines to be incompressible, (for hydraulic turbines and pumps), and since the acceleration due to gravity does not vary under this situation, the terms g and ρ are taken out from the expressions of K_{s_T} and K_{s_P}. The portions left as $NP^{1/2}/H^{5/4}$ and $NQ^{1/2}/H^{3/4}$ are termed for practical purposes, as the *specific speed* N_s for turbines or pumps. Therefore, we can write,

$$N_{s_T} \text{ (specific speed for turbines)} = NP^{1/2}/H^{5/4} \tag{15.18}$$

$$N_{s_P} \text{ (specific speed for pumps)} = NQ^{1/2}/H^{3/4} \tag{15.19}$$

The name specific speed for these expressions has little justification. However, a meaning can be attributed from the concept of a hypothetical machine. For a turbine, N_{sT} is the speed of a member of the same homologous series as the actual turbine, so reduced in size as to generate unit power under a unit head of the fluid. Similarly, for a pump, N_{sP} is the speed of a hypothetical pump with reduced size but representing a homologous series so that it delivers unit flow rate at a unit head. The specific speed N_s is, therefore, not a dimensionless quantity. The dimension of N_s can be found from their expressions given by Eqs (15.18) and (15.19). The dimensional formula and the unit of specific speed are given as follows:

Specific speed	*Dimensional formula*	*Unit (SI)*
N_{s_T} (turbine)	$M^{1/2}\ T^{-5/2}\ L^{-1/4}$	$kg^{1/2}/s^{5/2}m^{1/4}$
N_{s_P}(pump)	$L^{3/4}\ T^{-3/2}$	$m^{3/4}/s^{3/2}$

The dimensionless parameter K_s is often known as the dimensionless specific speed to distinguish it from N_s. The values of specific speed N_s (for maximum efficiencies) for different types of turbines and pumps will be discussed later.

Example 15.1

A radial flow hydraulic turbine is required to be designed to produce 20 MW under a head of 16 m at a speed of 90 rpm. A geometrically similar model with an output of 30 kW and a head of 4 m is to be tested under dynamically similar conditions. At what speed must the model be run? What is the required impeller diameter ratio between the model and the prototype and what is the volume flow rate through the model if its efficiency can be asumed to be 90%?

Solution

Equating the power coefficients (π term containing the power P) for the model and prototype, we can write

$$\frac{P_1}{\rho_1 N_1^3 D_1^5} = \frac{P_2}{\rho_2 N_2^3 D_2^5}$$

(where subscript 1 refers to the prototype and subscript 2 to the model)
Considering the fluids to be incompressible, and same for both the prototype and model, we have

$$\begin{aligned} D_2/D_1 &= [P_2/P_1]^{1/5}\,(N_1/N_2)^{3/5} \\ &= [0.03/20]^{1/5}\,[N_1/N_2]^{3/5} \\ &= 0.272\,[N_1/N_2]^{3/5} \end{aligned} \tag{15.20}$$

Equating the head coefficients (π term containing the head H)

$$\frac{gH_1}{(N_1 D_1)^2} = \frac{gH_2}{(N_2 D_2)^2}$$

Then,

$$D_2/D_1 = [H_2/H_1]^{1/2}\,[N_1/N_2] = [4/16]^{1/2}\,[N_1/N_2] \tag{15.21}$$

Therefore, equating the diameter ratios from Eqs (15.20) and (15.21), we have

$$0.272\,[N_1/N_2]^{3/5} = [4/16]^{1/2}\,[N_1/N_2]$$

or $$[N_2/N_1]^{2/5} = 1.84$$

Hence, $$\begin{aligned} N_2 &= N_1\,(1.84)^{5/2} = 90 \times (1.84)^{5/2} \\ &= 413.32 \text{ rpm} \end{aligned}$$

From Eq. (15.20)

$$D_2/D_1 = 0.272\,[90/413.32]^{3/5} = 0.11$$

$$\text{Model efficiency} = \frac{\text{Power output}}{\text{Water power input}}$$

Hence, $$0.9 = \frac{30 \times 10^3}{\rho Q g H}$$

or $$Q = \frac{30 \times 10^3}{0.9 \times 10^3 \times 9.81 \times 4} = 0.85 \text{ m}^3/\text{s}$$

Therefore, model volume flow rate = 0.85 m^3/s.

Example 15.2

A reservoir has a head of 40 m and a channel leading from the reservoir permits a flow rate of 34 m^3/s. If the rotational speed of the rotor is 150 rpm, determine the dimensionless specific speed of the turbine.

Solution

We have,

$$\text{Turbine power} = \rho g Q H = 1000 \times 9.81 \times 34 \times 40$$
$$= 13.34\ \text{MW}$$

The dimensionless specific speed of a turbine is given according to Eq. (15.16) by

$$K_{S_T} = \frac{N P^{1/2}}{\rho^{1/2}(gH)^{5/4}}$$

$$= \frac{150 \times (13.34 \times 10^6)^{1/2}}{60 \times (1000)^{1/2} \times (9.81 \times 40)^{5/4}}$$

$$= 0.165\ \text{rev}$$

$$= 1.037\ \text{rad}$$

Example 15.3

A centrifugal pump handles liquid whose kinematic viscosity is three times that of water. The dimensionless specific speed of the pump is 0.183 rev and it has to discharge 2 m^3/s of liquid against a total head of 15 m. Determine the speed, test head and flow rate for a one-quarter scale model investigation of the full size pump if the model uses water.

Solution

Since the viscosity of the liquid in the model and prototype vary significantly, equality of Reynolds number must apply for dynamic similarity. Let subscripts 1 and 2 refer to prototype and model respectively.

Equating Reynolds number,

$$N_1 D_1^2/\nu_1 = N_2 D_2^2/\nu_2$$

or

$$N_2/N_1 = (4)^2/3 = 5.333$$

Equating the flow coefficients,

$$Q_1/N_1 D_1^3 = Q_2/N_2 D_2^3$$

or

$$Q_2/Q_1 = (N_2/N_1)\,(D_2/D_1)^3$$
$$= 5.333/(4)^3 = 0.0833$$

Equating head coefficients,

$$H_1/(N_1 D_1)^2 = H_2/(N_2 D_2)^2$$

or

$$H_2/H_1 = (N_2/N_1)^2\,(D_2/D_1)^2$$
$$= (5.33/4)^2 = 1.776$$

Dimensionless specific speed of the pump can be written according to Eq. (15.17) as

$$K_{s_p} = \frac{N_1 Q_1^{1/2}}{(gH_1)^{3/4}}$$

or

$$N_1 = \frac{K_{s_p}(gH_1)^{3/4}}{Q_1^{1/2}}$$

$$= \frac{0.183(9.81 \times 15)^{3/4}}{2^{1/2}}$$

$$= 5.47 \text{ rev/s}$$

Therefore, model speed $N_2 = 5.47 \times 5.33 = 29.15$ rev/s

Model flow rate $= 0.0833 \times 2 = 0.166$ m^3/s

Model head $= 15 \times 1.776 = 26.64$ m

Example 15.4

Specifications for an axial flow coolant pump for one loop of a pressurised water nuclear reactor are as follows.

Head	85 m
Flow rate	10,000 m^3/hour
Speed	1490 rpm
Diameter	1200 mm
Water density	714 kg/m^3
Power	2 MW (electrical)

The manufacturer plans to build a model. Test conditions limit the available electric power to 250 kW and flow to 0.25 m^3/s of cold water. If the model and prototype efficiencies are assumed equal, find the head, speed and scale ratio of the model. Calculate the dimensionless specific speed of the prototype and confirm that it is identical with the model.

Solution

Let subscripts 1 and 2 represent prototype and model respectively. Equating the flow power and head coefficients for the model and prototype, we have

$$Q_1/Q_2 = (N_1/N_2)(D_1/D_2)^3$$

or

$$N_1/N_2 = \left(\frac{10000}{0.25 \times 3600}\right)(D_2/D_1)^3$$

$$= 11.11\,(D_2/D_1)^3$$

Also,

$$P_1/P_2 = (N_1/N_2)^3 (D_1/D_2)^5 (\rho_1/\rho_2)$$

Substituting for (N_1/N_2), we have

$$2/0.25 = (11.11)^3 (D_2/D_1)^9 (D_1/D_2)^5 (714/1000)$$

or $$(D_2/D_1)^4 = \frac{8}{(11.11)^3 \times 0.714}$$

which gives the scale ratio $D_2/D_1 = 0.3$

Then, $$N_1/N_2 = 11.11 \times (0.3)^3 = 0.3$$

or $$N_2/N_1 = 1/0.3 = 3.33$$

$$H_2/H_1 = (N_2/N_1)^2 (D_2/D_1)^2$$

$$= \left\{\left(\frac{N_2}{N_1}\right)\left(\frac{D_2}{D_1}\right)\right\}^2 = \left\{\frac{1}{0.3} \times 0.3\right\}^2 = 1.0$$

The dimensionless specific speed is given by

$$K_{s_p} = \frac{N Q^{1/2}}{(gH)^{3/4}}$$

For the prototype

$$K_{s_{p1}} = \frac{2\pi \times 1490}{60} (10000/3600)^{1/2} (1/9.81)^{3/4} (1/85)^{3/4}$$

$$= 1.67 \text{ rad}$$

For the model

$$K_{s_{p2}} = 2\pi \times \frac{1490}{60} \times 3.33 \times \frac{(0.25)^{1/2}}{(9.81 \times 85)^{3/4}}$$

$$= 1.67 \text{ rad}$$

Therefore we see that the dimensionless specific speeds of both model and prototype are the same.

15.4 DIFFERENT TYPES OF ROTODYNAMIC MACHINES

In this section we shall discuss the hydraulic machines which use water as the fluid in practice.

15.4.1 Impulse Hydraulic Turbine: The Pelton Wheel

The only hydraulic turbine of the impulse type in common use is named after an American engineer, Laster A Pelton, who contributed much to its development in about 1880. Therefore this machine is known as the Pelton turbine or Pelton wheel. It is an efficient machine particularly suited to high heads. The rotor consists of a large circular disc or wheel on which a number (seldom less than 15) of spoon-shaped buckets are spaced uniformly round its periphery,as shown in Fig. 15.4. The wheel is driven by jets of water being discharged at atmospheric pressure the wheel

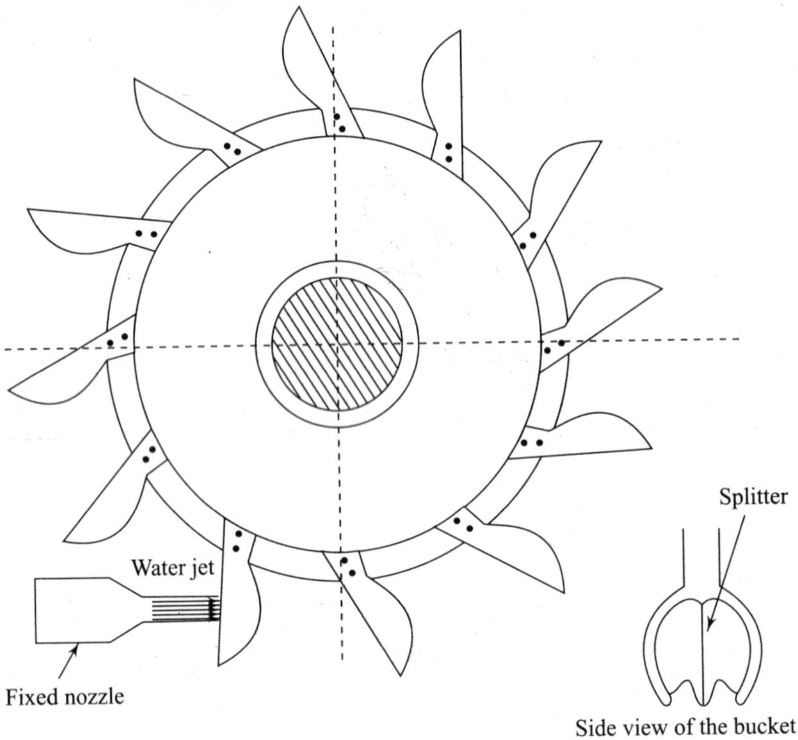

Fig. 15.4 A Pelton wheel

is driven by jets of water being discharged at atmospheric pressure from pressure nozzles. The nozzles are mounted so that each directs a jet along a tangent to the circle through the centres of the buckets. Down the centre of each bucket, there is a splitter ridge which divides the jet into two equal streams which flow round the smooth inner surface of the bucket and leaves the bucket with a relative velocity almost opposite in direction to the original jet.

For maximum change in momentum of the fluid and hence for the maximum driving force on the wheel, the deflection of the water jet should be 180°. In practice, however, the deflection is limited to about 165° so that the water leaving a bucket may not hit the back of the following bucket. Therefore, the camber angle of the buckets is made as 165° ($\theta = 165°$, Fig. 15.5(a)).

The number of jets is not more than two for horizontal shaft turbines and is limited to six for vertical shaft turbines. The flow partly fills the buckets and the fluid remains in contact with the atmosphere. Therefore, once the jet is produced by the nozzle, the static pressure of the fluid remains atmospheric throughout the machine. Because of the symmetry of the buckets, the side thrusts produced by the fluid in each half should balance each other.

Analysis of Force on the Bucket and Power Generation Figure 15.5(a) shows a section through a bucket which is being acted on by a jet. The plane of the section is parallel to the axis of the wheel and contains the axis of the jet. The absolute velocity of the jet V_1 with which it strikes the bucket is given by

$$V_1 = C_v\,[2gH]^{1/2}$$

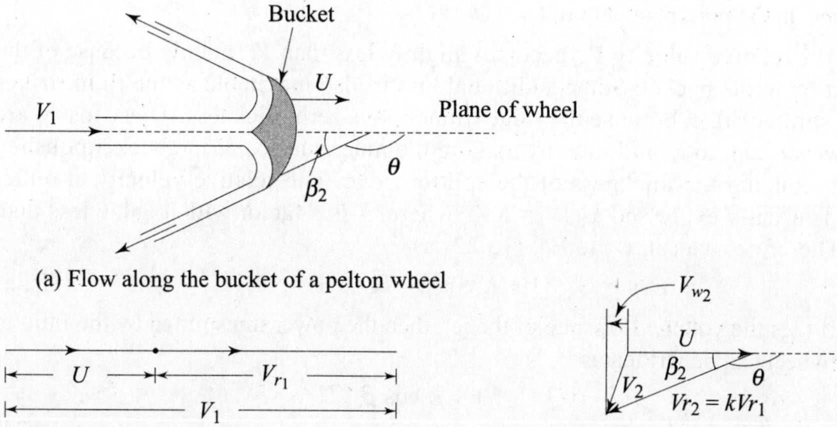

(a) Flow along the bucket of a pelton wheel

(b) Inlet velocity triangle

(c) Outlet velocity triangle

Fig. 15.5

where, C_v is the coefficient of velocity which takes care of the friction in the nozzle. H is the head at the entrance to the nozzle which is equal to the total or gross head of water stored at high altitudes minus the head lost due to friction in the long pipeline leading to the nozzle. Let the velocity of the bucket (due to the rotation of the wheel) at its centre where the jet strikes be U. Since the jet velocity V_1 is tangential, i.e., V_1 and U are colinear, the diagram of velocity vector at inlet (Fig. 15.5(b)) becomes simply a straight line and the relative velocity V_{r_1} is given by

$$V_{r_1} = V_1 - U$$

It is assumed that the flow of fluid is uniform and it glides the blade all along, including the entrance and exit sections, to avoid the unnecessary losses due to shock. Therefore, the direction of relative velocity at the entrance and the exit should match the inlet and outlet angles of the buckets respectively. The velocity triangle at the outlet is shown in Fig. 15.5(c). The bucket velocity U remains the same both at the inlet and outlet. With the direction of U being taken as positive, we can write:

The tangential component of inlet velocity (Fig. 15.5(b)),

$$V_{w_1} = V_1 = V_{r_1} + U$$

and the tangential component of outlet velocity (Fig. 15.5(c)),

$$V_{w_2} = -(V_{r_2}\cos\beta_2 - U)$$

where, V_{r_1} and V_{r_2} are the velocities of the jet relative to the bucket at its inlet and outlet and β_2 is the outlet angle of the bucket.

From Eq. (15.2) (Euler's equation for hydraulic machines), the energy delivered by the fluid per unit mass to the rotor can be written as

$$E/m = [V_{w_1} - V_{w_2}]U$$
$$= [V_{r_1} + V_{r_2} \cos \beta_2]\, U \qquad (15.22)$$

(since, in the present situation, $U_1 = U_2 = U$)

The relative velocity V_{r_2} becomes slightly less than V_{r_1} mainly because of the friction in the bucket. Some additional loss is also inevitable as the fluid strikes the splitter ridge, because the ridge cannot have zero thickness. These losses are however kept to a minimum by making the inner surface of the bucket polished and reducing the thickness of the splitter ridge. The relative velocity at outlet V_{r_2} is usually expressed as $V_{r_2} = KV_{r_1}$ where, K is a factor with a value less than 1. Therefore, we can write Eq. (15.22) as

$$E/m = V_{r_1} [1 + K \cos \beta_2]\, U \qquad (15.23)$$

If Q is the volume flow rate of the jet, then the power transmitted by the fluid to the wheel can be written as

$$P = \rho Q V_{r_1} [\, 1 + K \cos \beta_2]\, U$$
$$= \rho Q [1 + K \cos \beta_2]\, (V_1 - U)\, U \qquad (15.24)$$

The power input to the wheel is found from the kinetic energy of the jet arriving at the wheel and is given by $\frac{1}{2}\rho Q V_1^2$. Therefore, the wheel efficiency of a Pelton turbine can be written as

$$\eta_w = \frac{2\rho Q [1 + K \cos \beta_2](V_1 - U)U}{\rho Q V_1^2}$$

$$= 2[1 + K \cos \beta_2]\left[1 - \frac{U}{V_1}\right]\frac{U}{V_1} \qquad (15.25)$$

It is found from Eq. (15.25) that the efficiency η_w depends on K, β_2 and U/V_1. For a given design of the bucket, i.e., for constant values of β_2 and K, the efficiency η_w becomes a function of U/V_1 only, and we can determine the condition given by U/V_1 at which η_w becomes maximum.

For η_w to be maximum,

$$\frac{d\eta_w}{d(U/V_1)} = 2[1 + K \cos \beta_2]\left[1 - 2\frac{U}{V_1}\right] = 0$$

or

$$U/V_1 = \frac{1}{2} \qquad (15.26)$$

$d^2\eta_w/d(U/V_1)^2$ is always negative indicating that the Eq. (15.25) has only a maximum (not a minimum) value.

The condition given by Eq. (15.26) states that the efficiency of the wheel in converting the kinetic energy of the jet into mechanical energy of rotation becomes maximum when the wheel speed at the centre of the bucket becomes one-half of the incoming velocity of the jet. The overall efficiency η_0 will be less than η_w because of friction in bearing and windage, i.e., friction between the wheel and the atmosphere in which it rotates. Moreover, as the losses due to bearing friction and windage increase rapidly with speed, the overall efficiency reaches it peak when the ratio U/V_1 is slightly less than the theoretical value of 0.5. The value usually obtained in practice is about 0.46 (Fig. 15.6). An overall efficiency of 85–90% may usually be obtained in large machines. To obtain high values of wheel efficiency, the buckets should have a smooth surface and be properly designed. The length, width, and depth of the buckets are chosen about 2.5, 4 and 0.8 times the jet diameter. The buckets are notched for smooth entry of the jet.

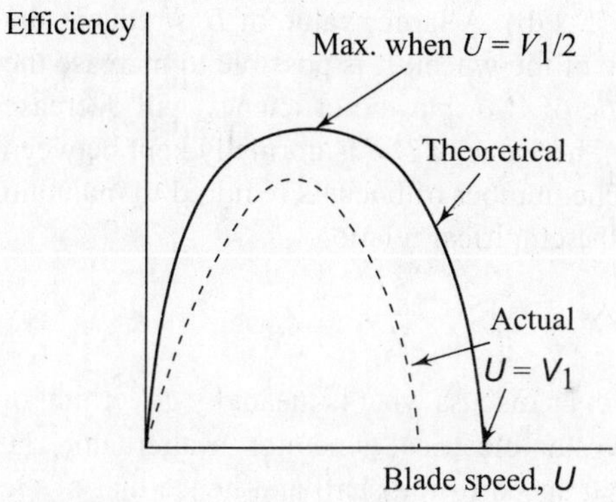

Fig. 15.6 Variation of wheel efficiency with blade speed

Specific Speed and Wheel Geometry The specific speed of a Pelton wheel depends on the ratio of jet diameter d and the wheel pitch diameter D (the diameter at the centre of the bucket). If the hydraulic efficiency of a Pelton wheel is defined as the ratio of the power delivered P to the wheel to the head available H at the nozzle entrance, then we can write

$$P = \rho\, Q\, g\, H\, \eta_h = \frac{\pi \rho d^2 V_1^3 \eta_h}{4 \times 2 C_v^2}$$

$$\left[\text{Since, } Q = \frac{\pi d^2}{4} V_1 \quad \text{and} \quad V_1 = C_v (2gH)^{1/2}\right]$$

The specific speed N_{s_T} [Eq. (15.18)] $= \dfrac{N P^{1/2}}{H^{5/4}}$

The rotational speed N can be written as

$$N = U/\pi D$$

Therefore, it becomes

$$N_{s_T} = \left(\frac{U}{\pi D}\right) \frac{(\pi)^{1/2} d V_1^{3/2} \eta_h^{1/2}}{(8)^{1/2} C_v} \left[\frac{2 g C_v^2}{V_1^2}\right]^{5/4} \rho^{1/2}$$

$$= \frac{g^{5/4}}{(\pi)^{1/2}\,2^{1/4}} C_v^{3/2} \frac{U}{V_1} \frac{d}{D} \eta_h^{1/2} \rho^{1/2} \tag{15.27a}$$

It may be concluded from Eq. (15.27a) that the specific speed N_{s_T} depends primarily on the ratio d/D as the quantities U/V_1, C_v and η_h vary only slightly. Using the typical values of $U/V_1 = 0.46$, $C_v = 0.97$ and $\eta_h = 0.85$, the approximate relation between the specific speed and diameter ratio is obtained as

$$N_{s_T} \cong 105\ (d/D)\ \text{kg}^{1/2}\text{s}^{-5/2}\text{m}^{-1/4} \tag{15.27b}$$

The optimum value of the overall efficiency of a Pelton turbine depends both on the values of the specific speed and the speed ratio. The Pelton wheels with a single jet operate in the specific speed range of 4–16, and therefore the ratio D/d lies between 6 and 26 as given by the Eq. (15.27b). A larger value of D/d reduces the rpm as well as the mechanical efficiency of the wheel. It is possible to increase the specific speed by choosing a lower value of D/d, but the efficiency will decrease because of the close spacing of buckets. The value of D/d is normally kept between 14 and 16 to maintain high efficiency. The number of buckets required to maintain optimum efficiency is usually fixed by the empirical relation

$$\text{Number of buckets,} = 15 + \frac{53}{N_{s_T}}$$

Governing of Pelton Turbine First let us discuss what is meant by governing of turbines in general. When a turbine drives an electrical generator or alternator, the primary requirement is that the rotational speed of the shaft and hence that of the turbine rotor has to be kept fixed. Otherwise the frequency of the electrical output will be altered. But when the electrical load changes depending upon the demand, the speed of the turbine changes automatically. This is because the external resisting torque on the shaft is altered while the driving torque due to change of momentum in the flow of fluid through the turbine remains the same. For example, when the load is increased, the speed of the turbine decreases and vice versa. A constancy in speed is therefore maintained by adjusting the rate of energy input to the turbine accordingly. This is usually accomplished by changing the rate of fluid flow through the turbine—the flow is increased when the load is increased and the flow is decreased when the load is decreased. This adjustment of flow with the load is known as *governing of turbines.*

In case of a Pelton turbine, an additional requirement for its operation at the condition of maximum efficiency is that the ratio of bucket to initial jet velocity U/V_1 has to be kept at its optimum value of about 0.46. Hence, when U is fixed, V_1 has to be fixed. Therefore the control must be made by a variation of the cross-sectional area, A, of the jet so that the flow rate changes in proportion to the change in the flow area keeping the jet velocity V_1 same. This is usually achieved by a spear valve in the nozzle (Fig. 15.7(a)). Movement of the spear along the axis of the nozzle changes the annular area between the spear and the housing. The shape of the spear is such, that the fluid coalesces into a circular jet and then the effect of the spear movement is to vary the diameter of the jet. Deflectors are often used (Fig. 15.7(b)) along with the spear valve to prevent the serious water hammer problem due to a sudden reduction in the rate of flow. These plates temporarily deflect the jet so that the entire flow does not reach the bucket; the spear valve may then be moved slowly to its new position to reduce the rate of flow in the pipeline gradually. If the bucket width is too small in relation to the jet diameter, the fluid is not smoothly deflected by the buckets and, in consequence, much energy is dissipated in turbulence and the efficiency drops considerably. On the other hand, if the buckets are unduly large, the effect of friction on the surfaces is unnecessarily high. The optimum value of the ratio of bucket width to jet diameter has been found to vary between 4 to 5.

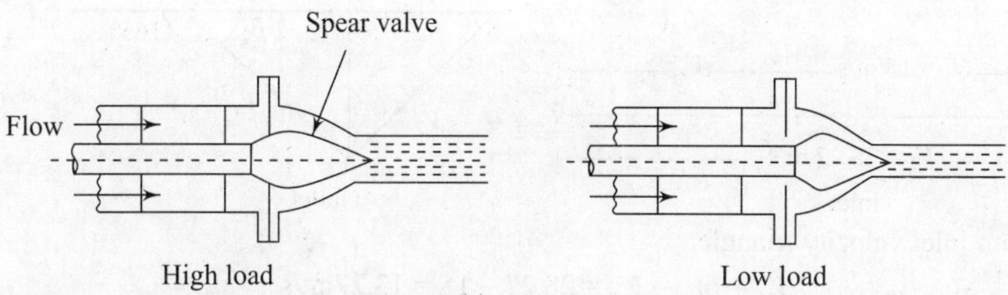

(a) Spear valve to alter jet area in a Pelton wheel

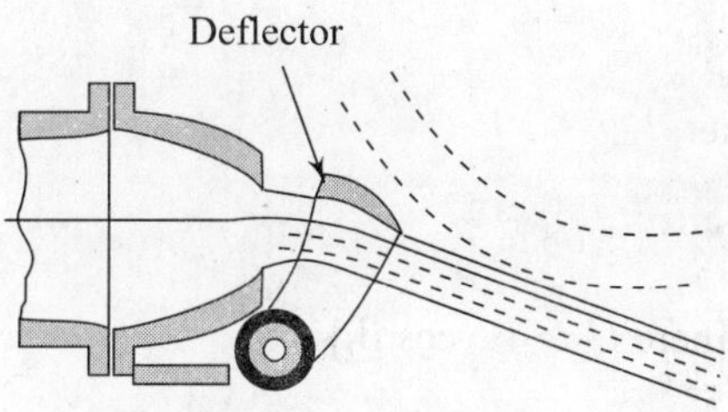

(b) Jet deflected from bucket

Fig. 15.7

Limitation of a Pelton Turbine The Pelton wheel is efficient and reliable when operating under large heads. To generate a given output power under a smaller head, the rate of flow through the turbine has to be higher which requires an increase in the jet diameter. The number of jets are usually limited to 4 or 6 per wheel. The increase in jet diameter in turn increases the wheel diameter. Therefore the machine becomes unduly large, bulky and slow-running. In practice, turbines of the reaction type are more suitable for lower heads.

Example 15.5

The mean bucket speed of a Pelton turbine is 15 m/s. The rate of flow of water supplied by the jet under a head of 42 m is 1 m^3/s. If the jet is deflected by the buckets at an angle of 165°, find the power and efficiency of the turbine. (Take coefficient of velocity C_v = 0.985).

Solution

Bucket speed is same at both inlet and outlet of the water jet.

Therefore, $U_1 = U_2 = 15$ m/s

Velocity of jet at inlet $V_1 = 0.985\ (2 \times 9.81 \times 42)^{1/2}$

$= 28.27$ m/s

Now the inlet and outlet velocity triangles are drawn as shown below:

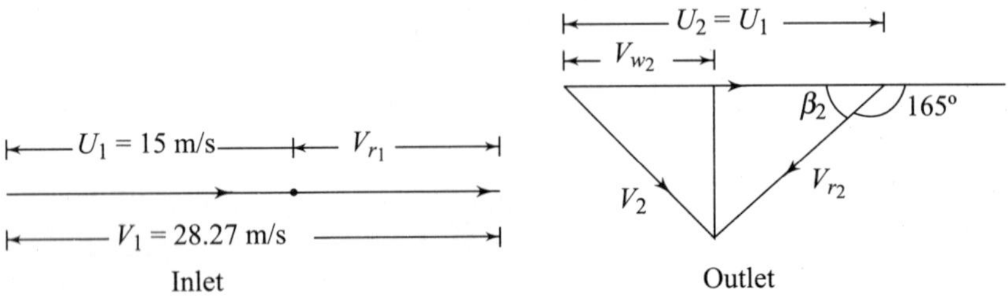

From inlet velocity triangle,

$$V_{r_1} = V_1 - U_1 = 28.27 - 15 = 13.27 \text{ m/s}$$

$$V_{w_1} = V_1 = 28.27 \text{ m/s}$$

The blade outlet angle is given by

$$\beta_2 = 180° - 165° = 15°$$

Neglecting the frictional losses in the bucket

$$V_{r_1} = V_{r_2} = 13.27 \text{ m/s}$$

From outlet velocity triangle

$$V_{w_2} = U_2 - V_{r_2} \cos \beta_2 \ [\text{here } U_2 > V_{r_2} \cos \beta_2]$$

$$= 15 - 13.27 \cos 15°$$

$$= 2.18 \text{ m/s}$$

Power developed, $P = \rho Q\,(V_{w_1} - V_{w_2})U_1$

$$= 10^3 \times 1 \times (28.27 - 2.18) \times 15$$

$$= 391.35 \text{ kW}$$

Turbine efficiency, $\eta = \dfrac{\text{Power developed}}{\text{Available power}}$

$$= \frac{391.35 \times 10^3}{10^3 \times 9.81 \times 1 \times 42}$$

$$= 0.95 = 95\%$$

Example 15.6

A single jet pelton turbine is required to drive a generator to develop 10 MW. The available head at the nozzle is 762 m. Assuming electric generator efficiency 95%, Pelton wheel efficiency 87%, coefficient of velocity for nozzle 0.97, mean bucket velocity 0.46 of jet velocity, outlet angle of the buckets 15° and the friction of the bucket reduces the relative velocity by 15 per cent, find (i) the diameter of the jet, (ii) the rate of flow of water through the turbine, and (iii) the force exerted by the jet on the buckets.

If the ratio of mean bucket circle diameter to the jet diameter is not to be less than 10, find the best synchronous speed for generation at 50 cycles per second and the corresponding mean diameter of the runner.

Solution

$$\text{Mechanical power output of the turbine} = \frac{\text{Electrical power output}}{\text{Generator efficiency}}$$

$$= \frac{10}{0.95}$$

$$= 10.53 \text{ MW}$$

Pelton wheel efficiency, $\eta = \dfrac{P}{\rho g Q H}$

where Q is the flow rate through the turbine.

Then, $Q = \dfrac{P}{\eta \times \rho g H}$

$$= \frac{10.53 \times 10^6}{0.87 \times 10^3 \times 9.81 \times 762} = 1.62 \text{ m}^3$$

If d_1 is the diameter of the jet, we can write

$$Q_1 = (\pi/4) \times d_1^2\, C_V\, (2gH)^{1/2}$$

where, C_V is the coefficient of velocity.

Then, $1.62 = (\pi/4) \times d_1^2 \times 0.97 \times (2 \times 9.81 \times 762)^{1/2}$

which gives $d_1 = 0.132 \text{ m} = 132 \text{ mm}$

The inlet and outlet velocity triangles are shown below:

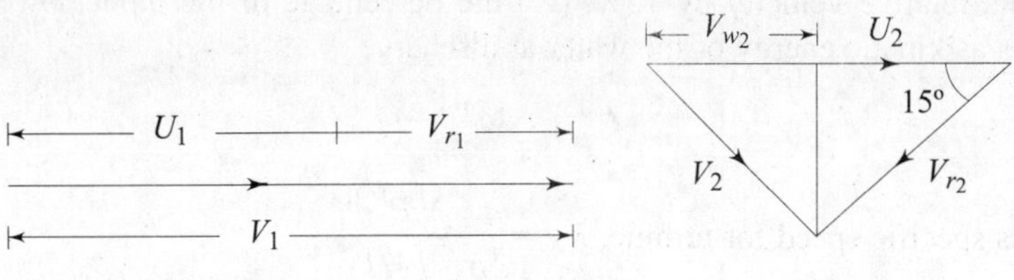

Jet velocity, $V_1 = C_v\, [2gH]^{1/2} = 0.97\, [2 \times 9.81 \times 762]^{1/2}$

$= 118.6 \text{ m/s}$

Mean bucket velocity, $U_1 = U_2 = 0.46 \times 118.6 = 54.56 \text{ m/s}$

From the inlet velocity triangle,

$$V_{w_1} = V_1 = 118.6 \text{ m/s}$$
$$V_{r_1} = V_1 - U_1 = 118.50 - 54.56 = 63.94 \text{ m/s}$$
$$V_{r_2} = 0.85 \times 63.94 = 54.35 \text{ m/s}$$

From the outlet velocity triangle,

$$V_{w_2} = U_2 - V_{r_2} \cos \beta_2 = 54.56 - 54.35 \times \cos 15°$$
$$= 2.06 \text{ m/s}$$

Therefore, the force exerted by the jet on the bucket is given by

$$F = \rho Q (V_{w_1} - V_{w_2}) = 10^3 \times 1.62 [118.5 - 2.06] \text{ N}$$
$$= 188.63 \text{ kN}$$

Considering the ratio of mean bucket circle diameter D to the jet diameter d as 10,

$$D = 10 \times 0.132 = 1.32 \text{ m}$$

Again, $U_1 = [\pi DN]/60$

Hence, $N = [54.56 \times 60] / (\pi \times 1.32] = 789.51 \text{ rpm}$

Frequency of generator $f = p \cdot N/60$

where p = Number of pair of poles

$p = 4$ gives $N_{syn} = [60 \times 50]/4$
$= 750$ rpm which is nearest to 789 rpm

Therefore, we choose $N_{syn} = 750$ rpm

Now, D (revised) $= [1.32 \times 789.51]/750 = 1.39$ m

Example 15.7

In a hydroelectric scheme, a number of Pelton wheels are to be used under the following conditions: total output required 30 MW; gross head 245 m; speed 6.25 rev/s; 2 jets per wheel; C_v of nozzles 0.97; maximum overall efficiency (based on conditions immediately before the nozzles) 81.5%; dimensionless specific speed not to exceed 0.022 rev. per jet; head lost to friction in pipeline is 12 m. Ratio of blade to jet speed is 0.46.

Calculate (i) the number of wheels required, (ii) the diameters of the jets and wheels, (iii) the hydraulic efficiency, if the blade deflects the water jet through 165° and reduces its relative velocity by 15%, (iv) the percentage of the input power which remains as kinetic energy of the water at discharge.

Solution

Dimensionless specific speed for turbine, $K_{s_T} = \dfrac{N P^{1/2}}{\rho^{1/2} (gH)^{5/4}}$

Here, $K_{s_T} = 0.022$ rev per jet.

The available head to the turbine (i.e., at the inlet to the nozzle)

$$H = 245 - 12 = 233 \text{ m}$$

Hence, power per jet, $P = [0.022 \times (10^3)^{1/2} \times (9.81 \times 233)^{5/4}/6.25]^2$

$= 3.09 \times 10^6 \text{ W} = 3.09 \text{ MW}$

(i) Therefore no. of wheels = 30/[3.09 × 2]

$$= 5 \text{ (no. of wheels would be an integer)}$$

(ii) If Q is the flow rate in m^3/s per jet, then,

$$10^3 \times Q \times 9.81 \times 233 \times 0.815 = 3.09 \times 10^6$$

which gives $Q = 1.66\ m^3/s$

Velocity of the jet, $V_1 = 0.97 \times [2 \times 9.81 \times 233]^{1/2}$

$$= 65.58 \text{ m}$$

Hence, $1.66 = \frac{\pi}{4} \times d^2 \times 65.58$ (where d is the diameter of the jet)

or $d = \sqrt{\frac{4 \times 1.66}{\pi \times 65.58}} = 0.179 \text{ m} = 179 \text{ mm}$

Blade or wheel speed, $U = 0.46 \times 65.58 = 30.17$ m/s

Therefore wheel diameter $D = \frac{30.17}{\pi \times 6.25} = 1.54$ m

(iii) The inlet and outlet velocity triangles are drawn as shown below:

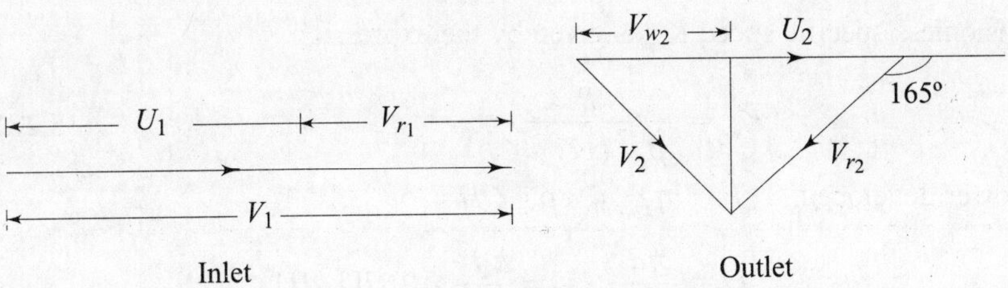

$$U_1 = U_2 = 30.17 \text{ m/s}$$

$$V_1 = V_{w_1} = 65.58 \text{ m/s}$$

$$V_{r_1} = V_1 - U_1 = 65.58 - 30.17 = 35.41 \text{ m/s}$$

The relative velocity at outlet, $V_{r_2} = 0.85 \times 35.41$

®= 30.1 m/s

From outlet velocity triangle, $V_{w_2} = 30.17 - 30.1 (\cos 15°)$

$$= 1.1 \text{ m/s}$$

Hydraulic efficiency, $\eta_h = \frac{(V_{w_1} - V_{w_2}) U_1}{g H}$

$$= \frac{(65.58 - 1.1) \times 30.17}{9.81 \times 233} = 0.851$$

$$= 85.1\%$$

The kinetic energy at the outlet/unit mass = $V_2^2/2$

Input power / unit mass = gH
where H is the net head to the turbine (at nozzle inlet).
Let x be the percentage of input power remaining as kinetic energy of water at discharge.

Then,
$$x = \frac{V_2^2}{2gH} \times 100$$

From outlet velocity triangle,
$$V_2^2 = [(30.1 \times \sin 15°)^2 + (1.1)^2]$$
$$= 61.90 \text{ m}^2/\text{s}^2$$

Therefore,
$$x = \frac{61.90}{2 \times 9.81 \times 233} \times 100 = 1.35\%$$

Example 15.8

The blading of a single jet Pelton wheel runs at its optimum speed which is 0.46 times the jet speed. The overall efficiency of the machine is 0.85. Show that the dimensionless specific speed is 0.192 d/D rev, where d represents the jet diameter and D the wheel diameter. For the nozzle, the velocity coefficient $C_v = 0.97$.

Solution

Dimensionless specific speed K_{s_T} is given by the expression

$$K_{s_T} = \frac{N P^{1/2}}{\rho^{1/2} (gH)^{5/4}} \tag{15.28}$$

The power developed,
$$P = \eta_{\text{overall}} \times \rho g QH$$

Again,
$$Q = \frac{\pi d^2}{4} \times V_1 = \frac{\pi d^2}{4} \times 0.97[2gH]^{1/2}$$
$$= 1.08\, d^2\, (gH)^{1/2}$$

Hence,
$$P = 0.85 \times \rho \times [1.08\, d^2\, (gH)^{1/2}]\, gH$$
$$= 0.92\, \rho\, d^2 (gH)^{3/2} \tag{15.29}$$

The rotational speed,
$$N = U/\pi d$$
again the wheel speed,
$$U = 0.46 \times V_1 = 0.46 \times 0.97\, (2\, gH)^{1/2}$$
$$= 0.63\, (gH)^{1/2}$$

Hence,
$$N = \frac{0.63 (gH)^{1/2}}{\pi D} = 0.2 \frac{(gH)^{1/2}}{D} \tag{15.30}$$

Substituting the values of P and N from Eqs (15.29) and (15.30) respectively into Eq. (15.28), we have

$$K_{s_T} = \left[0.2 \frac{(gH)^{1/2}}{D}\right] \left[0.92\, \rho d^2 (gH)^{3/2}\right]^{1/2} \frac{1}{\rho^{1/2} (gH)^{5/4}}$$
$$= 0.192\, \frac{d}{D}$$

15.4.2 Reaction Turbine

The principal feature of a reaction turbine that distinguishes it from an impulse turbine is that only a part of the total head available at the inlet to the turbine is converted to velocity head, before the runner is reached. Also in the reaction turbines, the working fluid, instead of engaging only one or two blades, completely fills the passages in the runner. The pressure or static head of the fluid changes gradually as it passes through the runner along with the change in its kinetic energy based on absolute velocity due to the impulse action between the fluid and the runner. Therefore the cross-sectional area of flow through the passages of the runner changes gradually to accommodate the variation in static pressure of the fluid. A reaction turbine is usually well suited for low heads. A radial flow hydraulic turbine of reaction type was first developed by an American engineer, James B. Francis (1815–92) and is named after him as the Francis turbine. The schematic diagram of a Francis turbine is shown in Fig. 15.8 (a).

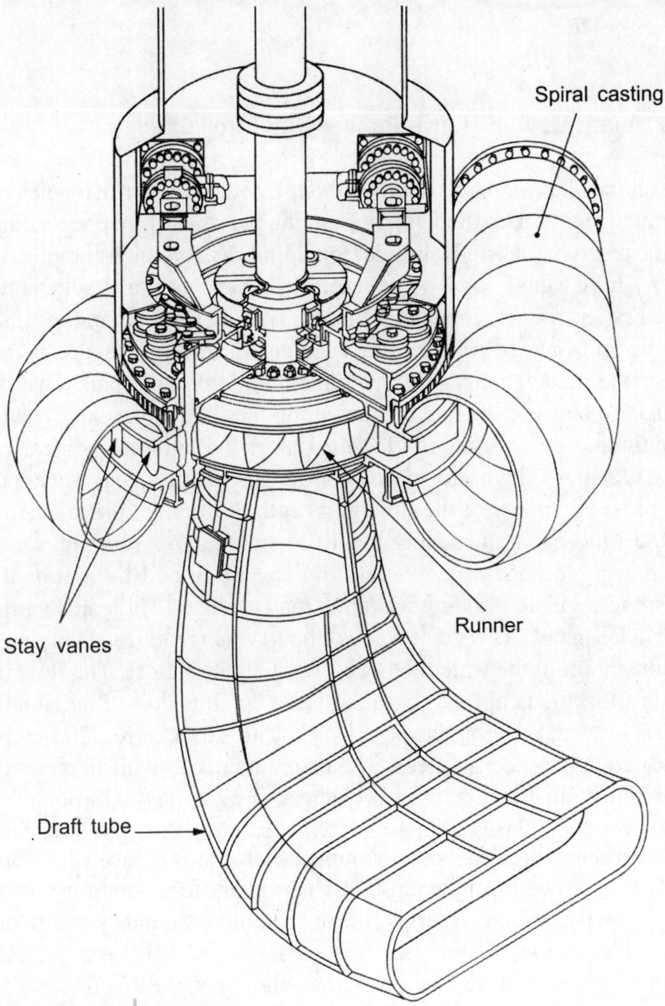

Fig. 15.8(a) A Francis turbine

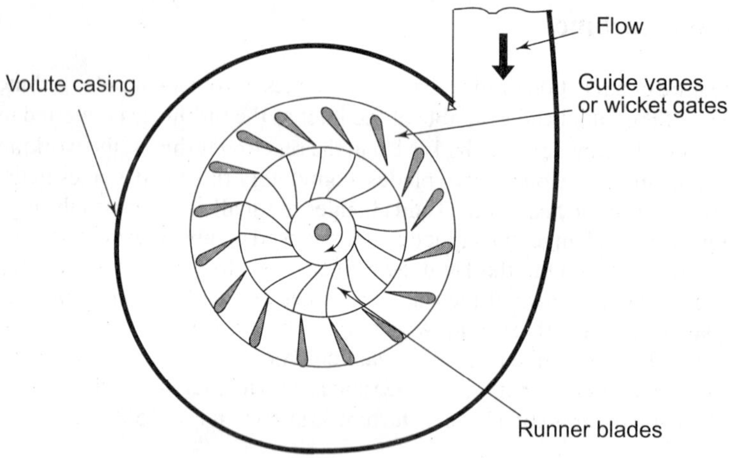

(b)

Fig. 15.8(b) A schematic view of Francis Runner with Scroll Casing

Most of these machines have vertical shafts although some smaller machines of this type have horizontal shafts. The fluid enters from the penstock (pipeline leading to the turbine from the reservoir at a high altitude) to a spiral casing which completely surrounds the runner. This casing is known as a scroll casing or a volute. A schematic sectional view of the Francis runner along with the scroll casing is shown in Fig. 15.8 (b). The cross-sectional area of this casing decreases uniformly along the circumference to keep the fluid velocity constant in magnitude along its path towards the guide vane. This is so because the rate of flow along the fluid path in the volute decreases due to continuous entry of the fluid to the runner through the openings of the guide vanes or stay vanes. The basic purpose of the guide vanes (stay vanes) is to convert a part of pressure energy of the fluid at its entrance to the kinetic energy and then to direct the fluid on to the runner blades at an angle appropriate to the design. Moreover, the guide vanes are pivoted and can be turned by a suitable governing mechanism to regulate the flow while the load changes. The guide vanes are also known as wicket gates. The guide vanes impart a tangential velocity and hence an angular momentum to the water before its entry to the runner. The flow in the runner of a Francis turbine is not purely radial but a combination of radial and tangential. The flow is inward, i.e., from the periphery towards the centre. The height of the runner depends upon the specific speed. The height increases with an increase in the specific speed. The main direction of flow changes as water passes through the runner and is finally turned into the axial direction while entering the draft tube.

The draft tube is a conduit which connects the runner exit to the tail race where the water is being finally discharged from the turbine. The primary function of the draft tube is to reduce the velocity of the discharged water to minimise the loss of kinetic energy at the outlet. This permits the turbine to be set above the tail water without any appreciable drop of available head. A clear understanding of the function of the draft tube in any reaction turbine, in fact, is very important for the purpose of its design. The purpose of providing a draft tube will be better understood if we carefully study the net available head across a reaction turbine.

Net Head Across a Reaction Turbine and the Purpose of Providing a Draft Tube The effective head across any turbine is the difference between the head at the inlet to the machine and the head at the outlet from it. A reaction turbine always runs completely filled with the working fluid. The tube that connects the end of the runner to the tail race is known as a draft tube and should completely be filled with the working fluid flowing through it. The kinetic energy of the fluid finally discharged into the tail race is wasted. A draft tube is made divergent so as to reduce the velocity at the outlet to a minimum. Therefore a draft tube is basically a diffuser and should be designed properly with the angle between the walls of the tube to be limited to about 8 degrees so as to prevent the flow separation from the wall and to reduce accordingly the loss of energy in the tube. Figure 15.9 shows a flow diagram from the reservoir via a reaction turbine to the tail race.

The total head H_1 at the entrance to the turbine can be found out by applying Bernoulli's equation between the free surface of the reservoir and the inlet to the turbine as

$$H_0 = \frac{p_1}{\rho g} + \frac{V_1^2}{2g} + z + h_f$$

or

$$H_1 = H_0 - h_f = \frac{p_1}{\rho g} + \frac{V_1^2}{2g} + z$$

where h_f is the head lost due to friction in the pipeline connecting the reservoir and the turbine. Since the draft tube is a part of the turbine, the net head across the turbine, for the conversion of mechanical work, is the difference of total head at inlet to the machine and the total head at discharge from the draft tube at tail race and is shown as H in Fig. 15.9.

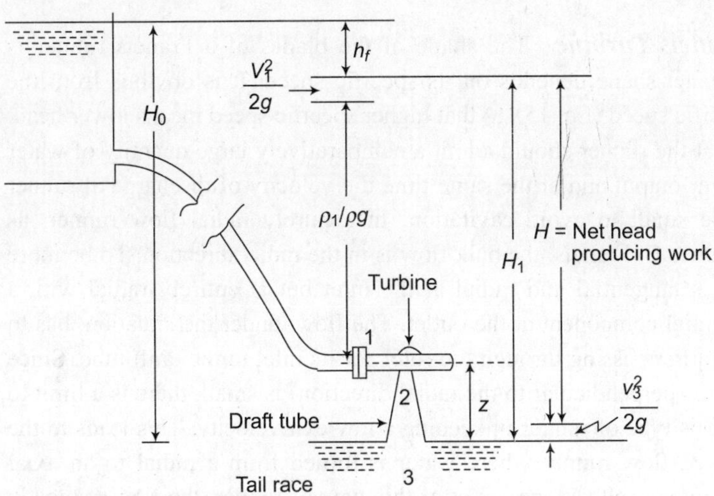

Fig. 15.9 Head across a reaction turbine®

Therefore, H = total head at inlet to machine (1) – total head at discharge (3)

$$= \frac{p_1}{\rho g} + \frac{V_1^2}{2g} + z - \frac{V_3^2}{2g} = H_1 - \frac{V_3^2}{2g}$$

$$= (H_0 - h_f) - \frac{V_3^2}{2g}$$

The pressures are defined in terms of their values above the atmospheric pressure. Sections 2 and 3 in Fig. 15.9 represent the exits from the runner and the draft tube respectively. If the losses in the draft tube are neglected, then the total head at 2 becomes equal to that at 3. Therefore, the net head across the machine is either $(H_1 - H_3)$ or $(H_1 - H_2)$. Applying Bernoulli's equation between 2 and 3 in consideration of flow, without losses, through the draft tube, we can write,

$$\frac{p_2}{\rho g} + \frac{V_2^2}{2g} + z = 0 + \frac{V_3^2}{2g} + 0$$

or
$$\frac{p_2}{\rho g} = -\left[z + \frac{V_2^2 - V_3^2}{2g}\right] \tag{15.31}$$

Since $V_3 < V_2$, both the terms in the bracket are positive and hence $p_2/\rho g$ is always negative, which implies that the static pressure at the outlet of the runner is always below the atmospheric pressure. Equation (15.31) also shows that the value of the suction pressure at runner outlet depends on z, the height of the runner above the tail race and $(V_2^2 - V_3^2)/2g$, the decrease in kinetic energy of the fluid in the draft tube. The value of this minimum pressure p_2 should never fall below the vapour pressure of the liquid at its operating temperature to avoid the problem of cavitation. Therefore, we find that the incorporation of a draft tube allows the turbine runner to be set above the tail race without any drop of available head by maintaining a vacuum pressure at the outlet of the runner.

Runner of Francis Turbine The shape of the blades of a Francis runner is complex. The exact shape depends on its specific speed. It is obvious from the equation of specific speed (Eq. 15.18) that higher specific speed means lower head. This requires that the runner should admit a comparatively large quantity of water for a given power output and at the same time the velocity of discharge at runner outlet should be small to avoid cavitation. In a purely radial flow runner, as developed by James B. Francis, the bulk flow is in the radial direction. To be more clear, the flow is tangential and radial at the inlet but is entirely radial with a negligible tangential component at the outlet. The flow, under the situation, has to make a 90° turn after passing through the rotor for its inlet to the draft tube. Since the flow area (area perpendicular to the radial direction) is small, there is a limit to the capacity of this type of runner in keeping a low exit velocity. This leads to the design of a mixed flow runner where water is turned from a radial to an axial direction in the rotor itself. At the outlet of this type of runner, the flow is mostly axial with negligible radial and tangential components. Because of a large discharge area (area perpendicular to the axial direction), this type of runner can pass a large amount of water with a low exit velocity from the runner. The blades for a reaction turbine are always so shaped that the tangential or whirling component of velocity at the outlet becomes zero ($V_{w_2} = 0$). This is made to keep the kinetic energy at outlet a minimum.

Figure 15.10 shows the velocity triangles at inlet and outlet of a typical blade of a Francis turbine. Usually the flow velocity (velocity perpendicular to the tangential direction) remains constant throughout, i.e. $V_{f1} = V_{f2}$ and is equal to that at the inlet to the draft tube.

The Euler's equation for turbine (Eq. (15.2)) in this case reduces to

$$E/m = e = V_{w_1} U_1 \tag{15.32}$$

where, e is the energy transfer to the rotor per unit mass of the fluid.

From the inlet velocity triangle shown in Fig. 15.10.

$$V_{w_1} = V_{f_1} \cot \alpha_1 \tag{15.33a}$$

and

$$U_1 = V_{f_1} (\cot \alpha_1 + \cot \beta_1) \tag{15.33b}$$

substituting the values of V_{w_1} and U_1 from Eqs (15.33a) and (15.33b) respectively into Eq. (15.32), we have

$$e = V_{f_1}^2 \cot \alpha_1 (\cot \alpha_1 + \cot \beta_1) \tag{15.34}$$

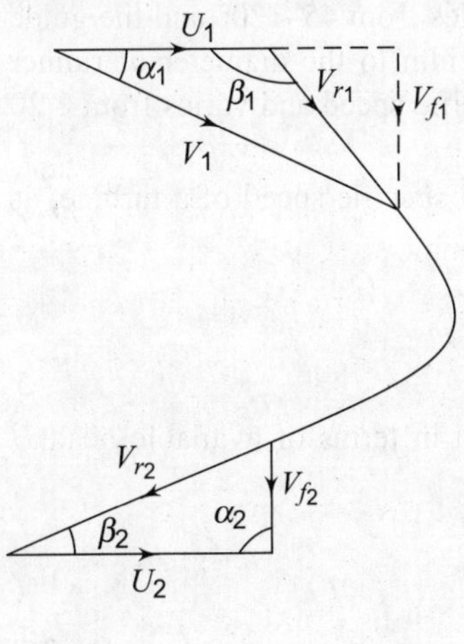

Fig. 15.10 Velocity triangle for a Francis runner

The loss of kinetic energy per unit mass becomes equal to $V_{f_2}^2/2$. Therefore, neglecting friction, the blade efficiency becomes

$$\eta_b = \frac{e}{e + (V_{f_2}^2/2)}$$

$$= \frac{2V_{f_1}^2 \cot\alpha_1 (\cot\alpha_1 + \cot\beta_1)}{V_{f_2}^2 + 2V_{f_1}^2 \cot\alpha_1 (\cot\alpha_1 + \cot\beta_1)}$$

Since, $V_{f_1} = V_{f_2}$, η_b can be written as

$$\eta_b = 1 - \frac{1}{1 + 2\cot\alpha_1 (\cot\alpha_1 + \cot\beta_1)}$$

The change in pressure energy of the fluid in the rotor can be found out by subtracting the change in its kinetic energy from the total energy released. Therefore, we can write for the degree of reaction

$$R = \frac{e - \frac{1}{2}(V_1^2 - V_{f_2}^2)}{e} = 1 - \frac{\frac{1}{2}V_{f_1}^2 \cot^2\alpha_1}{e}$$

$$(\text{Since } V_1^2 - V_{f_2}^2 = V_1^2 - V_{f_1}^2 = V_{f_1}^2 \cot^2\alpha_1)$$

using the expression of *e* from Eq. (15.29), we have

$$R = 1 - \frac{\cot\alpha_1}{2(\cot\alpha_1 + \cot\beta_1)} \tag{15.35}$$

The inlet blade angle β_1 of a Francis runner varies from 45–120° and the guide vane angle α_1 from 10–40°. The ratio of blade width to the diameter of runner B/D, at blade inlet, depends upon the required specific speed and varies from 1/20 to 2/3.

Expression for Specific Speed The dimensional specific speed of a turbine, as given by Eq. (15.18), can be written as

$$N_{s_T} = \frac{NP^{1/2}}{H^{5/4}}$$

Power generated P for a turbine can be expressed in terms of available head H and hydraulic efficiency η_h as

$$P = \rho Q g H \eta_h$$

Hence, it becomes

$$N_{s_T} = N(\rho Q g \eta_h)^{1/2} H^{-3/4} \tag{15.36}$$

Again, $N = U_1/\pi D_1$,

Substituting U_1 from Eq. (15.33b),

$$N = \frac{V_{f_1}(\cot\alpha_1 + \cot\beta_1)}{\pi D_1} \tag{15.37}$$

Available head H equals the head delivered by the turbine plus the head lost at the exit. Thus,

$$gH = e + (V_{f_2}^2/2)$$

Since
$$V_{f_1} = V_{f_2}$$

$$gH = e + (V_{f_1}^2/2)$$

with the help of Eq. (15.34), it becomes

$$gH = V_{f_1}^2 \cot\alpha_1 (\cot\alpha_1 + \cot\beta_1) + \frac{V_{f_1}^2}{2}$$

or
$$H = \frac{V_{f_1}^2}{2g}[1 + 2\cot\alpha_1(\cot\alpha_1 + \cot\beta_1)] \tag{15.38}$$

Substituting the values of H and N from Eqs (15.38) and (15.37) respectively into the expression of N_{s_T} given by Eq. (15.36), we get,

$$N_{s_T} = 2^{3/4} g^{5/4} (\rho\, \eta_h Q)^{1/2} \frac{V_{f_1}^{-1/2}}{\pi D_1} (\cot\alpha_1 + \cot\beta_1)\, [1+2\cot\alpha_1 (\cot\alpha_1 + \cot\beta_1)]^{-3/4}$$

Flow velocity at inlet V_{f_1} can be substituted from the equation of continuity as

$$V_{f_1} = \frac{Q}{\pi D_1 B}$$

where B is the width of the runner at its inlet.
Finally, the expression for N_{s_T} becomes,

$$N_{s_T} = 2^{3/4} g^{5/4} (\rho\, \eta_h)^{1/2} \left(\frac{B}{\pi D_1}\right)^{1/2} (\cot\alpha_1 + \cot\beta_1)\, [1+2\cot\alpha_1 (\cot\alpha_1 + \cot\beta_1)]^{-3/4} \tag{15.39}$$

For a Francis turbine, the variations of geometrical parameters like α_1, β_1 B/D have been described earlier. These variations cover a range of specific speed between 50 and 400. Higher specific speed corresponds to a lower head. This requires that runner should admit a comparatively large quantity of water. For a runner of given diameter, the maximum flow rate is achieved when the flow is parallel to the axis. Such a machine is known as an axial flow reaction turbine. It was first designed by an Austrian engineer, Viktor Kaplan and is therefore named after him as Kaplan turbine.

Example 15.9

A Francis turbine has a diameter of 1.4 m and rotates at 430 rpm. Water enters the runner without shock with a flow velocity of 9.5 m/s and leaves the runner without whirl with an absolute velocity of 7 m/s. The difference between the sum of the static and potential heads at the entrance to the runner and at the exit from the runner is 62 m. The turbine develops 12.25 MW. The flow rate through the turbine is 12 m^3/s for a net head of 115 m.

Find (i) absolute velocity of water at entry to the runner and the angle of the inlet guide vanes, (ii) entry angle of the runner blades, and (iii) loss of head in the runner.

Solution

The inlet and outlet velocity triangles are drawn as shown below:

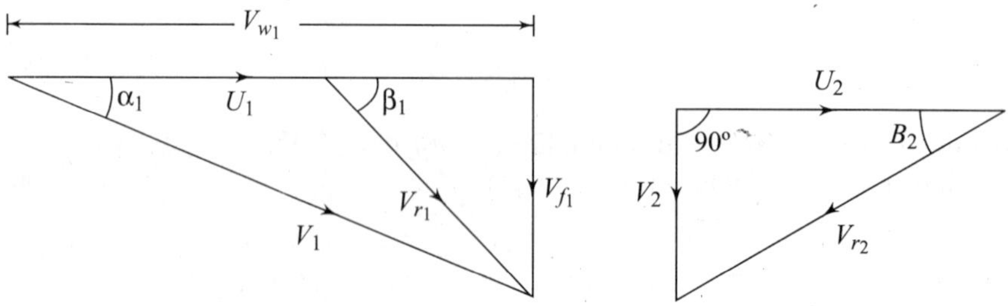

(i) Runner tip speed $U_1 = \dfrac{\pi ND}{60} = \dfrac{\pi \times 430 \times 1.4}{60} = 31.52$ m/s

Since, $V_{w_2} = 0$,

Power given to the runner by water $= \rho\, Q\, V_{w_1}\, U_1$

Hence, $12.25 \times 10^6 = 10^3 \times 12 \times V_{w_1} \times 31.52$

which gives $V_{w_1} = 32.39$ m/s

Inlet guide vane angle, α_1 is given by

$$\tan \alpha_1 = [9.5 / 32.29]$$

or

$$\alpha_1 = \tan^{-1} [9.5/32.39] = 16.35°$$

From the inlet velocity diagram, the absolute velocity at runner inlet

$$V_1 = [V_{f1}^2 + V_{w1}^2]^{1/2} = [(9.5)^2 + (32.39)^2]^{1/2} = 33.75 \text{ m/s}$$

(ii) Runner blade entry angle β_1 is given by

$$\tan \beta_1 = \frac{9.5}{32.39 - 31.52}$$

which gives $\beta_1 = 84.77°$

(iii) Total head across the runner

= Head transferred to the runner
+ Head lost in the runner

At the inlet, $H_1 = (p_1/\rho g) + (V_1^2/2g) + z_1$

At the outlet, $H_2 = (p_2/\rho g) + (V_2^2/2g) + z_2$

where, z_1 and z_2 are the elevations of runner inlet and outlet from a reference datum. For zero whirl at outlet, the work done per unit weight of the fluid = $[V_{w_1} U_1/g]$
Hence loss of head in the runner becomes

$$h_L = H_1 - H_2 - [V_{w_1} U_1/g]$$

$$= \left[\frac{p_1 - p_2}{\rho g}\right] + \left[\frac{V_1^2 - V_2^2}{2g}\right] + [z_1 - z_2] - [V_{w_1} U_1 / g]$$

It is given that

$$\left[\frac{p_1 - p_2}{\rho g}\right] + [z_1 - z_2] = 62 \text{ m}$$

Therefore,

$$h_L = 62 + \left[\frac{(33.75)^2 - (7)^2}{2 \times 9.81}\right] - \left[\frac{31.52 \times 32.39}{9.81}\right]$$

$$= 13.49 \text{ m}$$

Example 15.10

An inward flow vertical shaft reaction turbine runs at a speed of 375 rpm under an available total head of 62 m above the atmospheric pressure. The external diameter of the runner is 1.5 m and the dimensionless specific speed based on the power transferred to the runner is 0.14 rev. Water enters the turbine without shock with a flow velocity of 9 m/s and leaves the runner without whirl with an absolute velocity of 7 m/s. The discharge velocity of water at tailrace is 2.0 m/s. The mean height of the runner entry plane is 2 m above the tailrace level while the entrance to the draft tube is 1.7 m above the tailrace level. At the entrance to the runner, the static pressure head is 35 m above the atmospheric pressure, while at exit from the runner, the static pressure head is 2.2 m below the atmospheric pressure.

Assuming a hydraulic efficiency of 90 %, find (i) the runner blade entry angle, and (ii) the head loss in the guide vanes, in the runner and in the draft tube.

Solution

(i) Runner speed at the inlet,

$$U_1 = \frac{\pi N D}{60} = \frac{\pi \times 375 \times 1.5}{60}$$

$$= 29.45 \text{ m/s}$$

Since $V_{w_2} = 0$,
the power transferred to the runner per unit mass flow of water = $V_{w_1} U_1$

Hydraulic efficiency, $\eta_h = \dfrac{V_{w_1} U_1}{g H}$

Therefore,

$$0.9 = \frac{V_{w_1} \times 29.45}{9.81 \times 62}$$

or $$V_{w_1} = \frac{0.9 \times 9.81 \times 62}{29.45} = 18.59 \text{ m/s}$$

The inlet velocity triangle is shown below:

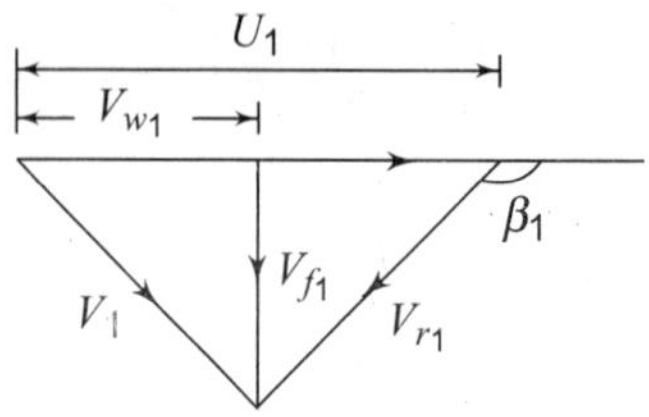

From the velocity triangle,

$$\tan(180° - \beta_1) = \frac{V_{f_1}}{U_1 - V_{w_1}} = \frac{9}{(29.45 - 18.59)} = 0.83$$

Hence, $$\beta_1 = 140.35°$$

(ii) Let the loss of head in the guide vanes be h_{1_g}. Then applying Bernoulli's equation between the inlet to guide vanes and exit from the guide vanes (i.e., the inlet to the runner), we have

$$\frac{p_0}{\rho g} + \frac{V_0^2}{2g} + z_0 = \frac{p_1}{\rho g} + \frac{V_1^2}{2g} + z_1 + h_{1_g}$$

(0 and 1 apply to inlet and exit of guide vanes, respectively)

From the velocity triangle at runner inlet,

$$V_1^2 = (18.59)^2 + (9)^2 = 426.59 \text{ m}^2/\text{s}^2$$

Again, $$\frac{p_0}{\rho g} + \frac{V_0^2}{2g} + z_0 = 62 \text{ m (total head to the turbine)}$$

Therefore, $$62 = \left(35 + \frac{426.59}{2 \times 9.81} + 2\right) + h_{1_g}$$

Hence, $$h_{1_g} = 62 - 58.74 = 3.26 \text{ m}$$

For the loss of head in the runner h_{1_r}, the application of Bernoulli's equation between points at the runner entry and runner exit gives

$$\frac{p_1}{\rho g} + \frac{V_1^2}{2g} + z_1 = \frac{p_2}{\rho g} + \frac{V_2^2}{2g} + z_2 + h_{1_r} + W$$

where W is the work head delivered by the fluid to the runner and is given by

$$W = \frac{V_{w_1} U_1}{g} = \frac{18.59 \times 29.45}{9.81} = 55.81 \text{ m}$$

Therefore,

$$h_{1_r} = \left[35+\frac{426.59}{2\times 9.81}+2\right]-\left[-2.2+\frac{7^2}{2\times 9.81}+1.7\right]-55.81$$

$$= 58.74 - 2.0 - 55.81 = 0.93 \text{ m}$$

For the losses of head h_{1_d} in the draft tube, the Bernoulli's equation between the points at entry and exit of the draft tube gives

$$\frac{p_2}{\rho g}+\frac{V_2^2}{2g}+z_2 = \frac{p_3}{\rho g}+\frac{V_3^2}{2g}+z_3+h_{1_d}$$

where subscript 2 represents the runner outlet, i.e., the inlet of draft tube, and subscript 3 represents the exit from draft tube. p_3 is the atmospheric pressure (zero gauge) and z_3 is the datum level.
Therefore,

$$\left[-2.2+\frac{49}{2\times 9.81}+1.7\right] = \left[0+\frac{4}{2\times 9.81}+0\right]+h_{1_d}$$

which gives, $h_{1_d} = 1.8$ m

Example 15.11

The diameter of the runner of a vertical-shaft turbine is 450 mm at the inlet. The width of the runner at the inlet is 50 mm. The diameter and width at the outlet are 300 mm and 75 mm, respectively. The blades occupy 8% of the circumference. The guide vane angle is 24°, the inlet angle of the runner blade is 95° and the outlet angle is 30°. The fluid leaves the runner without any whirl. The pressure head at the inlet is 55 mm above that at the exit from the runner. The fluid friction losses account for 18% of the pressure head at the inlet. Calculate the speed of the runner and the output power (use mechanical efficiency as 95%).

Solution

Applying Bernoulli's equation between the inlet and the outlet of the runner, we have

$$\frac{p_1}{\rho g}+\frac{V_1^2}{2g} = \frac{p_2}{\rho g}+\frac{V_2^2}{2g}+W+h_{1_r} \tag{15.40}$$

where W is the work head given by the fluid to runner and h_{1_r} is the head loss in the runner, subscript 1 represents the runner inlet while 2 represents the runner outlet.

$$\frac{p_1}{\rho g}-\frac{p_2}{\rho g} = 55 \text{ m (given in the problem)}$$

and $h_{1_r} = 0.18\times 55 = 9.9$ m (given in the problem)

The inlet and outlet velocity triangles are shown below:

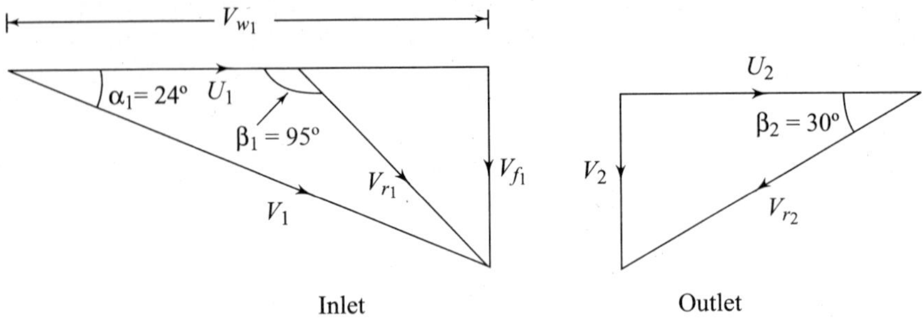

$$W = [V_{w_1} U_1]/g$$

From inlet velocity triangle,

$$V_{w_1} = V_1 \cos 24° = 0.913\ V_1$$

From continuity,

$$V_{f_1} D_1 B_1 = V_2 D_2 B_2$$

or $$V_{f_1} \times 450 \times 50 = V_2 \times 300 \times 75$$

which gives $$V_2 = V_{f_1}$$

Therefore, $$V_2 = V_{f_1} = V_1 \sin 24° = 0.406\ V_1$$

From the consideration of rotational speed,

$$U_1/D_1 = U_2/D_2$$

or $$U_1 = \frac{D_1}{D_2} U_2 = \frac{450}{300} U_2 = 1.5\ U_2$$

Again, from the outlet velocity triangle,

$$U_2 = \frac{V_2}{\tan 30°} = \frac{0.406\ V_1}{\tan 30°} = 0.703\ V_1$$

Hence, $$U_1 = 1.5 \times 0.703\ V_1 = 1.05\ V_1$$

Therefore, $$W = [V_{w_1} U_1]/g = \frac{0.913 \times 1.05}{g} V_1^2 = \frac{0.96\ V_1^2}{g}$$

Now Eq. (15.40) can be written as

$$55 - 9.9 = \frac{-V_1^2}{2g} + \frac{(0.406 V_1)^2}{2g} + \frac{0.96\ V_1^2}{g}$$

or $$45.1 = \frac{V_1^2}{2g} [-1 + (0.406)^2 + 2 \times 0.96] = 1.08\ \frac{V_1^2}{2g}$$

Hence, $$V_1 = [45.1 \times 2 \times 9.81/1.08]^{1/2} = 28.62 \text{ m/s}$$

$$U_1 = 1.05 \times 28.62 = 30.05 \text{ m/s}$$

Therefore, $N = 30.05/[\pi \times 0.45] = 21.26$ rev./s

Rate of flow, $Q = 0.92\,\pi D_1 B_1 \times V_{f_1}$

$$V_{f_1} = 0.406 \times 28.62 = 11.62 \text{ m/s}$$

Hence, $Q = 0.92 \times \pi \times 0.45 \times (0.05) \times 11.62 = 0.755 \text{ m}^3/\text{s}$

Therefore, power developed, $P = \rho Q V_{w_1} U_1$

$$= 10^3 \times 0.755 \times (0.96) \times (28.62)^2 = 593.60 \text{ kW}$$

15.4.3 Development of Kaplan Runner from the Change in the Shape of Francis Runner with Specific Speed

Figure 15.11 shows in stages the change in the shape of a Francis runner with the variation of specific speed. The first three types [Fig. 15.11(a), (b) and (c)] have, in order, the Francis runner (radial flow runner) at low, normal and high specific speeds. As the specific speed increases, discharge becomes more and more axial. The fourth type, as shown in Fig. 15.11(d), is a mixed flow runner (radial flow at inlet but axial flow at outlet) and is known as Dubs runner, which is mainly suited for high specific speeds. Figure 15.11(e) shows a propeller type runner with a less number of blades where the flow is entirely axial (both at inlet and outlet). This type of runner is the most suitable one for very high specific speeds and is known as Kaplan runner or axial flow runner.

From the inlet velocity triangle for each of the five runners, as shown in Figs 15.11(a) to 15.11(e), it is found that an increase in specific speed (or a decreased in head) is accompanied by a reduction in inlet velocity V_1. But the flow velocity V_{f1} at the inlet increases allowing a large amount of fluid to enter the turbine. The most important point to be noted in this context is that the flow at the inlet to all the runners, except the Kaplan one, is in radial and tangential directions. Therefore, the inlet velocity triangles of those turbines (Figs 15.11(a) to 15.11(d)) are shown in a plane containing, the radial and tangential directions, and hence the flow velocity V_{f1} represents the radial component of velocity.

In case of a Kaplan runner, the flow at inlet is in axial and tangential directions. Therefore, the inlet velocity triangle in this case (Fig. 15.11(e)) is shown in a plane containing the axial and tangential directions, and hence the flow velocity V_{f1} represents the axial component of velocity V_a. The tangential component of velocity is almost nil at the outlet of all runners. Therefore, the outlet velocity triangle (Fig. 15.11(f)) is identical in shape for all the runners. However, the exit velocity V_2 is axial in Kaplan and Dubs runner, while it is the radial one in all other runners.

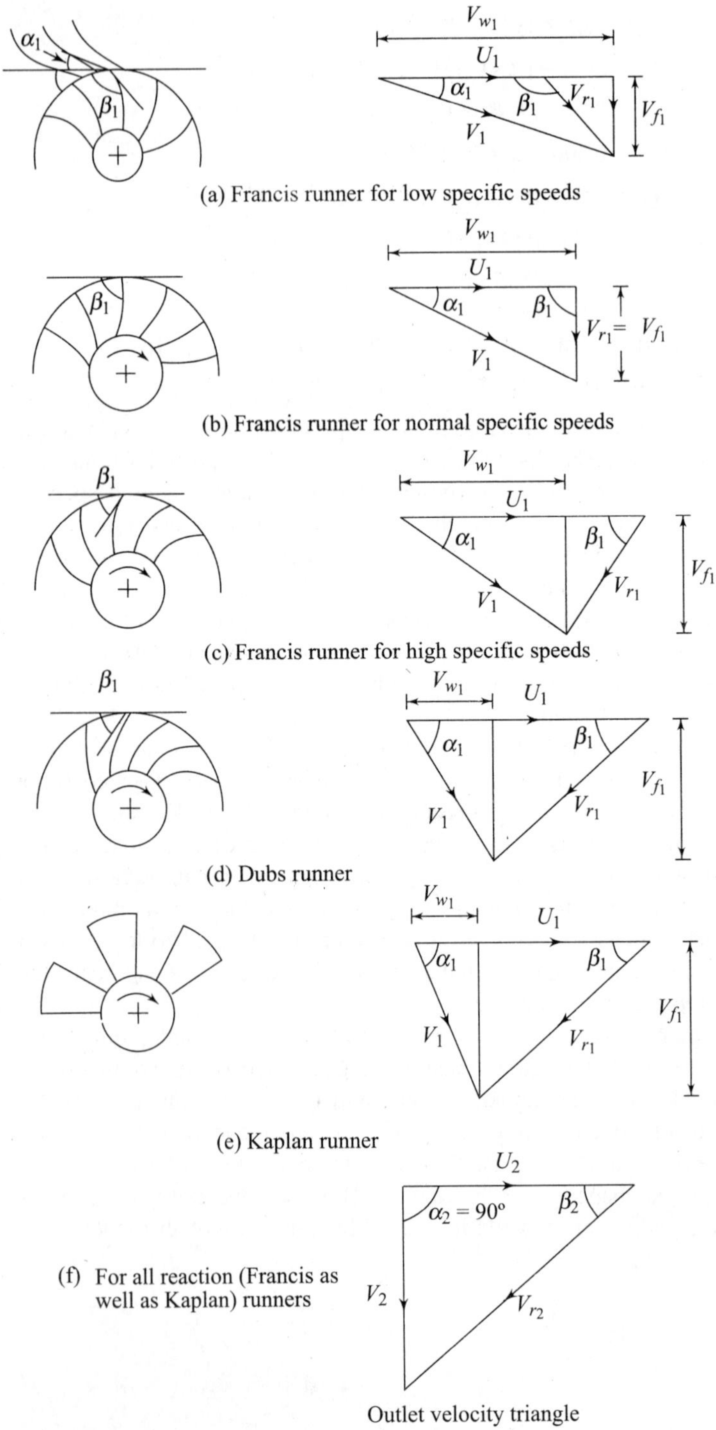

Fig. 15.11 Evolution of Kaplan runner from Francis one

Figure 15.12 shows a schematic diagram of a propeller or Kaplan turbine. The function of the guide vane is same as in case of a Francis turbine. Between the guide vanes and the runner, the fluid in a propeller turbine turns through a right-angle into the axial direction and then passes through the runner. The runner usually has four or six blades and closely resembles a ship's propeller. Neglecting the frictional effects, the flow approaching the runner blades can be considered to be a free vortex with whirl velocity being inversely proportional to the radius, while on the other hand, the blade velocity is directly proportional to the radius. To take care of this different relationship of the fluid velocity and the blade velocity with the change in radius, the blades are twisted. The angle with the axis is greater at the tip than at the root.

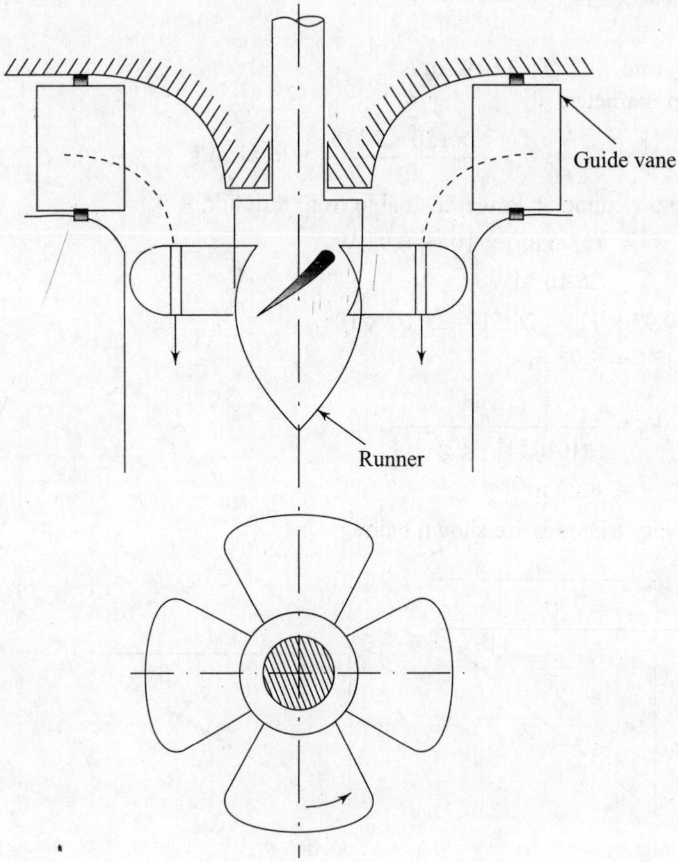

Fig. 15.12 A propeller or Kaplan turbine

Example 15.12

An axial flow hydraulic turbine has a net head of 23 m across it, and when running at a speed of 150 rpm, develops 23 MW. The blade tip and hub diameters are 4.75 and

2.0 m, respectively. If the hydraulic efficiency is 93 % and the overall efficiency 85 %, calculate the inlet and outlet blade angles at the mean radius, assuming axial flow at outlet.

Solution

Mean diameter, $d_m = (4.75+2)/2 = 3.375$ m

Power available from the fluid = (Power developed)/(Overall efficiency)

Hence, $$10^3 \times 9.81 \times 23 \times Q = \frac{23 \times 10^6}{0.85}$$

which gives the flow rate, $Q = 119.92$ m^3/s

Rotor speed at mean diameter,

$$U_m = \frac{\pi N d_m}{60} = \frac{\pi \times 150 \times 3.375}{60} = 26.51 \text{ m/s}$$

Power developed by the runner = Power available from the fluid × η_h

$$= (23/0.85) \times 10^6 \times 0.93 \text{ W}$$
$$= 25.16 \text{ MW}$$

Therefore, $10^3 \times 119.92 \times V_{w_1} \times 26.51 = 25.16 \times 10^6$

which gives $V_{w_1} = 7.92$ m/s.

Axial velocity, $$V_a = \frac{119.92}{\pi[(4.75)^2 - (2)^2]/4}$$
$$= 8.22 \text{ m/s}$$

Inlet and outlet velocity triangles are shown below:

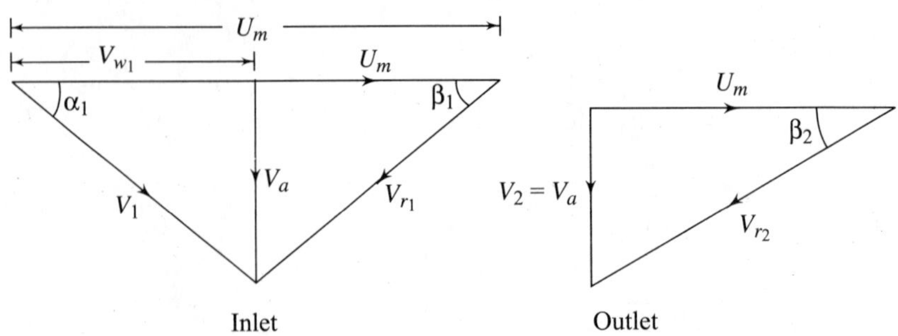

For the inlet velocity triangle,

$$\tan \beta_1 = \frac{V_a}{U_m - V_{w_1}} = \frac{8.22}{26.51 - 7.92}$$

which gives $\beta_1 = 23.85°$

At outlet, $\tan \beta_2 = V_a/U_m = 8.22/26.51$

which gives $\beta_2 = 17.23°$

Different Types of Draft Tubes incorporated in Reaction Turbines The draft tube is an integral part of a reaction turbine. Its principle has been explained earlier. The shape of the draft tube plays an important role especially for high specific speed turbines, since the efficient recovery of kinetic energy at the runner outlet depends mainly on it. Typical draft tubes, employed in practice, are discussed as follows.

Straight Divergent Tube (Fig. 15.13(a)) The shape of this tube is that of a frustrum of a cone. It is usually employed for low specific speed, vertical shaft Francis turbine. The cone angle is restricted to 8° to avoid the losses due to separation. The tube must discharge sufficiently low under tail water level. The maximum efficiency of this type of draft tube is 90%. This type of draft tube improves speed regulation on falling load.

Simple Elbow Type (Fig. 15.13(b)) The vertical length of the draft tube should be made small in order to keep down the cost of excavation, particularly in rock. The exit diameter of the draft tube should be as large as possible to recover kinetic energy at the runner's outlet. The cone angle of the tube is again fixed from the consideration of losses due to flow separation. Therefore, the draft tube must be bent to keep its definite length. Simple-elbow-type draft tube will serve such a purpose. Its efficiency is, however, low (about 60%). This type of draft tube turns the water from the vertical to the horizontal direction with a minimum depth of excavation. Sometimes, the transition from a circular section in the vertical portion to a rectangular section in the horizontal part (Fig. 15.13(c)) is incorporated in the design to have a higher efficiency of the draft tube. The horizontal portion of the draft tube is generally inclined upwards to lead the water gradually to the level of the tail race and to prevent entry of air from the exit end.

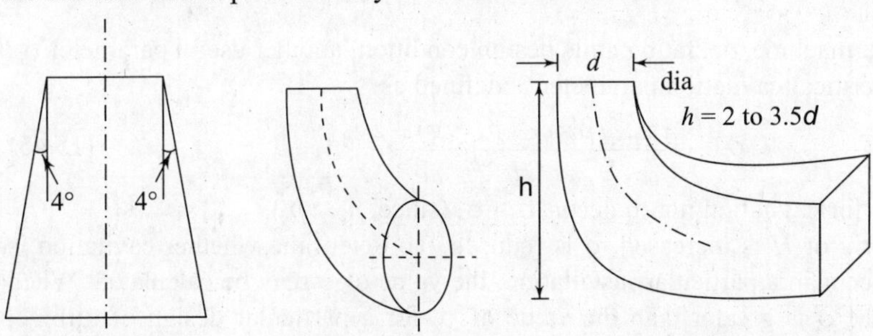

(a) Straight type (b) Simple elbow type (c) Elbow type with varying cross-section

Fig. 15.13 Different types of draft tubes

Cavitation in Reaction Turbines The phenomenon of cavitation has already been discussed in Section 5.5 in Chapter 5. To avoid cavitation, the minimum pressure in the passage of a liquid flow, should always be more than the vapour pressure of the liquid at the working temperature. In a reaction turbine, the point of minimum pressure is usually at the outlet end of the runner blades, i.e., at the inlet to the draft tube. For the flow between such a point and the final discharge into the tail race (where the pressure is atmospheric), the Bernoulli's equation can be written, in consideration of the velocity at the discharge from draft tube to be negligibly small, as

$$\frac{p_e}{\rho g} + \frac{V_e^2}{2g} + z = \frac{p_{\text{atm}}}{\rho g} + h_f \tag{15.41}$$

where, p_e and V_e represent the static pressure and velocity of the liquid at the outlet of the runner (or at the inlet to the draft tube). The larger the value of V_e, the smaller is the value of p_e and the cavitation is more likely to occur. The term h_f in Eq. (15.41) represents the loss of head due to friction in the draft tube and z is the height of the turbine runner above the tail water surface. For cavitation not to occur $p_e > p_v$, where p_v is the vapour pressure of the liquid at the working temperature.

An important parameter in the context of cavitation is the available suction head (inclusive of both static and dynamic heads) at exit from the turbine and is usually referred to as the net positive suction head 'NPSH' which is defined as

$$\text{NPSH} = \frac{p_e}{\rho g} + \frac{V_e^2}{2g} - \frac{p_v}{\rho g} \tag{15.42}$$

With the help of Eq. (15.41) and in consideration of negligible frictional losses in the draft tube ($h_f = 0$), Eq. (15,42) can be written as

$$\text{NPSH} = \frac{p_{\text{atm}}}{\rho g} - \frac{p_v}{\rho g} - z \tag{15.43}$$

A useful design parameter σ, known as Thoma's cavitation parameter (after the German engineer Dietrich Thoma, who first introduced the concept) is defined as

$$\sigma = \frac{\text{NPSH}}{H} = \frac{(p_{\text{atm}}/\rho g) - (p_v/\rho g) - z}{H} \tag{15.44}$$

For a given machine, operating at its design condition, another useful parameter σ_c, known as critical cavitation parameter is defined as

$$\sigma_c = \frac{(p_{\text{atm}}/\rho g) - (p_e/\rho g) - z}{H} \tag{15.45}$$

Therefore, for cavitation not to occur, $\sigma > \sigma_c$ (since, $p_e > p_v$).

If either z or H is increased, σ is reduced. To determine whether cavitation is likely to occur in a particular installation, the value of σ may be calculated. When the value of σ is greater than the value of σ_c for a particular design of turbine, cavitation is not expected to occur.

In practice, the value of σ_c is used to determine the maximum elevation of the turbine above tail water surface for cavitation to be avoided. The parameter σ_c increases with an increase in the specific speed of the turbine. Hence, turbines having higher specific speed must be installed closer to the tail water level.

15.4.4 Performance Characteristics of Reaction Turbines

It is not always possible in practice, although desirable, to run a machine at its maximum efficiency due to changes in operating parameters. Therefore, it becomes important to know the performance of the machine under conditions for which the efficiency is less than the maximum. It is more useful to plot the basic dimensionless performance parameters (Fig. 15.14) as derived earlier from the similarity principles of fluid machines. Thus one set of curves, as shown in Fig. 15.14, is applicable not just to the conditions of the test, but to any machine in the same homologous series, under any altered conditions.

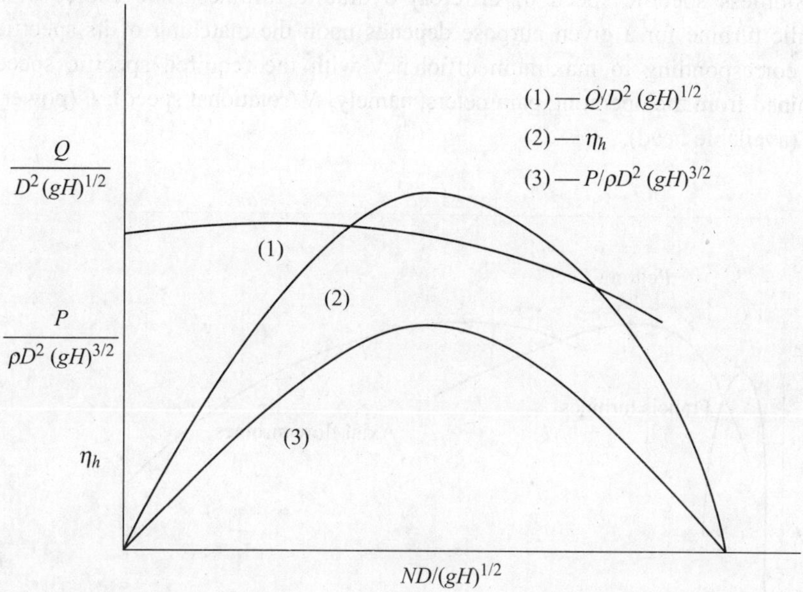

Fig. 15.14 Performance characteristics of a reaction turbine (in dimensionless parameters)

Figure 15.15 is one of the typical plots where variation in efficiency of different reaction turbines with the rated power is shown.

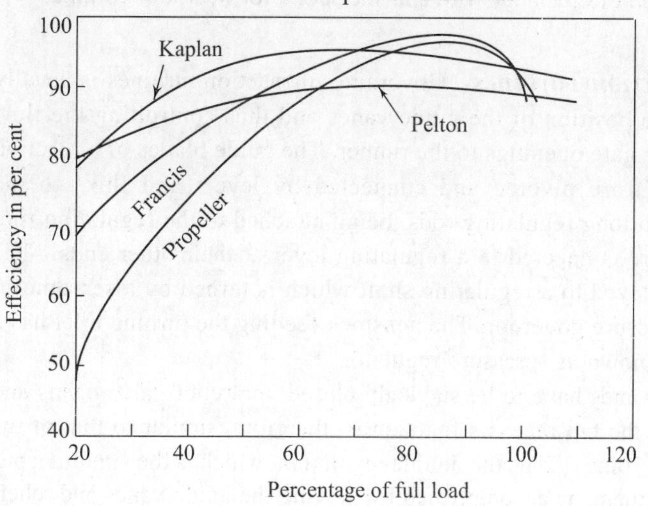

Fig. 15.15 Variation of efficiency with load

Comparison of Specific Speeds of Hydraulic Turbines Specific speeds and their ranges of variation for different types of hydraulic turbines have already been discussed earlier. Figure 15.16 shows the variation of efficiencies with the dimensionless specific speed of different hydraulic turbines. The choice of a hydraulic turbine for a given purpose depends upon the matching of its specific speed corresponding to maximum efficiency with the required specific speed determined from the operating parameters, namely, N (rotational speed), P (power) and H (available head).

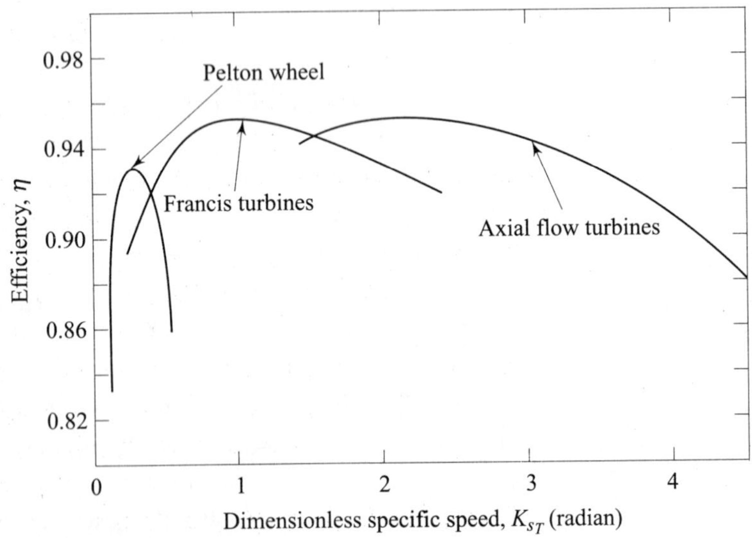

Fig. 15.16 Variation of efficiency with specific speed for hydraulic turbines

Governing of Reaction Turbines Governing of reaction turbines is usually done by altering the position of the guide vanes and thus controlling the flow rate by changing the gate openings to the runner. The guide blades of a reaction turbine (Fig. 15.17) are pivoted and connected by levers and links to the regulating ring. Two long regulating rods, being attached to the regulating ring at their one ends, are connected to a regulating lever at their other ends. The regulating lever is keyed to a regulating shaft which is turned by a servomotor piston of the oil pressure governor. The penstock feeding the turbine inlet has a relief valve better known as 'pressure regulator'.

When the guide vanes have to be suddenly closed, the relief valve opens and diverts the water to the tail race. Its function is, therefore, similar to that of the deflector in Pelton turbines. Thus the double regulation, which is the simultaneous operation of two elements is accomplished by moving the guide vanes and relief

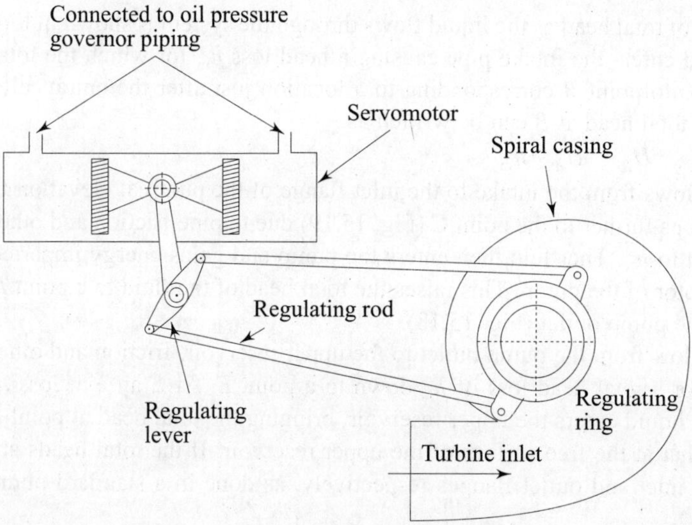

Fig. 15.17 Governing of reaction turbine

valve in Francis turbine by the governor. Double regulation system for Kaplan turbines comprises the movement of guide vanes as well as of runner vanes.

15.4.5 Rotodynamic Pumps

A rotodynamic pump is a device where mechanical energy is transferred from the rotor to the fluid by the principle of fluid motion through it. Therefore, it is essentially a turbine in reverse. Like turbines, pumps are classified according to the main direction of fluid path through them like (i) radial flow or centrifugal, (ii) axial flow and (iii) mixed flow types.

Centrifugal Pumps The centrifugal pump, by its principle, is converse of the Francis turbine. The flow is radially outward, and hence the fluid gains in centrifugal head while flowing through it. However, before considering the operation of a pump in detail, a general pumping system is discussed as follows.

General Pumping System and the Net Head Developed by a Pump The word pumping, referred to a hydraulic system commonly implies to convey liquid from a low to a high reservoir. Such a pumping system, in general, is shown in Fig. 15.18. At any point in the system, the elevation or potential head is measured from a fixed reference datum line. The total head at any point comprises the pressure head, the velocity head and the elevation head. For the lower reservoir, the total head at the free surface is H_A and is equal to the elevation of the free surface above the datum line since the velocity and static gauge pressure at A are zero. Similarly, the total head at the free surface in the higher reservoir is $(H_A + H_S)$ and is equal to the elevation of the free surface of the reservoir above the reference datum.

The variation of total head as the liquid flows through the system is shown in Fig. 15.19. The liquid enters the intake pipe causing a head loss h_{in} for which the total energy line drops to point B corresponding to a location just after the entrance to intake pipe. The total head at B can be written as

$$H_B = H_A - h_{\text{in}}$$

As the fluid flows from the intake to the inlet flange of the pump at elevation z_1 the total head drops further to the point C (Fig. 15.19) due to pipe friction and other losses equivalent to h_{f_1}. The fluid then enters the pump and gains energy imparted by the moving rotor of the pump. This raises the total head of the fluid to a point D (Fig. 15.19) at the pump outlet (Fig. 15.18).

In course of flow from the pump outlet to the upper reservoir, friction and other losses account for a total head loss of h_{f_2} down to a point E. At E an exit loss h_e occurs when the liquid enters the upper reservoir, bringing the total head at point F (Fig. 15.19) to that at the free surface of the upper reservoir. If the total heads are measured at the inlet and outlet flanges respectively, as done in a standard pump test, then

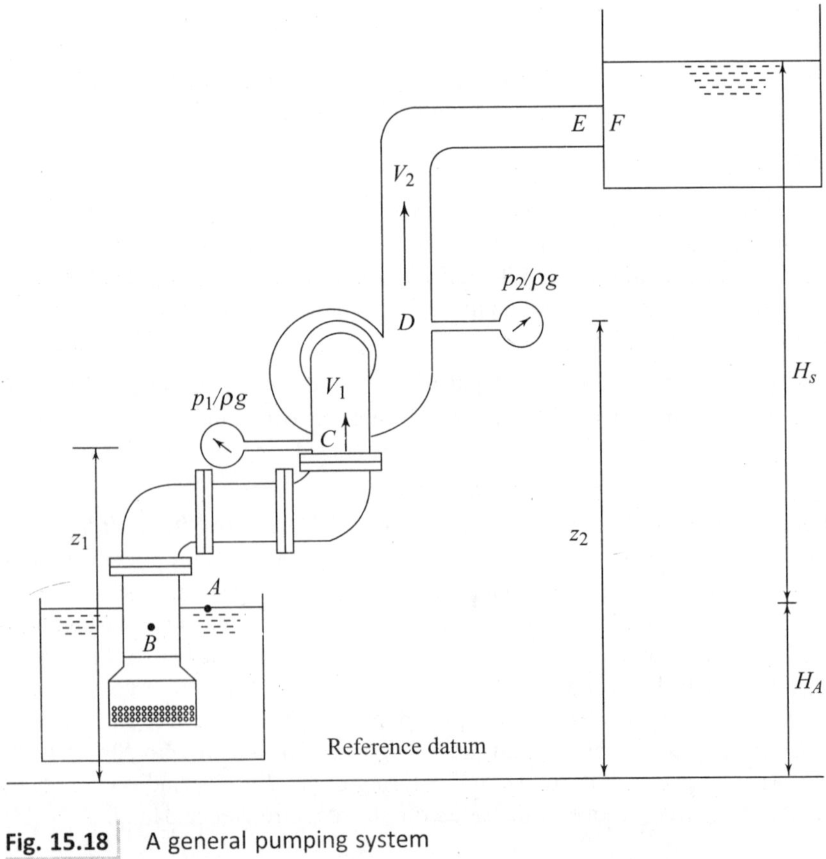

Fig. 15.18 A general pumping system

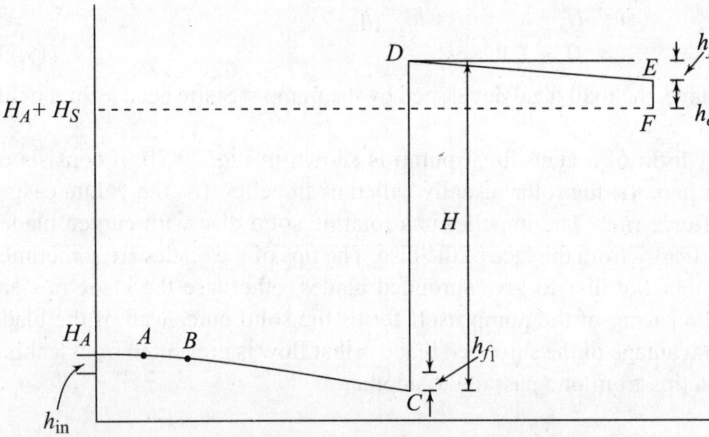

Fig. 15.19 Change of head in a pumping system

Total inlet head to the pump = $(p_1/\rho g) + (V_1^2/2g) + z_1$

Total outlet head of the pump = $(p_2/\rho g) + (V_2^2/2g) + z_2$

where V_1 and V_2 are the velocities in suction and delivery pipes respectively.

Therefore, the total head developed by the pump,

$$H = [(p_2 - p_1)/\rho g] + [(V_2^2 - V_1^2)/2g] + [z_2 - z_1] \quad (15.46)$$

The head developed H is termed as the *manometric head*. If the pipes connected to inlet and outlet of the pump are of the same diameter, $V_2 = V_1$, and therefore the head developed or manometric head H is simply the gain in Piezometric pressure head across the pump which could have been recorded by a manometer connected between the inlet and outlet flanges of the pump. In practice, $(z_2 - z_1)$ is so small in comparison to $(p_2 - p_1)/\rho g$ that it is ignored. It is therefore not surprising to find that the static pressure head across the pump is often used to describe the total head developed by the pump. The vertical distance between the two levels in the reservoirs H_s is known as the static head or static lift. Relationship between H_s, the static head and H, the head developed can be found out by applying Bernoulli's equation between A and C and between D and F (Fig. 15.18) as follows:

Between A and C,

$$0 + 0 + H_A = \frac{p_1}{\rho g} + \frac{V_1^2}{2g} + z_1 + h_{in} + h_{f_1} \quad (15.47)$$

Between D and F,

$$\frac{p_2}{\rho g} + \frac{V_2^2}{2g} + z_2 = 0 + 0 + H_s + H_A + h_{f_2} + h_e \quad (15.48)$$

substituting H_A from Eq. (15.47) into Eq. (15.48), and then with the help of Eq. (15.46), we can write

$$\begin{aligned} H &= H_s + h_{in} + h_{f_1} + h_{f_2} + h_e \\ &= H_s + \Sigma \text{ losses} \end{aligned} \tag{15.49}$$

Therefore, we have, the total head developed by the pump = Static head + Sum of all the losses.

The simplest form of a centrifugal pump is shown in Fig. 15.20. It consists of three important parts (i) the rotor, usually called as impeller, (ii) the volute casing and (iii) the diffuser ring. The impeller is a rotating solid disc with curved blades standing out vertically from the face of the disc. The tips of the blades are sometimes covered by another flat disc to give shrouded blades, otherwise the blade tips are left open and the casing of the pump itself forms the solid outer wall of the blade passages. The advantage of the shrouded blade is that flow is prevented from leaking across the blade tips from one passage to another.

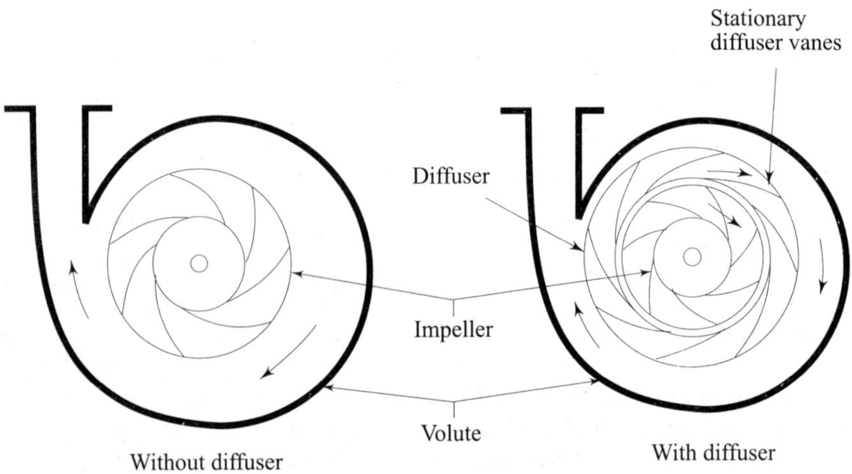

Fig. 15.20 A centrifugal pump

As the impeller rotates, the fluid is drawn into the blade passage at the impeller eye, the centre of the impeller. The inlet pipe is axial and therefore fluid enters the impeller with very little whirl or tangential component of velocity and flows outwards in the direction of the blades. The fluid receives energy from the impeller while flowing through it and is discharged with increased pressure and velocity into the casing. To convert the kinetic energy of fluid at the impeller outlet gradually into pressure energy, diffuser blades mounted on a diffuser ring are used.

The stationary blade passages so formed have an increasing cross-sectional area which reduces the flow velocity and hence increases the static pressure of the fluid. Finally, the fluid moves from the diffuser blades into the volute casing which is a passage of gradually increasing cross section and also serves to reduce the velocity of fluid and to convert some of the velocity head into the static head. Sometimes pumps have only volute casing without any diffuser.

Figure 15.21 shows an impeller of a centrifugal pump with the velocity triangles drawn at the inlet and the outlet. The blades are curved between the inlet and outlet radius. A particle of fluid moves along the broken curve shown in Fig.15.21.

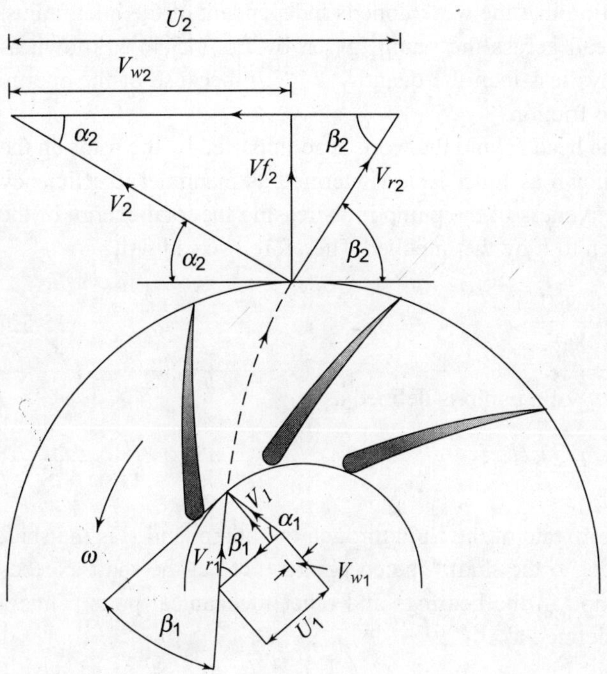

Fig. 15.21 Velocity triangles for centrifugal pump impeller

Let β_1 be the angle made by the blade at the inlet with the tangent to the inlet radius, while β_2 is the blade angle with the tangent at the outlet. V_1 and V_2 are the absolute velocities of fluid at inlet and outlet respectively, while V_{r_1} and V_{r_2} are the relative velocities (with respect to blade velocity) at inlet and outlet respectively. Therefore, according to Eq. (15.3),

Work done on the fluid per unit weight = $(V_{w_2} U_2 - V_{w_1} U_1)/g$ (15.50)

A centrifugal pump rarely has any sort of guide vanes at the inlet. The fluid therefore approaches the impeller without an appreciable whirl and so the inlet angle of the blades is designed to produce a right-angled velocity triangle at the inlet (as shown in Fig. 15.21). At conditions other than those for which the impeller was designed, the direction of relative velocity V_r does not coincide with that of a blade. Consequently, the fluid changes direction abruptly on entering the impeller. In addition, the eddies give rise to some back flow into the inlet pipe, thus causing fluid to have some whirl before entering the impeller. However, considering the operation under design conditions, the inlet whirl velocity

V_{w_1} and accordingly the inlet angular momentum of the fluid entering the impeller is set to zero. Therefore, Eq. (15.50) can be written as

Work done on the fluid per unit weight = $V_{w_2} U_2/g$ (15.51)

We see from this equation that the work done is independent of the inlet radius. The difference in total head across the pump [given by Eq. (15.46)], known as manometric head, is always less than the quantity $V_{w_2}U_2/g$ because of the energy dissipated in eddies due to friction.

The ratio of manometric head H, and the work head imparted by the rotor on the fluid $V_{w_2} U_2/g$ (usually known as Euler head) is termed as manometric efficiency η_m. It represents the effectiveness of the pump in increasing the total energy of the fluid from the energy given to it by the impeller. Therefore, we can write

$$\eta_m = \frac{gH}{V_{w_2}U_2} \tag{15.52}$$

The overall efficiency η_0 of a pump is defined as

$$\eta_0 = \frac{\rho Q g H}{P} \tag{15.53}$$

where, Q is the volume flow rate of the fluid through the pump, and P is the shaft power, i.e., the input power to the shaft. The energy required at the shaft exceeds $V_{w_2} U_2/g$ because of friction in the bearings and other mechanical parts. Thus a mechanical efficiency is defined as

$$\eta_{\text{mech}} = \frac{\rho Q V_{w_2} U_2}{P} \tag{15.54}$$

so that,
$$\eta_0 = \eta_m \times \eta_{\text{mech}} \tag{15.55}$$

Slip Factor Under certain circumstances, the angle at which the fluid leaves the impeller may not be the same as the actual blade angle. This is due a phenomenon known as fluid slip, which finally results in a reduction in V_{w2} the tangential component of fluid velocity at impeller outlet. One possible explanation for slip is given as follows.

In course of flow through the impeller passage, there occurs a difference in pressure and velocity between the leading and trailing faces of the impeller blades. On the leading face of a blade there is relatively a high pressure and low velocity, while on the trailing face, the pressure is lower and hence the velocity is higher. This results in a circulation around the blade and a non-uniform velocity distribution at any radius. The mean direction of flow at outlet, under this situation, changes from the blade angle at outlet β_2 to a different angle β_2' as shown in Fig. 15.22. Therefore the tangential velocity component at outlet V_{w_2} is reduced to V'_{w_2}, as shown by the velocity triangles in Fig. 15.22, and the difference ΔV_w is defined as the slip. The slip factor σ_s is defined as

$$\sigma_s = V'_{w_2}/V_{w_2}$$

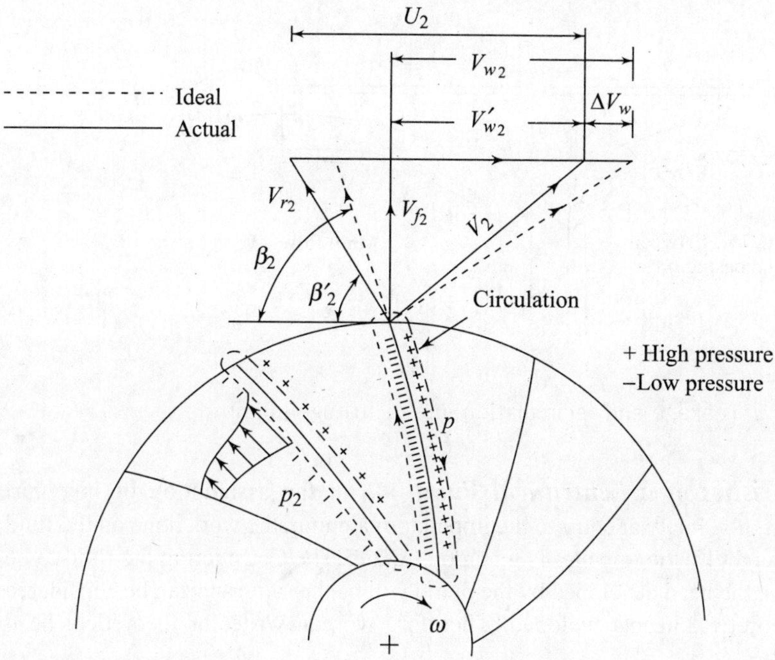

Fig. 15.22 Slip and velocity distribution in the impeller blade passage of a centrifugal pump

With the application of slip factor σ_s, the work head imparted to the fluid (Euler head) becomes $\sigma_s V_{w_2} U_2/g$. The typical values of slip factor lie in the region of 0.9.

Losses in a Centrifugal Pump It has been mentioned earlier that the shaft power P or energy that is supplied to the pump by the prime mover is not the same as the energy received by the liquid. Some energy is dissipated as the liquid passes through the machine. The losses can be divided into different categories as follows:

(a) Mechanical friction power loss due to friction between the fixed and rotating parts in the bearing and stuffing boxes.

(b) Disc friction power loss due to friction between the rotating faces of the impeller (or disc) and the liquid.

(c) Leakage and recirculation power loss. This is due to loss of liquid from the pump and recirculation of the liquid in the impeller. The pressure difference between impeller tip and eye can cause a recirculation of a small volume of liquid, thus reducing the flow rate at outlet of the impeller as shown in Fig. 15.23.

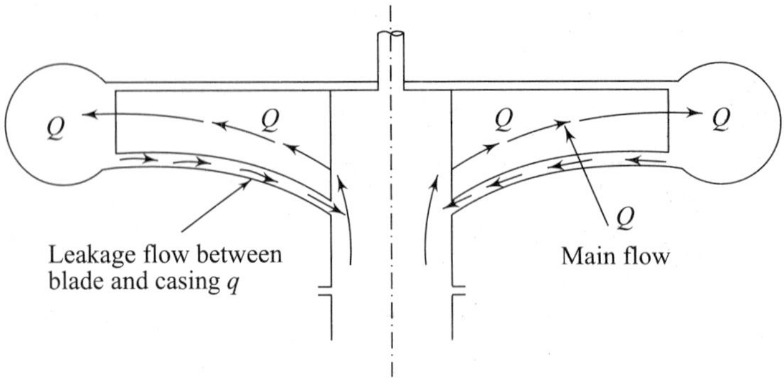

Fig. 15.23 Leakage and recirculation in a centrifugal pump

Characteristics of a Centrifugal Pump With the assumption of no whirl component of velocity at entry to the impeller of a pump, the work done on the fluid per unit weight by the impeller is given by Eq. (15.51). Considering the fluid to be frictionless, the head developed by the pump will be the same and can be considered as the theoretical head developed. Therefore we can write for theoretical head developed H_{theo} as

$$H_{\text{theo}} = \frac{V_{w_2} U_2}{g} \tag{15.56}$$

From the outlet velocity triangle (Fig. 15.21).

$$V_{w_2} = U_2 - V_{f_2} \cot \beta_2 = U_2 - (Q/A) \cot \beta_2 \tag{15.57}$$

where Q is rate of flow at impleller outlet and A is the flow area at the periphery of the impeller. The blade speed at outlet U_2 can be expressed in terms of rotational speed of the impeller N as

$$U_2 = \pi D N$$

Using this relation and the relation given by Eq. (15.57), the expression of theoretical head developed can be written from Eq. (15.56) as

$$\begin{aligned} H_{\text{theo}} &= \frac{\pi^2 D^2 N^2}{g} - \left[\frac{\pi DN}{gA} \cot \beta_2 \right] Q \\ &= K_1 - K_2 Q \end{aligned} \tag{15.58}$$

where, $K_1 = \dfrac{\pi^2 D^2 N^2}{g}$ and $K_2 = (\pi DN/gA) \cot \beta_2$

For a given impeller running at a constant rotational speed. K_1 and K_2 are constants, and therefore head and discharge bears a linear relationship as shown by Eq. (15.58). This linear variation of H_{theo} with Q is plotted as curve I in Fig. 15.24.

If slip is taken into account, the theoretical head will be reduced to $\sigma_s V_{w_2} U_2/g$. Moreover the slip will increase with the increase in flow rate Q. The effect of slip in head-dicharge relationship is shown by the curve II in Fig. 15.24. The loss due to slip can occur in both a real and an ideal fluid, but in a real fluid the shock losses at entry to the blades, and the friction losses in the flow passages have to be considered. At the design point the shock losses are zero since the fluid moves tangentially onto the blade, but on either side of the design point the head loss due to shock increases according to the relation

$$h_{\text{shock}} = K_3 (Q_f - Q)^2 \tag{15.59}$$

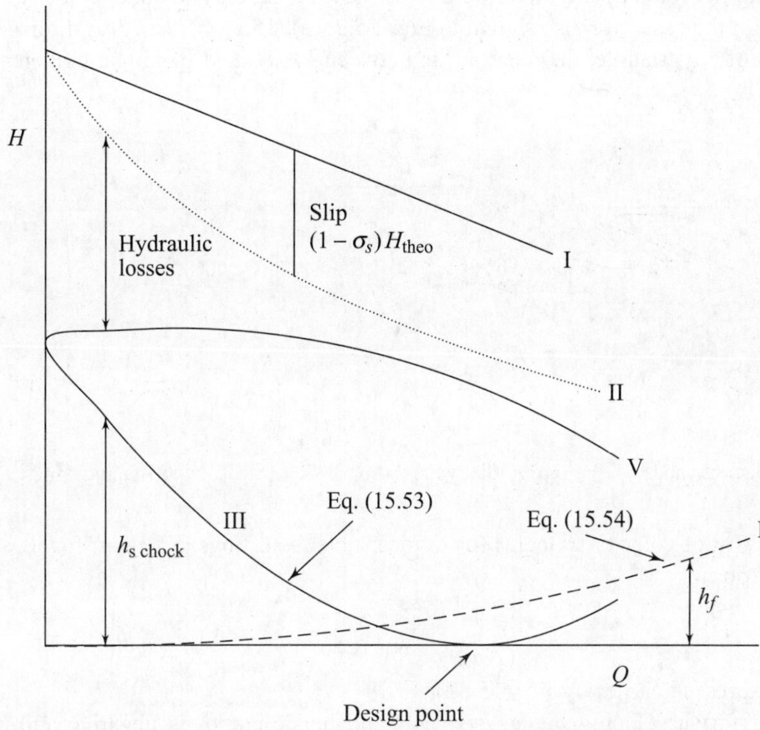

Fig. 15.24 Head-discharge characteristics of a centrifugal pump

where Q_f is the off design flow rate and K_3 is a constant. The losses due to friction can usually be expressed as

$$h_f = K_4 Q^2 \tag{15.60}$$

where, K_4 is a constant.

Equations (15.59) and (15.60) are also shown in Fig. 15.24 (curves III and IV) as the characteristics of losses in a centrifugal pump. By substracting the sum of the losses from the head in consideration of the slip, at any flow rate (by substracting the sum of ordinates of the curves III and IV from the ordinate of the curve II at all values of the abscissa), we get the curve V which represents the relationship of the actual head with the flow rate, and is known as the head-discharge characteristic curve of the pump.

Effect of Blade Outlet Angle The head-discharge characteristic of a centrifugal pump depends (among other things) on the outlet angle of the impeller blades which in turn depends on the blade settings. Three types of blade settings are possible (i) the forward facing for which the blade curvature is in the direction of rotation and, therefore, $\beta_2 > 90°$ (Fig. 15.25a), (ii) radial, when $\beta_2 = 90°$ (Fig. 15.25b), and (iii) backward facing for which the blade curvature is in a direction opposite to that of the impeller rotation and therefore $\beta_2 < 90°$ (Fig. 15.25c). The outlet velocity triangles for all the cases are also shown in Figs 15.25(a), 15.25(b), 15.25(c). From the geometry of any triangle, the relationship between V_w, U_2 and β_2 can be written as

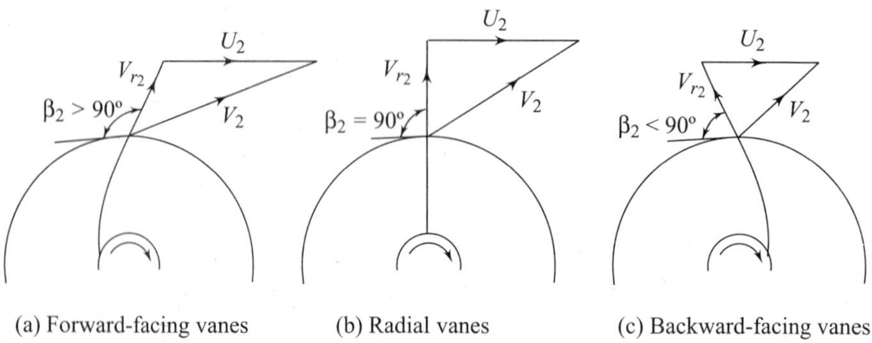

Fig. 15.25 Outlet velocity triangles for different blade settings in a centrifugal pump

$$V_{w_2} = U_2 - V_{f_2} \cot \beta_2$$

which was expressed earlier by Eq. (15.51).

In case of forward facing blade, $\beta_2 > 90°$ and hence cot β_2 is negative and therefore V_{w_2} is more than U_2. In case of radial blade, $\beta_2 = 90°$ and $V_{w_2} = U_2$. In case of backward facing blade, $\beta_2 < 90°$ and $V_{w_2} < U_2$. Therefore the sign of K_2, the constant in the theoretical head-discharge relationship given by the Eq. (15.58), depends accordingly on the type of blade setting as follows:

For forward curved blades, $K_2 < 0$

For radial blades, $K_2 = 0$

For backward curved blades, $K_2 > 0$

With the incorporation of above conditions, the relationship of head and discharge for three cases are shown in Fig. 15.26. These curves ultimately revert to their more recognised shapes as the actual head-discharge characteristics respectively after consideration of all the losses as explained earlier (Fig. 15.27).

For both radial and forward facing blades, the power is rising monotonically as the flow rate is increased. In the case of backward facing blades, the maximum efficiency occurs in the region of maximum power. If, for some reasons, Q increases beyond Q_D there occurs a decrease in power. Therefore the motor used to drive the pump at part load, but rated at the design point, may be safely used at the maximum power. This is known as self-limiting characteristic. In case of radial and forward-facing blades, if the pump motor is rated for maximum power, then it will be under-utilised most of the time, resulting in an increased cost for the extra rating. Whereas, if a smaller motor is employed, rated at the design point, then if Q increases above Q_D, the motor will be overloaded and may fail. It, therefore, becomes more difficult to decide on a choice of motor in these later cases (radial and forward-facing blades).

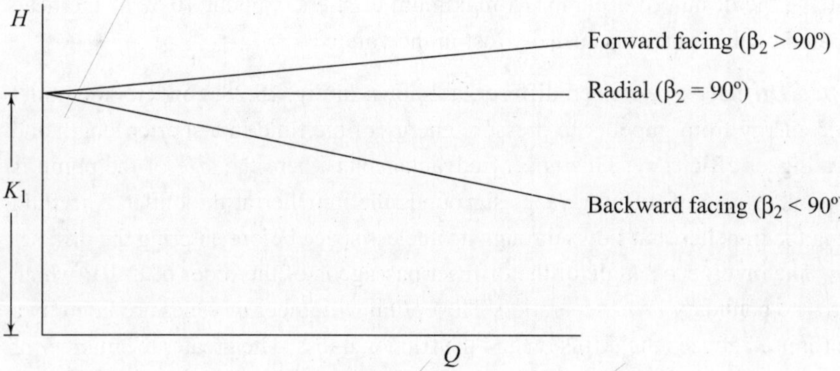

Fig. 15.26 Theoretical head-discharge characteristic curves of a centrifugal pump for different blade settings

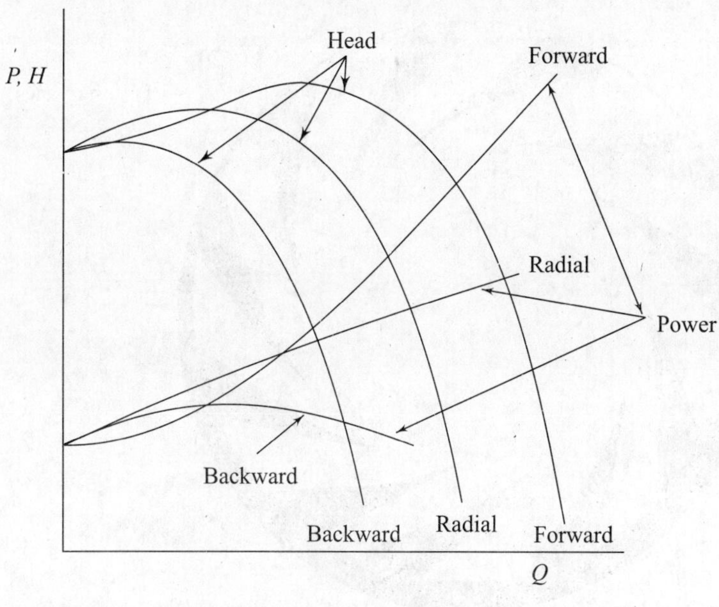

Fig. 15.27 Actual head-discharge and power-discharge characteristic curves of a centrifugal pump

Flow through Volute Chambers Apart from frictional effects, no torque is applied to a fluid particle once it has left the impeller. The angular momentum of fluid is therefore constant if friction is neglected. Thus the fluid particles follow the path of a free vortex. In an ideal case, the radial velocity at the impeller outlet remains constant round the circumference. The combination of uniform radial velocity with the free vortex ($V_w . r$ = constant) gives a pattern of spiral streamlines which should be matched by the shape of the volute. This is the most important feature of the design of a pump. At maximum efficiency, about 10 % of the head generated by the impeller is usually lost in the volute.

Vanned Diffuser A vanned diffuser, as shown in Fig. 15.28, converts the outlet kinetic energy from impeller to pressure energy of the fluid in a shorter length and with a higher efficiency. This is very advantageous where the size of the pump is important. A ring of diffuser vanes surrounds the impeller at the outlet. The fluid leaving the impeller first flows through a vaneless space before entering the diffuser vanes. The divergence angle of the diffuser passage is of the order of 8–10° which ensures no boundary layer separation. The optimum number of vanes are fixed by a compromise between the diffusion and the frictional loss. The greater the number of vanes, the better is the diffusion (rise in static pressure by the reduction in flow velocity) but greater is the frictional loss. The number of diffuser vanes should have no common factor with the number of impeller vanes to prevent resonant vibration.

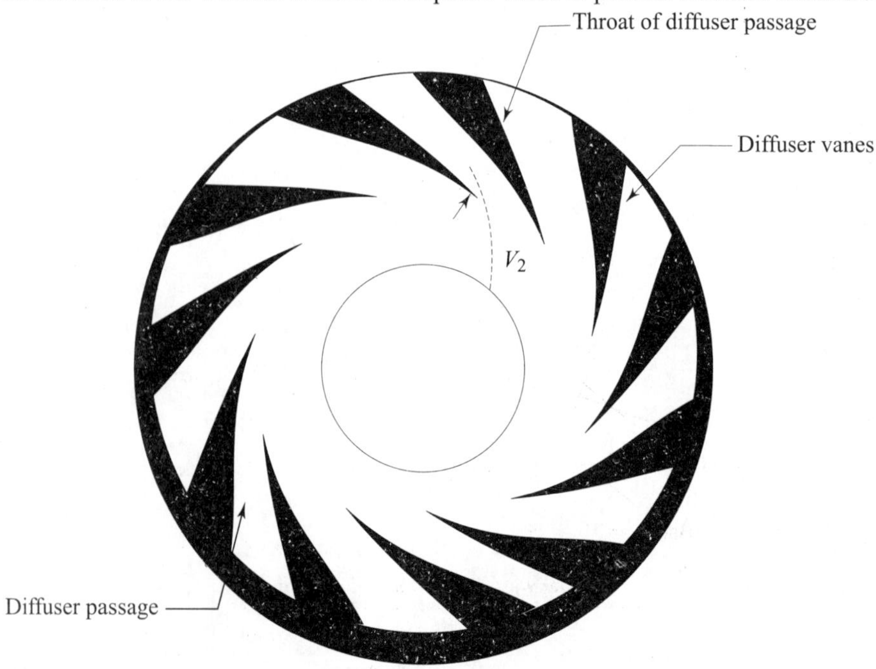

Fig. 15.28 A vanned diffuser of a centrifugal pump

Cavitation in Centrifugal Pumps Cavitation is likely to occur at the inlet to the pump, since the pressure there is the minimum and is lower than the atmospheric pressure by an amount that equals the vertical height above which the pump is situated from the supply reservoir (known as sump) plus the velocity head and frictional losses in the suction pipe. Applying the Bernoulli's equation between the surface of the liquid in the sump and the entry to the impeller, we have

$$\frac{p_i}{\rho g} + \frac{V_i^2}{2g} + z = \frac{p_A}{\rho g} - h_f \tag{15.61}$$

where, p_i is the pressure at the impeller inlet and p_A is the pressure at the liquid surface in the sump which is usually the atmospheric pressure, z is the vertical height of the impeller inlet from the liquid surface in the sump, h_f is the loss of head in the suction pipe. Strainers and non-return valves are commonly fitted to intake pipes. The term h_f must therefore include the losses occurring past these devices, in addition to losses caused by pipe friction and by bends in the pipe.

In a similar way as described in case of a reaction turbine, the net positive suction head 'NPSH' in case of a pump is defined as the available suction head (inclusive of both static and dynamic heads) at the pump inlet above the head corresponding to vapour pressure.
Therefore,

$$\text{NPSH} = \frac{p_i}{\rho g} + \frac{V_i^2}{2g} - \frac{p_v}{\rho g} \tag{15.62}$$

Again, with the help of Eq. (15.55), we can write

$$\text{NPSH} = \frac{p_A}{\rho g} - \frac{p_v}{\rho g} - z - h_f$$

The Thoma's cavitation parameter σ and critical cavitation parameter σ_c are defined accordingly (as done in case of reaction turbine) as

$$\sigma = \frac{\text{NPSH}}{H} = \frac{(p_A/\rho g) - (p_v/\rho g) - z - h_f}{H} \tag{15.63}$$

and
$$\sigma_c = \frac{(p_A/\rho g) - (p_i/\rho g) - z - h_f}{H} \tag{15.64}$$

We can say that for cavitation not to occur,

$$\sigma > \sigma_c \text{ (i.e., } p_i > p_v)$$

In order that σ should be as large as possible, z must be as small as possible. In some installations, it may even be necessary to set the pump below the liquid level at the sump (i.e., with a negative value of z) to avoid cavitation.

Axial Flow or Propeller Pump The axial flow or propeller pump is the converse of axial flow turbine and is very similar to it in appearance. The impeller consists of a central boss with a number of blades mounted on it. The impeller rotates within a cylindrical casing with fine clearance between the blade tips and the casing walls. Fluid particles, in course of their flow through the pump, do not change their radial locations. The inlet guide vanes are provided to properly direct the fluid to the rotor. The outlet guide vanes are provided to eliminate the whirling component of velocity at discharge. The usual number of impeller blades lies between 2 and 8, with a hub diameter to impeller diameter ratio of 0.3 to 0.6.

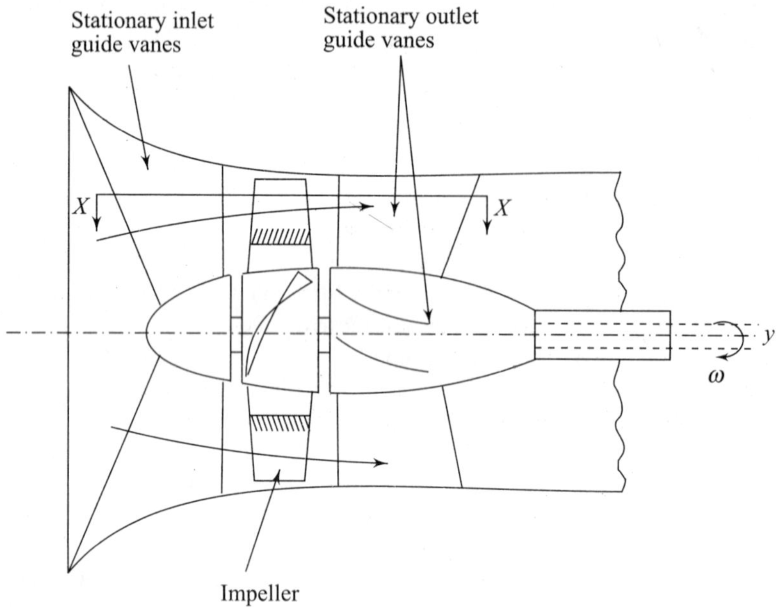

Fig. 15.29 A propeller of an axial flow pump

Matching of Pump and System Characteristics The design point of a hydraulic pump corresponds to a situation where the overall efficiency of operation is maximum. However the exact operating point of a pump, in practice, is determined from the matching of the pump characteristic with the headloss-flow, characteristic of the external system (i.e., pipe network, valve and so on) to which the pump is connected.

Let us consider the pump and the piping system as shown in Fig. 15.18. Since the flow is highly turbulent, the losses in pipe system are proportional to the square of flow velocities and can, therefore, be expressed in terms of constant loss coefficients. Therefore, the losses in both the suction and delivery sides can be written as

$$h_1 = fl_1 V_1^2/2gd_1 + K_1 V_1^2/2g \tag{15.65a}$$

$$h_2 = fl_2 V_2^2/2gd_2 + K_2 V_2^2/2g \tag{15.65b}$$

where, h_1 is the loss of head in suction side and h_2 is the loss of head in delivery side and f is the Darcy's friction factor, l_1, d_1 and l_2, d_2 are the lengths and diameters of the suction and delivery pipes respectively, while V_1 and V_2 are accordingly the average flow velocities. The first terms in Eqs (15.65a) and (15.65b) represent the major energy loss while the second terms represent the sum of all the minor losses through the loss coefficients K_1 and K_2 which include losses due to valves and pipe bends, entry and exit losses, etc. Therefore the total head the pump has to develop in order to supply the fluid from the lower to upper reservoir is

$$H = H_s + h_1 + h_2 \tag{15.66}$$

Now flow rate through the system is proportional to the flow velocity. Therefore resistance to flow in the form of losses is proportional to the square of the flow rate and is usually written as

$$h_1 + h_2 = \text{system resistance} = K\,Q^2 \tag{15.67}$$

where K is a constant which includes, the lengths and diameters of the pipes and the various loss coefficients. System resistance as expressed by Eq. (15.67), is a measure of the loss of head at any particular flow rate through the system. If any parameter in the system is changed, such as adjusting a valve opening, or inserting a new bend, etc., then K will change. Therefore, total head of Eq. (15.66) becomes,

$$H = H_s + KQ^2 \tag{15.68}$$

The head H can be considered as the total opposing head of the pumping system that must be overcome for the fluid to be pumped from the lower to the upper reservoir.

The Eq. (15.68) is the equation for system characteristic, and while plotted on H–Q plane (Fig. 15.30), represents the system characteristic curve. The point of interesection between the system characteristic and the pump characteristic on H–Q plane is the operating point which may or may not lie at the design point that corresponds to maximum efficiency of the pump. The closeness of the operating and design points depends on how good an estimate of the expected system losses has been made. It should be noted that if there is no rise in static head of the liquid (for example, pumping in a horizontal pipeline between two reservoirs at the same elevation), H_s is zero and the system curve passes through the origin.

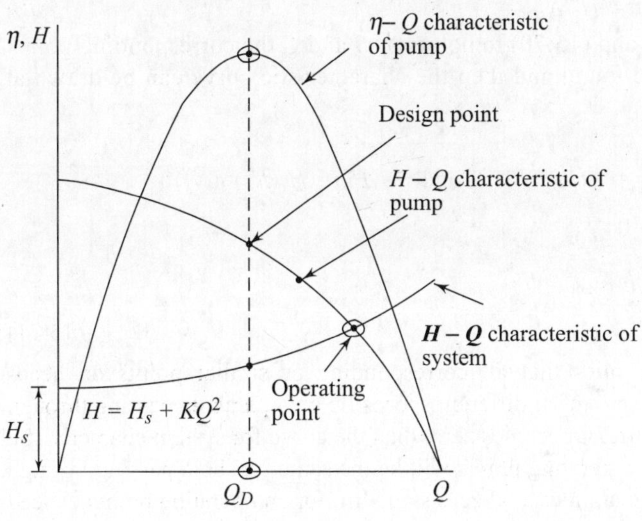

Fig. 15.30 H-Q Characteristics of pump and system

Effect of Speed Variation Head-discharge characteristic of a given pump is always referred to a constant speed. If such characteristic at one speed is known, it is possible to predict the characteristic at other speeds by using the principle of similarity. Let A, B, C are three points on the characteristic curve (Fig. 15.31) at speed N_1.

For points A, B and C, the corresponding heads and flows at a new speed N_2 are found as follows:

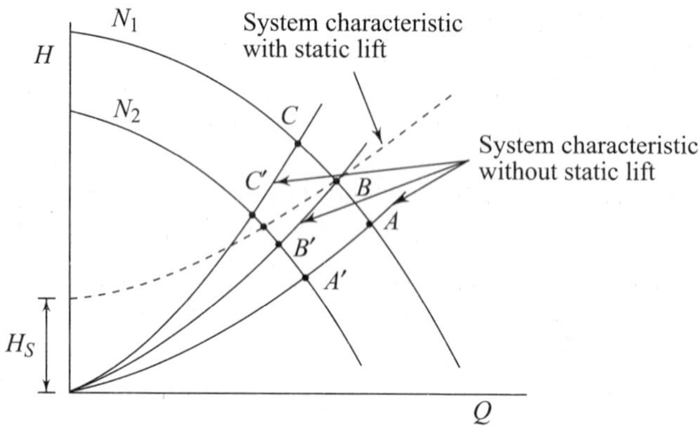

Fig. 15.31 Effect of speed variation on operating point of a centrifugal pump

From the equality of π_1 terms [Eq. (15.14)] gives

$$Q_1/N_1 = Q_2/N_2 \text{ (since for a given pump } D \text{ is constant)} \quad (15.69)$$

and similarly, equality of π_2 terms [Eq. (15.14)] gives

$$H_1/N_1^2 = H_2/N_2^2 \quad (15.70)$$

Applying Eqs (15.69) and (15.70) to points A, B and C the corresponding points A' B' and C' are found and then the characteristic curve can be drawn at the new speed N_2

Thus,

$$Q_2 = Q_1N_2/N_1 \quad \text{and} \quad H_2 = H_1(N_2)^2/(N_1)^2$$

which gives

$$\frac{H_2}{H_1} = \frac{Q_2^2}{Q_1^2}$$

or

$$H \propto Q^2 \quad (15.71)$$

Equation (15.71) implies that all corresponding or similar points on head-discharge characteristic curves at different speeds lie on a parabola passing through the origin. If the static lift H_s becomes zero, then the curve for system characteristic and the locus of similar operating points will be the same parabola passing through the origin. This means that, in case of zero static lift, for an operating point at speed N_1, it is only necessary to apply the similarity laws directly to find the corresponding operating point at the new speed since it will lie on the system curve itself (Fig. 15.31).

Variation of Pump Diameter A variation in pump diameter may also be examined through the similarity laws. For a constant speed,

$$Q_1/D_1^3 = Q_2/D_2^3$$

and

$$H_1/D_1^2 = H_2/D_2^2$$

or

$$H \propto Q^{2/3} \tag{15.72}$$

Pumps in Series and Parallel When the head or flow rate of a single pump is not sufficient for an application, pumps are combined in series or in parallel to meet the desired requirement. Pumps are combined in series to obtain an increase in head or in parallel for an increase in flow rate. The combined pumps need not be of the same design.

Figures 15.32 and 15.33 depict the combined H–Q characteristic for the cases of identical pumps connected in series and parallel respectively. It is found that the operating point changes in both cases. Figure 15.34 shows the combined characteristics of two different pumps connected in series and parallel.

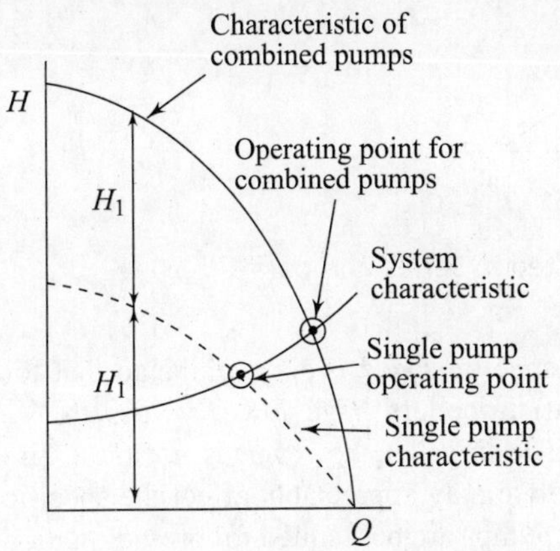

Fig. 15.32 Two similar pumps connected in series

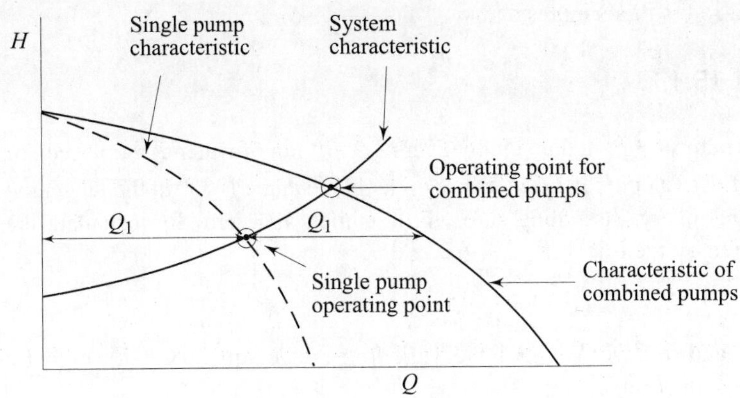

Fig. 15.33 Two similar pumps connected in parallel

Specific Speed of Centrifugal Pumps The concept of specific speed for a pump is same as that for a turbine. However, the quantities of interest are N, H and Q rather than N, H and P, like in the case of a turbines.
For pumps,

$$N_{s_p} = N\, Q^{1/2}/H^{3/4} \tag{15.73}$$

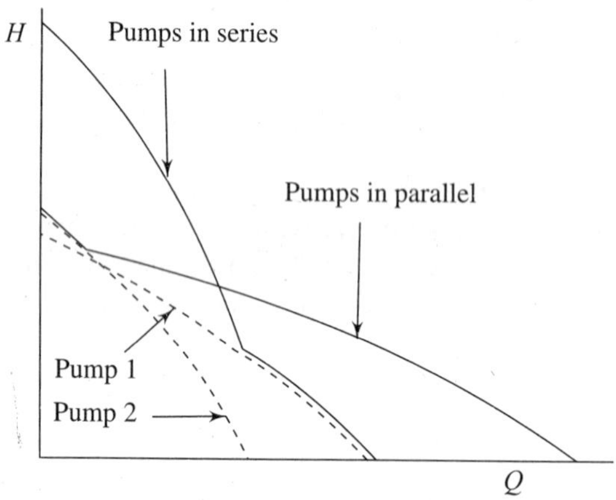

Fig. 15.34 Two different pumps connected in series and parallel

The effect of the shape of the rotor on specific speed is also similar to that for turbines. That is, radial flow (centrifugal) impellers have lower values of N_{s_p} compared to those of axial-flow designs. The impeller, however, is not the entire pump and in particular, the shape of the volute may appreciably affect the specific speed. Nevertheless, in general, centrifugal pumps are best suited for providing high heads at moderate rates of flow as compared to axial flow pumps which are suitable for large rates of flow at low heads. Similar to turbines, the higher the specific speed, the more compact is the machine for given requirements. For multistage pumps, the specific speed refers to a single stage.

Example 15.13

A centrifugal pump 1.3 m in diameter delivers 3.5 m^3/min of water at a tip speed of 10 m/s and a flow velocity of 1.6 m/s. The outlet blade angle is 30° to the tangent at the impeller periphery. Assuming zero whirl at inlet, and zero slip, calculate the torque delivered by the impeller.

Solution

With zero slip and zero whirl at inlet, the work done on the liquid per unit weight by the impeller can be written as

$$W = V_{w_2}\, U_2/g$$

Therefore, power supplied, $P = \rho Q V_{w_2} U_2$
(subscript 2 represents the outlet)
From the outlet velocity triangle shown below:

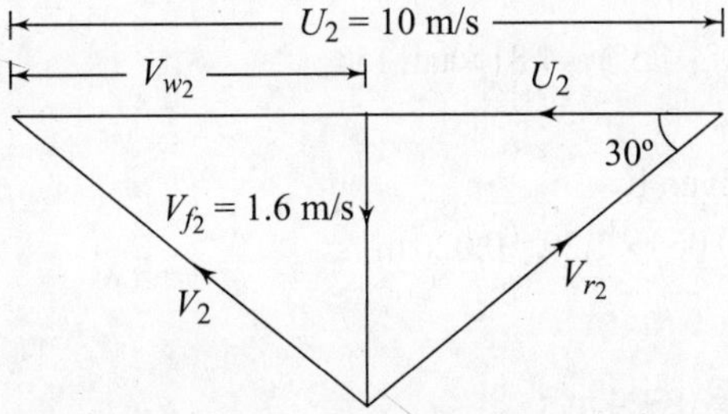

$$V_{w_2} = 10 - \frac{16}{\tan 30°} = 7.23 \text{ m}$$

Hence,
$$P = 10^3 \times \frac{3.5}{60} \times 7.23 \times 10 = 4217.5 \text{ W}$$

$$\text{Torque delivered} = \frac{\text{Power}}{\text{Angular velocity}} = \frac{4217.5 \times 0.65}{10}$$
$$= 274.14 \text{ Nm}$$

Example 15.14

An impeller with an eye radius of 51 mm and an outside diameter of 406 mm rotates at 900 rpm. The inlet and outlet blade angles measured from the radial flow direction are 75° and 83° respectively, while the depth of blade is 64 mm.

Assuming zero inlet whirl, zero slip and an hydraulic efficiency of 89%, calculate (i) the volume flow rate through the impeller, (ii) the stagnation and static pressure rise across the impeller, (iii) the power transferred to the fluid, and (iv) the input power to the impeller.

Solution

The inlet and outlet velocity triangles are shown below:

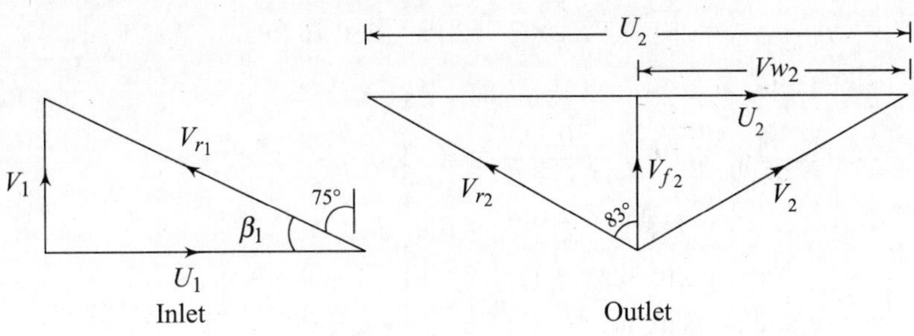

(i) At the inlet, the impeller blade velocity is

$$U_1 = \left(\frac{900 \times 2\pi}{60}\right) \times 0.051 = 4.81 \text{ m/s}$$

$$\tan \beta_1 = V_1/U_1$$

$$V_1 = 4.81 \times \tan(90° - 75°) = 4.81 \times \tan 15°$$

$$= 1.29 \text{ m/s}$$

volume flow rate through the pump is given by

$$Q = 2\pi \times 0.051 \times 0.064 \times 1.29 = 0.026 \text{ m}^3/\text{s}$$

(ii) From continuity,

$$V_{f_2} = \frac{0.051 \times 1.29}{0.203} = 0.324 \text{ m/s}$$

At the outlet, the velocity of the impeller blades is given by

$$U_2 = \left(\frac{900 \times 2\pi}{60}\right) \times 0.203 = 19.13 \text{ m/s}$$

Power transferred to the fluid per unit weight by the impeller can be written as

$$E = \frac{V_{w_2} U_2}{g} = \frac{\left(U_2 - \frac{V_{f_2}}{\tan 7°}\right) U_2}{g}$$

$$= \left(19.13 - \frac{0.324}{\tan 7°}\right) \frac{19.13}{9.81} = 32.16 \text{ m}$$

Therefore, total head developed by the pump $= H = 0.89 \times 32.16 = 28.62$ m

If the changes in potential head across the pump is neglected, the total head developed by the pump can be written as

$$H = \left[\frac{p_2 - p_1}{\rho g}\right] + \left[\frac{V_2^2 - V_1^2}{2g}\right]$$

Therefore, the rise in stagnation or total pressure becomes

$$\frac{p_{0_2} - p_{0_1}}{\rho g} = \left[\frac{p_2}{\rho g} + \frac{V_2^2}{2g}\right] - \left[\frac{p_1}{\rho g} + \frac{V_1^2}{2g}\right] = H$$

Hence, $\quad p_{02} - p_{01} = 10^3 \times 9.81 \times 28.62 \text{ Pa} = 280.76 \text{ kPa}.$

At the impeller exit,

$$V_{w_2} = 19.13 - \frac{0.324}{\tan 7°} = 16.49 \text{ m/s}$$

Therefore, $\quad V_2 = [V_{f_2}^2 + V_{w_2}^2]^{1/2}$

$$= [(0.324)^2 + (16.49)^2]^{1/2} = 16.49 \text{ m/s}$$

Solving for the static pressure head,

$$\frac{p_2 - p_1}{\rho g} = H - \left[\frac{V_2^2 - V_1^2}{2g}\right]$$

$$= 28.62 - \left[\frac{(16.49)^2 - (1.29)^2}{2 \times 9.81}\right] = 14.84 \text{ m}$$

$$p_2 - p_1 = 10^3 \times 9.81 \times 14.89 \text{ Pa} = 145.58 \text{ kPa}$$

(iii) Power given to the fluid $= \rho g Q H$

$$= 10^3 \times 9.81 \times 0.026 \times 28.62 \text{ W} = 7.30 \text{ kW}$$

(iv) Input power to the impeller = 7.30/0.89 = 8.20 kW.

Example 15.15

The basic design of a centrifugal pump has a dimensionless specific speed of 0.075 rev. The blades are forward facing on the impeller and the outlet angle is 120° to the tangent, with an impeller passage width at the outlet being equal to one-tenth of the diameter. The pump is to be used to raise water through a vertical distance of 35 m at a flow rate of 0.04 m^3/s. The suction and delivery pipes are each of 150 mm diameter and have a combined length of 40 m with a friction factor of 0.005. Other losses at the pipe entry, exit, bends, etc. are three times the velocity head in the pipes. If the blades occupy 6 % of the circumferential area and the hydraulic efficiency (neglecting slip) is 76 %, what will be the diameter of the pump impeller?

Solution

Velocity in the pipes $v = \dfrac{0.04 \times 4}{\pi \times (0.15)^2} = 2.26$ m/s

Total losses in the pipe

$$h_1 = \frac{4fl}{2gd}v^2 + \frac{3v^2}{2g} = \left[\frac{4 \times 0.005 \times 40}{0.15} + 3\right] \times \frac{(2.26)^2}{2 \times 9.81}$$

$$= 2.17 \text{ m}$$

Therefore, total head required to be developed = 35 + 2.17

= 37.17 m

The speed of the pump is determined from the consideration of specific speed as

$$0.075 = \frac{N(0.04)^{1/2}}{(9.81 \times 37.17)^{3/4}}$$

or

$$N = \frac{0.075\,(9.81 \times 37.17)^{3/4}}{(0.04)^{1/2}} = 31.29 \text{ rev/s}$$

Let the impeller diameter be D.

Flow area perpendicular to the impeller outlet periphery

$$= \pi D \times D/10 \times 0.94 = 0.295\,D^2$$

The inlet and outlet velocity triangles are drawn below:

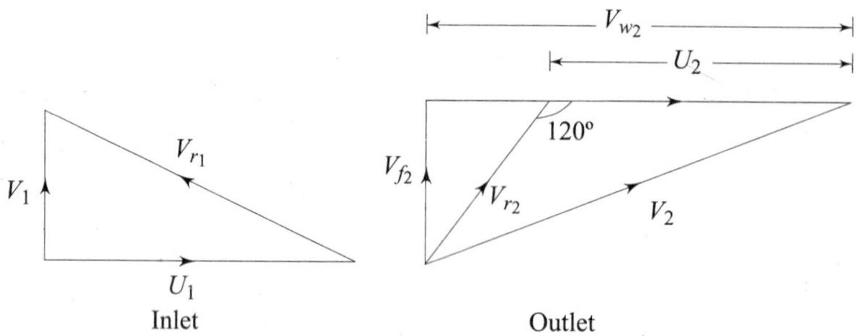

$$V_{f_2} = \frac{Q}{0.295\,D^2} = \frac{0.04}{0.295\,D^2} = \frac{0.135}{D^2}\ \text{m/s}$$

$$U_2 = \pi ND = 31.29 \times \pi \times D = 98.3\,D\ \text{m/s}$$

$$\eta_h\ \text{(Hydraulic efficiency)} = gH/(V_{w_2} U_2)$$

or $$0.76 = \frac{9.81 \times 37.17}{98.3\,D \times V_{w_2}}$$

which gives, $$V_{w_2} = \frac{4.88}{D}\ \text{m/s}$$

From the outlet velocity triangle,

$$\tan 60° = \frac{V_{f_2}}{V_{w_2} - U_2} = \frac{0.135}{D^2\,[4.88/D - 98.3\,D]}$$

or $$D^3 = 0.0496\,D - 0.0008$$

which gives $$D = 0.214\ \text{m}$$

Example 15.16

When a laboratory test was carried out on a pump, it was found that, for a pump total head of 36 m at a discharge of 0.05 m^3/s, cavitation began when the sum of the static pressure and the velocity head at the inlet was reduced to 3.5 m. The atmospheric pressure was 750 mm of Hg and the vapour pressure of water was 1.8 kPa. If the pump is to operate at a location where atmospheric pressure was reduced to 620 mm of Hg and the temperature is so reduced that the vapour pressure of water is 830 Pa, what is the value of the cavitation parameter when the pump develops the same total head and discharge? Is it necessary to reduce the height of the pump and if so by how much?

Solution

Cavitation began when, $\dfrac{p_1}{\rho g} + \dfrac{V_1^2}{2g} = 3.5$ m

(where subscript 1 refers to the condition at inlet to the pump)

and at this condition, $p_1 = p_{vap}$

Therefore, $$V_1^2/2g = 3.5 - \frac{1.8 \times 10^3}{9.81 \times 10^3}$$

$$= 3.32 \text{ m (net positive suction head)}$$

Hence, the cavitation parameter, $$\sigma = \frac{V_1^2}{2gH}$$

$$= 3.32/36 = 0.092$$

This dimensionless parameter will remain same for both the cases.

Applying Bernoulli's equation, between the liquid level at sump and the inlet to the pump (taking the sump level as datum), we can write for the first case,

$$\frac{p_1}{\rho g} + \frac{V_1^2}{2g} + z_1 = \frac{p_{atm}}{\rho g} - h_{f_1} \text{ (sum of head losses)}$$

or $$(z_1 + h_{f_1}) = \frac{p_{atm}}{\rho g} - \sigma \cdot H - \frac{p_1}{\rho g}$$

$$= (0.75 \times 13.6) - 3.32 - \frac{1.8}{9.81}$$

$$= 6.7 \text{ m}$$

For the second case,

$$\frac{p_1'}{\rho g} + \frac{V_1'^2}{2g} + z_1' = \frac{p_{atm}'}{\rho g} - h_{f_1}'$$

(Superscript′ refer to the second case)

or $$(z_1' + h_{f_1}') = \frac{p_{atm}'}{\rho g} - \sigma H - \frac{p_{vap}'}{\rho g}$$

$$= (0.62 \times 13.6) - 3.32 - \frac{830}{9.81 \times 10^3}$$

$$= 5.03 \text{ m}$$

Since the flow rate is same, $h_{f_1} = h_{f_1}'$

Therefore, the pump must be lowered a distance

$$(z_1 - z_1') = 6.7 - 5.03 = 1.67 \text{ m}$$

at the new location.

15.5 RECIPROCATING PUMP

We have described at the beginning of this chapter that the fluid machines can be divided into two categories depending upon their principle of operation, viz., the rotodynamic type and the positive displacement type. While the functioning of a rotodynamic machine depends on the hydrodynamic principles of continuous flow of a fluid through it, the working principle of a positive displacement machine is based on the change of volume occupied by a certain amount of fluid within the machine. The reciprocating pump is a positive displacement type of pump.

A reciprocating pump consists primarily of a piston or a plunger executing reciprocating motion inside a close fitting cylinder (Fig. 15.35). The motion of the piston outwards (i.e., towards the right in Fig. 15.35) causes a reduction of pressure in the cylinder, and therefore liquid flows into the cylinder through the inlet valve. The reverse movement of the piston (i.e., the motion of piston inside the cylinder) pushes the liquid and increases its pressure. Then the inlet valve closes and the outlet valve opens so that the high pressure liquid is discharged into the delivery pipe. Usually, the operation of the valve is controlled automatically by the pressure in the cylinder. In some designs, ports on the wall of the cylinder are provided instead of valves. These ports are covered and uncovered by the movement of the piston.

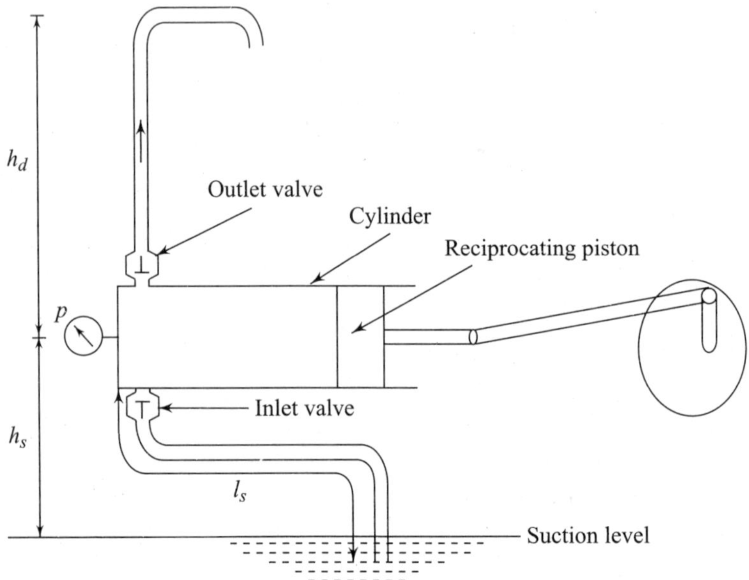

Fig. 15.35 A reciprocating pump

The axial force exerted by the piston on the fluid at any instant is pA, where p is the instantaneous pressure of the liquid in the cylinder and A is the cross-sectional area of the piston. Due to a motion of the piston through a small distance dx along the axis, the work done on the liquid becomes $pA\,\mathrm{d}x = p\mathrm{d}\forall$ where $\mathrm{d}\forall$ represents the volume swept by the piston due to its movement through a distance dx. Therefore the net work done by the pump is given by $\int p\,\mathrm{d}\forall$, calculated round the complete cycle. This can be represented by the area enclosed by a curve of pressure against volume. For an incompressible fluid, the ideal form of the diagram would be a simple rectangle *ABCD,* as shown in Fig. 15.36, since the rise or fall in pressure will not be associated with any change in volume. In practice, however, the acceleration and deceleration of the piston give rise to corresponding acceleration and deceleration of the liquid in the associated pipelines. At the beginning of the suction stroke, the liquid is accelerated, and hence an additional pressure difference is required. This makes the suction pressure at *A* to assume a lower valve at *E* (Fig. 15.36).

Similarly, due to deceleration of liquid at the end of the suction stroke, a rise of pressure in the cylinder is needed and therefore the end point *B* in the suction stroke gets shifted to *F*. Neglecting the frictional effect and considering the motion of the piston to be a simple harmonic one, the suction stroke is represented by a straight line *EF*. A further modification of the diagram results from the effect of friction and other losses in the suction pipe. The losses are zero at the ends of the stroke when the velocity is zero, and a maximum at mid-stroke (again for simple harmonic motion of the piston) when the velocity is at its maximum. The base of the diagram (Fig. 15.36) therefore becomes *ELF*. Inertia and friction in the delivery pipe cause similar modification of the ideal delivery stroke *DC* to *KPG*.

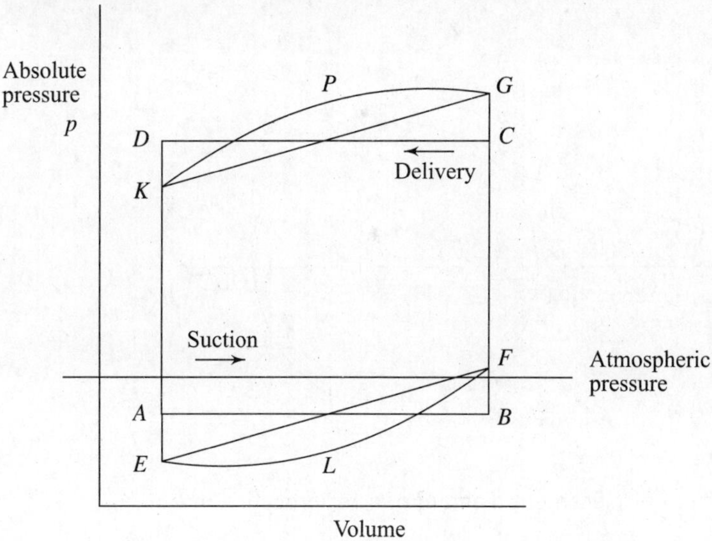

Fig. 15.36 Pressure-displacement diagram for a reciprocating pump

Finally, the actual shape of the pressure volume diagram becomes *ELFGPK*. The effects of inertia and friction in the cylinder are normally negligible as compared to those in the suction and delivery pipes. The speed of such a pump is usually restricted by the pressure corresponding to the point E of the diagram which is the minimum pressure point in a cycle. This pressure must not be allowed to fall below a pressure where the air cavitation (liberation of dissolved gases from the liquid) starts.

Analytical Expressions of Accelerating Heads During Suction and Delivery Strokes It has already been mentioned that the liquid mass in suction and delivery pipes gets accelerated and decelerated due to the typical accelerating and decelerating motion undergone by the piston during suction and delivery strokes. This causes a non-uniform additional head, known as acceleration head, which the pump has to develop during the suction and delivery strokes along with the constant theoretical suction and delivery head respectively. To obtain an expression of the acceleration head in each stroke, it is essential to determine first the velocity of the piston. This can be obtained from the consideration of crank revolution. The motion of the piston is usually considered to be a simple harmonic one with zero velocities at ends and maximum at the centre. However, this assumption is only true when the ratio of the length of the connecting rod to that of the crank is very large.

Let us consider the displacement of the piston, after a time t from its inner dead centre position (IDC) be x (Fig. 15.37). Then we can write

$$x = r - r\cos\theta$$

where r is the radius of the crank and θ is the angular displacement of the crank during the time interval t. If ω is the angular velocity of the crank, then we have $\theta = \omega t$.

and $$x = r - r\cos\omega t$$

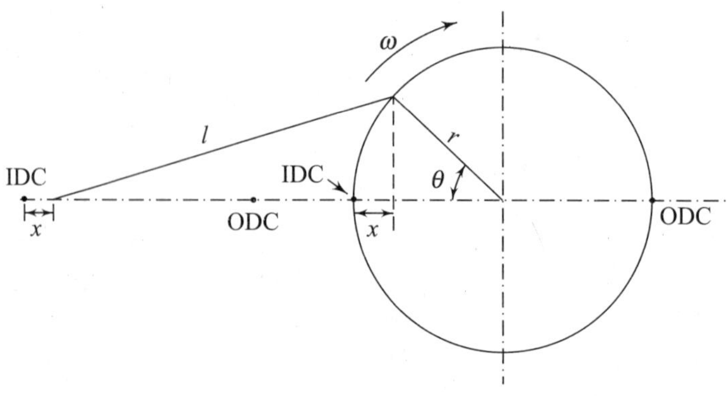

Fig. 15.37 Piston displacement diagram of a reciprocating machine

Hence, the instantaneous velocity of the piston $\frac{dx}{dt} = r\omega \sin \omega t$. Considering the liquid in the piston to be moving with the velocity of the piston, the velocity V of liquid in the pipeline can be written from the principle of continuity as

$$V = \frac{A}{a} r\,\omega \sin \omega t$$

where, A and a are the cross-sectional areas of the cylinder and pipeline respectively. The acceleration of liquid in the pipeline can be written as

$$\frac{dV}{dt} = \frac{d}{dt}\left(\frac{A}{a} r\,\omega \sin \omega t\right) = \frac{A}{a} r\,\omega^2 \cos \omega t$$

Therefore, the force F required to accelerate the liquid mass is given by

$$F = \rho\, a l\, \frac{A}{a} r\,\omega^2 \cos \omega t$$

$$= \rho\, l A r \omega^2 \cos \theta \;(\text{since } \theta = \omega t)$$

(l is the length of the pipeline)

The pressure head caused by the force F is given by

$$\frac{F}{a \rho g} = \frac{l}{g} \frac{A}{a} r\,\omega^2 \cos \theta$$

This is known as the acceleration head h_a. Using subscripts s and d to represent the quantities for suction and delivery sides, we can write

$$h_{a_s} = \frac{l_s}{g} \cdot \frac{A}{a_s} r\,\omega^2 \cos \theta$$

$$h_{a_d} = \frac{l_d}{g} \cdot \frac{A}{a_d} r\,\omega^2 \cos \theta$$

It is evident from these expressions and Fig. 15.37 that the maximum and minimum acceleration heads take place at the beginning and at the end of each stroke respectively with zero at the middle of the stroke. The magnitude of maximum acceleration head $= \frac{l}{g}\frac{A}{a} r\,\omega^2$. The pump is, therefore, required to develop an additional head of $\frac{l_s}{g}\frac{A}{a} r\,\omega^2$, at the beginning of the suction stroke, over the constant suction head determined by the height of the pump above the supply level. Similarly, an additional head of $\frac{l_d}{g}\frac{A}{a_d} r\,\omega^2$ is required to be developed by the pump, at the beginning of the delivery stroke, over the constant delivery head determined by the static lift of the pump. This has already been shown in Fig. 15.36.

Rate of Delivery

Single Acting Piston or Plunger Pump In a single acting piston pump the entrance and discharge of liquid takes place from one side of the piston only. Therefore one stroke is meant only for suction and the other stroke is meant only for discharge. Rate of delivery against crank angle for such type of a pump is shown in Fig. 15.38. During the first half revolution of the crank there is only suction and therefore the rate of delivery is zero. During the second half (corresponding to crank angles between 180° to 360°) of the crank revolution, the discharge takes place. Since the motion of the piston is approximately simple harmonic, the rate of delivery versus the crank angle curve will be a sine curve. Velocity of discharge of water at any instant is proportional to the velocity of the piston at that instant. Therefore the sine curve, shown in Fig. 15.38 also represents the velocity of discharge to some scale.

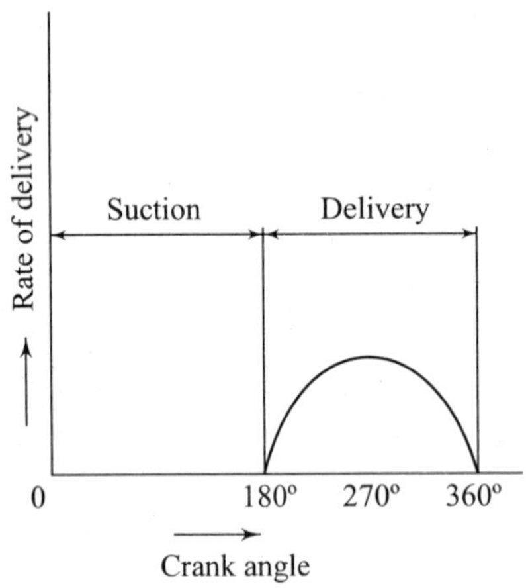

Fig. 15.38 Rate of delivery versus crank angle for a single acting reciprocating pump

Double Acting Piston or Plunger Pump In this type of pump provisions are made for the entry and discharge of the liquid from both the sides of the piston. Therefore each stroke is a suction-cum-delivery stroke. The curve of rate of delivery against the angle of rotation of the crank is therefore the two sine curves drawn at a phase difference of 180°.

Multi-Cylinder Pumps We observe that the rate of delivery from a single cylinder, whether single acting or double acting, is non-uniform. Multi-cylinder pumps are used to obtain a somewhat uniform discharge. In multi-cylinder pumps, a number of cylinders are connected in parallel, their cranks being equally spaced over 360°. The fluctuating discharge from the individual cylinders are thus added together resulting in an almost uniform total discharge. This is illustrated in Fig. 15.39 for a three-cylinder pump with the cranks at 120° to each other.

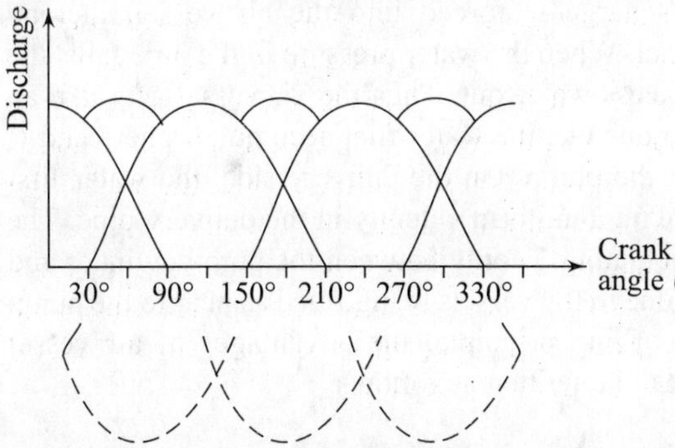

Fig. 15.39 Rate of delivery versus crank angle for a three cylinder reciprocating pump

Air Vessel The pulsation of pressure due to inertia or acceleration heads in suction and delivery pipe and the non-uniformity of discharge during the delivery stroke may largely be eliminated by connecting a large and closed chamber to both the suction and delivery pipes at points close to the pump cylinder as shown in Fig. 15.40. These vessels are known as air vessels.

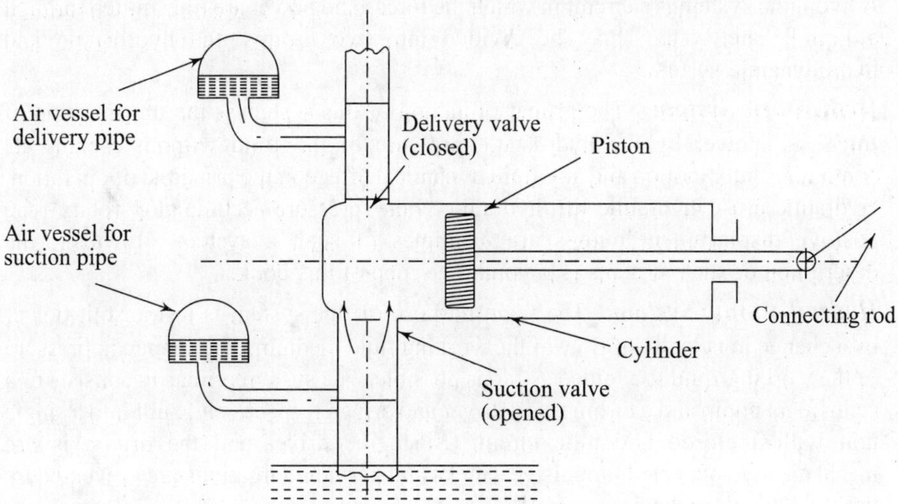

Fig. 15.40 Reciprocating pump connected with air vessels

Working Principle An air vessel in a reciprocating pump acts like a fly-wheel of an engine. The top of the vessel contains compressed air which can contract or expand to absorb most of the pressure fluctuations. Whenever the pressure rises, water in excess of the mean discharge is forced into the air vessel, thereby compressing the air within the vessel. When the water pressure in the pipe falls, the compressed air again ejects the excess water out. Thus the air vessel acts like an intermediate reservoir. On the suction side, the water first accumulates here and is then transferred to the cylinder of the pump. On the delivery side, the water first goes to the vessel and then flows with a uniform velocity in the delivery pipe. The column of water which is now fluctuating, is only between the pump cylinder and the air vessels which is very small due to the vessels being fitted as near to the pump cylinder as possible. From the working principle, the advantages of air vessel attached to a reciprocating pump can be written as follows:

(a) *Suction side*:
  (i) Reduces the possibility of cavitation
  (ii) The pump can be run at a higher speed
  (iii) The length of the suction pipe below the air vessel can be increased.

(b) *Delivery side:*
  (i) A large amount of power consumed in supplying the accelerating head can be saved.
  (ii) Maintains almost a constant rate of discharge

15.6 HYDRAULIC SYSTEM

A hydraulic system is a circuit in which the forces and power are transmitted through a liquid. The system may be divided into two groups, the hydrostatic and hydrodynamic system.

Hydrostatic System The primary function of this system is the transmission of force and power by the hydrostatic pressure of the fluid without causing its continuous bulk motion and any fluid dynamical effect on the principle of operation. Hydraulic press, hydraulic lift, hydraulic crane, pressure accumulator, rotary type positive displacement pumps are examples of such a system. However, the description of such systems is beyond the scope of this book.

Hydrodynamic System The main purpose of this system is to transmit power by a change in velocity of flow of the working fluid medium. The change in pressure of the working fluid is avoided as far as possible. The system primarily consists of a centrifugal pump and a turbine, as a driver and driven respectively, built into a single unit with a closed hydraulic circuit. Since the driver and the driven is not mechanically connected, impulsive shocks and periodic vibrations are prevented by the fluid coupling them.

The hydrodynamic transmission systems are of two types—hydraulic coupling and hydraulic torque convertor.

Hydraulic or Fluid Coupling The essential features of a fluid coupling are shown in Fig. 15.41. The primary function of the coupling is to transmit power with the same torque on the driving and driven shafts. It mainly consists of a radial pump impeller keyed to a driving shaft A, and a radial reaction turbine keyed to a driven

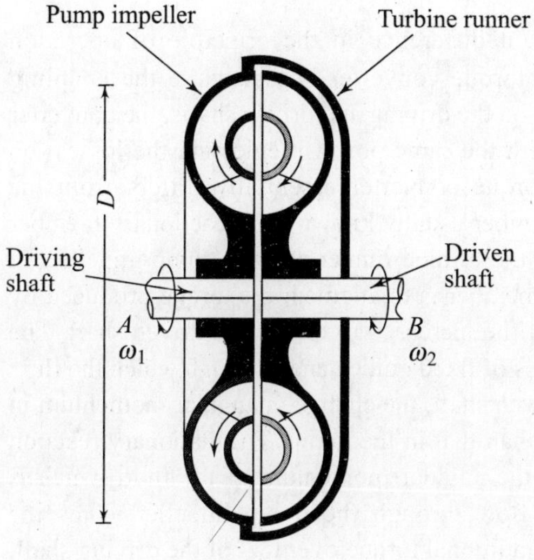

Fig. 15.41 Fluid or hydraulic coupling

shaft *B*. The blades of both the pump impeller and turbine runner are of straight radial type. There is no mechanical connection between the driving and the driven shafts. The two shafts together form a casing completely filled with the working fluid which is usually ordinary mineral lubricating oil. If shaft *A* is allowed to rotate, the oil will pass through the impeller blades and will flow radially outwards with higher energy. The fluid will then strike the turbine runners, and while flowing radially inwards, transfer power to the turbine blades. With the increase in the speed of shaft *A*, sufficient head is developed in the fluid at the outlet of the pump impeller so that the power transferred to the turbine rotor becomes high enough to set the driven shaft *B* in motion. Due to slip, the two shafts rotate at different speeds. If the driver and the follower rotate at the same speed, the circulation of oil cannot take place. This is because of the fact that the head produced by the pump should be greater than the centrifugal head resisting flow through the turbine. At equal speed of the shafts *A* and *B*, the heads would balance each other and then no flow would occur and no torque would be transmitted. If ω_1 and ω_2 are the angular velocities of driving and driven shafts respectively, then the slip is expressed as $(\omega_1 - \omega_2)/\omega_1$. Under usual operating conditions, the slip is about 2 to 3 per cent. From the dimensional analysis, the torque T can be expressed in terms of the pertinent controlling dimensionless variables as

$$\frac{T}{\rho\omega_1^2 D^5} = F\left(\frac{\omega_2}{\omega_1}, \frac{\rho\omega_1 D^2}{\mu}, \frac{\mathrm{V}}{D^3}\right)$$

The term $T/\rho\omega_1^2 D^5$ is known as the torque coefficient, and $\rho\omega_1 D^2/\mu$ corresponds to the Reynolds number of fluid flow. $\forall$ is the volume of the fluid in the coupling and D is the diameter of the impeller or the runner.

Fluid Torque Converter The main difference in the principle of operation between a fluid coupling and fluid torque converter is that while the coupling transmits power with the same torque on the driving and driven shafts, the converter provides for torque multiplication with the same power (neglecting the losses) on the driving and driven shafts. A torque converter essentially differs from the coupling in that a third stationary member usually known as a reactionary member (Fig. 15.42) is incorporated between the turbine runner and the pump impeller. In fact, the function of the reactionary member is to augment the torque produced by the driving shaft and then to transmit the increased torque to the driven shaft. The reactionary member consists of a series of fixed guide vanes through which the fluid flows. For a greater torque on the driven shaft, the change in angular momentum in the turbine runner should be greater than that in the pump. The stationary reaction blades are so shaped as to increase the angular momentum of the fluid which is further increased in the course of flow through the pump impeller. Thus the stationary members contribute to an additional torque over that of the driving shaft. The amplification of torque depends on the design of the stationary blades and the speed ratio (ratio of angular velocities of the driven and driving shafts).

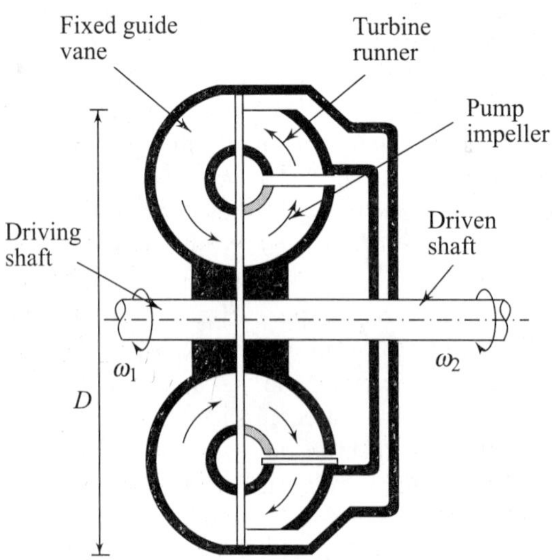

Fig. 15.42 Fluid or hydraulic torque converter

SUMMARY

- A fluid machine is termed as a turbine when the stored energy of a fluid is transferred to mechanical energy of the rotating member of the machine, and is termed as a pump or compressor when the mechanical energy of the moving parts of the machine is transfeired to increase the energy stored by the fluid. The machines for which the principle of operation depends on the theory of fluid dynamics are known as rotodynamic machines, while the machines which function on the principle of a change in volume of certain amount of fluid trapped in the machines are known as positive displacement machines.
- In a rotodynamic fluid machine, the head (energy per unit weight of the fluid) transferred by the fluid to the machine is given by $(V_{w_1} U_1 - V_{w_2} U_2)/g$. A negative sign of the expression implies the head transferred by the machine to the fluid. The above expression can be split up into three terms to show the three distinct components of energy transfer as

$$\frac{V_{w_1} U_1 - V_{w_2} U_2}{g} = \frac{1}{2g}\left[(V_1^2 - V_2^2) + (U_1^2 - U_2^2) - (V_{r_2}^2 - V_{r_1}^2)\right]$$

  The first term on the right-hand side represents the change in the absolute dynamic head of the fluid, the second and the third terms pertain to the change in the pressure head. For an axial flow machine, $U_1 = U_2$. The second term becomes positive for a radially inward flow machine like turbines while it becomes negative for a radially outward flow machine like centrifugal pump.
- The hydraulic efficiency of a turbine is defined as the ratio of mechanical energy delivered by the rotor to the energy available from fluid, while for a pump, it is defined as the ratio of useful energy gained by the fluid at final discharge and the mechanical energy supplied to the rotor. The pertinent dimensionless parameters governing the principle of operation of fluid machines are

$$\frac{Q}{ND^3}, \frac{gH}{N^2D^2}, \frac{\rho ND^2}{\mu}, \frac{P}{\rho N^3 D^5}, \frac{E/\rho}{N^2D^2}$$

  The dimensionless specific speed of a turbine is given by $NP^{1/2}/\rho^{1/2}\,(gH)^{5/4}$ and the corresponding dimensional version is $Np^{1/2}/H^{5/4}$. The dimensionless specific speed of a pump is given by $NQ^{1/2}/(gH)^{3/4}$, and the dimensionl version is $NQ^{1/2}/H^{3/4}$. The values of specific speed are quoted for maximum efficiency of the machine.
- The only hydraulic turbine of impulse type is the Pelton wheel. The buckets of the wheel in a Pelton turbine is exposed to the atmosphere, and the high pressure water expands to the atmospheric pressure in a nozzle and strikes the bucket as a water jet. A Pelton wheel runs at its maximum bucket efficiency (defined as the ratio of work developed by the buckets to the kinetic

energy of water available at the rotor inlet) when the ratio of the blade speed to the jet speed becomes 0.46. The governing of a Pelton turbine is made by changing the cross-sectional area of the water jet by a spear valve in the nozzle. The Pelton wheel is efficient under large heads, but unsuitable to smaller heads.

- The reaction turbines are efficient under smaller heads. In a reaction machine, there is a change in the pressure head of the fluid while flowing through the rotor. A Francis turbine is a radial flow reaction turbine. To keep the kinetic energy at the outlet a minimum, the tangential component of velocity at the outlet becomes zero. Therefore, the head developed is given by $V_{w_1}U_1/g$. With an increase in specific speed and decrease in head, the shape of the radial flow Francis runner changes to that of an axial flow machine known as Kaplan runner. The draft tube is a conduit which connects the runner exit to the tail race. The primary function of a draft tube is to reduce the discharge velocity of water to minimise the loss of kinetic energy at the outlet and to permit the turbine to be set above the tail race without any appreciable drop in available head. A draft tube has to be properly designed to avoid the phenomenon of cavitation which is likely to occur at the tube inlet. Governing of reaction turbines is usually done by altering the position of the guide vanes and thus controlling the flow rate by changing the gate openings to the runner.
- A centrifugal pump, by its principle, is converse of the Francis turbine. The flow is radially outward. The fluid enters the impeller eye with zero tangential velocity. Therefore, the head developed by the fluid is given by $\sigma V_{w_2}U_2/g$. The term σ is known as the slip factor which takes care of the deviation of actual tangential velocity component at outlet from the theoretical one due to the secondary flow within the blade passages resulting in a non-uniform velocity distribution at any radius. The actual operating point of a centrifugal pump is determined by the matching or intersection of head-discharge characteristic curve of the pump and the head loss-flow rate characteristic curve of the pipeline to which the pump is connected.
- A reciprocating pump is a positive displacement type of pump and works on the principle of forcing a definite amount of liquid in a cylinder by the reciprocating motion of a piston within it. The rate of discharge from a single cylinder pump is non-uniform. The delivery is made uniform by using multi-cylinder pumps in parallel with their cranks being equally spaced over 360°. Incorporation of air vessel at the suction side reduces the possibility of cavitation at higher speed, keeping a higher length of suction pipe below the air vessel. The introduction of an air vessel at the delivery side maintains almost a constant discharge with the saving of a large amount of power consumed in supplying the accelerating head.
- The primary function of a fluid coupling is to transmit power through the dynamic action of the fluid with the same torque on the driving and driven shafts, while a fluid torque converter transmits torque with amplification keeping the power on the driving and driven shafts the same.

EXERCISES

15.1 A quarter scale turbine model is tested under a head of 10.8 m. The full-scale turbine is required to work under a head of 30 m and to run at 7.14 rev/s. At what speed must the model be run? If it develops 100 kW and uses 1.085 m^3 of water per second at this speed, what power will be obtained from the full-scale turbine? The efficiency of the full-scale turbine being 3% greater than that of the model? What is the dimensionless specific speed of the full-scale turbine?

Ans. (17.14 rev/s, 7.66 MW, 0.513 rev/s)

15.2 A Pelton wheel operates with a jet of 150 mm diameter under the head of 500 m. Its mean runner diameter is 2.25 m and it rotates with a speed of 375 rpm. The angle of bucket tip at outlet as 15°, coefficient of velocity is 0.98, mechanical losses equal to 3% of power supplied and the reduction in relative velocity of water while passing through bucket is 15%. Find (i) the force of jet on the bucket, (ii) the power developed (iii) bucket efficiency and (iv) the overall efficiency.

Ans. ((i) 165.15 kN, (ii) 7.3 MW, (iii) 90.3%, (iv) 87.6%)

15.3 A Pelton wheel works at the foot of a dam because of which the head available at the nozzle is 400 m. The nozzle diameter is 160 mm and the coefficient of velocity is 0.98. The diameter of the wheel bucket circle is 1.75 m and the buckets deflect the jet by 150°. The wheel-to-jet speed ratio is 0.46. Neglecting friction, calculate (i) the power developed by the turbine, (ii) its speed and (iii) hydraulic efficiency.

Ans. ((i) 6.08 MW, (ii) 435.9 rpm, (iii) 89.05%)

15.4 A powerhouse is equipped with Pelton type impulse turbines . Each turbine delivers a power of 14 MW when working under a head of 900 m and running at 600 rpm. Find the diameter of the jet and mean diameter of the wheel. Assume that the overall efficiency is 89%, velocity coefficient of the jet 0.98, and speed ratio 0.46.

Ans. (132 mm, 1.91 m)

15.5 A Francis turbine has a wheel diameter of 1.2 m at the entrance and 0.6 m at the exit. The blade angle at the entrance is 90° and the guide vane angle is 15°. The water at the exit leaves the blades without any tangential velocity. The available head is 30 m and the radial component of flow velocity is constant. What would be the speed of the wheel in rpm and blade angle at the exit? Ignore friction.

Ans. (268 rpm, 28.2°)

15.6 In a vertical shaft inward-flow reaction turbine, the sum of the pressure and kinetic head at the entrance to the spiral casing is 120 m and the vertical distance between this section and the tail race level is 3 m. The peripheral velocity of the runner at the entry is 30 m/s, the radial velocity of water is constant at 9 m/s and discharge from the runner is without swirl. The estimated hydraulic losses are (i) between turbine entrance and exit from the guide vanes 4.8 m, (ii) in the runner 8.8 m, (iii) in the draft tube 0.79 m, and (iv) kinetic head rejected to the tail race 0.46 m. Calculate the guide vane angle and the runner blade angle at the inlet and the pressure heads at the entry to and the exit from the runner.

Ans. ((i) 14.28°, (ii) 120.78°, (iii) 47.34 m, (iv) –5.88 m)

15.7 A Kaplan turbine operating under a net head of 20 m develops 16 MW with an overall efficiency of 80 %. The diameter of the runner is 4.2 m, while the hub diameter is 2 m and the dimensionless specific speed is 3 rad. If the hydraulic efficiency is 90%, calculate the inlet and exit angles of the runner blades at the mean blade radius if the flow leaving the runner is purely axial.

Ans. (25°, 19.4°)

15.8 The following data refer to an elbow type draft tube:
Area of circular inlet = 25 m^2
Area of rectangular outlet = 116 m^2
Velocity of water at inlet to draft tube = 10 m/s
The frictional head loss in the draft tube equals to 10% of the inlet velocity head.
Elevation of inlet plane above tail race level = 0.6 m
Determine (i) Vacuum or negative head at the inlet, and (ii) Power thrown away in tail race.

Ans. (4.95 m vac, 578 kW)

15.9 Show that when runner blade angle at inlet of a Francis turbine is 90° and the velocity of flow is constant, the hydraulic efficiency is given by $2/(2 + \tan^2 \alpha)$, where α is the vane angle.

15.10 A Kaplan turbine develops 10 MW under a head of 4.3 m. Taking a speed ratio of 1.8, flow ratio of 0.5, boss diameter 0.35 times the outer diameter and overall efficiency of 90%, find the diameter and speed of the runner.

Ans. (9.12 m, 34.6 rpm)

15.11 A conical-type draft tube attached to a Francis turbine has an inlet diameter of 3 m and its area at outlet is 20 m^2. The velocity of water at inlet, which is 5 m above tail race level, is 5 m/s. Assuming the loss in draft tube equals to 50% of velocity head at outlet, find (i) the pressure head at the top of the draft tube, (ii) the total head at the top of the draft tube taking tail race level as datum, and (iii) power lost in draft tube.

Ans. ((i) 6.03 m vac, (ii) 0.24 m, (iii) 0.08 m)

15.12 Calculate the least diameter of impeller of a centrifugal pump to just start delivering water to a height of 30 m, if the inside diameter of impeller is half of the outside diameter and the manometric efficiency is 0.8. The pump runs at 1000 rpm.

Ans. (0.6 m)

15.13 The impeller of a centrifugal pump is 0.5 m in diameter and rotates at 1200 rpm. Blades are curved back to an angle of 30° to the tangent at outlet tip. If the measured velocity of flow at the outlet is 5 m/s, find the work input per kg of water per second. Find the theoretical maximum lift to which the water can be raised if the pump is provided with whirlpool chamber which reduces the velocity of water by 50%.

Ans. (72.78 m, 65.87 m)

15.14 The impeller of a centrifugal pump is 0.3 m in diameter and runs at 1450 rpm. The pressure gauges on suction and delivery sides show the difference of

25 m. The blades are curved back to an angle of 30°. The velocity of flow through impeller, being constant, equals to 2.5 m/s, find the manometric efficiency of the pump. If the frictional losses in impeller amounts to 2 m, find the fraction of total energy which is converted into pressure energy by impeller. Also find the pressure rise in pump casing.

Ans. (58.35%, 54.1%, 1.83 m of water)

15.15 A centrifugal pump is required to work against a head of 20 m while rotating at the speed of 700 rpm. If the blades are curved back to an angle of 30° to tangent at outlet tip and velocity of flow through impeller is 2 m/s, calculate the impeller diameter when (i) all the kinetic energy at impeller outlet is wasted and (ii) when 50% of this energy is converted into pressure energy in pump casing.

Ans. ((i) 0.55 m, (ii) 0.48 m)

15.16 During a laboratory test on a pump, appreciable cavitation began when the pressure plus the velocity head at inlet was reduced to 3.26 m while the change in total head across the pump was 36.5 m and the discharge was 48 litres/s. Barometric pressure was 750 mm of Hg and the vapour pressure of water 1.8 kPa. What is the value of σ_c? If the pump is to give the same total head and discharge in a location where the normal atmospheric pressure is 622 mm of Hg and the vapour pressure of water is 830 Pa, by how much must the height of the pump above the supply level be reduced?

Ans. (0.084, 1.65 m)

15.17 A single acting reciprocating pump having a cylinder diameter of 150 mm and stroke of 300 mm. is used to raise the water through a height of 20 m. Its crank rotates at 60 rpm. Find the theoretical power required to run the pump and the theoretical discharge. If actual discharge is 5 litres/s, find the percentage slip. If delivery pipe is 100 mm in diameter and is 15 m long, find the acceleration head at the beginning of the stroke.

Ans. (1.04 kW, 0.0053 m^3/s. 5.66, 20.37 m)

15.18 A reciprocating pump has a suction head of 6 m and delivery head of 15 m. It has a bore of 150 mm and stroke of 250 mm and piston makes 60 double strokes in a minute. Calculate the force required to move the piston during (i) suction stroke, and (ii) during the delivery stroke. Find also the power required to drive the pump.

Ans. ((i) 1.04 kN, (ii) 2.60 kN, 1.81 kW)

16

COMPRESSORS, FANS AND BLOWERS

In Chapter 15, we discussed the basic fluid mechanical principles governing energy transfer in a fluid machine. A brief description of different types of fluid machines using water as the working fluid was also given in Chapter 15. However, there exist a large number of fluid machines in practice, which use air, steam and gas (the mixture of air and products of burnt fuel) as working fluids. The density of fluids change with a change in pressure as well as in temperature as they pass through the machines. These machines are called 'compressible flow machines'or more popularly 'turbomachines'. Apart from the change in density with pressure, other features of compressible flow, depending upon the flow regimes, are also observed in course of flow of fluids through turbomachines. Therefore, the basic equation of energy transfer (Euler's equation, as discussed in Chapter 15) along with the equation of state relating the pressure, density and temperature of the working fluid and other necessary equations of compressible flow, (as discussed in Chapter 14), are needed to describe the performance of a turbomachine. However, a detailed discussion on all types of turbomachines is beyond the scope of this book. We shall present a very brief description of a few compressible flow machines, namely, compressors, fans and blowers in this chapter.

16.1 CENTRIFUGAL COMPRESSORS

A centrifugal compressor is a radial flow rotodynamic fluid machine that uses mostly air as the working fluid and utilises the mechanical energy imparted to the machine from outside to increase the total internal energy of the fluid mainly in the form of increased static pressure head.

During the Second World War most of the gas turbine units used centrifugal compressors. Attention was focused on the simple turbojet units where low power-plant weight was of great importance. Since the war, however, axial compressors have been developed to the point where it has an appreciably higher isentropic efficiency. Though centrifugal compressors are not that popular today, there is renewed interest in the centrifugal stage, used in conjunction with one or more axial stages, for small turbofan and turboprop aircraft engines.

A centrifugal compressor essentially consists of three components:

1. A **stationary casing**

2. A **rotating impeller,** as shown in Fig. 16.1(a), which imparts high velocity to the air. The impeller may be single or double sided, as shown in Fig. 16.1 (b) and (c) but the fundamental theory is the same for both of them.
3. A **diffuser** consists of a number of fixed diverging passages in which the air is decelerated with a consequent rise in static pressure.

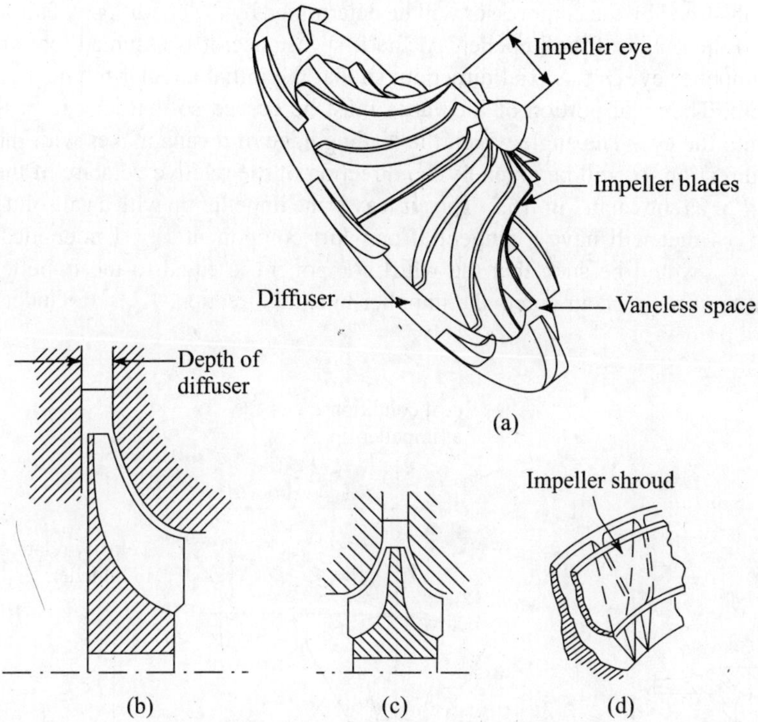

Fig. 16.1 Schematic views of a centrifugal compressor

Principle of Operation Air is sucked into the impeller eye and whirled outwards at high speed by the impeller disk. At any point in the flow of air through the impeller, the centripetal acceleration is obtained by a pressure head so that the static pressure of the air increases from the eye to the tip of the impeller. The remainder of the static pressure rise is obtained in the diffuser, where the very high velocity of air leaving the impeller tip is reduced to almost the velocity with which the air enters the impeller eye.

Usually, about half of the total pressure rise occurs in the impeller and the other half in the diffuser. Owing to the action of the vanes in carrying the air around with the impeller, there is a slightly higher static pressure on the forward side of the vane than on the trailing face. The air will thus tend to flow around the edge of the vanes in the clearing space between the impeller and casing. This results in a loss of efficiency and the clearance must be kept as small as possible. Sometimes, a shroud attached to the blades, as shown in Fig. 16.1(d), may eliminate such a loss, but it is avoided because of increased disc friction loss and of manufacturing difficulties.

The straight and radial blades are usually employed to avoid any undesirable bending stress to be set up in the blades. The choice of radial blades also determines that the total pressure rise is divided equally between the impeller and the diffuser.

Work Done and Pressure Rise Since no work is done on the air in the diffuser, the energy absorbed by the compressor will be determined by the conditions of air at the inlet and the outlet of the impeller. At the first instance, it is assumed that air enters the impeller eye in the axial direction, so that the initial angular momentum of air is zero. The axial portion of the vanes must be curved so that air can pass smoothly into the eye. The angle which the leading edge of a vane makes with the tangential direction, α, will be given by the direction of the relative velocity of the air at inlet, V_{r1}, as shown in Fig. 16.2. The air leaves the impeller tip with an absolute velocity of V_2 that will have a tangential or whirl component V_{w2}. Under ideal conditions, V_2, would be such that the whirl component is equal to the impeller speed U_2 at the tip. Since air enters the impeller in axial direction, $V_{w1} = 0$. Under

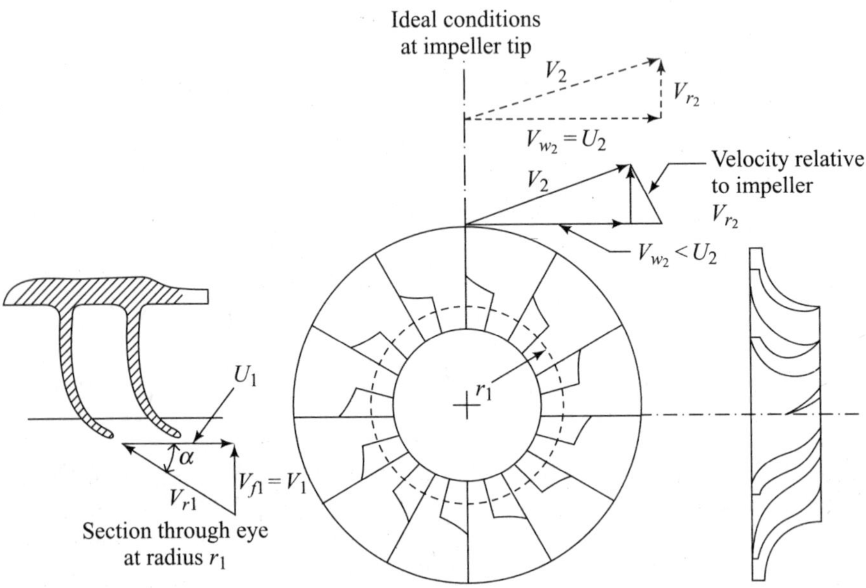

Fig. 16.2 Velocity triangles at inlet and outlet of impeller blades

the situation of $V_{w1} = 0$ and $V_{w2} = U_2$, we can derive from Eq. (15.2), the energy transfer per unit mass of air as

$$\frac{E}{m} = U_2^2 \tag{16.1}$$

Due to its inertia the air trapped between the impeller vanes is reluctant to move round with the impeller and we have already noted that this results in a higher static pressure on the leading face of a vane than on the trailing face. It also prevents the air from acquiring a whirl velocity equal to the impeller speed. This effect is known as slip. Because of slip, we obtain $V_{w2} < U_2$. The slip factor σ is defined in a similar way as in the case of a centrifugal pump as

$$\sigma = \frac{V_{w2}}{U_2}$$

The value of σ lies between 0.9 and 0.92. The energy transfer per unit mass in case of slip becomes

$$\frac{E}{m} = V_{w2}\,U_2 = \sigma U_2^2 \tag{16.2}$$

One of the widely used expressions for σ was suggested by Stanitz from the solution of potential flow through the impeller passages. It is given by

$$\sigma = 1 - \frac{0{\cdot}63\pi}{n}, \text{ where } n \text{ is the number of vanes.}$$

Power Input Factor The power input factor takes into account the effect of disk friction, windage, etc., for which a little more power has to be supplied than required by the theoretical expression. Considering all these losses, the actual work done (or energy input) on the air per unit mass becomes

$$w = \Psi\sigma U_2^2 \tag{16.3}$$

where Ψ is the power input factor.

From steady flow energy equation and in consideration of air as an ideal gas, one can write for adiabatic work w per unit mass of air flow as

$$w = c_p\,(T_{2t} - T_{1t}) \tag{16.4}$$

where T_{1t} and T_{2t} are the stagnation temperatures at inlet and outlet of the impeller, and c_p is the mean specific heat over the entire temperature range. With the help of Eq. (16.3), we can write

$$w = \Psi\,\sigma U_2^2 = c_p\,(T_{2t} - T_{1t}) \tag{16.5}$$

The stagnation temperature represents the total energy held by a fluid. Since no energy is added in the diffuser, the stagnation temperature rise across the impeller must be equal to that across the whole compressor. If the stagnation temperature at the outlet of the diffuser is designated by T_{3t} then $T_{3t} = T_{2t}$. One can write from Eq. (16.5)

$$\frac{T_{2t}}{T_{1t}} = \frac{T_{3t}}{T_{1t}} = 1 + \frac{\Psi\sigma U_2^2}{c_p\,T_{1t}} \tag{16.6}$$

The overall stagnation pressure ratio can be written as

$$\frac{p_{3t}}{p_{1t}} = \left(\frac{T'_{3t}}{T_{1t}}\right)^{\frac{\gamma}{\gamma-1}}$$

$$= \left[1 + \frac{\eta_c (T_{3t} - T_{1t})}{T_{1t}}\right]^{\frac{\gamma}{\gamma-1}} \tag{16.7}$$

where, T'_{3t} and T_{3t} are the stagnation temperatures at the end of an ideal (isentropic) and actual process of compression we know respectively (Fig. 16.3), and η_c is the isentropic efficiency defined as

$$\eta_c = \frac{T'_{3t} - T_{1t}}{T_{3t} - T_{1t}} \tag{16.8}$$

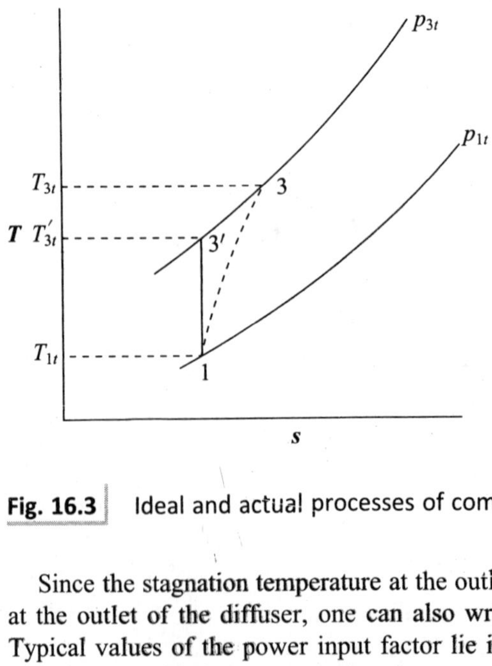

Fig. 16.3 Ideal and actual processes of compression on T-s plane

Since the stagnation temperature at the outlet of the impeller is the same as that at the outlet of the diffuser, one can also write T_{2t} in place of T_{3t} in Eq. (16.8). Typical values of the power input factor lie in the region of 1.035 to 1.04. If we know η_c, we will be able to calculate the stagnation pressure rise for a given impeller speed. The variation in stagnation pressure ratio across the impeller with the impeller speed is shown in Fig. 16.4. For common materials, U_2 is limited to 450 m/s.

Figure 16.5 shows the inducing section of a compressor. The relative velocity V_{r1} at the eye tip has to be held low otherwise the Mach number (based on V_{r1}) given by $M_{r1} = \dfrac{V_{r1}}{\sqrt{\gamma R T_1}}$ will be too high causing shock losses. Mach number M_{r1} should be in the range of 0.7–0.9. The typical inlet velocity triangles for large and medium or small eye tip diameter are shown in Fig. 16.6 (a) and (b) respectively.

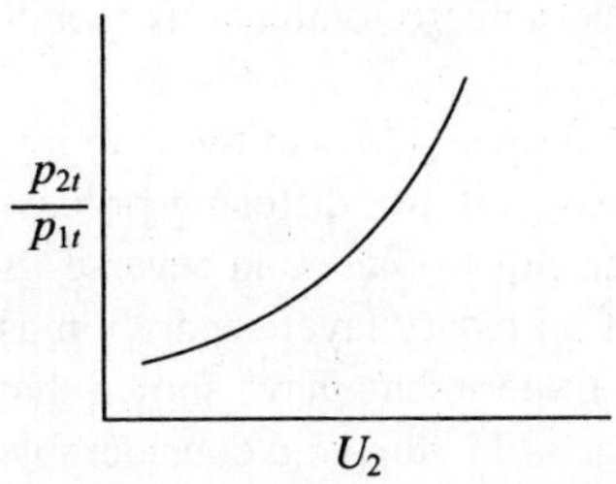

Fig. 16.4 Variation in stagnation pressure ratio with impeller tip speed

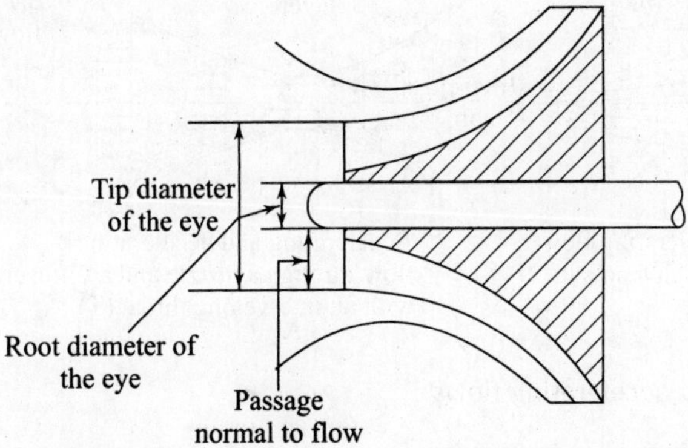

Fig. 16.5 Inducing section of a centrifugal compressor

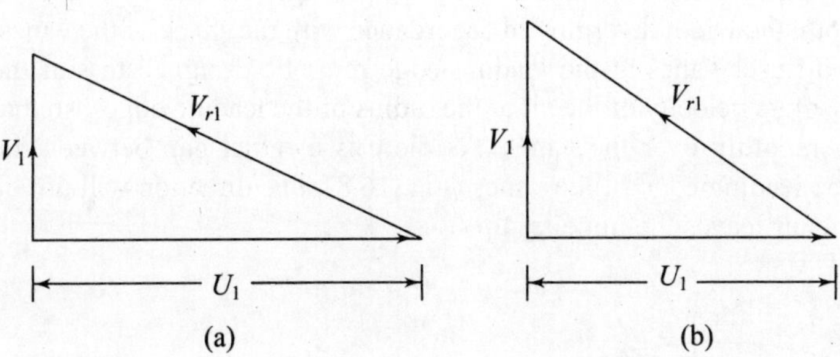

Fig. 16.6 Velocity triangles at the tip of the eye

16.1.1 Diffuser

The basic purpose of a compressor is to deliver air at high pressure required for burning fuel in a combustion chamber so that the burnt products of combustion at high pressure and temperature are used in turbines or propelling nozzles (in case of an aircraft engine) to develop mechanical power. The problem of designing an efficient combustion chamber is eased if velocity of the air entering the combustion chamber is as low as possible. It is necessary, therefore to design the diffuser so that only a small part of the stagnation temperature at the compressor outlet corresponds to kinetic energy.

It is much more diffucult to arrange for an efficient deceleration of flow than it is to obtain efficient acceleration. There is a natural tendency in a diffusing process for the air to break away from the walls of the diverging passage and reverse its direction. This is typically due to the phenomenon of boundary layer separation as explained in Section 9.6. This is shown in Fig. 16.7. Experiments have shown that the maximum permissible included angle of divergence is 11° to avoid considerable losses due to flow separation.

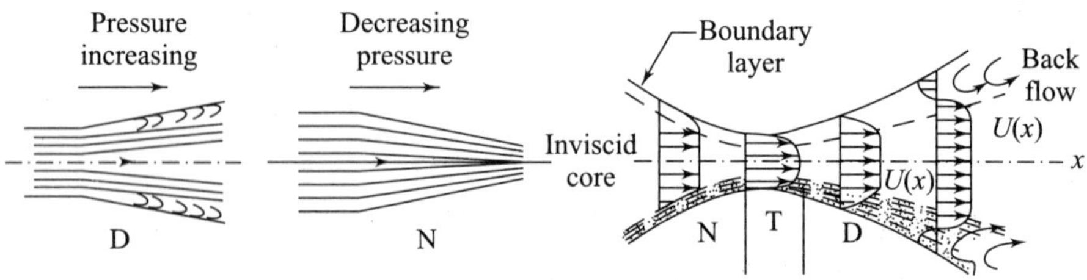

Decelerating flow through a diffuser (D)

Accelerating flow through a nozzle (N)

Accelerating and decelerating flow through a nozzle and a diffuser with an intervening throat (T)

Fig. 16.7 Accelerating and decelerating flows

In order to control the flow of air effectively and carry-out the diffusion process in as short a length as possible, the air leaving the impeller is divided into a number of separate streams by fixed diffuser vanes. Usually the passages formed by the vanes are of constant depth, the width diverging in accordance with the shape of the vanes. The angle of the diffuser vanes at the leading edge must be designed to suit the direction of the absolute velocity of the air at the radius of the leading edges, so that the air will flow smoothly over the vanes. As there is a radial gap between the impeller tip and the leading edge of the vanes (Fig. 16.8), this direction will not be that with which the air leaves the impeller tip.

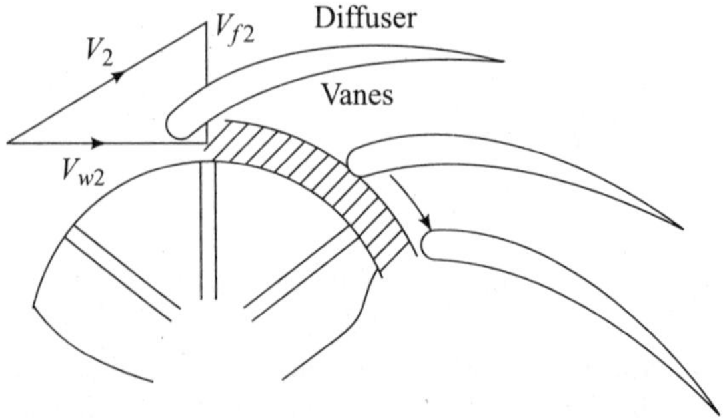

Fig. 16.8 Diffuser vanes

To find the correct angle for diffuser vanes, the flow in the vaneless space should be considered. No further energy is supplied to the air after it leaves the impeller. If we neglect the frictional losses, the angular momentum $V_w r$ remains consant. Hence V_w decreases from the impeller tip to the diffuser vane, in inverse proportion to the radius. For a channel of constant depth, the area of flow in the radial direction is directly proportional to the radius. The radial velocity V_f will therefore also decrease from the impeller tip to the diffuser vane, in accordance with the equation of continuity. If both V_f and V_w decrease from the impeller tip then the resulant velocity V decreases from the impeller tip and some diffusion takes place in the vaneless space. The consequent increase in density means that V_f will not decrease in inverse proportion to the radius as done by V_w, and the way V_f varies must be found from the equation of continuity.

16.1.2 Losses in a Centrifugal Compressor

The losses in a centrifugal compressor are almost of the same types as those in a centrifugal pump described in Section 15.3.3 of Chapter 15. However, the following features are to be noted.

Frictional Losses A major portion of the losses is due to fluid friction in stationary and rotating blade passages. The flow in impeller and diffuser is decelerating in nature. Therefore the frictional losses are due to both skin friction and boundary layer separation. The losses depend on the friction factor, length of the flow passage and square of the fluid velocity. The variation of frictional losses with mass flow is shown in Fig. 16.9.

Incidence Losses During the off-design conditions, the direction of relative velocity of fluid at the inlet does not match with the inlet blade angle and therefore the fluid cannot enter the blade passage smoothly by gliding along the blade surface. The loss in energy that takes place because of this is known as incidence loss. This is sometimes referred to as shock losses. However, the word shock in this context should not be confused with the aerodynamic sense of shock which is a sudden discontinuity in fluid properties and flow parameters that arises when a supersonic flow decelerates to a subsonic one as described in Chapter 14.

Clearance and Leakage Losses Certain minimum clearances are necessary between the impeller shaft and the casing and between the outer periphery of the impeller eye and the casing. The leakage of gas through the shaft clearance is minimised by the employing glands. The clearance losses depend upon the impeller diameter and the static pressure at the impeller tip. A larger diameter of the impeller is necessary for a higher peripheral speed (U_2) and it is very difficult in this situation to provide sealing between the casing and the impeller eye tip.

The variations of frictional losses, incidence losses and the total losses with mass flow rate are shown in Fig. 16.9. The leakage losses comprise a small fraction of the total loss. The incidence losses attain the minimum value at the designed mass flow rate. The shock losses are, in fact, zero at the designed flow rate. However, the incidence losses, as shown in Fig. 16.9, comprise both shock losses and impeller

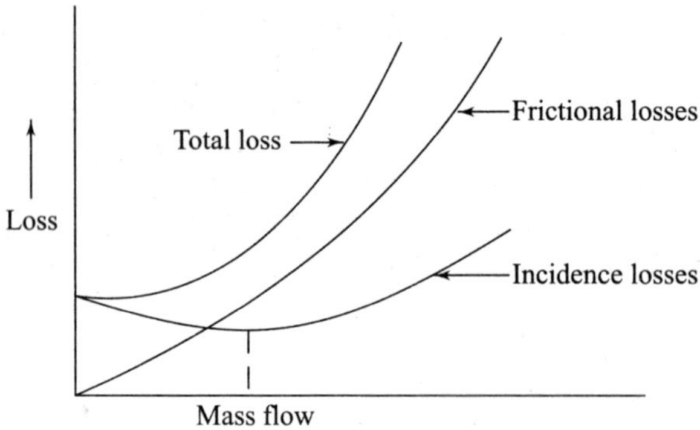

Fig. 16.9 Dependence of various losses with mass flow in a centrifugal compressor

entry loss due to a change in the direction of fluid flow from the axial to the radial direction in the vaneless space before entering the impeller blades. The impeller entry loss is similar to that in a pipe bend and is very small compared to other losses. This is why incidence losses show a non-zero minimum value (Fig. 16.9) at the designed flow rate.

16.1.3 Compressor Characteristics

The theoretical and actual head-discharge relationships of a centrifugal compressor are the same as those of a centrifugal pump, as described in Chapter 15. Therefore the curves of *H–Q* are similar to those of Figs 15.26 and 15.27. However, the performance of a compressor is usually specified by curves of delivery pressure and temperature against mass flow rate for various fixed values of rotational speed at given values of inlet pressure and temperature. It is always advisable to plot such performance characteristic curves with dimensionless variables. To find these dimensionless variables, we start with an implicit functional relationship of all the variables as

$$F(D, N, m, p_{1t}, p_{2t}, RT_{1t}, RT_{2t}, = 0 \tag{16.9}$$

where D = characteristic linear dimension of the machine, N = rotational speed, m = mass flow rate, p_{1t} = stagnation pressure at compressor inlet, p_{2t} = stagnation pressure at compressor outlet, T_{1t} = stagnation temperature at compressor inlet, T_{2t} = stagnation temperature at compressor outlet, and R = characteristic gas constant.

By making use of Buckingham's π theorem, we obtain the non-dimensional groups (π terms) as

$$\frac{p_{2t}}{p_{1t}}, \frac{T_{2t}}{T_{1t}}, \frac{m\sqrt{RT_{1t}}}{D^2\, p_{1t}}, \frac{ND}{\sqrt{RT_{1t}}}$$

The third and fourth non-dimensional groups are defined as 'non-dimensional mass flow' and non-dimensional rotational speed' respectively. The physical interpretation of these two non-dimensional groups can be ascertained as follows.

$$\frac{m\sqrt{RT}}{D^2 p} = \frac{\rho AV\sqrt{RT}}{D^2 p} = \frac{p}{RT}\frac{AV\sqrt{RT}}{D^2 p} \propto \frac{V}{\sqrt{RT}} \propto M_F$$

$$\frac{ND}{\sqrt{RT}} = \frac{U}{\sqrt{RT}} \propto M_R$$

Therefore, the 'non-dimensional mass flow' and 'non-dimensional rotational speed' can be regarded as flow Mach number, M_F and rotational speed Mach number, M_R.

When we are concerned with the performance of a machine of fixed size compressing a specified gas, R and D may be omitted from the groups and we can write

$$\text{Function}\left(\frac{p_{2t}}{p_{1t}}, \frac{T_{2t}}{T_{1t}}, \frac{m\sqrt{T_{1t}}}{p_{1t}}, \frac{N}{\sqrt{T_{1t}}}\right) = 0 \tag{16.10}$$

Though the terms $m\sqrt{T_{1t}}/p_{1t}$ and $N/\sqrt{T_{1t}}$ are truly not dimensionless, they are referred to as 'non-dimensional mass flow' and 'non-dimensional rotational speed' for practical purposes. The stagnation pressure and temperature ratios p_{2t}/p_{1t} and T_{2t}/T_{1t} are plotted against $m\sqrt{T_{1t}}/p_{1t}$ in the form of two families of curves, each curve of a family being drawn for fixed values of $N/\sqrt{T_{1t}}$. The two families of curves represent the compressor characteristics. From these curves, it is possible to draw the curves of isentropic efficiency η_c *vs* $m\sqrt{T_{1t}}/p_{1t}$ for fixed values of $N/\sqrt{T_{1t}}$. We can recall, in this context, the definition of the isentropic efficiency as

$$\eta_c = \frac{T'_{2t} - T_{1t}}{T_{2t} - T_{1t}} = \frac{(p_{2t}/p_{1t})^{\frac{\gamma-1}{\gamma}} - 1}{(T_{2t}/T_{1t}) - 1} \tag{16.11}$$

Before describing a typical set of characteristics, it is desirable to consider what might be expected to occur when a valve placed in the delivery line of the compressor running at a constant speed, is slowly opened. When the valve is shut and the mass flow rate is zero, the pressure ratio will have some value A (Fig. 16.10), corresponding to the centrifugal pressure head produced by the action

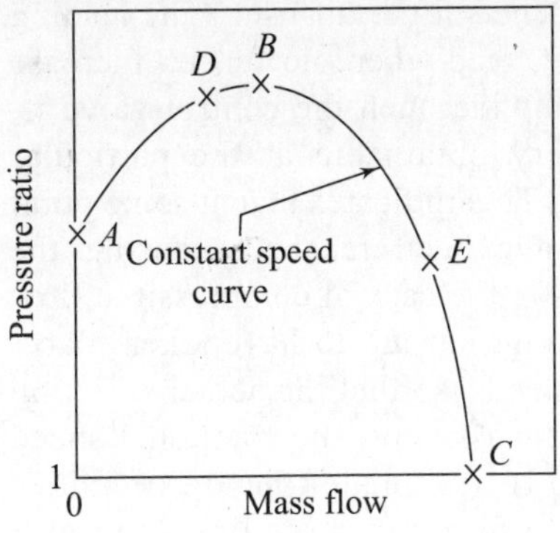

Fig. 16.10 The theoretical characteristic curve, after Cohen et al. [1]

of the impeller on the air trapped between the vanes. As the valve is opened, flow commences and the diffuser begins to influence the pressure rise, for which the pressure ratio increases. At some point *B*, efficiency approaches its maximum and the pressure ratio also reaches its maximum. Further increase of mass flow will result in a fall of pressure ratio. For mass flows greatly in excess of that corresponding to the design mass flow, the air angles will be widely different from the vane angles and breakaway of the air will occur. In this hypothetical case, the pressure ratio drops to unity at *C*, when the valve is fully open and all the power is absorbed in overcoming internal frictional resistances.

In practice, the operating point *A* could be obtained if desired but a part of the curve between *A* and *B* could not be obtained due to surging. It may be explained in the following way. If we suppose that the compressor is operating at a point *D* on the part of characteristic curve (Fig. 16.10) having a positive slope, then a decrease in mass flow will be accompanied by a fall in delivery pressure. If the pressure of the air downstream of the compressor does not fall quickly enough, the air will tend to reverse its direction and will flow back in the direction of the resulting pressure gradient. When this occurs, the pressure ratio drops rapidly causing a further drop in mass flow until the point *A* is reached, where the mass flow is zero. When the pressure downstream of the compressor has reduced sufficiently due to reduced mass flow rate, the positive flow becomes established again and the compressor picks up repeat the cycle of events which occurs at high frequency.

This surging of air may not happen immediately as the operating point moves to the left of *B* because the pressure downstream of the compressor may at first fall at a greater rate than the delivery pressure. As the mass flow is reduced, the reverse will apply and the conditions are unstable between *A* and *B*. As long as the operating point is on the part of the characteristic having a negative slope, decrease in mass flow is accompanied by a rise in delivery pressure and the operation is stable.

Let us consider the constant speed curve *ABC* in Fig. 16.10. There is an additional limitation to the operating range, between *B* and *C*. As the mass flow increases and the pressure decreases, the density is reduced and the radial component of velocity must increase. At constant rotational speed this means an increase in resultant velocity and hence in angle of incidence at the diffuser vane leading edge. At some point, say *E*, the position is reached where no further increase in mass flow can be obtained no matter how wide open the control valve is. This point represents the maximum delivery obtainable at the particular rotational speed for which the curve is drawn. This indicates that at some point within the compressor sonic conditions have been reached, causing the limiting maximum mass flow rate to be set as in the case of compressible flow through a converging diverging nozzle. Choking is said to have taken place. Other curves may be obtained for different speeds, so that the actual variation of pressure ratio over the complete range of mass flow and rotational speed will be shown by curves such as those in Fig. 16.11. The left-hand extremities of the constant speed curves may be joined to form the surge line, the right-hand extremities indicate choking(Fig. 16.11).

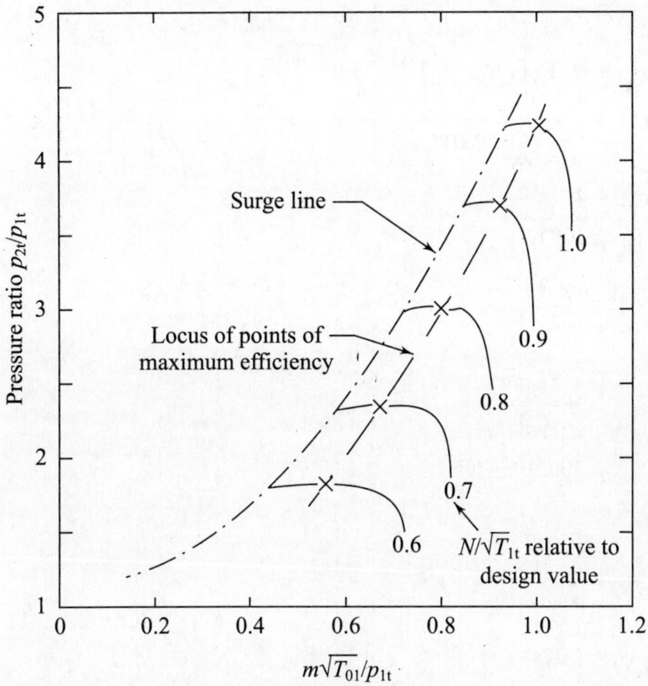

Fig. 16.11 Variations of pressure ratio over the complete range of mass flow for different rotational speeds, after Cohen et al. [1]

Example 16.1

Air at a stagnation temperature of 27° C enters the impeller of a centrifugal compressor in the axial direction. The rotor which has 15 radial vanes, rotates at 20000 rpm. The stagnation pressure ratio between the diffuser outlet and the impeller inlet is 4 and the isentropic efficiency is 85%. Determine (i) the impeller tip radius and (ii) power input to the compressor when the mass flow rate is 2 kg/s. Assume a power input factor of 1.05 and a slip factor $\sigma = 1 - 2/n$, where n is the number of vanes. For air, take $\gamma = 1.4$, $R = 287$ J/kg K.

Solution

(i) From Eq (16.7), we can write

$$T_{3t} - T_{1t} = \frac{T_{1t}\left[(p_{3t}/p_{1t})^{\frac{\gamma-1}{\gamma}} - 1\right]}{\eta_c}$$

again with the help of Eq (16.5) and $T_{2t} = T_{3t}$ it becomes

$$U_2^2 = \frac{c_p T_{1t}\left[(p_{3t}/p_{1t})^{\frac{\gamma-1}{\gamma}} - 1\right]}{\eta_c \sigma \psi}$$

Here,

$$p_{3t}/p_{1t} = 4$$

$$T_{1t} = 300 \text{ K}$$

$$c_p = \frac{\gamma R}{\gamma - 1}$$

$$= \frac{1.4 \times 287}{0.4}$$

$$= 1005 \text{ J/kg K.}$$

$$\sigma = 1 - \frac{2}{15}$$

$$= 0.867$$

$$\psi = 1.05$$

Therefore,

$$U_2^2 = \frac{1005 \times 300 \times \left(4^{\frac{0.4}{1.4}} - 1\right)}{0.85 \times 0.867 \times 1.05}$$

which gives $U_2 = 435$ m/s

Thus the impeller tip radius is

$$r_2 = \frac{435 \times 60}{2\pi \times 20000}$$

$$= 0.21 \text{ m}$$

(ii) Power input to the air $= \dfrac{2 \times 1.05 \times 0.867 \times (435)^2}{1000}$ kW

$$= 344.52 \text{ kW}$$

Example 16.2

Determine the pressure ratio developed and the specific work input to drive a centrifugal air compressor of an impeller diameter of 0.5 m and running at 7000 rpm. Assume zero whirl at the entry and $T_{1t} = 290$ K. The slip factor and power input factor to be unity, the process of compression is isentropic and for air $c_p = 1005$ J/kg K, $\gamma = 1.4$.

Solution

The impeller tip speed

$$U_2 = \frac{\pi \times 0.5 \times 7000}{60}$$

$$= 183.26 \text{ m/s}$$

With the help of Eqs (16.6) and (16.7), we can write

$$\text{Pressure ratio} = \left[1 + \frac{U_2^2}{c_p T_{1t}}\right]^{\frac{\gamma}{\gamma-1}}$$

$$= \left[1 + \frac{(183.26)^2}{1005 \times 290}\right]^{\frac{1.4}{0.4}}$$

$$= 1.46$$

From Eq (16.3), specific work input $= U_2^2 = \dfrac{(183.26)^2}{1000} = 33.58$ kJ/kg

Example 16.3

A centrifugal compressor has an impeller tip speed of 360 m/s. Determine (i) the absolute Mach number of flow leaving the radial vanes of the impeller and (ii) the mass flow rate. The following data are given:

Impeller Tip speed	360 m/s
Radial component of flow velocity at impeller exit	30 m/s
Slip factor	0.9
Flow area at impeller exit	0.1 m^2
Power input factor	1.0
Isentropic efficiency	0.9
Inlet stagnation temperature	300 K
Inlet stagnation pressure	100 kN/m^2
R (for air)	287 J/kg K
γ (for air)	1.4

Solution

The absolute Mach number is the Mach number based on absolute velocity.

Therefore, $$M_2 = \frac{V_2}{\sqrt{\gamma R T_2}}$$

Now V_2 and T_2 have to be determined.

From the velocity triangle at the impeller exit,

$$V_2 = \sqrt{V_{w2}^2 + V_{f2}^2}$$

In case of slip, $V_{w2} = \sigma U_2$

Hence,

$$V_2 = \sqrt{(\sigma U_2)^2 + V_{f2}^2}$$

$$= \sqrt{(0.9 \times 360)^2 + (30)^2}$$

$$= 325.38 \text{ m/s}$$

From Eq. (16.5)

$$T_{2t} = T_{1t} + \frac{\psi \sigma U_2^2}{c_p}$$

$$\left[c_p = \frac{\gamma R}{\gamma - 1} = \frac{1.4 \times 287}{0.4} = 1005 \text{ J/kg K} \right]$$

$$T_{2t} = 300 + \frac{0.9 \times (360)^2}{1005}$$

$$= 416 \text{ K.}$$

$$T_2 = T_{2t} - \frac{V_2^2}{2c_p}$$

$$= 416 - \frac{(325.38)^2}{2 \times 1005}$$

$$= 363.33 \text{ K.}$$

Therefore,

$$M_2 = \frac{325.28}{\sqrt{1.4 \times 287 \times 363.33}}$$

$$= 0.85$$

Mass flow rate $\dot{m} = \rho_2 A_2 V_{f2}$

We have to find out ρ_2

With the help of Eq. (16.7), we can write

$$\frac{p_{2t}}{p_{1t}} = \left[1+\frac{0.9\times(416-300)}{300}\right]^{\frac{1.4}{0.4}}$$

$$= 2.84$$

Again, $$\frac{p_2}{p_{2t}} = \left(\frac{T_2}{T_{2t}}\right)^{\frac{1.4}{0.4}} = \left(\frac{363.33}{416}\right)^{\frac{1.4}{0.4}} = 0.623$$

Hence,

$$p_2 = 0.623\, p_{2t}$$

$$= 0.623 \times 2.84\, p_{1t}$$

$$= 0.623 \times 2.84 \times 100 \text{ kPa}$$

$$= 176.93 \text{ kPa}$$

Therefore, $$\dot{m} = \left(\frac{p_2}{RT_2}\right)\cdot A_2V_{f2}$$

$$= \frac{176.93\times10^3}{287\times363.33}\times0.1\times30$$

$$= 5.09 \text{ kg/s}$$

Example 16.4

Air at a temperature of 27°C flows into a centrifugal compressor running at 20,000 rpm. The following data are given:

Slip factor	0.80
Power input factor	1
Isentropic efficiency	80%
Outer diameter of blade tip	0.5 m

Assuming the absolute velocities of air entering and leaving the compressor are same, find (i) static temperature rise of air passing through the compressor, and (ii) the static pressure ratio.

Solution

Velocity of the blade tip,

$$U_2 = \frac{\pi\times0.5\times20{,}000}{60}$$

$$= 523.6 \text{ m/s}$$

From Eq. (16.5),

$$\text{Stagnation temperature rise } (T_{2t} - T_{1t}) = \frac{\psi \sigma U_2^2}{c_p}$$

$$= \frac{0.80 \times (523.6)^2}{1005} = 218.23^\circ \text{ C}$$

(c_p of air has been taken as 1005 J/kg K)

Since the absolute velocities at the inlet and the outlet of the stage are the same, the rise in stagnation temperature equals to that in static temperature.

The static pressure ratio can be written as

$$\frac{p_2}{p_1} = \left(\frac{T_2'}{T_1}\right)^{\frac{\gamma}{\gamma-1}}$$

$$= \left[1 + \frac{\eta_c (T_2 - T_1)}{T_1}\right]^{\frac{\gamma}{\gamma-1}}$$

$$= \left[1 + \frac{0.8 \times 218.23}{300}\right]^{\frac{1.4}{0.4}}$$

$$= 4.98$$

16.2 AXIAL FLOW COMPRESSORS

The basic components of an axial flow compressor are a rotor and a stator; the former carrying the moving blades and the latter the stationary rows of blades. The stationary blades convert the kinetic energy of the fluid into pressure energy, and also redirect the flow into an angle suitable for entry to the next row of moving blades. Each stage will consist of one rotor row followed by a stator row but it is usual to provide a row of so-called inlet guide vanes. This is an additional stator row upstream of the first stage in the compressor and serves to direct the axially approaching flow correctly into the first row of the rotating blades. Two forms of rotor have been taken up, namely, drum type and disk type. A disk-type rotor is illustrated in Fig. 16.12. The disk type is used where consideration of low weight is most important. There is a contraction of the flow annulus from the low to the high-pressure end of the compressor. This is necessary to maintain the axial velocity at a reasonably constant level throughout the length of the compressor despite the increase in density of air. Figure 16.13 illustrates this flow through compressor stages.

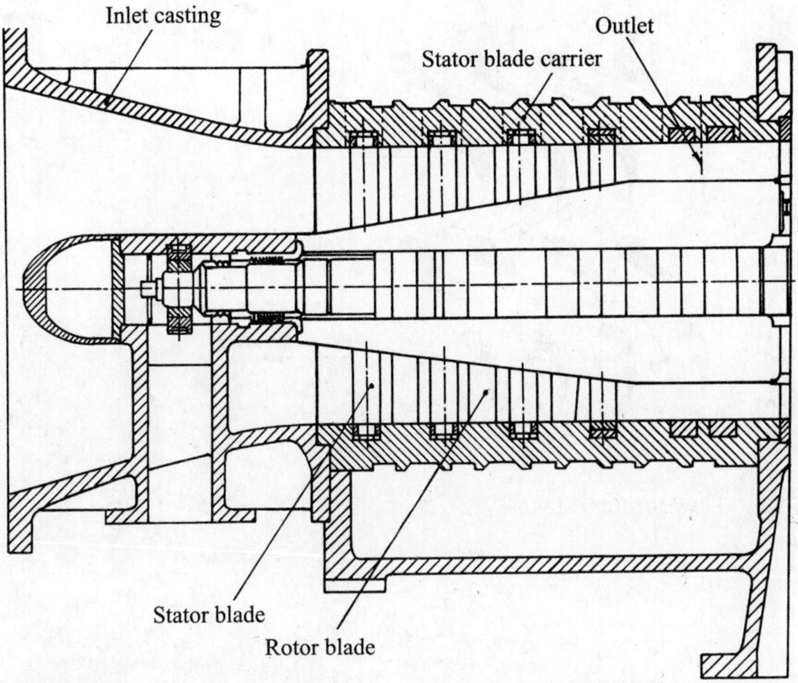

Fig. 16.12 Disk type axial flow compressor

The basic principle of acceleration of the working fluid, followed by diffusion to convert acquired kinetic energy into a pressure rise, is applied in the axial compressor. The flow is considered as occurring in a tangential plane at the mean blade height where the blade peripheral velocity is U. This two-dimensional approach means that, in general, the flow velocity will have two components, one axial and one peripheral denoted by subscript w, implying a whirl velocity. It is first assumed that the air approaches the rotor blades with an absolute velocity V_1, at an angle α_1 to the axial direction. In combination with the peripheral velocity U of the blades, its relative velocity will be V_{r1} at an angle β_1 as shown in the upper velocity triangle (Fig. 16.14). After passing through the diverging passages formed between the rotor blades which do work on the air and increase its absolue velocity, the air will emerge with the relative velocity of V_{r2} at angle β_2 which is less than β_1. This turning of air towards the axial direction is, as previously mentioned, necessary to provide an increase in the effective flow area and is brought about by the camber of the blades. Since V_{r2} is less than V_{r1} due to diffusion, some pressure rise has been accomplished in the rotor. The velocity V_{r2} in combination with U gives the absolute velocity V_2 at the exit from the rotor at an angle angle α_2 to the axial direction. The air then passes through the passages formed by the stator blades where it is further diffused to velocity V_3 at an angle α_3 which in most designs equals to α_1 so that it is prepared for entry to the next stage. Here again, the turning of the air towards the axial direction is brought about by the camber of the blades.

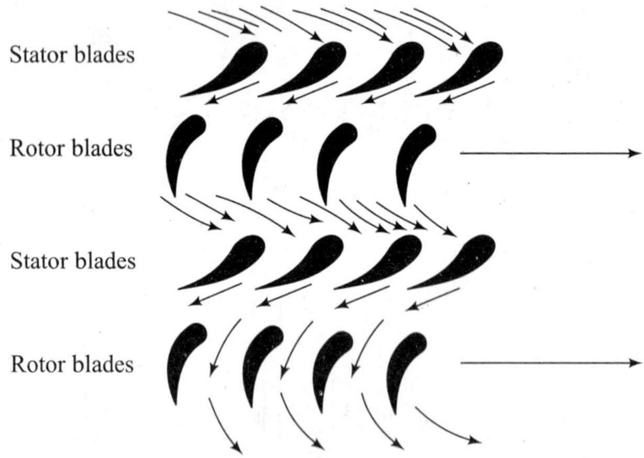

Fig. 16.13 Flow through stages

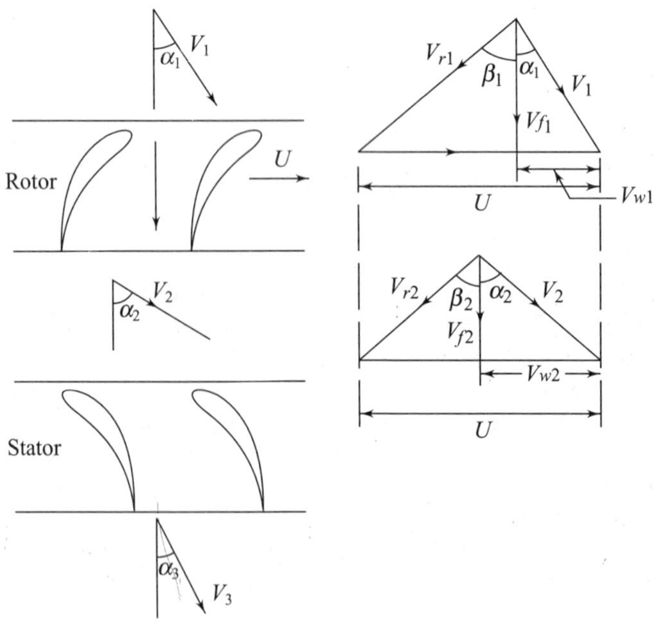

Fig. 16.14 Velocity triangles of a stage of an axial flow compressor

Two basic equations follow immediately from the geometry of the velocity triangles. These are as follows:

$$\frac{U}{V_f} = \tan \alpha_1 + \tan \beta_1 \tag{16.12}$$

$$\frac{U}{V_f} = \tan \alpha_2 + \tan \beta_2 \tag{16.13}$$

in which $V_f = V_{f1} = V_{f2}$ is the axial velocity, assumed consant through the stage. The work done per unit mass or specific work input, w being given by

$$w = U(V_{w2} - V_{w1}) \tag{16.14}$$

This expression can be put in terms of the axial velocity and air angles to give

$$w = UV_f(\tan \alpha_2 - \tan \alpha_1) \tag{16.15}$$

or by using Eqs (16.12) and (16.13)

$$w = UV_f(\tan \beta_1 - \tan \beta_2) \tag{16.16}$$

This input energy will be absorbed usefully in raising the pressure and velocity of the air. A part of it will be spent in overcoming various frictional losses. Regardless of the losses, the input will reveal itself as a rise in the stagnation temperature of the air ΔT_{st}. If the absolute velocity of the air leaving the stage V_3 is made equal to that at the entry V_1, the stagnation temperature rise ΔT_{st} will also be the static temperature rise of the stage, ΔT_s, so that

$$\Delta T_{st} = \Delta T_s = \frac{UV_f}{c_p}(\tan \beta_1 - \tan \beta_2) \tag{16.17}$$

In fact, the stage temperature rise will be less than that given in Eq. (16.17) owing to three-dimensional effects in the compressor annulus. Experiments show that it is necessary to multiply the right-hand side of Eq. (16.17) by a work-done factor λ which is a number less than unity. This is a measure of the ratio of actual work-absorbing capacity of the stage to its ideal value.

The radial distribution of axial velocity is not constant across the annulus but becomes increasingly peaky (Fig. 16.15) as the flow proceeds, setting down to a fixed profile at about the fourth stage. Equation (16.16) can be written with the help of Eq. (16.12) as

$$\begin{aligned} w &= U[(U - V_f \tan \alpha_1) - V_f \tan \beta_2] \\ &= U(U - V_f(\tan \alpha_1 + \tan \beta_2)) \end{aligned} \tag{16.18}$$

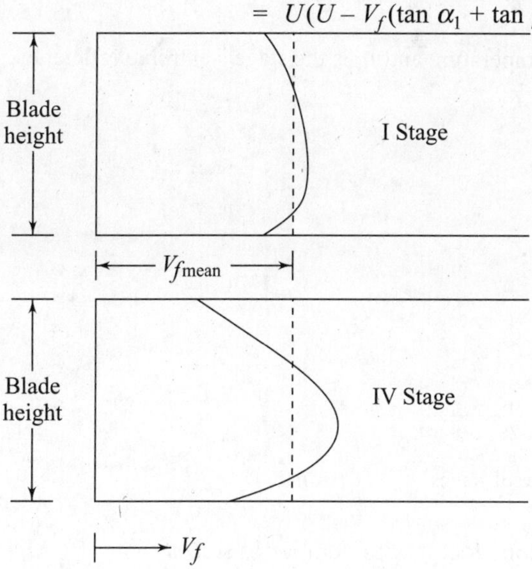

Fig. 16.15 Axial velocity distributions

Since the outlet angles of the stator and the rotor blades fix the value of α_1 and β_2 and hence the value of ($\tan \alpha_1 + \tan \beta_2$). Any increase in V_f will result in a decrease in w and vice versa. If the compressor is designed for constant radial distribution of V_f as shown by the dotted line in Fig. 16.15, the effect of an increase in V_f in the central region of the annulus will be to reduce the work capacity of blading in that area. However, this reduction is somewhat compensated by an increase in w in the regions of the root and tip of the blading because of the reduction of V_f at these parts of the annulus. However, the net result is a loss in total work capacity because of the adverse effects of blade tip clearance and boundary layers on the annulus walls. This effect becomes more pronounced as the number of stages is increased and the way in which the mean value varies with the number of stages. The variation of λ with the number of stages is shown in Fig. 16.16. Care should be taken to avoid confusion of this factor with the idea of an efficiency. If w is the expression for the specific work input (Eq. 16.14), then λw is the actual amount of work which can be supplied to the stage. The application of an isentropic efficiency to the resulting temperature rise will yield the equivalent isentropic temperature rise from which the stage pressure ratio may be calculated. Thus, the actual stage temperature rise is given by

$$\Delta T_{st} = \frac{\lambda U V_f}{c_p} (\tan \beta_1 - \tan \beta_2) \tag{16.19}$$

and the pressure ratio R_s by

$$R_s = \left[1 + \frac{\eta_s \Delta T_{st}}{T_{1t}}\right]^{\frac{\gamma}{\gamma - 1}} \tag{16.20}$$

where T_{1t} is the inlet stagnation temperature and η_s is the stage isentropic efficiency.

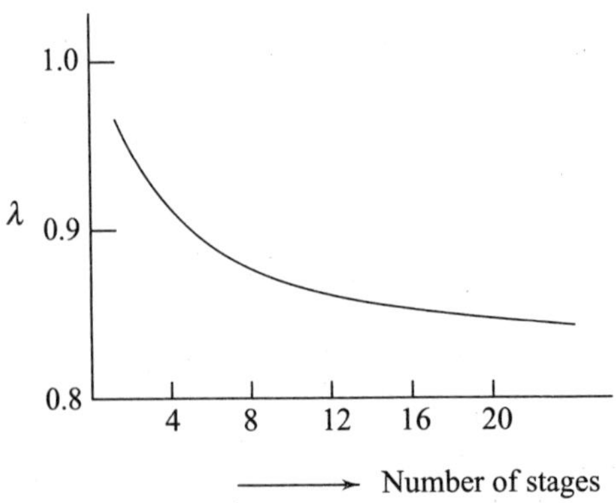

Fig. 16.16 Variation of work-done factor with number of stages

Example 16.5

At the mean diameter, $U = 200$ m/s, $V_f = 180$ m/s, $\beta_1 = 43.9°$ and $\beta_2 = 13.5°$. The factor $\lambda = 0.86$ and $\eta_s = 0.85$ and inlet temperature T_{1t} is 288 K. Calculate the pressure ratio.

Solution

$$\Delta T_{st} = \frac{0{\cdot}86 \times 200 \times 180}{1{\cdot}005 \times 10^3} \text{ (tan 43.9° – tan 13.5°)}$$

$$= 22.24 \text{ K}$$

and
$$R_s = \left[1 + \frac{0{\cdot}85 \times 22{\cdot}24}{288}\right]^{3.5} = 1.25$$

[c_p of air has been taken as 1005 J/kg K]

Example 16.6

The conditions of air at the entry of an axial flow compressor stage are $p_1 = 100 \text{kN/m}^2$ and $T_1 = 300$ K. The air angles are $\beta_1 = 51°$, $\beta_2 = 10°$, $\alpha_1 = \alpha_3 = 8°$.

The mean diameter and peripheral speed are 0.5 m and 150 m/s respectively. Mass flow rate through the stage is 30 kg/s; the work done factor is 0.95 and mechanical efficiency is 90%. Assuming an isentropic stage efficiency of 85%, determine (i) blade height at entry (ii) stage pressure ratio, and (iii) the power required to drive the stage (for air, $R = 287$ J/kg K, $\gamma = 1.4$)

Solution

(i) $$\rho_1 = \frac{p_1}{RT_1} = \frac{100 \times 10^3}{287 \times 300} = 1.16 \text{ kg/m}^3$$

From Eq. (16.12),

$$\frac{U}{V_f} = \tan\alpha_1 + \tan\beta_1$$

Hence,
$$V_f = \frac{150}{\tan 8° + \tan 51°}$$

$$= 109.06 \text{ m/s}$$

$$\dot{m} = V_f \rho_1 (\pi d h_1)$$

$$30 = 109.06 \times 1.16 \times \pi \times 0.5 h_1$$

which gives $h_1 = 0.15$ m

(ii) From Eq. (16.19)

$$\Delta T_{st} = \frac{\lambda U V_f}{c_p}(\tan\beta_1 - \tan\beta_2)$$

Again, $$c_p = \frac{1.4}{(1.4-1)} \times 287 = 1005 \text{ J/kg K}$$

Hence, $$\Delta T_{st} = \frac{0.95\times150\times190.06}{1005}(\tan 51^\circ - \tan 10^\circ)$$

$$= 16.37^\circ \text{ C}$$

With the help of Eq. (16.20) we can write

$$\text{Pressure ratio, } R_s = \left[1+\frac{0.85\times16.37}{300}\right]^{\frac{1.4}{0.4}}$$

$$= 1.17$$

(iii) $$P = \frac{\dot{m}w}{\eta_m} = \frac{\dot{m}c_p\,\Delta T_{st}}{\eta_m}$$

$$= \frac{30\times1005\times16.37}{0.9\times1000}\text{ kW} = 548.39 \text{ kW}$$

16.2.1 Degree of Reaction

A certain amount of diffusion (a rise in static pressure) takes place as the air passes through the rotor as well as the stator; the rise in pressure through the stage is in general, attributed to both blade rows. The term degree of reaction is a measure of the extent to which the rotor itself contributes to the increase in the static head of fluid. It is defined as the ratio of the static enthalpy rise in the rotor to that in the whole stage. Variation of c_p over the relevent temperature range will be negligibly small and hence this ratio of enthalpy rise will be equal to the corresponding temperature rise.

It is useful to obtain a formula for the degree of reaction in terms of the various velocities and air angles associated with the stage. This will be done for the most common case in which it is assumed that the air leaves the stage with the same velocity (absolute) with which it enters ($V_1 = V_3$).

This leads to $\Delta T_s = \Delta T_{st}$. If ΔT_A and ΔT_B are the static temperature rises in the rotor and the stator respectively, then from Eqs (16.15), (16.16) and (16.17),

$$\begin{aligned} w &= c_p(\Delta T_A + \Delta T_B) = c_p\Delta T_s \\ &= UV_f(\tan\beta_1 - \tan\beta_2) \\ &= UV_f(\tan\alpha_2 - \tan\alpha_1) \end{aligned} \qquad (16.21)$$

Since all the work input to the stage is transferred to air by means of the rotor, the steady flow energy equation yields

$$w = c_p\, \Delta T_A + \frac{1}{2}(V_2^2 - V_1^2)$$

With the help of Eq. (16.21), it becomes

$$c_p \Delta T_A = UV_f(\tan \alpha_2 - \tan \alpha_1) - \frac{1}{2}(V_2^2 - V_1^2)$$

But $V_2 = V_f \sec \alpha_2$ and $V_1 = V_f \sec \alpha_1$, and hence

$$c_p \Delta T_A = UV_f(\tan \alpha_2 - \tan \alpha_1) - \frac{1}{2}\, V_f^2(\sec^2 \alpha_2 - \sec^2 \alpha_1)$$

$$= UV_f(\tan \alpha_2 - \tan \alpha_1) - \frac{1}{2}\, V_f^2(\tan^2 \alpha_2 - \tan^2 \alpha_1) \qquad (16.22)$$

The degree of reaction

$$\Lambda = \frac{\Delta T_A}{\Delta T_A + \Delta T_B} \qquad (16.23)$$

With the help of Eq. (16.22), it becomes

$$\Lambda = \frac{UV_f(\tan \alpha_2 - \tan \alpha_1) - \frac{1}{2} V_f^2(\tan^2 \alpha_2 - \tan^2 \alpha_1)}{UV_f(\tan \alpha_2 - \tan \alpha_1)}$$

By adding up Eq. (16.12) and Eq. (16.13), we get

$$\frac{2U}{V_f} = \tan \alpha_1 + \tan \beta_1 + \tan \alpha_2 + \tan \beta_2$$

Hence,

$$\Lambda = 1 - \frac{V_f}{2U}(\tan \alpha_2 + \tan \alpha_1)$$

or

$$\Lambda = \frac{V_f}{2U}\left(\frac{2U}{V_f} - \frac{2U}{V_f} + \tan \beta_1 + \tan \beta_2\right)$$

or

$$\Lambda = \frac{V_f}{2U}(\tan \beta_1 + \tan \beta_2) \qquad (16.24)$$

As the case of 50% reaction blading is important in design, it is of interest to see the result for $\Lambda = 0.5$

$$\tan \beta_1 + \tan \beta_2 = \frac{U}{V_f}$$

and it follows from Eqs (16.12) and (16.13) that

$$\tan \alpha_1 = \tan \beta_2, \text{ i.e. } \alpha_1 = \beta_2 \qquad (16.25a)$$

$$\tan \beta_1 = \tan \alpha_2, \text{ i.e. } \beta_1 = \alpha_2 \qquad (16.25b)$$

Furthermore, since V_f is constant through the stage.

$$V_f = V_1 \cos \alpha_1 = V_3 \cos \alpha_3$$

And since we have initially assumed that $V_3 = V_1$, it follows that $\alpha_1 = \alpha_3$. Because of this equality of angles, namely, $\alpha_1 = \beta_2 = \alpha_3$ and $\beta_1 = \alpha_2$, blading designed on this basis is sometimes referred to as *symmetrical blading*.

It is to be remembered that in deriving Eq. (16.24) for Λ we have implicitly assumed a work done factor λ of unity in making use of Eq. (16.22). A stage designed with symmetrical blading is referred to as 50% reaction stage although Λ will differ slightly for λ.

Example 16.7

The preliminary design of an axial flow compressor is to be based upon a simplified consideration of the mean diameter conditions. Suppose that the characteristics of a repeating stage of such a design are as follows:

Stagnation temperature rise (ΔT_{st})	30 K
Degree of reaction (Λ)	0.6
Flow coefficienty (V_f/U)	0.5
Blade speed (U)	300 m/s

Assuming constant axial velocity across the stage and equal absolute velocities at inlet and outlet, determine the blade angles of the rotor for a shock free flow. (c_p for air = 1005 J/kg K).

Solution

Specific work input $\quad w = 1005 \times 30$ J/kg

From Eq. (16.17)

$$1005 \times 30 = (300)^2 \times (0.5)(\tan \beta_1 - \tan \beta_2)$$

or

$$\tan \beta_1 - \tan \beta_2 = 0.67$$

Again from Eq. (16.24),

$$0.6 = \frac{0.5}{2}(\tan \beta_1 + \tan \beta_2)$$

$$\tan \beta_1 + \tan \beta_2 = 2.4$$

The above two equations give

$$\beta_1 = 56.92°, \quad \beta_2 = 40.86°$$

16.3 FANS AND BLOWERS

Fans and blowers (Fig. 16.17) are turbomachines which deliver air at a desired high velocity (and accordingly at a high mass flow rate) but at a relatively low static pressure. The total pressure rise across a fan is extremely low and is of the order of a few millimeters of water gauge. The rise in static pressure across a blower is relatively higher and is more than 1000 mm of water guage that is required to overcome the pressure losses of the gas during its flow through various passages.

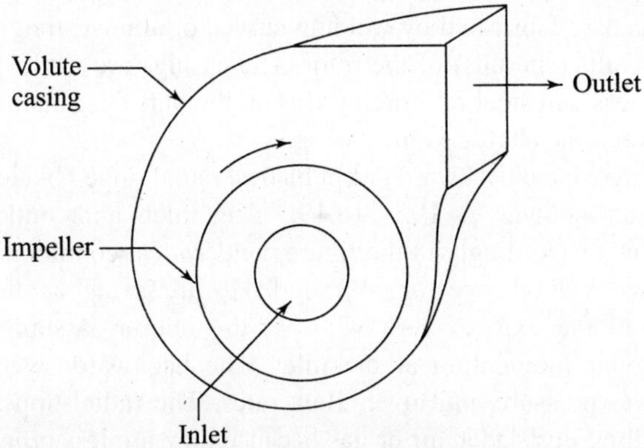

Fig. 16.17 A centrifugal fan or blower

A large number of fans and blowers for relatively high pressure applications are of centrifugal type. The main components of a centrifugal blower are shown in Fig. 16.18. It consists of an impeller which has blades fixed between the inner and outer diameters. The impeller can be mounted either directly on the shaft extension of the prime mover or separately on a shaft supported between two additional bearings. Air or gas enters the impeller axially through the inlet nozzle which provides slight acceleration to the air before its entry to the impeller. The action of the impeller swings the gas from a smaller to a larger radius and delivers the gas at a high pressure and velocity to the casing. The flow from the impeller blades is collected by a spiral-shaped casing known as *volute casing* or *spiral casing*. The casing can further increase the static pressure of the air and it finally delivers the air to the exit of the blower.

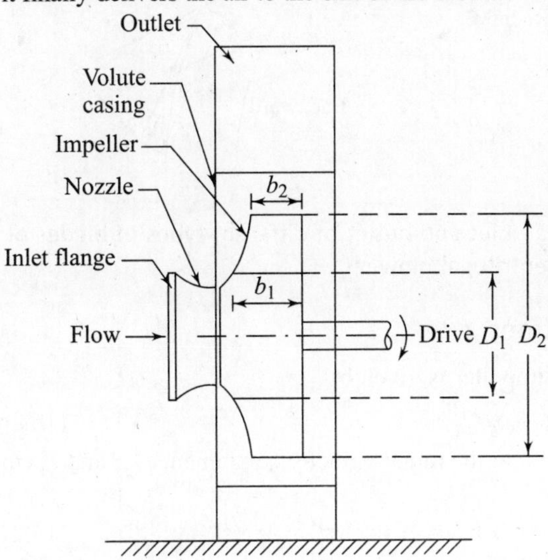

Fig. 16.18 Main components of a centrifugal blower

The centrifugal fan impeller can be fabricated by welding curved or almost straight metal blades to the two side walls (shrouds) of the rotor. The casings are made of sheet metal of different thickness and steel reinforcing ribs on the outside. Suitable sealing devices are used between the shaft and the casing.

A centrifugal fan impeller may have backward swept blades, radial tipped blades or forward swept blades as shown in Fig. 16.19. The inlet and outlet velocity triangles are also shown accordingly in the figure. Under ideal conditions, the directions of the relative velocity vectors V_{r1} and V_{r2} are same as the blade angles at the entry and the exit. A zero whirl at the inlet is assumed which results in a zero angular momentum at the inlet. The backward swept blades are employed for lower pressure and lower flow rates. The radial-tipped blades are employed for handling dust-laden air or gas because they are less prone to blockage, dust erosion and failure. The radial-tipped blades in practice are of forward-swept type at the inlet, as shown in Fig. 16.19. The forward-swept blades are widely used in practice. On account of the forward-swept blade tips at the exit, the whirl component of exit velocity (V_{w2}) is large which results in a higher stage pressure rise.

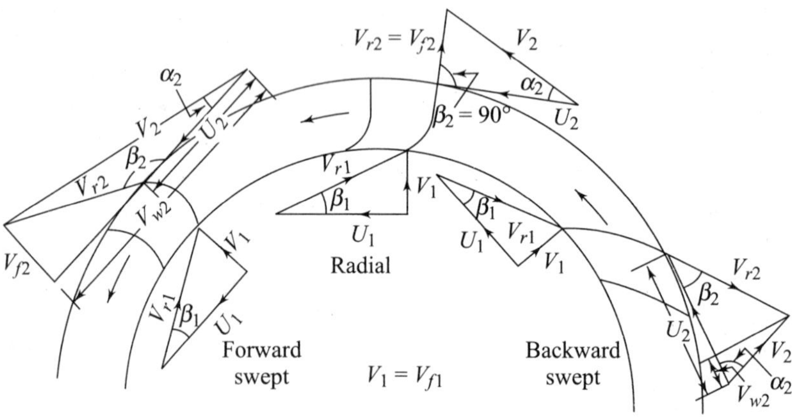

Fig. 16.19 Velocity triangles at inlet and outlet of different types of blades of an impeller of a centrifugal blower

16.3.1 Parametric Calculations

The mass flow rate through the impeller is given by

$$\dot{m} = \rho_1 Q_1 = \rho_2 Q_2 \tag{16.26}$$

The areas of cross sections normal to the radial velocity components V_{f1} and V_{f2} are $A_1 = \pi D_1 b_1$ and $A_2 = \pi D_2 b_2$

$$m = \rho_1 V_{f1} (\pi D_1 b_1) = \rho_2 V_{f2} (\pi D_2 b_2) \tag{16.27}$$

The radial component of velocities at the impeller entry and exit depend on its width at these sections. For small pressure rise through the impeller stage, the density change in the flow is negligible and the flow can be assumed to be almost incompressible. For constant radial velocity

$$V_{f1} = V_{f2} = V_f \tag{16.28}$$

Eqs (16.27) and (16.28) give

$$b_1/b_2 = D_2/D_1 \tag{16.29}$$

16.3.2 Work

The work done is given by Euler's Equation (Eq. 15.2) as

$$w = U_2V_{w2} - U_1V_{w1} \tag{16.30}$$

It is reasonable to assume zero whirl at the entry. This condition gives

$$\alpha_1 = 90°,\ V_{w1} = 0 \text{ and hence, } U_1V_{w1} = 0$$

Therefore we can write,

$$V_1 = V_{f1} = V_{f2} = U_1 \tan \beta_1 \tag{16.31}$$

Equation (16.30) gives

$$w = U_2V_{w2} = U_2^2\left(\frac{V_{w2}}{U_2}\right) \tag{16.32}$$

For any of the exit velocity triangles (Fig. 16.19)

$$U_2 - V_{w2} = V_{f2} \cot \beta_2$$

$$\frac{V_{w_2}}{U_2} = \left[1 - \frac{V_{f2} \cot \beta_2}{U_2}\right] \tag{16.33}$$

Eqs (16.32) and (16.33) yield

$$w = U_2^2\,[1 - \varphi \cot \beta_2] \tag{16.34}$$

where $\varphi\ (= V_{f2}/U_2)$ is known as the flow coefficient

$$\text{Head developed in metres of air} = H_a = \frac{U_2V_{w2}}{g} \tag{16.35}$$

$$\text{Equivalent head in metres of water} = H_w = \frac{\rho_a H_a}{\rho_w} \tag{16.36}$$

where ρ_a and ρ_w are the densities of air and water respectively.

Assuming that the flow fully obeys the geometry of the impeller blades, the specific work done in an isentropic process is given by

$$(\Delta h_0) = U_2\,(1 - \varphi \cot\beta_2) \tag{16.37}$$

The power required to drive the fan is

$$P = m\,(\Delta h_0) = mU_2V_{w2} = mU_2^2\,(1 - \varphi \cot \beta_2)$$

$$= mc_p\,(\Delta T_0) \tag{16.38}$$

The static pressure rise through the impeller is due to the change in centrifugal energy and the diffusion of relative velocity component. Therefore, it can be written as

$$p_2 - p_1 = (\Delta p) = \frac{1}{2}\rho\,(U_2^2 - U_1^2) + \frac{1}{2}\rho\,(V_{r1}^2 - V_{r2}^2) \tag{16.39}$$

The stagnation pressure rise through the stage can also be obtained as

$$(\Delta p_0) = \frac{1}{2}\rho\,(U_2^2 - U_1^2) + \frac{1}{2}\rho\,(V_{r1}^2 - V_{r2}^2) + \frac{1}{2}\rho\,(V_2^2 - V_1^2) \tag{16.40}$$

From (16.39) and (16.40), we get

$$(\Delta p_0) = (\Delta p) + \frac{1}{2}\rho\,(V_2^2 - V_1^2) \tag{16.41}$$

From any of the outlet velocity triangles (Fig. 16.19),

$$\frac{V_2}{\sin\beta_2} = \frac{U_2}{\sin\{\pi - (\alpha_2 + \beta_2)\}}$$

or

$$\frac{V_2}{\sin\beta_2} = \frac{U_2}{\sin(\alpha_2 + \beta_2)} \tag{16.42}$$

or

$$V_{w2} = V_2\cos\alpha_2 = \frac{U_2\sin\beta_2\cos\alpha_2}{\sin(\alpha_2 + \beta_2)}$$

or

$$\frac{V_{w2}}{U_2} = \frac{\sin\beta_2\cos\alpha_2}{\sin\alpha_2\cos\beta_2 + \cos\alpha_2\sin\beta_2}$$

or

$$\frac{V_{w2}}{U_2} = \frac{\tan\beta_2}{\tan\alpha_2 + \tan\beta_2} \tag{16.43}$$

work done per unit mass is also given by (from (16.32) and (16.43))

$$w = U_2^2\left(\frac{\tan\beta_2}{\tan\alpha_2 + \tan\beta_2}\right) \tag{16.44}$$

16.3.3 Efficiency

On account of losses, the isentropic work $\frac{1}{\rho}$ (Δp_0) is less than the actual work.

Therefore the stage efficiency is defined by

$$\eta_s = \frac{(\Delta p_0)}{\rho U_2 V_{w2}} \tag{16.45}$$

16.3.4 Number of Blades

Too few blades are unable to fully impose their geometry on the flow, whereas too many of them restrict the flow passage and lead to higher losses. Most of the efforts to determine the optimum number of blades have resulted in only empirical relations given below

(i) $$n = \frac{8{\cdot}5 \sin \beta_2}{1 - D_1/D_2} \tag{16.46}$$

(ii) $$n = 6{\cdot}5 \left(\frac{D_2 + D_1}{D_2 - D_1}\right) \sin \frac{1}{2} (\beta_1 + \beta_2) \tag{16.47}$$

(iii) $$n = \frac{1}{3} \beta_2 \tag{16.48}$$

For a detailed procedure on design, please refer to Stepanoff [2].

16.3.5 Impeller Size

The diameter ratio (D_1/D_2) of the impeller determines the length of the blade passages. The smaller the ratio, the longer is the blade passage. The following value for the diameter ratio is often used by the designers:

$$\frac{D_1}{D_2} = 1{\cdot}2(\varphi)^{1/3} \tag{16.49}$$

where $\varphi = V_{f2}/U_2$

The following relation for the blade width to diameter ratio is recommended:

$$b_1/D_1 \approx 0{\cdot}2 \tag{16.50}$$

If the rate of diffusion in a parallel wall impeller is too high, the tapered shape towards the outer periphery, is preferable.

The typical performance curves describing the variations of head, power and efficiency with discharge of a centrifugal blower or fan are shown in Fig. 16.20.

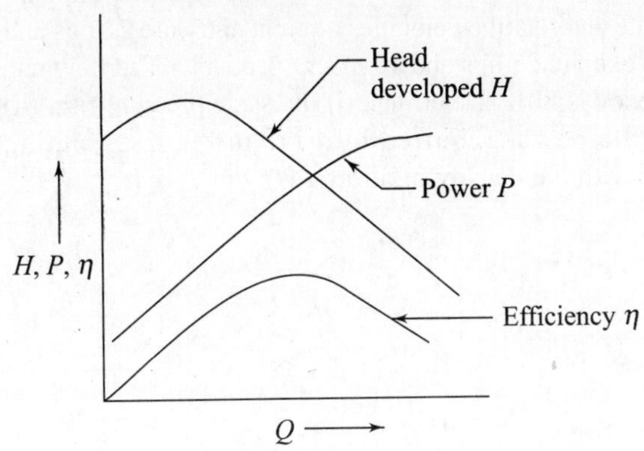

Fig. 16.20 Performance characteristic curves of a centrifugal blower or fan

16.3.6 Fan Laws

The relationships of discharge Q, head H and Power P with the diameter D and rotational speed N of a centrifugal fan can easily be expressed from the diamensionless performance parameters determined from the principle of similarity of rotodynamic machines as described in Section 15.3.2. These relationships are known as Fan Laws and described as follows:

$$Q = K_q D^3 N \tag{16.51}$$

$$H = \frac{K_h D^2 N^2 \rho}{g} \tag{16.52}$$

$$P = \frac{K_p D^5 N^3 \rho}{g}$$

where K_q, K_h, and K_p are constants.

For the same fan, the dimensions get fixed and the laws are

$$\frac{Q_1}{Q_2} = \frac{N_1}{N_2}$$

$$\frac{H_1}{H_2} = \left(\frac{N_1}{N_2}\right)^2 \text{ and } \frac{P_1}{P_2} = \left(\frac{N_1}{N_2}\right)^3$$

For a different size and other conditions remaining same, the laws are

$$\frac{Q_1}{Q_2} = \left(\frac{D_1}{D_2}\right)^3, \frac{H_1}{H_2} = \left(\frac{D_1}{D_2}\right)^2 \text{ and } \frac{P_1}{P_2} = \left(\frac{D_1}{D_2}\right)^5$$

Example 16.8

A centrifugal fan running at 1500 rpm has inner and outer diameter of the impeller as 0.2 m and 0.24 m. The absolute and relative velocities of air at entry are 21 m/s and 20 m/s respectively and those at exit are 25 m/s and 18 m/s respectively. The flow rate is 0.6 kg/s and the motor efficiency is 80%. Determine (i) the stage pressure rise, (ii) degree of reaction and (iii) the power required to drive the fan. Assuming the flow to be incompressible with the density of air as 1.2 kg/m^3.

Solution

(i)

$$U_1 = \frac{\pi \times 0.20 \times 1500}{60}$$

$$= 15.71 \text{ m/s}$$

$$U_2 = \frac{\pi \times 0.24 \times 1500}{60}$$

$$= 18.85 \text{ m/s}$$

From Eq. (16.40), the total pressure rise across the stage is

$$(\Delta p_t)_{\text{stage}} = \frac{1}{2}\rho\,[(V_2^2 - V_1^2) + (U_2^2 - U_1^2) + (V_{r1}^2 - V_{r1}^2)]$$

$$= \frac{1}{2} \times 1.2\,[(25^2 - 21^2) + (18.85^2 - 15.71^2) + (20^2 - 18^2)]$$

$$= 221.11 \text{ N/m}^2$$

(ii) The static pressure rise across the stage is

$$(\Delta p_s)_{\text{stage}} = \frac{1}{2}\rho\,[(U_2^2 - U_1^2) + (V_{r1}^2 - V_{r2}^2)]$$

$$= \frac{1}{2} \times 1.2\,[(18.85^2 - 15.71^2) + (20^2 - 18^2)]$$

$$= 110.71 \text{ N/m}^2$$

$$\text{The degree of reaction} = \frac{110.71}{221.11}$$

$$= 0.5$$

(iii) The specific power input to the stage is

$$w = \frac{(\Delta p_0)_{\text{stage}}}{\rho}$$

$$= \frac{221.11}{1.2}$$

$$= 184.26 \text{ J/kg}$$

Therefore, the power required to drive the fan is

$$P = \frac{\dot{m}\, w}{\eta_m}$$

$$= \frac{0{\cdot}6 \times 184{\cdot}26}{0{\cdot}8}$$

$$= 138.19 \text{ W}$$

SUMMARY

- A centrifugal compressor is a radial flow machine which utilises the mechanical energy imparted to the machine from outside to increase the internal energy of the fluid mainly in the form of increased static pressure.
- A centrifugal compressor mainly consists of a rotating impeller which imparts energy to the fluid flowing past the impeller blades and a diffuser comprising a number of fixed diverging passages in which the fluid is decelerated with a consequent rise in static pressure. Usually, about half of the total pressure rise occurs in the impeller and the other half in the diffuser.
- The losses in a centrifugal compressor are due to (i) fluid friction in stationary and rotating blade passages, (ii) leakage through the clearances between the impeller shaft and casing and between the outer periphery of the impeller eye and the casing, (iii) incidence of fluid with shock during off design conditions.
- The performance characteristics of a compressor are usually specified by curves of stagnation pressure ratio (p_{2t}/p_{1t}) and stagnation temperature ratio (T_{2t}/T_{1t}) against non-dimensional mass flow $\left(m\sqrt{RT_{1t}}/D^2 p_{1t}\right)$ and non-dimensional rotational speed $\left(ND/\sqrt{RT_{1t}}\right)$.
- Most of the positive slope part of the characteristic curve $\left(p_{2t}/p_{1t} \text{ vs } m\sqrt{RT_{1t}}/D^2 p_{1t}\right)$ of a centrifugal compressor cannot be obtained in practice because of the phenomenon of **surging** which is an unstable operation of the compressor manifested by a cyclic reversal of pressure gradient and flow in the delivery pipe.
- An axial flow compressor consists of several stages. Each stage has a rotor carrying the moving blades and a stator comprising the stationary blades. While the rotor of a stage imparts energy to the fluid, the stator serves the process of diffusion to increase the static pressure. The bulk flow is in the axial direction and brought about by the camber of the blades in the stator.
- An important parameter in the design of an axial flow compressor is the degree of reaction which is defined as the ratio of the static enthalpy rise in the rotor to that in the whole stage. A 50% degree of reaction results in a symmetrical blading which means the inlet and outlet angles of a rotor blade are equal to those of a stator blade.
- Fans and blowers are turbomachines which deliver air at a desired high velocity but at a relativelely low static pressure. A large number of fans and blowers for relatively high pressure applications are of centrifugal type. A centrifugal blower consists of an impeller which has blades fixed between inner and outer diameters and a spiral-shaped volute casing.
- The relationships of discharge (Q), head (H) and power (P) with the

diameter (D) and rotational speed (N) of a fan are known as **Fan Laws**. For the same fan, $Q \propto N$, $H \propto N^2$ and $P \propto N^3$. For fans of different sizes, $Q \propto D^3$, $H \propto D^2$ and $P \propto D^5$.

REFERENCES

1. Cohen, H., Rogers, G. F. C., and Saravanamuttoo, H.I.H., *Gas Turbine Theory*, Longman, 1996.
2. Stepanoff, A.J., *Centrifugal and Axial Flow Pumps: Theory, Design and Application,* John Wiley and Sons, New York, 1967.

EXERCISES

[Note: For the Problems, assume $R = 287$ J/kg K and $\gamma = 1.4$ and $C_p = 1005$ J/kg K for air]

16.1 Determine the pressure ratio developed and the specific work input to drive a centrifugal air compressor having an impeller diameter of 0.5 m and running at 7000 rpm. Assume zero whirl at the entry and $T_{1t} = 288$ K.

Ans. (1.47, 33.58 kJ/kg)

16.2 A centrifugal compressor develops a pressure ratio of 4 : 1. The inlet eye of the compressor impeller is 0.3 m in diameter. The axial velocity at inlet is 120 m/s and the mass flow rate is 10 kg/s. The velocity in the delivery duct is 110 m/s. The tip speed of the impeller is 450 m/s and runs at 16,000 rpm with a total head isentropic efficiency of 80%. The inlet stagnation pressure and temperature are 101 kN/m^2 and 300 K. Calculate (i) the static temperatures and pressures at inlet and outlet of the compressor, (ii) the static pressure ratio, (iii) the power required to drive the compressor.

Ans. ($T_1 = 292.8$ K, $T_2 = 476.45$ K, $p_1 = 93$ kN/m^2, $p_2 = 386.9$ kN/m^2, $p_2/p_1 = 4.16$, $P = 1.83$ MW)

16.3 The following results were obtained from a test on a small single-sided centrifugals compressor:

Compressor delivery stagnation pressure	2.97 bar
Compressor delivery stagnation temperature	429 K
Static pressure at impeller tip	1.92 bar
Mass flow	0.60 kg/s
Rotational speed	766 rev/s
Ambient conditions	0.99 bar 288 K

Determine the isentropic efficiency of the compressor.

The diameter of the impeller is 0.165 m, the axial depth of the vaneless diffuser is 0.01 m and the number of impeller vanes is 17. Making use of the Stanitz equation for slip factor, calculate the stagnation pressure at the impeller tip.

Ans. (0.75, 3.13 bar)

16.4 A single-sided centrifugal compressor is to deliver 14 kg/s of air when operating at a pressure ratio of 4:1 and a speed of 200 rev/s. The inlet stagnation conditions are 288 K and 1.0 bar. The slip factor and power input factor may be taken as 0.9 and 1.04 respectively. The overall isentropic efficiency is 0.80. Determine the overall diameter of the impeller.

Ans. (0.69 m)

16.5 Each stage of an axial flow compressor is of 50% degree of reaction and has the same mean blade speed and the same value of outlet relative velocity angle $\beta_2 = 30°$. The mean flow coefficient (V_f/U) is constant for all stages at 0.5. At the entry to the first stage, the stagnation temperature is 290 K, the stagnation pressure is 101 kPa. The static pressure is 87 kPa and the flow area is 0.38 m^2. Determine the axial velocity, the mass flow rate and the shaft power needed to drive the compressor when there are 6 stages and the mechanical efficiency is 0.98.

Ans. (135.51 m/s, 56.20 kg/s, 10.68 MW)

16.6 An axial flow compressor stage has blade root, mean and tip velocities of 150, 200 and 250 m/s respectively. The stage is to be designed for a stagnation temperature rise of 20 K and an axial velocity of 150 m/s, both constant from root to tip. The work done factor is 0.93. Assuming degree of reaction is 0.5 at the mean radius, determine the stage air angles at root mean and tip for a free vortex design where the whirl component of velocity varies inversely with the radius

Ans. (α_1= 17.04° (= β_2), β_1 = 45.75° (= α_2) at mean radius;
α_1 = 13.77°, β_1 = 54.88°, β_2 = 40.36°, α_2 = 39.34° at tip;
α_1 = 22.10°, β_1 = 30.71°, β_2 = – 19.95°, α_2 = 53.74° at root)

16.7 An axial compressor has the following data:

Temperature and pressure at entry	300 K, 1.0 bar
Degree of reaction	50%
Mean blade diameter	0.4 m
Rotational speed	15,000 rpm
Blade height at entry	0.08 m
Air angles at rotor and stator exit	25°
Axial velocity	150 m/s
Work done factor	0.90
Isentropic stage efficiency	85%
Mechanical efficiency	97%

Determine (i) air angles at the rotor and stator entry (ii) the mass flow rate of air (iii) the power required to drive the compressor, (iv) the pressure ratio developed by the stage (v) The mach number (based on relative velocities) at the rotor entry.

Ans. ((i) 25°, 58.44° (ii) 17.51 kg/s, (iii) 0.89 MW, (iv) 1.58, (v) 0.83)

16.8 An axial flow compressor stage has a mean diameter of 0.6 m and runs at 15,000 rpm. If the actual temperature rise and pressure ratio developed are 30° C and 1.36 respectively, determine (i) the power required to drive the compressor while delivering 57 kg/s of air. Assume mechanical efficiency of

86% and an initial temperature of 35° C, (ii) the isentropic efficiency of the stage and (iii) the degree of reaction if the temperature at the rotor exit is 55° C.

Ans. ((i) 2 MW, (ii) 94.2%, (iii) 66.6%)

16.9 A centrifugal blower takes in 200 m^3/min of air at a pressure and the temperature of 101 kN/m^2 and 45° C and delivers it at a pressure of 750 mm of water gauge. Assuming the efficiencies of the blower and drive as 80% and 82% respectively, determine (i) the power required to drive the blower and (ii) the pressure and temperature of air at blower exit.

Ans. (37.38 kW, 108.36 kN/m^2, 326.06 K)

16.10 A backward-swept centrifugal fan develops a pressure of 80 mm of water guage. It has an impeller diameter of 0.89 m and runs at 720 rpm. The blade angle at tip is 39° and the width of the impeller is 0.1 m. Assuming a constant radial velocity of 9.15 m/s and density of air as 1.2 kg/m^3, determine the fan efficiency, discharge and power required.

Ans. (87.61%, 2.56 m^3/s, 2.29 kW)

Appendix A

PHYSICAL PROPERTIES OF FLUIDS

Table A.1 Physical Properties of Some Common Liquids at 20°C and 101.325 kNm^2

Liquid	*Density, ρ (kg/m^3)*	*Isentropic bulk modulus of elasticity, E_s (GN/m^2)*	*Surface tension incontact with air, σ × 10^2 (N/m)*
Benzene	879	1.48	2.89
Carbon tetrachloride	1595	1.36	2.70
Castor oil	969	2.11	–
Glycerine	1260	4.59	6.30
Kerosene	820	1.43	2.68
Lubricating oil	880	1.44	–
Mercury	13550	28.50	48.40
Sea water	1025	2.42	7.00
Water	998	2.24	7.28

Table A.2 International Standard Atmosphere

Altitude above sea level (m)	*Temperature (K)*	*Absolute pressure (kN/m^2)*	*Density (kg/m^3)*
0*	288.15*	101.325*	1.2250*
1000	281.7	89.88	1.1117
2000	275.2	79.50	1.0066
4000	262.2	61.66	0.8194
6000	249.2	47.22	0.6602
8000	236.2	35.65	0.5258
10000	223.3	26.50	0.4134
11500	216.7	20.98	0.3375
14000	216.7	14.17	0.2279
16000	216.7	10.35	0.1665
18000	216.7	7.565	0.1216
20000	216.7	5.529	0.08892
22000	218.6	4.097	0.06451
24000	220.6	2.972	0.04694
26000	222.5	2.188	0.03426
28000	224.5	1.616	0.02508
30000	226.5	1.197	0.01848
32000	228.5	0.889	0.01356

* STP conditions

Dynamic viscosity of common fluids has been shown in Fig. 1.13 in Chapter 1. Figure A.1 shows the kinematic viscosity of common fluids.

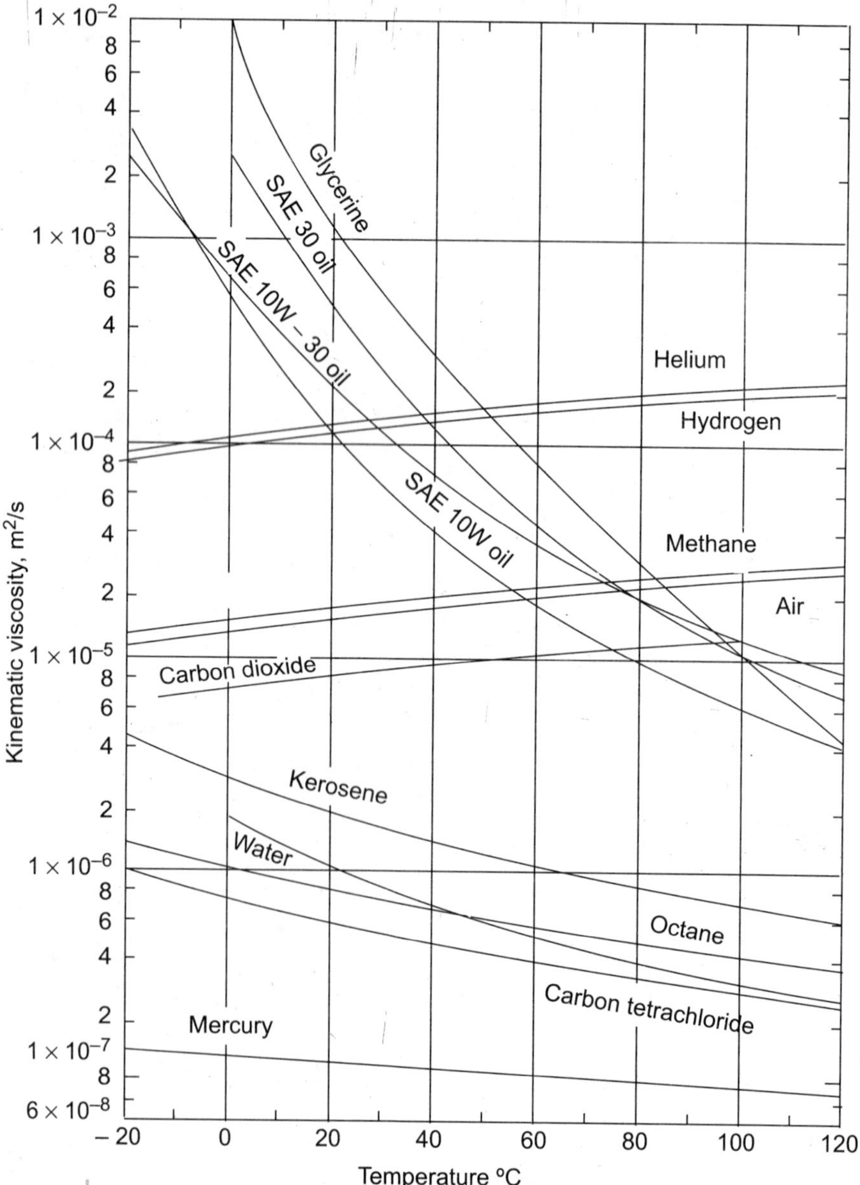

Fig. A.1 Kinematic viscosity of common fluids (at 101.325 kN/m^2) as a function of temperature

Appendix B

REVIEW OF PRELIMINARY CONCEPTS IN VECTORS AND THEIR OPERATIONS

B.1 DEFINITION OF VECTOR

Definition of scalar and vector quantities has been provided in Section 3.2. Vector quantities are denoted by symbols either with an arrow or a cap at the top, like $\vec{A}\,\vec{B}\,\vec{C}$, etc. or $\hat{A}\,\hat{B}\,\hat{C}$, etc. A vector quantity $\vec{A}$ is written in terms of its components in a rectangular cartesian coordinates system (Fig. B.1) as

$$\vec{A} = \hat{i}\,A_x + \hat{j}\,A_y + \hat{k}\,A_z$$

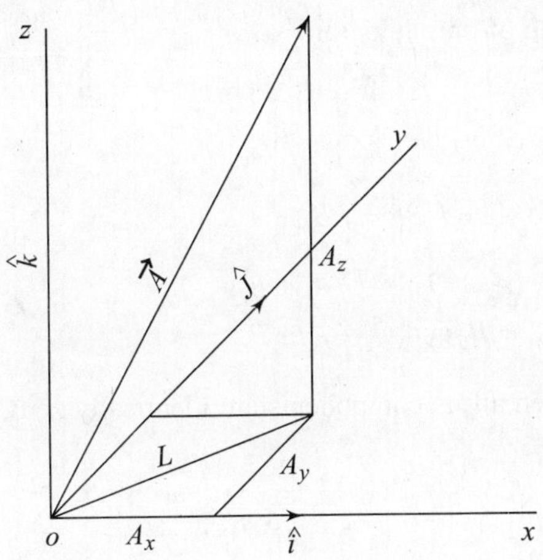

Fig. B.1 Magnitude and components of a vector

where $\hat{i}$, $\hat{j}$ and $\hat{k}$ are the unit vectors and A_x, A_y and A_z are the components of $\vec{A}$ in x, y, z directions respectively. $|\vec{A}|$ is the magnitude of $\vec{A}$. From Fig. B.1.

$$[|\vec{A}|]^2 = L^2 + A_z^2 = A_x^2 + A_y^2 + A_z^2$$

Therefore, $$|\vec{A}| = \sqrt{A_x^2 + A_y^2 + A_z^2}$$

B.2 ADDITION OF VECTORS

Vector quantities are added in consideration of both magnitude and direction. Thus, for addition of two vectors $\vec{A}$ and $\vec{B}$, we have from the rule of parallelogram (Fig. B.2).

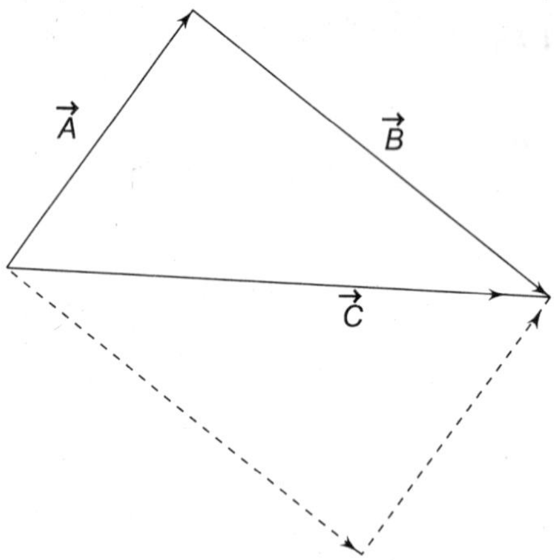

Fig. B.2 Addition of vectors by the rule of parallelogram

$$\vec{C} = \vec{A} + \vec{B} \text{ and } \vec{C} = \hat{i}\,C_x + \hat{j}\,C_y + \hat{k}\,C_z$$

$$= \hat{i}\,A_x + \hat{j}\,A_y + \hat{k}\,A_z + \hat{i}\,B_x + \hat{j}\,B_y + \hat{k}\,B_z$$

$$= \hat{i}\,(A_x + B_x) + \hat{j}\,(A_y + B_y) + \hat{k}\,(A_z + B_z)$$

Hence, $$C_x = A_x + B_x,\ C_y = A_y + B_y \text{ and } C_z = A_z + B_z$$

If a vector $\vec{D}$ equals to zero, then all its components are identically zero, i.e., $D_x = D_y = D_z = 0$.

B.3 PRODUCT OF VECTORS

B.3.1 The Dot Product (or Scalar Product)

The dot product of two vector quantities $\vec{A}$ and $\vec{B}$ is defined as

$$\vec{A} \cdot \vec{B} = |\vec{A}|\,|\vec{B}| \cos \theta_{AB}$$

where θ_{AB} is the angle between the vectors. The dot product is a scalar quantity which physically represents the product of $|\vec{A}|$ with the component of $|\vec{B}|$ in the direction of $\vec{A}$. If $\theta_{AB} < \pi/2$, its magnitude is positive, while for $\theta_{AB} > \pi/2$, it is negative. If $\theta_{AB} = \pi/2$, $\vec{A} \cdot \vec{B} = 0$. The dot products of unit vectors in a Cartesian coordinate system are :

$$\hat{i} \cdot \hat{i} = 1 \qquad \hat{j} \cdot \hat{i} = 0 \qquad \hat{k} \cdot \hat{i} = 0$$

$$\hat{i} \cdot \hat{j} = 0 \qquad \hat{j} \cdot \hat{j} = 1 \qquad \hat{k} \cdot \hat{j} = 0$$

$$\hat{i} \cdot \hat{k} = 0 \qquad \hat{j} \cdot \hat{k} = 0 \qquad \hat{k} \cdot \hat{k} = 1$$

Therefore,
$$\vec{A} \cdot \vec{B} = (\hat{i} A_x + \hat{j} A_y + \hat{k} A_z) \cdot (\hat{i} B_x + \hat{j} B_y + \hat{k} B_z)$$
$$= A_x B_x + A_y B_y + A_z B_z$$

The following rules apply for the dot product of vectors:

(i) The dot product is commutative, i.e., $\vec{A} \cdot \vec{B} = \vec{B} \cdot \vec{A}$

(ii) The dot product is distributive, i.e., $\vec{A} \cdot (\vec{B} + \vec{C}) = \vec{A} \cdot \vec{B} + \vec{A} \cdot \vec{C}$

(iii) The dot product is not associative, i.e., $\vec{A}(\vec{B} \cdot \vec{C}) \neq (\vec{A} \cdot \vec{B})\vec{C}$.

B.3.2 Cross Product of Vectors

The cross product of two vector quantities $\vec{A}$ and $\vec{B}$ is written as $\vec{A} \times \vec{B}$. It is a vector quantity whose magnitude is given by $|\vec{A} \times \vec{B}| = |\vec{A}|\,|\vec{B}| \sin \theta_{AB}$ and is perpendicular to both $\vec{A}$ and $\vec{B}$. The sense of $\vec{A} \times \vec{B}$ is given by the right-hand rule, that is, as $\vec{A}$ is rotated into $\vec{B}$, then $\vec{A} \times \vec{B}$ points in the direction of the right thumb. This is shown in Fig. B.3.

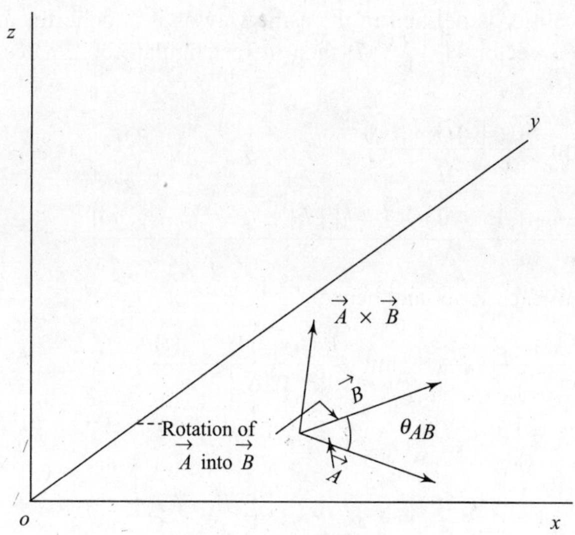

Fig. B.3 Cross product of vectors

If $\vec{A}$ and $\vec{B}$ are parallel, then $\sin \theta_{AB} = 0$ and $\vec{A} \times \vec{B} = 0$. The cross products among unit vectors in a Cartesian coordinate system are

$$\hat{i} \times \hat{i} = 0 \qquad \hat{j} \times \hat{i} = -\hat{k} \qquad \hat{k} \times \hat{i} = \hat{j}$$

$$\hat{i} \times \hat{j} = \hat{k} \qquad \hat{j} \times \hat{j} = 0 \qquad \hat{k} \times \hat{j} = -\hat{i}$$

$$\hat{i} \times \hat{k} = -\hat{j} \qquad \hat{j} \times \hat{k} = \hat{i} \qquad \hat{k} \times \hat{k} = 0$$

The cross product $\vec{A} \times \vec{B}$ is usually written in a determinant form as

$$\vec{A} \times \vec{B} = \begin{vmatrix} \hat{i} & \hat{j} & \hat{k} \\ A_x & A_y & A_z \\ B_x & B_y & B_z \end{vmatrix}$$

By expanding the determinant we have

$$\vec{A} \times \vec{B} = \hat{i}\,(A_y B_z - A_z B_y) + \hat{j}\,(A_z B_x - A_x B_z) + \hat{k}\,(A_x B_y - A_y B_x)$$

The following properties apply for the cross product of vectors:

(i) The cross product is not commutative, i.e., $\vec{A} \times \vec{B} \neq \vec{B} \times \vec{A}$ since the interchange of two rows changes the sign of a determinant, $\vec{A} \times \vec{B} = -\vec{B} \times \vec{A}$.

(ii) The cross product is distributive, i.e., $\vec{A} \times (\vec{B} + \vec{C}) = \vec{A} \times \vec{B} + \vec{A} + \vec{C}$

(iii) The cross product is not associative, i.e., $\vec{A} \times (\vec{B} \times \vec{C}) \neq (\vec{A} \times \vec{B}) \times \vec{C}$

B.4 DIFFERENTIATION OF VECTORS

The derivative of a vector quantity is defined in the same way as it is done for a scalar quantity. Let there be a vector $\vec{A} = \vec{A}(t)$ then in rectangular coordinates $A_x = A_x(t)$, $A_y = A_y(t)$, $A_z = A_z(t)$

$$\frac{d\vec{A}}{dt} = \lim_{\Delta t \to 0} \frac{\vec{A}(t + \Delta t) - \vec{A}(t)}{\Delta t}$$

$$= \lim_{\Delta t \to 0} \frac{\hat{i}\,[A_x(t + \Delta t) - A_x(t)] + \hat{j}\,[A_y(t + \Delta t) - A_y(t)] + \hat{k}\,[A_z(t + \Delta t) - A_z(t)]}{\Delta t}$$

The limiting process applies to each term, and hence

$$\frac{d\vec{A}}{dt} = \lim_{\Delta t \to 0} \frac{A_x(t + \Delta t) - A_x(t)}{\Delta t}\hat{i} + \lim_{\Delta t \to 0} \frac{A_y(t + \Delta t) - A_y(t)}{\Delta t}\hat{j}$$

$$+ \lim_{\Delta t \to 0} \frac{A_z(t + \Delta t) - A_z(t)}{\Delta t}\hat{k}$$

$$= \frac{dA_x}{dt}\hat{i} + \frac{dA_y}{dt}\hat{j} + \frac{dA_z}{dt}\hat{k}$$

Similarly, if $\vec{A} = \vec{A}\,(x, y, z)$ that is, $A_x = A_x\,(x, y, z)$, etc. then

$$\frac{\partial \vec{A}}{\partial x} = \hat{i}\frac{\partial A_x}{\partial x} + \hat{j}\frac{\partial A_y}{\partial x} + \hat{k}\frac{\partial A_z}{\partial x}$$

$$\frac{\partial \vec{A}}{\partial y} = \hat{i}\frac{\partial A_x}{\partial y} + \hat{j}\frac{\partial A_y}{\partial y} + \hat{k}\frac{\partial A_z}{\partial y}$$

$$\frac{\partial \vec{A}}{\partial z} = \hat{i}\frac{\partial A_x}{\partial z} + \hat{j}\frac{\partial A_y}{\partial z} + \hat{k}\frac{\partial A_z}{\partial z}$$

B.5 VECTOR OPERATOR ∇

B.5.1 Definition of ∇

The vector operator del, ∇, is defined as

$$\nabla = \hat{i}\frac{\partial}{\partial x} + \hat{j}\frac{\partial}{\partial y} + \hat{k}\frac{\partial}{\partial z} \qquad \text{(Cartesian coordinates)}$$

$$\nabla = \hat{i}_r\frac{\partial}{\partial r} + \hat{i}_\theta\frac{\partial}{r\partial \theta} + \hat{i}_z\frac{\partial}{\partial z} \qquad \text{(Cylindrical coordinates)}$$

where, $\hat{i}_r$, $\hat{i}_\theta$ and $\hat{i}_z$ are the unit vectors in r, θ and z directions respectively in a cylindrical coordinate system. Three possible products and other functions can be formed with the operator ∇ as follows:

B.5.2 Gradient

When ∇ operates on a differentiable scalar function, the resulting term is known as the gradient of the scalar function. Let $\psi\,(x, y, z)$ be a scalar function,

Then, $$\nabla\psi = \text{gradient } \psi = \text{grad } \psi = \hat{i}\frac{\partial \psi}{\partial x} + \hat{j}\frac{\partial \psi}{\partial y} + \hat{k}\frac{\partial \psi}{\partial z}$$

It has to be noted that though ψ is a scalar function, $\nabla\psi$ is a vector function (or field).

B.5.3 Divergence

The dot product of ∇ and a vector function (or field) results in a scalar function (or field) known as divergence. For a vector field $\vec{A}\,(x, y, z)$ in a rectangular Cartesian coordinate system,

$$\nabla \cdot \vec{A} = \text{divergence } \vec{A} \text{ (or div } \vec{A}\text{)}$$

$$= \left(\hat{i}\frac{\partial}{\partial x} + \hat{j}\frac{\partial}{\partial y} + \hat{k}\frac{\partial}{\partial z}\right).(\hat{i}A_x + \hat{j}A_y + \hat{k}A_z)$$

$$= \frac{\partial A_x}{\partial x} + \frac{\partial A_y}{\partial y} + \frac{\partial A_z}{\partial z}$$

$$\left(\text{Since, } \frac{\partial \hat{i}}{\partial x} = \frac{\partial \hat{j}}{\partial y} = \frac{\partial \hat{k}}{\partial z} = 0\right)$$

In cylindrical coordinates, if $\vec{A} = \vec{A}\,(r, \theta, z)$, then

$$\nabla \cdot \vec{A} = \left(\hat{i}_r \frac{\partial}{\partial r} + \hat{i}_\theta \frac{1}{r}\frac{\partial}{\partial \theta} + \hat{i}_z \frac{\partial}{\partial z}\right).(\hat{i}_r A_r + \hat{i}_\theta A_\theta + \hat{i}_z A_z)$$

$$= \frac{\partial A_r}{\partial r} + \frac{A_r}{r} + \frac{1}{r}\frac{\partial A_r}{\partial \theta} + \frac{\partial A_z}{\partial z}$$

$$\left(\text{Since, } \frac{\partial \hat{i}_r}{\partial r} = \frac{\partial \hat{i}_z}{\partial z} = 0 \text{ and } \frac{\partial \hat{i}_r}{\partial \theta} = \hat{i}_\theta, \frac{\partial \hat{i}_\theta}{\partial r} = 0,\right.$$

$$\left.\frac{\partial \hat{i}_\theta}{\partial \theta} = -\hat{i}_r \frac{\partial \hat{i}_\theta}{\partial \theta} = -\hat{i}_r, \hat{i}_r.\hat{i}_r = \hat{i}_\theta.\hat{i}_\theta = \hat{i}_z.\hat{i}_z = 1\right)$$

The divergence of velocity vector $\vec{V}$, i.e., $\nabla \cdot \vec{V}$ was used to describe the continuity equation [Eq. (3.6)] in Section 3.4.

B.5.4 Curl

The cross product between ∇ and a vector function (or field) results in a vector function (or field) known as curl. For a vector field $\vec{A} = \vec{A}\,(x, y, z)$ in Cartesian coordinates,

$$\nabla \times \vec{A} = \text{Curl } \vec{A} = \begin{vmatrix} \hat{i} & \hat{j} & \hat{k} \\ \dfrac{\partial}{\partial x} & \dfrac{\partial}{\partial y} & \dfrac{\partial}{\partial z} \\ A_x & A_y & A_z \end{vmatrix}$$

or

$$\nabla \times \vec{A} = \hat{i}\left(\frac{\partial A_z}{\partial y} - \frac{\partial A_y}{\partial z}\right) + \hat{j}\left(\frac{\partial A_x}{\partial z} - \frac{\partial A_z}{\partial x}\right) + \hat{k}\left(\frac{\partial A_y}{\partial x} - \frac{\partial A_x}{\partial y}\right)$$

In cylindrical coordinates, $\vec{A} = \vec{A}\,(r, \theta, z)$. Then

$$\nabla \times \vec{A} = \left(\hat{i}_r \frac{\partial}{\partial r} + \hat{i}_\theta \frac{1}{r}\frac{\partial}{\partial \theta} + \hat{i}_z \frac{\partial}{\partial z}\right) \times (\hat{i}_r A_r + \hat{i}_\theta A_\theta + \hat{i}_z A_z)$$

$$= \hat{i}_r \frac{\partial}{\partial r} \times (\hat{i}_r A_r) + \hat{i}_r \frac{\partial}{\partial r} \times (\hat{i}_\theta A_\theta) + \hat{i}_r \frac{\partial}{\partial r} \times (\hat{i}_z A_z)$$

$$+ \hat{i}_\theta \frac{1}{r}\frac{\partial}{\partial \theta} \times (\hat{i}_r A_r) + \hat{i}_\theta \frac{1}{r}\frac{\partial}{\partial \theta} \times (\hat{i}_\theta A_\theta) + \hat{i}_\theta \frac{1}{r}\frac{\partial}{\partial \theta} \times (\hat{i}_z A_z)$$

$$+\hat{i}_z \frac{\partial}{\partial z}\times(\hat{i}_r A_r) + \hat{i}_z \frac{\partial}{\partial z}\times(\hat{i}_\theta A_\theta)+\hat{i}_z\frac{\partial}{\partial z}\times(\hat{i}_z A_z)$$

$$= \hat{i}_r\times\hat{i}_r \cancelto{0}{\frac{\partial A_r}{\partial r}} + \hat{i}_r \times\left(A_r \cancelto{0}{\frac{\partial \hat{i}_r}{\partial r}}\right) + \hat{i}_r\times\hat{i}_\theta \frac{\partial A_\theta}{\partial r} + \hat{i}_r\times\left(A_\theta \cancelto{0}{\frac{\partial \hat{i}_\theta}{\partial r}}\right)$$

$$+\hat{i}_r\times\hat{i}_z\frac{\partial A_z}{\partial r} + \hat{i}_r\times\left(\cancelto{0}{\hat{i}_\theta\times\hat{i}_\theta}\,\frac{1}{r}\frac{\partial A_\theta}{\partial\theta}\right)+\hat{i}_\theta\times\hat{i}_r\frac{1}{r}\frac{\partial A_r}{\partial\theta}$$

$$+\hat{i}_\theta\times\left(A_r\cancelto{=i_\theta}{\frac{1}{r}\frac{\partial \hat{i}_r}{\partial\theta}}\right)+\hat{i}_\theta\times\hat{i}_\theta\,\frac{1}{r}^{=0}\,\frac{\partial A_\theta}{\partial\theta}$$

$$+\hat{i}_\theta\times\left(A_\theta\cancelto{=-i_r}{\frac{1}{r}\frac{\partial\hat{i}_\theta}{\partial\theta}}\right)+\hat{i}_\theta\times\left(A_r\cancelto{=0}{\frac{\partial\hat{i}_r}{\partial z}}\right)+\hat{i}_\theta\times\left(A_z\cancelto{=0}{\frac{1}{r}\frac{\partial\hat{i}_z}{\partial\theta}}\right)$$

$$+\hat{i}_z\times\hat{i}_r\frac{\partial A_r}{\partial z}+\hat{i}_z\times\left(A_r\frac{\partial\hat{i}_r}{\partial z}^{=0}\right)+\hat{i}_z\times i_\theta\frac{\partial A_\theta}{\partial z}+\hat{i}_z\times\left(\cancelto{=0}{A_\theta\frac{\partial\hat{i}_\theta}{\partial z}}\right)$$

$$+\cancelto{=0}{\hat{i}_z\times\hat{i}_z}\,\frac{\partial A_z}{\partial z}+\hat{i}_z\times\left(\cancelto{=0}{A_z\frac{\partial\hat{i}_z}{\partial z}}\right)$$

or,

$$\nabla\times\vec{A} = \hat{i}_z\frac{\partial A_\theta}{\partial r}-\hat{i}_\theta\frac{\partial A_z}{\partial r}-\hat{i}_z\frac{1}{r}\frac{\partial A_r}{\partial\theta}+\hat{i}_z\frac{A_\theta}{r}+\hat{i}_r\frac{1}{r}\frac{\partial A_z}{\partial\theta}$$

$$+\hat{i}_\theta\frac{\partial A_r}{\partial z}-\hat{i}_r\frac{\partial A_\theta}{\partial z}$$

and finally,

$$\nabla\times\vec{A} = \text{curl }\vec{A} = \hat{i}_r\left(\frac{1}{r}\frac{\partial A_z}{\partial\theta}-\frac{\partial A_\theta}{\partial z}\right)+\hat{i}_\theta\left(\frac{\partial A_r}{\partial z}-\frac{\partial A_z}{\partial r}\right)$$

$$+\hat{i}_z\frac{1}{r}\left(\frac{\partial}{\partial r}(r A_\theta)-\frac{\partial A_r}{\partial\theta}\right)$$

The curl of velocity vector $\vec{V}$, i.e., $\nabla\times\vec{V}$ was used in describing the rotation of a fluid element in Sec. 3.2.5.

B.5.5 Laplacian

The scalar function obtained by the dot product $\nabla\cdot\nabla$ is known as the Laplacian and is denoted by the symbol ∇^2.

Therefore, in Cartesian coordinates,

$$\nabla^2 = \nabla\cdot\nabla = \left(\hat{i}\frac{\partial}{\partial x}+\hat{j}\frac{\partial}{\partial y}+\hat{k}\frac{\partial}{\partial z}\right)\cdot\left(\hat{i}\frac{\partial}{\partial x}+\hat{j}\frac{\partial}{\partial y}+\hat{k}\frac{\partial}{\partial z}\right)$$

$$= \frac{\partial^2}{\partial x^2} + \frac{\partial^2}{\partial y^2} + \frac{\partial^2}{\partial z^2}$$

In cylindrical coordinates,

$$\nabla^2 = \nabla \cdot \nabla = \left(\hat{i}_r \frac{\partial}{\partial r} + \hat{i}_\theta \frac{1}{r}\frac{\partial}{\partial \theta} + \hat{i}_z \frac{\partial}{\partial z}\right) \cdot \left(\hat{i}_r \frac{\partial}{\partial r} + \hat{i}_\theta \frac{\partial}{r\partial \theta} + \hat{i}_z \frac{\partial}{\partial z}\right)$$

$$= \frac{\partial^2}{\partial r^2} + \hat{i}_\theta \frac{1}{r}\left(\frac{\partial \hat{i}_r}{\partial \theta}\frac{\partial}{\partial r} + \hat{i}_r \frac{\partial^2}{\partial \theta \partial r} + \frac{\partial \hat{i}_\theta}{\partial \theta}\frac{1}{r}\frac{\partial}{\partial \theta} + \hat{i}_\theta \frac{1}{r}\frac{\partial^2}{\partial \theta^2}\right) + \frac{\partial^2}{\partial z^2}$$

$$= \frac{\partial^2}{\partial r^2} + \frac{1}{r}\frac{\partial}{\partial r} + \frac{1}{r^2}\frac{\partial^2}{\partial \theta^2} + \frac{\partial^2}{\partial z^2}$$

$$= \frac{1}{r}\frac{\partial}{\partial r}\left(r\frac{\partial}{\partial r}\right) + \frac{1}{r^2}\frac{\partial^2}{\partial \theta^2} + \frac{\partial^2}{\partial z^2}$$

B.6 VECTOR IDENTITIES

B.6.1 $\nabla \times \nabla\theta = 0$, where θ is Any Scalar Function

This relation may be verified by expanding it into components. Therefore, in Cartesian coordinates,

$$\nabla \times \nabla\theta = \nabla \times \left(\hat{i}\frac{\partial \theta}{\partial x} + \hat{j}\frac{\partial \theta}{\partial y} + \hat{k}\frac{\partial \theta}{\partial z}\right)$$

$$\begin{vmatrix} \hat{i} & \hat{j} & \hat{k} \\ \partial/\partial x & \partial/\partial y & \partial/\partial z \\ \partial\theta/\partial x & \partial\theta/\partial y & \partial\theta/\partial z \end{vmatrix}$$

$$= \hat{i}\left(\frac{\partial^2\theta}{\partial y\partial z} - \frac{\partial^2\theta}{\partial z\partial y}\right) + \hat{j}\left(\frac{\partial^2\theta}{\partial z\partial x}\frac{\partial^2\theta}{\partial x\partial z}\right) + \hat{k}\left(\frac{\partial^2\theta}{\partial x\partial y} - \frac{\partial^2\theta}{\partial y\partial x}\right)$$

If $\theta = \theta(x, y, z)$ is a continuous, differentiable function, then

$$\frac{\partial^2\theta}{\partial x\partial y} = \frac{\partial^2\theta}{\partial y\partial x}; \frac{\partial^2\theta}{\partial x\partial z} = \frac{\partial^2\theta}{\partial z\partial x} \text{ and } \frac{\partial^2\theta}{\partial y\partial z} = \frac{\partial^2\theta}{\partial z\partial y}$$

consequently, $\nabla \times \nabla\theta = 0$

The proof of the identity in cylindrical coordinates is a more lengthy process and is left as an exercise for the readers.

B.6.2 For Two Vector Functions $\vec{A}$ and $\vec{B}$

$$\nabla\left(\vec{A}\cdot\vec{B}\right) = \left(\vec{A}\cdot\nabla\right)\vec{B}+\left(\vec{B}\cdot\nabla\right)\vec{A}+\vec{A}\times\left(\nabla\times\vec{B}\right)+\vec{B}\times\left(\nabla\times\vec{A}\right)$$

In Cartesian coordinates, we can write

$$\vec{A}\cdot\vec{B} = A_x B_x + A_y B_y + A_z B_z$$

and so,

$$\begin{aligned}\nabla\left(\vec{A}\cdot\vec{B}\right) = {} & \hat{i}\left\{\frac{\partial}{\partial x}(A_x B_x)+\frac{\partial}{\partial x}\left(A_y B_y\right)+\frac{\partial}{\partial x}(A_z B_z)\right\} \\ & +\hat{j}\left\{\frac{\partial}{\partial y}(A_x B_x)+\frac{\partial}{\partial y}\left(A_y B_y\right)+\frac{\partial}{\partial y}(A_z B_z)\right\} \\ & +\hat{k}\left\{\frac{\partial}{\partial z}(A_x B_x)+\frac{\partial}{\partial z}\left(A_y B_y\right)+\frac{\partial}{\partial z}(A_z B_z)\right\}\end{aligned} \tag{B.1}$$

Again,

$$\vec{A}\cdot\nabla = A_x\frac{\partial}{\partial x}+A_y\frac{\partial}{\partial y}+A_z\frac{\partial}{\partial z}$$

and so,

$$\begin{aligned}\left(\vec{A}\cdot\nabla\right)\vec{B} = {} & \hat{i}\left(A_x\frac{\partial B_x}{\partial x}+A_y\frac{\partial B_x}{\partial y}+A_z\frac{\partial B_x}{\partial z}\right) \\ & +\hat{j}\left(A_x\frac{\partial B_y}{\partial x}+A_y\frac{\partial B_y}{\partial y}+A_z\frac{\partial B_y}{\partial z}\right) \\ & +\hat{k}\left(A_x\frac{\partial B_z}{\partial x}+A_y\frac{\partial B_z}{\partial y}+A_z\frac{\partial B_z}{\partial z}\right)\end{aligned} \tag{B.2}$$

In a similar, way, we can write

$$\begin{aligned}\left(\vec{B}\cdot\nabla\right)\vec{A} = {} & \hat{i}\left(B_x\frac{\partial A_x}{\partial x}+B_y\frac{\partial A_x}{\partial y}+B_z\frac{\partial A_x}{\partial z}\right) \\ & +\hat{j}\left(B_x\frac{\partial A_y}{\partial x}+B_y\frac{\partial A_y}{\partial y}+B_z\frac{\partial A_y}{\partial z}\right) \\ & +\hat{k}\left(B_x\frac{\partial A_z}{\partial x}+B_y\frac{\partial A_z}{\partial y}+B_z\frac{\partial A_z}{\partial z}\right)\end{aligned} \tag{B.3}$$

and

$$\vec{A}\times\left(\nabla\times\vec{B}\right) = \begin{vmatrix}\hat{i} & \hat{j} & \hat{k} \\ A_x & A_y & A_z \\ \left(\frac{\partial B_z}{\partial y}-\frac{\partial B_y}{\partial z}\right) & \left(\frac{\partial B_x}{\partial z}-\frac{\partial B_z}{\partial x}\right) & \left(\frac{\partial B_y}{\partial x}-\frac{\partial B_x}{\partial y}\right)\end{vmatrix}$$

$$= \hat{i}\left(A_y\frac{\partial B_y}{\partial x} + A_z\frac{\partial B_z}{\partial x} - A_y\frac{\partial B_x}{\partial y} - A_z\frac{\partial B_x}{\partial z}\right)$$

$$- \hat{j}\left(A_x\frac{\partial B_y}{\partial x} + A_z\frac{\partial B_y}{\partial z} - A_x\frac{\partial B_x}{\partial y} - A_z\frac{\partial B_z}{\partial y}\right)$$

$$+ \hat{k}\left(A_x\frac{\partial B_x}{\partial z} + A_y\frac{\partial B_y}{\partial z} - A_x\frac{\partial B_z}{\partial x} - A_y\frac{\partial B_z}{\partial y}\right) \quad \text{(B.4)}$$

Similarly, $\vec{B}\times(\nabla\times\vec{A}) = \hat{i}\left(B_y\frac{\partial A_y}{\partial x} + B_z\frac{\partial A_z}{\partial x} - B_y\frac{\partial A_x}{\partial y} - B_z\frac{\partial A_x}{\partial z}\right)$

$$- \hat{j}\left(B_x\frac{\partial A_y}{\partial x} + B_z\frac{\partial A_y}{\partial z} - B_x\frac{\partial A_x}{\partial y} - B_z\frac{\partial A_z}{\partial y}\right)$$

$$+ \hat{k}\left(B_x\frac{\partial A_x}{\partial z} + B_y\frac{\partial A_y}{\partial z} - B_x\frac{\partial A_z}{\partial x} - B_y\frac{\partial A_z}{\partial y}\right) \quad \text{(B.5)}$$

Adding Eqs (B.2), (B.3), (B.4) and (B.5) we have

$$(\vec{A}\cdot\nabla)\vec{B} + (\vec{B}\cdot\nabla) + \vec{A}\times(\nabla\times\vec{B}) + \vec{B}\times(\nabla\times\vec{A})$$

$$= \hat{i}\left\{\frac{\partial}{\partial x}(A_x B_x) + \frac{\partial}{\partial x}(A_y B_y) + \frac{\partial}{\partial x}(A_z B_z)\right\}$$

$$+ \hat{j}\left\{\frac{\partial}{\partial y}(A_x B_x) + \frac{\partial}{\partial y}(A_y B_y) + \frac{\partial}{\partial y}(A_z B_z)\right\}$$

$$+ \hat{k}\left\{\frac{\partial}{\partial z}(A_x B_x) + \frac{\partial}{\partial z}(A_y B_y) + \frac{\partial}{\partial z}(A_z B_z)\right\} \quad \text{(B.6)}$$

Comparison of Eqs (B.1) and (B.6) proves that

$$\nabla(\vec{A}\cdot\vec{B}) = (\vec{A}\cdot\nabla)\vec{B} + (\vec{B}\cdot\nabla)\vec{A} + \vec{A}\times(\nabla\times\vec{B}) + \vec{B}\times(\nabla\times\vec{A})$$

INDEX